HEINEMANN BIOLOGY 2

6TH EDITION

Zoë Armstrong
Christina Bliss
Erin Bruns
Naomi Campanale
Elaine Georges
Caryn Hellberg
Carina Jansen
Jacqueline Kerr
Katherine McMahon
Heather Maginn
Jonathan Meddings
Sue Siwinski
Natasha Ward

VCE UNITS 3 AND 4 • 2022-2026

Pearson Australia
(a division of Pearson Australia Group Pty Ltd)
707 Collins Street, Melbourne, Victoria 3008
PO Box 23360, Melbourne, Victoria 8012
www.pearson.com.au

First published 2021 by Pearson Australia
2024 2023 2022 2021
10 9 8 7 6 5 4 3 2 1

Project Leads: Misal Belvedere, Malcolm Parsons
Lead Publisher: Malcolm Parsons
Content Developer: Rebecca Wood
Lead Development Editors: Fiona Cooke, Zoe Hamilton
Project Manager: Michelle Thomas
Production Editors: Natalie Lincoln, Virginia O'Brien
Series Designer: Anne Donald
Rights and Permissions Editors: Siân Human, Madeleine Roberts
Publishing Services Analyst: JitPin Chong
Illustrators: Bruce Rankin, DiacriTech, Anne Donald, QBS
Editor: Lorna Hendry
Proofreader: Jane Fitzpatrick
Indexer: Ann Philpott

Printed in Malaysia by Vivar Printing

National Library of Australia Cataloguing-in-Publication entry

A catalogue record for this book is available from the National Library of Australia

ISBN: 9780655700081

Pearson Australia Group Pty Ltd ABN 40 004 245 943

Indigenous Australians
Some of the images used in *Heinemann Biology 2* 6e might have associations with deceased Indigenous Australians.
Please be aware that these images might cause sadness or distress in Aboriginal or Torres Strait Islander communities.

HEINEMANN BIOLOGY 2
6TH EDITION

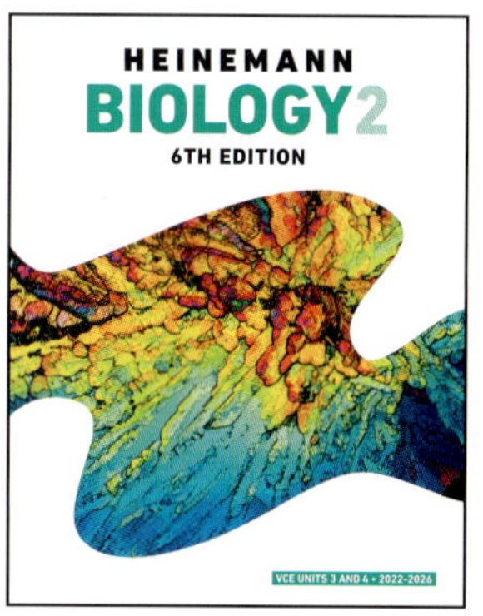

Writing and Development Team

We are grateful to the following people for their time and expertise in contributing to the **Heinemann Biology 2** project.

Zoë Armstrong
Science communicator and conservation biologist
Author, reviewer and digital question author

Doug Bail
Education consultant
SPARKlab developer

Christina Bliss
Teacher
Author and digital question author

Reuben Bolt
Pro Vice Chancellor Indigenous Leadership, CDU
Reviewer

Erin Bruns
Teacher, Head of Department
Author and reviewer

Naomi Campanale
Medical research scientist
Author and digital question author

Donna Chapman
Laboratory technician
Reviewer

Elaine Georges
Teacher
Author and reviewer

Martin Hall
Teacher
Series reviewer and teacher support author

Caryn Hellberg
Science communicator, former teacher and genetic counsellor
Author and digital question author

Neil van Herk
Teacher
Lead reviewer and digital question author

Carina Jansen
Teacher
Author, reviewer and digital question content lead

Jacqueline Kerr
Teacher
Author, reviewer and digital question author

Katherine McMahon
Teacher
Author, reviewer and digital question author

Heather Maginn
Educational travel program development manager
Author, reviewer and digital question author

Jonathan Meddings
Science and medical writer
Author and digital question author

Yvonne Sanders
Teacher, former Head of Department
Skills and Assessment author

Sue Siwinski
Teacher, former Head of Department, education consultant
Author, reviewer and teacher support author

Natasha Ward
Education officer
Author

Laurence Wooding
Laboratory technician
Teacher support author

The Publisher wishes to thank and acknowledge Pauline Ladiges and Barbara Evans for their contribution in creating the original works of the series and longstanding dedicated work with Pearson and Heinemann.

Contents

Unit 4 How does life change and respond to challenges?

AREA OF STUDY 3

Heinemann Biology 2 6th edition includes a comprehensive set of resources to support Area of Study 3 via your Pearson Places bookshelf.

How to use this book Heinemann Biology 2 6th edition

Heinemann Biology 2 6th edition has been written to the new VCE Biology Study Design 2022–2026. The book covers Units 3 and 4. Explore how to use this book below.

Chapter opener

Chapter opening pages link the study design to the chapter content. Key knowledge addressed in the chapter is clearly listed. To help you find where each outcome is covered in the chapter, the relevant section numbers are written in bold.

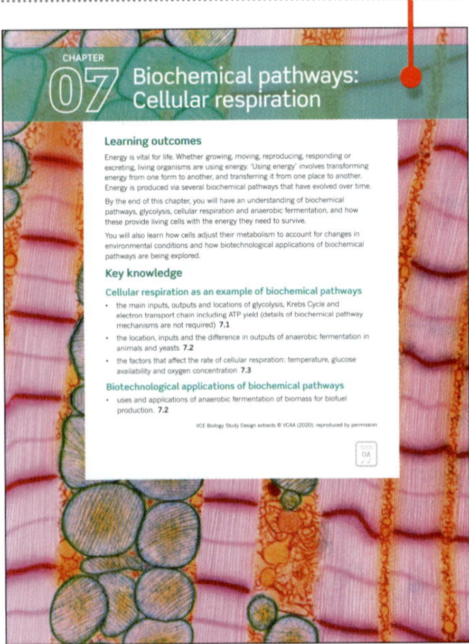

Case study

Case studies place biology in an applied situation or relevant context. Text and artwork refer to the nature and practice of biology, applications of biology and associated issues, and the historical development of biological concepts and ideas.

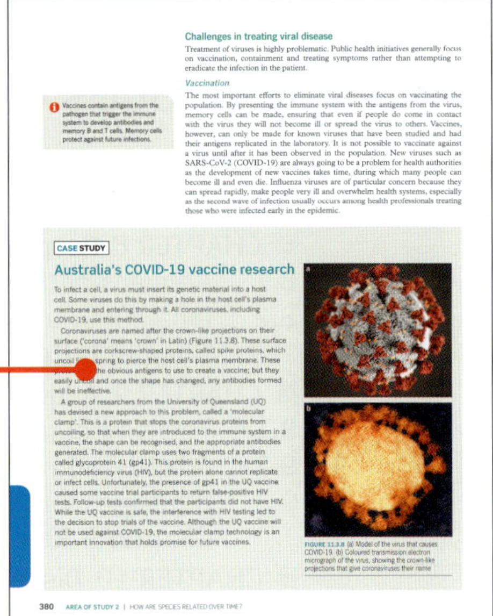

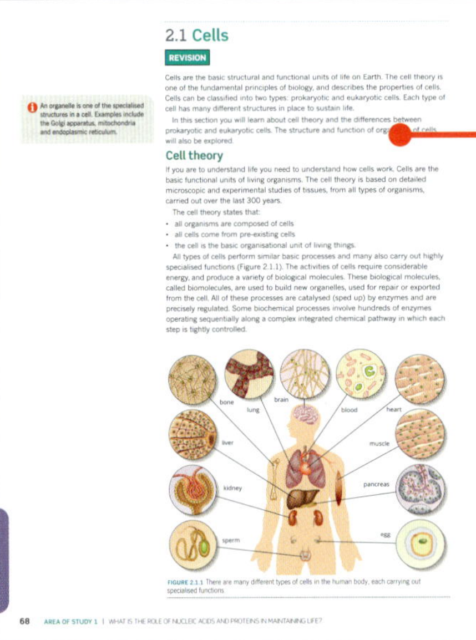

Revision

Revision sections contain vital information from Year 11.

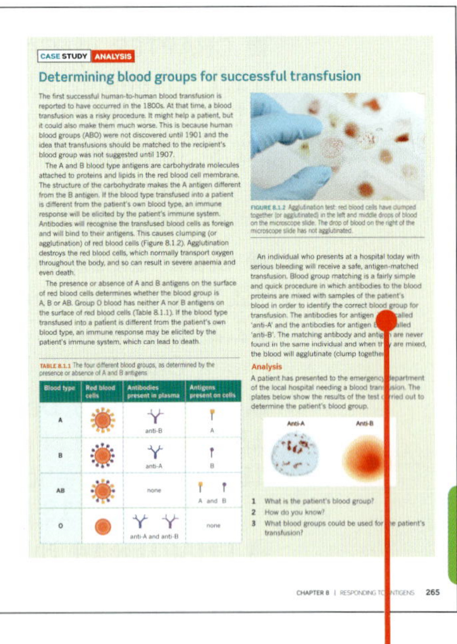

Highlight

Focus on important information such as key definitions and summary points.

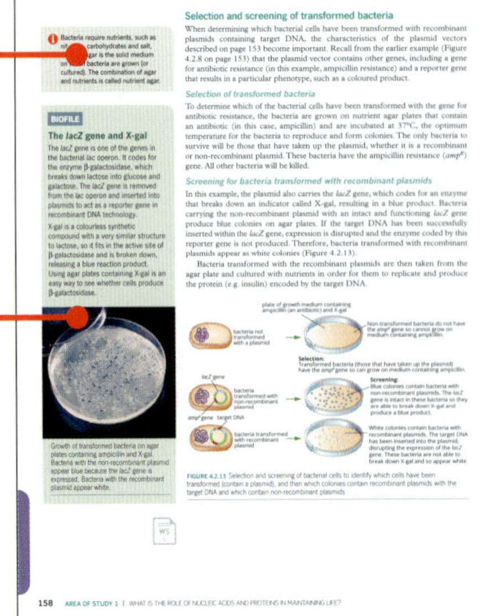

BioFile

BioFiles include interesting information and real world examples.

Case study: Analysis

These case studies include real-world data that can be analysed and evaluated.

Area of Study review

Each area of study concludes with a comprehensive set of exam-style questions, including multiple choice and short answer, to support you in your exam preparation.

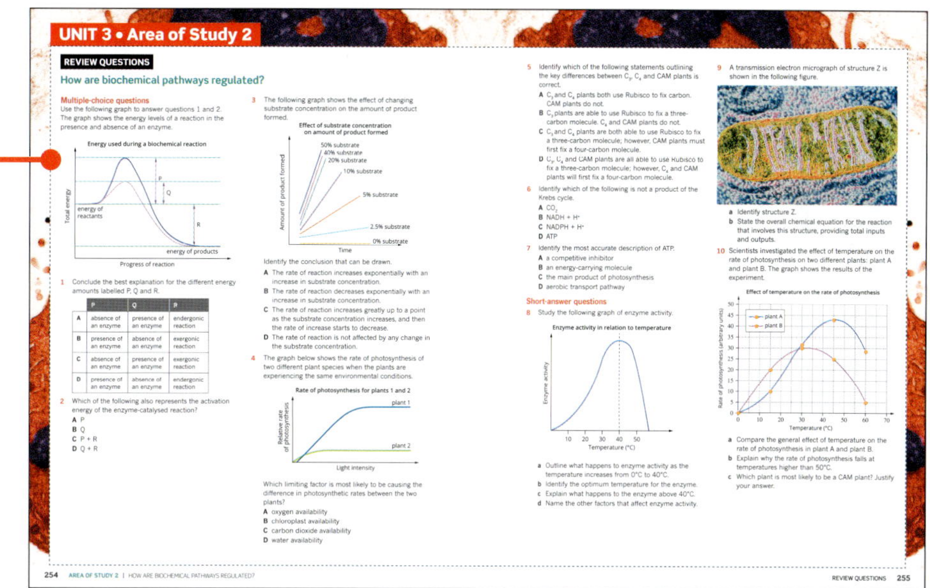

Section summary

Each section includes a summary to help you consolidate key points and concepts.

Section review

Each section concludes with questions that test your ability to recall, explain and apply key concepts.

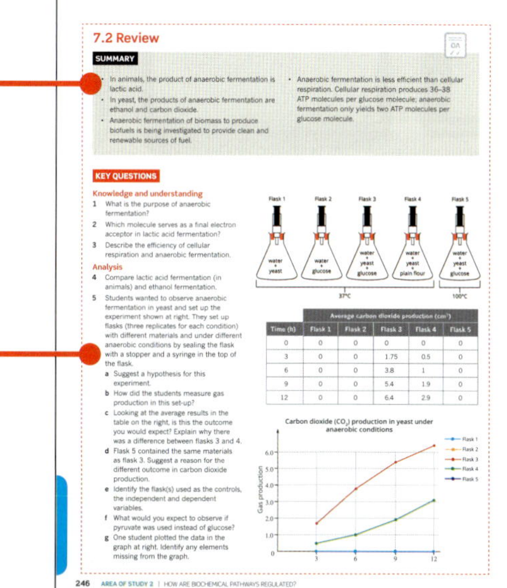

Chapter review

Each chapter concludes with a list of key terms and questions that test your understanding of the key knowledge covered in the chapter.

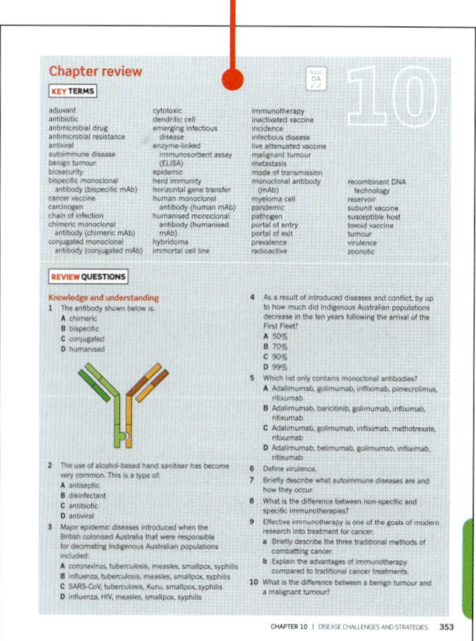

Icons

This icon is used to alert you to engage with auto-corrected questions through Pearson Places.

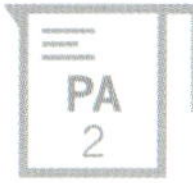

These icons indicate when it is the best time to engage with a worksheet (WS), a practical activity (PA) or exam questions (EQ) in *Heinemann Biology 2 Skills and Assessment* book.

Glossary

Key terms are shown in **bold** throughout, and listed at the end of each chapter. A comprehensive glossary at the end of the book defines all key terms.

Answers

Comprehensive answers for all section review, chapter review and Area of Study review questions are provided via *Heinemann Biology 2 eBook* 6th edition.

CHAPTER 01

Scientific investigation

Learning outcomes

The development of a set of key science skills is a core component of the study of VCE Biology and applies across Units 1 to 4 in all areas of study. Chapter 1 scaffolds the development of these skills. The opportunity to develop, use and demonstrate these skills in a variety of contexts is important ahead of undertaking investigations and when evaluating the research of others.

Although this chapter can be read as a whole, it is best to refer to it and use it when the need arises as you work through other chapters. For example, you may need a refresher on the process of the scientific method. It also contains useful checklists to assist when drawing scientific diagrams, graphing and completing aspects of your report. Similarly, when performing a practical investigation, refer to this chapter to make sure you collect data properly and that your data is of high quality.

Key science skills

Develop aims and questions, formulate hypotheses and make predictions

- identify, research and construct aims and questions for investigation **1.1, 1.2**
- identify independent, dependent and controlled variables in controlled experiments **1.1, 1.2**
- formulate hypotheses to focus investigation **1.1, 1.2**
- predict possible outcomes **1.1, 1.2**

Plan and conduct investigations

- determine appropriate investigation methodology: case study; classification and identification; controlled experiment; correlational study; fieldwork; literature review; modelling; product, process or system development; simulation **1.1, 1.2**
- design and conduct investigations; select and use methods appropriate to the investigation, including consideration of sampling technique and size, equipment and procedures, taking into account potential sources of error and uncertainty; determine the type and amount of qualitative and/or quantitative data to be generated or collated **1.1, 1.3**
- work independently and collaboratively as appropriate and within identified research constraints, adapting or extending processes as required and recording such modifications **1.1, 1.2**

Comply with safety and ethical guidelines

- demonstrate safe laboratory practices when planning and conducting investigations by using risk assessments that are informed by safety data sheets (SDS), and accounting for risks **1.2**
- apply relevant occupational health and safety guidelines while undertaking practical investigations **1.2**
- demonstrate ethical conduct when undertaking and reporting investigations **1.2**

Generate, collate and record data

- systematically generate and record primary data, and collate secondary data, appropriate to the investigation, including use of databases and reputable online data sources **1.3, 1.4**
- record and summarise both qualitative and quantitative data, including use of a logbook as an authentication of generated or collated data **1.4**
- organise and present data in useful and meaningful ways, including schematic diagrams, flow charts, tables, bar charts and line graphs **1.5, 1.6**
- plot graphs involving two variables that show linear and non-linear relationships **1.5, 1.6**

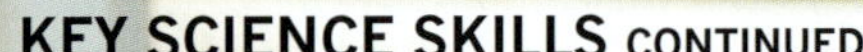

Analyse and evaluate data and investigation methods

- process quantitative data using appropriate mathematical relationships and units, including calculations of ratios, percentages, percentage change and mean **1.5**
- identify and analyse experimental data qualitatively, handling where appropriate concepts of: accuracy, precision, repeatability, reproducibility and validity of measurements; errors (random and systematic); and certainty in data, including effects of sample size in obtaining reliable data **1.4, 1.5**
- identify outliers, contradictory or provisional data **1.4, 1.5**
- repeat experiments to ensure findings are robust **1.4**
- evaluate investigation methods and possible sources of personal errors/mistakes or bias, and suggest improvements to increase accuracy and precision and to reduce the likelihood of errors **1.4, 1.6**

Construct evidence-based arguments and draw conclusions

- distinguish between opinion, anecdote and evidence, and scientific and non-scientific ideas **1.2**
- evaluate data to determine the degree to which the evidence supports the aim of the investigation, and make recommendations, as appropriate, for modifying or extending the investigation **1.4, 1.6**
- evaluate data to determine the degree to which the evidence supports or refutes the initial prediction or hypothesis **1.4, 1.6**
- use reasoning to construct scientific arguments, and to draw and justify conclusions consistent with the evidence and relevant to the question under investigation **1.6**
- identify, describe and explain the limitations of conclusions, including identification of further evidence required **1.6**
- discuss the implications of research findings and proposals **1.6**

Analyse, evaluate and communicate scientific ideas

- use appropriate biological terminology, representations and conventions, including standard abbreviations, graphing conventions and units of measurement **1.4, 1.5, 1.6**
- discuss relevant biological information, ideas, concepts, theories and models and the connections between them **1.1, 1.2, 1.6**
- analyse and explain how models and theories are used to organise and understand observed phenomena and concepts related to biology, identifying limitations of selected models/theories **1.1, 1.6**
- critically evaluate and interpret a range of scientific and media texts (including journal articles, mass media communications and opinions in the public domain), processes, claims and conclusions related to biology by considering the quality of available evidence **1.2, 1.4**
- analyse and evaluate bioethical issues using relevant approaches to bioethics and ethical concepts, including the influence of social, economic, legal and political factors relevant to the selected issue **1.2**
- use clear, coherent and concise expression to communicate to specific audiences for specific purposes in appropriate scientific genres, including scientific reports and posters **1.6**
- acknowledge sources of information and assistance and use standard scientific referencing conventions **1.6**

OA
✓✓

1.1 The scientific method

Biology is the study of living organisms. As scientists, biologists extend their understanding using the scientific method, which involves investigations that are carefully designed, conducted and reported (Figure 1.1.1). Well-designed research is based on a sound knowledge of what is already understood about a subject, as well as careful preparation and observation.

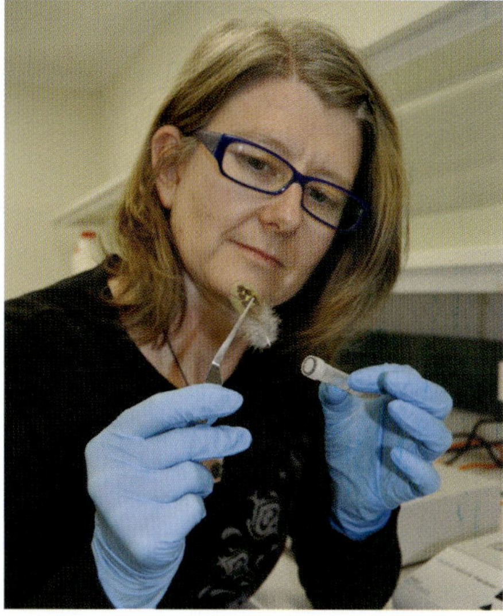

FIGURE 1.1.1 Biological research uses a variety of methodologies and methods. Analysis of DNA extracted from feathers by scientists at the Museum of Western Australia has confirmed that the night parrot (*Pezoporus occidentalis*) is not extinct, as previously thought.

OBSERVATION

Observation includes using all your senses and a wide variety of instruments and laboratory techniques to allow closer observation. Through careful inquiry and observation you can learn a lot about organisms, the ways they function, and their interactions with each other and the environment. For example, animals clearly function very differently from plants. Animals usually move around, take in nutrients and water, and often interact with each other in groups. Plants, however, are stationary, turn their leaves towards the light and grow. Many other distinguishing macroscopic structures and behaviours can be discerned from simple observation. Microscopic observation of cells reveals similarities and differences in the cellular structure of plant and animal cells, as well as the specialisations in the cells of a particular organism.

Practical investigations

The idea for a practical investigation of a complex problem arises from prior learning and observations that raise further questions. For example, indoor plants do not grow well in the long term without artificial lighting, which suggests light is required for photosynthesis in plants (Figure 1.1.2). This aspect of photosynthesis can be researched and the new knowledge applied to other applications, such as methods for growing plants in the laboratory for genetic selection and modification for crop improvement.

FIGURE 1.1.2 Laboratory methods such as plant tissue culture rely on careful observations and data collection about the requirements for growth of plants in natural conditions. Laboratory investigations then provide new information that can be applied to plants growing in the field.

Interpreting observations

How observations are interpreted depends on past experiences and knowledge, but to enquiring minds they will usually provoke further questions such as:

- How do organisms gain and expend energy?
- How do multicellular organisms develop specialised tissues?
- What are the molecular building blocks of cells?
- How do species change and evolve over time?
- How do cells communicate with each other?

Many of these questions cannot be answered by observation alone, but they can be answered through scientific investigations. Good scientists have acute powers of observation and enquiring minds, and they make the most of these chance opportunities, like Alexander Fleming did when he discovered penicillin.

● *You will now be able to answer key question 1.*

CASE STUDY

Observation and discovery

Scottish physician Alexander Fleming was growing cultures of *Staphylococcus* bacteria in his laboratory in the 1920s (Figure 1.1.3a). Some of the agar plates he was growing the bacteria on became contaminated with a fungus called *Penicillium notatum*. From his observation that the bacteria were unable to grow in the region around the contaminating fungus, Fleming inferred that the fungus was releasing a substance that killed the bacteria. Experiments followed that used extracts from the fungus. When a paper disc was soaked in this extract and applied to an agar plate culture of *Staphylococcus*, a clear zone appeared around the disc (Figure 1.1.3b). The bacteria could not grow in this area, demonstrating the antibacterial properties of this substance. Fleming named it penicillin after the type of fungus producing the chemical.

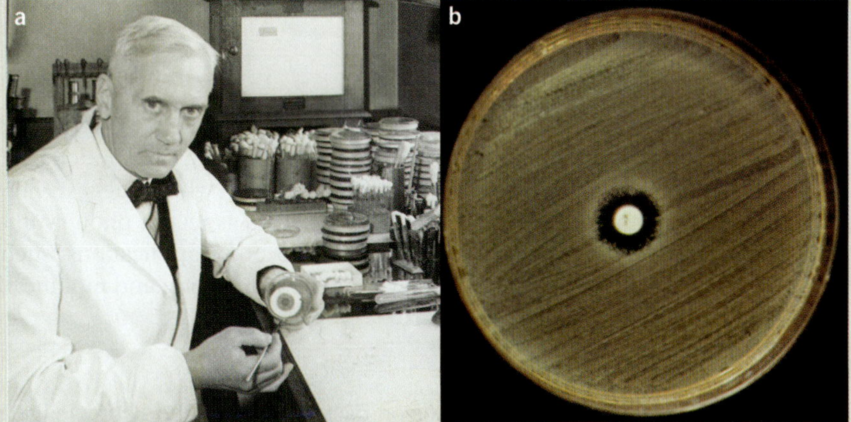

FIGURE 1.1.3 (a) Scottish biologist Alexander Fleming. (b) A culture of *Staphylococcus aureus* bacteria with a white disc containing penicillin placed at the centre. *S. aureus* has not been able to grow near the penicillin disc.

After Fleming made the initial key observation that led to the discovery of naturally occurring antibiotics, the Australian scientist Howard Florey (then working at Oxford, England) and his colleagues further developed the methods for extracting penicillin on a large scale, and showed it was effective against staphylococcal and pneumococcal infections. Following the success of penicillin, pharmaceutical companies searched for other naturally occurring antibiotics, many of which were found in fungi (Figure 1.1.4).

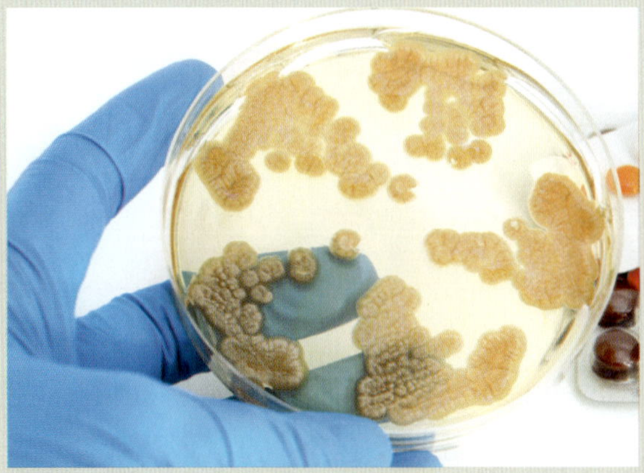

FIGURE 1.1.4 Agar plate with fungal colonies

THE SCIENTIFIC PROCESS

Scientists observe, study what is already known, and then ask questions. Using their knowledge and experience, scientists suggest possible explanations for the things they observe. A **hypothesis** is a prediction based on scientific reasoning that can be tested experimentally. This is the basis of the **scientific method** (Figure 1.1.5).

A hypothesis is a prediction based on scientific reasoning about what an investigator might expect to see in the results of their experiment.

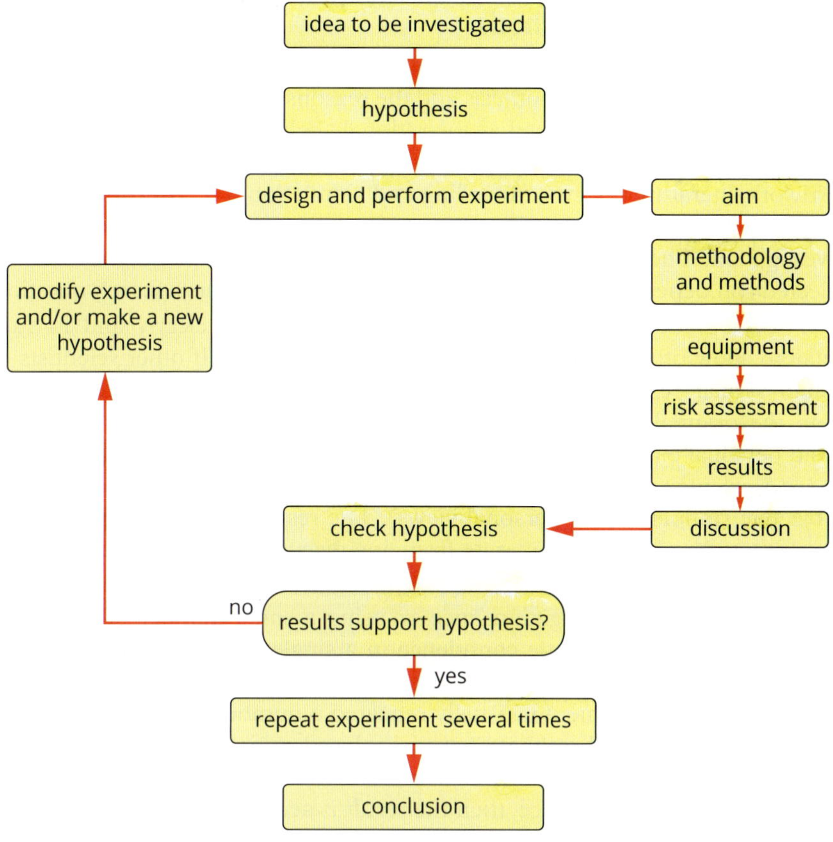

FIGURE 1.1.5 The scientific method

Carefully designed experiments are conducted to determine whether the predictions are accurate or not. If the results of an experiment do not fall within an acceptable range, the hypothesis is rejected. If the predictions are found to be accurate, the hypothesis is supported. If, after many different experiments, one hypothesis is supported by all the results obtained so far, then this explanation can be given the status of a **theory** or **principle**.

There is nothing mysterious about the scientific method. You might use the same process to find out how an unfamiliar machine works if you had no instructions. Careful observation is usually the first step.

● *You will now be able to answer key question 2.*

Research questions

In science, there is little value in asking questions that cannot be answered. A hypothesis must be testable, but your inability to test a particular hypothesis does not mean that the hypothesis cannot be supported.

Your ability to test a hypothesis may be limited by the resources and equipment you have available. If you ask a research question, form and test your hypothesis, and find your hypothesis is supported, that does not mean it is true in all circumstances. Likewise, if your hypothesis is not supported, that does not mean it is never true.

For example, you might hypothesise 'If hydrogen peroxide is a toxic by-product of cellular respiration that is broken down by catalase, then all eukaryotes will contain catalase'. However, there may be a eukaryote that lacks catalase, but testing every eukaryotic organism would be impossible, and just because a eukaryote without catalase hasn't been identified does not mean none exist.

● *You will now be able to answer key question 3.*

Methodology and methods

Scientific investigations must be able to be repeated by other scientists to be considered reliable.

The **methodology** is a brief description of the general approach taken to investigate the research question or hypothesis and the reasons why this approach is taken. The methodology can be described as the rationale behind your investigative methods. Examples of scientific investigation methodologies are controlled experiments, fieldwork, literature reviews, modelling and simulation. The **methods** (also known as procedures) are the specific steps that are taken to collect data during the investigation. The type of scientific investigation methodology and the methods selected will depend on the aim of the investigation and the research question.

The methodology and methods must be described clearly and in sufficient detail to allow other scientists to repeat the investigation. If other scientists cannot obtain similar results using the same methods and conditions, then the results from the original investigation are considered unreliable. It is also important to avoid personal bias that might affect the collection of data or the analysis of results. A good scientist works hard to be objective (free of personal bias) rather than subjective (influenced by personal views). The results of an investigation must be clearly stated and must be separate from any discussion of the conclusions that are drawn from the results.

Conducting an investigation once or using a small number of samples is not sufficient. You can have little confidence in a single result because you cannot be sure that the result was not due to some unusual circumstance that occurred at the time. The same experiment is usually repeated a number of times over a period of time and the combined results are then analysed statistically. If the statistics show that there is a low probability (usually less than 5%) that the results could have occurred as a result of chance, then the result is accepted as being significant.

● *You will now be able to answer key question 4.*

Experimental controls

The experimental conditions of the control group are identical to those of the experimental group, except that the independent variable is also kept constant.

The independent variable is the only variable that the experimenter changes in a controlled experiment. The dependent variable is measured to determine the effect of changing the independent variable.

It is difficult—sometimes impossible—to eliminate all **variables** that might affect the outcome of an investigation. In biology, time of day, temperature, amount of light, humidity and unidentified infections in organisms are examples of such variables. A way to eliminate the possibility that random factors affect results is to set up a second group within the experiment (called a **control group**) that is identical in every way to the first group (the **experimental group**) except for the single experimental variable that is being tested. This is a controlled experiment, because it allows you to examine one variable at a time. Controlled experiments are an important way of testing your hypothesis.

The variable that the experimenter is manipulating is the **independent variable**.

The **dependent variable** is what is measured when the independent variable changes. All of the other factors that could vary but must be kept the same in all experimental groups are called **controlled variables**.

● *You will now be able to answer key question 5.*

When investigating antibacterial activity of compounds extracted from fungi or other sources, the variables to consider include the source, purity and concentration of the extract, the composition and consistency of the agar plates, the type of bacteria tested, the amount of substance on the test disc, the thickness of the discs and the incubation temperature. The independent variable would be the extract being tested. The dependent variable would be the presence and size of the zone of inhibition around the disc. The other variables listed above all need to be controlled. In Section 1.4 you will learn about setting up an investigation with controls.

In a controlled experiment, controlled (fixed) variables are kept constant.

Forming conclusions

Conclusions are evidence-based statements that are developed from the analysis of results. When drawing conclusions from the results of an investigation, the quality of the data needs to be considered—the data should be accurate, reliable and valid. A conclusion is valid if it provides a response to the research question that the investigation set out to answer. Conclusions should summarise and explain the results of the investigation, and identify the extent to which the investigation addressed the research question or hypothesis.

Speculation involves going beyond the results to make suggestions about what might be occurring. Conclusions are necessary, but speculation is interesting and thought-provoking. Both concluding and speculating are worthwhile, but you must be careful to keep them separate. It is also the usual practice of scientists to accept the simplest hypothesis that accounts for all the evidence available.

The conclusion made by Fleming, that *Penicillium notatum* produced a substance that can kill bacteria, was evidence-based. It has been repeated many times and the principle has been generalised to the search for other antibiotics in a range of fungi and other organisms, including bacteria and plants.

LIMITATIONS OF THE SCIENTIFIC METHOD

The scientific method is not perfect; however, it remains the best way to understand our surroundings, and to constantly improve on that understanding. Even when the scientific method is strictly adhered to, there is still an element of chance in scientific discovery.

The scientific method can be applied only to hypotheses that can be tested, or to questions that can be answered. A hypothesis that is not testable can be neither supported nor disproved by the scientific method. Such hypotheses therefore remain as possible explanations. For example, Fleming's observation led to the hypothesis that certain fungi can produce chemicals that inhibit the growth of certain bacteria. This was testable for *Penicillium* and other fungi that can be grown on agar plates in the laboratory. If the hypothesis was broadened to 'All fungi produce antibiotics', this might not be testable, as testing it would depend on being able to grow all fungi and all potential bacterial targets in the laboratory.

A hypothesis can never be proven by a scientific study. It can only be supported under the conditions that have been tested.

It is also important to understand that although a hypothesis may be supported by experimental data, the same hypothesis may not be supported in all circumstances—it has only been found to be true under the conditions that have been tested.

The scientific method cannot be used to test morality or ethics. These judgements belong to the fields of philosophy, history, politics and law. Science can, however, provide valuable information that people can take into account when making these judgements. For example, science can be used to predict the environmental consequences of pollution and the medical consequences of chemical weapons, but it cannot itself make value or moral judgements about either.

DETERMINING APPROPRIATE INVESTIGATION METHODOLOGY

When it comes to beginning your scientific investigation, you will need to think about the best way to address your research question. For some investigations, setting up a controlled experiment may require using equipment that is not readily available to you in the school setting. This may mean you need to look at a computer simulation to model the outcomes of the investigation. Other approaches could include a literature review of other studies focused on a similar research question. The different approaches that you could use are outlined in Table 1.1.1.

TABLE 1.1.1 Scientific investigation methodologies

Type of methodology	Explanation	Example
case study	investigation of a real or hypothetical situation, such as an activity, event, problem or behaviour, often involving analysis of data within a real-world context	looking at the impact of an oil spill in one part of the world and using this analysis to prepare, hypothesise and plan for the impact of an oil spill of similar magnitude in another part of the world
classification and identification	arranging objects, events or organisms into manageable groups by identifying shared or similar features	using morphology (physical features of organisms) to group them into taxonomic groups based on shared characteristics
controlled experiment	experimental investigation that involves formulating a hypothesis and testing the effect of an independent variable on the dependent variable, while controlling all other variables in the experiment	investigating the impact of a change in temperature on the activity of an enzyme
correlational study	making observations and recording events and behaviours to investigate the relationship or association between variables	investigating the correlation between body mass index and the incidence of coronary heart disease
fieldwork	observing and interacting with particular environments to determine if a relationship exists between organisms or environmental factors and organisms; often involves observations and sampling of organisms and environments	chi-square test to investigate whether a relationship exists between two different species of marine molluscs in an intertidal zone
literature review	critical analysis of what has already been investigated and published, using secondary data from other people's investigations or from experimental research to explain events or propose new ideas or relationships	analysis of data looking at the impact of smoking on lung cancer in a variety of research papers to support, refute or develop new hypotheses
modelling	using models as representations of objects, systems or processes to aid understanding or make predictions	model of the connections between neurons in the human brain constructed from brain-scanning technology
product, process or system development	using scientific understanding and advances in technsology to design a new tool, method or process to meet the demands or needs of society	developing a new biodegradable packaging material
simulation	using mathematical models or simulations to test hypotheses, conduct virtual experiments or model the complexity of whole cells, systems, organs or organisms	computer simulation of immune cells attacking other cells

EXPERIMENTATION

Once you have a testable hypothesis, you are ready to conduct an experiment to test it. Every experiment has to be designed and planned carefully. You need to be sure that someone else can repeat your experiment exactly the way you did it and get similar results. In Section 1.2 you will learn how to formulate your hypothesis and design an experiment to test it.

● *You will now be able to answer key questions 6 and 7.*

MODELS

Scientific models are used to create and test theories and explain concepts. They may also be developed as prototypes for functional devices such as replacement organs. The introduction of computer technology, including two- and three-dimensional animations, has helped to create more detailed and realistic representations of biological processes. Different types of models can be used, but each model has limitations in the type of information it can provide.

Modelling concepts

Models are created to answer specific questions or demonstrate specific processes. How a model is designed will depend on its purpose. The two most familiar types of models are visual models and physical models, but mathematical models and computational models are also common and increasingly important in the biosciences. Models help to make sense of ideas by visualising:

- objects that are difficult to see because of their size (too big or too small) or position, such as ecosystems, organs such as the heart and pancreas, cells, molecules and atoms
- processes that cannot easily be seen directly, such as digestion, feedback loops, biochemical reactions, gene expression and protein folding
- abstract ideas, such as energy transfer and the particulate nature of matter
- complex processes, such as networks of biochemical reactions, genome organisation and regulation, evolution, and brain connectivity and function.

For example, models of all the connections between neurons in the human brain have been constructed from brain scanning technology. The models are used to predict and test signalling and communication between neurons (Figure 1.1.6).

A deeper understanding of concepts can be developed through models. However, you need to identify the benefits and limitations of using a particular model to represent a concept. Furthermore, the quality and validity of a model is limited by the depth and accuracy of the information used to construct the model.

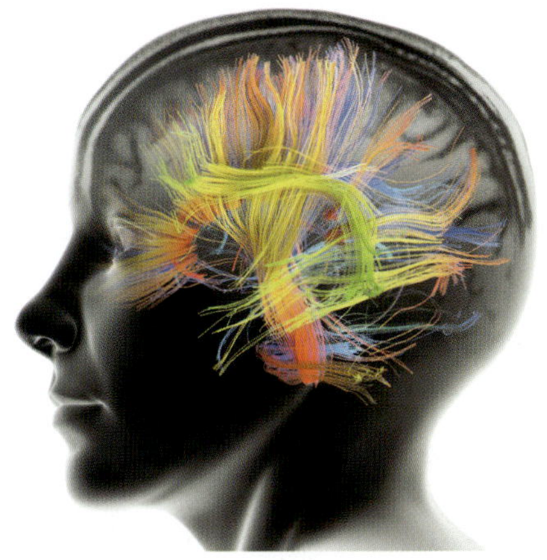

FIGURE 1.1.6 A model of the brain's wiring pattern explored in the Human Connectome Project

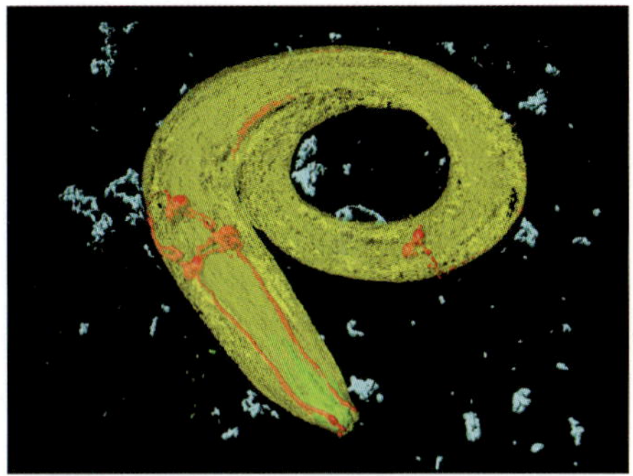

FIGURE 1.1.7 Model organism *Caenorhabditis elegans*, a nematode (roundworm). Confocal laser scanning micrograph of *C. elegans* with neurons stained green and the digestive tract stained red. *C. elegans* is a soil-dwelling nematode worm about 1 mm long and one of the most studied animals in biological and genetic research.

Model organisms

Biologists use live bacteria, animals and plants as model organisms for the investigation of cells and systems in situ and in vivo. It is possible to test in animals hypotheses that cannot be tested in humans for ethical reasons. Most of the advances in understanding animal and plant biology, genetics, pathology and medicine result from the use of model organisms. These organisms include the bacterium *Escherichia coli*, the nematode *Caenorhabditis elegans* (Figure 1.1.7), rats and mice, the plant *Arabidopsis thaliana* and the fruit fly *Drosophila melanogaster*.

Efforts are being made to reduce the number of animals used in research, and strict ethical guidelines must be followed in their use. Studies performed in vitro, and advances in computer simulation and 'virtual' cells and organisms that have made *in silico* studies possible, allow for a reduced reliance on live animals. But keep in mind that the value and validity of a virtual model or simulation is only as good as the data and information used to construct the model. This ultimately comes from living cells and organisms.

● *You will now be able to answer key questions 8–10.*

> Studies that are in situ are 'in position' or 'in place', such as when studying cells functioning within an intact organ, or molecules in their normal cellular location.

> Studies that are in vivo are 'within the living', such as when cells are studied in a living organism.

> Studies that are in vitro are 'in glass' or in a dish or test tube, such as when cells are removed from the organism and studied in a culture dish (it doesn't have to be glass).

> Studies that are *in silico* are 'in silicon', which refers to the silicon chips used in computers for computer simulations.

1.1 Review

OA ✓✓

SUMMARY

- Well-designed experiments are based on a sound knowledge of what is already understood or known and careful observation.
- The scientific method is an accepted procedure for conducting investigations.
- A hypothesis is a possible explanation for a set of observations that can be used to make predictions, which can then be tested experimentally.
- Controlled experiments allow us to examine one factor at a time; they are a commonly used methodology for testing hypotheses.
- Scientific investigations are undertaken to test hypotheses. The results of an investigation may support or reject a hypothesis, but cannot show it to be true in all circumstances.
- Science cannot be used to evaluate hypotheses that are not testable, nor can it make value or moral judgements.
- Models are useful tools that can be created and used to assist in a deeper understanding of concepts.

KEY QUESTIONS

Knowledge and understanding

1 The scientific method is a multistep process. Which two of the following are important parts of the method?

A observations made by eye and with instrumentation
B subjective decisions based on data collected
C careful manipulation of results to fit your ideas
D the use of prior knowledge to help objectively interpret new data

2 The following steps of the scientific method are out of order. Place a number (1–7) to the left of each point to indicate the correct sequence.

	form a hypothesis
	collect results
	plan experiment and equipment
	draw conclusions
	question whether results support hypothesis
	state the biological question to be investigated
	perform experiment

3 Scientists make observations and ask questions from which a testable hypothesis is formed.

a Define hypothesis.

b Three statements are given below. One is a theory, one is a hypothesis and one is an observation. Identify which is which.

i If skin cells are exposed to UV light, then cells will be damaged.
ii The skin burned when exposed to UV light.
iii Skin is formed from units called cells.

4 **a** What do 'objective' and 'subjective' mean?

b Why must experiments be conducted objectively?

5 Define 'independent', 'controlled' and 'dependent' variables.

6 **a** Explain what is meant by the term 'controlled experiment'.

b A student conducted an experiment to find out whether a bacterial species could use sucrose (cane sugar) as an energy source for growth. She already knew that these bacteria could use glucose for energy. Three components of the experiment are listed. Next to each one, indicate the type of variable described.

i presence or absence of sucrose
ii measurement of cell density after 24 h
iii incubation temperature, volume of culture, size of flask

Analysis

7 A scientist conducts a set of experiments, analyses the results and publishes them in a scientific journal. Other scientists in different laboratories repeat the experiment, but do not get the same results as the original scientist. Suggest several possible reasons that could explain this.

8 The following diagram illustrates a body function involving a feedback loop. Describe what the model shows, and discuss the benefits and limitations of this diagram as a visual model of biological feedback.

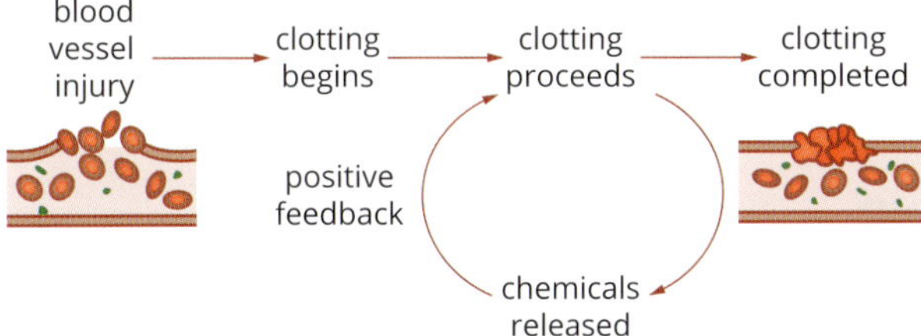

9 Below is a molecular model of the enzyme catalase, which converts hydrogen peroxide to water and oxygen. Suggest reasons why scientists construct molecular models in addition to simple diagrams or a written description of its molecular composition.

10 Discuss how computer modelling could assist in representing scientific concepts and advancing scientific knowledge.

1.2 Planning investigations

FIGURE 1.2.1 Scientists collecting grape vine samples for genetic research on the geographical origins of vines in the Mediterranean Basin

Practical investigations are those for which you gather the raw data yourself. These often take the form of experiments, activities, field trips or surveys (Figure 1.2.1). There are many elements to this type of practical investigation. A step-by-step approach will help you through the process and assist you in completing a solid and worthwhile investigation.

Taking the time to carefully plan and design an investigation before you begin will help you maintain a clear and concise focus throughout. Preparation is essential. In this section you will learn about some of the key steps to take when planning investigations:

- choosing a topic
- defining key terms
- sourcing information
- obtaining ethics approval
- ensuring occupational health and safety
- writing a protocol and schedule.

CHOOSING A TOPIC

Throughout this course you will conduct practical work (laboratory or fieldwork) on a range of topics. For Unit 4 Area of Study 3 you are required to design and conduct a scientific investigation related to cellular processes and/or how life changes and responds to challenges.

'Cellular processes' are any of the cell processes and biochemical pathways covered in Unit 3, such as polypeptide synthesis, gene expression, enzyme regulation, cellular respiration and photosynthesis.

'How life changes and responds to challenges' covers topics in Unit 4, such as immune responses to pathogens, changes in allele frequencies in populations, manipulation of gene pools through selective breeding, evolutionary changes in species over time and trends in hominin evolution.

When you choose a topic, consider the following:

- Choose a research question you find interesting.
- Start with a topic about which you already have some background information, or some clues about how to perform the experiments.
- Check that your school laboratory has the resources for you to perform the experiments or investigate the topic.
- Choose a topic that can provide clear, measurable data.

A number of topics that may be addressed in the course are suggested in Table 1.2.1. You will learn more about useful research techniques for topics like these in Section 1.3.

Before you start

The topics in Table 1.2.1 are only suggestions. Select your topic based on what resources are available to you. Before commencing your investigation, check that you have:

- the materials required to grow or culture an organism (e.g. plants, bacteria, yeast, protists or invertebrates)
- equipment such as microscopes, pH meters, spectrophotometers, centrifuges, and data loggers
- the materials needed to perform the experiments, such as biochemical test strips (for glucose, protein), enzymes and substrates, acids and bases.

Also ensure that you:

- can order any materials needed that are not on hand
- have a solid understanding of the theory behind your investigation
- are trained to use the required equipment
- have a detailed plan for the practical components of your investigation
- are able to access the school laboratory when you need to.

TABLE 1.2.1 Potential areas for investigation in Units 3 and 4

Laboratory experiments may be used to investigate factors affecting cellular and/or biochemical processes.	Possible topics for laboratory investigation include: • phagocytosis or endocytosis in living cells • photosynthesis in plants, algae or cyanobacteria • cellular respiration in plants, algae, bacteria, fungi or yeast • comparison of photosynthetic pigments by chromatography • enzyme activity in living cells or tissues, or purified enzymes • plant and animal responses to infection • antibiotics—mode of action and biological effectiveness • enzymes and electrophoresis for DNA manipulation and analysis • transformation of bacteria by plasmid transfer.
Fieldwork may be used for an investigation on cellular processes or for investigating biological change over time.	Possible topics for fieldwork investigation include: • collecting samples (e.g. for photosynthetic pigment extraction) • surveying populations for phenotypes and phenotypic change • assessing impacts of selective breeding programs • investigating the role of geological change on populations and evolutionary processes.
The use of data from online databases may facilitate, or be central to, your investigation.	Possible uses of online databases include: • bioinformatics using DNA sequence data • comparison of protein structures with digital 3D protein models • global statistics on disease incidence and vaccination • species distribution • characteristics and images of hominin and other fossils • geological sites of fossil evidence.

DEFINING KEY TERMS

When you begin a scientific investigation, you first have to develop and evaluate a research question, determine the associated variables, formulate a hypothesis and define the aims. It is important to understand that each of these can be refined as the planning of your investigation continues.

- The **research question** defines what is being investigated. For example: Is the rate of photosynthesis in plants dependent on temperature?
- The variables are the factors that change during your experiment. For example: Temperature is a variable for the photosynthesis example given earlier.
- The hypothesis is a statement that can be tested and is based on previous knowledge, evidence or observations, and that attempts to answer the research question. For example: If the temperature increases from 20°C to 40°C, then the rate of photosynthesis will increase.
- The **aim** is a statement describing in detail what will be investigated. For example: To investigate the effect of temperature on the rate of photosynthesis in plants at 20°C, 30°C and 40°C.

Determining your research question

Before conducting an experimental investigation you need a research question to address. Once you have come up with a topic or idea of interest, the first thing you need to do is conduct a search of the relevant literature; that is, you must read scientific reports and other articles on the topic to find out what is already known, and what is not known or not yet agreed upon. The literature also gives you important information for the introduction to your report and ideas for experimental methods. Use this information to generate questions.

When you have defined the research question, you are able to formulate a hypothesis, identify the measurable variables, proceed with designing your investigation and suggest a possible outcome of the experiment.

When writing a research question, it is advisable to include the independent and dependent variables. For example, what is the effect of [the independent variable] on [the dependent variable]?

Stop to evaluate the question before you progress; it may need further refinement or even further investigation before it is suitable as a basis for an achievable and worthwhile investigation. Consider the following checklist:

- ☐ relevance—Make sure your question is related to your chosen topic. For your practical investigation decide whether your question will relate to cellular structure or organisation, or to structural, physiological or behavioural adaptations of an organism to an environment.
- ☐ clarity and measurability—Make sure your question can be framed as a clear hypothesis. If the question cannot be stated as a specific hypothesis, then it is going to be very difficult to complete your research.
- ☐ time frame—Make sure your question can be answered within a reasonable period of time. Ensure your question isn't too broad.
- ☐ knowledge and skills—Make sure you have a level of knowledge and a level of laboratory skills that will allow you to explore the question. Keep the question simple and achievable.
- ☐ safety and ethics—Consider the safety and ethical issues associated with the question you will be investigating. If there are issues, determine if these need to be addressed.
- ☐ advice—Seek advice from your teacher on your question. Their input may prove very useful. Their experience may lead them to consider aspects of the question that you have not thought about.

Defining your variables

The factors that can change during your experiment or investigation are called the variables. An experiment or investigation determines the relationship between variables. There are three categories of variables:

- independent—a variable that is controlled by the researcher (the one that is selected and manipulated)
- dependent—a variable that may change in response to a change in the independent variable, and is measured or observed
- controlled variables—the variables that are kept constant during the investigation.

You should have only one independent variable. Otherwise you could not be sure which independent variable was responsible for changes in the dependent variable.

Making predictions and constructing a hypothesis

The hypothesis is a prediction of what you think will happen during a scientific investigation. It is a statement that can be tested (based on evidence and prior knowledge) to answer your research question. It defines a proposed relationship between two variables. To do this, you will need to identify the dependent and independent variables.

A good hypothesis is written in terms of the dependent and independent variables:

If *x* happens, then *y* will happen. The 'if ' part of the hypothesis refers to the independent variable—the variable you alter in the experiment. The 'then' part relates to the dependent variable—the variable you measure or observe.

For example:

If *yeast is grown in acidic conditions,* ***then*** *the rate of cellular respiration will decrease.*

A hypothesis does not need to include 'if ' and 'then' in its wording. For example, the previous hypothesis could also be stated the following way:

The rate of cellular respiration in yeast will decrease when yeast cells are grown in acidic conditions.

A good hypothesis can be tested to determine whether it is supported (verified), or not supported (falsified) by the investigation. To be testable, your hypothesis should include variables that are measurable.

When you evaluate your research question, consider the variables, and think about different potential hypotheses; it helps to create a table that outlines them. For example, Table 1.2.2 outlines a research question, the variables, and a potential hypothesis that relates to the effect of glucose on the rate of cellular respiration in yeast.

TABLE 1.2.2 Example of research question, variables and potential hypothesis

Research question	Will the rate of cellular respiration in yeast cells be faster if the cells are exposed to higher amounts of glucose?
Independent variable	glucose concentration
Dependent variable	rate of cellular respiration measured as change in CO_2 released over time
Controlled variables	yeast culture volume, temperature, light conditions
Potential hypothesis	The rate of cellular respiration in yeast will increase as glucose concentration increases.

Determining your aim

The aim is the key step required to test your hypothesis. The aim should directly relate to the variables in the hypothesis, describing how each will be studied or measured. The aim does not need to include the details of the method.

For example:

- Hypothesis: If algae are exposed to low light levels, then the rate of photosynthesis will decrease.
- Aim: To compare the rates of photosynthesis in algae at different distances from a light source.
- Variables: distance from light source, i.e. light intensity (independent) and rate of photosynthesis (dependent).

● *You will now be able to answer key questions 1–3, 7 and 8.*

SOURCING INFORMATION

When you are sourcing information during your search of the literature, researching experimental methods and investigating a broader issue, consider whether that information is from primary or secondary sources. You should also consider the advantages and disadvantages of using resources such as books or the internet.

Primary and secondary sources

Primary and secondary sources provide valuable information for research.

Primary sources of information are created by a person directly involved in an investigation. Examples of primary sources are results from research and peer-reviewed scientific articles. **Secondary sources** of information are a synthesis, review or interpretation of primary sources. Examples of secondary sources are textbooks, newspaper articles and websites.

Sometimes the same type of resource may be classified as both a primary and a secondary source, depending on when and by whom it was written. For example, a scientist's journal article on a clinical trial of treatments for teenage obesity is a primary source, while a general magazine article about teenage obesity written by a journalist and referring to the scientific study is a secondary source. Table 1.2.3 on page 16 compares primary and secondary sources.

Secondary sources of information may have a bias, so you need to determine if they are reliable sources of information. You will learn about assessing the accuracy, reliability and validity of data in Section 1.4.

TABLE 1.2.3 Summary of primary and secondary sources

	Primary sources	Secondary sources
Characteristics	• first-hand records of events or experiences • written at the time the event happened • original documents	• interpretations of primary sources • written by people who did not see or experience the event • use information from original documents but rework it
Examples	• results of experiments • scientific journal/magazine articles • reports of scientific discoveries • photographs, specimens, maps and artefacts • interviews with experts • websites (if they meet the criteria above)	• textbooks • biographies • newspaper articles • magazine articles • radio and television documentaries • websites that interpret the scientific work of others • podcasts

Using books and the internet

Peer-reviewed scientific journals are the best sources of information, but you are unlikely to have access to many of them, and much of the information is difficult to interpret if you are not an expert in the field.

As books, magazines and internet searches will be your most commonly used resources for information, you should be aware of their limitations (Table 1.2.4). Reputable science magazines you might find in your school library include *New Scientist*, *Cosmos*, *Scientific American* and *Double Helix* (Figure 1.2.2).

FIGURE 1.2.2 A reputable science magazine you might find in your school library

TABLE 1.2.4 Advantages and disadvantages of book and internet resources

	Book resources	Internet resources
Advantages	• written by experts • authoritative information • reviewed to ensure information is accurate • logical, organised layout • content is relevant to the topic • contain a table of contents and index to help find relevant information	• quick and easy to access • allow access to hard-to-find information • access to the whole world; millions of websites • up-to-date information • may be interactive and use animations to enhance understanding
Disadvantages	• may not have been published recently • usable by only one person at a time	• time-consuming looking for relevant information • a lot of 'junk' sites and biased material • search engines may not display the most useful sites • cannot always tell how up-to-date information is • difficult to tell if information is accurate • hard to tell who has responsibility for authorship • information may not be well ordered • less than 10% of sites are educational

Evaluating books and journals

Your textbook should be your first source of reliable information. Other information should be consistent with it. Articles published in journals and magazines often present findings of new research, which may or may not be confirmed later, so be careful not to treat such sources of information as established fact. Scientific journals are **peer-reviewed** (critically reviewed by other specialist scientists), which gives them more credibility than other sources.

Evaluating websites

Remember that anyone can publish anything on the internet, so it is important to evaluate the credibility, currency and content of online information. To evaluate online information, follow this checklist:

- ☐ credibility—Consider who the author is, their qualifications and expertise; check for their contact information and for a trusted abbreviation in the web address, such as .gov or .edu; websites using .com may have a bias towards selling a product (but this product could be a reputable science magazine or journal), and .org sites might have a bias towards one point of view (although these sites can be a good starting point for general information).
- ☐ currency—Check the date the information you are using was last revised.
- ☐ content—Consider whether the information presented is fact or opinion; check for properly referenced sources; compare information to other reputable sources, including books and science journals.

● *You will now be able to answer key questions 9 and 10.*

ETHICS

Ethics is a set of moral principles by which your actions can be judged as right or wrong. Every society or group of people has its own principles or rules of conduct. Scientists have to obtain approval from an ethics committee and follow ethical guidelines when conducting research that involves animals, including, and especially, humans.

Applying ethical principles means:

- considering the implications of investigations of organisms and the environment—you should aim to maximise benefit while minimising harm and risk
- recognising the intrinsic value of life and respecting the welfare, autonomy, beliefs, perceptions and customs of others
- using integrity when recording and reporting the outcomes of your investigation, and when using other people's data (such as in a literature review)
- forming a conclusion about science-related ethical issues using scientific knowledge and skills, while also considering the needs of all parties involved
- recognising the importance of social, economic and political values when forming conclusions using scientific understanding.

Ethics approval

If you work with animals as part of your studies, you may need to obtain a licence. Check with your school, teacher or laboratory technician. All animal use should follow the Victorian Government's guidelines for the care and use of animals in schools. These guidelines recommend that schools consider the '3Rs rule':

- Replace the use of animals with other methods where possible.
- Reduce the number of animals used.
- Refine techniques to reduce the impact on animals.

You should treat animals with respect and care. The welfare of the animal must be the most important factor to consider when determining the use of animals in experiments. If at any time the animal being used in your experiment is distressed or injured, the experiment must stop.

If human volunteers are needed, then the participants need to be fully briefed on the aim of, and methods involved in, the study, and they must give informed consent. They should also be given the opportunity to see the results of the study and their potential impact on the science community.

OCCUPATIONAL HEALTH AND SAFETY

While planning for an investigation in the laboratory or outside in the field, it is important for your safety and the safety of others that you consider the potential risks.

Everything we do has some risk involved. **Risk assessments** are performed to identify, assess and control hazards. A risk assessment should be performed for any situation, whether in the laboratory or out in the field, that could cause harm to people or animals. Always identify the risks and control them to keep everyone safe.

To identify risks, think about:

- the activity that you will be conducting
- where in the environment you will be working (e.g. in a laboratory, the school grounds, or a natural environment)
- how you will use equipment, chemicals, organisms or parts of organisms that you will be handling
- what clothing you should wear.

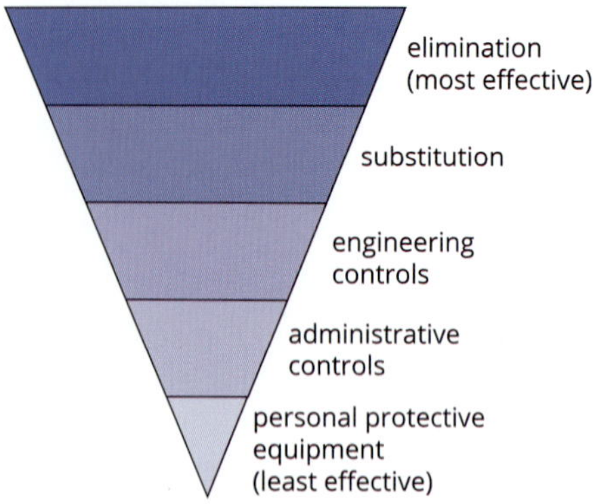

The hierarchy of risk control (Figure 1.2.3) is organised from the most effective risk management measures at the top of the pyramid to the least effective at the bottom of the pyramid.

elimination (most effective)
substitution
engineering controls
administrative controls
personal protective equipment (least effective)

FIGURE 1.2.3 The hierarchy of risk control in this pyramid is shown from top to bottom in order of decreasing effectiveness.

Take the following steps to manage risks when planning and conducting an investigation:

- Elimination—eliminate dangerous equipment, methods or substances.
- Substitution—find different equipment, methods or substances to use that will achieve the same result, but have less risk associated.
- Engineering—modify equipment to reduce risks. Ensure there is a barrier between the person and the hazard. Examples include physical barriers, such as guards in machines, or fume hoods when working with volatile substances.
- Administration—provide guidelines, special procedures, warning signs and safe behaviours for any participants.
- Personal protective equipment (PPE)—wear safety glasses, lab coats, gloves, respirators and any other necessary safety equipment where appropriate, and provide these to other participants. As PPE can be damaged, it is considered the least effective control measure, but it remains an essential safety feature after other control measures are in place (Figure 1.2.4).

FIGURE 1.2.4 A lab coat, gloves and safety glasses are essential items of personal protective equipment in the laboratory.

Science outdoors

Your investigation may involve outdoor fieldwork (Figure 1.2.5). All the potential risks, and ways to minimise them, must be considered when planning fieldwork. Ways to reduce risk include use of suitable protective clothing, knowledge of the terrain, having up-to-date maps, and checking predicted weather and fire risk.

FIGURE 1.2.5 Researchers excavating human fossils at a cave in Atapuerca, Spain. Hard hats, ropes, harnesses, strong clothing and footwear are essential during fossil research in the field.

Chemical safety

Some chemicals used in laboratories are harmful. When you are working with chemicals in the laboratory or at home, it is important to keep them away from your body. Laboratory chemicals can enter the body in three ways:

- ingestion—Chemicals that have been ingested (eaten) may be absorbed across cells lining the mouth or enter the stomach, and may then be absorbed into the bloodstream.
- inhalation—Chemicals that are breathed in (inhaled) can cross the thin cell layer of the alveoli in the lungs and enter the bloodstream.
- absorption—Some chemicals are able to pass through the skin in a process called absorption.

When working with any type of chemical you should:

- identify the chemical codes and be aware of the dangers they are warning about
- become familiar with the relevant safety data sheet, formerly known as the material safety data sheet
- use personal protective equipment
- wipe up any spills
- wash your hands thoroughly after use.

Chemical codes

The chemicals in laboratories, supermarkets, pharmacies and hardware shops have warning symbols on their labels. These are a chemical code indicating the nature of the contents (Table 1.2.5). From 1 January 2017, the Globally Harmonised System of Classification and Labelling of Chemicals (GHS) pictograms were introduced into Australia. This system is used for labelling containers and in safety data sheets. Some of the pictograms that you may see denote chemicals that are corrosive, pose a health hazard or are flammable. These chemical codes will need to be analysed and addressed when you are planning and conducting scientific investigations. You will perform a risk assessment in which these chemical codes will be provided, then, after analysing them, you may need to modify your experimental plan so that safety is improved.

TABLE 1.2.5 GHS pictograms used as warning symbols on chemical labels

GHS pictogram	Use	GHS pictogram	Use	GHS pictogram	Use
	flammable liquids, solids and gases; including self-heating and self-igniting substances		oxidising liquids, solids and gases, may cause or intensify fire		explosion, blast or projection hazard
	corrosive chemicals; may cause severe skin and eye damage and may be corrosive to metals		gases under pressure		fatal or toxic if swallowed, inhaled or in contact with skin
	low level toxicity; this includes respiratory, skin and eye irritation, skin sensitisers and chemicals harmful if swallowed, inhaled or in contact with skin		hazardous to aquatic life and the environment		chronic health hazards; this includes aspiratory and respiratory hazards, carcinogenicity, mutagenicity and reproductive toxicity

Safety data sheets

Every chemical substance used in a laboratory has a **safety data sheet (SDS)**. This contains important information about the possible hazards in using the substance and how it should be handled and stored. An SDS states:

- the name of the hazardous substance
- the chemical and generic names of certain ingredients
- the chemical and physical properties of the hazardous substance
- health hazard information
- how to store the chemical safely
- precautions for safe use and handling
- how to dispose of the chemical safely
- the name of the manufacturer or importer, including an Australian address and telephone number.

An SDS contains important safety and first aid information for teachers and technicians about each chemical you commonly use in the laboratory.

The SDS provides employers, workers and emergency crews with the necessary information to safely manage the risk of hazardous substance exposure.

● *You will now be able to answer key questions 4–6.*

1.2 Review

OA ✓✓

SUMMARY

- A research question is a statement that broadly defines what is being investigated.
- A hypothesis:
  - is a statement that can be tested and is based on previous knowledge and evidence or observations, and addresses the research question
  - often takes the form of a proposed relationship between two or more variables in a cause and effect relationship
  - must be testable; that is, able to be supported (verified) or not supported (falsified) by investigation.
- The three types of variables are:
  - independent—a variable that is controlled by the researcher (the one that is selected and manipulated)
  - dependent—a variable that may change in response to a change in the independent variable, and is measured or observed
  - controlled variables—variables that are kept constant during the investigation.
- An aim is a statement describing in detail what will be investigated.
- Primary sources of information are created by a person directly involved in an investigation. Secondary sources of information are a synthesis, review or interpretation of primary sources.
- Ethical and safety considerations must be of the highest priority at all times during a scientific investigation.

KEY QUESTIONS

Knowledge and understanding

1 Write each of the three inferences below as an 'if… then…' hypothesis that could be tested in an experiment.

  a Fungi produce compounds that kill bacteria.

  b An enzyme in stomach fluid causes meat to be digested.

  c Acidic conditions are not good for cellular respiration in eukaryotic cells.

2 Write a hypothesis for each of the following scenarios:

  a A student investigating algal blooms wondered whether *Chlorella*, a unicellular eukaryotic alga, carries out photosynthesis faster than *Anabaena*, a cyanobacterium.

  b A student on work placement at a dairy research station wondered whether dairy cattle with mastitis (a bacterial infection of the udder) would have more white blood cells such as neutrophils in their blood to fight the infection.

3 Which of these hypotheses is written in the correct format? Explain why the other options are not good hypotheses.

  A If light and temperature increase, then the rate of photosynthesis increases.

  B Cellular respiration is affected by temperature.

  C Light is related to the rate of photosynthesis.

  D Light triggers a response in motile algae to move towards the light source.

4 Complete the following table to list and describe the three ways a laboratory chemical could enter the body and how you might prevent this occurring.

Mode of entry	How the substance enters	Prevention

5 If you spilled a chemical substance with the following label on yourself, what would be the appropriate thing to do?

6 If you spilled a live bacterial culture on the lab bench, you would use paper towel to soak up the liquid.

  a Who would you consult about proper clean-up procedures?

  b What personal protective equipment (PPE) would you wear during this clean up?

  c What would you use to clean the bench top?

continued over page

1.2 Review *continued*

Analysis

7 Identify the independent, dependent and controlled variables that would be needed to investigate each of the following hypotheses.

a If a gene for salt tolerance is delivered to wheat plants, then the modified wheat will be able to grow in saline soils.

b Algae that live in shaded water have photosynthetic pigments that are different from those of algae that live in water exposed to full sunlight.

8 Consider the planning process for the following investigation of an enzyme that breaks down cellulose:

Aim:

Method:

1. Set up 6 equal-sized test tubes in a test tube rack.
2. Label 2 test tubes pH 5. Add 5 mL of pH 5 buffer to each tube.
3. Label 2 tubes pH 7. Add 5 mL of pH 7 buffer to each tube.
4. Label 2 tubes pH 9. Add 5 mL of pH 9 buffer to each tube.
5. Add 0.1 mL of cellulase enzyme solution to one tube at each pH.
6. Add 0.1 mL of the appropriate buffer (pH 5, 7 or 9) to the other tube at each pH.
7. Place the test tube rack, with all tubes, in a 37°C water bath.
8. Add 0.1 g shredded cellulose paper to each of the test tubes.
9. Incubate for 24 h.
10. Take 1 mL of solution from each tube and test for presence of glucose.

Experimental design:

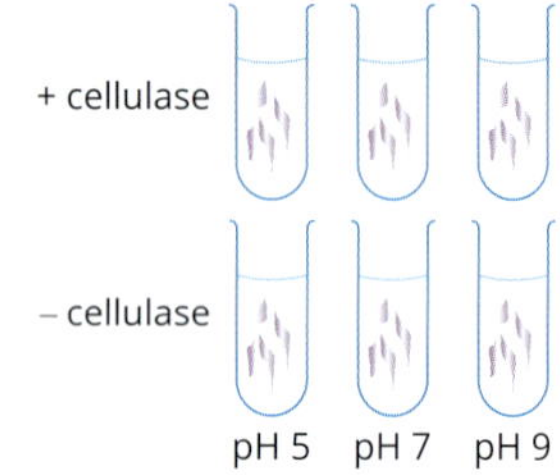

a Before conducting this experiment, what information would be researched as background for the introduction of the practical report?

b What information would you need to find out to conduct the experiment effectively?

c Identify the independent variable for the experiment.

d Identify the dependent variable for the experiment.

e List the controlled variables stated in the method.

f Write an aim for this experiment.

g Why was it important to use the set of test tubes without cellulase?

h Suggest improvements to the design of this experiment.

9 Decide whether each of the following is a primary or a secondary source of information.

a a newspaper article about genetically edited human embryos

b an experiment to investigate molecular changes within cells treated with hormones

c an interview with a fisheries molecular scientist about using DNA analysis for tracking tiger sharks

d a website with information about genetic engineering

10 You are learning about genetically inherited diseases and are searching for facts about cystic fibrosis. From the list below, which would be the best resource to use? Give reasons for your choice.

A the book *Cystic Fibrosis*, published in 1997

B the article 'Living with cystic fibrosis' published in the *Daily Mail* on 23 February 2008

C the Cystic Fibrosis Australia website accessed on 30 August 2020

1.3 Techniques used in scientific investigations

In this section you will learn about designing and selecting methods to use in scientific investigations. You will be introduced to different techniques, and understand how selecting appropriate equipment and methods will allow you to obtain accurate and precise measurements. The choice of techniques, sample size and data collection will also be discussed.

MICROSCOPY

Your practical investigation may involve the study of live cells or prepared slides using microscopic methods. You may need to include cell size, number and cellular behaviour as part of your experimental evidence. You probably have access to light microscopes with magnifications up to 400× and possibly 1000× (oil immersion).

Field of view and size of specimens

Biological drawings should include a scale. Calculating the field of view under the microscope is required for estimating the size of specimens viewed. To calculate the field of view you use a minigrid. This is a 1 mm × 1 mm grid with a smaller microgrid of 100 μm × 100 μm in the centre (used with the 40× objective) (Figure 1.3.1).

a

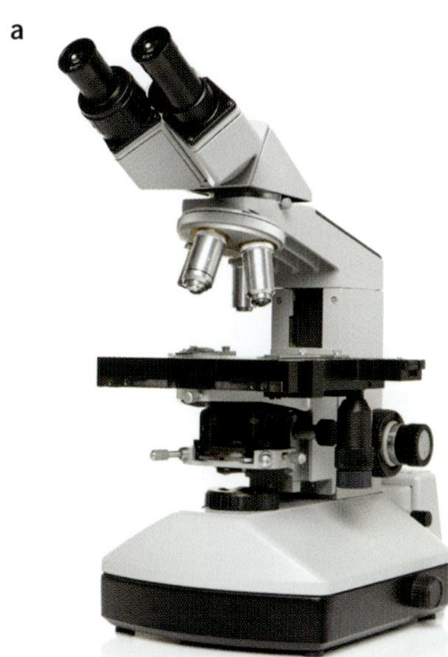

b

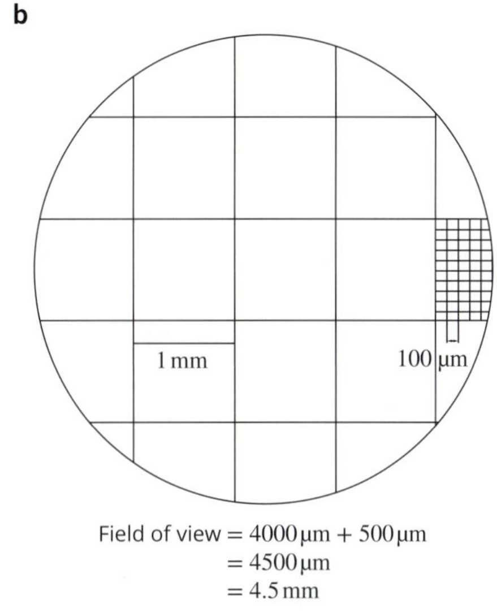

Field of view = 4000 μm + 500 μm
= 4500 μm
= 4.5 mm

FIGURE 1.3.1 (a) Light microscopes are used extensively in biology. (b) Using a minigrid allows you to measure the field of view and calculate cell size. This is a view of a minigrid at 40× magnification. Each large square is 1 mm^2, so the field of view is 4.5 mm (or 4500 μm).

Once you have calculated your field of view for each lens, you can estimate the size of the cells. For example, you may be studying the processes of phagocytosis and lysosome action in *Amoeba* or *Paramecium* (Figure 1.3.2). If at 400× magnification you estimate that one cell occupies half the diameter of the field of view (or two cells span the field of view) and the field of view is 450 μm, then the estimated size of each cell is 450 μm ÷ 2 cells = 225 μm.

Typical magnifications and fields of view in a school light microscope are listed. Microscopes usually have 10× eyepieces. The total magnification is the product of the eyepiece (ocular lens) and the objective lens.

Objective lens	Total magnification	Field of view
4×	40×	4.5 mm
10×	100×	1.5 mm
40×	400×	450 μm
100×	1000×	150 μm

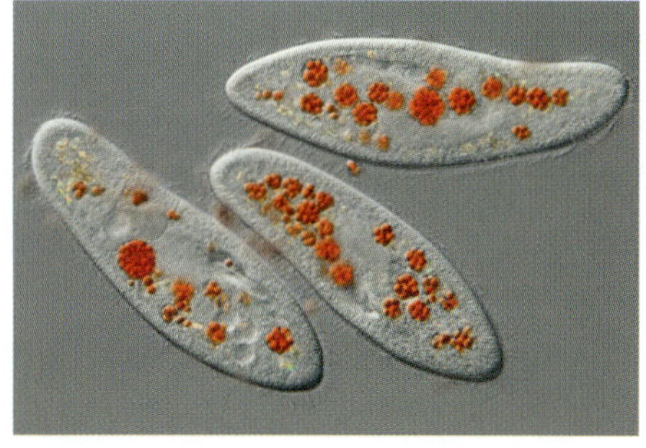

FIGURE 1.3.2 *Paramecium caudatum* viewed through a light microscope. Yeast cells (stained red) that have been engulfed by the *Paramecium* can be seen in the vacuoles.

If using microscopy, it is important that a large sample size is used. For example, in an investigation into the effect of an acidic environment on bacterial cell growth, a sample of a culture could be taken and placed on a slide to be stained and viewed using oil immersion microscopy. If only one slide was used, it would be recommended to look at multiple fields of view. The number of bacteria within each field of view could then be calculated and an average number of cells on the slide determined. As the sample size increases, the results will more accurately represent the overall population and the effect of errors and uncertainty in the method will be reduced. In this example, more slides could be prepared from the same bacterial cultures, and multiple fields of view from each slide calculated to obtain an even larger sample size.

● *You will now be able to answer key question 1.*

CELL CULTURE

Cell culture is a core technique in biological sciences; unfortunately, animal cell culture is restricted to laboratories that have specialised equipment and training, as well as ethics and safety approvals. However, in your school lab you are able to grow cultures of eukaryotic cells including unicellular algae (e.g. *Chlorella*), protists (e.g. *Paramecium caudatum* and *Amoeba proteus*) and yeast (e.g. *Saccharomyces cerevisiae*, baker's yeast) (Table 1.3.1). Keep in mind that cells take time to grow, so plan early. You can also grow cultures of bacteria (low risk category 1) such as *Escherichia coli*, *Staphylococcus epidermidis* and *Bacillus subtilis* on agar plates or in broth cultures. Live cell cultures can be used to investigate factors affecting cellular processes that may be reflected in cell growth rates, cell responses and other cellular processes.

TABLE 1.3.1 Growing cells for biology investigations

	Bacteria and yeast are cultured in appropriate liquid nutrient broth or nutrient agar plates.
	Algae and protists can be grown in suitable protist medium in sterile glassware. Algae are grown in good light conditions. Protists prefer the dark.
	Plant tissue culture. Small segments of stem or leaf are surface sterilised to remove contaminants. Explants (any sample taken from the organism, like a cutting) are cultured on plant nutrient agar over days or weeks.

CASE STUDY

Growing body parts

The study of cell biology makes extensive use of cells grown in culture (Figure 1.3.3). Cells are removed from an animal or plant and grown in vitro, in a dish bathed in nutrient medium under sterile conditions. Cells are treated for investigations of cellular processes and analysed by methods such as immunofluorescence microscopy, and analysis of biochemical pathways and gene expression.

FIGURE 1.3.3 Cells attach to the surface of the culture dish. A medium containing nutrients such as glucose, amino acids, vitamins and proteins covers the cells. The red colour of the medium is a pH indicator.

Cell culture is essential for stem cell research and exploration of stem cells as a therapeutic tool. Stem cells differentiate, so they can be identified with fluorescent tags and sorted by a technique called flow cytometry (Figure 1.3.4).

FIGURE 1.3.4 A scientist analyses cells with a flow cytometer. Cells are tagged with a fluorescent marker, sorted in the flow cytometer and visualised on a computer screen.

Replacement of cells and tissues lost through disease or injury is a challenge for medical research. Advances in cell culture methods, including stem cell technology, provide an opportunity for meeting this challenge. Tissue engineering combines cell culture and biopolymer scaffolds for the growth of specialised cells into the shape of body tissues or organs. Adult cells with the ability to replicate can be cultured directly in vitro; for example, skin cells are grown as sheets to replace skin damaged by burns. Alternatively, stem cells are cultured and differentiated into the cell types needed to reconstruct tissues. Skin, cartilage, heart valves and corneas are examples of tissues grown in the lab (Figure 1.3.5).

FIGURE 1.3.5 Corneal tissue grown in the laboratory. It was cultured from human epithelial cells that line the cornea of the eye.

Growing complex organs such as a urinary bladder has been done in the laboratory; however, achieving a functional organ in the body has so far proven challenging due to the difficulty of connecting all the blood vessels and nerves needed for a functional organ.

INVESTIGATING CELLULAR PROCESSES

Cellular respiration and photosynthesis can be detected in several ways, some of which provide qualitative results, and others which provide quantitative data. A few examples of materials you might use in a laboratory to conduct an investigation are described here.

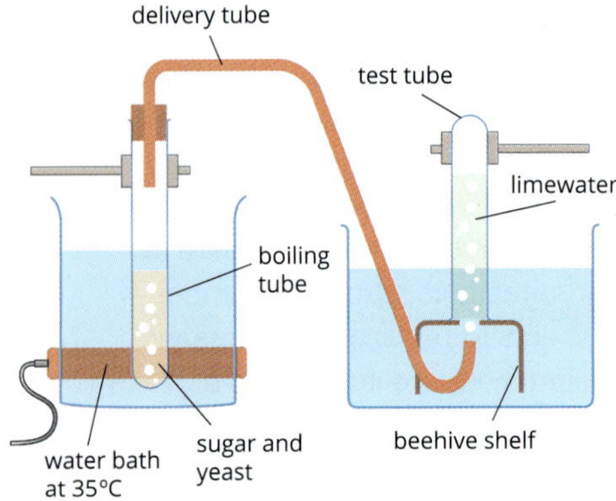

FIGURE 1.3.6 Yeast carbon dioxide test: A water bath being used to heat a stoppered test tube containing a yeast and sucrose solution (left) is connected by a delivery tube to a test tube of limewater (right). Cloudiness in limewater is a positive test for carbon dioxide, indicating fermentation in the yeast.

Cellular respiration

Cellular respiration may be studied in plant seedlings, yeast cultures, or insect populations. Carbon dioxide (CO_2) is a product of cellular respiration. Carbon dioxide can be detected directly using a data logger with a carbon dioxide sensor, or indirectly by mixing the air from a growth chamber with calcium carbonate solution, commonly known as limewater (Figure 1.3.6). Carbon dioxide dissolved in water forms carbonic acid, causing an acidic pH change, so cellular respiration in water plants, algae and yeast cultures can be detected with a pH indicator, pH test strips or a pH meter. Yeast in broth culture or immobilised in alginate balls may be investigated for factors affecting cellular respiration, such as temperature, nutrient concentration and inhibitors.

Photosynthesis

Photosynthesis can be studied in plants, growing seedlings, algae and cyanobacteria. In water, photosynthesis of plants and algae can be measured by oxygen production using a photosynthometer, which is a syringe connected by tubing to pond water surrounding a water plant, such as *Elodea*. Oxygen is collected and measured in the syringe (Figure 1.3.7a). Another approach is to measure the change in pH of the water as carbon dioxide is removed for photosynthesis; for example, by using hydrogen carbonate pH indicator (Figure 1.3.7b). This method of measuring photosynthesis is another indirect measurement. Carbon dioxide dissolves in water to form carbonic acid. This lowers the pH, which is why a pH indicator can be used to indirectly measure photosynthetic rate.

Photosynthesis can also be investigated in small leaf discs; the discs trap oxygen gas as they photosynthesise, become buoyant and float. Conditions that may affect the rate of photosynthesis, such as light intensity, temperature and chlorophyll concentration, can be investigated using these methods.

> Direct measurements involve measuring the exact product of a reaction, whereas indirect measurements involve measuring the change in an indicator (such as pH, colour or turbidity) that would be affected by the products of the reaction.

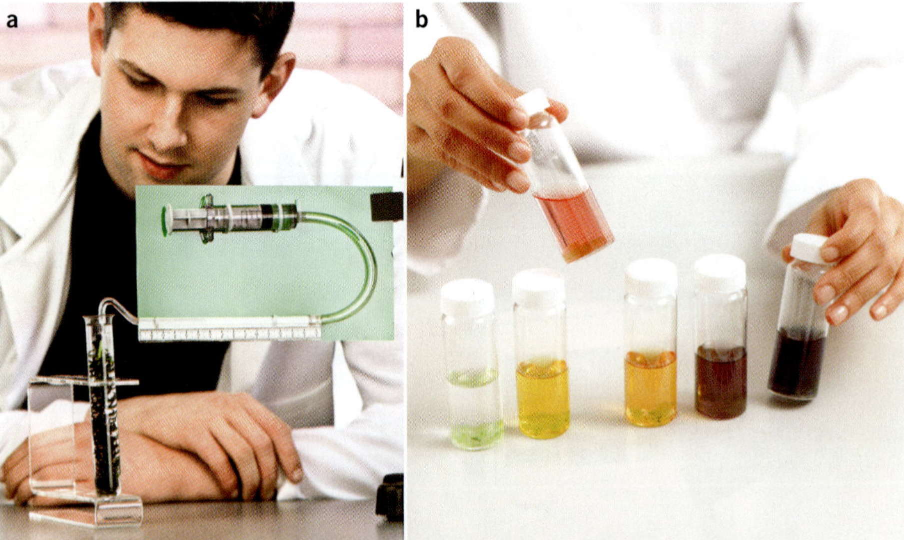

FIGURE 1.3.7 Students investigating photosynthesis by different methods: (a) Measuring oxygen produced by pond weed (in the test tube). As the pond weed photosynthesises it produces oxygen, which is collected and has its volume measured in the photosynthometer (syringe connected by tubing to pond water). (b) Measuring pH change with hydrogen carbonate indicator. Algae are immobilised in small alginate balls in each tube. pH change reflects carbon dioxide intake for photosynthesis.

You will now be able to answer key questions 2 and 6.

TOOLS TO SUPPORT YOUR PRACTICAL INVESTIGATIONS

A variety of tools can be used when conducting scientific investigations. The choice of equipment is important to ensure that your measurements are accurate (to minimise error), and that your results are reproducible and reliable (minimising uncertainty). Equipment that might be of use when conducting investigations is outlined in Table 1.3.2.

TABLE 1.3.2 Tools that may be available for practical investigations

Simple indicator of pH	Measuring pH or temperature	Measuring solutes
Tool: A dipstick test for the full pH range. A strip with pH-sensitive coloured pads is dipped into a solution then read against a reference colour chart after a defined time. **Purpose:** To measure the pH of a solution.	**Tool:** Electronic meters and probes. **Purpose:** To measure pH or temperature.	**Tool:** Strip tests for measuring glucose, protein and other solutes: • Multistix tests for several substances • Uriscan strips test glucose and protein • Glucostix tests glucose only **Purpose:** Usually designed to test urine. Coloured pads on the strip are dipped into urine or other solutions; colour develops and is read against a reference chart. Detection is often based on an enzyme reaction within the pad.
	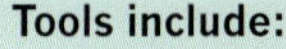	
Data loggers for a range of measurements	**Biochemical/chemical tests to detect molecules**	**Measuring absorbance, optical density or turbidity**
Tool: Common types of probes and capabilities in data have the ability to measure: • pH • temperature • oxygen concentration • carbon dioxide concentration • absorption colorimeter • concentration of various compounds. **Purpose:** To enable data collection over significant time periods.	**Tools include:** **a** biuret reagent* for detecting protein (colour change from blue to purple) **b** Benedict's reagent* for detecting reducing sugars such as glucose, maltose, fructose; not sucrose (colour change from blue to orange/red) **c** iodine–potassium iodide (IKI)* reagent for detecting starch (colour change from yellow/orange to deep blue). **Purpose:** To detect different biochemical reactions.	**Tool:** Colorimeter or spectrophotometer. **Purpose:** To quantify colour reactions, or turbidity for monitoring cell growth.
	a b c * Some tests are qualitative; quantitative or 'semi-quantitative' results may be achieved if combined with standards and absorbance readings.	

● *You will now be able to answer key questions 3 and 4.*

INVESTIGATING ENZYMATIC REACTIONS

Enzymes are biological catalysts that regulate biochemical processes in cells. Enzymes speed up the conversion of substrates to products. They are not consumed in the reaction. The general equation for enzymatic reactions is:

$$\text{substrate(s)} \xrightarrow{\text{enzyme}} \text{product(s)}$$

To measure enzymatic reactions, you can either measure the decrease in the amount of substrate or the increase in the amount of product. You will learn more about enzymes in Chapter 5.

Enzymes that could be used for your practical investigation may be present in biological material or purchased from a commercial supplier. To study the factors regulating an enzyme reaction you need an enzyme, its substrate, and a method to measure the change in amount of the substrate or product or both. Table 1.3.3 outlines some examples.

TABLE 1.3.3 Useful enzymes for practical investigations

	Sources	Examples
The enzyme	present in fresh cells and tissues	• catalase in some bacteria, liver or potato • amylase in seeds
	present in commercial products	• proteases in meat tenderiser • lipases and proteases in washing powders • amylases in some wallpaper strippers
	purchased in purified form from a biochemicals supplier	• trypsin and pepsin • amylase • catalase • cellulase • restriction enzymes
The substrate	present in biological tissues	• protein in egg white or meat • starch in grain-based foods • cellulose in plant material
	purchased in purified form from a biochemicals supplier	• starch • albumin • hydrogen peroxide • DNA from plasmid or bacteriophage

FIGURE 1.3.8 (a) Some bacteria produce catalase. A bacterial colony is placed on a slide with a drop of hydrogen peroxide. If catalase is produced by the bacteria, oxygen is released and seen as bubbles. (b) Measuring the volume of oxygen released from a yeast culture after catalase enzyme is added. Catalase reacts with hydrogen peroxide, a by-product of cellular reactions. (c) Investigating the effect of pH on digestive proteases acting on muscle tissue.

Some reactions can be detected by visible changes in the mass of the starting material, a change in physical appearance or the production of bubbles from a gaseous product. Other reactions require a biochemical detection test that gives an obvious colour reaction, or a spectrophotometer to detect a colourless product. Colour reactions without measurement are examples of qualitative results. If you have equipment to measure absorbance of a colour reaction, such as a colorimeter, then it may be possible obtain quantitative results for more accurate and reliable data.

Examples of enzyme experiments are shown in Figure 1.3.8.

Measuring absorbance for quantifying reactions

If your school has a colorimeter or spectrophotometer, then you may be able to get quantitative results when conducting experiments that use colour-based reactions, such as the detection of protein or starch. You can also use this instrument to measure the turbidity and optical density of bacterial cultures or yeast broths, providing a quantitative measure of their growth rates and an ability to account for sources of error and uncertainty.

A sample is placed in a special tube called a cuvette, which is placed in the instrument. Light of a particular wavelength is shone through the sample, which absorbs some of the light (Figure 1.3.9). The appropriate wavelength of light to select is the one that is maximally absorbed by the sample, and this differs for each substance measured. For example, blue solutions absorb light around 600 nm, and red solutions absorb light around 490 nm (wavelength is measured in nanometres, nm, and represented by the symbol λ). The meter reads the amount of light absorbed by the sample. A sample with a high concentration of the substance will absorb more light and therefore give a higher absorbance reading.

- *You will now be able to answer key question 5.*

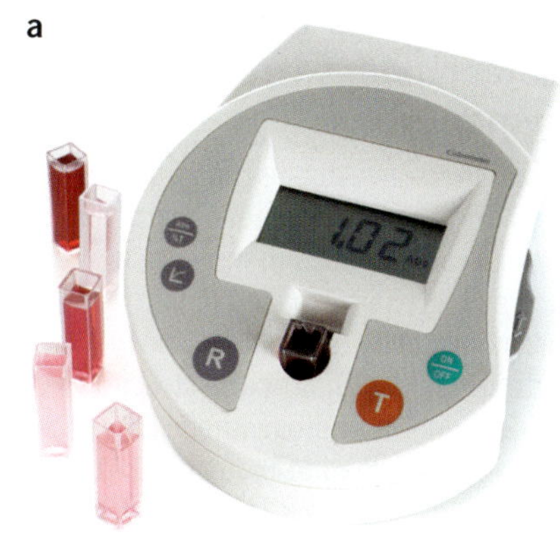

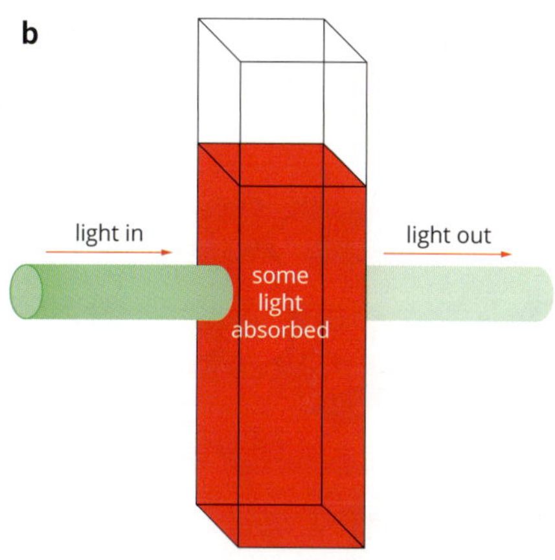

FIGURE 1.3.9 (a) A colorimeter or spectrophotometer reads absorbance of light. A sample is placed in a cuvette and placed in the instrument. (b) Light of a particular wavelength is shone through the sample. The meter reads the amount of light absorbed by the sample. A sample with higher concentration gives a higher absorbance reading.

CHROMATOGRAPHY

Chromatography methods available in your school laboratory may include paper or thin layer chromatography (Figure 1.3.10). Photosynthetic pigments vary in different organisms, such as different plants, algae and cyanobacteria. Different photosynthetic pigments have different properties of light absorbance, so are relevant to the rates of photosynthesis in different conditions.

Amino acids, the building block of proteins, can also be investigated by chromatography with the detection agent ninhydrin, which must be used safely in a fume hood.

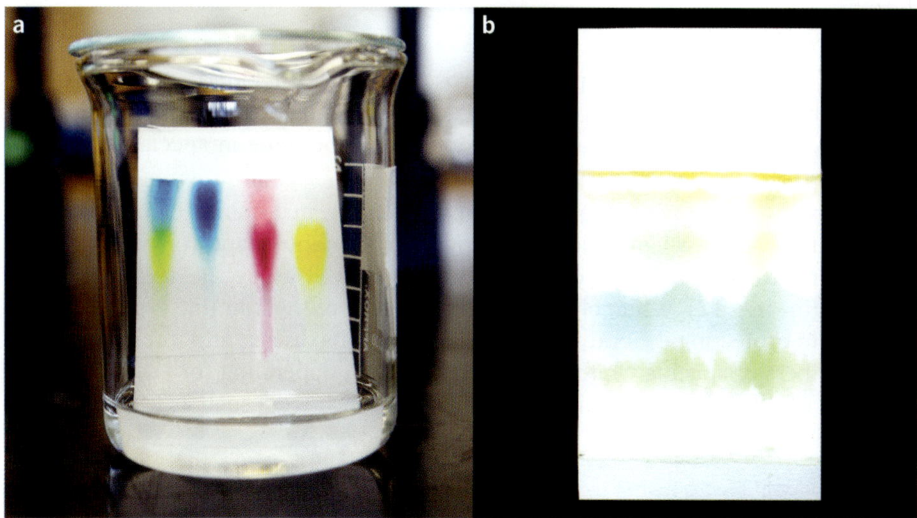

FIGURE 1.3.10 (a) Thin-layer chromatography (TLC) plate in a beaker, showing separated components (colours). TLC is performed on a sheet of glass, plastic or foil coated in a thin layer of adsorbent material. (b) An example of plant pigment molecules separated by paper chromatography. The sample is applied to the plate or paper and a solvent is drawn up the plate or paper via capillary action. Different components move up at different rates, causing them to separate.

BIOFILE

Biotechnology from plant pigments

Researchers investigate the structure of the different photosynthetic pigments in plants and algae with instruments such as high performance liquid chromatography (see figure below). Better understanding of the structure and function of photosynthetic pigments may lead to biotechnology applications such as silicon-based artificial photosynthesis systems for CO_2 capture, and genetic enhancement of photosynthetic organisms used as food sources and for pharmaceutical production.

Leaf pigment chromatography. A plant physiology researcher extracts a sample of pigments from leaf tissue (green liquid, front right) for analysis in the high-performance liquid chromatography machine in the background. Photographed at the ARS (Agricultural Research Service) Natural Products Utilization Research Unit in Oxford, Mississippi, USA, which conducts research for the US Department of Agriculture.

ELECTROPHORESIS

Gel electrophoresis is a technique for separating proteins and DNA according to their size. Electrophoresis uses specialised equipment, which may not be available at your school. For DNA analysis, a gel with DNA samples loaded in small wells is placed in an electrophoresis apparatus (Figure 1.3.11a). An electric current is applied, causing DNA fragments of different sizes to separate (Figure 1.3.11b). This method is used to investigate how different restriction enzymes cut DNA and to analyse plasmids for gene cloning. It is also used to view the products of a polymerase chain reaction (PCR), which is used for many applications, such as DNA barcoding.

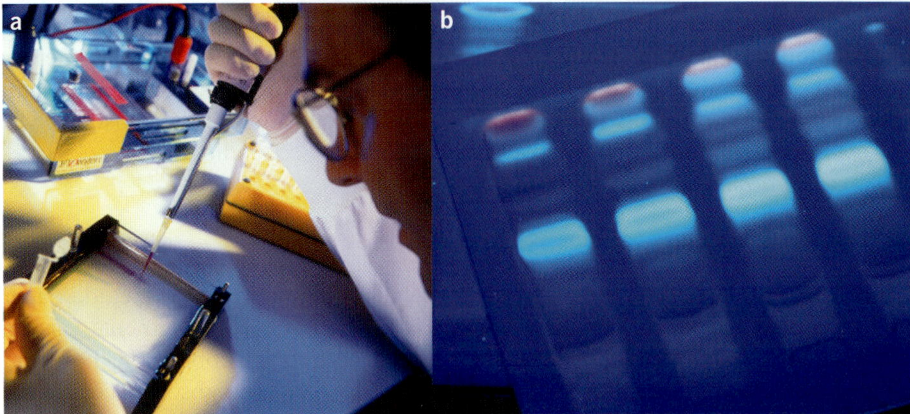

FIGURE 1.3.11 (a) A scientist loads DNA samples into the wells of an agarose gel for electrophoresis. Upper portion of photograph shows an operating electrophoresis chamber, with electrodes connected to red and black power cables. (b) DNA bands form when an electric current is applied because smaller fragments of DNA travel faster and therefore further than larger fragments.

IMMUNOLOGICAL INVESTIGATIONS

You may have access to prepared microscope slides of blood smears that can be used to investigate human responses to invading pathogens. (It is not safe to prepare your own blood smears. This should only be done in specialised labs or clinical settings by trained personnel.)

Different types of white blood cells (or leukocytes) in a stained blood smear can be identified under a light microscope by their size and shape (or morphology), particularly the morphology of their nucleus, as well as by the colour the cells stain (Figure 1.3.12). White blood cells are the cells of the immune system that are involved in protecting the body against both infectious disease and foreign invaders. The numbers of specific white blood cells can vary, depending for example on the presence of different types of infection, or in blood cancers such as leukaemia and myeloma.

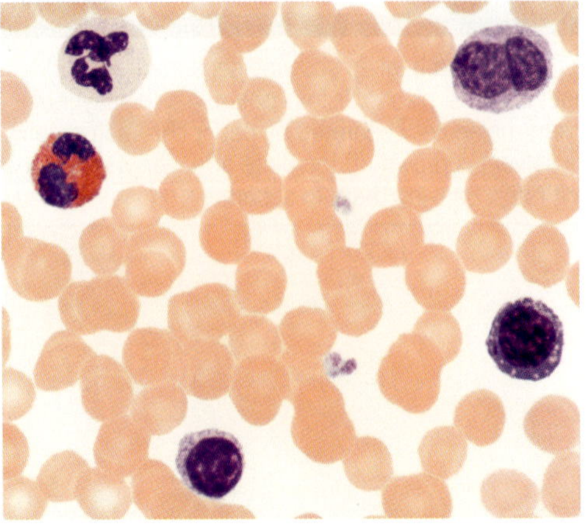

FIGURE 1.3.12 Light micrograph of a normal human blood smear showing the different types of white blood cells (neutrophil, monocyte, eosinophil, lymphocyte and basophil) and red blood cells (or erythrocytes). The mature human red blood cell is small, round, biconcave, and lacks a nucleus and organelles.

COLLATING SECONDARY DATA FROM DATABASES

Many databases of biological information in addition to gene and protein sequences are available. They include databases for biochemical pathways and cellular signalling. Other open-access databases provide a large body of information for investigating the living world, biosciences and molecular biology. They include databases from museums and research institutions and include the records of specimens, fauna and flora, biodiversity and fossil collections (Table 1.3.4). They may include images, physical data and information about the geographic distribution of samples that can be used in scientific investigations (Figure 1.3.13).

TABLE 1.3.4 Useful databases for investigating biodiversity

Bioinformatics database	Type of data, information or applications
Encyclopedia of Life Tree of Life Web Project	species information, biodiversity, taxonomy, phylogeny
Museums Victoria	species data, classification, geographic distribution over time, skull image databases, biological data, fossils
Australia Museum—Learning Resources	evolution and extinction of Australian mammals; human evolution with 3D virtual skulls
American Museum of Natural History Smithsonian Museum of Natural History	research and collections with links to various resources, e.g. palaeobiology, bioinformatics
The Paleobiology Database Fossilworks	databases of fossils, geological distribution, timescales, analysis tools, construct maps

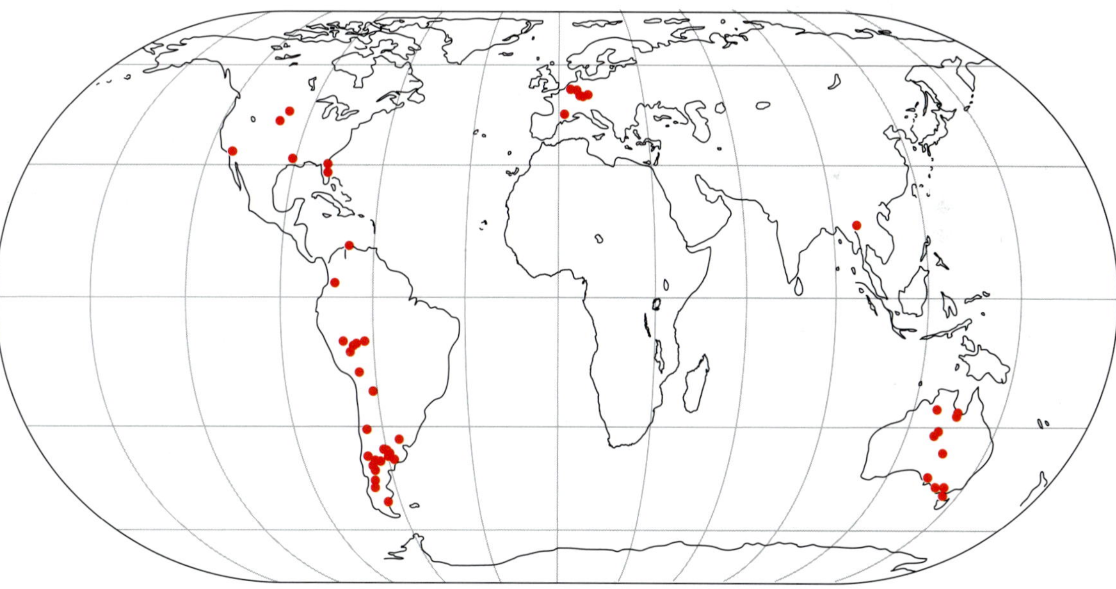

FIGURE 1.3.13 Map showing the distribution of marsupials in the Miocene geological period, constructed using a palaeontology database with search and mapping tools

CASE STUDY **ANALYSIS**

The rise of bioinformatics

Bioinformatics is the use of mathematics, statistics and computer science to analyse and understand biological data. Bioinformatics computer programs can be used to organise raw biological data to visualise patterns, identify genes, model protein structures (Figure 1.3.14), compare DNA sequences (Figure 1.3.15), predict evolutionary relationships and discover and design drugs, along with many other applications.

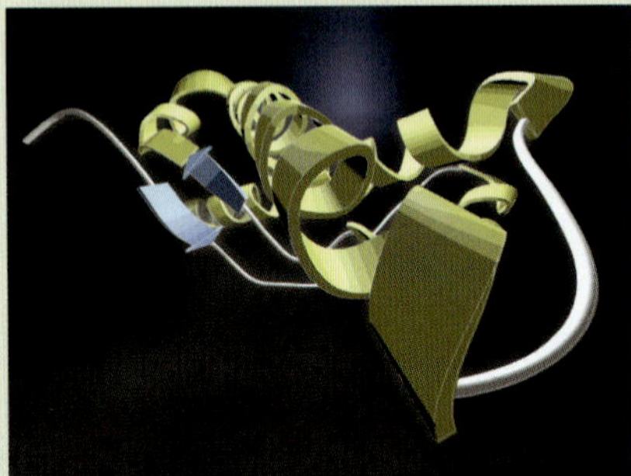

FIGURE 1.3.14 Bioinformatics tools enable you to view the secondary structure of proteins such as this bovine prion protein, which misfolds and forms clumps in the brains of animals with 'mad cow disease'.

Bioinformatics is one of the fastest growing areas of biological science and is now integral to most research and development in biology. The global bioinformatics market is predicted to grow from US$4.1 billion in 2014 to US$18.9 billion by 2026. This rapid growth is primarily driven by the medical biotechnology sector, with demand for more comprehensive and efficient storage and access to personal medical data, and the development of personalised medicine.

One of the best known applications of bioinformatics is whole genome sequencing. The Human Genome Project is still known as the world's largest collaborative biological project. It began in 1990 and by 2003 the three billion nucleotide bases of the human genome had been sequenced. With the technology available at the time, this was an enormous undertaking, costing approximately US$3 billion dollars. With rapid advances in sequencing technology, the output of genome sequencing has skyrocketed, while the cost to sequence a genome has plummeted. It now costs just over $1000 to have your entire genome sequenced, making it affordable for many people.

Although sequencing technology is becoming more accessible, the pool of raw data for analysis is growing, requiring efficient data storage and management systems and the processing power to analyse it. The potential applications of whole genome data are vast but are limited by the bioinformatics tools, computational power and specialised knowledge of bioinformatics currently available to most biologists. As the demand for and capabilities of this technology grows, the scope of biological research is also shifting. Biologists are increasingly required to hone interdisciplinary skills in computer science, mathematics and statistics in order to keep pace with the rapid rise of bioinformatics.

A sample of online bioinformatics resources is listed in Table 1.3.5.

TABLE 1.3.5 Bioinformatics resources

Centres providing bioinformatics databases	Type of data or information; applications
Biology Workbench, San Diego Supercomputer Center	search DNA and protein sequences; sequence alignment, construct evolutionary trees
US National Center for Biotechnology Information (NCBI) (GenBank) European Bioinformatics Institute (EMBL)	nucleotide (gene) sequences; protein sequences and protein structures, chromosome maps, genome maps, SNPs, epigenetics, molecular homology
NCBI–Cn3D OpenScience—Jmol	3D protein structure viewing—free downloads
Sanger Institute	bacterial, protozoan, virus and helminth (worm) genomes

```
B4F917.1   13 SIKLWPPSESTRIMLVDRMTNNLST..ESIFSRK..YRLLGKQEAHENAKTIEELCFALADE.....HFREEPDGDGSSAVQLYAKETSKMMLEVLK 100
A9S1V2.1   23 VFKLWPPSQGTREAVRQKMALKLSS..ACFESQS..FARIELADAQEHARAIEEVAFGAAQE......ADSGGDKTGSAVVMVYAKHASKLMLETLR 109
B9GSN7.1   13 SVKLWPPGQSTRLMLVERMTKNFIT..PSFISRK..YGLLSKEEAEEDAKKIEEVAFAAANQ.....HYEKQPDGDGSSAVQIYAKESSRLMLEVLK 100
Q8H056.1   30 SFSIWPPTQRTRDAVVRRLVDTLGG..DTILCKR..YGAVPAADAEPAARGIEAEAFDAAAA..SGEAAATASVEEGIKALQLYSKEVSRRLLDFVK 120
Q0D4Z3.2   44 SLSIWPPSQRTRDAVVRRLVQTLVA..PSILSQR..YGAVPEAEAGRAAAAVEAEAYAAVTES.SSAAAAPASVEDGIEVLQAYSKEVSRRLLELAK 135
B9MVW8.1   56 SFSIWPPTQRTRDAIISRLIETLST..TSVLSKR..YGTIPKEEASEASRRIEEEAFSGAST.......VASSEKDGLEVLQLYSKEISKRMLETVK 141
Q0IYC5.1   29 SFAVWPPTRRTRDAVVRRLVAVLSGDTTTALRKRYRYGAVPAADAERAARAVEAQAFDAASA....SSSSSSSVEDGIETLQLYSREVSNRLLAFVR 121
A9NW46.1   13 SIKLWPPSESTRLMLVERMTDNLSS..VSFFSRK..YGLLSKEEAAENAKRIEETAFLAAND.....HEAKEPNLDDSSVVQFYAREASKLMLEALK 100
Q9C500.1   57 SLRIWPPTQKTRDAVLNRLIETLST..ESILSKR..YGTLKSDDATTVAKLIEEEAYGVASN.......AVSSDDDGIKILELYSKEISKRMLESVK 142
Q2HRI7.1   25 NYSIWPPKQRTRDAVKNRLIETLST..PSVLTKR..YGTMSADEASAAAIQIEDEAFSVANA.......SSSTSNDNVTILEVYSKEISKRMIETVK 110
Q9M7N3.1   28 SFKIWPPTQRTREAVVRRLVETLTS..QSVLSKR..YGVIPEEDATSAARIIEEEAFSVASV.ASAASTGGRPEDEWIEVLHIYSQEIXQRVVESAK 119
Q9M7N6.1   25 SFSIWPPTQRTRDAVINRLIESLST..PSILSKR..YGTLPQDEASETARLIEEEAFAAAGS.......TASDADDGIEILQVYSKEISKRMIDTVK 110
Q9LE82.1   14 SVKMWPPSKSTRLMLVERMTKNITT..PSIFSRK..YGLLSVEEAEQDAKRIEDLAFATANK.....HFQNEPDGDGTSAVHVYAKESSKLMLDVIK 101
```

FIGURE 1.3.15 Bioinformatics tools allow comparison of many gene or protein sequences at once.

Analysis

A student was aiming to compare the genomes of four different organisms to determine their evolutionary relationships. The more closely related two organisms are, the more similar their DNA will be. A small section of the gene for cytochrome c (a protein involved in cellular respiration) is shown below for the organisms (labelled A to D).

Organism A	AGCCTATTTACGCAGTACGTAAACCCTATATACTATGCA	
Organism B	AGCCTATTTACGGACTACGTTAACACTATATACTATGCA	A/B = 4
Organism C	ACCCTATTTACGCAGTACGTAAACACTATATACTATGCA	A/C = 1, B/C = 3
Organism D	ACCCTATTTACGCAGTACGTAAACCCTATATACTATGGA	A/D = 2, B/D = 6, C/D = 2

a When comparing the four sequences, which two organisms seem to be most closely related? Explain.

b Explain whether electrophoresis could be used to determine the similarities and differences between these organisms.

1.3 Review

SUMMARY

- Biologists use a range of techniques in scientific investigations.
- Cellular respiration and photosynthesis can be detected in several ways, some of which provide qualitative results, and others, quantitative data.
- When investigating cellular respiration, you might perform a yeast carbon dioxide test to detect the presence of carbon dioxide.
- When investigating photosynthesis, you might use a photosynthometer to measure the amount of oxygen produced by a pond weed.
- Bioinformatics is the use of mathematics, statistics and computer science to analyse and understand biological data.
- Online databases such as the Encyclopedia of Life and Fossilworks are useful for investigations of biological change over time.

KEY QUESTIONS

Knowledge and understanding

1 In a cell experiment you were viewing filamentous algae under the microscope. The following diagram represents what you saw in the field of view with the 40× objective lens. The eyepiece lens on your microscope is 10×.

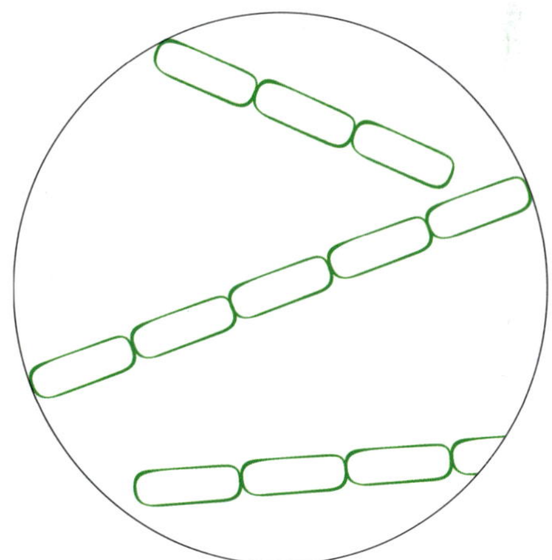

You have previously calculated that the field of view when using this lens is 450 μm.

a What is the total magnification?

b What is the length of each cell?

continued over page

1.3 Review *continued*

2 Recall two methods you could use for detecting CO_2 generation during cellular respiration in yeast, water plants or algae.

3 Recall two methods you could use for detecting photosynthesis in plants or algae.

4 Which materials or method(s) from list A–I could you use for the experiments listed in the table? Copy and complete the table by writing the letter(s) into the right-hand column.

A biochemical test
B bacterial culture
C glucose test strip
D pH meter, indicator or pH stick
E data logger—temperature probe
F plant tissue culture
G data logger—oxygen probe
H staining and microscopy
I spectrophotometer/colorimeter

	Materials or method(s)
i measure oxygen released in photosynthesis	
ii test the effectiveness of antibiotics on rate of bacterial growth	
iii quantitative measure of protein concentration in an enzymatic reaction	
iv identify phagocytosis in ciliate protozoa	
v measure glucose in an enzyme experiment	

Analysis

5 The general formula for an enzymatic reaction is

$$\text{substrate(s)} \xrightarrow{\text{enzyme}} \text{product(s)}$$

a If you measured the amount of substrate, what would you expect to see if the reaction continued?

b Suggest another way to measure the progress of the reaction, and the direction of change expected if the reaction occurs.

6 A student investigated the effect of changing temperature on the rate of photosynthesis in *Elodea* (a freshwater plant) by measuring the pH change in the water over time. Explain why this is considered an indirect measurement of photosynthetic rate.

1.4 Data collection and quality

In this section you will learn about data collection, and how to identify and reduce sources of error that can affect data quality. You will learn about generating primary data and how to record both qualitative and quantitative data. You will also learn about the various factors that contribute to data quality, and the importance of controlled experiments in producing valid results.

FIGURE 1.4.1 A student recording the results of a photosynthesis experiment directly into a logbook

KEEPING A LOGBOOK

Throughout Units 3 and 4, and during your practical investigation for Unit 4 Area of Study 3, you must keep a logbook that includes every detail of your research (Figure 1.4.1). The following checklist will help you remember what to include in your logbook:

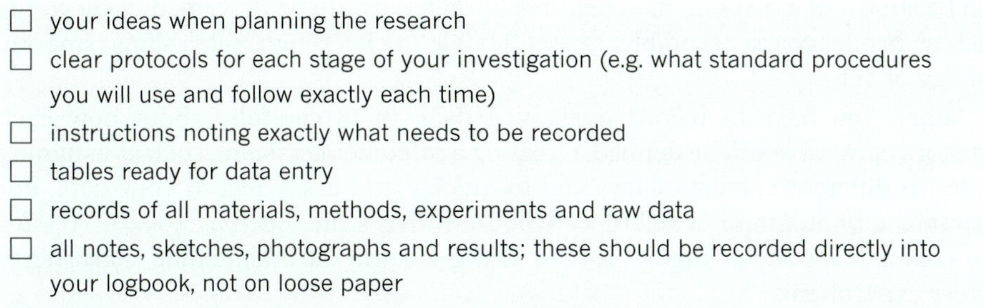

- ☐ your ideas when planning the research
- ☐ clear protocols for each stage of your investigation (e.g. what standard procedures you will use and follow exactly each time)
- ☐ instructions noting exactly what needs to be recorded
- ☐ tables ready for data entry
- ☐ records of all materials, methods, experiments and raw data
- ☐ all notes, sketches, photographs and results; these should be recorded directly into your logbook, not on loose paper
- ☐ records of any incidents or errors that may influence the results.

DATA COLLECTION

The measurements or observations that *you* collect during *your* investigation are your **primary data**. Keep in mind there are different types of data that can be collected in a scientific investigation, including **secondary data** (data you have not collected yourself), so when planning your investigation, consider the type of data you will collect and how best to record it. Data can be raw or processed, and qualitative or quantitative.

> Primary data is data you collect yourself. Secondary data is data that someone else has collected.

Raw and processed data

The data you record in your logbook is **raw data**. This data often needs to be processed or analysed before it can be presented. **Processed data** is raw data that has been organised, altered or analysed to produce meaningful information. If an error occurs in processing the data, or you decide to present the data in a different format, you will always have the recorded raw data to refer back to.

Raw data that should be recorded includes:

- ☐ tables of results
- ☐ all observations and other notes
- ☐ diagrams and/or photographs of results.

> Raw data is the data you record directly into your logbook.

> Processed data is data obtained by applying a calculation or formula to raw data.

For example, you might want to study the effect of glucose concentration on cellular respiration in yeast. To do this you might record two sets of raw data: the concentration of glucose added to each yeast culture flask and the amount of carbon dioxide produced by each culture (Figure 1.4.2).

Table 1 • Carbon dioxide released by yeast cells in different glucose concentrations

Culture flask	Glucose concentration (g/L)	Amount of CO_2 released (ppm)	Amount of CO_2 released (ppm per 10^6 cells)
1	0.0	5	0.5
2	1.0	50	5.0
3	5.0	210	21.0
4	10.0	250	25.0

FIGURE 1.4.2 An example of a table that you might include in your logbook for primary data collection. Data tables should have a title and headings for each column and row, including units as required.

You can then process this data further. For example, to compare across different experiments in which the cell number may vary, it is useful to perform a cell count and express the result per cell (or per million cells). This value, shown in the last column in Figure 1.4.2 on page 35, is processed data.

● *You will now be able to answer key question 1.*

Qualitative data

Data collected about categorical variables is known as **qualitative data**. Categorical variables can be counted but not measured, and relate to a type or category, such as colour, or states, such as on/off or wet/dry.

FIGURE 1.4.3 (a) Colour-based biochemical reactions and (b) structural features are examples of qualitative data.

Recording qualitative data

Qualitative data can be represented as names, symbols or numbers. Observations of categorical variables can be descriptions or images. For example, dog breeds can be shown in a diagram, and textures of materials can be described using words such as brittle, coarse, crumbly, dense, flexible, rocky, rough, silky, slimy, smooth, spongy or velvety.

When you have to record qualitative data, think carefully about how each categorical variable will be defined. Creating a referencing system, such as assigning codes to different colours, allows you to quickly and easily record your data. For example, a photograph of reference colours with a scale (such as +++, ++, + for the colour reactions in Figure 1.4.3a) is a good way of maintaining consistency across experiments.

If you are recording details of structural features, such as when comparing variations in the patterns on turtle shells (Figure 1.4.3b), make a key with diagrams to define your criteria for recording each feature. Samples may have both qualitative data, such as a particular pattern on the top of the shell, and quantitative data, such as the number of clearly defined sections on the shell.

Quantitative data

Data collected about numeric variables is **quantitative data**. Like categorical variables, numeric variables can be counted. Unlike categorical variables, numeric variables can also be measured, because they have a measurable quantity, such as length, mass or time. Numeric variables can be discrete or continuous:

- **Discrete variables** are values that can be counted or measured, but which can only have certain values. Examples are number of chromosomes in a karyotype, number of white blood cells on a slide, or the number of times a lever is pressed.
- **Continuous variables** may be any number value within a given range that can be measured. Examples are age, temperature, length, mass and wavelength.

Recording quantitative data

When you record quantitative data, remember to use SI (International System of Units) units, such as grams, centimetres or millimetres.

Sometimes qualitative data can become quantitative if accurate and consistent measurement is applied. For example, biochemical reactions based on a colour change can be prepared with known concentrations. If a colorimeter or spectrophotometer is available to read absorbance values, then you obtain quantitative data. A calibration curve or standard curve can then be prepared or reading the experimental values. You will learn about standard curves in Section 1.5.

Figure 1.4.4 summarises the different types of data and their variables.

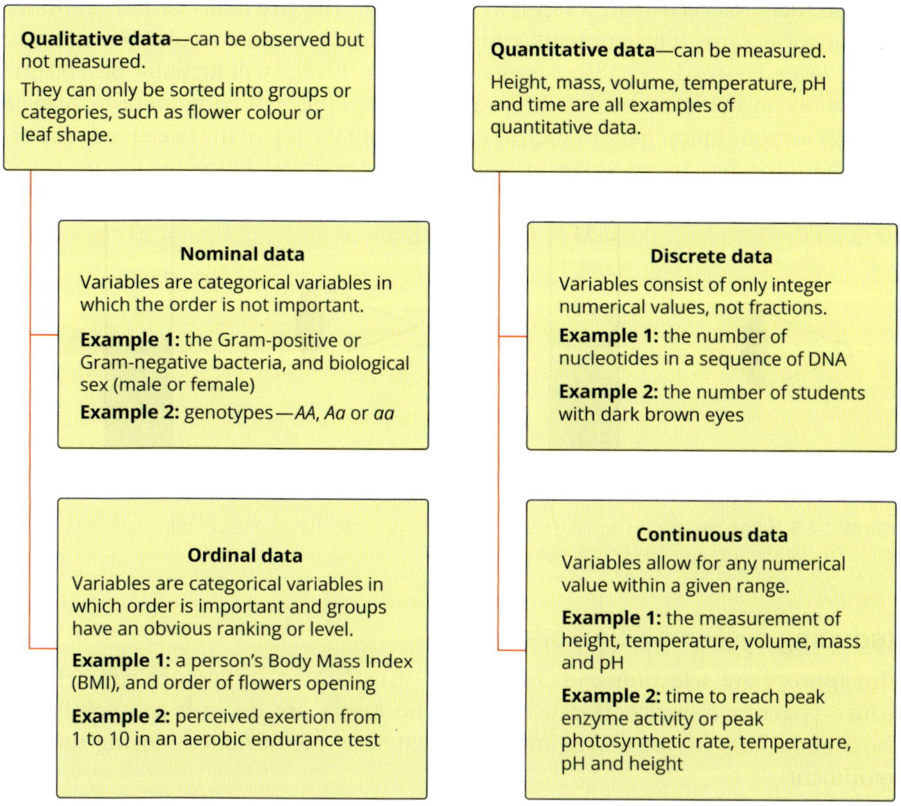

Qualitative data—can be observed but not measured.

They can only be sorted into groups or categories, such as flower colour or leaf shape.

Nominal data

Variables are categorical variables in which the order is not important.

Example 1: the Gram-positive or Gram-negative bacteria, and biological sex (male or female)

Example 2: genotypes—*AA*, *Aa* or *aa*

Ordinal data

Variables are categorical variables in which order is important and groups have an obvious ranking or level.

Example 1: a person's Body Mass Index (BMI), and order of flowers opening

Example 2: perceived exertion from 1 to 10 in an aerobic endurance test

Quantitative data—can be measured.

Height, mass, volume, temperature, pH and time are all examples of quantitative data.

Discrete data

Variables consist of only integer numerical values, not fractions.

Example 1: the number of nucleotides in a sequence of DNA

Example 2: the number of students with dark brown eyes

Continuous data

Variables allow for any numerical value within a given range.

Example 1: the measurement of height, temperature, volume, mass and pH

Example 2: time to reach peak enzyme activity or peak photosynthetic rate, temperature, pH and height

FIGURE 1.4.4 Qualitative and quantitative variables

● *You will now be able to answer key questions 2–4.*

IDENTIFYING AND REDUCING ERRORS

When an instrument is used to measure a physical quantity and obtain a numerical value, the aim is to determine the true value. The **true value** is the value, or range of values, that would be obtained if the variable could be measured perfectly. However, for a number of reasons the measured value is often not the true value. The difference between the true value and the measured value is called the error. This error in the measured value is the result of errors in the experiment. **Personal errors** are often mistakes or miscalculations. If you have made a personal error, the data from the trial should be ignored and the trial should be repeated. The two types of experimental errors are systematic errors and random errors.

Systematic errors

A **systematic error** (or bias) is a consistent error that occurs every time you take a measurement and affects the accuracy of a measurement. Systematic errors are not easy to spot, because they do not appear as a single difference in the data set. Instead, repeated measurements give results that differ by the same amount from the true value. There are many different types of systematic errors, but the most common types are selection bias and measurement bias.

> A systematic error is an error that affects every result in the data set by the same amount. An example is a temperature probe that measures 0.2°C higher than the actual value.

Selection bias

Selection bias occurs when your sample is not representative of the population being studied. This can have a number of different causes, including sampling bias, which is when your sample has not been selected randomly, and time-interval bias, which is when you stop your study too early because the results support your hypothesis.

Measurement bias

Measurement bias is usually a result of instruments that are faulty or not calibrated, or the incorrect use of instruments, which produces inaccurate results. For example, if a scale under-reads by 1%, a measurement of 99 mm will actually be 100 mm. Another example would be if you repeatedly used a piece of equipment incorrectly throughout your investigation, such as reading from the top of the **meniscus** instead of the bottom when using a measuring cylinder or graduated pipette (Figure 1.4.5).

A meniscus is the curved upper surface of liquid in a tube.

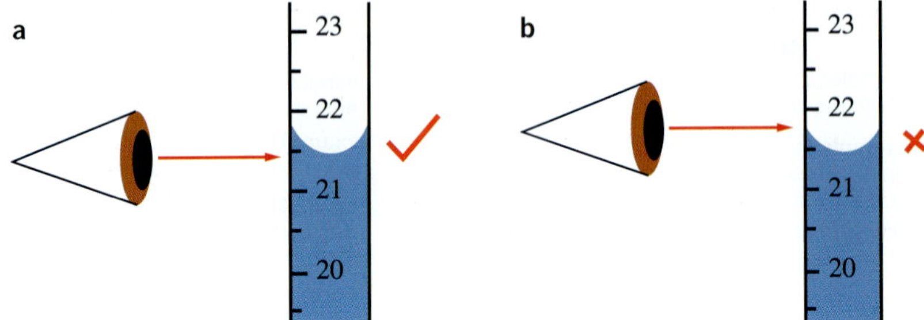

FIGURE 1.4.5 When measuring liquid levels in cylinders and pipettes, measure the value at (a) the bottom of the meniscus of the liquid, not (b) the top.

Reducing systematic errors

The appropriate selection and correct use of calibrated equipment will help you reduce systematic errors. Because systematic errors are difficult to identify, it is also a good idea (if you have time) to repeat your measurements using different equipment.

Appropriate equipment

Use the equipment best suited to the data you need to collect. Determining the units and scale of the data you are collecting will help you to select the correct equipment. For example, if you need to measure 10 mL of a liquid, using a 10 mL graduated pipette or a 20 mL measuring cylinder will give more accurate readings than when using a 200 mL measuring cylinder, because the pipette or 20 mL cylinder will have a finer scale.

Calibrated equipment

Accurate measurement requires properly calibrated equipment. Before you conduct your investigation, make sure your instruments or measuring devices are properly calibrated and functioning correctly (Figure 1.4.6). Your school laboratory may have a set of standard masses that can be used to calibrate a balance or scale. A pH meter should have a set of standard pH solutions (e.g. at pH 4, pH 7 and pH 9) to check the meter readings and adjust the meter if necessary.

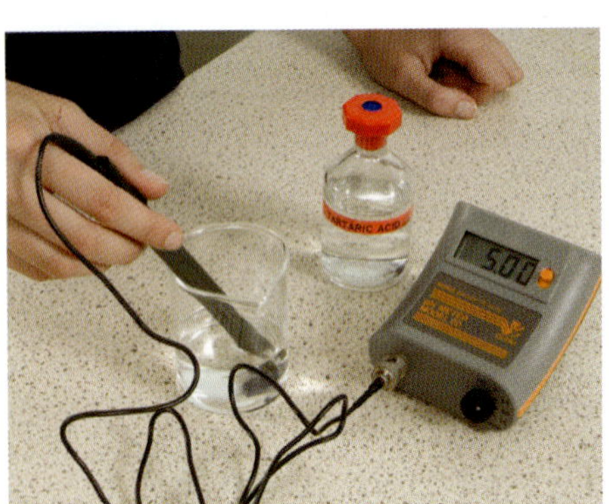

FIGURE 1.4.6 Measuring the pH level of tartaric acid with a pH meter. To ensure an accurate reading, the student would first have calibrated the meter using standard solutions of known pH.

Random errors

Random errors (also called variability) are unpredictable variations that can occur with each measurement. Random errors affect precision and can occur because instruments are affected by small variations in their surroundings, such as changes in temperature. All instruments have a limited precision, so the results they produce will always fall within a range of values.

Reducing random errors

To reduce random errors you need to take more measurements or increase your sample size. You can then calculate the average (the mean), which should be close to the true value.

More measurements

The impact of random errors can be minimised by taking more measurements and then calculating the average value. In general, more measurements will improve the accuracy of the processed data (calculated values). The minimum number of measurements you should make is three. If one reading differs greatly from the rest, mention this in your results and discuss possible reasons for the difference.

Sample size

Increasing the sample size reduces the effect of random errors, which in turn makes your data more reliable. For example, if you are conducting an investigation into the effects of light intensity on the rate of photosynthesis in *Elodea*, do not test your hypothesis on just one stem. Test several stems (minimum three). If two stems photosynthesise and one does not, it is reasonable to conclude that one stem was unhealthy or the conditions incorrect. **Provisional data** is data that is subject to revision. If there are significant errors present, or results that you identify as outliers, you may wish to conduct another measurement under the same conditions. Using a large number of samples will reduce the likelihood of your results being skewed.

DATA QUALITY

The results of your data analysis will only be as good as the quality of the data. A well-designed scientific investigation should produce accurate, precise, reliable and valid results. You should consider all of these factors when collecting primary data in your investigations, and also when you evaluate the quality of secondary data from other sources. Being able to discuss systematic and random errors, and their effect on accuracy and precision, strengthens your written evaluation once your results have been obtained and analysed.

Accuracy and precision

In science and statistics the terms 'accuracy' and 'precision' have very specific and different meanings:

- **Accuracy** is the ability to obtain the true value of the variable being measured. To obtain accurate results, you must minimise systematic errors.
- **Precision** is how closely a set of measurements agree with each other. Precision is different from accuracy in that it does not indicate how close the measurements are to the true value. To obtain precise results, you must minimise random errors.

To understand more clearly the difference between accuracy and precision, think about firing arrows at an archery target (Figure 1.4.7). Accuracy is being able to hit the bullseye, whereas precision is being able to hit the same spot every time you shoot. If you hit the bullseye every time you shoot, you are both accurate and precise (Figure 1.4.7a). If you hit the same area of the target every time but not the bullseye, you are precise but not accurate (Figure 1.4.7b). If you hit the area around the bullseye each time but don't always hit the bullseye, you are accurate but not precise (Figure 1.4.7c). If you hit a different part of the target every time you shoot, you are neither accurate nor precise (Figure 1.4.7d).

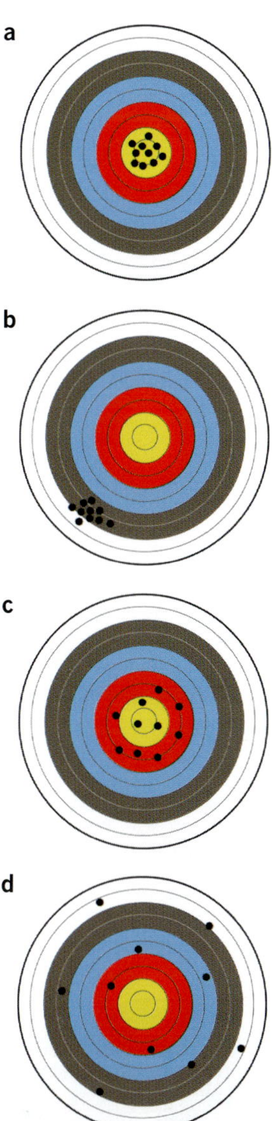

FIGURE 1.4.7 Examples of accuracy and precision: (a) both accurate and precise, (b) precise but not accurate, (c) accurate but not precise, and (d) neither accurate nor precise

Recording numerical data

When using measuring instruments, the number of significant figures (or digits) and decimal places you use is determined by how precise your measurements are.

This depends on the scale, accuracy and precision of the instrument and technique you are using (Figure 1.4.8). For example, a beaker is used to measure volumes approximately and has limited accuracy, for example ±5% (meaning the actual value could be in the range of 5% above or below the value shown). A graduated pipette is more accurate, with accuracies of ±0.1% or ±0.2%. Your pipette may be accurate but if your technique using the pipette is variable, the overall accuracy and precision will be limited.

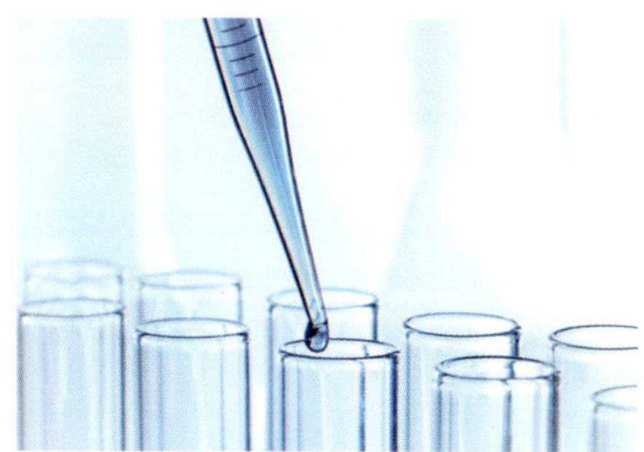

FIGURE 1.4.8 A 5 mL graduated pipette can measure volumes to an accuracy of one hundredth of a millilitre, or 5 mL ± 0.01 mL. The pipette has major divisions of 1 mL and minor divisions of 0.1 mL. You can estimate to 0.01 mL and record volumes to 2 decimal places, for example 3.80 mL or 4.52 mL.

When you record raw data and report processed data, use the number of significant figures available from your equipment or observation. Using either a greater or smaller number of significant figures can be misleading. For example, Table 1.4.1 shows measurements of five tissue samples taken using an electronic balance accurate to two decimal places. The data was entered into a spreadsheet to calculate the mean, which was presented with four decimal places. You would record the mean as 20.83 g, not 20.8260 g, because two decimal places is the precision limit of the instrument. Recording 20.8260 g would be an example of false precision.

TABLE 1.4.1 An example of false precision in a data calculation

Sample	1	2	3	4	5	mean
Mass (g)	20.13	20.62	21.22	20.99	21.17	20.8260

● *You will now be able to answer key questions 5 and 8.*

Repeatability

Repeatability (sometimes called reliability) is the ability to obtain the same results if an experiment is repeated under the exact same set of experimental conditions (Figure 1.4.9). Because a single measurement or experimental result could be affected by errors, **replication** of samples within an experiment and **repeat trials** are key components of repeatability. To improve repeatability you should:

- specify the materials and methods in detail
- include replicate (several) samples within each experiment
- take repeat readings of each sample
- run the experiment or trial more than once.

Conducting the experiment more than once allows the researcher to determine if their results are reproducible. **Reproducibility** is the ability to obtain the same results if an experiment is repeated under different conditions. Different conditions might include a different researcher conducting the experiment, the use of different equipment or instruments, or conducting the experiment at a different time or location.

FIGURE 1.4.9 If you can reproduce your results, they are reliable.

● *You will now be able to answer key questions 6 and 7.*

Validity

Validity refers to whether your results measure what the investigation set out to measure. Results are invalid, for example, if you think you have measured a variable but have actually measured something else. Factors influencing validity include:

- whether your experiment measures what it claims to measure. In other words, your experiment should test your hypothesis.
- the certainty that something observed in your experiment was the result of your experimental conditions and not some other cause that you did not consider. In other words, whether the independent variable influenced the dependent variable in the way you have concluded.
- the degree to which your findings can be generalised to the wider population from which your sample is taken, or to a different population, place or time.

Controls

To ensure your results are valid, carefully determine:

- the independent variable (the variable that you will change) and how you will change it
- the dependent variable (the variable that you will measure)
- the controlled variables (the variables that must remain constant) and how you will maintain them (see Section 1.2).

Your experiment should be designed so that only one independent variable is changed at a time. The remaining variables must be kept constant (or controlled) so that meaningful conclusions can be drawn in turn about the effect of each independent variable on the dependent variable you are measuring.

A control group is a comparison group. To have a control group, you need to set up two groups side by side within an experiment. Both groups are the same, except for the variable you are testing. This is the independent variable in the hypothesis, and is applied to your experimental group but not your control group. All the other variables have to stay the same. You do not want them to change, as these may affect the result of your experiment. For example, when testing a new medication (the variable being tested—the independent variable), two groups of patients are involved. The control group of patients is given a **placebo** (a blank capsule). The other group is given the actual medication, and the data collected from this group is compared to the data from the control group to see the effects of the medication.

> A placebo simulates the same treatment but contains no active ingredient. In this way it acts as a control group.

Randomisation

Random selection of your sample reduces selection bias and improves validity. Selection bias occurs when your sample doesn't reflect the wider population you wish to generalise your results to. For example, if you were scoring phenotypes in large trials of genetically selected or genetically modified crop plants, a study design in which you choose locations at random throughout the plot, rather than choosing locations only at the edge of a plot, would reduce selection bias and improve the validity of the investigation.

USING AND EVALUATING SECONDARY DATA

In researching your topic for various investigations (e.g. case study, correlational study, controlled experiment, literature review, modelling) you will find a range of sources of information. Not all will have reliable information suitable for a scientific investigation. Determine whether the information source is reputable, such as a university, research and education organisation, or peer-reviewed journal. Other sources of information, such as interest groups or companies, may have a specific bias (as outlined in Table 1.4.2 on page 42). Current secondary school, university and specialist area textbooks are good places to start. The best source of experimental data and up-to-date information comes from peer-reviewed scientific journals that have been recently published.

> Peer-reviewed means that other scientists have checked the information and have agreed that it is appropriate for publication.

As many peer-reviewed journals require a subscription, you may not have access to the original articles in full, but you can probably find the abstract (a summary of the study). Also, these original articles can be very complex and hard to interpret if you are not an expert in the field, so an alternative way to access current information about a topic is through print and online science magazines, such as *New Scientist* and *The Scientist*. Good science magazines and journalists provide the background, the results and the relevance of the study in a way that non-experts can understand. However, the methods will be in the original peer-reviewed report.

Your investigation may use scientific data such as protein structures, DNA sequences, or fossil and biogeographic data from open-access databases. Most of these databases are linked to large research centres and are usually reliable sources of information.

DATA QUALITY SUMMARY

Table 1.4.2 summarises factors to consider when evaluating and using primary and secondary data. Make sure you consider all the factors that might affect the quality of the data when you are doing your research and when you write a report of your investigation.

TABLE 1.4.2 Summary of factors impacting quality of primary and secondary data

	Primary data	Secondary data
Accuracy	• Use appropriate and calibrated instruments. • Address systematic errors.	• Use reputable sources such as peer-reviewed journals and books. • Check that systematic errors have been addressed.
Precision	• Use an appropriate number of significant figures. • Address random errors.	• Check that random errors were addressed. • Check that any data analysis was appropriate.
Repeatability	• Specify the materials and methods in detail. • Use replicates within the experiment. • Perform repeat readings. • Repeat your experiment.	• Check that the experimental method was relevant. • Were the results analysed and statistically significant? • Check that information is consistent with other reputable sources.
Validity	• Ensure your experiment tests your hypothesis. • Randomise your sample and use one or more controls.	• Check the study and information is current. • Check the information is based on scientific investigation, controlled trials or research. • Determine if the source is unbiased, or from a particular interest group, e.g. pharmaceutical company, religious group. • Check that the results relate to the hypothesis and aims.

● *You will now be able to answer key questions 9–11.*

1.4 Review

SUMMARY

- Record all information objectively in your logbook including your data and method of investigation.
- Raw data is the data you collect in your logbook.
- Processed data is raw data that has been mathematically manipulated.
- Beware of potential errors when conducting an investigation:
  - Systematic errors are consistent errors that reduce accuracy.
  - Random errors are unpredictable errors that reduce precision.
- Reduce systematic errors by:
  - selecting appropriate equipment
  - properly calibrating equipment
  - using equipment correctly
  - repeating experiments.
- Reduce random errors by:
  - having a large sample size
  - repeating measurements.
- Accuracy is the ability to obtain the true value of the variable being measured.
- Precision is how closely a set of measurements agree with each other.
- Repeatability is the ability to reproduce your results under the exact same set of experimental conditions.
- Reproducibility is the ability to obtain the same results if an experiment is repeated under different conditions.
- Validity refers to whether your results measure what the investigation set out to measure.
- Controlled experiments are important for obtaining valid results.

KEY QUESTIONS

Knowledge and understanding

1 Explain the difference between raw and processed data.

2 a Describe quantitative data.
b Describe qualitative data.

3 Identify which of the following pieces of information about plant material used in a plant hormone experiment are qualitative and which are quantitative. Place a tick in the appropriate column.

Information	Qualitative	Quantitative
leaf colour		
leaf smoothness		
length of stem		
number of leaves		
presence of roots		
change in internode length		

4 Using a Venn diagram, present the differences and similarities between discrete and continuous data.

5 What type of error is associated with:
a inaccurate measurements
b imprecise measurements

6 Describe the difference between replication and repeat trials.

7 Explain why replication and repeat trials are necessary.

Analysis

8 Two sets of data are given below. Both sets contain errors. Identify which set is more likely to contain a systematic error and which is more likely to contain a random error.

Data set A: 11.4, 10.9, 11.8, 10.6, 1.5, 11.1

Data set B: 25, 27, 22, 26, 28, 23, 25, 27

9 A student conducted the following experiment on *Chlorella*, a unicellular freshwater alga. The student's research found a method to put the alga into small, jelly-like alginate balls. Due to equipment limitations (only two oxygen sensors were available), only two sets of data could be collected at one time (see student's report below).

Aim: To investigate the effect of light intensity on photosynthesis in *Chlorella*.

Hypothesis:

Materials:

- freshly prepared alginate balls with *Chlorella* (unicellular alga)
- two equal-sized tubes
- two dissolved oxygen sensors and data logger
- bright fibre optic light

Method:

1. Add 10 mL tap water to each tube.
2. Add 10 algae balls to each tube.
3. Place a probe for detecting dissolved oxygen into each tube and connect to the data logger.
4. Place tube one 30 cm from the light source. Place tube two 5 cm from the light source.
5. Turn on the light and measure the O_2 concentration over 24 h.

Experimental design:

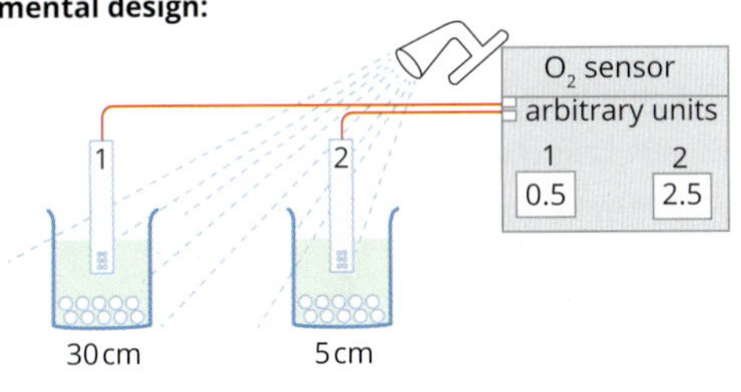

a Suggest a hypothesis for this experiment.

b Identify the independent variable.

c Identify the dependent variable.

d Identify the controlled variables.

e Will the results be objective or subjective? Explain.

f Suggest ways to improve the repeatability and validity of this experiment.

g Based on the results of the oxygen sensors shown in the diagram, what conclusion would you draw from this experiment?

h Will the conclusion be valid for all algae? Explain.

10 Consider the following seedling growth investigation.

a State the independent variable for the experiment.

b State the dependent variable for the experiment.

c List the controlled variables stated in the method.

d Explain the importance of controlling all variables except the dependent variable.

e Identify three variables that could be used to modify this experiment and describe a modification for each variable.

f Write a research question for each variable used to modify the experiment in question 10e.

g Refine each research question from 10f.

Aim:
To investigate the effect of pH on seedling growth.

Hypothesis:
If the soil pH is increased, then seedling growth will increase.

Method:

1. Germinate twenty pea seeds on damp cotton wool and choose twelve with a height of about 12 mm.
2. Plant a seedling in each of twelve pots of the same size. For each pot, use 80 g of quality potting mix, and water with 10 mL of tap water. Safety note: ensure that gloves and a mask are worn when handling potting mix, as it may contain harmful microbes.
3. Label each pot with the pH treatment the soil will receive: four pots at pH 5, four pots at pH 7 and four pots at pH 9.
4. Weigh each pot to the nearest 0.1 g. Draw up a data table and record the results for each pot in the column for day 0.
5. Reweigh the seedlings in their pots 2 days later. Record the results for each pot in the column for day 2.
6. Immediately after weighing, give each plant 10 mL of water at the appropriate pH according to the label on the pot.
7. Repeat steps 5 and 6 every 2 days for the next 10 days.
8. Keep plants in the same position where light is available to maintain lighting conditions.
9. Repeat steps 1–8 twice to reduce the chance of variability between trials.

11 Researchers were investigating the effect of a new oral vitamin D drug. A group of 50 patients were administered with the vitamin D therapy, and another 50 patients were given a placebo.

a When developing the protocols and patients involved in the clinical trial, it was ensured there was an equal number of male and female participants. Explain why this was necessary.

b What would have been in the tablets the placebo group received?

c Would a sample size of 50 people in each group be considered a large sample size? Explain.

1.5 Data analysis and presentation

In Section 1.4 you learnt about different types of data and the factors that affect data quality. In this section you will learn about different descriptive statistics you can use to analyse your data, as well as how you can present your data in tables and graphs.

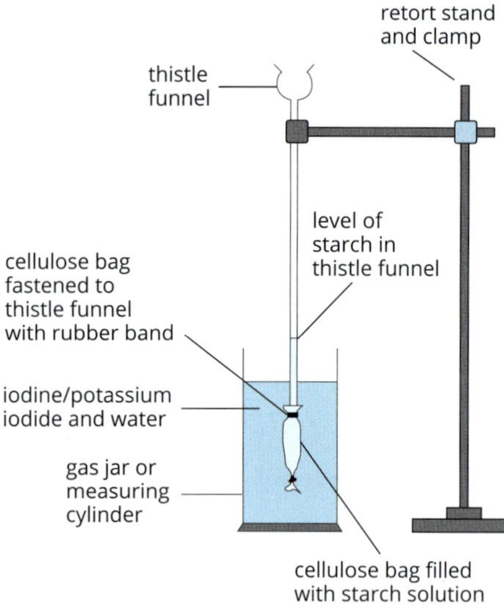

FIGURE 1.5.1 Diagram showing a dialysis tubing arrangement. Note the straight lines for labels that are horizontal where possible, and the realistic proportions of different parts in relation to each other.

PREPARING DIAGRAMS

It is important to learn how to draw and label diagrams of equipment and biological specimens in your studies of biology. There are certain rules you should follow in order to produce a diagram that will be acceptable in your reports and exams.

When drawing scientific equipment, diagrams should:

- ☐ be large, simple, two-dimensional pencil drawings
- ☐ have ruled lines where possible
- ☐ keep proportions realistic (Figure 1.5.1).

When drawing biological specimens follow these guidelines:

- ☐ Draw the whole diagram (including labels, lines, magnification, heading and scale if possible) in pencil.
- ☐ A diagram of microscopic objects does not require a circle representing the field of view.
- ☐ Draw one or a few cells to represent a sample; there is no need to try and draw every cell in a field of view.
- ☐ Draw your diagram with simple and clear lines (do not sketch).
- ☐ Use stippling (small dots to represent shading) rather than shading to indicate depth.
- ☐ Make your diagram as large as possible (at least 10 × 10 cm).
- ☐ Draw only the structures that you see, not things you think you should see, such as mitochondria (Figure 1.5.2).
- ☐ If there are many features to show, it is useful to pair a photo with a detailed supporting diagram that shows cellular detail (Figure 1.5.3).
- ☐ Include clear labels for the features you want to highlight.
- ☐ Place labels outside the drawing.
- ☐ Make sure label pointers do not cross over each other.
- ☐ Line up labels on either side of the diagram where possible.
- ☐ Use straight lines without arrowheads that meet the features being labelled.
- ☐ Include a scale bar or scale (e.g. 1:100) in the diagram, or state the magnification (e.g. 400×) in the caption.

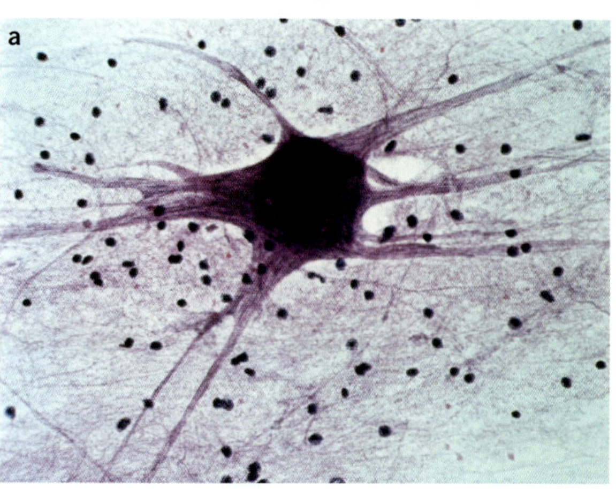

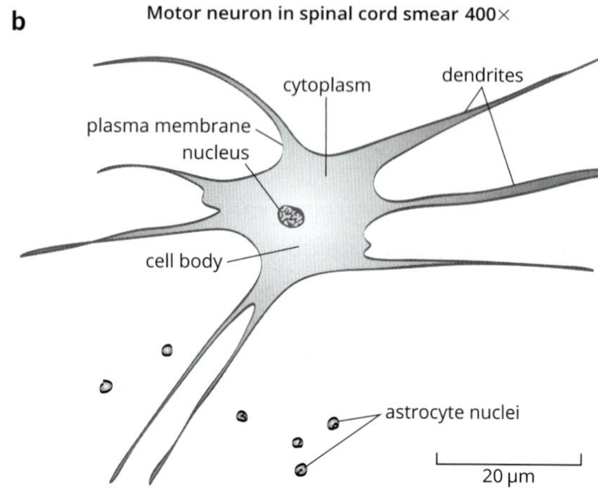

FIGURE 1.5.2 (a) Photomicrograph and (b) scientific diagram showing a motor neuron in a spinal cord smear

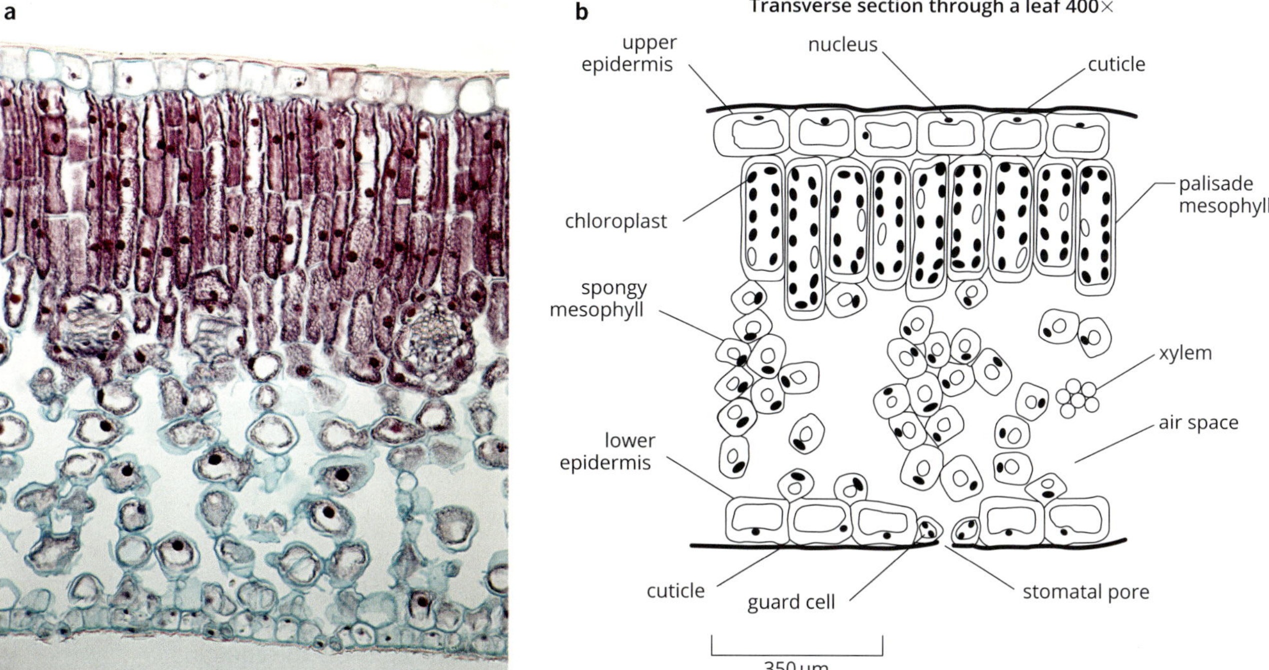

FIGURE 1.5.3 (a) Photomicrograph and (b) diagram of a transverse section through a leaf

DESCRIPTIVE STATISTICS

Descriptive statistics can be used to analyse both quantitative and qualitative data. An important type of descriptive statistic is the measure of central tendency. It is good practice to use a measure of central tendency to provide a clearer understanding of the data.

Measures of central tendency

Measures of central tendency are single values that allow you to describe the central position in a set of data. Measures of central tendency are sometimes also called measures of central location. The mean, median and mode are all measures of central tendency.

Consider the following data set: 3, 5, 7, 8, 8, 8, 10.

- The **mean** (or average) is the sum of the values divided by the number of values, which in this case is $\frac{(3 + 5 + 7 + 8 + 8 + 8 + 10)}{7} = 7$.
- The **median** is the 'middle' value in an ordered list of values, which in this case is the fourth value, which is 8.
- The **mode** is the value that occurs most often in a list of values, which in this case is 8. This measure is particularly useful for describing qualitative or discrete data.

The appropriate measure of central tendency to use depends on the type of data you are working with (Table 1.5.1).

> The mean, median and mode are all measures of central tendency. The mean can be used for nominal, ordinal, and discrete or continuous data so it is common to see the mean used in a variety of investigations.

TABLE 1.5.1 When to use the different measures of central tendency

Type of data	Mode	Median	Mean
nominal (qualitative)	✔	✘	✘
ordinal (qualitative)	✔	✔	maybe
discrete or continuous (quantitative)	✔	✔	✔

Percentage change

Calculating the change in a variable is a helpful statistic because it provides a general trend or pattern, rather than listing a specific value, which will vary depending on the sample being studied. Percentage change applies to increases and decreases relative to the control or the starting point of the measurement.

For example, Table 1.5.2 shows data collected in an experiment that investigated the osmotic strength of different solutions. Four sets of dialysis tubing (a semipermeable membrane), each containing a different solution, were suspended in a beaker of physiological saline solution. The mass was measured at the start and after 24 hours.

TABLE 1.5.2 Percentage change in mass of dialysis tubing over 24 h

	Mass (g) at 0 h	Mass (g) at 24 h	% change
sample 1	20.55	20.89	1.65
sample 2	20.01	21.94	9.65
sample 3	21.25	22.09	3.95
sample 4	20.55	20.32	-1.12

The percentage change in mass is calculated using the equation:

$$\text{percentage mass change} = \frac{\text{final mass} - \text{original mass}}{\text{original mass}} \times 100$$

Calculating percentage change accounts for variation and/or errors in the replicates within your experiment, or for the same experiment repeated by others. In Table 1.5.2, the starting mass is not identical in each sample, perhaps due to errors in measuring the volume put into the tubing. Although the final mass for sample 3 is the greatest, the percentage change is less than for sample 2 because the original mass was higher. Calculating percentage mass change shows that sample 2 has the greatest osmotic effect.

Percentage difference

The percentage difference (also often expressed as a fraction) is a measure of the precision of two measurements. It is calculated by working out the difference between the two measurements and dividing by the average of the two measurements:

$$\text{percentage difference} = \frac{\text{measurement 1} - \text{measurement 2}}{\text{average of measurements}} \times 100$$

For example, if your two measurements were 25 cm and 24 cm, you would calculate percentage difference as follows:

$$\text{percentage difference} = \frac{(25 - 24)}{\frac{(25 + 24)}{2}} \times 100 = \frac{1}{24.5} = 0.041 \times 100 = 4.1\%$$

The range of a set of values can be found by subtracting the lowest value in the data set from the highest value.

Range

The **range** is simply the difference between the highest and lowest values in a data set. Table 1.5.3 shows the measurements taken for five different plants after treatment with a plant hormone.

TABLE 1.5.3 Plant height in a hormone treatment experiment

Plant	1	2	3	4	5	Mean	Range
hormone-treated plants (mm)	158	378	320	377	363	319.2	378 – 158 = 220
untreated control plants (mm)	140	135	170	171	193	161.8	193 – 135 = 58

To determine the range for values in Table 1.5.3 you would subtract the smallest value from the largest value. Notice how an abnormally large or abnormally small value in the data set makes the variability appear high. If one value appears way out of range, such as plant 1 in the hormone-treated group, it is considered an **outlier** and can be deleted from the calculations. The range for the hormone-treated plants would then be 378 – 320 = 58. This illustrates the importance of having a sample size that is large enough to limit the impact of anomalies in the data set.

Uncertainty in measurement

When averaging repeat measurements, the **uncertainty** should be reported alongside your average. Uncertainty results from errors and represents a realistic range within which the true value is likely to be. A simple way to calculate the uncertainty is the range divided by 2:

$$\text{uncertainty} = \pm \frac{(\text{maximum value} - \text{minimum value})}{2}$$

For example, if an experiment were conducted to measure the length of time it takes to convert a substrate to a product in an enzymatic reaction, and three replications of the experiment produced the times 2.50, 3.47 and 2.81 seconds, the average time taken would be 2.93 seconds. The uncertainty would be calculated as follows:

$$\text{uncertainty} = \pm \frac{(3.47 - 2.50)}{2} = \pm 0.48$$

The result showing the mean and uncertainty is expressed as mean = 2.93 ± 0.48 seconds.

For the data set in Table 1.5.3, in which the range was calculated, the uncertainties are as follows:

- control plants 161.8 ± 29.0
- hormone-treated plants 359.5 ± 29.0 (with the outlier removed).

The higher the uncertainty, the less reliable your data may be (see Section 1.4).

● *You will now be able to answer key questions 1–3.*

PRESENTING DATA

When you have completed your experiment, you will need to organise and present the data. This makes it much easier to identify trends or patterns in the data. It also helps to identify any relationships that result from cause and effect between the independent and dependent variables, and helps you see if one variable has had any effect on another variable.

There are a number of ways to present data, including tables, graphs, flow charts and diagrams. The best way of visualising your data depends on its nature. Try several formats before you make a final decision to create the best possible presentation.

Presenting data in tables

Tables record number values and allow you to organise your data.

Presenting raw data in tables

Tables organise data into rows and columns, and vary in complexity according to the nature of your data. Tables can be used to organise raw data and processed data, or to summarise results.

The simplest form of a table is a two-column chart. The first column should contain the independent variable (the one you manipulate) and the second column should contain the dependent variable (the one that may change in response to a change in the independent variable).

As you can see in Figure 1.5.4, tables should have the following features:

- a descriptive title
- column headings (including the units)
- aligned figures (align the decimal points)
- the independent variable placed in the left column
- the dependent variable/s placed in the right column/s.

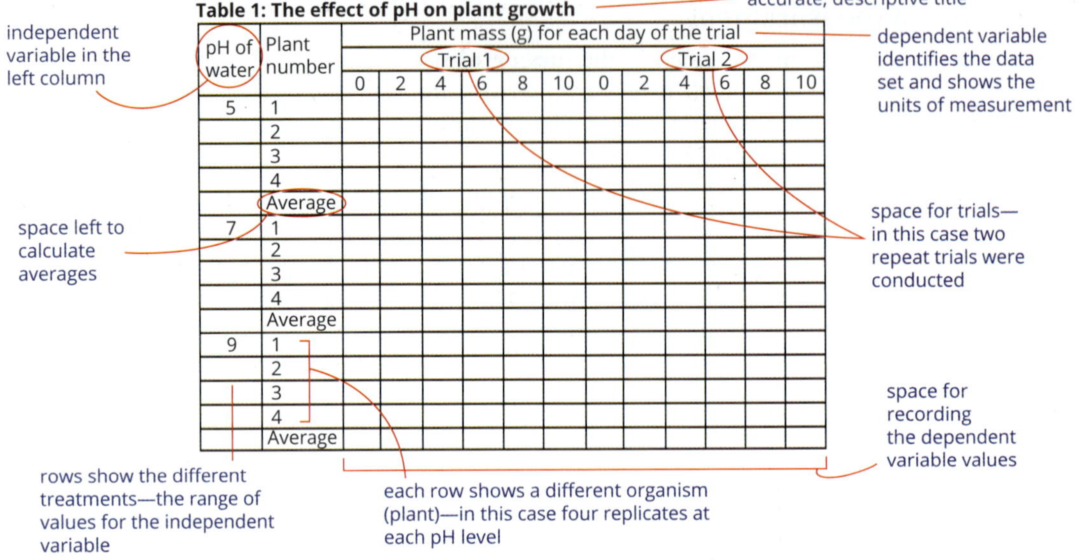

FIGURE 1.5.4 Features of a good table

You should tailor the layout of your data table to suit your experiment. Table 1.5.4 is an example of a raw data table. It contains data from an experiment on the effect of temperature on the activity of enzyme X. A reaction between the enzyme and substrate was conducted for 10 minutes and the reaction product was measured. Three trials were performed.

TABLE 1.5.4 Raw data table for the effect of temperature on reaction rate of enzyme X; measurement of reaction product

	Product released (μg)		
Temperature (°C)	**Trial 1**	**Trial 2**	**Trial 3**
10	100	120	120
20	850	790	820
40	1350	1420	1390
60	1250	1210	1150
80	200	220	230

Presenting processed data in tables

Table 1.5.5 also contains data on the relationship between temperature and mean enzyme reaction rate. It presents the data in a processed format; that is, the replicate values from Table 1.5.4 have been averaged to calculate the mean. The mean reaction rate per minute was also calculated using the equation $\frac{\text{mean}}{10 \text{ min}}$. The mean of the reaction rate per minute and its uncertainty are listed in Table 1.5.6.

TABLE 1.5.5 Processed data table for the effect of temperature on enzyme X reaction rate; calculation of the mean product (μg) and rate (μg/min)

Temperature (°C)	Mean (μg)	Mean rate (μg/min)
10	113.3	11.3
20	820.0	82.0
40	1386.7	138.7
60	1203.3	120.3
80	216.7	21.7

TABLE 1.5.6 Processed data table for the effect of temperature on enzyme X reaction rate; calculation of mean and uncertainty

Temperature (°C)	Mean rate (μg/min)	Uncertainty
10	11.3	± 1.0
20	82.0	± 3.0
40	138.7	± 3.5
60	120.3	± 5.0
80	21.7	± 1.5

Presenting data in graphs

In general, tables provide more detailed data than graphs, but it is easier to observe trends and patterns in data in graph form than in table form. Graphs are used when two variables are being considered and one variable is dependent on the other. Table 1.5.7 on page 50 summarises suitable graphs for qualitative and quantitative data.

There are several types of graphs, including line graphs, bar graphs, scatterplots and pie charts. The best one to use will depend on the nature of the data.

General rules to follow when preparing a graph include the following:

- ☐ Keep the graph simple and uncluttered.
- ☐ Use a descriptive title.
- ☐ Represent the independent variable on the *x*-axis.
- ☐ Represent the dependent variable on the *y*-axis.
- ☐ Start each axis at zero.
- ☐ Match the length of the axes to the data.
- ☐ Clearly label axes with both the variable and the unit in which it is measured.
- ☐ Use small symbols such as circles or squares for data points.
- ☐ Use different symbols for different data sets.

TABLE 1.5.7 Summary of suitable graphs for discrete and continuous data

Type of data	Appropriate type of graph
discrete	bar graph, histogram or pie chart

Examples

Bar graph showing the turbidity of river water at four locations:

Water turbidity at various locations along the Murrumbidgee River

Turbidity (NTU)

3500, 3000, 2500, 2000, 1500, 1000, 500, 0

stormwater outlet; stagnant water; turbulent water under bridge; 1 m from river edge

Location of water sampled, Murrumbidgee River

Histogram showing the height distribution of students in class 12A:

Student heights in Class 12A

Frequency

12, 10, 8, 6, 4, 2, 0

136–140; 141–145; 146–150; 151–155; 156–160

Height (cm)

Pie chart representing the length of time a population of mammalian cells spends in each stage of the cell cycle:

Proportion of time spent in each stage of the cell cycle

M, G2, G1, S

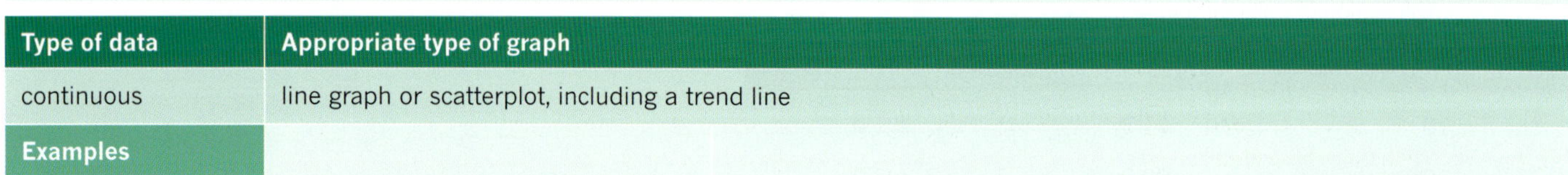

Type of data	Appropriate type of graph
continuous	line graph or scatterplot, including a trend line

Examples

Line graph showing absorbance of sodium in a sports drink:

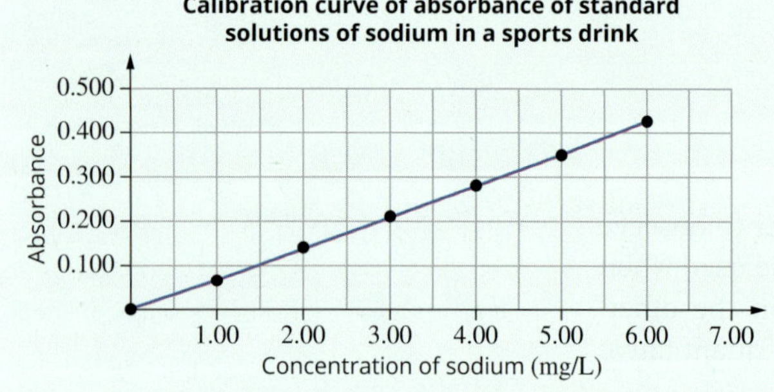

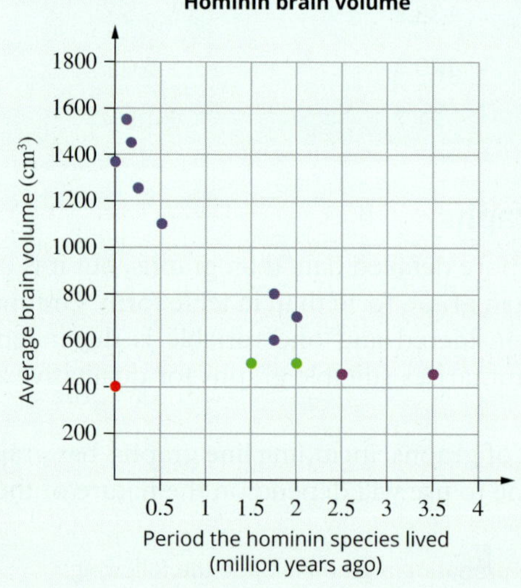

Scatterplot of brain volume in a range of hominin species—colour code for different genera: *Homo*, blue; *Paranthropus*, green; *Australopithecus*, pink; a modern chimpanzee included for comparison, red

Missing data

When you have missing data, leave a gap for it, as shown in Figure 1.5.5. Ensure that the axes are complete (do not skip values) and do not join data points that have data missing between them. Joining points could be misleading. For example, if the data in Figure 1.5.5 was collected to determine the need for a pertussis (whooping cough) booster vaccination program, it is important to know which age groups in the population are most at risk, so that the right age groups are targeted and public health funds are well-directed. Try to predict the result for the missing data in the 25–59 year old age group. The actual value of 9000 pertussis notifications was recorded by the Australian Government Department of Health. This would mean a significant change in the approach to vaccination programs for individuals within this age group.

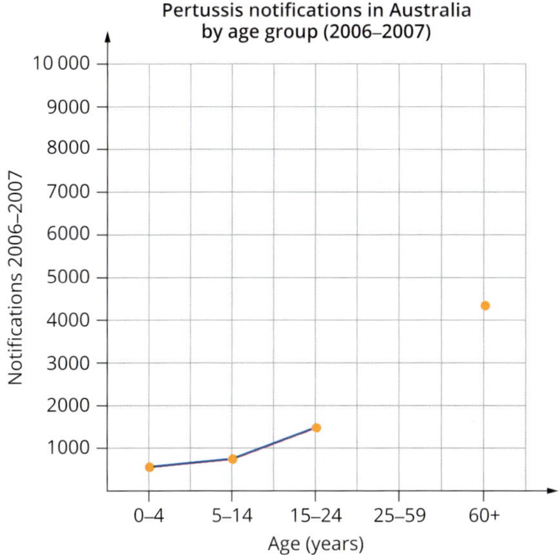

FIGURE 1.5.5 Line graph with missing data

Outliers

Sometimes when you plot data, there may be one point that does not fit the trend and is clearly an error. This is called an outlier. An outlier is often caused by a mistake made in measuring or recording data, or from a random error in the measuring equipment. You should plot all of the data points on your graph, but if a data point in a continuous data series is clearly outside the trend line (an outlier) you can ignore it when drawing the line of best fit (Figure 1.5.6). Outliers can also be calculated mathematically.

● *You will now be able to answer key questions 4–8.*

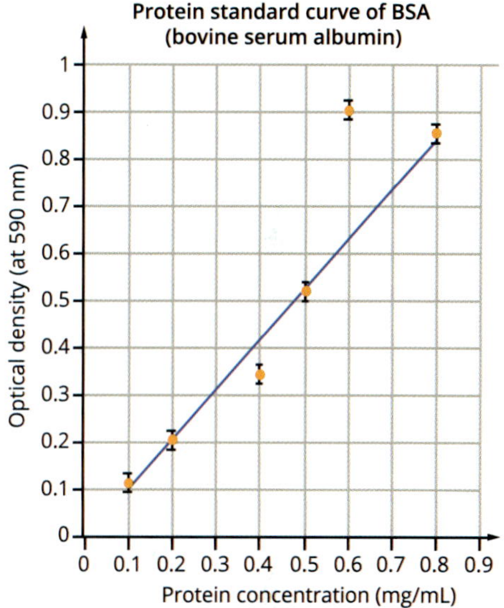

FIGURE 1.5.6 Line graph showing an outlier, which has been ignored when adding the line of best fit

Distorting the truth

Poorly constructed graphs can distort the truth. For example, in Figure 1.5.7 you can see two graphs that show the same data—the test results of two groups of students. One group of students did not eat breakfast before doing the test, and scored an average of 42 marks out of 50. The other group of students did eat breakfast and scored an average of 48 marks out of 50. One graph distorts the difference in marks between the two groups by using a scale from 40 to 50 marks on the *y*-axis. It is important to make sure the graphs you create do not distort your data. You should also be wary of distorted data when interpreting graphs in other publications.

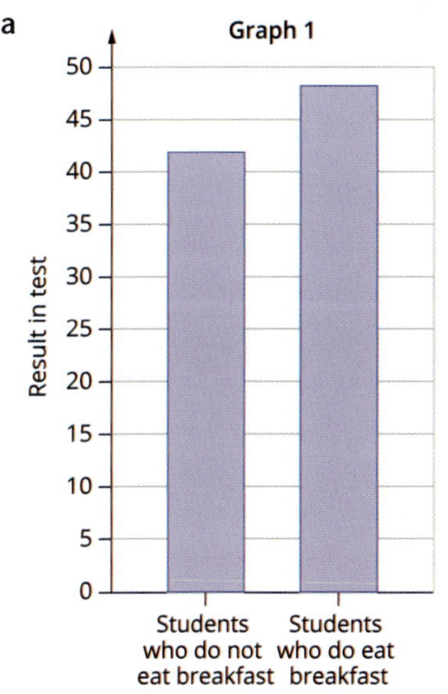

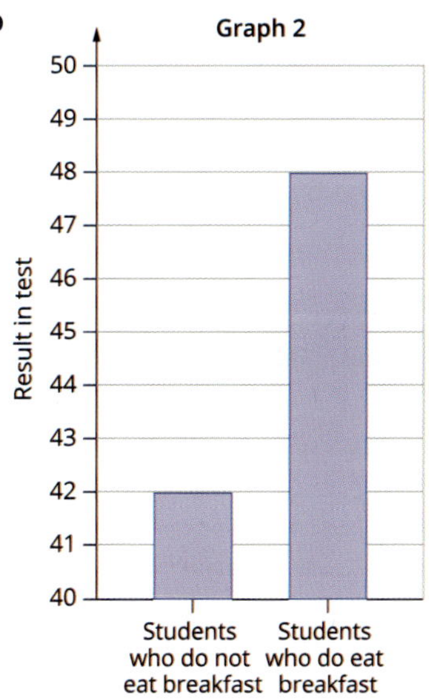

FIGURE 1.5.7 (a) Graph showing the difference between the test marks of two groups of students out of the total 50 marks on the *y*-axis. (b) Graph showing the difference between the test marks of the two groups with only a narrow range of marks on the *y*-axis, which distorts the difference and makes it appear larger than it really is

1.5 Review

OA ✓✓

SUMMARY

- Descriptive statistics can be used for qualitative and quantitative data.
- Descriptive statistics include three measures of central tendency:
  - the mean, which is the sum of the values divided by the number of values
  - the median, which is the 'middle' value in an ordered list of values
  - the mode, which is the value that occurs most often in a list of values.
- Other helpful descriptive statistics include:
  - percentage change, which applies to increases and decreases relative to the control or the starting point of the measurement
  - percentage difference, which is a measure of the precision of two measurements
  - range, which is simply the difference between the highest and lowest values in a data set
  - uncertainty, which results from errors and represents a realistic range within which the true value is likely to be.
- Tables are used to record raw and processed data.
- Tables allow the presentation of more detail, while graphs allow trends to be shown more clearly.
- When presenting the results of an investigation, do not distort the truth—for example, this means you must select appropriate scales on graph axes, include outliers in graphs, and include and explain all errors.

KEY QUESTIONS

Knowledge and understanding

1. For the following data set, calculate and record
   - **a** the median
   - **b** the mode
   - **c** the mean and uncertainty.

   Show your calculations.

   Data: 21, 28, 19, 19, 25, 24, 20
2. Calculate the percentage change in mass for these plants exposed to light of different intensity.

Plant	Mass on day 1 (g)	Mass on day 2 (g)	% change
plant 1 (control)	12.3	12.5	
plant 2 (intense light)	12.4	12.7	
plant 3 (low light)	12.1	12.0	

3. Describe the advantages of calculating percentage change for the results of an experiment repeated by different groups of scientists.
4. Describe at least four ways the graph below could be improved.

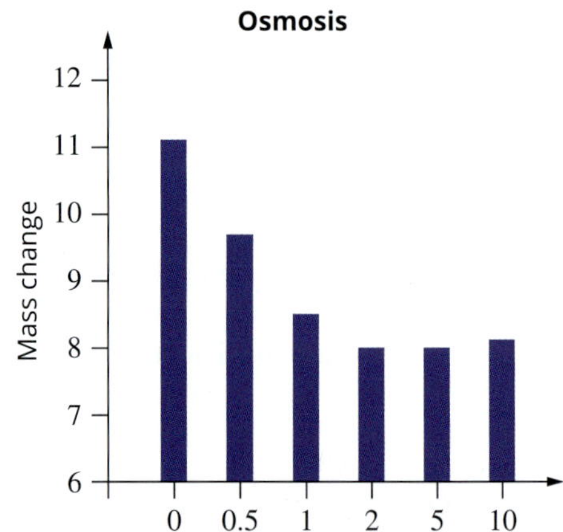

5. When should you draw a line of best fit on a graph, and when should you rule lines from point to point?
6. What are outliers, and what is the statistical measurement most affected by them?

Analysis

7 Immunologists have measured the levels of antibodies in blood serum to gather background data on population responses to infection. They collected the following data on the concentration of two different types of antibody, IgG and IgA, from subjects ranging in age from 6 months to 20 years (the antibody levels are listed in order of increasing age of subject).
- Age of subject: 6 months, 1, 2, 4, 10, 20 years
- Concentration of IgG (mg/100 mL): 300, 600, 800, 1000, 1500, 1500
- Concentration of IgA (mg/100 mL): 50, 100, 100, 150, 200, 400

a Prepare a data table.

b Prepare a graph of the data.

8 A student wants an estimate of the protein concentration in a sample that she isolated. The student prepares a standard curve of known protein concentrations. The following data was obtained.

Protein standard (mg/mL)	Absorbance (at 590 nm)
0.00	0.00
0.10	0.20
0.20	0.18
0.40	0.34
0.50	0.45
0.60	0.52
0.75	0.58
0.90	0.72
1.00	0.86
Unknown	0.70

a Plot the data on a scatterplot, using graph paper or a spreadsheet program.

b Identify any outliers in this set of data.

c Draw a trend line.

d Use the graph to determine the protein concentration of the unknown solution.

1.6 Reporting investigations

FIGURE 1.6.1 Posters at a scientific conference

Now that you have thoroughly researched your topic, formulated a research question and hypothesis, conducted experiments, and collected data, it is time to bring it all together. The final part of an investigation involves summarising the findings in an objective, clear and concise manner for your audience.

Scientists report their findings in a number of ways: written peer-reviewed journal articles, on web pages, and at scientific conferences with short oral presentations or scientific posters (Figure 1.6.1). Regardless of the reporting and presentation method, the same key information is presented in the same order.

The coursework for Unit 3 will include your completed practical logbook (which you should be completing as you conduct practical activities) and either written reports or multimodal presentations of scientific investigations. Upon completion of the scientific investigation for Unit 4, Area of Study 3, you are required to present your methodology, methods, results and conclusions as a scientific poster (Figure 1.6.2). In this section you will learn how to present your findings effectively, discuss your investigation, and draw evidence-based conclusions in relation to your hypothesis and research question.

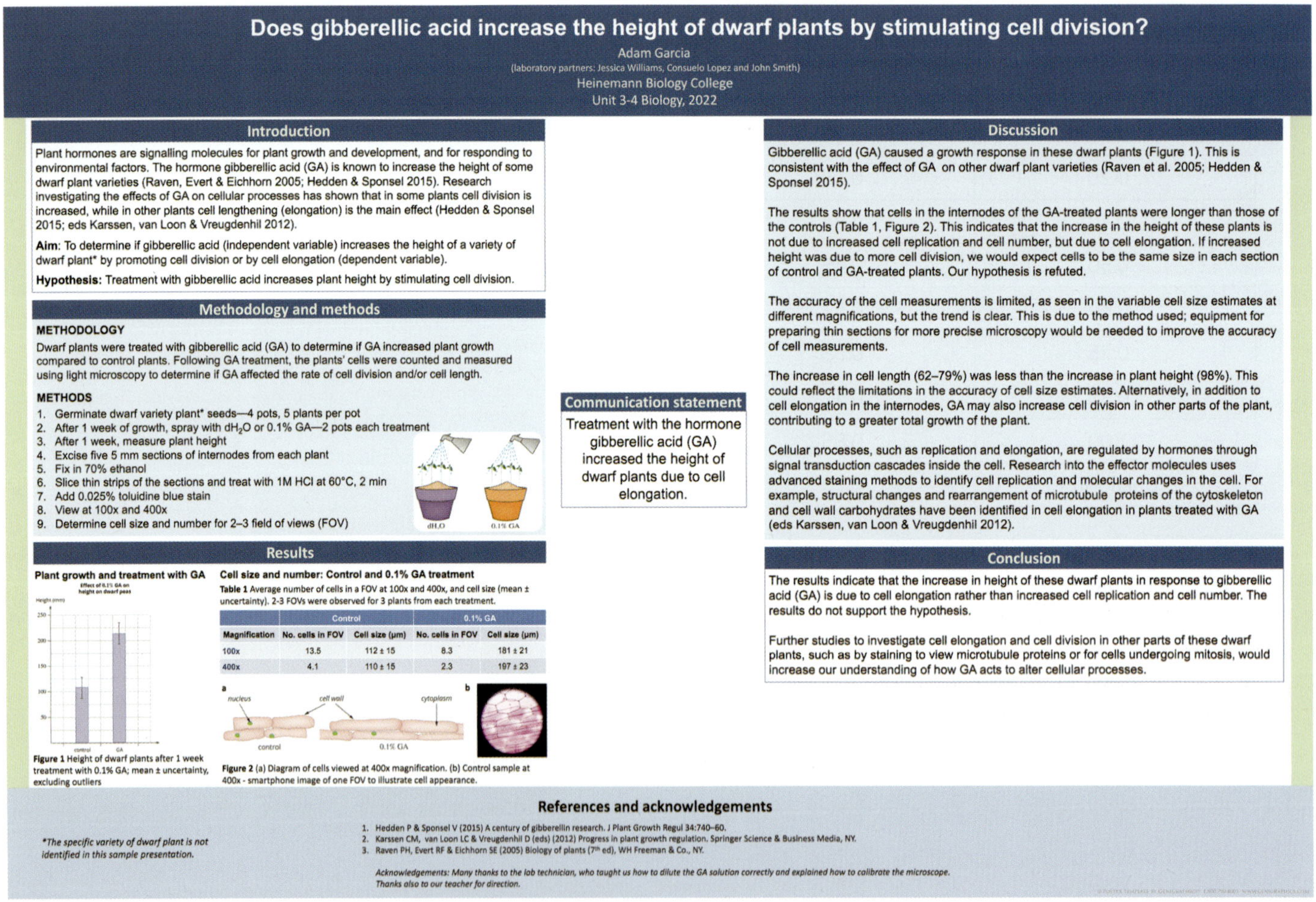

Does gibberellic acid increase the height of dwarf plants by stimulating cell division?

Adam Garcia

(laboratory partners: Jessica Williams, Consuelo Lopez and John Smith)

Heinemann Biology College

Unit 3-4 Biology, 2022

Introduction

Plant hormones are signalling molecules for plant growth and development, and for responding to environmental factors. The hormone gibberellic acid (GA) is known to increase the height of some dwarf plant varieties (Raven, Evert & Eichhorn 2005; Hedden & Sponsel 2015). Research investigating the effects of GA on cellular processes has shown that in some plants cell division is increased, while in other plants cell lengthening (elongation) is the main effect (Hedden & Sponsel 2015; eds Karssen, van Loon & Vreugdenhil 2012).

Aim: To determine if gibberellic acid (independent variable) increases the height of a variety of dwarf plant* by promoting cell division or by cell elongation (dependent variable).

Hypothesis: Treatment with gibberellic acid increases plant height by stimulating cell division.

Methodology and methods

METHODOLOGY

Dwarf plants were treated with gibberellic acid (GA) to determine if GA increased plant growth compared to control plants. Following GA treatment, the plants' cells were counted and measured using light microscopy to determine if GA affected the rate of cell division and/or cell length.

METHODS

1. Germinate dwarf variety plant* seeds—4 pots, 5 plants per pot
2. After 1 week of growth, spray with dH_2O or 0.1% GA—2 pots each treatment
3. After 1 week, measure plant height
4. Excise five 5 mm sections of internodes from each plant
5. Fix in 70% ethanol
6. Slice thin strips of the sections and treat with 1M HCl at 60°C, 2 min
7. Add 0.025% toluidine blue stain
8. View at 100x and 400x
9. Determine cell size and number for 2–3 field of views (FOV)

dH_2O 0.1% GA

Results

Plant growth and treatment with GA

Effect of 0.1% GA on height on dwarf peas

Height (mm)

250 200 150 100 50

control GA

Figure 1 Height of dwarf plants after 1 week treatment with 0.1% GA; mean ± uncertainty, excluding outliers

Cell size and number: Control and 0.1% GA treatment

Table 1 Average number of cells in a FOV at 100x and 400x, and cell size (mean ± uncertainty). 2-3 FOVs were observed for 3 plants from each treatment.

	Control		0.1% GA	
Magnification	No. cells in FOV	Cell size (μm)	No. cells in FOV	Cell size (μm)
100x	13.5	112 ± 15	8.3	181 ± 21
400x	4.1	110 ± 15	2.3	197 ± 23

a

nucleus

cell wall

cytoplasm

control

0.1% GA

b

Figure 2 (a) Diagram of cells viewed at 400x magnification. (b) Control sample at 400x - smartphone image of one FOV to illustrate cell appearance.

Communication statement

Treatment with the hormone gibberellic acid (GA) increased the height of dwarf plants due to cell elongation.

Discussion

Gibberellic acid (GA) caused a growth response in these dwarf plants (Figure 1). This is consistent with the effect of GA on other dwarf plant varieties (Raven et al. 2005; Hedden & Sponsel 2015).

The results show that cells in the internodes of the GA-treated plants were longer than those of the controls (Table 1, Figure 2). This indicates that the increase in the height of these plants is not due to increased cell replication and cell number, but due to cell elongation. If increased height was due to more cell division, we would expect cells to be the same size in each section of control and GA-treated plants. Our hypothesis is refuted.

The accuracy of the cell measurements is limited, as seen in the variable cell size estimates at different magnifications, but the trend is clear. This is due to the method used; equipment for preparing thin sections for more precise microscopy would be needed to improve the accuracy of cell measurements.

The increase in cell length (62–79%) was less than the increase in plant height (98%). This could reflect the limitations in the accuracy of cell size estimates. Alternatively, in addition to cell elongation in the internodes, GA may also increase cell division in other parts of the plant, contributing to a greater total growth of the plant.

Cellular processes, such as replication and elongation, are regulated by hormones through signal transduction cascades inside the cell. Research into the effector molecules uses advanced staining methods to identify cell replication and molecular changes in the cell. For example, structural changes and rearrangement of microtubule proteins of the cytoskeleton and cell wall carbohydrates have been identified in cell elongation in plants treated with GA (eds Karssen, van Loon & Vreugdenhil 2012).

Conclusion

The results indicate that the increase in height of these dwarf plants in response to gibberellic acid (GA) is due to cell elongation rather than increased cell replication and cell number. The results do not support the hypothesis.

Further studies to investigate cell elongation and cell division in other parts of these dwarf plants, such as by staining to view microtubule proteins or for cells undergoing mitosis, would increase our understanding of how GA acts to alter cellular processes.

The specific variety of dwarf plant is not identified in this sample presentation.

References and acknowledgements

1. Hedden P & Sponsel V (2015) A century of gibberellin research. J Plant Growth Regul 34:740–60.
2. Karssen CM, van Loon LC & Vreugdenhil D (eds) (2012) Progress in plant growth regulation. Springer Science & Business Media, NY.
3. Raven PH, Evert RF & Eichhorn SE (2005) Biology of plants (7th ed), WH Freeman & Co., NY.

Acknowledgements: Many thanks to the lab technician, who taught us how to dilute the GA solution correctly and explained how to calibrate the microscope. Thanks also to our teacher for direction.

FIGURE 1.6.2 Example of a scientific poster

PRESENTATION FORMATS

All modes of presenting scientific investigations have the same elements, but with different emphases on visual or textual components depending on the mode of delivery. Table 1.6.1 provides some guidelines for different presentation formats.

TABLE 1.6.1 Characteristics of the main formats for presenting research work

Format	Characteristics	General guidelines for the presentation format
poster presentation	• concise visual display of information • suitable for presenting information to many people • summary of ideas	• title that attracts attention • large headings that stand out • subheadings of a smaller size • attractive presentation • balance of written material and visual material such as diagrams, photographs, tables, graphs • writing large enough to be read from a distance
written report of a practical activity	• presents clear and detailed information on a topic • suitable for providing detailed and more comprehensive background information	• appropriate written style for introduction, materials, methodology, methods, results, discussion and conclusion • use subheadings to organise sections • text should be supported by tables, graphs, diagrams or photographs
oral presentation with supporting slides and/ or handouts	• easy-to-follow format • good for presenting to a large audience • supporting slides can be printed as notes to be given to the audience • opportunity to answer questions from the audience	• brief oral descriptions • use clear visuals that complement what is spoken • minimal text on each slide • consistent format on all slides—background, colours and text • images, diagrams and graphs are clear and large
online presentation e.g. website, blog	• can present visual and written information • accessible to a worldwide audience • easy to follow • easy to update with new information	• include hyperlinks to related information • include multimedia, such as video clips and audio, if appropriate • use the same format throughout—font, background, colours • use clear headings • list all hyperlinks on the main page • include your name, credentials and date of publication

EFFECTIVE SCIENCE WRITING

Effective science writing is objective, clear and concise, and has a consistent narrative and visual support. If you have time, it is a good idea to put your finished writing aside for a few days and then go back and read it over again, fixing anything that is incorrect or poorly written. Checking the spelling is also an essential part of editing your writing. Do not rely only on computer programs to check spelling; they can make mistakes too, and often do not recognise scientific words. Make sure the spellchecker is set to Australian English; the default setting is usually American English.

Objective writing

Scientific reports should be written in an objective (unbiased) style. This is in contrast to literary writing, which often uses subjective (biased) techniques of persuasion (Table 1.6.2).

TABLE 1.6.2 Examples of unscientific and scientific writing

Unscientific writing examples	Scientific writing examples
Examples of biased and subjective language: • The results were weird/bad/atrocious/wonderful... • This produced a disgusting odour... • This is a major health crisis... • This breathtakingly beautiful golden bowerbird...	Examples of unbiased and objective language: • The results showed... • This produced a pungent odour... • This is a serious health issue... • The golden bowerbird...
Examples of exaggeration: • The object weighed a colossal amount... • No one has ever seen this phenomenon... • The magnesium exploded into flames... • Millions of ants swarmed over the nest...	Examples of accurate language: • The object weighed about 250 kg... • This phenomenon has not been reported previously... • The magnesium burnt vigorously... • Ants swarmed over the nest...
Examples of everyday language: • The bacteria passed away... • The results don't... • We guessed that... • Previous researchers were slack and missed...	Examples of formal language: • The bacteria died... • The results do not... • It was predicted/hypothesised... • Previous researchers did not report...

Qualified writing

It is best to avoid words that are absolute, such as always, never, shall, will, or proven. Instead, qualify your writing using words such as may, might, possible, probably, likely, suggests, indicates, appears, tends, can and could.

Concise writing

To be concise use short sentences with a simple structure. The opposite of being concise is being verbose (wordy). When editing your writing consider how you could say the same thing using fewer words (Table 1.6.3).

TABLE 1.6.3 Examples of verbose writing and concise alternatives

Verbose	Concise
due to the fact that	because
Carlos undertook an investigation into	Carlos investigated
It is possible that the cause could be	the cause may be
a total of five experiments	five experiments
the end result	the result
in the event that	if
at the time of writing	today

Verbose	Concise
is well known to be	is
on an annual basis	yearly
until such time as	until
in the vicinity of	near
while in the process of preparation	while preparing
I am of the opinion that	I think that
we took measurement readings	we measured

Voice

'Voice' means whether the subject of the sentence is the 'doer' or 'receiver' of the action. In the active voice the subject is the doer; for example, 'We added 20 mL of sodium chloride to the beaker.' In the passive voice the subject is the receiver; for example, '20 mL of water was added to the solution.' Scientific writing regularly avoids using the active voice and use of personal pronouns (Table 1.6.4).

It is common practice in scientific report writing to write in the passive voice.

TABLE 1.6.4 Examples of active and passive voice

Active voice	Passive voice
We recorded oxygen levels hourly.	The oxygen concentration was recorded every 60 minutes.
We used a pH meter to measure pH.	The pH was recorded with a pH meter.
A thermostat controlled the temperature in the water bath.	The temperature in the water bath was controlled by a thermostat.
We placed 50 g of solute in a conical flask containing distilled water and then slowly added 1 mol L^{-1} hydrochloric acid drop by drop.	Fifty grams of solute was placed in a conical flask containing distilled water, and then 1 mol L^{-1} hydrochloric acid was added dropwise.

Tense

Use the past tense when describing your research, including the planning, the experiments and the results, as well as the work of previous researchers. For everything else (including describing facts and theories) you should use the present tense. Avoid using conditional verbs (could or would) and the future tense (unless you are talking about something that has not yet happened). Table 1.6.5 shows some examples of the incorrect and correct use of tenses in scientific writing.

TABLE 1.6.5 Examples of correct and incorrect use of tense

Incorrect tense	Correct tense
Zhu (2013) describes a similar phenomenon.	Zhu (2013) described a similar phenomenon.
Hormone will then be added to the tips of coleoptiles.	Hormone was added to the tips of coleoptiles.
Enzyme B reacts best at pH 9.	Enzyme B reacted best at pH 9.
The DNA sequence comparison supports the conclusion that species A and B share a common ancestor.	The DNA sequence comparison supported the conclusion that species A and B share a common ancestor.

Visual support

Use graphs or diagrams to present complex concepts or information. This will reduce the number of words you need, and also make your research more accessible for your audience. Details of experimental methods can be presented as a diagram or flow chart. This can make it easier to see the methods than to read through a series of steps. Flow charts use simple diagrams, small text boxes and connecting lines to represent the methods and sequence of steps in a scientific method. Diagrams should use clear outlines and labels—they are not works of art.

WRITING A SCIENTIFIC REPORT

Whether the investigation is presented as a poster, written report or oral presentation, the same key elements are included in the same sequence, as summarised in Figure 1.6.3.

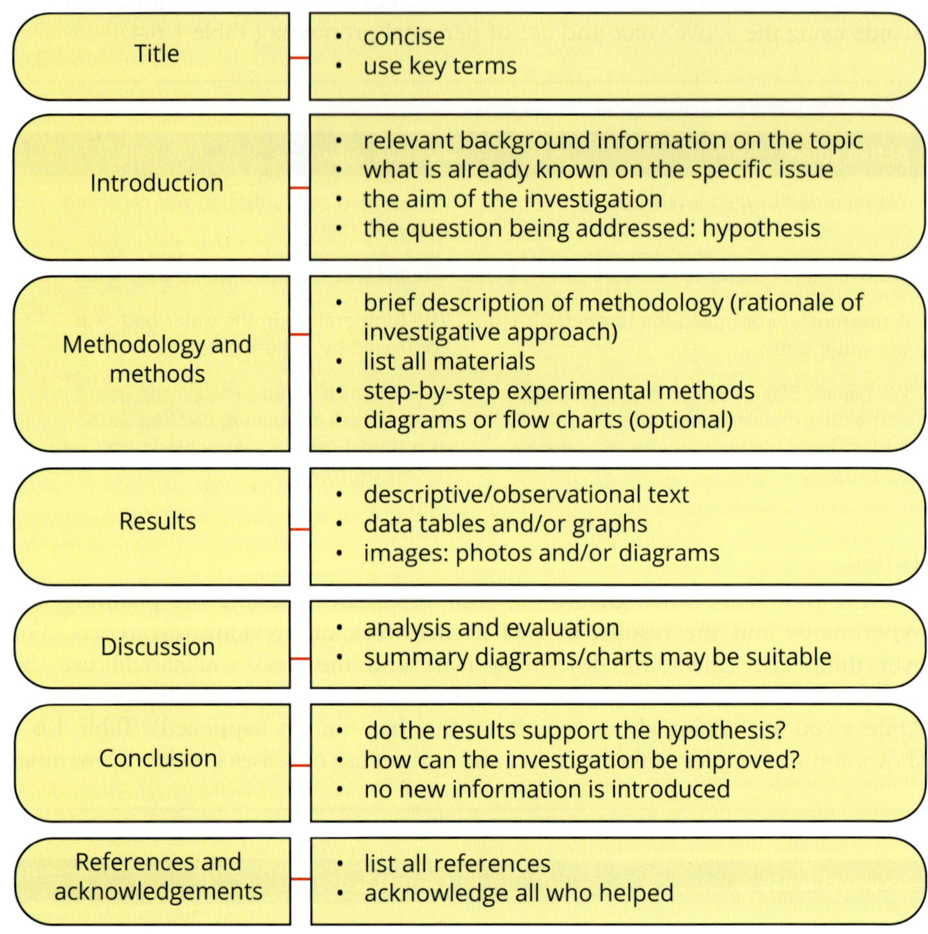

FIGURE 1.6.3 Elements of a scientific report or presentation

Title

The title should give a clear idea of what the report is about, without being too long. It should include key terms that tell the reader what your study is about.

Introduction

The introduction sets the context of your report. It should outline relevant biological ideas, concepts, theories and models, and how they relate to your specific research question and hypothesis. It introduces the key terms, the specific question to be addressed, and states your hypothesis. Any references used in the introduction should be correctly cited. This section should also identify the independent, dependent and controlled variables (see Section 1.2).

For example, consider a student investigating the cellular processes affected by a growth-promoting plant hormone. The research and introduction for this investigation might include the following points:

- the name and chemical nature of the hormone
- where the hormone is found (natural or synthetic)
- what is currently known about the actions of the hormone
- the specific question being addressed, identifying the independent and dependent variables
- the hypothesis.

While researching this topic, the student found prior evidence that suggests this hormone increases the height of some dwarf plants, but the mechanism for this effect was not clear. There were some reports of increased cell division, while other studies reported a change in cell length. The student's hypothesis was that the hormone would increase the growth of dwarf peas by increasing the cell number.

Methodology and methods

The methodology and methods section outlines the rationale of the investigative approach and describes in detail all the steps that were undertaken during the investigation, including a list of the materials used. For a poster presentation use step-wise lists, diagrams of specific methods, and/or flow charts of the overall experimental design. There should be enough detail for someone else to replicate your experiments. Therefore, your method needs to be in the correct sequence and include how you observed, measured, recorded and analysed the results.

Here is an example of a methodology and methods section for an experiment on plant hormone action as it might be presented in a written report. For a poster presentation, the methods may be easier to follow in a step-wise list accompanied by large, clearly labelled diagrams. Alternatively, flow charts are a good way to clearly present experimental designs.

Materials:

- 20 dwarf pea seeds
- 3 pots and potting mix
- plant hormone—gibberellic acid (GA) solutions, 0.01% and 0.1%, diluted from 1% stock solution in distilled water (dH_2O)
- small spray bottles
- scalpel blade and forceps
- toluidine blue stain (0.025%)
- 1 mol L^{-1} HCl
- microscope slides and coverslips
- compound light microscope

Methodology:

A controlled experiment was conducted in which dwarf plants were treated with gibberellic acid (GA) to determine if GA increased plant growth compared to control plants. Following GA treatment, the plants' cells were counted and measured using light microscopy to determine if GA affected the rate of cell division and/or cell length.

Methods:

Example experiment 1: Plant growth and treatment with GA

Dwarf pea seeds were germinated and transferred into 3 pots with potting mix, 5 plants per pot. After 1 week, when the seedlings were approximately 20 mm tall, plants were sprayed with either dH_2O, 0.01% GA or 0.1% GA (Figure 1). Plant height was measured 1 and 2 weeks later.

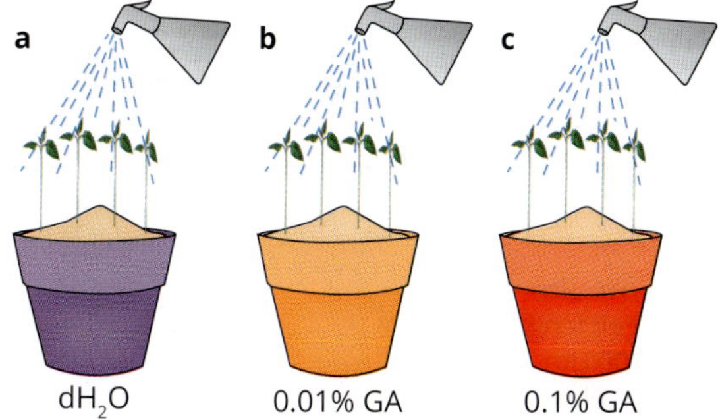

Figure 1: Seedlings were sprayed with (a) dH_2O, (b) 0.01% GA or (c) 0.1% GA.

Example experiment 2: Microscopic analysis of internode cells

At week 3, a 5 mm section of internode was cut from the stem, placed on a microscope slide and sliced lengthwise. Three drops of 1 mol L^{-1} HCl were added to the tissue and the slide placed on a 60°C hotplate for 2 minutes. Excess HCl was soaked up with paper towel; 2 drops of toluidine blue stain were added for 2 minutes, then a coverslip was placed on the tissue and gently pressed down (Figure 2). The slide was viewed under the microscope at 100× magnification. Two stems from each pot were stained and viewed in this manner.

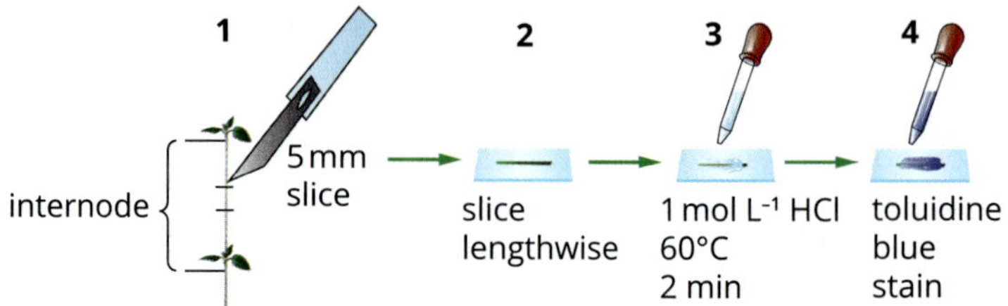

Figure 2: (1) A 5 mm section of internode was cut from the stem and (2) placed lengthwise on a microscope slide, then (3) three drops of 1 mol L^{-1} HCl were added and the slide heated at 60°C for 2 minutes. (4) Finally, 2 drops of toluidine blue stain were added for 2 minutes, then a coverslip placed on top before viewing by light microscopy.

Results

The results section is a record of your observations. It is where you present your data using graphs, diagrams, tables or photographs. In Section 1.5 you learnt tips on using graphs and tables appropriately.

For the plant hormone experiment described above, the results section might include the following table and figures.

Results:

Example experiment 1: Effect of hormone on plant growth

Table 1: Results of plant height (mm) at week 3 of GA treatment

	Plant height (mm) at different GA concentrations		
Plant no.	0	0.01%	0.10%
1	23	117	158
2	20	210	378
3	22	240	320
4	30	211	377
5	31	198	363
mean	25	195	319

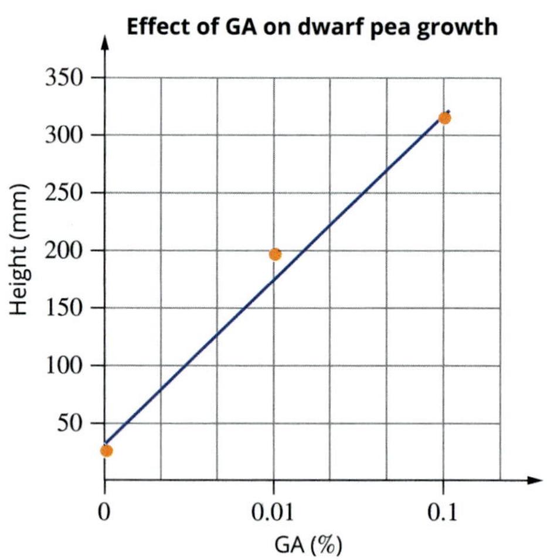

Figure 3: Graph of average plant height at week 3 (2 weeks after GA treatment)

Example experiment 2: Microscopy. The effect of GA on cell growth

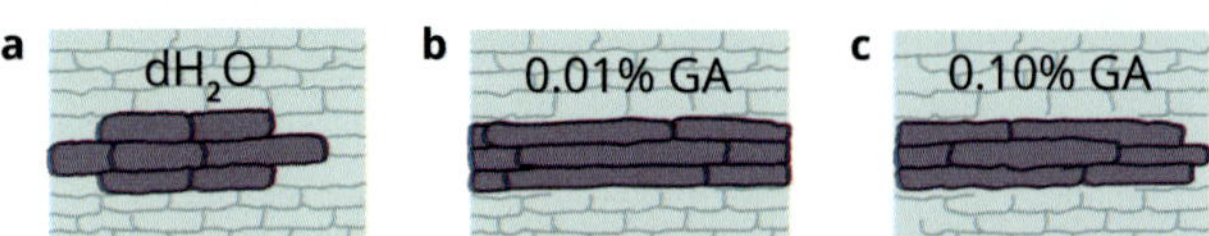

Figure 4: Diagrams of representative samples at week 3 (2 weeks after GA treatment), viewed at 100× magnification. Estimated average cell length in (a) control dH_2O, 60 μm; (b) 0.01% GA, 100 μm; (c) 0.10% GA, 100 μm.

Discussion

In the discussion you interpret your results, and discuss how your findings relate to your initial question and hypothesis, the research of others, and the biological concepts outlined in your introduction. It is also important to evaluate the methods used and the impact of any errors on the results and conclusions that can be formed.

Interpret the results

When you interpret your results, you need to state clearly whether a pattern, trend or relationship was observed between the independent and dependent variables, describe what kind of pattern it was, and specify under what conditions it was observed.

In experiments with continuous variables, such as a range of concentrations, temperatures or pH, the types of relationships that may occur between variables are:

- **linear relationship**—variables that change in linear or direct proportion to each other produce a straight trend line (Figure 1.6.4a)
- **exponential relationship**—variables that change exponentially in proportion to each other produce a curved trend line (Figure 1.6.4b, c)
- **inverse relationship**—when there is an inverse relationship, one variable increases as the other variable decreases; this relationship may be linear or exponential (Figure 1.6.4d, e)
- none—when there is no relationship between two variables, one variable will not change even if the other does (Figure 1.6.4f).

More complex relationships might have to be evaluated mathematically to obtain a formula that describes the trend line.

Interpreting the result for the plant hormone experiment previously described, Figure 3 on page 60 shows that as the GA concentration increases, the height of the peas increases.

● *You will now be able to answer key question 1.*

Evaluate investigative methodology and methods

Your discussion should evaluate your investigative methods and identify any issues that could have affected the validity, reliability, accuracy or precision of the data. Any possible sources of error in your experiment should be stated. Remember that controls are essential to the reliability and validity of your investigation, so if you have overlooked or were unable to control a variable that should have been controlled, this may explain unexpected results.

Make recommendations for modifying or extending the investigation. In the example plant hormone experiment, the sources of error the experimenter should consider include whether there were enough replicates to obtain reliable data, whether microscopy was an appropriate method for determining cell number and cell length, whether the microscope was calibrated, and whether enough cells were viewed. When writing your report, provide specific suggestions for improvements to the methodology based on what you have learnt.

It is also important to acknowledge contradictions in data and information. Again consider the example plant hormone experiment, in which the results of Experiment 2 indicated an increase in cell length in the GA-treated plants compared to the controls, but both GA concentrations had the same effect. This is not consistent with the concentration effect on plant height. So this raises several questions. Is it a limitation of the experimental design or methods? Are there more biological effects that are not being detected or measured? In your discussion, acknowledge these sorts of issues and make suggestions for further experiments to address them.

a Direct or linear proportional relationship
- Variables change at the same rate (graph line is straight, slope is constant)
- Positive relationship—as *x* increases, *y* increases

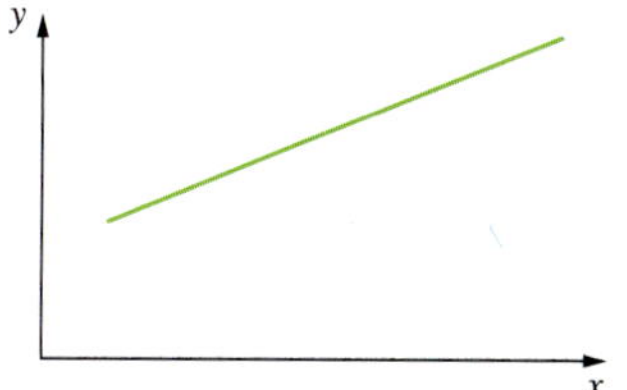

b Exponential relationship
- As *x* increases, *y* increases slowly, then more rapidly

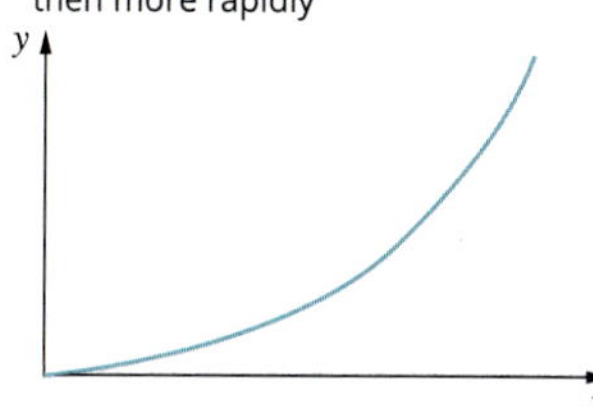

c Exponential rise, then levels off or plateaus (stops rising)
- As *x* increases, *y* increases rapidly at first, then slows, then finally does not increase at all—*y* reaches a maximum value

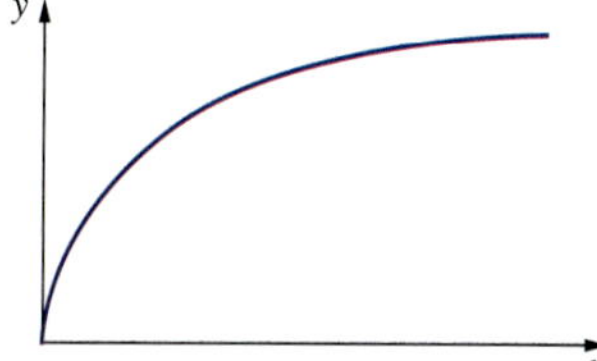

d Inverse direct or linear proportional relationship
- Variables change at the same rate (graph line is straight, slope is constant)
- Negative relationship—as *x* increases, *y* decreases

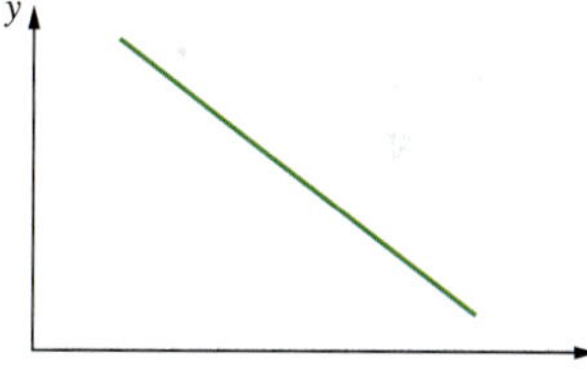

e Inverse exponential relationship
- As *x* increases, *y* decreases rapidly, then more slowly, until a minimum *y* value is reached

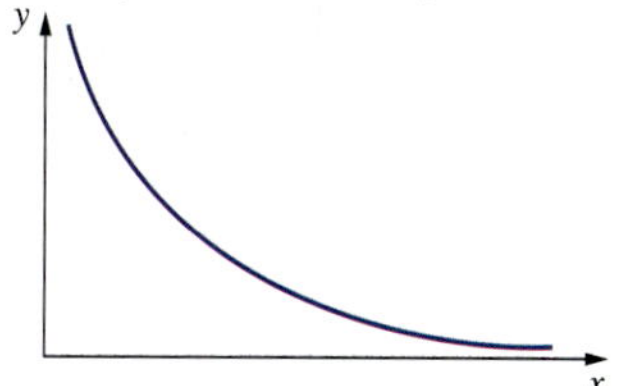

f No relationship between *x* and *y*
- As *x* increases, *y* remains the same

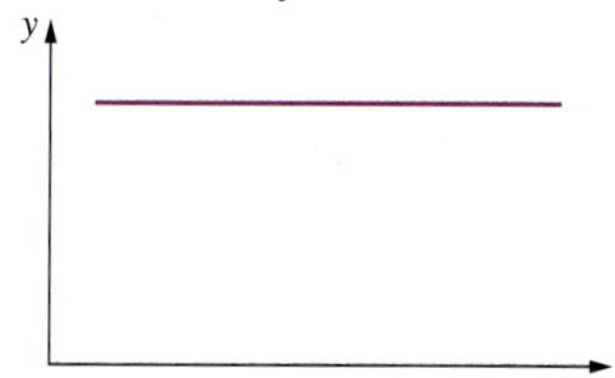

FIGURE 1.6.4 Line graphs illustrating common relationships between variables: (a) direct linear relationship, (b, c) exponential relationships, (d, e) inverse relationships, (f) no relationship

Some experimental findings may lead you to formulate new research questions and develop new hypotheses. An extension of the experiment may be to make an alteration that will enable further investigation. For example, if the effect of temperature has been investigated, further understanding of temperature could be determined by using a different temperature range in a modification of the original method.

Some experimental findings may raise questions about what to do with the new information. This is of particular concern if animal studies have been undertaken. As stated in Section 1.2, research that involves animals, including humans, needs to have obtained approval from an ethics committee. If there are questions raised about the application of any findings from an experiment to animal or human models, the ethical implications of reporting on this data must recognise the potential sociocultural, economic and political impact these results may have.

Relate findings to biological concepts

In your introduction you established a context. Now you have a framework in which to discuss whether your data supports or refutes your hypothesis. Providing context also enables you to compare your results with existing research and knowledge. Use the points in the Figure 1.6.5 to help frame your discussion.

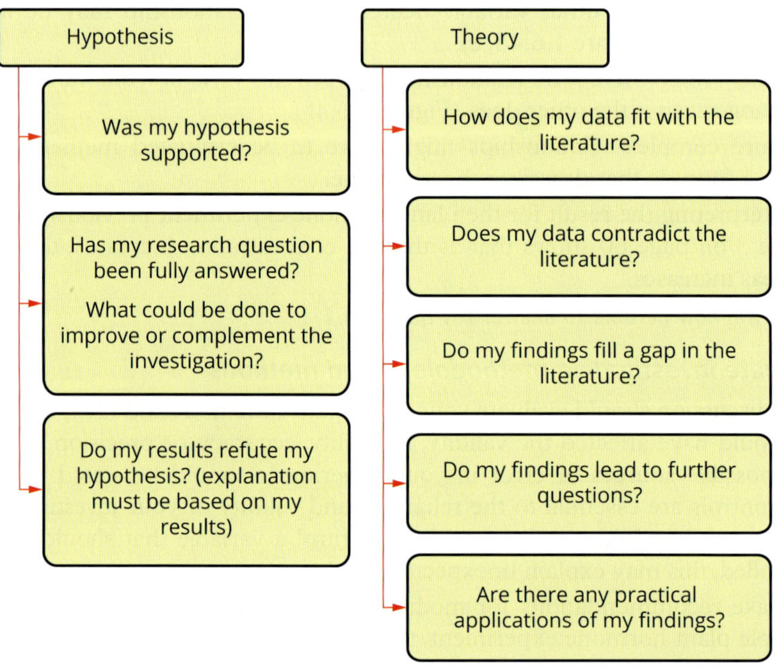

FIGURE 1.6.5 Points to help frame your discussion

Conclusion

Your conclusion should be one or two paragraphs that uses your evidence (data) to support or refute the hypothesis. It should provide a carefully considered response to your research question based on your results and discussion. You should clearly state whether your hypothesis was supported or not. Draw your conclusions by identifying trends, patterns and relationships in the data.

It is important to recognise the limitations of your data and the limitations of the scientific method. Be careful not to overstate your conclusion. Your results will support or refute the hypothesis. They will not 'prove' something is true, as you can only ever provide evidence that indicates the probability of something being true.

Do not provide irrelevant information or introduce new information in your conclusion. Refer to the specifics of your hypothesis and research question, and do not make generalisations.

- *You will now be able to answer key questions 2–4.*

References

All the scientific papers and other sources that are mentioned in the report are to be listed at the end of your report. Cite the source of any information you obtained from secondary sources in the text of your report whenever it is used and referred to, and provide a list of references at the end of your report. This demonstrates that you are aware of previous work in the area, and allows readers to locate sources of information if they want to study them further.

The usual approach is to give a short reference in the text, such as 'Hedden and Sponsel (2015)', and give the full reference in the reference list. If you are stating factual information from another source, you can either quote it word-for-word, or rewrite it in your own words. However, if you rewrite it you must make it clear that the information is not your own. Plagiarism (claiming that another person's work is your own) is not tolerated in scientific research.

Table 1.6.6 shows examples of ways to reference the three most common sources of information: journal articles, books and web pages. Use a consistent format for all references. The examples in Table 1.6.6 follow the American Psychological Association (APA) (seventh edition) referencing system, but there are other referencing systems that you might be required to use during your scientific career.

TABLE 1.6.6 Examples of references in APA style seventh edition for three common information sources

Source of information and example of reference in text	Format for listing references and example of a reference as written in the reference list
Research article or review article in a scientific journal	**Author, initials. (year). Title of article. Journal title, volume number(issue number), page numbers. Digital object identifier (doi) or URL**
GA is well established as a naturally occurring plant growth regulator with effects on… (Hedden & Sponsel, 2015)	Hedden, P. & Sponsel, V. (2015). A century of Gibberellin Research. *Journal of Plant Growth Regulation, 34*, 740–760. doi:10.1007/s00344-015-9546-1
Book	**Author, initials. (year). Title of book (edition, if not first). Publisher.**
The molecular mechanisms of plant growth regulators are being studied by many groups (Karssen et al. 2012)	Karssen, C. M., van Loon, L. C., & Vreugdenhil, D. (Eds). (2012). *Progress in Plant Growth Regulation.* Springer Science & Business Media.
Online article or page	**Author, initials/name of organisation. (year). Title of webpage or web document. URL**
Plant hormones play many roles in plant growth and development, and sensing and responding to environment, possibly even by 'hearing' (Coghlan, 1998)	Coghlan, A. (1998). Sensitive Flower. *New Scientist,* (2153). https://www.newscientist.com/article/mg15921534-900-sensitive-flower/

Acknowledgements

Finally, it is important to acknowledge anyone who has assisted you in your investigation. This includes people who helped you find appropriate literature and references, learn to use equipment, prepare solutions, set up the experiments, find and navigate online databases, edit your report or prepare graphs and images. For example, statements similar to the following could be used:

- This research was supported by the staff of the Science Faculty at Western High School in Melbourne.
- Special thanks to Ms Smith for preparing stock solutions.

1.6 Review

OA ✓✓

SUMMARY

- Your reports should include the following sections:
  - title
  - introduction
  - materials and methods
  - results
  - discussion
  - conclusion
  - references
  - acknowledgements.
- The title should give a clear idea of what the report is about, without being too long.
- The introduction sets the context of your report. It should outline relevant biological ideas, concepts, theories and models, and how they relate to your specific question and hypothesis.
- The methodology and methods section should:
  - outline the methodology used and the rationale for using this approach
  - clearly state the materials required and the methods used to collect data during your investigation
  - be presented in a clear, logical order that accurately reflects how you conducted your study.
- The results section should state your results and present them using graphs, figures and tables, but not interpret the results.
- The discussion should:
  - interpret data
  - evaluate the investigative method and make recommendations for improving the method
  - explain the link between investigation findings and relevant biological concepts.
- The conclusion should succinctly link the evidence collected to the hypothesis and research question, indicating whether the hypothesis was supported or refuted.
- References and acknowledgements should be presented in an appropriate format.

KEY QUESTIONS

Knowledge and understanding

1 a Which of the graphs A–D shows that the rate of cellular respiration increases in direct proportion to an increase in temperature?

b Which of the graphs A–D depicts the results of a mammalian cell culture experiment in which a hormone stimulates cells to multiply exponentially, and then slow down when the nutrient supply is depleted?

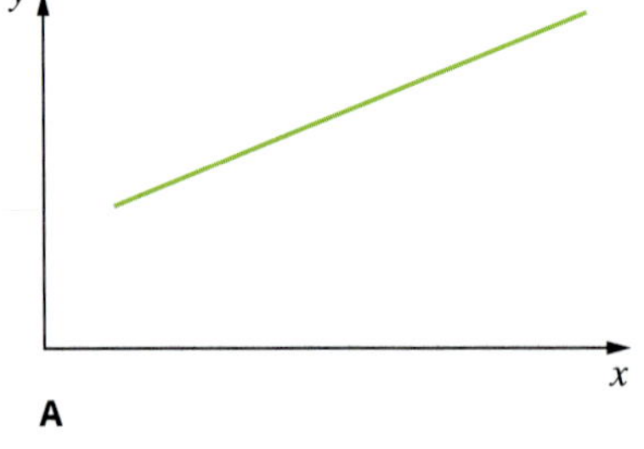

A

B

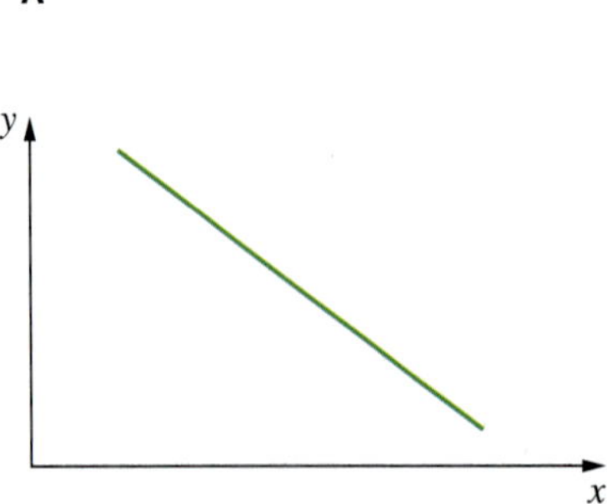

C

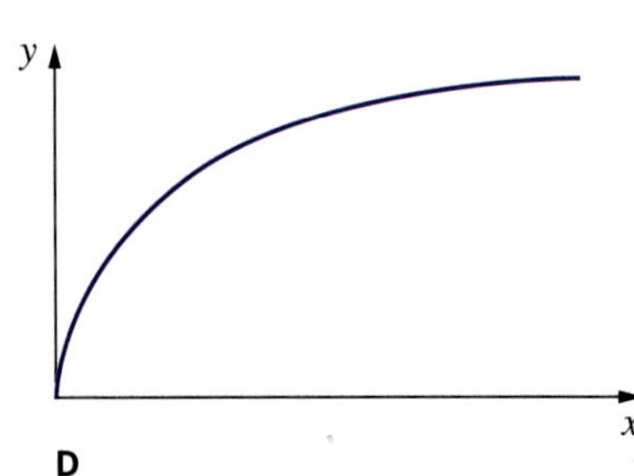

D

Analysis

2 A scientist designed and conducted an experiment to test the following hypothesis: An increased consumption of fast food causes a decrease in the function of the liver.

   a The discussion section of the scientist's report included comments on the repeatability and validity of the investigation. Read each of the following statements and determine whether they relate to repeatability or validity.

      i Only teenage boys were tested.

      ii Six boys were tested.

   b The scientist then conducted the fast food study with 50 people in the experimental group and 50 people in the control group. In the experimental group, all 50 people gained weight. The scientist concluded all the subjects gained weight as a result of the experiment. Is this conclusion valid? Explain why or why not.

   c What recommendations would you make to the scientist to improve the investigation?

3 Review the plant hormone experiment outlined in this section and answer the following questions.

   a Discuss whether the experimental design, materials and methods were described clearly enough. For example, are there any missing experimental details needed to repeat the experiment? Would you suggest a different layout?

   b How would you interpret the results?

   c Write a conclusion for the experiment. Remember to state whether or not the results support the hypothesis.

4 A scientist designed and completed an experiment to test the following hypothesis: 'If there is a negative correlation between water temperature and pH, then water that is heated to 100°C will have a lower pH than water that is cooled to 5°C'.

   a Write a possible aim for this scientist's experiment.

   b What would be the independent, dependent and controlled variables in this investigation?

   c What kind of data would be collected? Would it be qualitative or quantitative?

   d List the equipment that could be used and the type of precision expected for each item.

   e What would you expect the graph of the results to look like if the scientist's hypothesis was correct?

01

KEY TERMS

accuracy
aim
conclusion
continuous variable
control group
controlled variable
dependent variable
discrete variable
experimental group
exponential relationship
hypothesis
independent variable
inverse relationship
linear relationship
mean
median
meniscus
method
methodology
mode
observation
outlier
peer-reviewed
personal error
placebo
precision
primary data
primary source
principle
processed data
provisional data
qualitative data
quantitative data
random error
random selection
range
raw data
repeat trials
repeatability
replication
reproducibility
research question
risk assessment
safety data sheet (SDS)
scientific method
secondary data
secondary source
systematic error
theory
true value
uncertainty
validity
variable

UNIT 3 How do cells maintain life?

To achieve the outcomes in Unit 3, you will draw on key knowledge outlined in each area of study and the related key science skills on pages 7–9 of the study design. The key science skills are discussed in Chapter 1 of this book.

AREA OF STUDY 1

What is the role of nucleic acids and proteins in maintaining life?

Outcome 1: On completion of this unit the student should be able to analyse the relationship between nucleic acids and proteins, and evaluate how tools and techniques can be used and applied in the manipulation of DNA.

AREA OF STUDY 2

How are biochemical pathways regulated?

Outcome 2: On completion of this unit the student should be able to analyse the structure and regulation of biochemical pathways in photosynthesis and cellular respiration, and evaluate how biotechnology can be used to solve problems related to the regulation of biochemical pathways.

CHAPTER 02

Cells and the composition of cells

Learning outcomes

By the end of this chapter, you will be able to describe the various functions that proteins have in cells and the nature of the proteome. You will also be able to explain how proteins are synthesised and to outline the importance of the four hierarchical levels of protein structure. Finally, you will be able to describe the role of various organelles involved in the protein secretory pathway.

Key knowledge

- amino acids as the monomers of a polypeptide chain and the resultant hierarchical levels of structure that give rise to a functional protein **2.3**
- proteins as a diverse group of molecules that collectively make an organism's proteome, including enzymes as catalysts in biochemical pathways **2.3**
- the role of rough endoplasmic reticulum, Golgi apparatus and associated vesicles in the export of proteins from a cell via the protein secretory pathway. **2.3**

WS 1

OA ✓✓

2.1 Cells

REVISION

> An organelle is one of the specialised structures in a cell. Examples include the Golgi apparatus, mitochondria and endoplasmic reticulum.

Cells are the basic structural and functional units of life on Earth. The cell theory is one of the fundamental principles of biology, and describes the properties of cells. Cells can be classified into two types: prokaryotic and eukaryotic cells. Each type of cell has many different structures in place to sustain life.

In this section you will learn about cell theory and the differences between prokaryotic and eukaryotic cells. The structure and function of organelles of cells will also be explored.

Cell theory

If you are to understand life you need to understand how cells work. Cells are the basic functional units of living organisms. The cell theory is based on detailed microscopic and experimental studies of tissues, from all types of organisms, carried out over the last 300 years.

The cell theory states that:

- all organisms are composed of cells
- all cells come from pre-existing cells
- the cell is the basic organisational unit of living things.

All types of cells perform similar basic processes and many also carry out highly specialised functions (Figure 2.1.1). The activities of cells require considerable energy, and produce a variety of biological molecules. These biological molecules, called biomolecules, are used to build new organelles, used for repair or exported from the cell. All of these processes are catalysed (sped up) by enzymes and are precisely regulated. Some biochemical processes involve hundreds of enzymes operating sequentially along a complex integrated chemical pathway in which each step is tightly controlled.

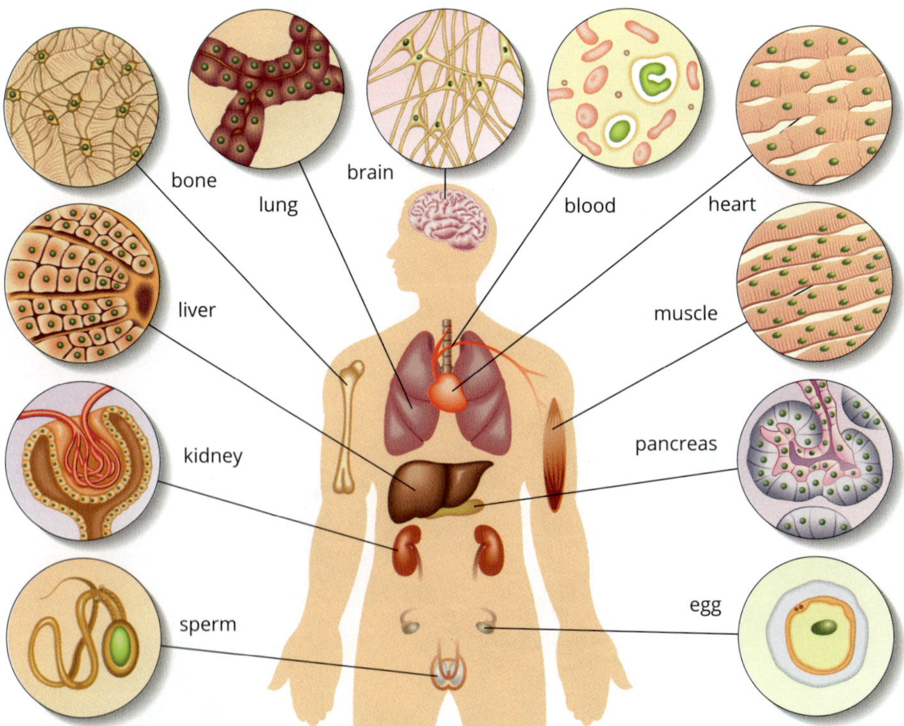

FIGURE 2.1.1 There are many different types of cells in the human body, each carrying out specialised functions.

Types of cells

There are two fundamentally different types of cells. Organisms are classified according to the type of cell they are composed of:

- Prokaryotes are composed of prokaryotic cells and include bacteria and archaea. Prokaryotic cells are usually unicellular and are generally smaller and less complex than eukaryotic cells. They do not contain membrane-bound organelles or a nucleus.
- Eukaryotes are composed of eukaryotic cells and include plants, animals, fungi and protists. Eukaryotic cells contain a membrane-bound nucleus and membrane-bound organelles.

> Protists are a large group of mostly single-celled organisms that cannot be classified as a plant, animal or fungus, but that have plant, animal and fungus-like characteristics.

Common features of cells

There is really no such thing as a typical cell. Cells are specialised for many different purposes and their structures reflect those purposes. However, there are some features that are shared by all cells:

- Plasma membrane (also called the cell membrane) is a semipermeable structure that separates the interior of the cell from the exterior environment.
- Cytoplasm comprises the cytosol (gel-like substance). In eukaryotes, the cytoplasm includes the cytosol and the organelles present in the cell except the nucleus.
- Deoxyribonucleic acid (DNA) carries hereditary information, directs the cell's activities and is passed accurately from generation to generation.
- Ribosomes are structures that assist in the synthesis of proteins. Although they are not membrane-bound, ribosomes are often grouped with organelles.

Table 2.1.1 on page 70 shows some examples of the structures found inside different types of eukaryotic and prokaryotic cells.

Prokaryotic cells

Prokaryotic organisms are unicellular and have a simple cell structure. Prokaryotic organisms can be found everywhere, even in extreme environments such as volcanoes.

The structure of a typical prokaryotic cell is shown in Figure 2.1.2. Prokaryotic cells are small and lack membrane-bound organelles, including a distinct nucleus. Their cytoplasm contains scattered ribosomes that are involved in the synthesis of proteins. The genetic material of prokaryotic cells is usually a single, circular DNA chromosome called the genophore, which is contained in an irregularly shaped region called the nucleoid. The nucleoid does not have a nuclear membrane, unlike the nucleus of eukaryotes.

In addition to this chromosomal DNA, many prokaryotic cells also contain small rings of double-stranded DNA called plasmids.

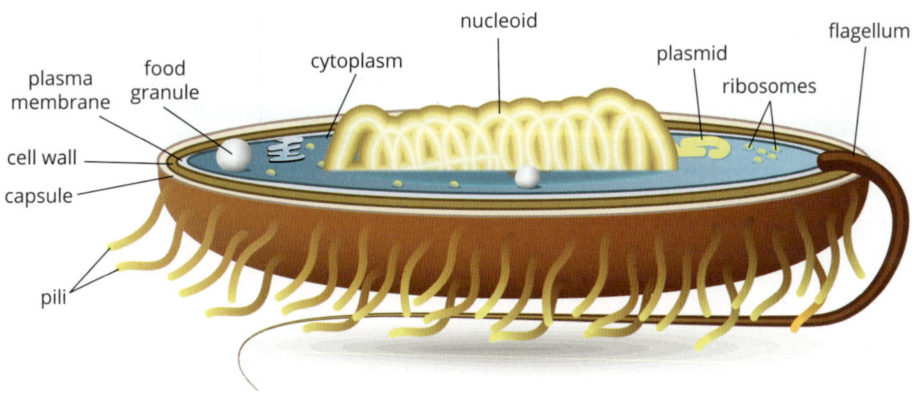

FIGURE 2.1.2 Cross-sectional diagram of a typical prokaryotic bacterial cell

TABLE 2.1.1 These representative diagrams of bacterial, fungal, protistan, plant and animal cells show the organelles that are visible using the electron microscope.

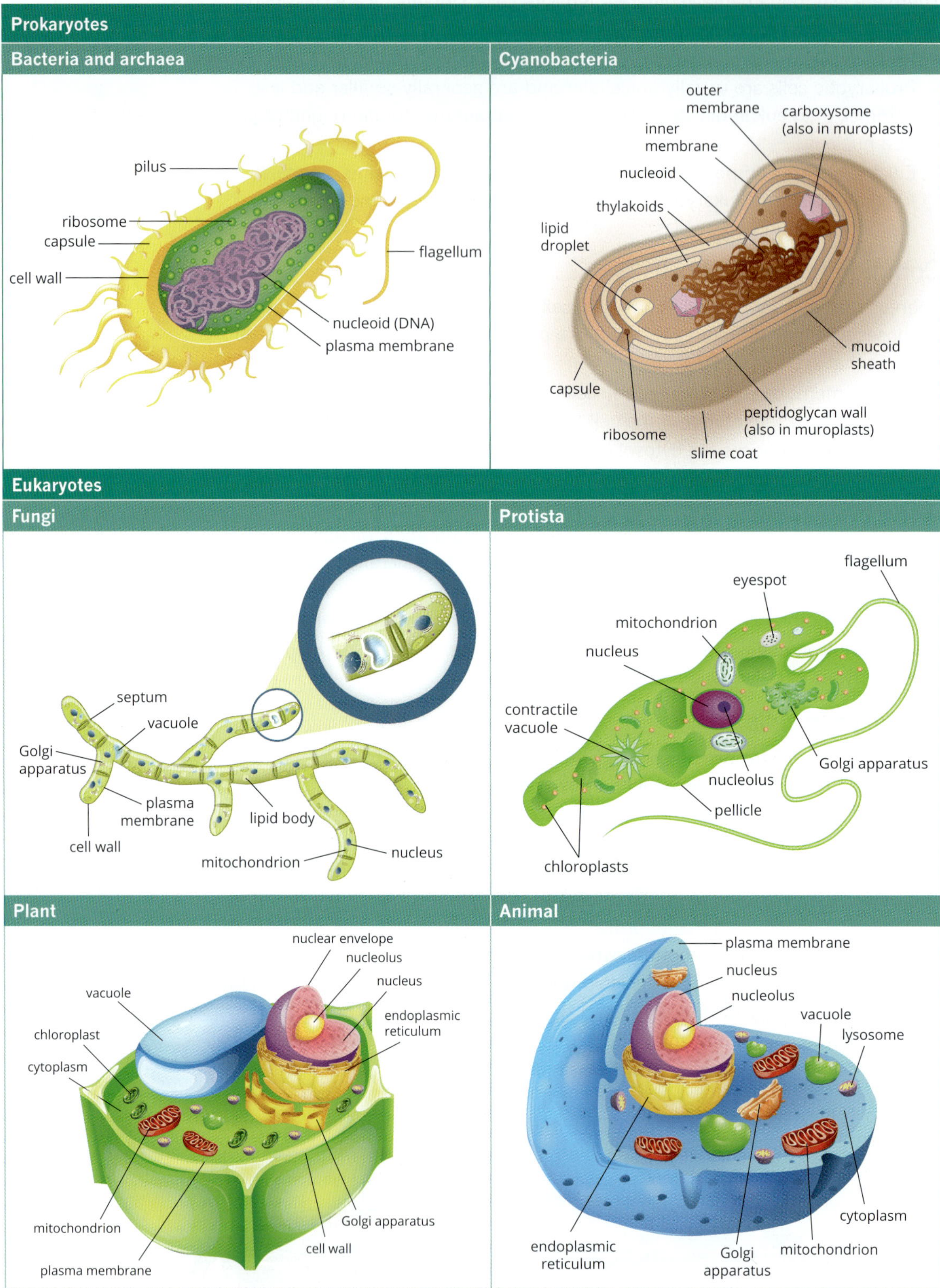

The plasma membrane of most prokaryotic cells is surrounded by an outer cell wall of protein and complex carbohydrate (peptidoglycan). Many bacteria also have a capsule outside the cell wall, which protects the cell from damage and dehydration. Many prokaryotes also have flagella (singular flagellum) that enable them to move freely, and small hair-like projections called pili (singular pilus), which are involved in the transfer of DNA between organisms and which can also help generate movement. Specialised pili that can attach to surfaces are called fimbriae.

Eukaryotic cells

Eukaryotic cells not only have a plasma membrane surrounding the cytoplasm, but also have internal (non-plasma) membranes that form specialised membrane-bound compartments within the cell. This is known as cell compartmentalisation. The membrane-bound structures are called organelles.

Organelles of eukaryotic cells

Organelles are subcellular structures that have a specific function. Because they have a specific function, their presence depends on the needs of the cell. In eukaryotes, some organelles are membrane-bound and some are not (Figure 2.1.3).

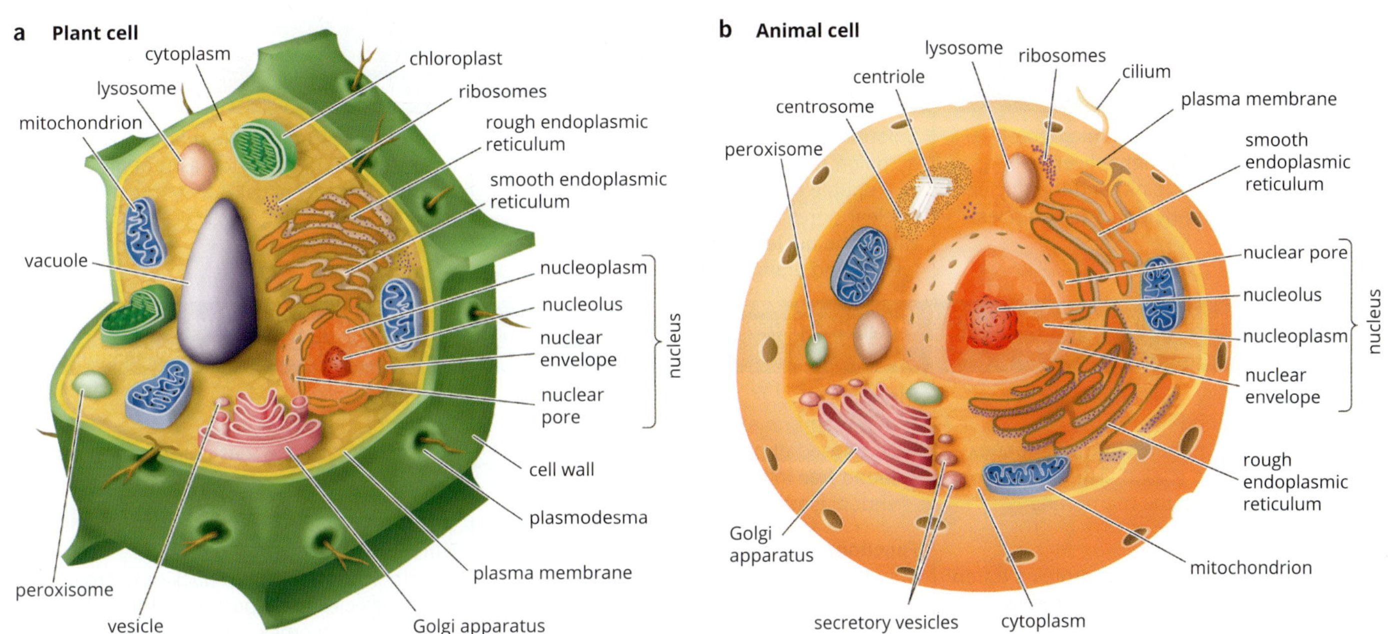

FIGURE 2.1.3 The many organelles of eukaryotic cells can be seen in these illustrations of (a) a plant and (b) an animal cell.

Ribosomes, rough and smooth endoplasmic reticilum, Golgi apparatus and vesicles are all important organelles involved in protein manufacture and movement within a cell. Their roles are outlined in Table 2.1.2 on pages 72 and 73, and will be discussed in more detail in Section 2.3.

TABLE 2.1.2 Structure and function of eukaryotic organelles

<table>
<tr><th colspan="2">Organelle: nucleus</th><th>Organelle: mitochondrion</th></tr>
<tr><td colspan="2">Structure:
• membrane-bound: double membrane
• contains DNA</td><td>Structure:
• membrane-bound: double membrane; the inner membrane is highly folded
• contains DNA</td></tr>
<tr><td colspan="2">Function:
• contains genetic information (used for the synthesis of proteins)
• directs activities of the cell</td><td>Function:
• release energy from organic compounds (cellular respiration)</td></tr>
<tr><td colspan="2">Present in plants: Yes</td><td>Present in plants: Yes</td></tr>
<tr><td colspan="2">Present in animals: Yes</td><td>Present in animals: Yes</td></tr>
<tr><td>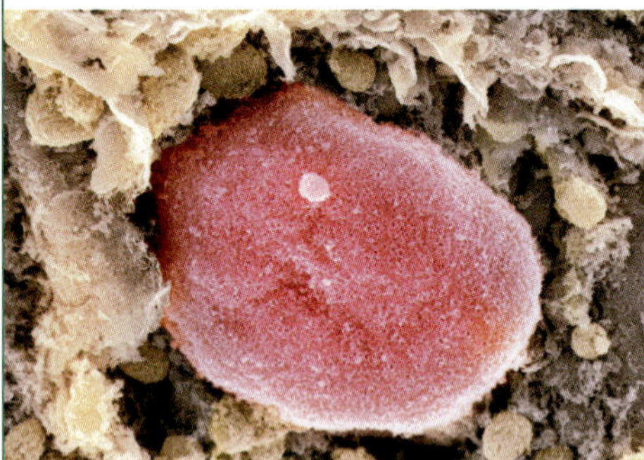
Coloured scanning electron micrograph (SEM) of a section through a liver cell showing the nucleus (pink) and the nuclear envelope with its many pores (tiny circles).</td><td>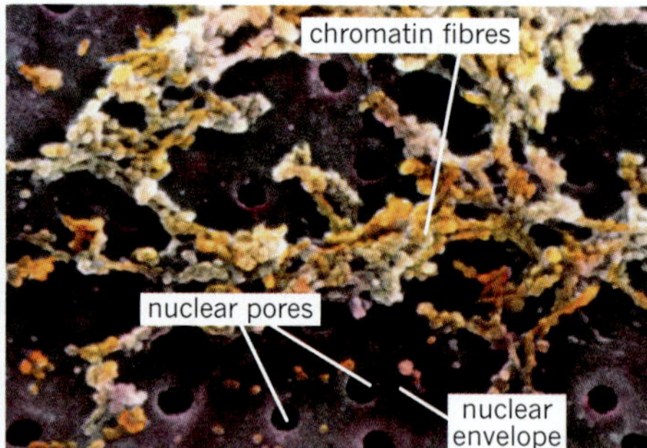

Coloured SEM of the nuclear envelope in an onion root tip cell. The envelope consists of a double membrane (purple), with nuclear pores (black circles). Contained within the nucleus are the chromatin fibres (yellow and orange).</td><td>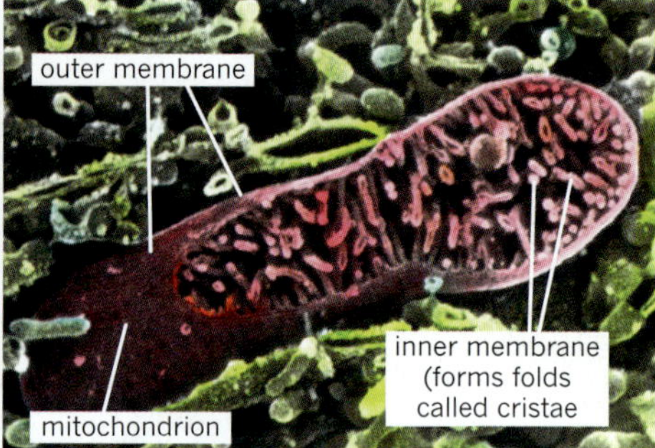

Coloured SEM of a single mitochondrion (pink, centre) in the cytoplasm of an intestinal epithelial cell.</td></tr>
<tr><th>Organelle: rough endoplasmic reticulum (RER)</th><th>Organelle: ribosome</th><th>Organelle: Golgi apparatus (also known as Golgi body, Golgi complex)</th></tr>
<tr><td>Structure:
• membrane-bound
• composed of a network of membranous tubules and sacs (called cisternae)
• ribosomes bind to the membrane</td><td>Structure:
• non-membrane-bound
• composed of proteins and ribosomal RNA
• found free in the cytosol or attached to endoplasmic reticulum (ER)</td><td>Structure:
• membrane-bound
• stack of cisternae that are not connected to each other</td></tr>
<tr><td>Function:
• synthesises and processes proteins (often by adding carbohydrates to proteins produced by the ribosomes to form glycoproteins)</td><td>Function:
• synthesises proteins (translate messenger RNA into proteins)
• RER-bound ribosomes synthesise proteins for export from the cell</td><td>Function:
• further processes and packages proteins into vesicles for export from the cell (except lysosomes, which remain in the cell)</td></tr>
<tr><td>Present in plants: Yes</td><td>Present in plants: Yes</td><td>Present in plants: Yes</td></tr>
<tr><td>Present in animals: Yes</td><td>Present in animals: Yes</td><td>Present in animals: Yes</td></tr>
<tr><td>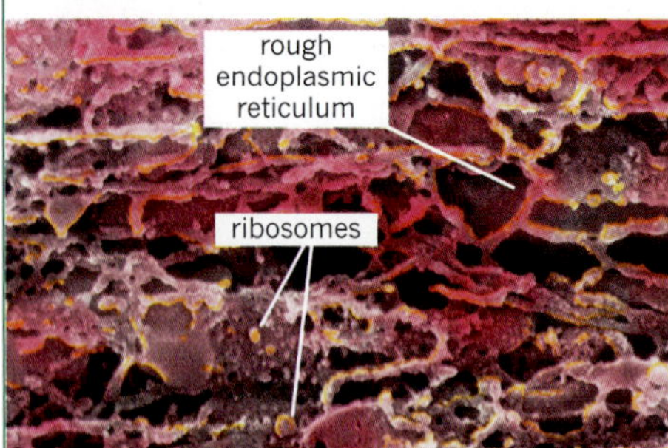

Coloured SEM of endoplasmic reticulum in an olfactory epithelium supporting cell. On the surface of some of the ER membranes are ribosomes (yellow spheres).</td><td>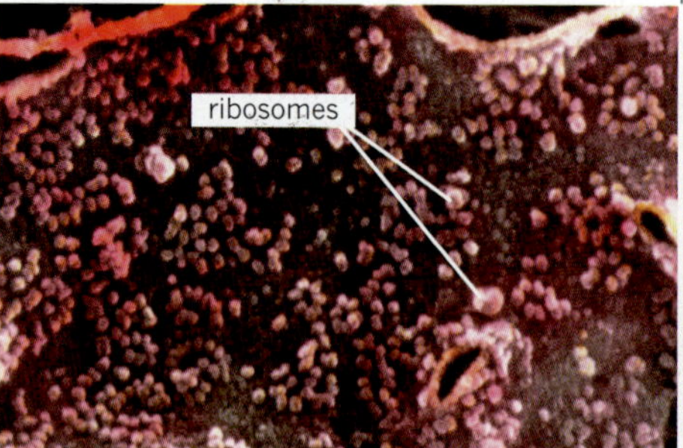

Coloured SEM of rough endoplasmic reticulum in an olfactory bulb mitral cell. On the surface of the ER membrane are numerous ribosomes (small spheres).</td><td>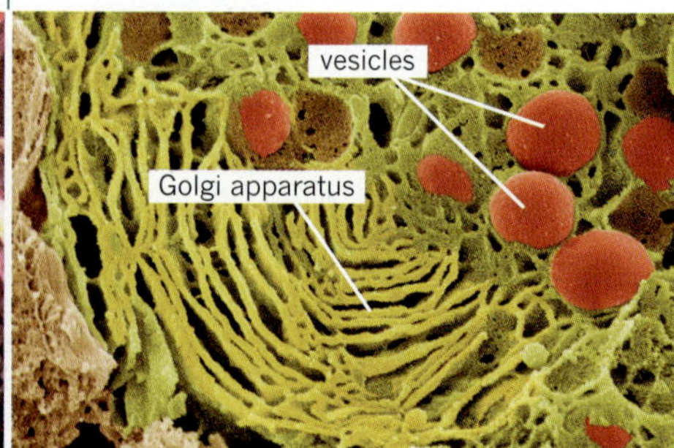

Coloured SEM of a pancreatic cell, showing the Golgi apparatus and vesicles.</td></tr>
</table>

TABLE 2.1.2 Structure and function of eukaryotic organelles (continued)

Organelle: chloroplast	Organelle: smooth endoplasmic reticulum (SER)
Structure: • membrane-bound: double membrane • contains thylakoids (disc-shaped, membranous sacs) • contains DNA	**Structure**: • membrane-bound • network of cisternae
Function: • uses light energy, carbon dioxide and water to produce glucose (photosynthesis)	**Function:** • synthesises lipids
Present in plants: Yes	**Present in plants:** Yes
Present in animals: No	**Present in animals:** Yes
Coloured transmission electron micrograph (TEM) of two chloroplasts seen in the leaf of a pea plant *Pisum sativum*. Each chloroplast is seen cut lengthways and contains stacks of flattened membranes (yellow) known as grana.	Coloured SEM showing smooth (top right) and rough (bottom centre) endoplasmic reticulum (light pink) inside a cell. Lipid droplets (round blue structures) and mitochondria can also be seen in this image.

2.1 Review

SUMMARY

- Cells are the basic structural units of organisms.
- The cell theory states that:
  - all organisms are composed of cells
  - all cells come from pre-existing cells
  - the cell is the smallest living organisational unit.
- There are two fundamentally different types of cells: prokaryotic and eukaryotic.
- Prokaryotes include bacteria and archaea.
- Eukaryotes include protists, fungi, plants and animals.
- Some common features shared by all cells are the plasma membrane, cytoplasm, genetic material in the form of DNA, and ribosomes.
- Prokaryotic cells have a simple structure, with a nucleoid lacking a membrane, scattered ribosomes, and DNA mainly in a single-stranded loop in the nucleoid.
- Eukaryotic cells have a complex structure, a membrane-bound nucleus, many organelles in the cell cytoplasm, and DNA mainly in chromosomes in the nucleus.
- The main structures in a plant cell include the nucleus, vacuole, Golgi apparatus, rough and smooth endoplasmic reticulum, ribosomes, mitochondria and cell wall.
- The main structures in an animal cell include the nucleus, ribosomes, Golgi apparatus, rough and smooth endoplasmic reticulum, vacuoles, mitochondria, lysosomes, vesicles and centrioles.

2.2 Molecular composition of organisms

REVISION

Elements are substances consisting of one type of atom.

Compounds are made up of two or more different types of atoms.

Molecules are made up of two or more atoms of the same or different types.

All life is composed of the same few elements. There are 92 naturally occurring elements. Only 11 of these are found in organisms in more than trace amounts, and four of these—carbon (C), hydrogen (H), oxygen (O) and nitrogen (N)—make up 99% of organisms by mass. The same elements are also found in rocks, soil and air; however, there is a difference in the way that these atoms are organised into larger compounds in living organisms (Figure 2.2.1). Organisms produce compounds that contain carbon and hydrogen known as organic compounds. All other compounds, whether in living or non-living things, are called inorganic compounds.

In this section, you will learn about the difference between organic and inorganic compounds. In addition, the four main types of organic molecules—nucleic acids, carbohydrates, lipids and proteins—will also be explored.

FIGURE 2.2.1 Aerial view of the Murray River wetlands lagoons and basin. The plants, water and animals all contain molecules. Only the plants and animals produce organic molecules.

Organic molecules

Organisms produce characteristic complex compounds that contain carbon and hydrogen (Figure 2.2.2). These are called organic compounds because the first of these compounds to be discovered were produced by or found in organisms. Most large organic molecules are composed of many smaller organic molecules linked together.

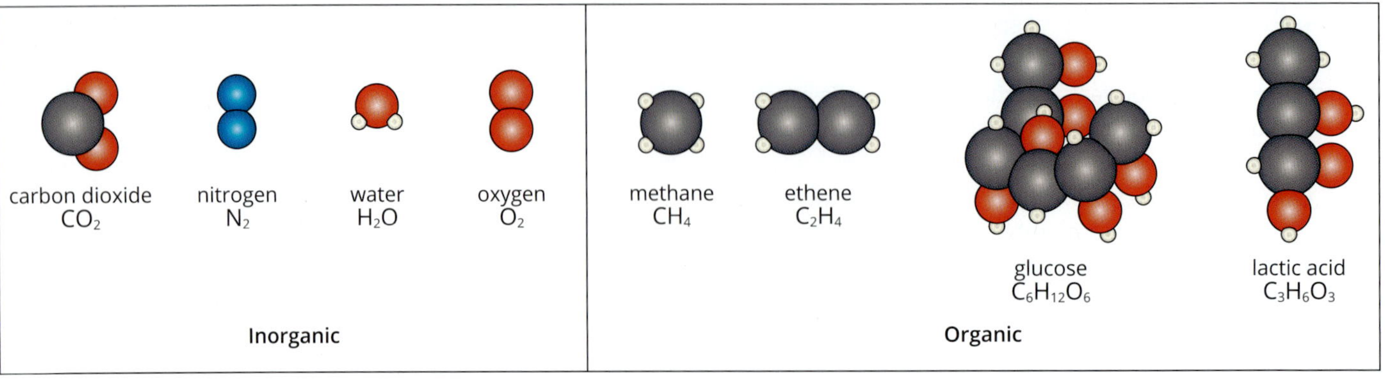

FIGURE 2.2.2 Some common molecules in organisms. Carbon atoms are coloured black, oxygen red, hydrogen white and nitrogen blue.

All other elements and compounds, whether in living or non-living things, are referred to as inorganic. Inorganic substances that are important for living organisms include water, oxygen, carbon dioxide, nitrogen and minerals.

The four main types of organic molecules are carbohydrates, proteins, nucleic acids and lipids (Figure 2.2.3). Carbohydrates, proteins and nucleic acids are huge and are also known as biomacromolecules. Biomacromolecules are chain-like molecules called polymers (poly meaning 'many' and mero meaning 'part'). Polymers are formed by joining together many smaller units (monomers; mono meaning 'one or single') to form a chain.

In organisms, organic molecules can be converted from one form into another. Units may be linked together to form larger molecules. For example, glucose units may be linked together to form larger carbohydrates such as starch, glycogen or cellulose (Figure 2.2.3a). Other chemical groups may be attached to form molecules such as glycoproteins (proteins with sugars attached, Figure 2.2.4) and phospholipids (lipids with phosphate attached, Figure 2.2.5). When food is plentiful, carbohydrates are converted into fats for storage; when it is scarce, the reverse will occur. Even proteins can be converted into small molecules to use for energy.

> A molecule is made up of two or more atoms that are held together by chemical bonds. Any molecule that is found in a living organism is called a biomolecule. Examples of biomolecules are fatty acids, carbohydrates and hormones.

> Large biomolecules are called biomacromolecules. Biomacromolecules can be made up of thousands of atoms and include proteins and nucleic acids.

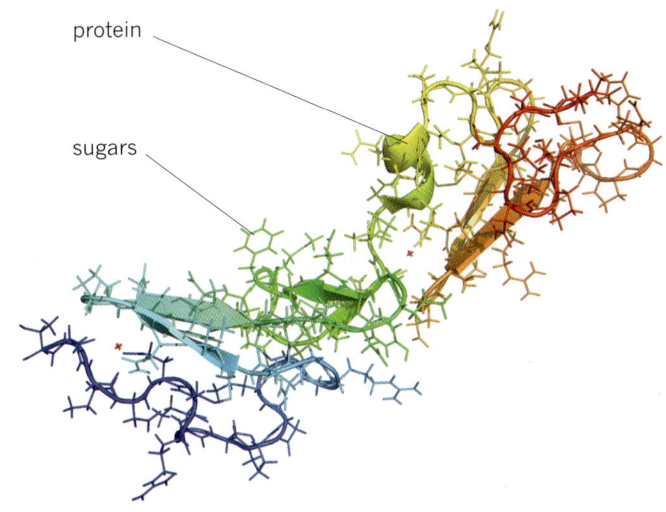

FIGURE 2.2.4 Computer model of a fibrillin glycoprotein. The protein component is represented by ribbons and the sugars are represented by the ring structures.

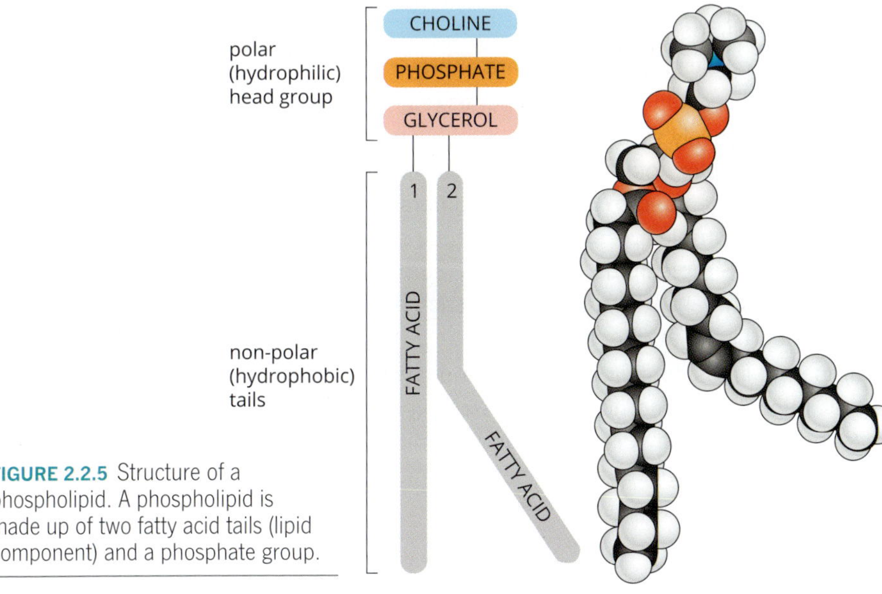

FIGURE 2.2.5 Structure of a phospholipid. A phospholipid is made up of two fatty acid tails (lipid component) and a phosphate group.

a Carbohydrates

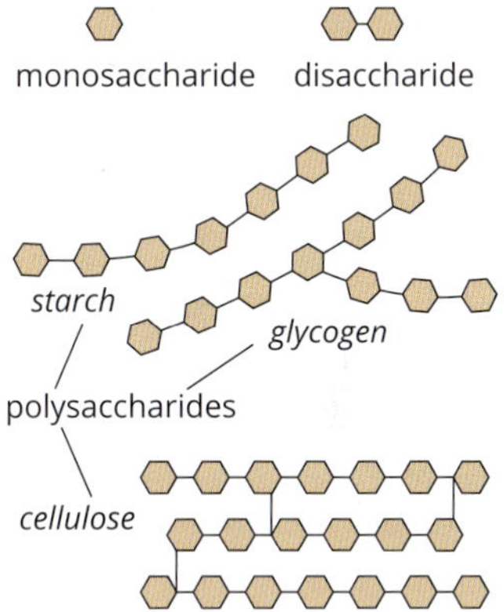

b Proteins

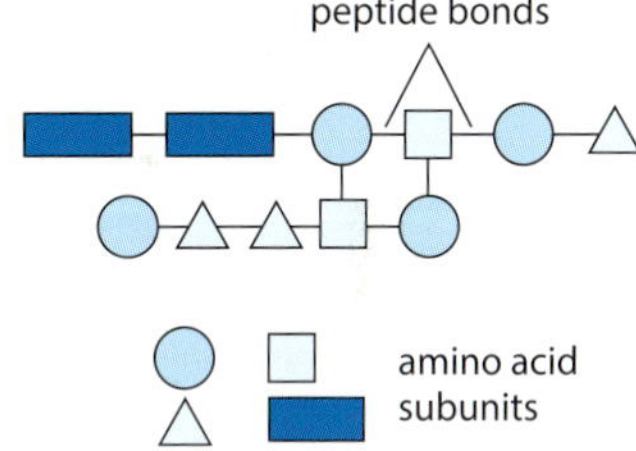

c Nucleic acids

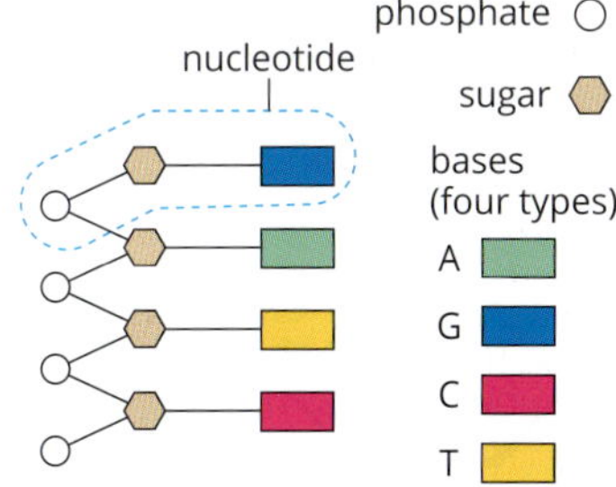

d Lipids

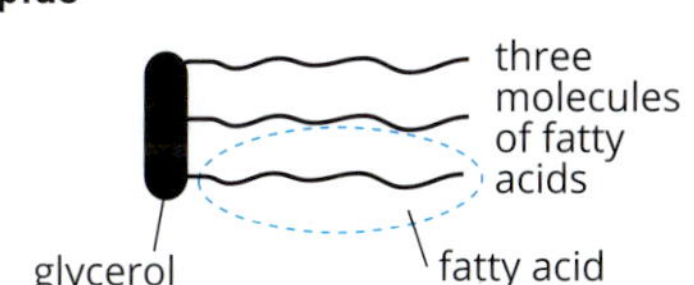

FIGURE 2.2.3 Structures of some organic molecules: (a) carbohydrates, (b) proteins, (c) nucleic acids and (d) lipids

a **Glucose**

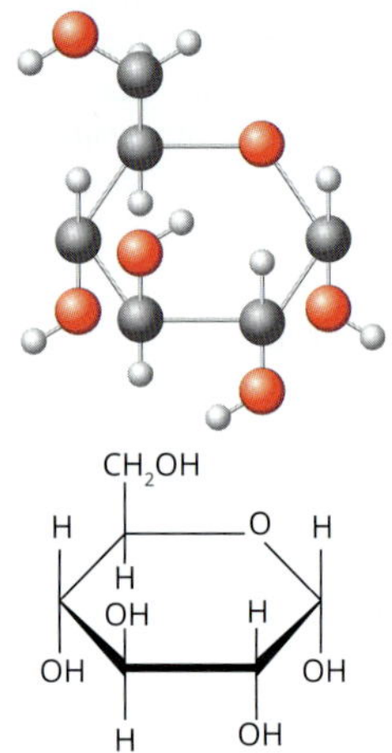

b **Fructose**

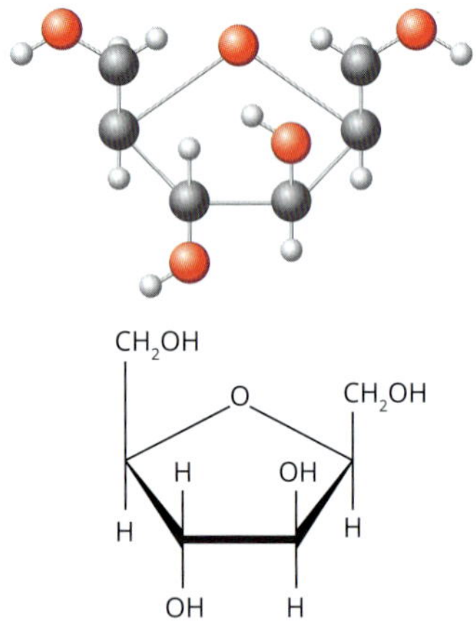

c **Galactose**

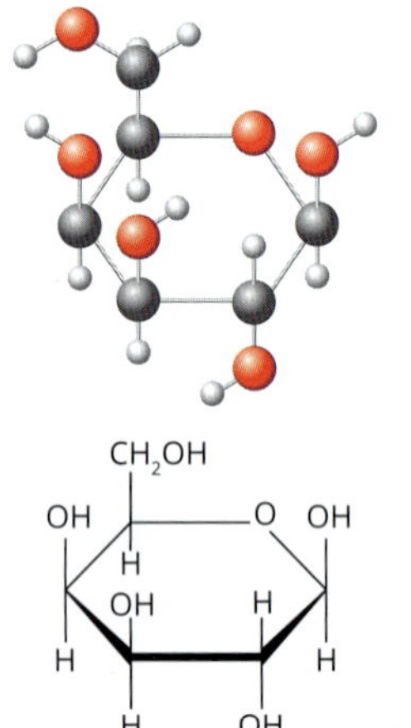

FIGURE 2.2.6 Structural chemical formula and model of three monosaccharides: (a) glucose, (b) fructose and (c) galactose. In the models, grey spheres represent carbon atoms, white spheres represent hydrogen atoms and red spheres represent oxygen atoms.

Carbohydrates

Carbohydrates are the most abundant organic molecules in nature.

- They are an important source of chemical energy for living organisms (e.g. glucose).
- They are used as energy reserves in plants (e.g. starch) and animals (e.g. glycogen).
- They form structural components such as cell walls (e.g. cellulose in plants).
- They form part of both DNA and RNA.
- They combine with proteins and lipids to form glycoproteins and glycolipids, as in plasma membranes.

Carbohydrates are compounds made of carbon, hydrogen and oxygen. There are three main groups of carbohydrates: monosaccharides, disaccharides and polysaccharides. The basic subunits of carbohydrates are the simple sugars, called monosaccharides, meaning 'single sugar' (Figure 2.2.3a on page 75). Examples of monosaccharides include glucose, fructose and galactose (Figure 2.2.6).

Carbohydrates are organic compounds, such as sugars, starch and cellulose, that are made of carbon, hydrogen and oxygen. Carbohydrates include monosaccharides, disaccharides and polysaccharides (complex carbohydrates). Only polysaccharides are polymers and hence also fall under the category of biomacromolecules.

When many sugars are joined together they form biomacromolecules called polysaccharides ('many sugars'). Cellulose, the major component of plant cell walls, is the most abundant organic molecule on Earth. Starch is the polysaccharide used for energy storage in plants. In animals, the polysaccharide glycogen is used for energy storage. These three polysaccharides are each composed of glucose subunits, but they differ in a number of ways (Figure 2.2.3a on page 75). Starch is a long chain molecule, glycogen has a branching structure and cellulose has additional bonds forming cross-linking between the subunits of the chain.

Lipids

Lipids are 'fatty' substances that are composed of non-polar hydrophobic molecules. This means that lipids are insoluble in water, giving rise to their critical role in living organisms—they can form an effective barrier between two watery environments.

The roles of lipids in organisms include:

- as the main component of plasma membranes and organelle membranes (as phospholipids)
- storing energy (e.g. fats and oils are energy-storing molecules)
- playing an important role as hormones (e.g. steroids).

Lipids are relatively small molecules and vary widely in structure. There are two general forms of lipids—simple and compound. Simple lipids are composed of carbon, hydrogen and oxygen, but in different proportions to those of carbohydrates. Simple lipids contain a much smaller proportion of oxygen than do carbohydrates. Simple lipids include fats (composed of fatty acids and glycerol, see Figure 2.2.3d on page 75), and steroids such as cholesterol (Figure 2.2.7) and the hormones cortisone and testosterone.

Compound lipids contain fatty acids and glycerol, as well as other elements such as phosphorus and nitrogen. Phospholipids have a hydrophilic end (the 'phospho' end) and a hydrophobic end (the lipid end). Phospholipids are the main components of biological membranes and their fundamental role in membrane function is described in Section 2.3.

Nucleic acids

Nucleic acids are the genetic material of all organisms, and they determine many of the features of an organism. Nucleic acids are biomacromolecules composed of long chains of monomers called nucleotides. A nucleotide consists of a phosphate, a sugar and a nitrogenous base (Figure 2.2.8).

There are two types of nucleic acids:

- Deoxyribonucleic acid (DNA)—DNA carries the 'instructions' required to assemble proteins from amino acid monomers using a genetic code. It is passed accurately from cell to cell during cell division. The four bases in DNA are adenine (A), thymine (T), guanine (G) and cytosine (C).
- Ribonucleic acid (RNA)—RNA plays a major role in the manufacture of proteins within cells. The four bases in RNA are adenine (A), uracil (U), guanine (G) and cytosine (C).

Proteins

Proteins are more complex than carbohydrates or lipids. Proteins make up over 50% of the dry weight of cells. There are thousands of different kinds of proteins, and their functions vary widely. Although carbohydrates and lipids are similar in all plants and animals, organisms can have a variety of unique proteins that are specific to a particular species.

Protein functions vary widely. Proteins can:

- catalyse cellular reactions (e.g. enzymes such as amylase)
- play an important role as hormones (e.g. insulin)
- act as carrier molecules (e.g. haemoglobin)
- form structural components in organisms (e.g. collagen)
- play an important role in the immune system (e.g. antibodies and antigens).

All proteins contain carbon, hydrogen, oxygen and nitrogen; many also contain sulfur, and often phosphorus and other elements. Proteins are biomacromolecules composed of chains of monomers called amino acids (Figure 2.2.9). Amino acids are linked by a particular kind of chemical bond, called a peptide bond, and form polypeptides or polypeptide chains. (Polypeptide means 'many peptide bonds'.) A protein is formed by one or more polypeptides arranged in a biologically functional way. In other words, 'protein' is the term used for a fully functioning molecule, while 'polypeptide' refers to a non-functioning version.

There are 20 different amino acids commonly found in proteins. Nine of these are known as essential amino acids because they cannot be produced by humans. So, humans must obtain these amino acids by consuming other organisms.

The properties of many proteins are determined by their shape, which is determined by their amino acid sequence. You will learn more about proteins, including their shapes and functions, in Section 2.3.

Cholesterol

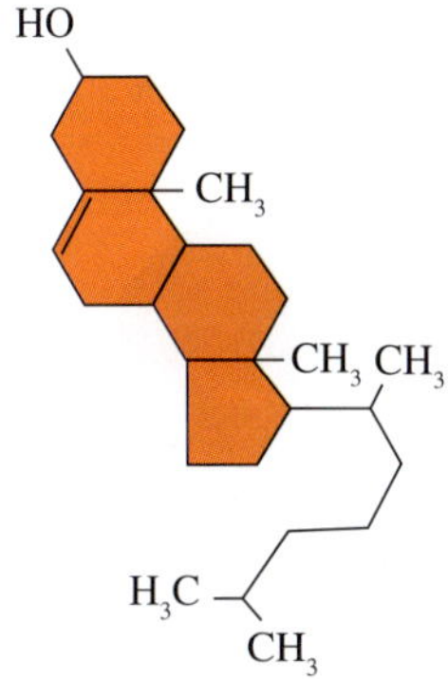

FIGURE 2.2.7 Structure of cholesterol

Nucleic acids

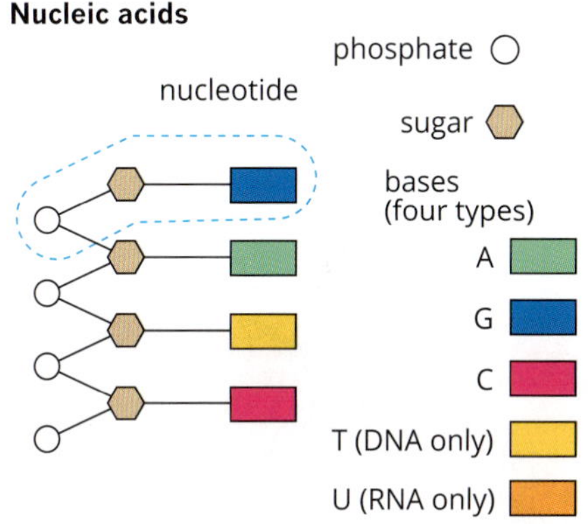

FIGURE 2.2.8 Structure of a nucleic acid

Proteins

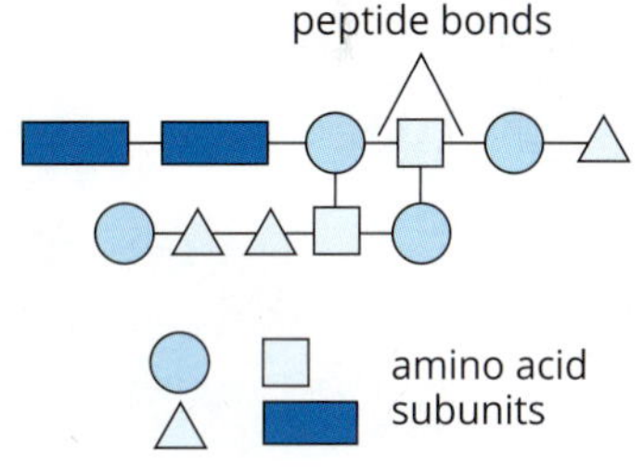

FIGURE 2.2.9 Structure of proteins

Enzymes are a type of protein that speed up the rate of cellular reactions without being consumed in the reaction.

Inorganic substances

Oxygen and carbon dioxide

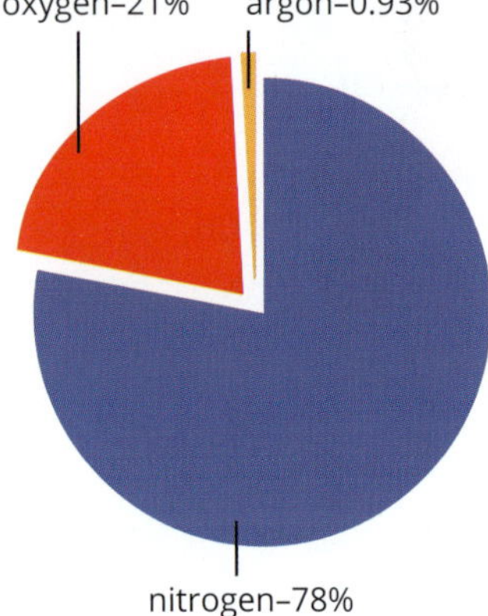

FIGURE 2.2.10 Air is a mixture that is mainly made up of nitrogen, oxygen and argon.

In most cells, oxygen is needed to release energy from food molecules in processes known collectively as cellular respiration. A constant supply of oxygen is therefore necessary to maintain the activity of these cells. This is usually easy for organisms that get their oxygen from air, because the atmosphere is 21% oxygen (Figure 2.2.10). However, oxygen is not very soluble in water, so organisms that get their oxygen from water are either small, flat and relatively inactive, or they have very efficient ventilation systems with large surface area for gaseous exchange, such as fish gills (Figure 2.2.11).

FIGURE 2.2.11 Coloured SEM of mackerel (*Scomber scombrus*) gills showing the large surface area for gaseous exchange

Carbon is the key atom in organic molecules. Carbon dioxide is taken from the atmosphere (which contains approximately 0.035% by volume of carbon dioxide, Figure 2.2.10) by plants, some bacteria and some protists. It is used in the process of photosynthesis to make sugars, some of which are eaten by animals. Carbon dioxide is returned to the atmosphere mainly by the decay of organic material and as an end-product of cellular respiration. This cycling of carbon through organisms and the atmosphere is critical to the survival of all organisms.

Nitrogen

Nitrogen is required by organisms in relatively large amounts because it is a key component of all proteins. There is plenty of nitrogen around because the atmosphere is about 78% nitrogen gas (Figure 2.2.10); however, most organisms are unable to use nitrogen in this form. Atmospheric nitrogen is converted by certain bacteria and cyanobacteria into compounds that can be used by plants in a process known as nitrogen fixation. The most important nitrogen-fixing bacteria are the symbiotic bacteria found in the roots of plants, including legumes, casuarinas and acacias (Figure 2.2.12). Nitrogen compounds produced by the bacteria in the soil are absorbed by plants and used to make amino acids. Heterotrophs obtain their amino acids by consuming plants and other organisms. They also produce nitrogen-rich waste (manure), which has traditionally been used as a plant fertiliser.

FIGURE 2.2.12 (a) Nodules containing nitrogen-fixing bacteria on the roots of a garden pea (*Pisum sativum*). (b) Coloured SEM of nitrogen-fixing soil bacteria (*Rhizobium* species) in a root nodule of a bean plant. These bacteria (green) have a symbiotic relationship with the plant.

Minerals

Mineral salts are naturally occurring inorganic compounds produced by the weathering of rocks. The water-soluble mineral salts produced are absorbed as ions into the roots of plants (Figure 2.2.13), making them available to be eaten by animals. Humans require more than 20 minerals. Biologically important minerals include phosphorus, potassium, calcium, magnesium, iron, sodium, iodine and sulfur. Many others are needed in small (trace) amounts.

Mineral ions are found in the cytosol of cells, in structural components (such as bone), and in the molecules of many enzymes and vitamins. They may also be incorporated into other important organic compounds in cells. Phosphorus is present in the phospholipids of plasma membranes and in ATP (adenosine triphosphate—an important energy carrier in cells, see Chapter 5). Magnesium is an important constituent of chlorophyll, and iron is the central atom in every haemoglobin molecule in red blood cells (Figure 2.2.14). Calcium, potassium and sodium ions are important for the normal performance of cardiac muscle cells, and calcium and phosphorus are found in bones and teeth (Figure 2.2.15).

FIGURE 2.2.13 Soil profile showing the horizons (layers), which vary in colour depending on the mineral content in the soil. Plants absorb these minerals when they draw water out of the soil.

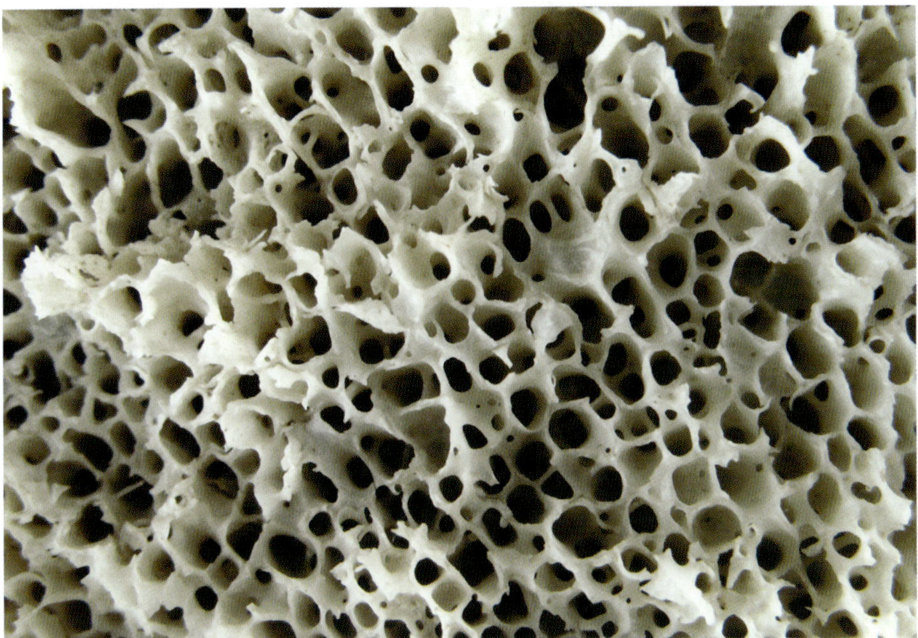

FIGURE 2.2.15 Bone matrix is made up of inorganic components including salts of calcium and phosphorus.

> Molecules of ATP provide energy for immediate use by the cell and are produced during glycolysis and cellular respiration.

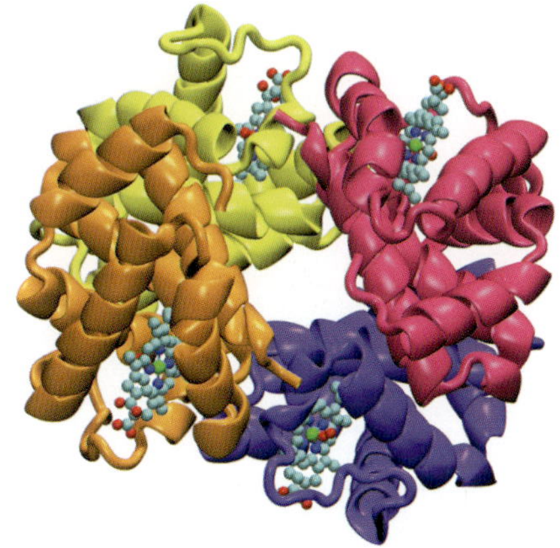

FIGURE 2.2.14 Haemoglobin is made up of four protein subunits (coloured ribbon structures). Each subunit has an oxygen-binding site, or haem group (turquoise). Within each haem group there is one atom of iron (green). Oxygen molecules are shown as paired red spheres.

2.2 Review

OA ✓✓

SUMMARY

- Organic compounds contain carbon and hydrogen and are found in living things.
- Inorganic substances are all elements and compounds other than organic compounds (e.g. oxygen, water and carbon dioxide).
- There are four main types of organic compounds: carbohydrates, proteins, nucleic acids and lipids.
- Biomacromolecules are large organic molecules formed by joining together many smaller units (monomers) to form a chain or polymer.
- Plants convert inorganic molecules into carbohydrates.
- Carbohydrates can be used as a source of energy and as energy stores. They also form structural components of cells and make up part of the structure of nucleic acids.
- Lipids are fatty substances that are not soluble in water. They can be used in plasma membranes, store energy and act as hormones.
- Nucleic acids contain the genetic material of all organisms. The two types of nucleic acids are RNA and DNA.
- The functions of proteins vary widely. Proteins can:
  - catalyse cellular reactions
  - play an important role as hormones
  - act as carrier molecules
  - form structural components in organisms
  - play an important role in the immune system.
- Oxygen is needed in most organisms to release energy from food molecules.
- Atmospheric carbon dioxide is the main source of carbon, which is the key atom in organic molecules.
- Nitrogen is 'fixed' from the atmosphere by certain bacteria.

2.3 Proteins

Nearly every function of a living organism depends on proteins. Proteins have a large range of functions in living organisms including speeding up chemical reactions, playing a role in cell-to-cell recognition and cellular communication, movement, storage and even structural support. A human has tens of thousands of different proteins, and each protein has a specific sequence of amino acids, giving it a unique shape that enables it to carry out a particular function.

In this section, you will learn about the nature of the proteome and the diversity in the functions of proteins. You will also learn about the synthesis of a polypeptide chain from amino acid monomers and the functional importance of the four hierarchical levels of protein structure.

THE NATURE OF THE PROTEOME

A **protein** is an organic compound consisting of one or more long chains of amino acids connected by peptide bonds. Proteins are present in every living organism and are essential to their structure and function.

The **proteome** is the complete set of proteins expressed by the **genome**—the complete set of genes or genetic material—of an individual cell or organism at a given time. The proteome varies between cell type, developmental stage and environmental conditions. Although a cell may contain the entire genome, only specific genes will be expressed, or 'switched on', at any given time. This ensures a cell produces only the proteins required for the specific functions it carries out.

For example, all human somatic cells (any body cell of an organism, apart from cells that give rise to eggs and sperm) in an individual contain identical genomes, but the array of proteins produced by a fibroblast cell is different from the array of proteins produced by a B lymphocyte. Fibroblasts are found in connective tissue and produce collagen to give the tissue strength and elasticity. B lymphocytes are found in the blood circulation and lymphatic system, and produce antibodies to help defend the body against infection or foreign, non-self, materials. A fibroblast does not produce antibodies and a B lymphocyte does not produce collagen. Each cell type only produces the proteins required to carry out its own specific functions.

There are many similarities between human and other proteomes, reflecting their common evolutionary origins. The human proteome contains proteins related by evolutionary descent (homologous) with 61% of the fruit fly proteome, 43% of the worm proteome and 46% of the proteome of baker's yeast.

Proteomics

Proteomics is the large-scale study of the structure, function and interactions of proteins. Proteomics is essential as it is proteins that actually carry out most of the activities of the cell, not the genes that encode them. By knowing when and where proteins are produced in an organism, as well as how proteins interact, we can better understand the functioning of cells and organisms.

One of the ways to determine changes in the proteome is by comparing the proteomes of cells under different conditions. For example, by comparing the protein expression of a diseased cell and a healthy cell, the proteins affected by the disease can be determined.

The research from proteomics can lead to the creation of protein biomarkers that can be used for screening individuals and populations for early detection of disease. The study of proteomics is also important in the production of drugs that interact with proteins involved in disease and alter their function.

BIOFILE

Rational drug design

Rational drug design uses high-speed computers to compare the three-dimensional structure of a faulty protein with a database containing many different chemical compounds. The compounds most likely to interact with the faulty protein are identified, and the interactions between these compounds and the faulty protein can then be tested in the laboratory to design drugs.

An example of a drug that has been created using rational drug design is Gleevec. This drug has been designed to interact with a faulty enzyme that leads to an overproduction of abnormal white blood cells in a rare type of leukaemia. Gleevec binds to the active site of the enzyme, altering its shape and preventing it from functioning.

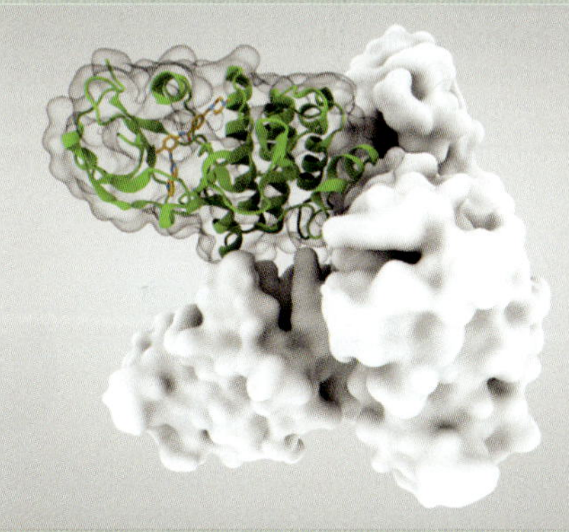

Illustration of Gleevec, a drug created using rational drug design.

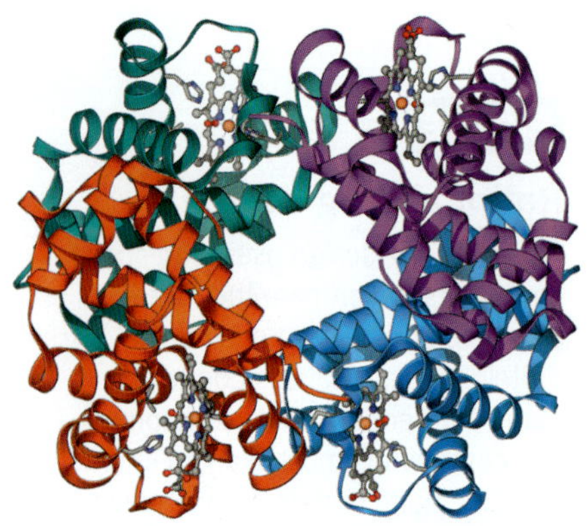

FIGURE 2.3.1 Ribbon diagram showing the three-dimensional structure of haemoglobin, a protein made up of four polypeptide chains

THE FUNCTIONAL DIVERSITY OF PROTEINS

Proteins have a large variety of functions, so they vary extensively in structure, with each type of protein having a unique three-dimensional shape (Figure 2.3.1). There are many different types of proteins in every organism. Each protein has a different function, and each plays a vital role in the regulation, functioning and maintenance of both individual cells and entire organisms. In fact, almost every function of living organisms depends on proteins. The specific structure of each protein enables it to carry out its function.

Some of the functional types of proteins are described in Table 2.3.1.

Enzymes

One of the most important groups of proteins is enzymes. **Enzymes** act as biological **catalysts** in metabolic reactions. This means that enzymes speed up the rate of biochemical reactions without being consumed within the reaction. They are large globular structures that act within specific reactions to either speed up **anabolic reactions** (reactions that make larger molecules) or **catabolic reactions** (reactions that break down larger molecules into smaller molecules). For example, lipase is an enzyme involved in the breakdown of the lipids during digestion. Without enzymes, many reactions within cells would be too slow to sustain life. You will learn more about the action of enzymes in Chapter 5.

TABLE 2.3.1 An overview of protein function

Function: enzymatic proteins	Function: hormonal proteins
Description: act as catalysts in biochemical pathways (enzymes) **Examples:** • catabolic enzymes, such as amylase, that catalyse the breakdown of bonds (also known as hydrolysis) (shown below) • anabolic enzymes, such as DNA polymerase, that catalyse the formation of bonds (also known as condensation reactions)	**Description:** coordinate an organism's activities by triggering a response **Examples:** insulin, glucagon
Catabolic enzymes, such as amylase, catalyse (speed up) reactions in which their specific substrate is broken into smaller products.	high blood sugar insulin secreted normal blood sugar The hormone insulin regulates blood sugar levels.
Function: immunological proteins	**Function: contractile and motor proteins**
Description: • protect against disease by recognising foreign bodies and microbes • activate immune cells **Examples:** immunoglobulins (antibodies), complement, major histocompatibility complex proteins	**Description:** • contractile proteins aid muscle contraction • motor proteins are responsible for the movement of cilia and flagella **Examples:** myosin, actin, kinesin , dynein
antibodies virus bacterium Antibodies help destroy viruses and bacteria.	actin myosin muscle tissue 30 μm Actin and myosin are responsible for muscle contraction.

TABLE 2.3.1 An overview of protein function (continued)

<table>
<tr><th>Function: structural proteins</th><th>Function: transport proteins</th></tr>
<tr><td>Description:
• provide support by forming the structural components of cells and organs
• assist in contractile functions in tissue such as muscle
Examples: collagen, keratin, actin, cytoskeleton</td><td>Description:
• transport of substances by acting as carrier molecules within or between cells
• act as membrane channel proteins
Examples: haemoglobin, sodium–potassium pump, calcium channel</td></tr>
<tr><td>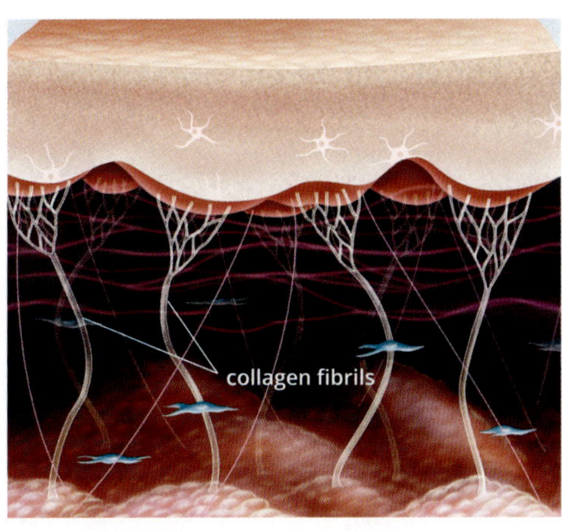

Collagen fibrils provide elasticity and support to the skin.</td><td>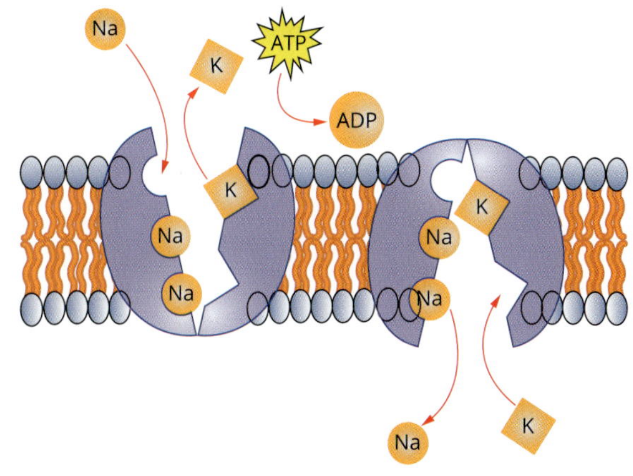

Sodium–potassium pump uses ATP to transport sodium ions and potassium ions across the plasma membrane.</td></tr>
<tr><th>Function: receptor proteins</th><th>Function: storage proteins</th></tr>
<tr><td>Description: assist the cell in responding to a chemical stimuli
Examples: neurotransmitter receptors, hormone receptors</td><td>Description: storage of metal ions and amino acids
Examples: ovalbumin and casein (to store amino acids), and ferritin (to store iron)</td></tr>
<tr><td>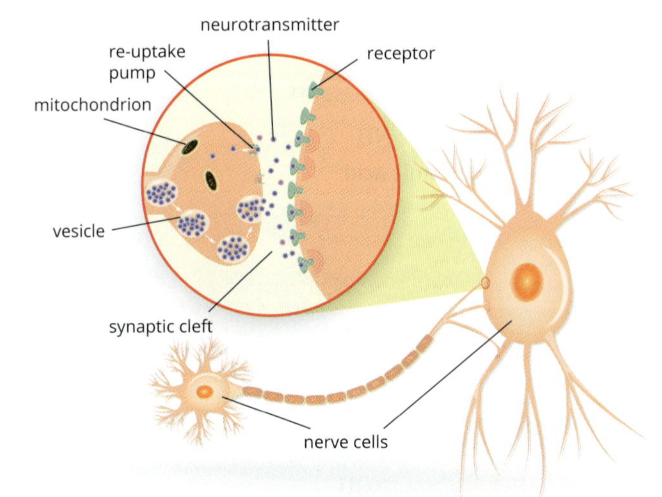

Receptors built into the plasma membrane of nerve cells detect neurotransmitters secreted by neighbouring nerve cells.</td><td>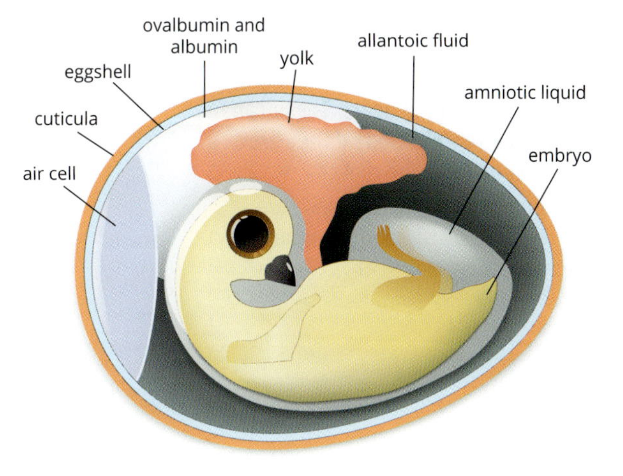

Ovalbumin is a protein found in egg white, used as an amino acid source for the developing embryo.</td></tr>
</table>

SYNTHESIS OF PROTEINS

There are many steps involved in producing a functional protein. As you will recall from Section 2.2, although protein structure, size and function are quite diverse, all proteins are made up of amino acids. **Amino acids** are smaller subunits (known as **monomers**) that are joined together in a particular order to form **polypeptide chains**. The polypeptide chains are then folded and coiled into proteins.

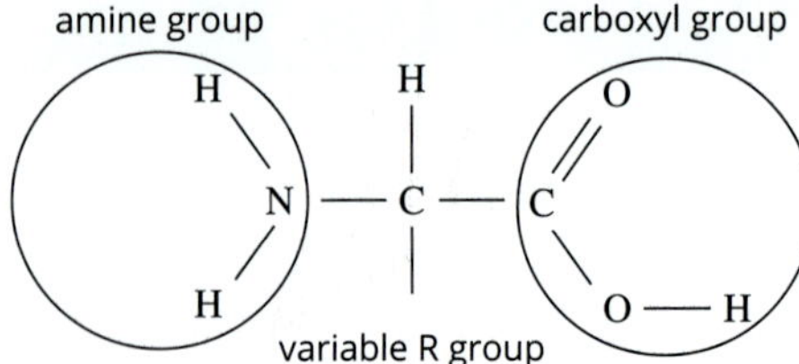

FIGURE 2.3.2 Basic structure of an amino acid, showing an amine, carboxyl and variable R groups

Amino acid structure

All amino acids have the same basic structure (Figure 2.3.2):

- an **amine group** (NH_2)
- a **carboxyl group** (COOH)
- a **variable R group** (or side chain).

In the synthesis of proteins in organisms, there are 20 different standard (or canonical) amino acids, and each has a different R group (Figure 2.3.3). The variable properties of the R group (e.g. charged or uncharged, polar or non-polar, hydrophobic or hydrophilic) determine the type of protein that the amino acid will form. R groups can be as simple as a hydrogen atom (as in the amino acid glycine) or more complex; for example, $-CH(CH_3)_2$ in the amino acid valine.

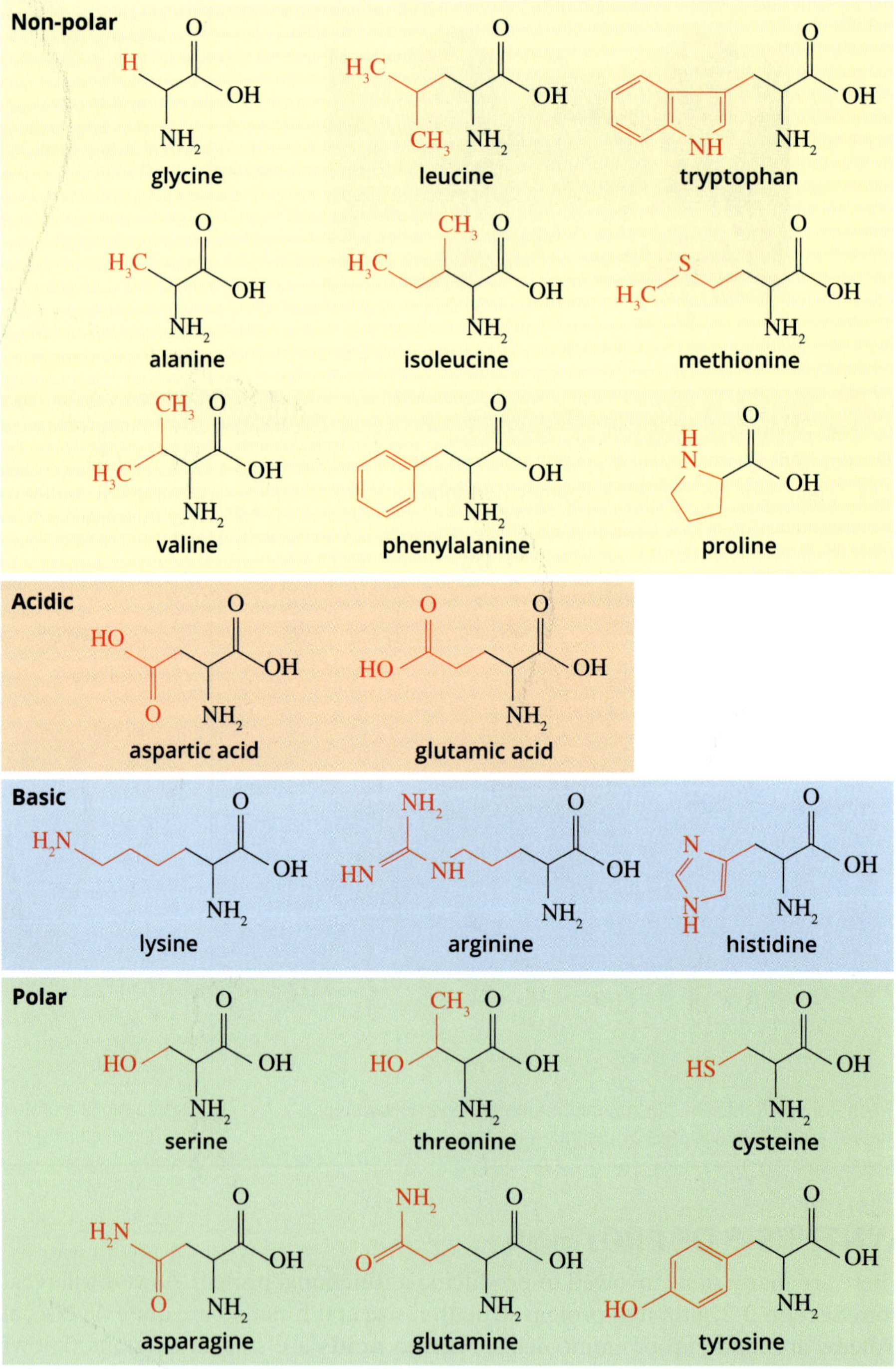

Non-polar

glycine, leucine, tryptophan, alanine, isoleucine, methionine, valine, phenylalanine, proline

Acidic

aspartic acid, glutamic acid

Basic

lysine, arginine, histidine

Polar

serine, threonine, cysteine, asparagine, glutamine, tyrosine

FIGURE 2.3.3 Chemical structure of the 20 standard amino acids. The R groups are coloured in red. Amino acids can be classified according to their chemical nature as non-polar, acidic, basic or polar.

BIOFILE

Amino acids in the human diet

In the human diet, amino acids can be classified into three main groups:

- Essential amino acids: These cannot be synthesised by the body and must be obtained from our diet. The nine essential amino acids are histidine, isoleucine, leucine, lysine, methionine, phenylalanine, threonine, tryptophan and valine.
- Non-essential amino acids: These can be produced by the body if they are not obtained from the diet.
- Conditional amino acids: These are only required by the body in times of illness or stress.

Polypeptide chains

Amino acids are joined by **peptide bonds**. A chain of amino acids joined by peptide bonds is known as a polypeptide chain. The backbone of the polypeptide chain is formed by the repeats of the carboxyl and amine groups, with the R groups forming the side chains of the polypeptide chain (Figure 2.3.4).

A polypeptide chain forms the **primary structure** of a protein. With further folding and modification, a fully functional protein can be formed.

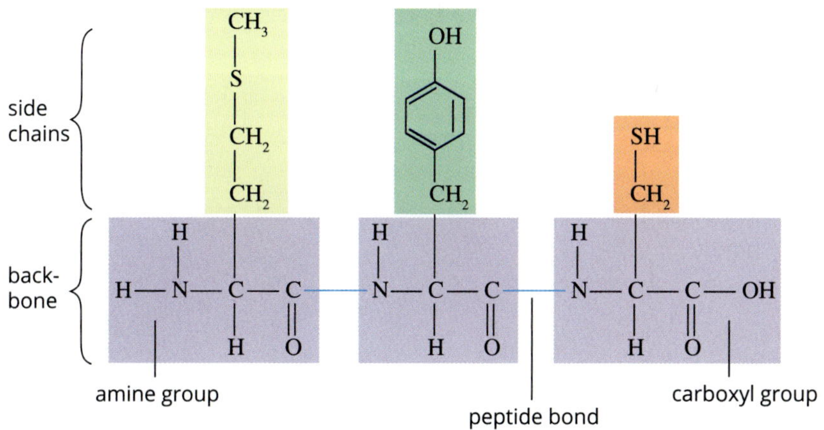

FIGURE 2.3.4 Formation of a polypeptide chain

BIOFILE

Protein size

Proteins come in vastly different sizes. The peptide hormones oxytocin and antidiuretic hormone are only nine amino acids long. Dystrophin is a large protein that is important for muscle cell structure. It has about 3600 amino acids. Defects in this protein lead to muscular dystrophy. Another muscle protein, titin, is involved in the elasticity of the muscle and is the largest known protein. It has about 30 000 amino acids!

A chain of amino acids joined by peptide bonds is known as a polypeptide chain.

PROTEIN STRUCTURE

Proteins are large biomolecules that can contain thousands of amino acids and may be synthesised as one or several polypeptide chains. These polypeptide chains are folded and organised into specific shapes that are vital to the correct functioning of the protein. Most proteins are required to bind to other molecules. A single change to one amino acid within the sequence can alter the shape, and consequently the function, of a protein.

There are four different levels of organisation when describing protein structure (Figure 2.3.5 on page 86):

- primary structure
- secondary structure
- tertiary structure
- quaternary structure.

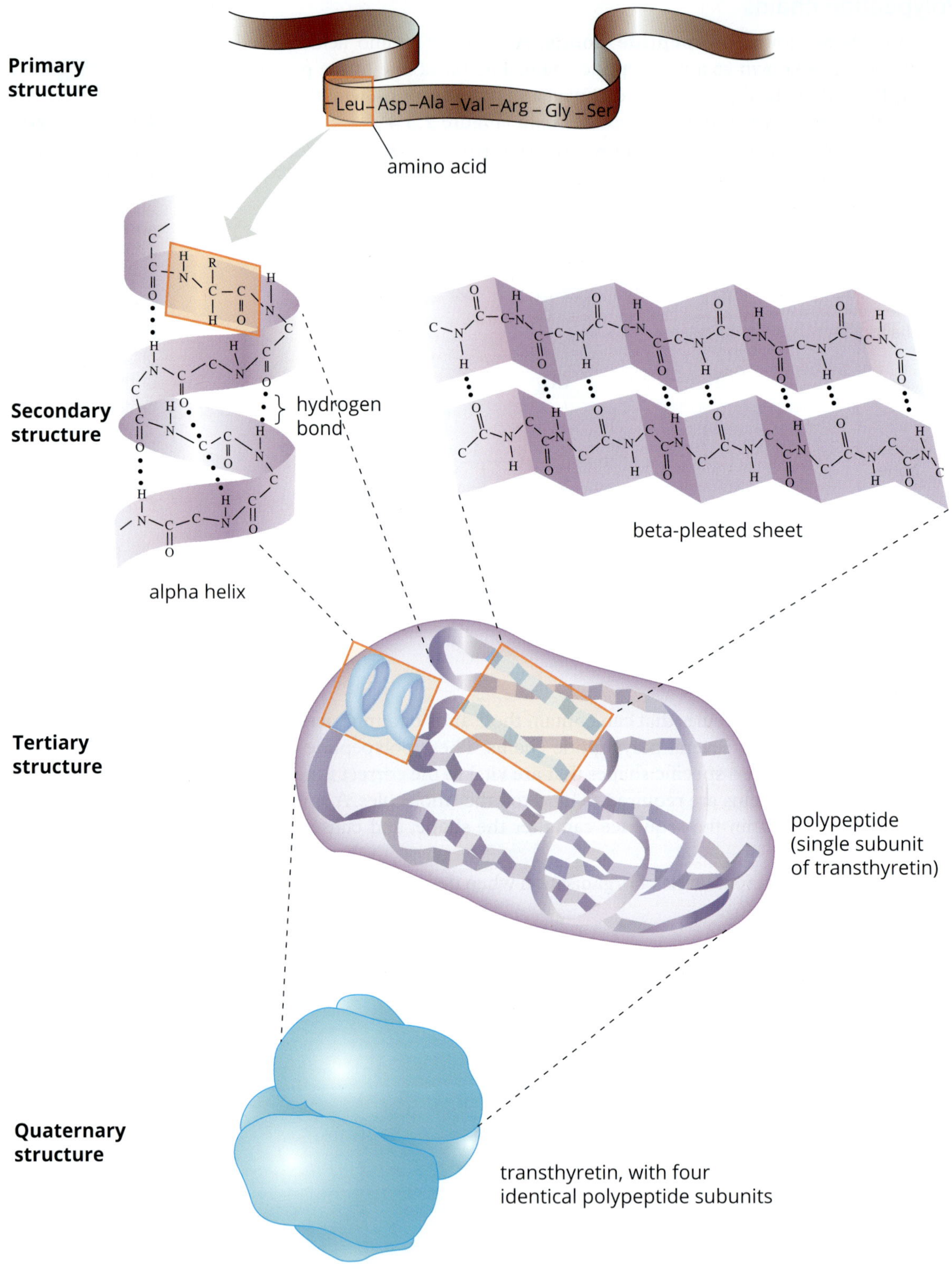

FIGURE 2.3.5 Overview of the four levels of protein structure—primary, secondary, tertiary and quaternary

Primary structure

The linear sequence of amino acids in the polypeptide chain is referred to as the primary structure of a protein (Figure 2.3.6). It is unique to each protein. The primary structure of one polypeptide may consist of only 50 amino acids, whereas another may consist of 103 amino acids. Linear sequences shorter than 50 amino acids are known as **peptides**.

The linear sequence:

- provides information on how proteins will fold
- of functional and non-functional proteins can be compared to identify what changes to the sequence render the protein non-functional
- can be compared between proteins to determine the evolutionary history of a protein.

Secondary structure

The next step in the formation of a functional protein is the folding or coiling of the polypeptide chain—its **secondary structure**. Folding or coiling occurs due to the formation of hydrogen bonds between the amine and carboxyl groups of amino acids within a polypeptide chain that have come in close proximity to each other. This results in the formation of secondary structures. There are three types of secondary structures:

- **Alpha helix**—Hydrogen bonds form between amine and carboxyl groups from non-adjacent peptide bonds within the polypeptide chain and the chain coils to form a helical shape (Figure 2.3.7a).
- **Beta-pleated sheets**—Hydrogen bonds form between amine and carboxyl groups in different parts of adjacent polypeptide chains, causing the chains to fold back on each other (Figure 2.3.7b).
- **Random coil**—Although parts of the polypeptide chain appear to have a random structure, the folding is not in fact random and the same folding occurs in all molecules of the same protein. For example, all insulin molecules will have the same random coils within their structure.

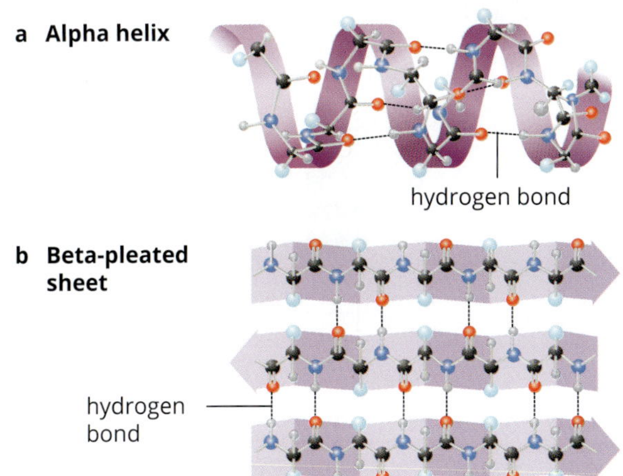

FIGURE 2.3.7 Regions stabilised by hydrogen bonds between atoms of the polypeptide chain result in the secondary structures (a) alpha helix and (b) beta-pleated sheets.

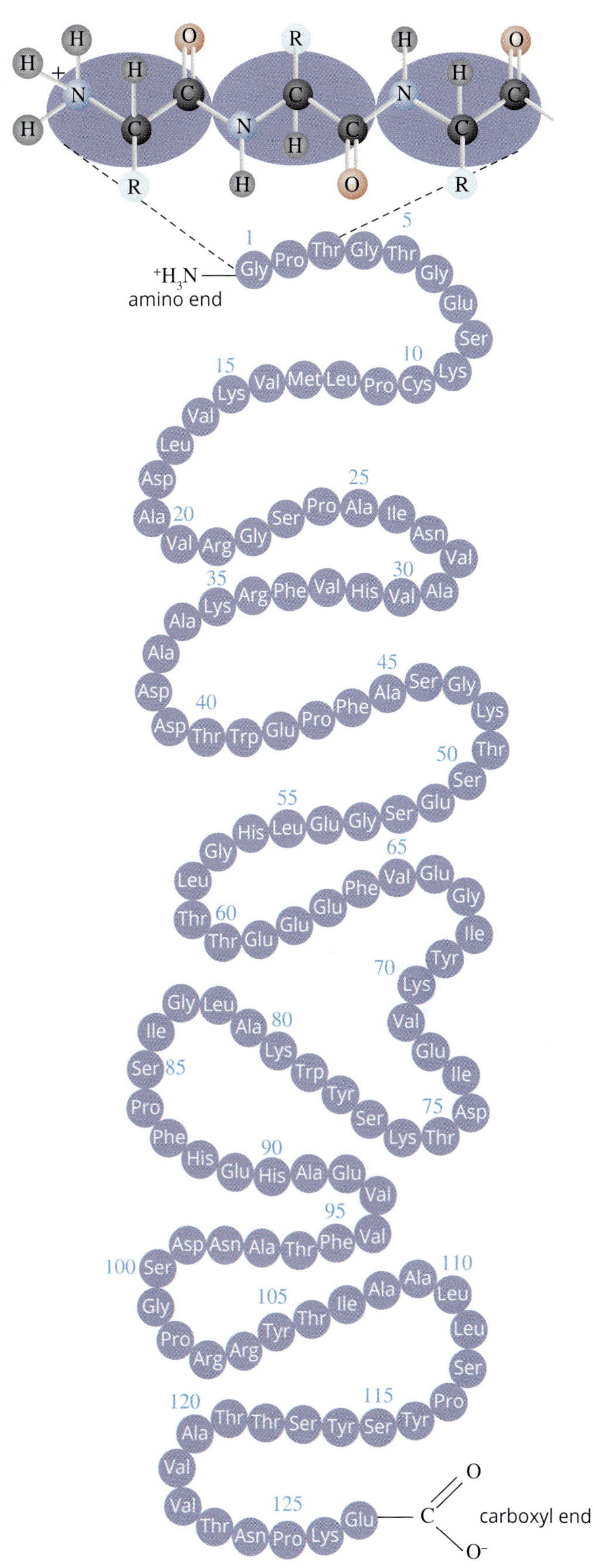

FIGURE 2.3.6 Primary structure of a protein showing the linear amino acid sequence of a polypeptide chain

Tertiary structure

Polypeptides also fold further, forming more stable globular or fibrous three-dimensional shapes (Figure 2.3.8). This is known as the **tertiary structure**, and is usually the result of a combination of alpha helices and beta-pleated sheets along with other folded areas. The tertiary structure occurs due to different types of bonds, such as the disulfide bridge and the hydrogen bridge, between the R groups (side chains) of the amino acids (Figure 2.3.9).

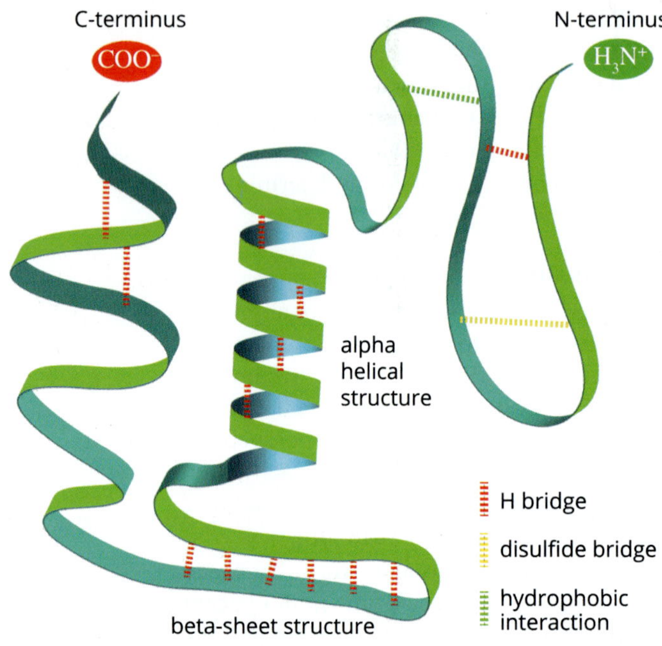

FIGURE 2.3.8 The tertiary structure of a protein is stabilised by the presence of different types of bonds.

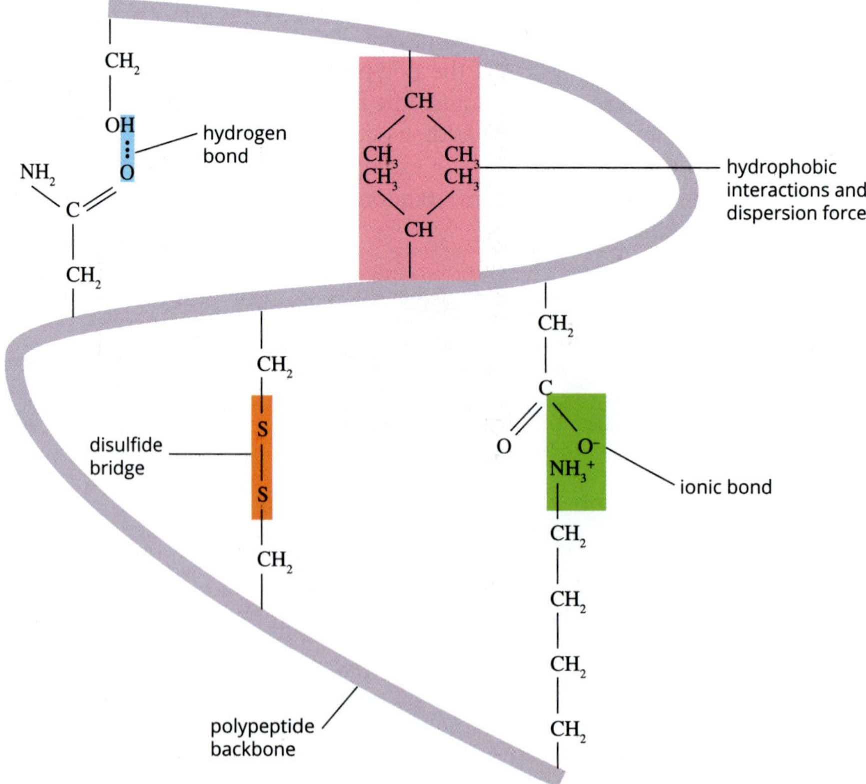

FIGURE 2.3.9 Different types of bonds between the R groups of the amino acids and weak dispersion forces

The three-dimensional structure of a protein is critical to its function. In some smaller polypeptides, this folding process occurs spontaneously due to its chemical environment. However, larger, more complex proteins require specialised proteins to help them fold correctly and, in some cases, to refold if they unravel and lose their native shape (**denature**).

The tertiary structure is the final structure for some proteins.

Quaternary structure

A **quaternary structure** is formed when two or more polypeptide chains or **prosthetic groups** (an inorganic compound that is involved in protein structure or function) join together to create a single functional protein. The polypeptides may be identical or different. Some proteins will not become active until they achieve their quaternary structure. A protein with a prosthetic group is known as a **conjugated protein**. Haemoglobin is an example of a conjugated protein with a quaternary structure (Figure 2.3.10).

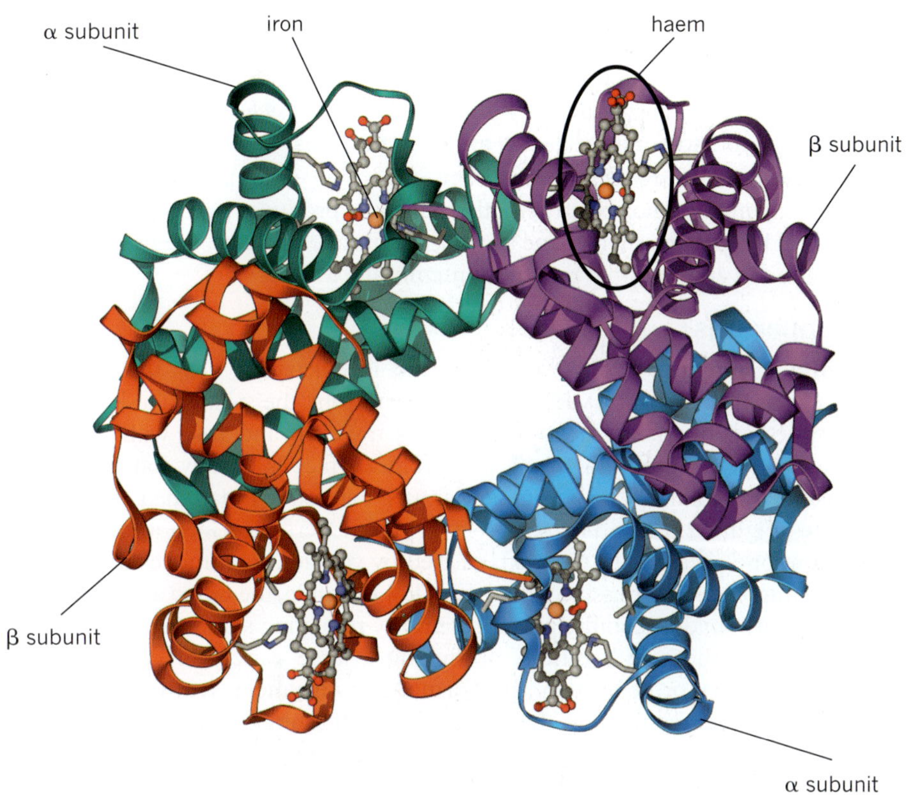

FIGURE 2.3.10 Quaternary structure of haemoglobin. Four polypeptides (two alpha (α) subunits of 141 amino acids and two beta (β) subunits of 146 amino acids) join together with haem prosthetic groups to form the functional haemoglobin molecule. Haemoglobin is a conjugated protein.

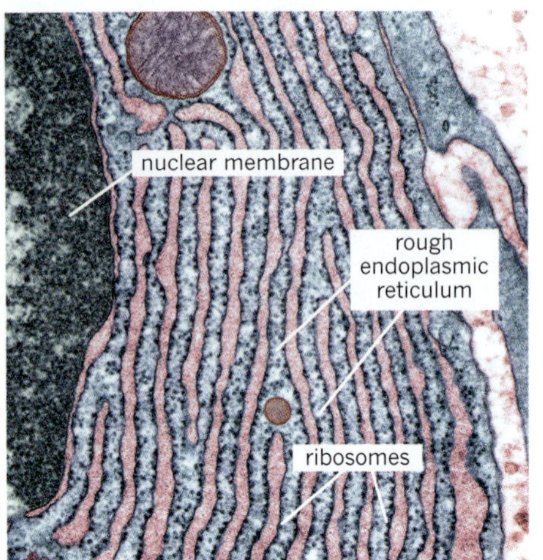

FIGURE 2.3.11 Coloured TEM of rough endoplasmic reticulum (blue) with ribosomes (black dots) on the outer surface. The outer layer of the nuclear membrane can be seen at the left edge (dark grey).

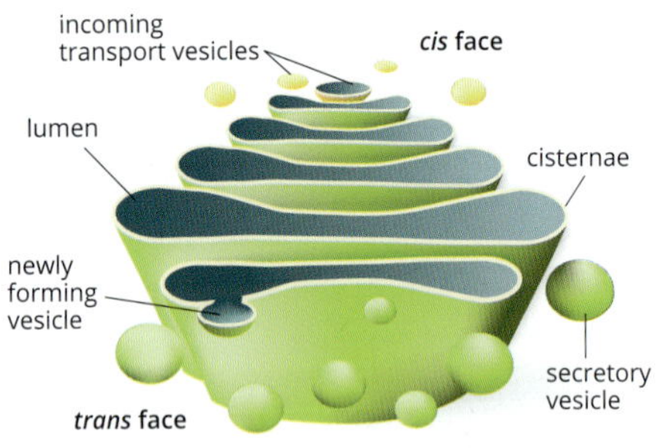

FIGURE 2.3.12 Structure of the Golgi apparatus

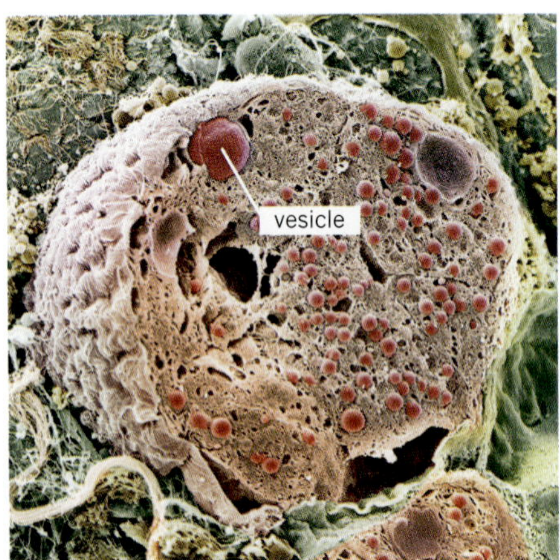

FIGURE 2.3.13 Coloured SEM of a pancreatic cell. Pancreatic cells produce and excrete digestive enzymes in vesicles.

PROTEIN SECRETORY PATHWAY

Secretory proteins are proteins that are produced to be exported out of a cell. The movement of secretory proteins occurs by **exocytosis**, also known as the **protein secretory pathway**. Before reaching the plasma membrane for exocytosis, secretory proteins must first be synthesised and modified.

Ribosomes and endoplasmic reticulum

Proteins destined for use within the cell are synthesised by free **ribosomes** that are found in the **cytosol**. Proteins that are to be secreted are synthesised by ribosomes that stud the outer surface of the **rough endoplasmic reticulum** (Figure 2.3.11).

As it is produced by the ribosome, the polypeptide chain is inserted into the lumen (the fluid-filled space between the membranes) of the rough endoplasmic reticulum through a pore in the membrane. Once the secretory protein has been synthesised, it is transported through the tubules of the rough endoplasmic reticulum, where it is modified. For example, if the secretory protein is a glycoprotein (most are), carbohydrates are attached to the protein in the endoplasmic reticulum by enzymes that are present on its membranes.

When the secretory proteins reach the end of the tubules, they are wrapped in the membranes of **vesicles** that bud off from the endoplasmic reticulum. The vesicles are then transported to other parts of the cell. Vesicles that move from one part of the cell to another are called **transport vesicles**.

In addition to making secretory proteins, the rough endoplasmic reticulum also produces transmembrane proteins. As the ribosome produces polypeptides that are to be part of the plasma membrane as transmembrane proteins, the polypeptide is inserted into the endoplasmic reticulum membrane itself.

Golgi apparatus

The **Golgi apparatus** is an organelle found in eukaryotic cells that processes and packages proteins into vesicles for export from the cell. After leaving the endoplasmic reticulum, the transport vesicles travel to and fuse with the Golgi apparatus at the ***cis* face** (Figure 2.3.12). The *cis* face is usually found near the endoplasmic reticulum. The Golgi apparatus consists of flattened sacs called **cisternae** (Figure 2.3.12). The secretory protein enters the Golgi apparatus and moves from one cisternae to the next, carried by vesicles. As it moves through the Golgi apparatus it is progressively modified. For example, the Golgi apparatus may modify the carbohydrate on the glycoproteins by removing some sugar monomers and substituting them with others, producing a large variety of carbohydrates. When the secretory protein, such as a hormone or enzyme, is ready for secretion, **secretory vesicles** containing the protein bud off from the ***trans* face** end of the Golgi apparatus and move to the plasma membrane, where the product for secretion is released out of the cell via exocytosis.

Sometimes secretory vesicles are not transported to the plasma membrane. Instead, the vesicles are stored in the Golgi apparatus until the secretory protein is needed. For example, in pancreatic cells, digestive enzymes are stored in secretory vesicles within the Golgi apparatus until the presence of food in the stomach triggers a signal for their secretion (Figure 2.3.13).

Exocytosis

When the secretory vesicle membrane and plasma membrane come into contact, specific proteins alter the arrangement of the **phospholipids** in the phospholipid bilayer of the plasma membrane, enabling the fusion of the two membranes. The fluid and dynamic nature of the plasma membrane enables this membrane fusion. Once the two membranes are fused, the contents of the secretory vesicle are released out of the cell. This is called exocytosis. The vesicle membrane becomes a permanent part of the plasma membrane (Figure 2.3.14). The plasma membrane is continually recycled as vesicles fuse during exocytosis and are conversely formed and released during **endocytosis**. A summary of the roles of the rough endoplasmic reticulum and Golgi apparatus in the exocytosis of proteins can be seen in Figure 2.3.15.

phospholipid bilayer
hydrophilic phosphate head
hydrophobic lipid tails

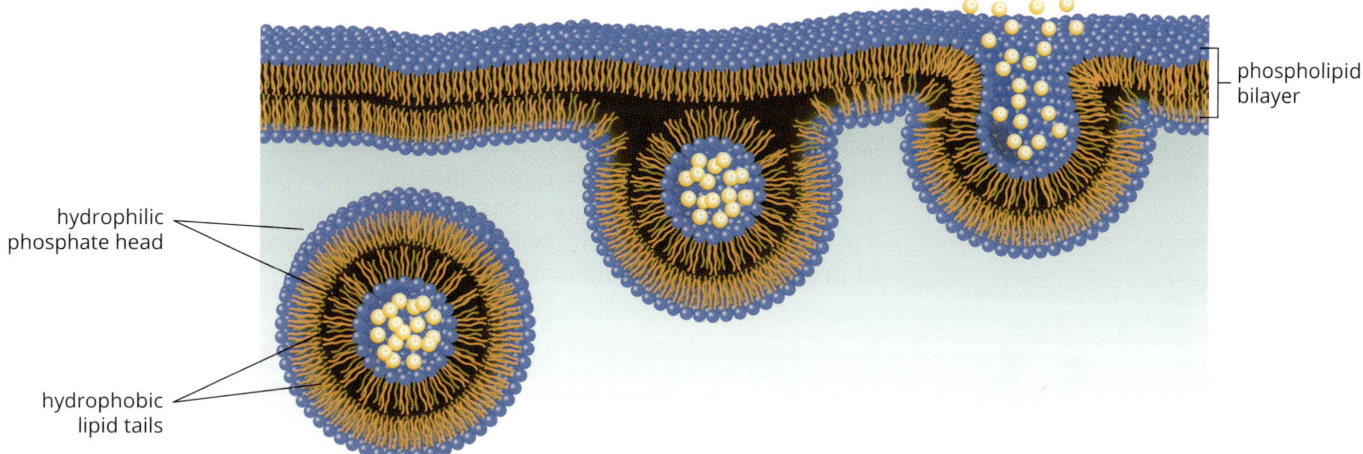

FIGURE 2.3.14 Exocytosis is the movement of a secretory vesicle towards the plasma membrane and the release of its contents.

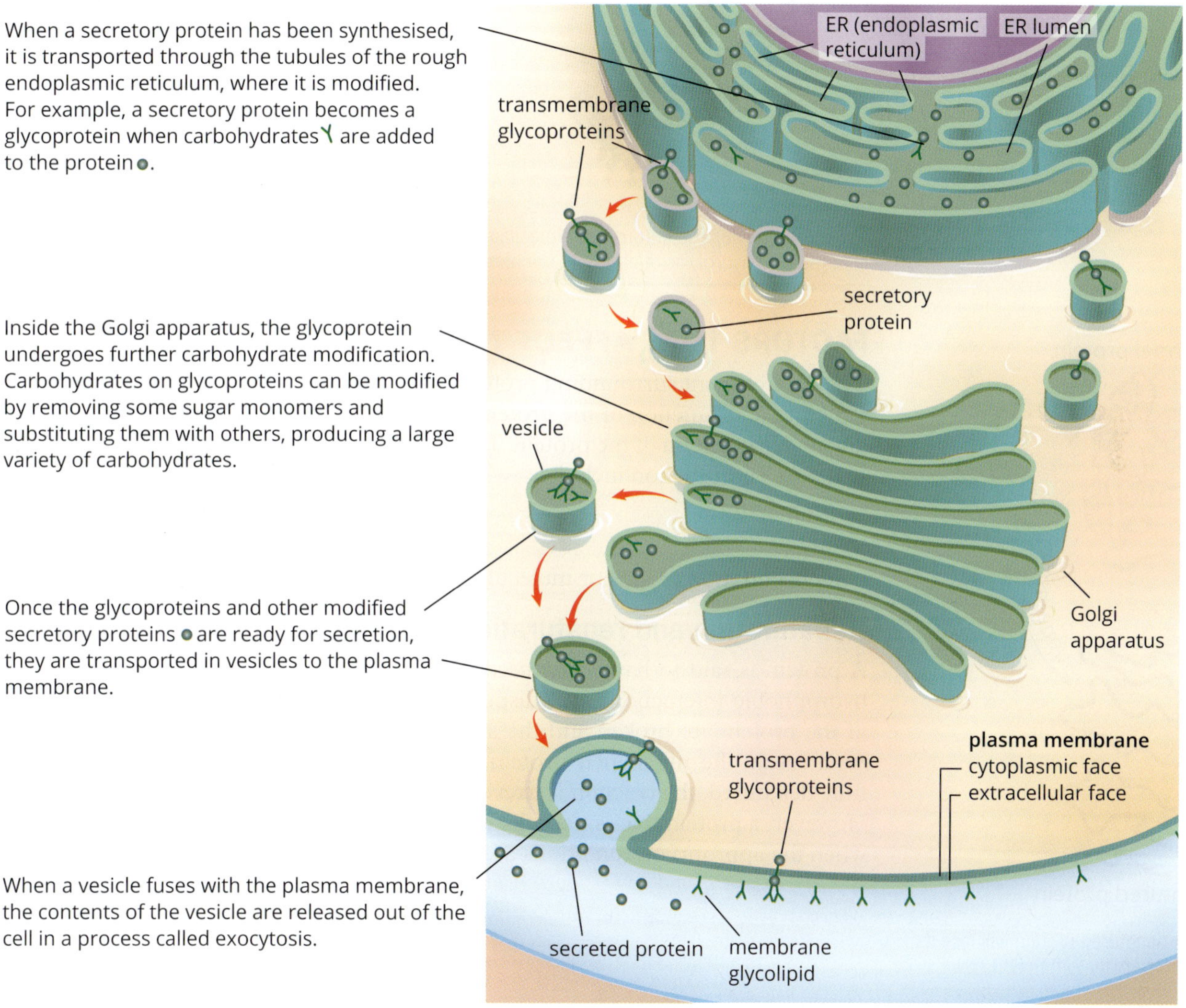

When a secretory protein has been synthesised, it is transported through the tubules of the rough endoplasmic reticulum, where it is modified. For example, a secretory protein becomes a glycoprotein when carbohydrates are added to the protein.
Inside the Golgi apparatus, the glycoprotein undergoes further carbohydrate modification. Carbohydrates on glycoproteins can be modified by removing some sugar monomers and substituting them with others, producing a large variety of carbohydrates.
Once the glycoproteins and other modified secretory proteins are ready for secretion, they are transported in vesicles to the plasma membrane.
When a vesicle fuses with the plasma membrane, the contents of the vesicle are released out of the cell in a process called exocytosis.
ER (endoplasmic reticulum)
ER lumen
transmembrane glycoproteins
secretory protein
vesicle
Golgi apparatus
plasma membrane
cytoplasmic face
extracellular face
transmembrane glycoproteins
secreted protein
membrane glycolipid

FIGURE 2.3.15 Roles of rough endoplasmic reticulum and Golgi apparatus in the protein secretory pathway

PROTEIN CLASSIFICATION

Proteins can be classed as one of two types depending on their shapes:

- **Fibrous proteins** are typically elongated and insoluble (Figure 2.3.16a). Many have structural roles and have little or no tertiary folding (e.g. collagen found in connective tissue and keratin found in hair and nails).
- **Globular proteins** are compactly folded and coiled into spherical tertiary and quaternary structures (Figure 2.3.16b). Globular proteins are generally soluble. They have a core with hydrophobic properties and an outer hydrophilic region. Most enzymes and hormones are globular proteins.

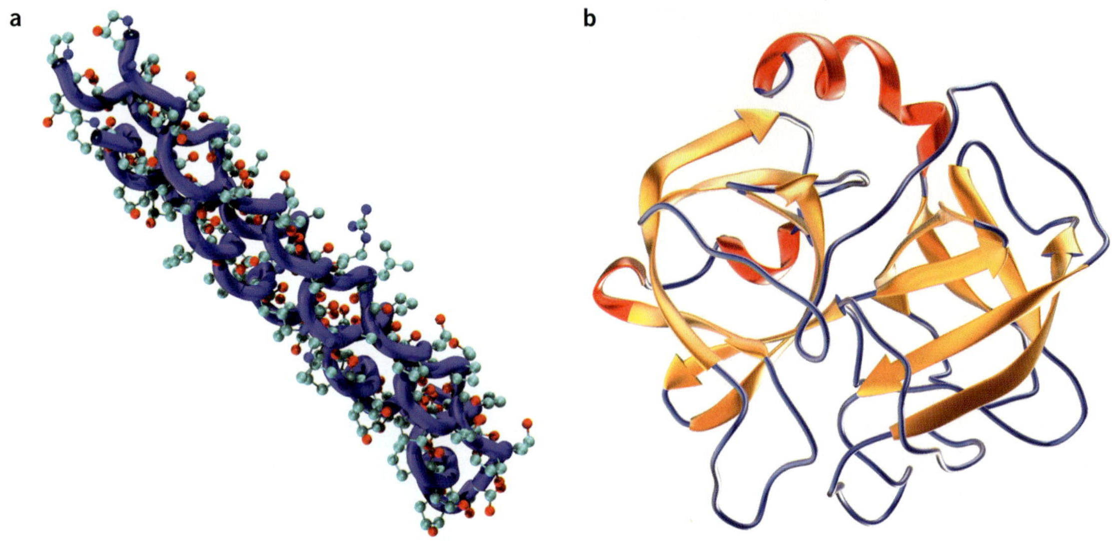

FIGURE 2.3.16 (a) The fibrous protein collagen. (b) The globular protein elastase (an enzyme that catalyses the hydrolysis of elastin in the pancreas)

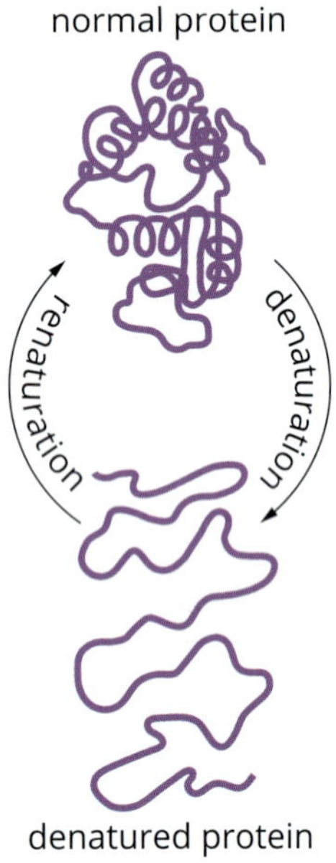

FACTORS THAT AFFECT THE FUNCTION OF A PROTEIN

The environment surrounding proteins plays an important role in maintaining the structure and function of the protein. Usually the loss of function of the protein is due to denaturation of the protein. The factors in the environment affecting protein structure and function include:

- temperature
- pH
- concentration of ions or molecules that act as cofactors.

Denaturation and renaturation of proteins

A protein is said to have denatured when the hydrogen bonds, disulfide bridges, hydrophobic interactions and dispersion forces that create the tertiary structure of the protein are broken and the shape of the protein is altered (Figure 2.3.17). As a result, the misshapen protein is biologically inactive. If a protein becomes fully denatured, the reaction is irreversible and the protein remains non-functional. However, a protein that is partially denatured may be able to fold again (renature) when the appropriate conditions are present.

FIGURE 2.3.17 A denatured protein will lose its shape and hence its ability to function.

The effect of temperature on protein function

Proteins can be denatured at high temperatures due to the breaking of bonds. For example, hydrogen bonds break at temperatures above 40°C. However, at temperatures below 35°C, the bonds are not flexible enough to allow the necessary conformational changes.

The optimal temperature for proteins varies with the organism and its environment. In humans, the optimum temperature for proteins is 37°C, but proteins found in organisms living in extreme environments, such as hot springs or icy environments, tend to have different optimal temperatures.

CASE STUDY ANALYSIS

Interaction of proteins with phospholipids in the plasma membrane

Many different proteins make up the plasma membrane. Figure 2.3.18 shows a channel protein spanning the phospholipid bilayer.

The blue molecules shown in Figure 2.3.18 are polar amino acids. They are unable to cross directly through the phospholipids as the phospholipids tails are hydrophobic. Hydrophobic substances are non-polar and do not like interacting with polar molecules. The protein channel provides an environment in which the amino acid can pass without repulsion.

The interaction of hydrophobic and hydrophilic molecules with each other provides information about the nature of various amino acids that make up the actual channel protein.

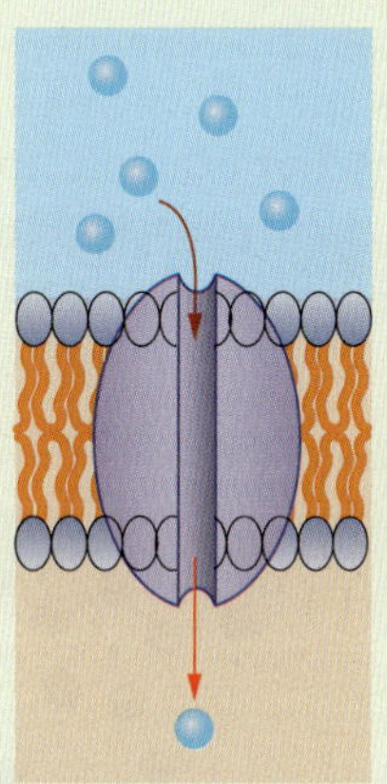

FIGURE 2.3.18 Channel protein spanning the phospholipid bilayer of the plasma membrane

Analysis

1. Given the position of the phospholipids in the plasma membrane, what types of amino acids must make up the part of the protein channel labelled X in the figure to the right?
2. What types of amino acids make up the region labelled Y in the figure to the right? Use Figure 2.3.3 on page 84 to list amino acids that may provide this chemical environment.

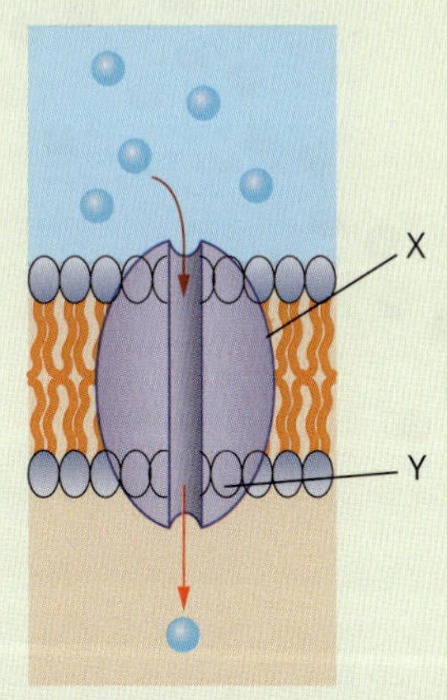

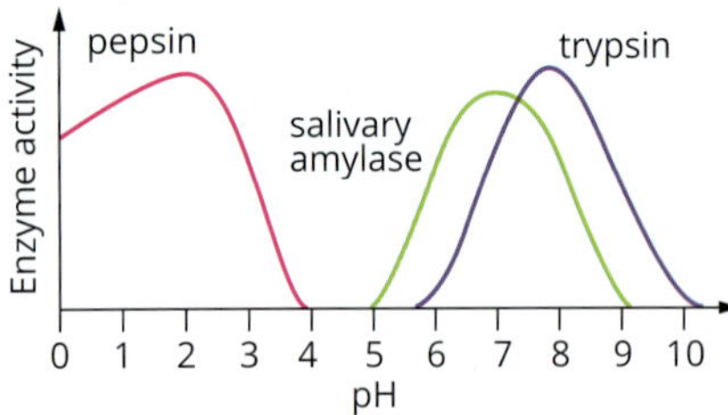

FIGURE 2.3.19 Graph showing the optimum pH for different digestive enzymes

The effect of pH on protein function

Most proteins have a specific pH range in which their function is optimal, but this range can be quite different for each specific protein (whereas the range of optimal temperatures is similar for most proteins within an organism). In humans, for example, the enzyme salivary amylase (which starts the digestion of starch in the mouth) has an optimum pH of about 7, the enzyme pepsin (a digestive enzyme found in the stomach) has an optimal pH of about 2, and the enzyme trypsin (a digestive enzyme found in the small intestines) has an optimal pH of about 8 (Figure 2.3.19).

If the pH reaches too far above or falls too far below the optimal pH, then the tertiary structure is affected. The interactions between the R groups of different amino acids are altered and the bonds between them are broken. As a result, the protein may be denatured, and in the case of enzymes, the enzyme activity will decrease.

The effect of cofactors on protein function

Some proteins require non-protein chemical compounds known as **cofactors** for their biological function. The presence and concentration of cofactors such as salts, specific elements such as iron, magnesium and calcium ions, or organic molecules such as vitamins, can play a significant role in the folding and function of proteins. For example, magnesium is essential for chlorophyll function in plants (Figure 2.3.20). A lack of magnesium ions causes the yellowing of leaves due to the plant's inability to synthesise chlorophyll.

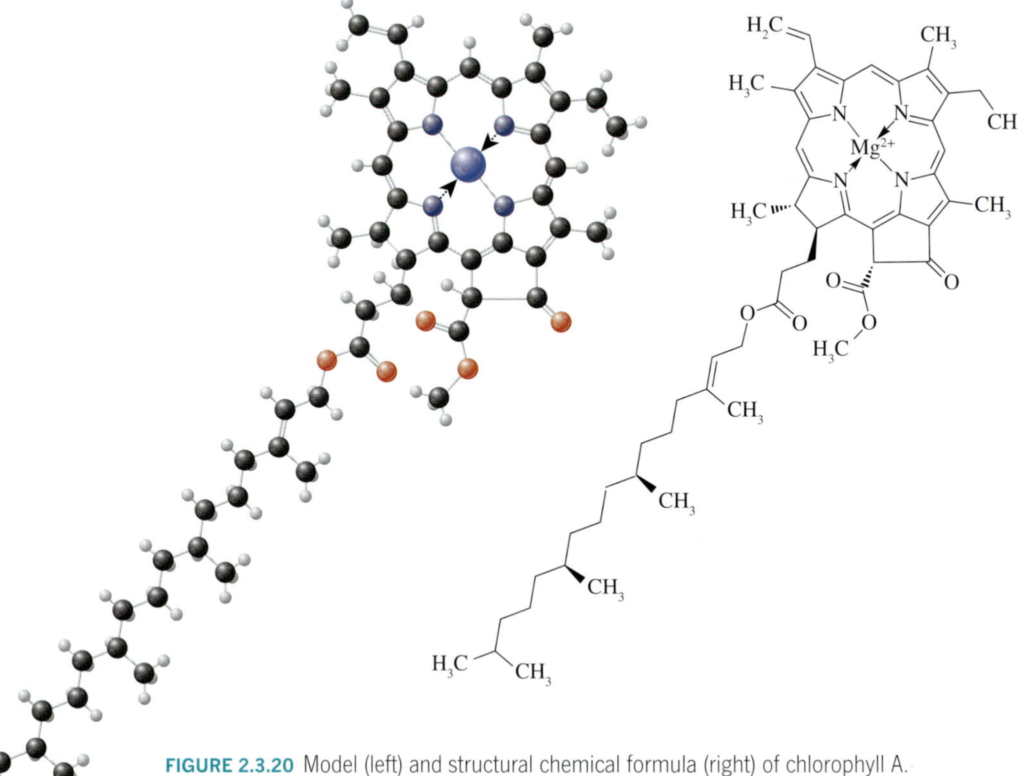

FIGURE 2.3.20 Model (left) and structural chemical formula (right) of chlorophyll A. The magnesium ion is coloured in blue in the model.

2.3 Review

OA ✓✓

SUMMARY

- Proteins have very diverse functions. The specific folding and final structure of proteins relate directly to their function.
- Functional types of proteins include:
  - enzymatic proteins
  - structural proteins
  - transport proteins
  - hormonal proteins
  - receptor proteins
  - immunological proteins
  - contractile and motor proteins
  - storage proteins.
- The proteome is the complete set of proteins expressed by the genome.
- Proteomics is the study of proteomes, including protein structure and function.
- Amino acids have an amine group, a carboxyl group and an R group. There are 20 standard amino acids. All have the same amine and carboxyl group, but differ in their R group.
- Amino acids are joined by peptide bonds to form polypeptide chains.
- Proteins are made up of one or more polypeptide chains, which are folded and organised into specific shapes that relate to their specific function.
  - Primary structure of a protein is the linear sequence of amino acids in the polypeptide chain.
  - Secondary structure of a protein is achieved with the folding or coiling of the polypeptide chains due to hydrogen bonds.
  - Tertiary structure of a protein is achieved by further folding, which creates more stable shapes. This structure occurs as a result of bonds forming between the R groups of the amino acids.
  - Quaternary structure of a protein is achieved when two or more polypeptide chains join to create a single functional protein.
- Proteins can be either fibrous or globular.
  - Fibrous proteins are elongated and insoluble.
  - Globular proteins are spherical and compact.
- Factors within the environment can have an impact on the structure and function of a protein, and can also lead to denaturation. These factors include temperature, pH, concentration of ions and molecules that act as cofactors.
- Proteins have optimal temperature and pH ranges within which they function most effectively.

KEY QUESTIONS

Knowledge and understanding

1 Proteins are key components of cells. Outline, with examples, at least five different roles carried out by proteins.

2 Draw the structure of an amino acid, and label the groups that are used in peptide bond formation.

Analysis

3 The enzyme amylase speeds up the breakdown of starch to simple sugar in both the mouth and stomach. Do new enzymes need to be made after each reaction to break down a starch molecule? Explain your answer.

4 Distinguish between peptides and polypeptides.

5 Use a single sentence and a simple diagram to explain what is meant by the following structures of a protein:

a primary
b secondary
c tertiary
d quaternary

6 Distinguish between fibrous and globular proteins.

7 **a** Explain what is meant by a protein becoming 'denatured'.
b Discuss the factors that can cause a protein to become denatured.

8 **a** What are cofactors?
b How do cofactors affect protein function?

Chapter review

02

KEY TERMS

alpha helix
amine group (amino group)
amino acid
anabolic reaction
beta-pleated sheet
carboxyl group
catabolic reaction
catalyst
cis face
cisterna (pl. cisternae)
cofactor
conjugated protein
cytosol
denature (denaturation)
endocytosis
enzyme
exocytosis
fibrous protein
genome
globular protein
Golgi apparatus
monomer
peptide
peptide bond
phospholipid
polypeptide chain
primary structure
prosthetic group
protein
protein secretory pathway
proteome
proteomics
quaternary structure
random coil
ribosome
rough endoplasmic reticulum (RER)
secondary structure
secretory protein
secretory vesicle
tertiary structure
trans face
transport vesicle
variable R group
vesicle

REVIEW QUESTIONS

Knowledge and understanding

1 Which of the following shows a beta-pleated sheet?

A
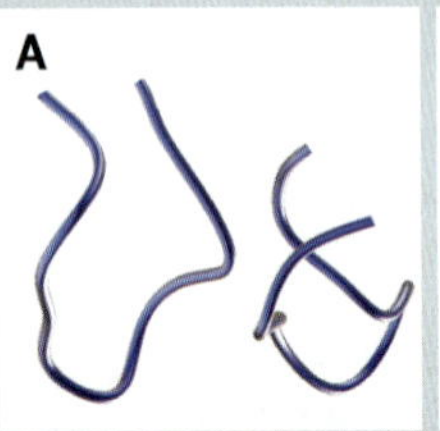

B
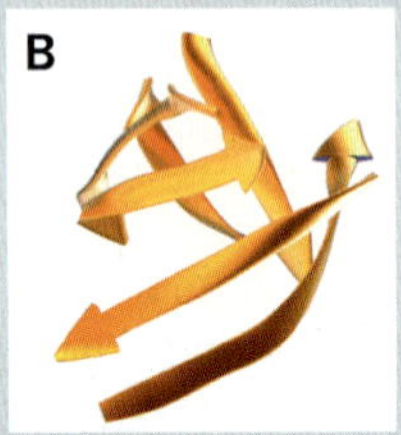

C
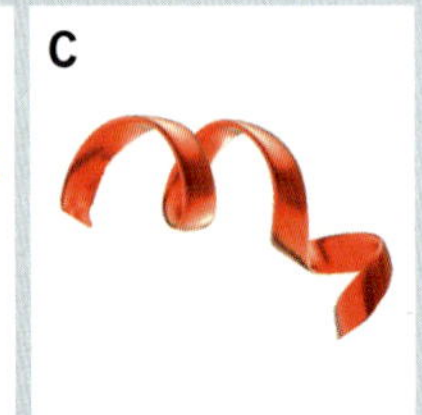

D None of the above

2 Consider the following diagrams of the four levels of protein structure.

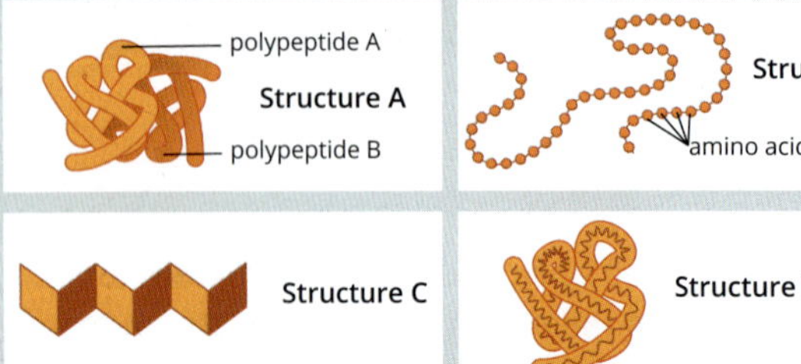

The diagram showing a quaternary structure is:

A structure A
B structure B
C structure C
D structure D

3 **a** Define the term 'proteome'.
b Outline the relationship between the genome and the proteome of an individual organism.

4 List the parts of an amino acid that are involved in the bonds formed in the following structures.
a primary structure
b secondary structure
c tertiary structure

5 In polypeptide synthesis, the function of the ribosome is to:
A synthesise the required amino acids
B ensure that the DNA base sequence is complete
C provide the energy needed for polypeptide synthesis
D provide the site for polypeptide synthesis

6 Look carefully at the diagrams in boxes P, Q, R and S, and the text in boxes W, X, Y and Z below. Place the letter for each into the table so that they correspond to the correct level of protein structure.

P	Q	R	S
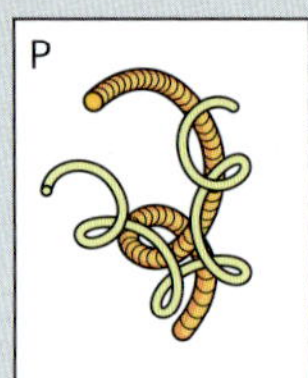	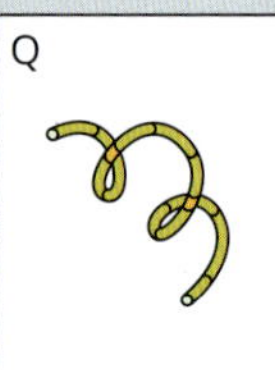	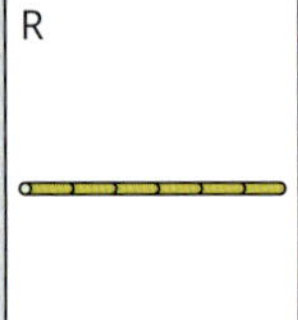	

W	X	Y	Z
Polypeptide chain becomes coiled or pleated.	Amino acids become joined by peptide bonds to form a polypeptide.	Two or more polypeptide chains become entwined and chemically bonded together.	Polypeptide chain folds on itself to form a 3D structure.

Level of structure	Diagram (P–Q)	Description (W–Z)
primary		
secondary		
tertiary		
quaternary		

Application and analysis

7 Distinguish between free ribosomes and ribosomes bound to the rough endoplasmic reticulum in relation to protein manufacture.

8 A scientist was interested in how serine (an amino acid) enters cells. The scientist established that serine is a polar molecule.

   a The polar nature of the amino acid rules out one method of entry for serine. Name and explain this method of entry.

   In order to further investigate the entry of serine, the scientist takes three cell cultures (A, B and C), adds radioactively labelled serine molecules in solution to each and then monitors the cells to determine whether the serine enters the cells. The results of the experiment are then tabulated.

Results of the ability of cell cultures to absorb serine following various treatments

Culture	Treatment	Result
A	radioactive serine solution only	radioactive serine found in the cells
B	mercury solution (damages protein carriers and channels) and radioactive serine solution	no radioactive serine in the cells
C	ATPase inhibitor and radioactive serine solution (ATPase catalyses the formation of ATP from ADP and inorganic phosphate (P_i))	radioactive serine found in the cells

   b What is the function of culture A in the experiment?

   c What do the results of the experiment suggest about the method used by the cell to take in serine?

   d What is the independent variable in this experiment?

9 Examine the cell below.

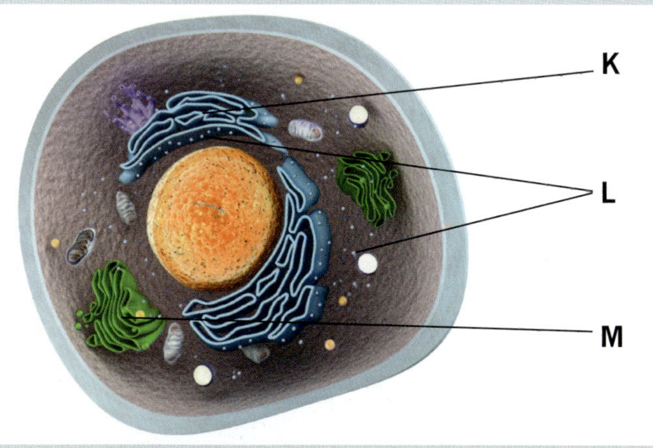

   a Identify the organelles K, L and M.

   b i Describe one similarity in function between K and M.

   ii Describe one difference in function between K and M.

   c A cell contains large numbers of organelle M. Suggest a possible function for the cell, giving a reason for your suggestion.

   d How might organelle L be involved in the specialised function of the cell?

10 Distinguish between the *cis* and *trans* face of the Golgi apparatus.

11 The pancreas is responsible for making two very important protein hormones (insulin and glucagon) involved in the regulation of blood glucose levels. Given hormones travel in the bloodstream to take effect on target cells in other parts of the body:

   a Discuss where in the cell the proteins will be produced.

   b Outline the protein secretory pathway, including the names of organelles, from manufacture to export of proteins from the cell.

12 Salivary amylase is a protein manufactured in the mouth to help speed up the digestion of starch. Explain whether the optimum pH of this enzyme would differ to that of an enzyme found in the stomach.

13 The following metabolic pathway involves the action of many enzymes within a cell.

   substrate 1 → substrate 2 → substrate 3 → product

   Explain what would happen to the amount of product produced if the enzyme that acts on substrate 2 was not produced in the cell. All other enzymes at each step of the pathway are present.

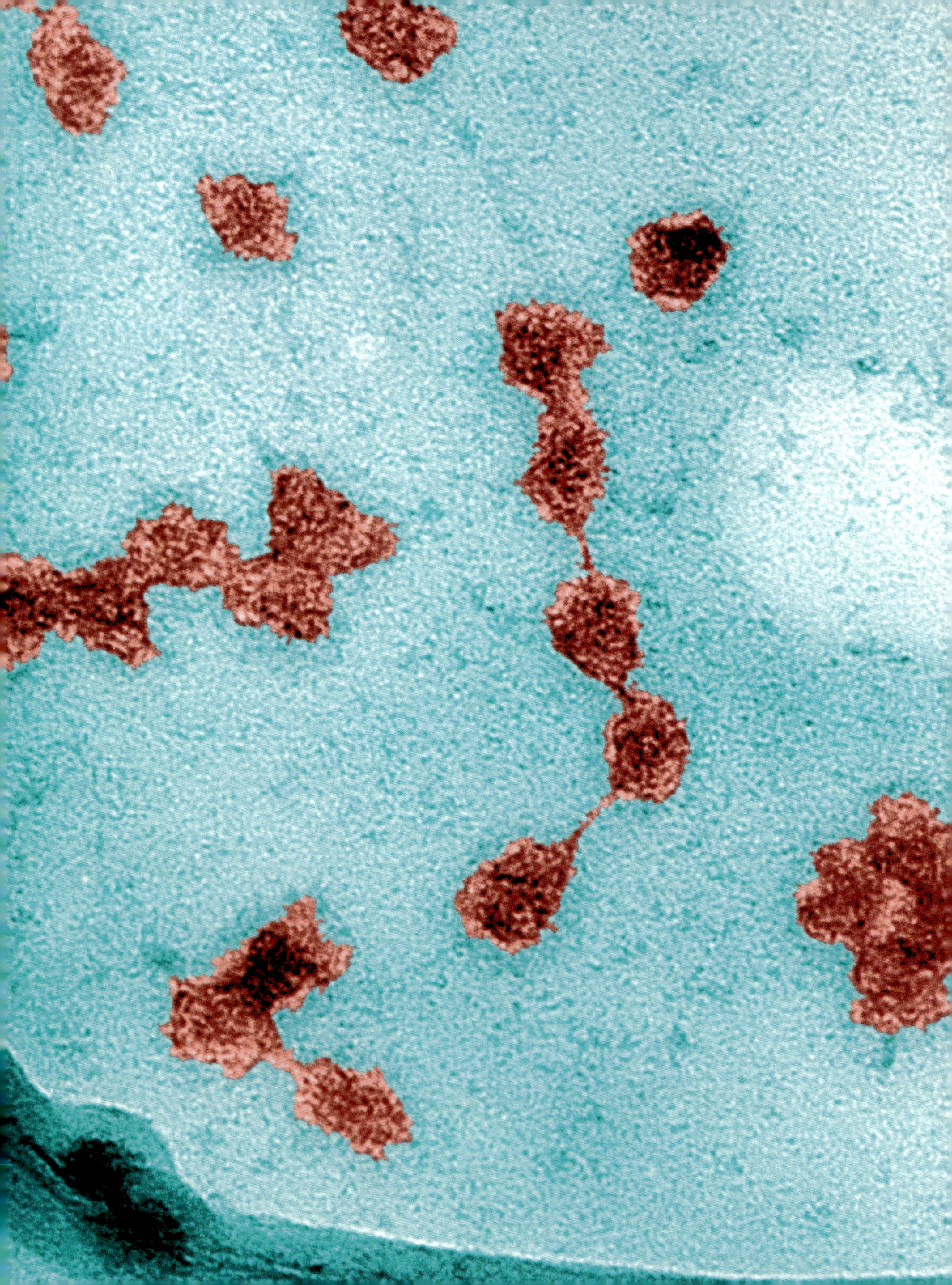

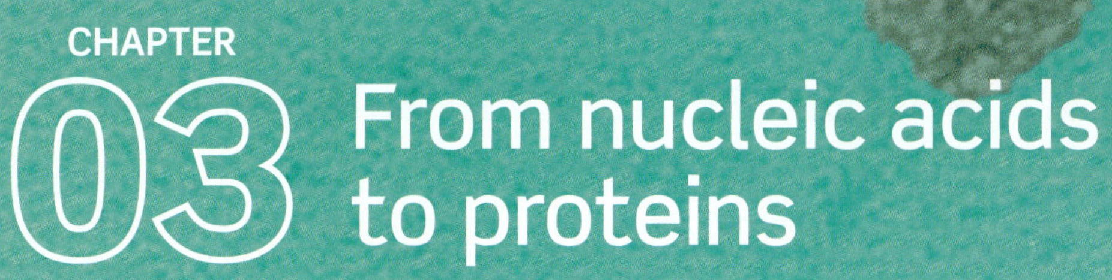

From nucleic acids to proteins

Learning outcomes

By the end of this chapter, you will be able to describe the structure and synthesis of the nucleic acids DNA and RNA. You will also understand the role of these nucleic acids as information molecules that encode instructions for protein synthesis, and the steps in eukaryotic gene expression: transcription, RNA processing and translation.

You will have explored gene structure and analysed the distinction between structural and regulatory genes and have an understanding of the regulation of gene transcription by transcriptional factors.

Key knowledge

- nucleic acids as information molecules that encode instructions for the synthesis of proteins: the structure of DNA, the three main forms of RNA (mRNA, rRNA and tRNA) and a comparison of their respective nucleotides **3.1, 3.2**
- the genetic code as a universal triplet code that is degenerate and the steps in gene expression, including transcription, RNA processing in eukaryotic cells and translation by ribosomes **3.2**
- the structure of genes: exons, introns and promoter and operator regions **3.2, 3.3**
- the basic elements of gene regulation: prokaryotic *trp* operon as a simplified example of a regulatory process **3.3**

3.1 Nucleic acids: DNA and RNA

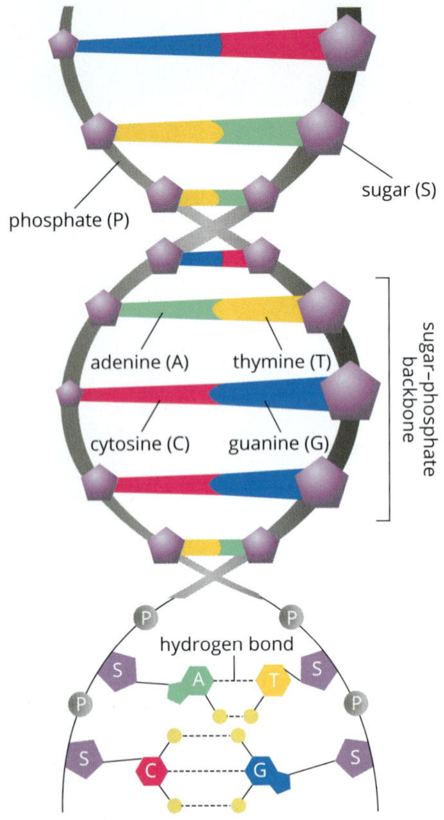

FIGURE 3.1.1 Structure of the DNA double helix biomolecule

> **i** Hereditary information refers to genetic material that is passed on from parent to offspring.

Nucleic acids are organic biomolecules that store and transmit inherited characteristics of organisms. There are two types of nucleic acids: deoxyribonucleic acid (DNA) and ribonucleic acid (RNA). Both DNA and RNA are made up of nitrogenous bases and a sugar–phosphate backbone. RNA is usually single-stranded and DNA has a double-stranded helix structure with complementary pairing of its nitrogenous bases (Figure 3.1.1).

In this section, you will learn about the structure of DNA and RNA.

NUCLEIC ACIDS

Nucleic acids are large **biomolecules** that store and transmit hereditary information. Specifically, nucleic acids encode instructions for the synthesis of proteins.

- **Deoxyribonucleic acid (DNA)** carries the instructions that code for the production of RNA, which may be functional (such as transfer RNA) or contain information for protein synthesis (messenger RNA). DNA is able to self-replicate.
- **Ribonucleic acid (RNA)** has different forms that perform different functions. It can carry a copy of a DNA sequence (messenger RNA) and it also has the ability to 'read' and translate the DNA information (transfer RNA). RNA plays a major role in the process of **protein synthesis**.

Nucleic acids are polymers, made up of repeated subunit monomers called **nucleotides**.

Nucleotides

A single nucleotide consists of three basic units:

- a **phosphate group**—the same in all nucleotides
- a five-carbon (pentose) sugar
  - **deoxyribose** in DNA nucleotides
  - **ribose** in RNA nucleotides
- a nitrogenous (nitrogen-containing) **base**.

The five carbon atoms in a pentose sugar molecule are labelled 1' to 5'. The phosphate is always attached to the 5' carbon and the base to the 1' carbon in a single nucleotide. You can see this structure in Figure 3.1.2.

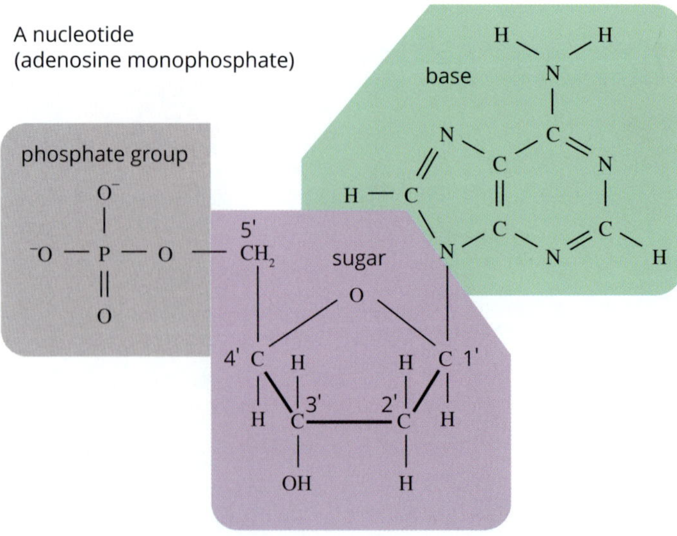

FIGURE 3.1.2 Basic structure of a DNA nucleotide, showing the phosphate group, the five-carbon sugar and the nitrogenous base adenine (A)

The nitrogenous bases

There are five different nitrogenous bases:

- **adenine (A)**
- **guanine (G)**
- **cytosine (C)**
- **thymine (T)**—in DNA only
- **uracil (U)**—in RNA only.

These five nitrogenous bases can be categorised into one of two groups based on their structure:

- **Purines** (A and G) have two rings in their structure.
- **Pyrimidines** (T, U and C) have one ring in their structure (Figure 3.1.3).

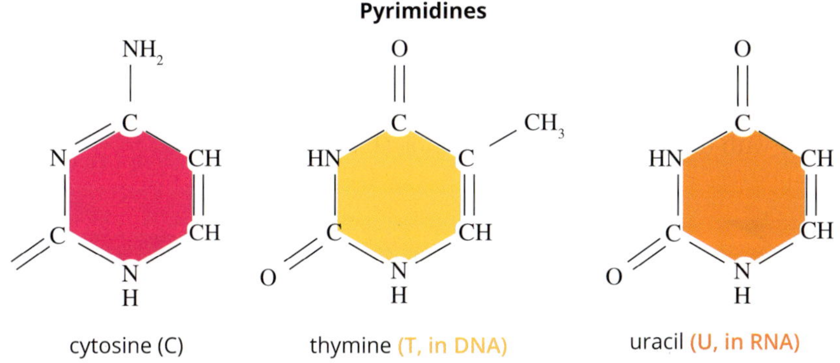

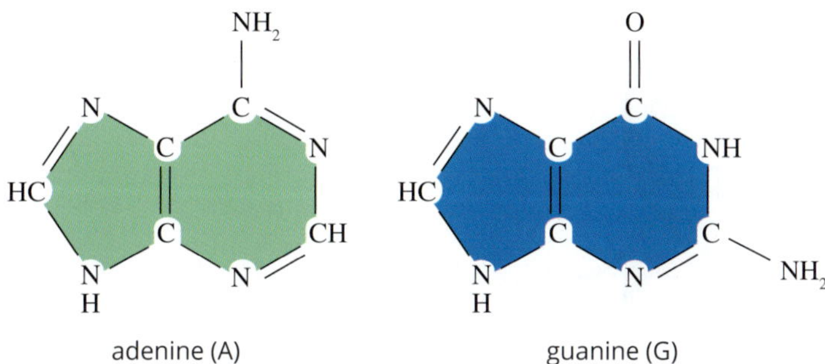

FIGURE 3.1.3 Structure of pyrimidines (cytosine, thymine and uracil) and purines (adenine and guanine). Pyrimidines have one ring and purines have two rings.

BIOFILE

Joining nucleotides

Free nucleotides link together to form strands through a condensation polymerisation reaction. The hydroxyl group (OH) on the 3' carbon atom of the sugar of one nucleotide joins with the phosphate on the fifth carbon of the pentose sugar (5') of the other nucleotide to form water, which is released. Free nucleotides can then be continuously added to the 3' carbons in this way, forming a long sugar–phosphate–sugar–phosphate backbone strand. The nucleotides in the sugar–phosphate chain are joined covalently with a strong phosphodiester bond. In comparison, the hydrogen bonds that hold the two strands together are much weaker.

In polynucleotide strands, one end has a free phosphate group on the 5' carbon; this is called the 5' end (five prime end). The other end of the strand has a free hydroxyl on the 3' carbon; this is called the 3' end (three prime end). Free nucleotides are always added to the hydroxyl group, meaning that DNA and RNA are synthesised in the 5' to 3' direction.

DNA STRUCTURE

DNA is double-stranded, as it consists of two chains of nucleotides, **polynucleotides** or 'strands' twisted into a **double helix** structure (Figure 3.1.4). The primary structure of DNA is the single strand of polynucleotides, which consists of a specific sequence of nitrogenous bases (A, T, C and G). The hydrogen bonds between pairs of nitrogenous bases stabilise the secondary structure of the DNA and form the double helix.

When DNA is uncoiled, the two strands can be represented as a ladder. The sides of the ladder consist of the sugar–phosphate backbones. The two strands of a DNA molecule are **antiparallel**, meaning that they run in opposite directions, with the 3' end of one strand matching with the 5' end of the other strand. The rungs of the ladder are the nitrogenous bases of each nucleotide.

Complementary base pairing occurs between the nitrogenous bases, forming the double-stranded DNA molecule. In complementary base pairing:

- the purine adenine (A) always pairs with the pyrimidine thymine (T), held together with two weak hydrogen bonds
- the purine guanine (G) always pairs with the pyrimidine cytosine (C), held together with three weak hydrogen bonds (Figure 3.1.4).

The diameter of the DNA double helix is approximately 2.0 nanometres and there are 10–10.5 pairs of nucleotide bases in each twist of the helix.

In the DNA double helix, hydrogen bonds hold the pairs of polynucleotides together. Note that there are two hydrogen bonds between adenine (A) and thymine (T), and three hydrogen bonds between cytosine (C) and guanine (G).

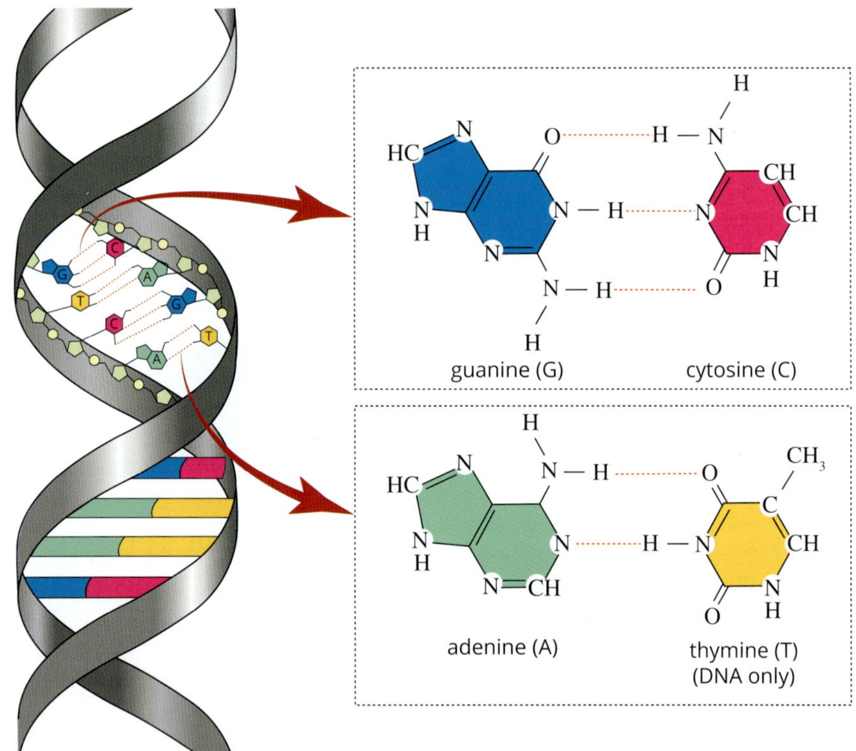

FIGURE 3.1.4 Helical structure of DNA. Two complementary strands form a double helix joined by base pairs guanine (G) and cytosine (C), and adenine (A) and thymine (T).

RNA STRUCTURE

Unlike DNA, RNA is usually found as a single strand, sometimes folded onto itself. RNA molecules are usually much shorter than DNA molecules. The nucleotides of RNA have the same basic structure as those of DNA, with a few differences. DNA contains the sugar deoxyribose, while RNA contains ribose. The nitrogenous base thymine in DNA is replaced by uracil in RNA, both of which pair with adenine (Figure 3.1.5). Uracil is more stable in single-stranded polynucleotides.

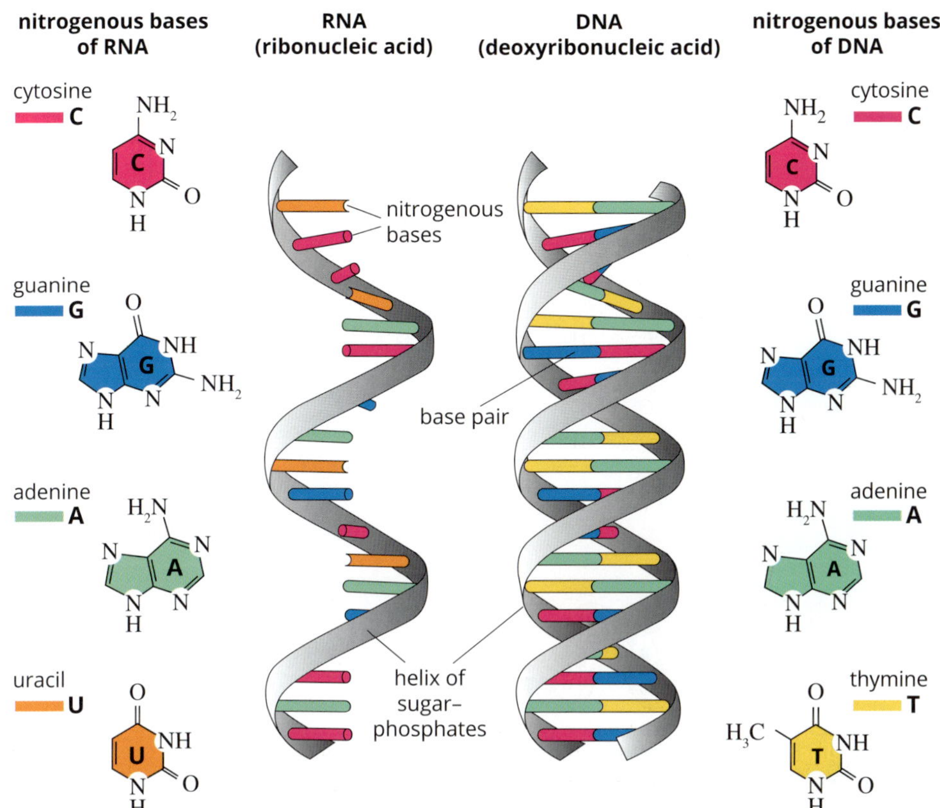

FIGURE 3.1.5 Comparison of the structures of RNA and DNA

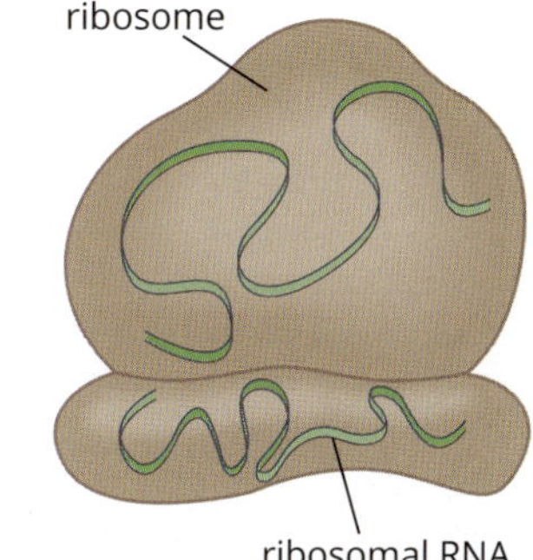

FIGURE 3.1.7 Ribosomal RNA (rRNA) forms ribosomes, the site of translation of the mRNA into proteins.

There are three main forms of RNA: messenger RNA (mRNA) (Figure 3.1.6), ribosomal RNA (rRNA) (Figure 3.1.7) and transfer RNA (tRNA) (Figure 3.1.8). The roles of each type of RNA are described on page 112.

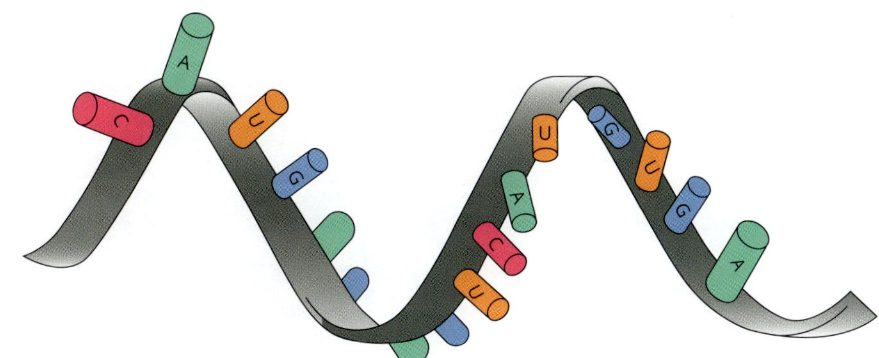

FIGURE 3.1.6 Messenger RNA (mRNA) carries a copy of the DNA's nucleotide sequence to be translated into proteins.

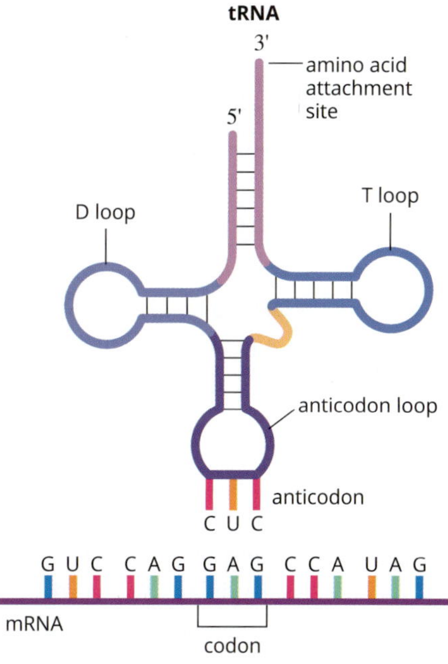

FIGURE 3.1.8 There are multiple transfer RNA (tRNA) molecules with anticodons to complement the different possible codons in mRNA. Each tRNA carries a specific amino acid.

Table 3.1.1 summarises and highlights the differences between the structure of DNA and RNA.

TABLE 3.1.1 Summary of the differences between DNA and RNA

	DNA	RNA
Relative length	long	short
Sugar	deoxyribose deoxyribose in DNA	ribose ribose in RNA
Bases	adenine cytosine guanine thymine thymine in DNA	adenine cytosine guanine uracil uracil in RNA
Strands	double	usually single
Base pairing	adenine–thymine cytosine–guanine	adenine–uracil cytosine–guanine

CASE STUDY ANALYSIS

Viral RNA

RNA is not always single-stranded. *Rotavirus* (Figure 3.1.9), for example, is a genus of double-stranded RNA (dsRNA) viruses. *Rotavirus* causes gastroenteritis in several animals including humans, with high mortality rates in young children and infants.

Rotavirus dsRNA exists as separate segments. Each segment is a small double helix containing a single gene. One of these genes codes for a protein that is thought to disrupt carbohydrate digestion and ion transport in cells of the small intestine. Water is drawn out of the cells via osmosis, leading to watery diarrhoea.

Since the eukaryotic host cells only contain single-stranded RNA, they produce Dicer enzymes that cut dsRNA. This is a protective mechanism that stops dsRNA being translated into dangerous proteins. *Rotavirus* gets around this problem by creating single-stranded mRNA from the dsRNA, which then invades the host cells. There are no chemical markers on the invading mRNA to stop the host cells from translating it into the diarrhoea-causing protein.

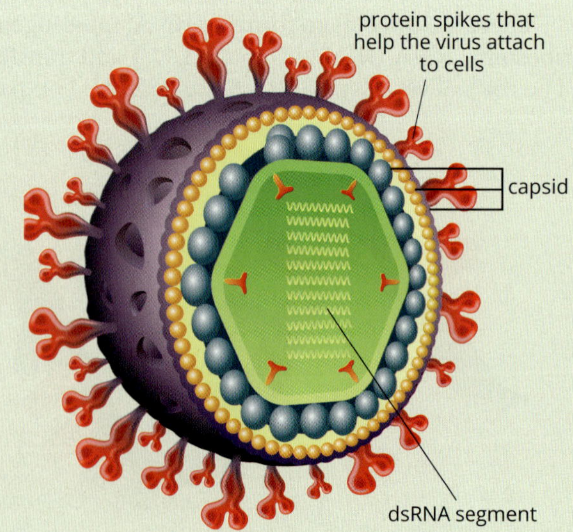

FIGURE 3.1.9 Double-stranded RNA exists in some viruses, including *Rotavirus* species.

EUKARYOTIC VS PROKARYOTIC GENETIC MATERIAL

Prokaryotic and eukaryotic genes are structurally different in many ways. These differences affect the way in which genetic information is transcribed, translated and expressed. Table 3.1.2 summarises the major differences between the structure of the genetic material of prokaryotes and eukaryotes.

TABLE 3.1.2 Summary of the major differences between the structure of the genetic material of prokaryotes and eukaryotes

Prokaryote	Eukaryote
There is one chromosome per cell.	There are multiple chromosomes per cell.
It has a circular chromosome without ends (no telomeres).	It has linear chromosomes with ends (with telomeres).
Contains plasmids—small, circular DNA.	Contains no plasmids but there are other sources of DNA apart from chromosomes—mitochondrial DNA and chloroplast DNA.
There is much less DNA than in eukaryotes (thousands to millions of bases, depending on species).	There is much more DNA than in prokaryotes (millions to billions of bases, depending on species).
There are fewer genes than in eukaryotes (thousands).	There are more genes than in prokaryotes (tens of thousands).
There is less non-coding DNA than in eukaryotes (greater number of genes per number of bases).	There is more non-coding DNA than in prokaryotes (fewer genes per number of bases).
DNA is not packaged into an organelle (less DNA to fit into the cell) (Figure 3.1.10, page 106).	DNA is tightly packaged—coiled around histones, which form nucleosomes, which are condensed into chromatin and packed as chromosomes into the nucleus (a lot of DNA to fit into a small space) (Figure 3.1.11, page 106).
Genes cluster into functional groups, known as operon regions (e.g. genes that code for enzymes in the same biochemical pathway are next to each other on the chromosome and are controlled by the same promoter, so all the genes for the pathway can be transcribed and expressed at once).	Operon regions are rare. (Genes that code for functionally similar enzymes can be physically far apart or located on different chromosomes. Eukaryotes have mechanisms to express these genes at the same time.)

In fact, the invading mRNA lacks some of the extra structures found on the host mRNA, which are usually used to regulate translation. The unregulated viral mRNA is translated very efficiently while the host's own proteins are prevented from being produced.

Another variation of RNA in viruses is that single-stranded RNA (ssRNA) can be classified as positive-sense or negative-sense, depending on the direction of the strand. Within the positive-sense ssRNA group is the subfamily *Orthocoronavirinae* (the coronaviruses), including the virus responsible for the disease COVID-19. The COVID-19 virus primarily affects a person's lungs, taking over the function of cells that normally produce surfactants to stop the alveoli (air sacs) collapsing. When its RNA enters a host cell, it is directly translated into polypeptides as though it were the host's mRNA, so the cells exhaust their resources making new viruses. When the cells die, fluids and white blood cells flood the area, causing the disease.

Negative-sense ssRNA viruses, on the other hand, have RNA that cannot be directly read as mRNA. Instead, their RNA strand is complementary to the strand the host cell translates, so these viruses need an enzyme to transcribe their genome. *Alphainfluenzavirus* is the genus responsible for influenza A, one of the groups of viruses that causes widespread outbreaks of serious cases of 'the flu'. *Alphainfluenzavirus* has a protein that binds to RNAs in the host cell, effectively recycling the cap of the host RNA and using it as a primer to transcribe its own RNA in the direction needed for translation. Influenza A, like COVID-19, also begins in the lungs, although different cells are targeted.

Analysis

1 Of the three types of viruses mentioned in the data, identify which two require their own RNA polymerase (also known as RNA replicase) to successfully infect host cells. Explain your answer.

2 Propose, with a reason, whether or not a damaged coronavirus could infect a host cell.

3 *Rotavirus* infects not just humans, but some other mammals and birds also. Would you expect the same proteins to be translated in these other animals? Explain your answer.

> Histones are proteins found in eukaryotic cells that tightly package DNA into structures called nucleosomes.

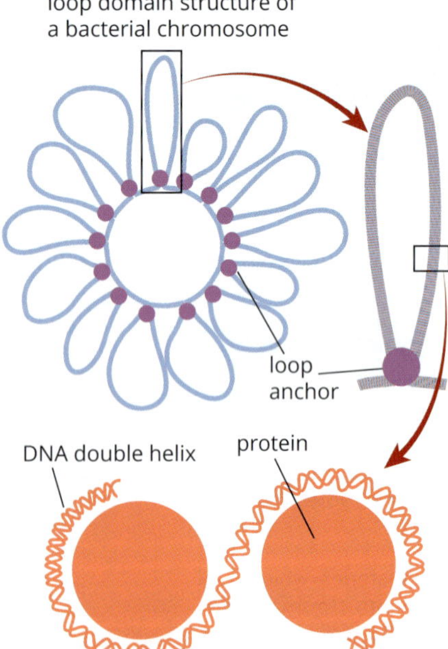

FIGURE 3.1.10 DNA is packaged into loop structures in prokaryotes. The DNA of prokaryotes does not have to be as tightly packaged as in eukaryotes because there is much less genetic material.

> Other proteins, called chaperones, assist with the assembly of nucleosomes.

Although there are many differences between the gene structures of prokaryotes and eukaryotes, shared evolutionary history means that there are also many fundamental similarities.

- Both prokaryotes and eukaryotes have double-stranded DNA that is made up of the nitrogenous bases adenine (A), thymine (T), cytosine (C) and guanine (G).
- Both have mRNA, which acts as an intermediate code to building proteins, with uracil (U) replacing thymine (T).
- Because the genetic information of prokaryotes and eukaryotes is composed of the same code, the way mRNA codons are translated into amino acids and proteins is also much the same.

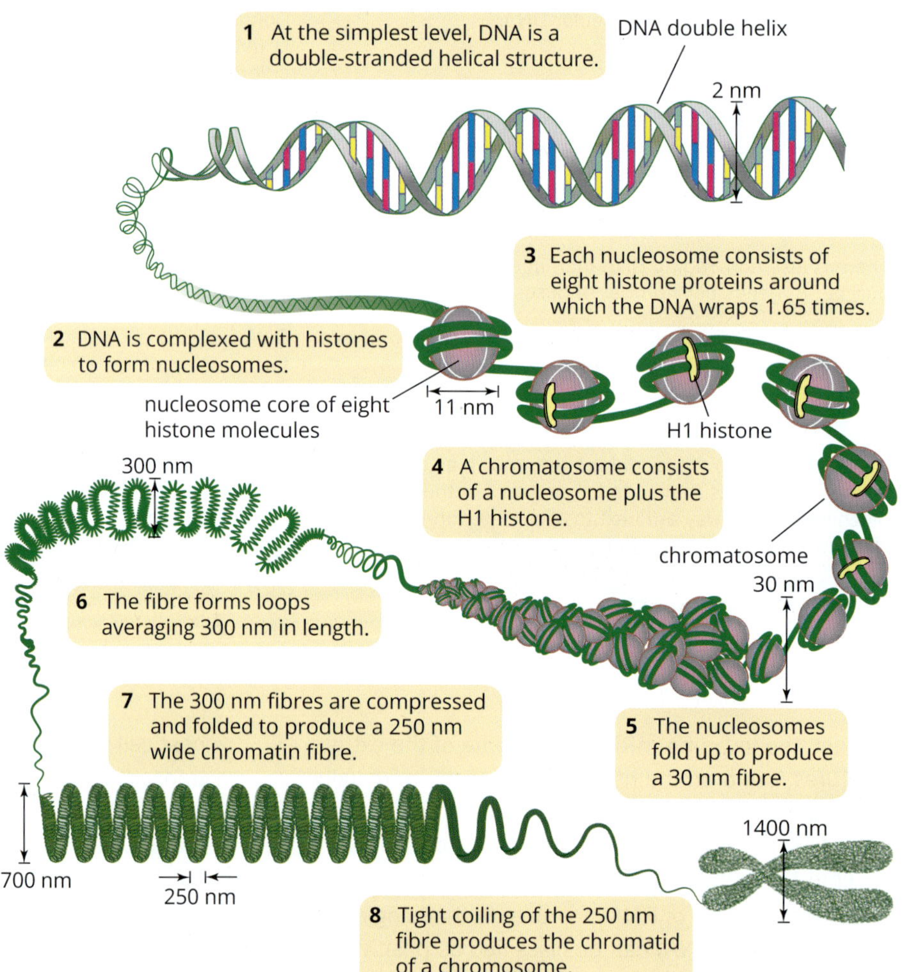

FIGURE 3.1.11 Packaging of DNA in eukaryotic cells. Because eukaryotes have large quantities of DNA to fit into a small space, the DNA needs to be tightly and efficiently packaged. (1) Double-stranded DNA is (2 and 3) tightly coiled around histones to form nucleosomes. (4) Nucleosomes and histones together form chromatosomes. The nucleosomes (5) fold, (6) loop and (7) compress into chromatin. (8) Tight coiling of the chromatin produces the chromatids of a chromosome.

3.1 Review

OA ✓✓

SUMMARY

- Nucleic acids are polymers of nucleotides (polynucleotides). There are two types of nucleic acids: deoxyribonucleic acid (DNA) and ribonucleic acid (RNA).
- Nucleotides are made up of a five-carbon sugar, a phosphate and a nitrogenous base.
- The nitrogenous bases adenine and guanine are purines; cytosine, thymine (present in DNA only) and uracil (RNA only) are pyrimidines.
- DNA is a long, coiled, double-stranded nucleic acid that forms a double helix.
  - The two strands of DNA are joined by complementary base pairing between the nitrogenous bases. Adenine joins with thymine by two weak hydrogen bonds, while cytosine joins with guanine by three hydrogen bonds.
  - The two strands of DNA are antiparallel (oriented in opposite directions).
- RNA is a short, usually single-stranded nucleic acid.

KEY QUESTIONS

Knowledge and understanding

1 What are the functions of DNA?

2 What are the three components of a nucleotide?

Analysis

3 Complete the table by identifying which polynucleotide(s) each nitrogenous base is found in (DNA, RNA or both), assigning the structure purine or pyrimidine, and filling in the name and structure of its complementary base(s).

Nitrogenous base			Complementary base(s)	
Name	DNA and/or RNA	Purine or pyrimidine	Name(s)	Purine or pyrimidine
adenine (A)				
guanine (G)				
cytosine (C)				
thymine (T)				
uracil (U)				

4 Construct a table or diagram to compare the structures of RNA and DNA. Include at least three distinct characteristics.

3.2 Gene structure and expression

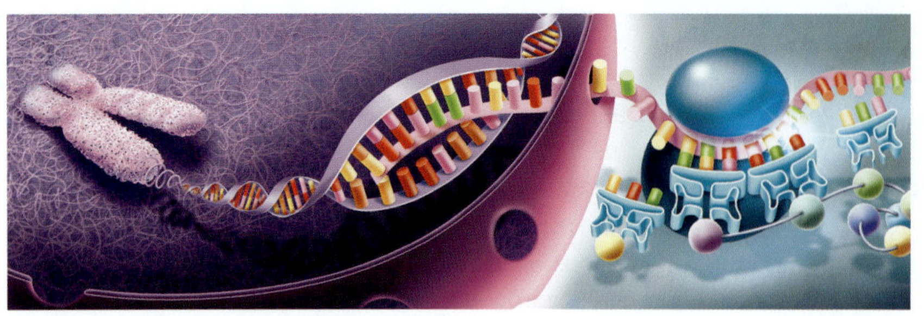

FIGURE 3.2.1 Overview of gene expression, showing mRNA being produced during transcription and a polypeptide being synthesised during translation

The genetic code represents the information stored in DNA as a triplet code within sections called genes. This information is used to synthesise the amino acid sequences that form proteins through a process called gene expression (Figure 3.2.1).

In this section, you will learn about the roles of DNA and RNA in protein synthesis and the different steps of gene expression.

GENES AND THE GENETIC CODE

A **gene** is a region of DNA that may be translated into a polypeptide or an RNA molecule that can be functional, such as tRNA. When coding for a polypeptide, it is the sequence of nucleotides within a gene that contains the information for the protein to be synthesised. Genes can be millions of nucleotides in length. For example, the longest known gene, which codes for the protein dystrophin, is 2.5 megabase pairs long (2 500 000 base pairs).

The genetic code

The **genetic code** is a set of rules that defines how the information in nucleic acids (DNA and RNA) is translated into proteins and functional RNA molecules. The information in DNA and RNA is stored as a three-letter code of nucleotides. In DNA, this three-letter code is called a **triplet**. When a DNA triplet is transcribed into mature **messenger RNA (mRNA)** through the process of **transcription**, the triplet is then called a **codon**. Most codons code for an amino acid (Figure 3.2.2). This includes AUG, which also has the role of initiating **translation**. The **stop codons**, however, end translation without attaching an amino acid. With the amino acid sequence determined by codons, a **polypeptide chain** is formed, which can then be modified into its tertiary and possibly quaternary structures to become a functional protein. The genetic code is almost universal—nearly every cell on Earth follows the same rules to translate codons into the same amino acids. A gene transferred into another species will therefore usually result in the synthesis of an identical protein.

Transcription is the production of single-stranded mRNA from DNA.

Translation is the process in which the sequence of an mRNA molecule is used to produce the amino acid sequence of a polypeptide.

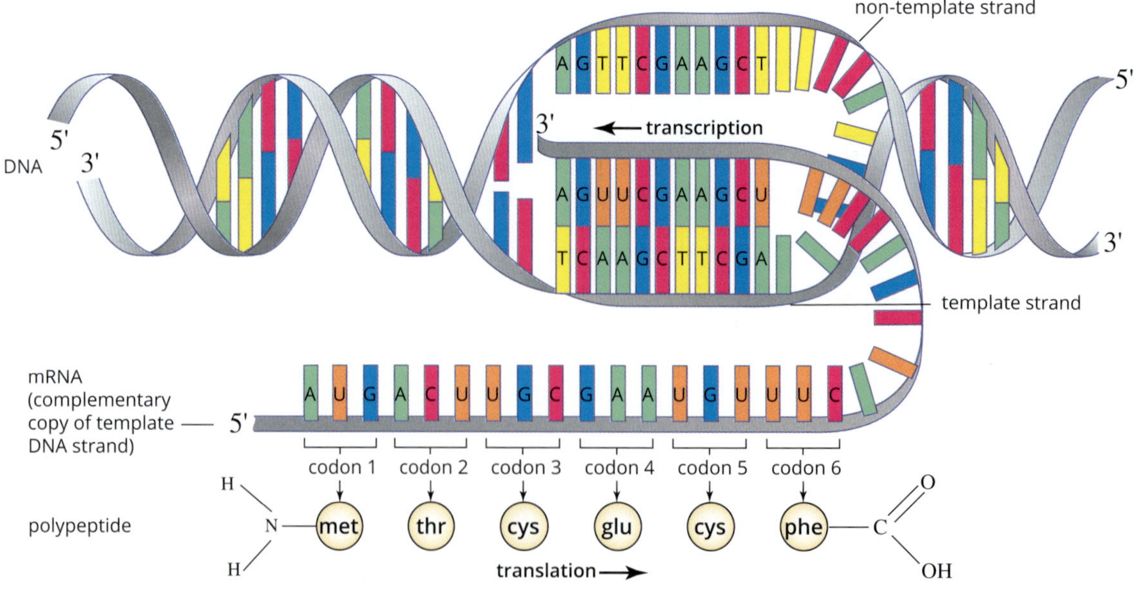

FIGURE 3.2.2 DNA is transcribed into mRNA (messenger RNA). mRNA is read as codons.

Degeneracy

The genetic code is said to be **degenerate** because more than one codon can code for the same amino acid (Figure 3.2.3) at the third base. As the genetic code uses four nucleotides, and three nucleotides code for an amino acid, the combinations of these nucleotides make a total of 64 possible codons ($4^3 = 64$), to code for the total 20 amino acids (Figure 3.2.3). The degeneracy of the code acts as a buffer for genetic mutations in that a single change in one base may not necessarily lead to a change in the amino acid produced and therefore may not cause a change in the structure of the protein produced.

> Amino acids all contain an amine ($-NH_2$) and a carboxyl ($-COOH$) group, but have a unique side chain.

Second base of codon

First base of codon	U		C		A		G		Third base of codon
U	UUU	phenylalanine (phe)	UCU	serine (ser)	UAU	tyrosine (tyr)	UGU	cysteine (cys)	U
	UUC		UCC		UAC		UGC		C
	UUA	leucine (leu)	UCA		UAA	STOP	UGA	STOP	A
	UUG		UCG		UAG		UGG	tryptophan (trp)	G
C	CUU	leucine (leu)	CCU	proline (pro)	CAU	histidine (his)	CGU	arginine (arg)	U
	CUC		CCC		CAC		CGC		C
	CUA		CCA		CAA	glutamine (gln)	CGA		A
	CUG		CCG		CAG		CGG		G
A	AUU	isoleucine (ile)	ACU	threonine (thr)	AAU	asparagine (asn)	AGU	serine (ser)	U
	AUC		ACC		AAC		AGC		C
	AUA		ACA		AAA	lysine (lys)	AGA	arginine (arg)	A
	AUG	methionine (met) START	ACG		AAG		AGG		G
G	GUU	valine (val)	GCU	alanine (ala)	GAU	aspartic acid (asp)	GGU	glycine (gly)	U
	GUC		GCC		GAC		GGC		C
	GUA		GCA		GAA	glutamic acid (glu)	GGA		A
	GUG		GCG		GAG		GGG		G

FIGURE 3.2.3 The genetic code for the 20 amino acids and stop codons

THE STRUCTURE OF GENES

Eukaryotic genes all have a number of structural features in common, including:

- **promoter regions**—an upstream binding region for the enzyme that is involved in the encoding process (which is RNA polymerase)
- **exons**—DNA regions that are the coding segments
- **introns** (or spacer DNA)—DNA regions that are non-coding segments.

These structural features are illustrated in Figure 3.2.4.

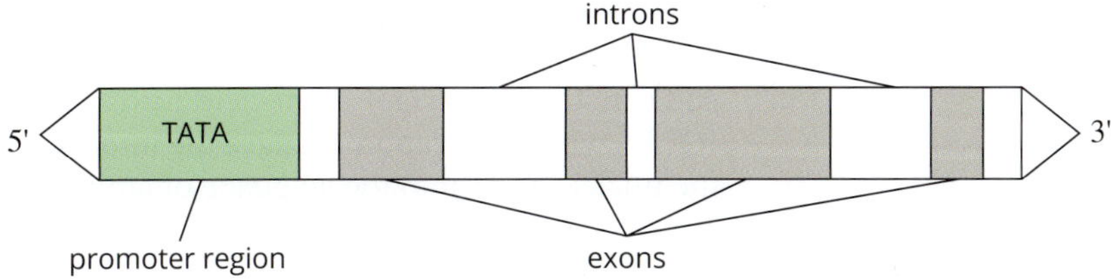

FIGURE 3.2.4 Eukaryotic genes contain common features: a promoter region, coding exons and non-coding introns.

RNA polymerase is needed to separate the double helix in a small section of DNA. The exposed nucleotides on one of the DNA strands can then act as a template for mRNA.

Promoter region

Promoter regions are sections of a gene that are found before the first triplet, at the 5' end of the site where transcription will begin. A promoter region:

- is the location where the RNA polymerase (the enzyme that initiates transcription) attaches to the gene
- identifies which DNA strand will be transcribed
- identifies where transcription of the gene will start
- identifies in which direction transcription will occur.

In many eukaryotic genes, the promoter region includes a characteristic sequence of repeating T and A bases, known as the **TATA box**.

Introns and exons

In eukaryotes, not all sections of a gene are translated:

- Exons are regions of a gene that are usually expressed as proteins or RNA. Exons come together to make up mRNA, which is then translated into proteins.
- Introns are non-coding regions of a gene. Introns are spliced out of the mRNA during the stage of gene expression called **RNA processing**.

There are no rules about the number of exons and introns in a gene. In the dystrophin gene, for example, 99% of its length is made up of introns.

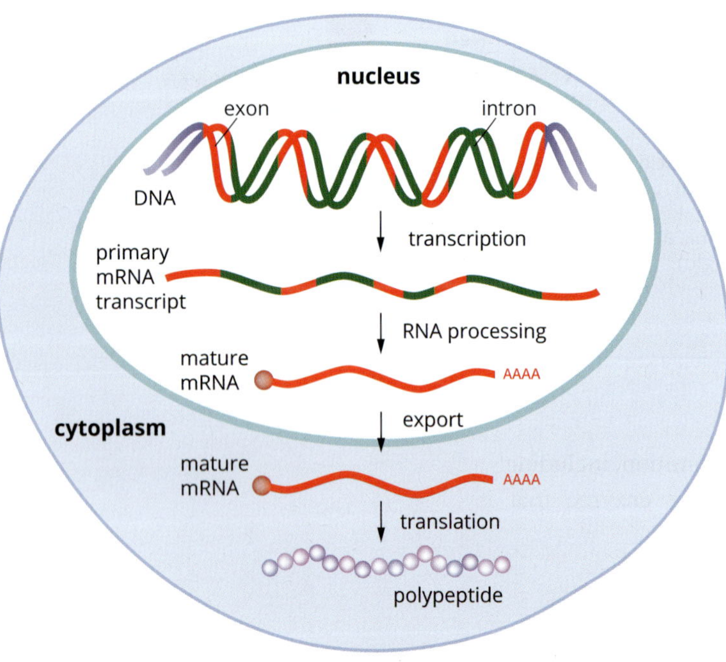

FIGURE 3.2.5 Transcription creates a primary RNA transcript from DNA. The introns are then spliced (cut out) during RNA processing to create a mature strand of mRNA. The mRNA exits the nucleus via a nuclear pore. A ribosome translates the mRNA into a polypeptide chain during translation.

GENE EXPRESSION

Gene expression is the process by which the information stored in a gene is used to synthesise a functional gene product (protein or RNA) (Figure 3.2.5). This process is highly regulated so that proteins or RNA molecules are only produced if and when they are required by a cell. Multicellular organisms, in particular, can have specialised cells that require a specific set of proteins. For example, in humans, the cells in connective tissue and bone require the protein fibrillin to form elastic fibres, and skin cells require the enzyme tyrosinase to produce melanin and other pigments. The ability to regulate gene expression conserves energy and materials (nucleotides and amino acids) in the cell.

Gene expression leading to protein synthesis in eukaryotic cells occurs in three stages:

- transcription
- RNA processing
- translation.

Transcription, in a broader context, means to copy text in the same language. Translation means changing text from one language into another. In a biological context, transcription keeps the genetic code in the 'language' of nucleic acids (DNA to RNA), while translation changes the information in the nucleic acids into the language of amino acids.

CASE STUDY **ANALYSIS**

Gene switching

Trypanosomes (*Trypanosoma brucei*) are parasites that cause African sleeping sickness (Figure 3.2.6). These protozoan parasites spend part of their life cycle in the tsetse fly (*Glossina fuscipes fuscipes*) (Figure 3.2.7), where they have adaptations to living in the fly's gut. The trypanosomes then migrate to the fly's salivary glands and are transmitted to humans and livestock when the fly feeds on mammalian blood. The trypanosomes undergo morphological changes in response to the environment of their new host. African sleeping sickness occurs in Sub-Saharan African, across 36 countries, and threatens the lives of millions of people. The disease causes about 9000 deaths per year. It causes damage to the central nervous system, leading to behavioural changes, confusion, loss of coordination and disturbance to sleep. Without treatment, the disease can be fatal.

T. brucei have up to 1000 genes, which code for proteins that will be positioned on their cell surface, but they can only express one of these genes at a time. When this occurs, one gene is transcribed and the rest are repressed. When a human is infected by a parasite, their immune system will usually recognise the proteins on the cell surface of the parasite and respond by producing antibodies as a defence. However, parasites such as trypanosomes can switch from expressing one gene to another, thereby overcoming the human defence response and evading detection.

Analysis

If *T. brucei* can alter its gene expression to evade human defences, would you expect it to express all of its proteins throughout its time in a human host? Use evidence in the data to explain your answer.

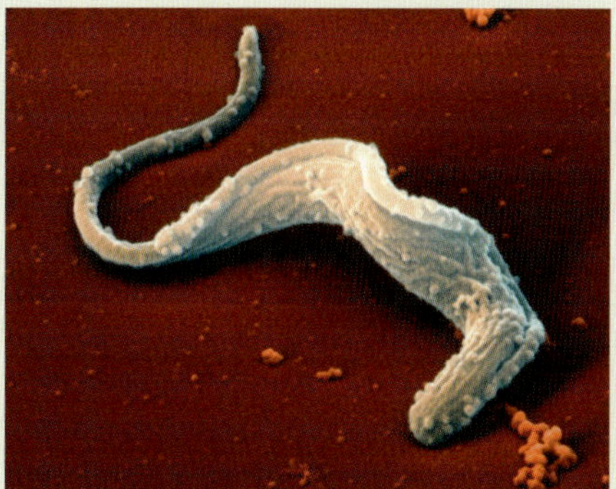

FIGURE 3.2.6 Coloured scanning electron micrograph of the parasite that causes African sleeping sickness, *Trypanosoma brucei*

FIGURE 3.2.7 A tsetse fly (*Glossina fuscipes fuscipes*) feeding on human blood

Roles of DNA and RNA in protein synthesis

DNA and RNA both play vital roles in protein synthesis. DNA provides the instructions, which are translated by RNA into proteins that carry out all of the functions that are essential to life.

DNA does not code for all molecules in the body, such as carbohydrates. However, it does code for the enzymes needed to assemble them.

DNA and protein synthesis

DNA stores and transmits hereditary information. The sequence of nucleotides is the code from which biological products are synthesised. These products are mainly proteins, but they may also be functional RNA molecules. A gene is a region of DNA that contains the information to produce a protein or a functional RNA molecule.

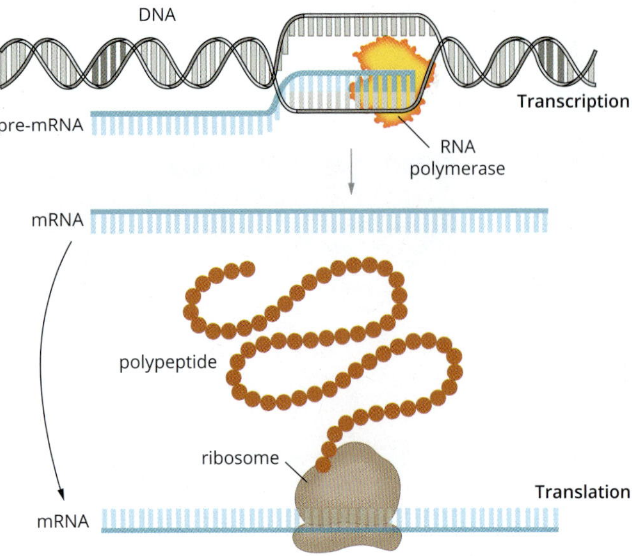

FIGURE 3.2.8 The nucleotide sequence is first transcribed to mRNA, which is then translated into a chain of amino acids.

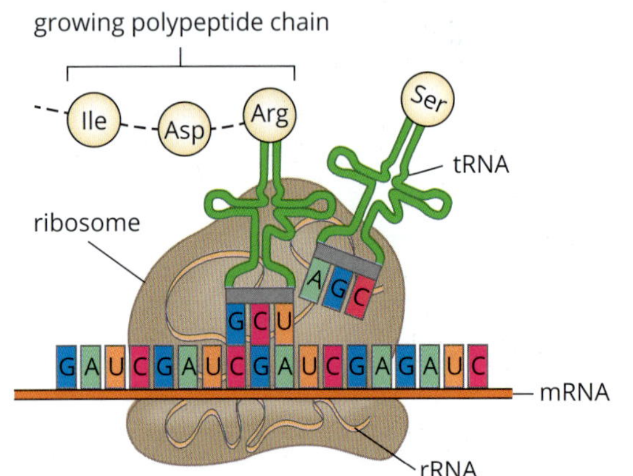

FIGURE 3.2.9 The three different types of RNA work together to use the information contained in a gene to synthesise a protein.

RNA and protein synthesis

RNA plays an important role in expressing the information contained in coding DNA (genes) to synthesise proteins. Each type of RNA has a different role in the process of protein synthesis:

- Messenger RNA (mRNA) is formed in the nucleus by the process of transcription. mRNA carries a copy of the nucleotide sequence of DNA that specifies the amino acid sequence for a particular protein. During transcription, pre-mRNA is first formed by the enzyme **RNA polymerase**. Pre-mRNA is then processed (post-transcriptional modification) to form mature mRNA, which is a single-stranded copy of the coding DNA (gene) (Figure 3.2.8). The mature mRNA travels from the nucleus to the cytoplasm where it binds to ribosomes ready for translation.
- **Ribosomal RNA (rRNA)** is synthesised in the nucleolus of the cell nucleus and is based on the nucleotide sequence of the DNA. Together with proteins, rRNA forms a small organelle called a ribosome. Ribosomes are the sites where the information in the mRNA is translated into a chain of amino acids (Figure 3.2.9).
- **Transfer RNA (tRNA)** molecules are the links between amino acids and mRNA. They have a region called an **anticodon** which is a sequence of nucleotides complementary to the codons found on mRNA, and a binding site for one type of amino acid (which freely attach themselves in the cytoplasm). Ribosomes have three sites where tRNA can bind: aminoacyl (A), peptidyl (P) and exit (E) as shown in Figure 3.2.10. A tRNA with the right anticodon will form a complex with the ribosome. As the next tRNA enters, a peptide bond is made between the amino acids. The ribosome moves along the mRNA and the tRNAs are shifted towards the exit site, while the polypeptide chain grows. You might expect there to be 64 types of tRNA based on the possible combinations of nucleotides in codons (Figure 3.2.3 on page 109). However, three combinations are stop codons (no tRNAs match them, so translation is terminated). There is also a so-called 'wobble position' in the first nucleotide of an anticodon, which allows for some flexibility in the base used, so that the maximum number of tRNAs found in any cell is only 41.

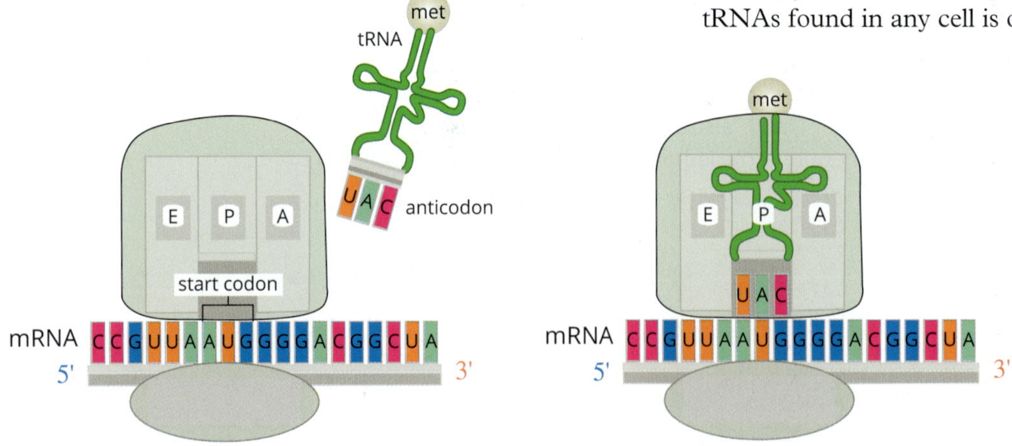

FIGURE 3.2.10 A tRNA molecule carrying the amino acid methionine (met) has an anticodon sequence that is complementary to the codon of the mRNA in the ribosome.

CASE STUDY **ANALYSIS**

Protein folding and the need for speed

The folding configuration of proteins gives rise to their three-dimensional structure, which is essential to their function. The diversity of shapes and sizes among proteins is quite extraordinary, as the examples in Figure 3.2.11 show.

Polypeptide chains fold at certain points because of the interactions between the different amino acids present. Particular sequences of amino acids can be identified as a folding point. The shorter the distance between two folding points, the faster the protein can be assembled and start carrying out its task. Fast folding is also thought to be advantageous, as a rapidly synthesised protein is less likely to get aggregated (clumped in the wrong shape). It is hypothesised that proteins need to be folded in under 10 milliseconds to minimise aggregation, yet some proteins average only 1.2 folds per second!

Using computers to analyse a dataset of 92000 proteins and 989 genomes from different organisms, researchers have shown that folding speed has tended to increase as proteins have evolved over time. They have also been able to apply mathematical models to their data to estimate when certain proteins started being synthesised in a range of organisms.

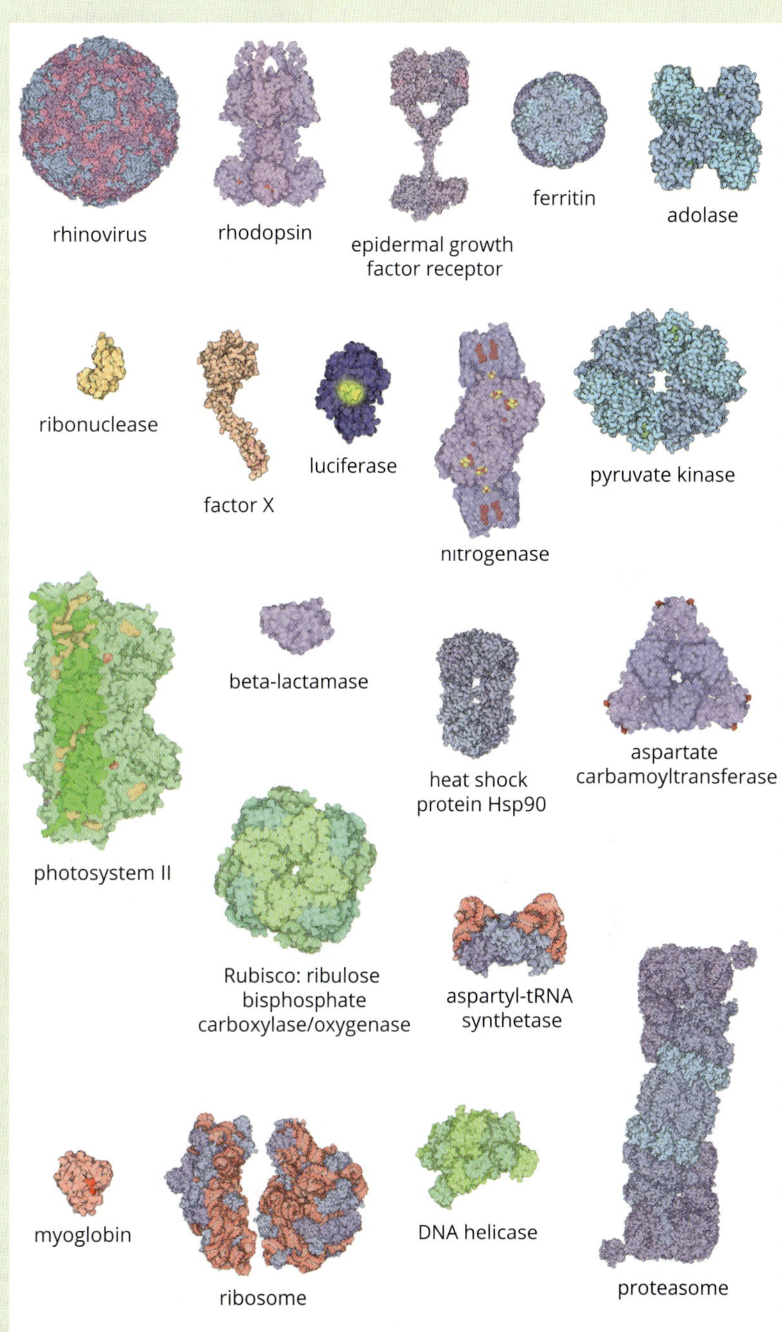

FIGURE 3.2.11 Examples of the diversity of structures and folding configurations of different proteins from the Worldwide Protein Data Bank

Analysis

1 For some of the organisms investigated in the dataset, not all of their protein structures were known. Explain how the researchers could still determine distances between protein folding points by examining the organisms' genomes.

2 a i How long would it take a small protein to fold if it required eight folds averaging one fold every 1.2 seconds?

ii How many times slower is this than the hypothesised time required for a protein to minimise aggregation?

b Suggest two reasons why protein aggregation (clumping) would be detrimental to an organism.

Transcription

The production of single-stranded mRNA from DNA is called transcription and occurs within the nucleus of eukaryotic cells. The DNA segment that undergoes transcription is known as the transcription unit.

Transcription occurs in three steps:

1 Initiation: Transcription factors combine with the region at the start of the gene, known as the promoter. The promoter region contains specific nucleotide sequences (TATA box) that are recognised by an appropriate subunit of the enzyme RNA polymerase. In eukaryotic cells, transcription factors are required for RNA polymerase to attach to the DNA. RNA polymerase then attaches to the promoter, unwinding and unzipping the DNA molecule by breaking the weak hydrogen bonds between the two strands to expose the bases (Figure 3.2.12a).

2 Elongation: During transcription, the RNA polymerase molecule covers a region of approximately 30 base pairs. Within this region, a segment of about 15 base pairs is uncoiled. This results in the formation of a transcription bubble. As the RNA polymerase moves along the gene, DNA strands located behind the transcription bubble are coiled again. The RNA polymerase moves along the DNA molecule, producing a strand of mRNA. It uses a strand of DNA as a template, attaching nucleotides (A, U, G, C) by complementary base pairing. mRNA is always synthesised in the 5' to 3' direction, with new nucleotides added to the 3' end. The initial mRNA molecule transcribed is called a primary RNA transcript (Figure 3.2.12b). The primary RNA transcript will then be processed into mature mRNA.

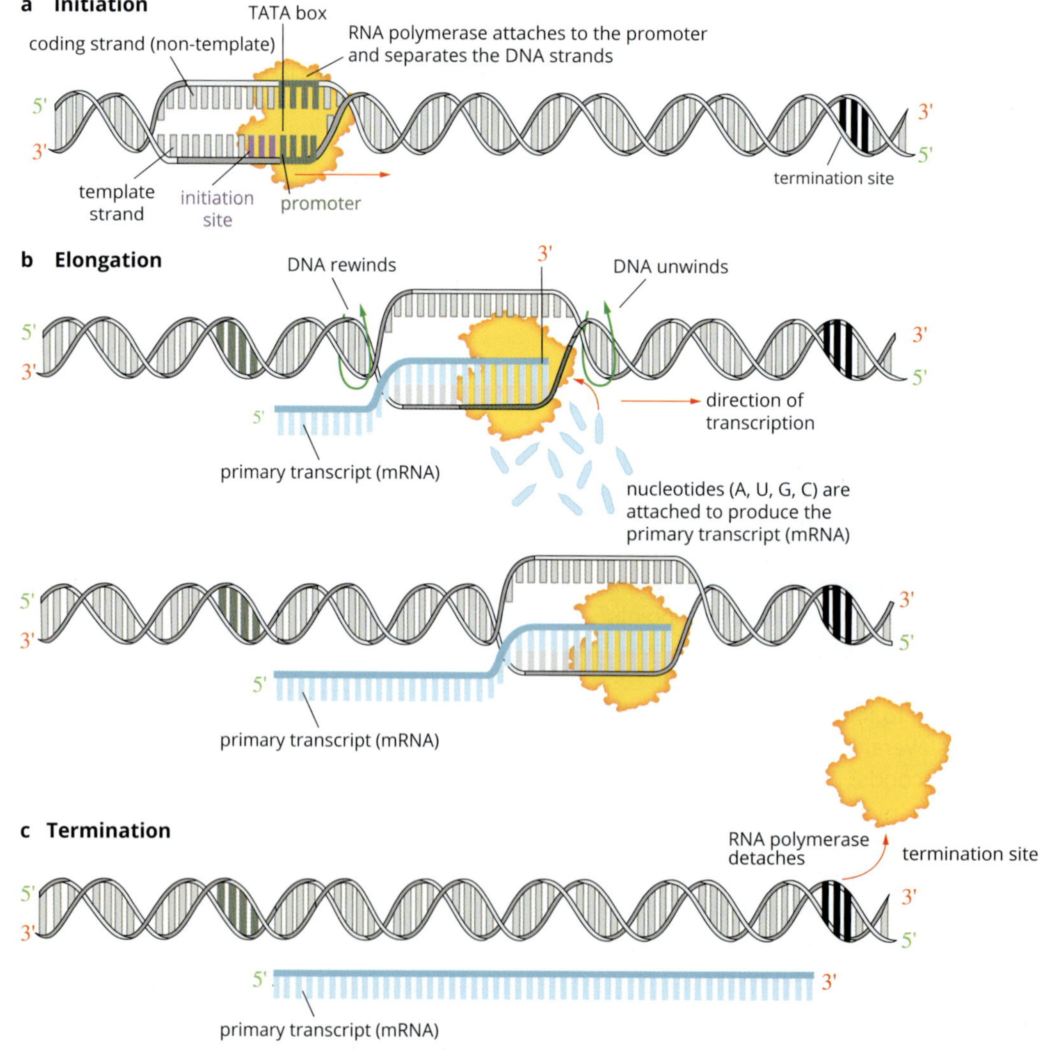

FIGURE 3.2.12 Transcription occurs in three stages: (a) initiation, (b) elongation and (c) termination.

3 Termination: Transcription ends when RNA polymerase reaches the termination site of the gene. This region contains a poly-A signal, which, once transcribed, triggers a set of proteins to cleave the mRNA. The RNA polymerase continues a little further downstream before detaching. The DNA strands join together, reforming the double helix (Figure 3.2.12c).

Many RNA polymerase molecules may attach to the gene being transcribed, producing many of the same mRNA molecules. The strand of DNA that is transcribed to the mRNA is known as the **template strand**, and the other complementary strand is known as the **coding strand**. The mRNA carries the same base sequence as the coding strand, except it contains uracil in place of thymine.

> The separated strands of DNA reform their hydrogen bonds spontaneously, without the need for additional enzymes

RNA processing

After transcription, the primary RNA transcript undergoes processing before it is translated. This stage of gene expression is called RNA processing and includes:

- the addition of a **5' cap**
- the addition of a **poly-A tail**
- **splicing** (removal) of the introns (mRNA maturation).

5' cap and poly-A tail

A cap consisting of a methylguanosine triphosphate molecule, called a 5' cap, is added to the 5' end of the primary RNA transcript while it is being synthesised during transcription. Once transcription has finished, a chain of up to 250 adenine nucleotides is added to the 3' end of the primary RNA transcript. This chain is called a poly-A tail.

These modifications to either end of the primary RNA transcript increase its stability and prevent it from degrading. Additionally, the 5' cap aids the binding of the ribosome to the mRNA at the beginning of translation.

Splicing

In eukaryotes, not every part of the original DNA sequence is incorporated into the mature mRNA. The included parts are knowns as exons, short for 'expressed regions'. Most exons directly code for a polypeptide, although a small number are used as signals needed to start and stop translation. The sequences in between exons are called introns, short for 'intragenic regions'. Most prokaryotes contain only exons and therefore the RNA processing described in this section does not occur in prokaryotes.

> Introns are also found in tRNA and rRNA, although their removal requires a slightly different process than the removal of introns in mRNA.

In eukaryotes, before a protein can be produced, the introns must be cut out of the primary RNA transcript to form the mature mRNA molecule. This process is known as splicing. During splicing, a complex molecule composed of protein and RNA molecules, called a **spliceosome**, removes the introns from the primary RNA transcript and joins the exon sections together to make mature mRNA (Figure 3.2.13). (Not all of the exons will necessarily be included, as you will see in the next section.) The single-stranded mature mRNA molecule then exits the nucleus via a nuclear pore.

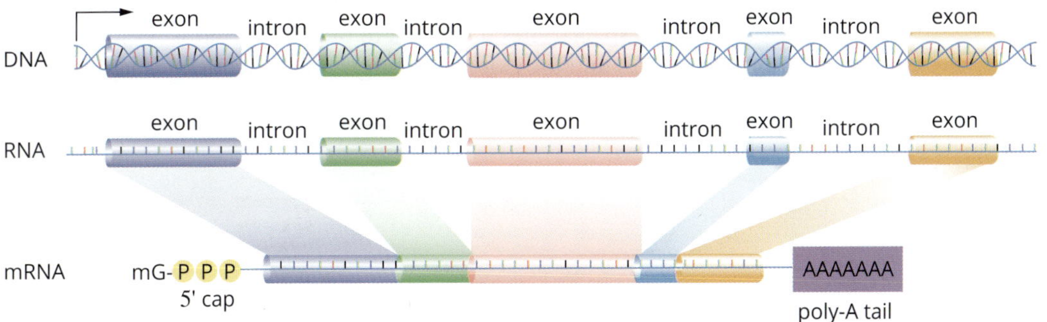

FIGURE 3.2.13 During RNA processing, the introns are spliced from the primary RNA transcript, resulting in mature messenger RNA, which consists of only exons.

Alternative splicing

A primary transcript can be spliced in many different ways, resulting in alternative mature mRNA strands from a single gene and, thus, different proteins. This is the result of some exons being removed along with the introns during RNA processing. For example, a particular gene may result in a mature mRNA that contains all exons 1–5, but the same gene may result in another mature mRNA that contains only exons 1, 2, 4 and 5 (Figure 3.2.14). Alternative splicing is one reason why the 21 000 genes of humans can produce so many more than 21 000 proteins.

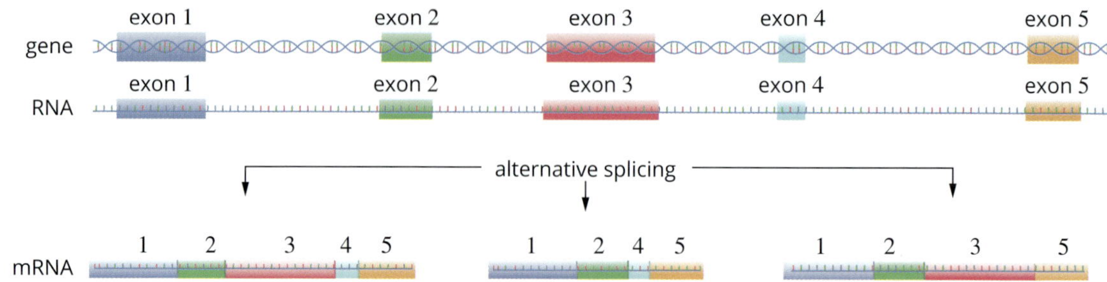

FIGURE 3.2.14 Alternative splicing of a single gene gives rise to alternative mRNA molecules, resulting in many different proteins.

In early research on gene structure, introns were called 'junk DNA' because it was believed they had no role in protein production. It is now known that gene expression is much more complex than first thought. Introns are essential to the process of alternative splicing as they contain nucleotide sequences needed for the spliceosome complex to form. Some introns also have additional regulatory functions, such as assisting mRNA export and protecting DNA from damage during transcription.

Translation

Translation is the process in which the codons on mRNA are translated into a sequence of amino acids, resulting in a polypeptide. This process occurs on ribosomes. Ribosomes bind to an mRNA molecule and act as docking stations for the tRNAs to deposit specific amino acids. A part of the tRNA, called an anticodon, recognises and binds to the codon on the mRNA by complementary base pairing. Each tRNA carries a specific amino acid related to the codon to which it binds.

Translation occurs in a series of steps, as outlined below.

1 Initiation: To begin protein synthesis, a small ribosomal subunit attaches to the 5' end of an mRNA strand. It then moves along the mRNA until it reaches a **start codon** AUG. The sequence AUG, which codes for the amino acid methionine, is the most common initiation codon of mRNA (there are some rare exceptions). A tRNA molecule with the anticodon UAC then brings the amino acid methionine to the mRNA. The tRNA molecule joins to the mRNA start codon, attaching by complementary base pairing between the codon and anticodon (Figure 3.2.15). A large ribosomal subunit also attaches to the tRNA and the small ribosomal subunit. The binding of both ribosomal subunits causes the formation of three special sites for tRNA to bind: the aminoacyl site (A site), the peptidyl site (P site) and the exit site (E site). The attachment of amino acids to their corresponding tRNA molecules occurs in the cytoplasm—a process that is catalysed by enzymes.

2 Elongation: Following the attachment of the amino acid methionine, another tRNA, with a complementary anticodon to the next codon on the mRNA attaches and adds its specific amino acid to the growing polypeptide chain (Figure 3.2.15). The deposited amino acid joins by a peptide bond to the first amino acid through a condensation polymerisation reaction. The ribosome then releases the tRNA and moves further along the mRNA strand. At each codon a new tRNA binds and adds another amino acid. The tRNA molecules can be reused, allowing them to pick up more of their specific amino acids and return to the mRNA molecule.

Ribosomes in eukaryotic cells can carry out translation (build polypeptides) at a rate of around 6–9 amino acids per second.

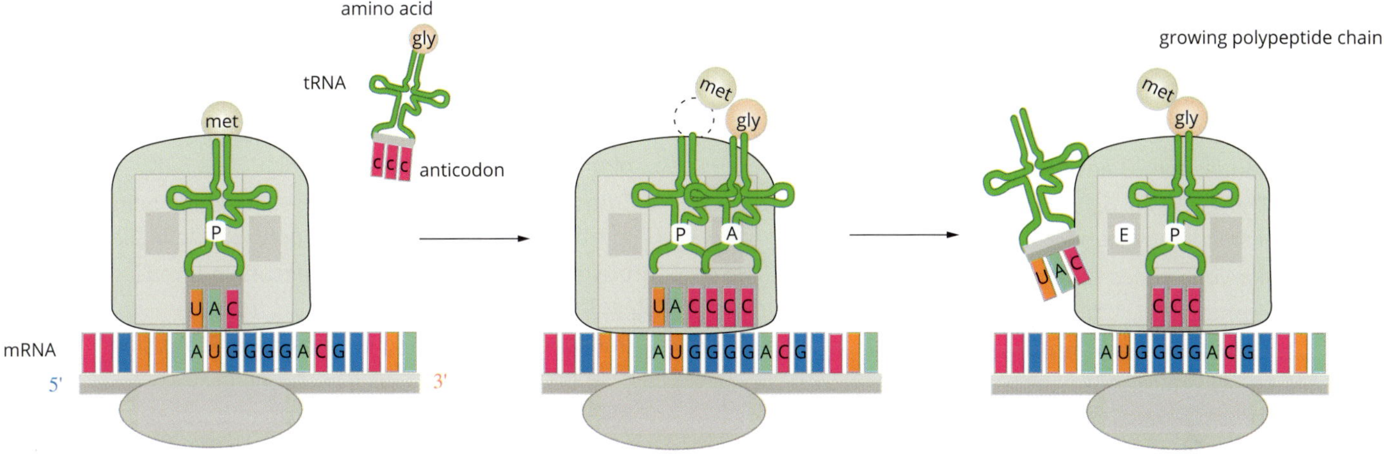

FIGURE 3.2.15 Process of translation on a ribosome. The ribosome moves along the mRNA one codon at a time, and tRNA molecules bring their specific amino acids to their complementary mRNA codon.

3 Termination: Attachment of amino acids continues until a stop codon is reached, which is a codon that does not code for an amino acid. The polypeptide chain is then released from the ribosome into the cytoplasm or the endoplasmic reticulum. Some proteins consist of more than one polypeptide; the polypeptides of these proteins associate in the cytoplasm or the Golgi apparatus to form the fully functional protein.

Many ribosomes can translate the same single strand of mRNA, enabling many polypeptide chains to be produced at the same time. Once the polypeptides are fully functional, they either remain in the cell for use, or are exported from the cell via exocytosis (also known as the protein secretory pathway) for use elsewhere in the organism. You learnt about the protein secretory pathway in Chapter 2.

BIOFILE

Epigenetics

Epigenetics is the study of the molecular changes that affect gene expression without altering the DNA sequence. Epigenetic changes therefore affect an organism's phenotype rather than genotype.

Epigenetic modification can occur directly to DNA in the form of methylation (the addition of a methyl group, which is a carbon with three hydrogens). This often represses gene expression. Modifications can also occur on histones, the proteins that package DNA in eukaryotic chromosomes. This affects the coiling of chromatin so that genes cannot be accessed to be transcribed. In females, a whole X chromosome is silenced so that excess gene products are not made (a process known as X-inactivation).

Epigenetic changes are passed down when cells divide, which is an important feature of differentiation in stem cells. During gamete formation, epigenetic tags are usually removed in a process called reprogramming. Research has found that some changes do remain, however, and can span multiple generations.

3.2 Review

OA ✓✓

SUMMARY

- The genetic code is the rules and information used by cells to synthesise proteins and functional RNA molecules from nucleic acids. It is almost universal; the rules of the code are the same in nearly all cells.
- DNA triplets (groups of three nucleotides) are transcribed to mRNA where they are called codons.
  - There are 64 possible codons (3^4).
  - The code is degenerate; there are 61 possible codons for 20 amino acids.
  - Three codons do not code for an amino acid; instead, they stop translation.
- Eukaryotic genes have common structures:
  - promotor, including the TATA box
  - exons (expressed regions)
  - introns (intragenic regions).
- DNA stores hereditary information. A gene is a region of DNA that codes for a protein or a functional RNA molecule.
- The role of RNA is to express the information contained in the nucleotide sequence of a gene.
  - Messenger RNA (mRNA) carries a copy of the genetic sequence in DNA, specifying the amino acid sequence for a particular protein.
  - Ribosomal RNA (rRNA) makes up part of a ribosome. Ribosomes are the sites where the information in the mRNA is translated into a chain of amino acids.
  - Transfer RNA (tRNA) carries specific amino acids to ribosomes in order to form polypeptide chains.
- Protein synthesis in eukaryotes occurs in three stages:
  - transcription of DNA to pre-mature mRNA using RNA polymerase
  - RNA processing, in which a 5' cap and poly-A tail are added, and introns are spliced so that the mature mRNA contains only exons and the 5' cap and poly-A tail
  - translation, which occurs on a ribosome where mRNA is translated into a sequence of amino acids, delivered by their specific tRNA molecules.

KEY QUESTIONS

Knowledge and understanding

1 What is the genetic code?

2 Describe these structural features of eukaryotic genes and their functions:
  a promoter regions
  b exons
  c introns

3 What are the three stages of protein synthesis?

4 What are splicing and alternative splicing?

5 There are many steps that occur during transcription and translation. These steps are provided below in random order.
  a Rewrite the steps of transcription in the correct order. Group the steps into the correct stage: initiation, elongation or termination.
  - The RNA polymerase moves along the DNA molecule, producing a strand of mRNA.
  - The RNA polymerase detaches, releasing the mRNA and allowing the DNA molecule to reform.
  - RNA polymerase uses a strand of DNA as a template, attaching nucleotides (A, U, G, C) by complementary base pairing.
  - Transcription factors combine with the region at the start of the gene, known as the promoter.
  - RNA polymerase reaches the termination site of the gene.
  - RNA polymerase attaches to the promoter, unwinding and unzipping the DNA molecule by breaking the weak hydrogen bonds between the strands to expose the bases.

b Rewrite the steps of translation in the correct order. Group the steps into the correct stage: initiation, elongation or termination.

- Following the attachment of the amino acid methionine, another tRNA with a complementary anticodon to the next codon on the mRNA attaches and adds its specific amino acid to the growing polypeptide chain.
- A stop codon is reached. The tRNA with the complementary anticodon has no amino acid bound to it.
- A small ribosomal subunit attaches to the 5' end of an mRNA strand. It then moves along the mRNA until it reaches a start codon (AUG).
- The polypeptide chain is released from the ribosome into the cytoplasm or the endoplasmic reticulum.
- The ribosome then releases the tRNA and moves along the mRNA strand. At each codon a new tRNA binds and adds another amino acid.
- A tRNA molecule with an anticodon (UAC) brings the amino acid methionine to the mRNA. The tRNA molecule joins to the mRNA start codon, attaching by complementary base pairing to its anticodon.

Analysis

Refer to Figure 3.2.3 on page 109 to complete question 6.

6 **a** The genetic code is degenerate. Complete the table by entering all of the possible mRNA codons for the amino acids listed, and determine the complementary triplets that would be in the corresponding template strand of DNA.

Amino acid	mRNA codons	Complementary DNA triplets
methionine (met)		
arginine (arg)		
cysteine (cys)		
leucine (leu)		
valine (val)		

b Give an example of where a change to one nucleotide in the template DNA would alter the polypeptide synthesised, and an example of a nucleotide change where the polypeptide would remain the same.

c In a codon, do any of the bases (first, second or third) have a greater role in determining the amino acid than others? Use examples in your answer.

3.3 Gene regulation

As you have learnt, the genome consists of many thousands of genes. A cell is able to express a selection of these genes at a given time. The genes expressed determine which proteins are produced, giving the cell its functionality and characteristics. Gene expression is the process through which information from a gene is used to synthesise a functional gene product—a protein or RNA. Gene expression in eukaryotes is tightly regulated by multiple mechanisms, at different points.

Although gene expression is controlled at many points, this section focuses on gene regulation at the transcription stage. You will learn the difference between structural and regulatory genes and understand how some proteins called transcription factors are able to regulate transcription by 'switching genes on and off' (Figure 3.3.1).

FIGURE 3.3.1 Transcription repressor protein (pink) bound to DNA (red and blue). The repressor protein physically blocks access to the DNA, preventing transcription of the underlying gene.

GENE REGULATION IN EUKARYOTES

Gene regulation is tightly controlled in both eukaryotes and prokaryotes. However, the process of gene expression in eukaryotes is more complex, since gene regulation occurs at a greater number of stages in eukaryotes than in prokaryotes.

In Section 3.2 you learnt that gene expression in eukaryotes comprises the processes of transcription, RNA processing and translation. Gene expression is highly controlled, and can be regulated at any of these stages (Figure 3.3.2). In eukaryotic cells, transcription and RNA processing occur within the nucleus, and translation occurs in the cytoplasm.

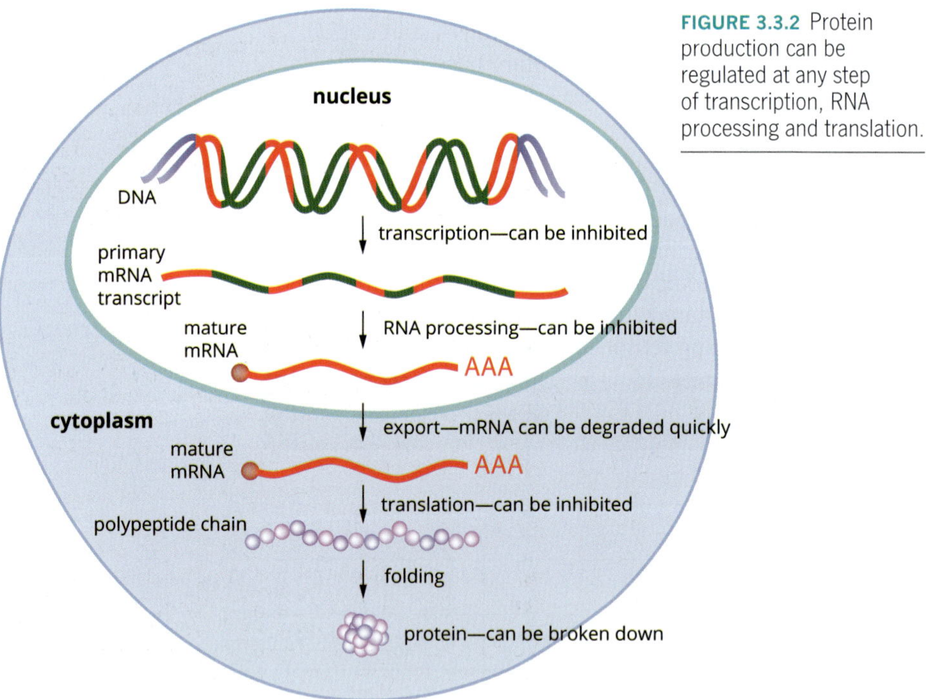

FIGURE 3.3.2 Protein production can be regulated at any step of transcription, RNA processing and translation.

GENE REGULATION IN PROKARYOTES

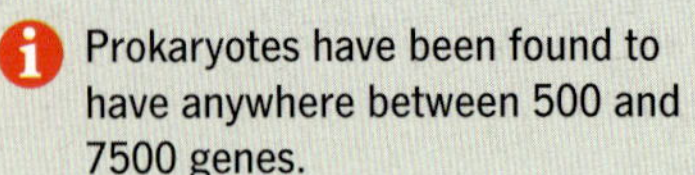

Prokaryotes have been found to have anywhere between 500 and 7500 genes.

In prokaryotic cells, gene expression consists only of transcription and translation and occurs in the cytoplasm of cells. (Prokaryotic cells do not have a nucleus or any other membrane-bound organelles.) Here, transcription and translation occur at almost the same time. Gene expression in prokaryotes is regulated during transcription, which will be the focus of this section.

REGULATORY AND STRUCTURAL GENES

Some genes are expressed constitutively (continually), while other genes can be **induced** or **repressed**.

- **Constitutive genes** are always switched on: they are transcribed continually.
- For other genes, transcription may be induced or repressed by transcription factors as needed, depending on the cell type, stage or environmental conditions.

Regulatory genes code for transcription factors. **Transcription factors** are proteins that control gene expression at the transcription stage. They bind to DNA sequences close to the promoter region of a gene or to RNA polymerase to induce or repress the expression of specific genes (Figure 3.3.3).

Structural genes code for proteins and RNAs that are not involved in gene regulation. For example, they can code for enzymes, protein channels, protein components for the cytoplasmic skeleton, or tRNA, among others.

Examples of constitutive genes include genes needed for cellular respiration, regulation of the cell cycle and the synthesis of tRNA.

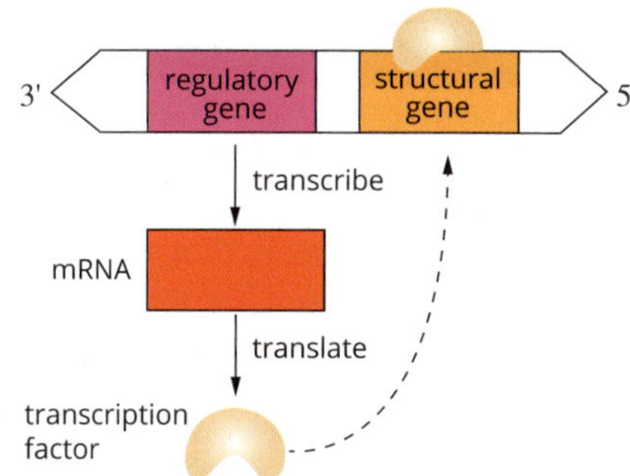

FIGURE 3.3.3 Regulatory genes code for transcription factors that induce or repress structural genes. This diagram shows an example of a transcription factor acting as an inhibitor of transcription.

OPERONS—REGULATING A CLUSTER OF GENES TOGETHER

In some instances, multiple structural genes are transcribed together and are controlled by a single promoter. These functional units of DNA are called **operons** and they mostly occur in prokaryotes. Operons have common features (Figure 3.3.4), including:

- a promoter region—the binding site of the RNA polymerase
- an **operator region**—the binding site of a transcription factor
- structural genes—the code that is transcribed and then translated into the products needed by the cell.

Repressor proteins bind to the operator, preventing transcription of the structural genes in the operon. In some operons, the repressor is bound to the operator most of the time, but can be removed to allow the operon to be switched on or induced. In other operons, the repressor is inactive most of the time, but can be activated when needed so that the operon is switched off or repressed.

One example of an operon that has been well studied in *Escherichia coli* (*E. coli*) is the *trp* operon.

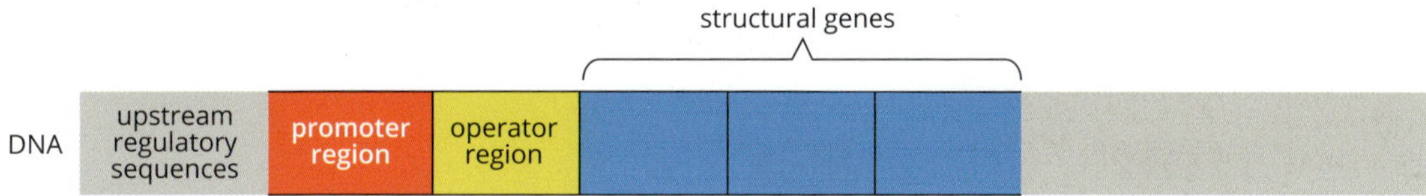

FIGURE 3.3.4 Operons contain common features: a promoter region, an operator region and structural genes.

BIOFILE

Tumour-suppressor proteins

Tumour-suppressor proteins function to prevent cancer. These proteins are produced in response to exposure to radiation and chemicals that cause damage to the structure of DNA.

An important tumour-suppressor protein is p53. If DNA damage has occurred, the p53 protein binds to specific sites on the DNA to repress genes that play a role in the continuation of the cell cycle. This inhibits cell division and prevents damaged DNA from replicating. If the damage is minor, p53 activates genes that repair DNA. In cases where the damage cannot be repaired, p53 will initiate cell death (apoptosis).

p53 plays a major role in the prevention of cancers. If the gene coding for the p53 protein is deleted or mutated, the risk of cancer is greatly increased. In 50% of all cancers, p53 has been found to be inactive.

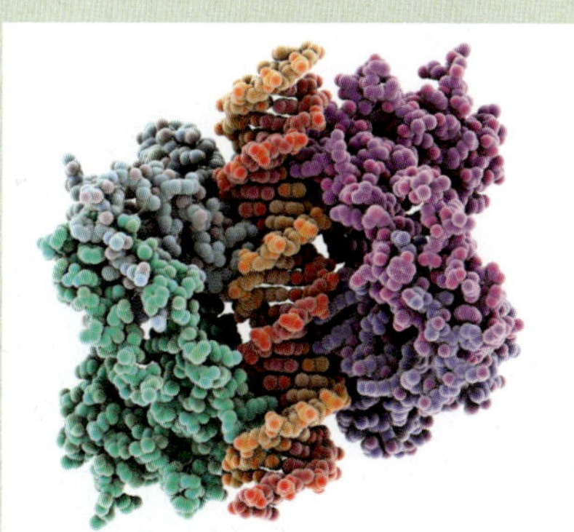

Molecular model of the tumour suppressor protein p53 (left and right) bound to a DNA molecule (centre)

The *trp* operon

Tryptophan is an amino acid used to build proteins and is often found in the environment where *E. coli* lives. When in short supply, however, *E. coli* and many other prokaryotes can produce their own tryptophan in an enzyme-controlled pathway. The DNA unit responsible for coding and regulating the production of these enzymes is called the ***trp* operon**. It is an example of a **repressible operon**, which means that it is switched on by default, but can be turned off.

The *trp* operon consists of a promoter, operator and five structural genes that code for either whole enzymes or subunits that combine to make enzymes. Upstream of the operon is the gene ***trpR***, which codes for a transcription factor called the ***trp* repressor**. The *trpR* gene is expressed constitutively, so the *trp* repressor is always present. On its own, the *trp* repressor is unable to bind to the operator. When tryptophan is available, however, it acts as a **corepressor**. The tryptophan binds to the *trp* repressor and causes it to change shape. This makes the repressor active, allowing it to block RNA polymerase and prevent the structural genes from being transcribed (Figure 3.3.5).

If tryptophan levels in the cell decrease, the tryptophan bound to the repressor detaches. The inactive repressor does not prevent transcription and the enzymes needed to synthesise tryptophan are produced.

a When no tryptophan is present:

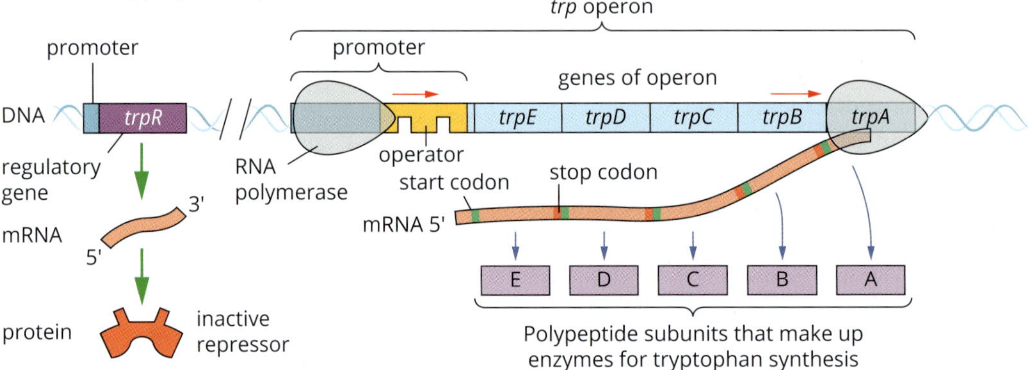

b When tryptophan is present:

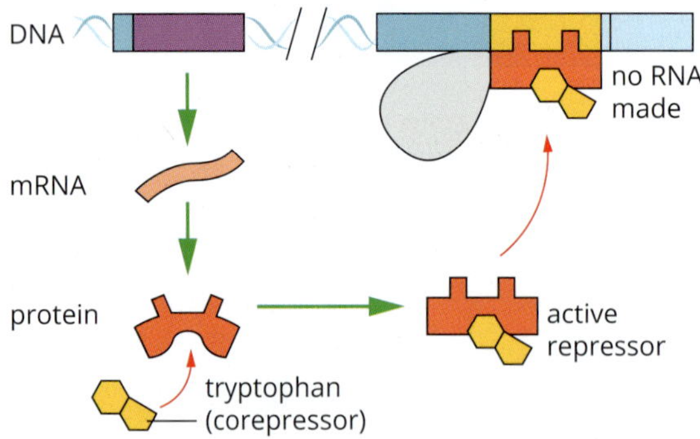

FIGURE 3.3.5 (a) The *trp* repressor is inactive by default. (b) Tryptophan binds to the *trp* repressor, altering its shape and allowing it to block transcription.

CASE STUDY ANALYSIS

Master regulatory genes

The development of a complex, trillion-celled adult organism from a single fertilised cell occurs in a series of steps. Master regulatory genes code for transcription factors that turn genes on and off in different cells in the developing embryo. They can also start a sequence of events by turning other regulatory genes on and off, leading to the production of transcription factors that will, in turn, regulate other genes, and so on. A single master regulatory gene can, in this way, control the development of a complex structure such as an eye or nervous system, or a whole organism.

Some of the most important master regulatory genes are the *Hox* genes. These genes control the structure and organisation of body segments during embryonic development. The proteins of *Hox* genes bind to regulatory regions of target genes, which then activate or repress cellular activity, directing the development of the organism. An error in one gene can result in minor malfunctions or even a body part growing in the wrong place.

Interestingly, the order in which these genes are located on their chromosomes corresponds to the order in which they are expressed, and the part of the body plan they control. The genes are shown as coloured squares in Figure 3.3.6. *Hox* genes in humans are found in clusters named A, B, C and D, each on a different chromosome.

This important gene family has been highly conserved throughout animal evolution. Because of the high degree of genetic similarity in these genes across animal species, researchers can use model organisms, such as *Drosophila*, to investigate human birth defects and diseases.

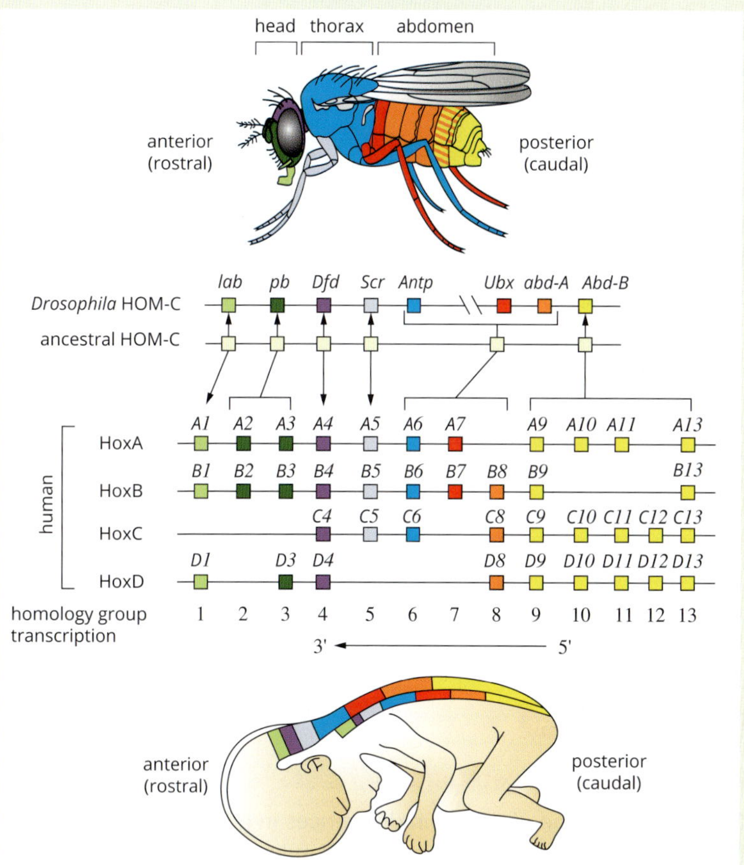

FIGURE 3.3.6 The master regulators of embryonic development in all animals are *Hox* genes. The colour coding shows the pattern of the *Hox* complex shared between *Drosophila* and humans as well as the body parts they control.

Analysis

1 How many *Hox* genes are found in *Drosophila*?

2 There are four clusters of human *Hox* genes. Explain which cluster you think is least likely to affect brain development in an embryo.

3 Consider the music played by an orchestra. This could be used as an analogy (comparison) for the events that occur when master genes regulate an organism's body plan.

a Complete the table by suggesting how each event in an orchestra is similar to the events in embryonic development.

Orchestra	Embryonic development
The conductor is responsible for controlling when musicians stop and start playing.	
The piece of music must be read in the correct order.	
The conductor does not play an instrument.	
Many musicians play at the same time.	
The conductor drops a sheet of music and skips ahead. One group of instruments is instructed to play too early.	
Mistakes can still be made by a musician even if the conductor does their job properly.	

b Identify two limitations of this analogy.

3.3 Review

OA ✓✓

SUMMARY

- Gene expression in eukaryotes can be regulated at any of the three stages: transcription, RNA processing and translation.
- Gene regulation in prokaryotes occurs during transcription.
- Constitutive genes are expressed continually.
- Regulatory genes code for the production of transcription factors, which induce or repress target genes by binding close to their promoter region or to RNA polymerase.
- Structural genes code for proteins and RNAs that are not involved in gene regulation; for example, enzymes.
- Operons are genes that are grouped together with a single promotor and an operator, which allows the grouped genes to be switched on or off together.
- The *trp* operon is an example of a repressible operon. It is switched off by the presence of tryptophan which acts as a corepressor, activating the repressor so that no more tryptophan is synthesised.
- Eukaryotes have much more DNA than prokaryotes, which requires them to package it more tightly.
- Despite differences in how nucleic acids are packaged in prokaryotes and eukaryotes, both types of organisms use the same fundamental processes to synthesise proteins from the same genetic code.

KEY QUESTIONS

Knowledge and understanding

1 What are some of the main differences in gene regulation between prokaryotic and eukaryotic cells?

2 Define the following terms:

a transcription factors
b operator
c regulatory gene
d structural gene
e constitutive gene
f induced gene
g repressed gene

3 Complete the diagram of this prokaryotic operon by inserting the names of the gene regions in the correct locations.
structural genes, termination site, promoter, operator

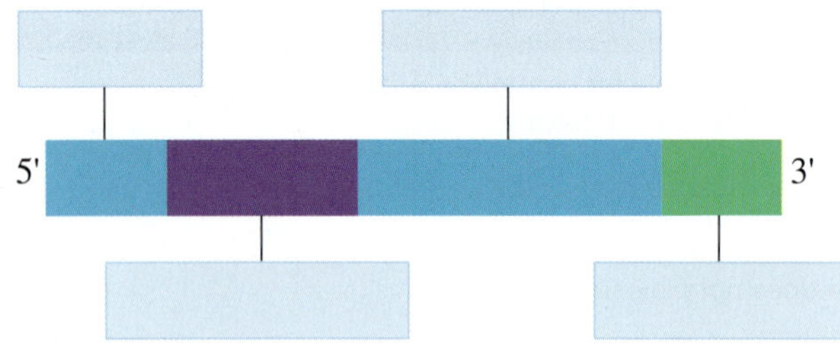

4 List five basic differences between prokaryotic and eukaryotic gene structure.

Analysis

5 **a** Complete the table by inserting in the correct order the processes that occur in *E. coli* in the presence and absence of tryptophan in the environmnent. Options can be used more than once.

Presence of tryptophan	Absence of tryptophan

- tryptophan-producing enzymes are synthesised
- expression of *trpR* gene
- corepressor binds to repressor
- repressor blocks RNA polymerase
- structural genes are transcribed
- structural genes are not expressed
- tryptophan detaches from repressor

b Explain why it is beneficial for *E. coli* to be able to repress the *trp* operon.

Chapter review

KEY TERMS

5' cap
adenine (A)
anticodon
antiparallel
base
biomolecule
coding strand
codon
complementary base pairing
constitutive gene
corepressor
cytosine (C)
degenerate
deoxyribonucleic acid (DNA)
deoxyribose
double helix
exon
gene
gene expression
gene regulation
genetic code
guanine (G)
induced
intron
messenger RNA (mRNA)
nucleic acid
nucleotide
operator region
operon
phosphate group
poly-A tail
polynucleotide
polypeptide chain
promoter region
protein synthesis
purine
pyrimidine
regulatory gene
repressed
repressible operon
repressor protein
ribonucleic acid (RNA)
ribose
ribosomal RNA (rRNA)
RNA polymerase
RNA processing
spliceosome
splicing
start codon
stop codon
structural gene
TATA box
template strand
thymine (T)
transcription
transcription factor
transfer RNA (tRNA)
translation
triplet
trp operon
trp repressor
trpR
tryptophan
uracil (U)

REVIEW QUESTIONS

Knowledge and understanding

1 What are the sequences that are included in the final mRNA product called?
A introns
B termons
C exons
D spliced codons

2 In polypeptide synthesis, the function of the ribosome is to:
A synthesise the required amino acids
B ensure that the DNA base sequence is complete
C provide the energy needed for the synthesis
D provide the site for the synthesis

3 DNA provides the code for the synthesis of polypeptides. Which one of the following statements is true?
A Every codon codes for its own exclusive amino acid.
B The code is read as sets of three bases called triplets.
C Each triplet codes for at least two different amino acids.
D There are 20 different amino acids therefore there are 20 different codons.

4 Geneticists used DNA sequencing to discover the base sequence of a plasmid. If 27% of the bases in the plasmid were cytosine then the plasmid consisted of:
A 27% cytosine–guanine pairs
B 46% adenine–thymine pairs
C 46% cytosine–guanine pairs
D 23% adenine–thymine pairs

5 The nucleotide sequence A G U G A C C A A could represent:
A part of the DNA template of a particular gene
B the amino acid chain of a polypeptide
C a sequence of mRNA
D a section of double helix

6 Why do eukaryotic cells contain histones?

7 Name the three main types of RNA involved in protein synthesis and outline their functions.

8 What is a promoter?

Application and analysis

9 Explain why the chromosomes of eukaryotic cells can be seen under a light microscope but those of prokaryotes cannot.

To answer questions 10–12, refer to the genetic code in the table below. With a few rare exceptions, the genetic code is accepted as being universal.

First position (5' end)	Second position				Third position (3' end)
	U	C	A	G	
U	phe phe leu leu	ser ser ser ser	tyr tyr STOP STOP	cys cys STOP trp	U C A G
C	leu leu leu leu	pro pro pro pro	his his gin gin	arg arg arg arg	U C A G
A	ile ile ile met	thr thr thr thr	asn asn lys lys	ser ser arg arg	U C A G
G	val val val val	ala ala ala ala	asp asp glu glu	gly gly gly gly	U C A G

10 a The following flow chart represents the production of a polypeptide chain as directed by the DNA template.

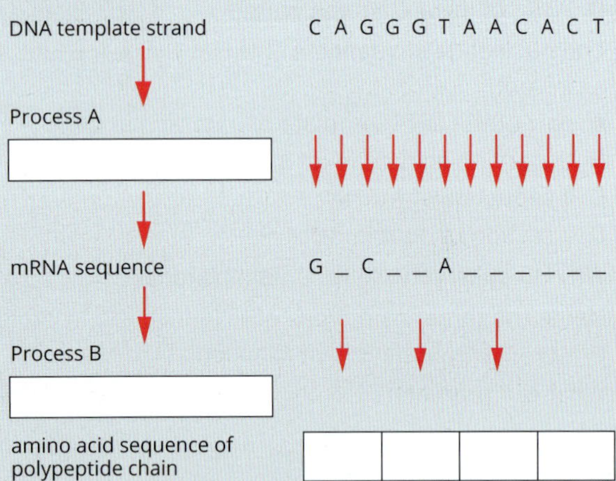

i Name processes A and B.
ii Complete the mRNA sequence.
iii Use the genetic code to determine the correct sequence of amino acids.

b i How is the fourth codon different from all of the others in this sequence?
ii Outline the significance of this codon in terms of polypeptide production.

11 The DNA sequence of a particular gene is shown below.

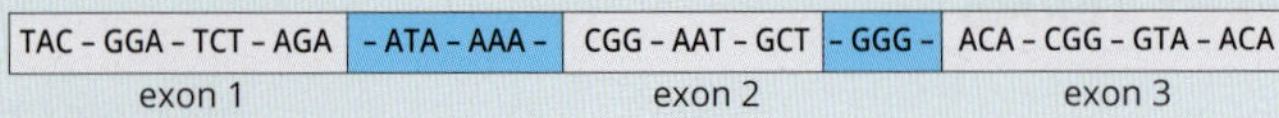

a What do the bright blue sections represent?
b Show the strand of mature mRNA that would be produced using this gene.
c Write the sequence of amino acids that would be translated from the mRNA.
d What amino acid sequence would be formed if an alternative splice, skipping exon 2, were to occur?

12 The DNA sequence of the promoter and first exon of a gene are shown below.

G G G C T C T A T A A A T A C C A C T T C A A T G C T

a Identify the TATA box.
b Explain where RNA polymerase will start transcription.
c The mRNA strand produced is shown below:
A U G G U G A A G U U A C G A
i Explain whether the DNA strand shown is the coding strand or the template strand.
ii Decode the strand.
d Suggest the likely consequence for the organism of a mutation that changes the sequence of the TATA box.

13 a Use the following terms to label the structures A–I on the diagram below:

amino acid	pre-mRNA (primary mRNA transcript)
anticodon	ribosome
loaded tRNA	RNA polymerase
mRNA	unloaded tRNA
polypeptide	

b Describe the role of structure B in protein synthesis.

c Describe the process occurring in structure H. Ensure you include the significance of structure I in your discussion.

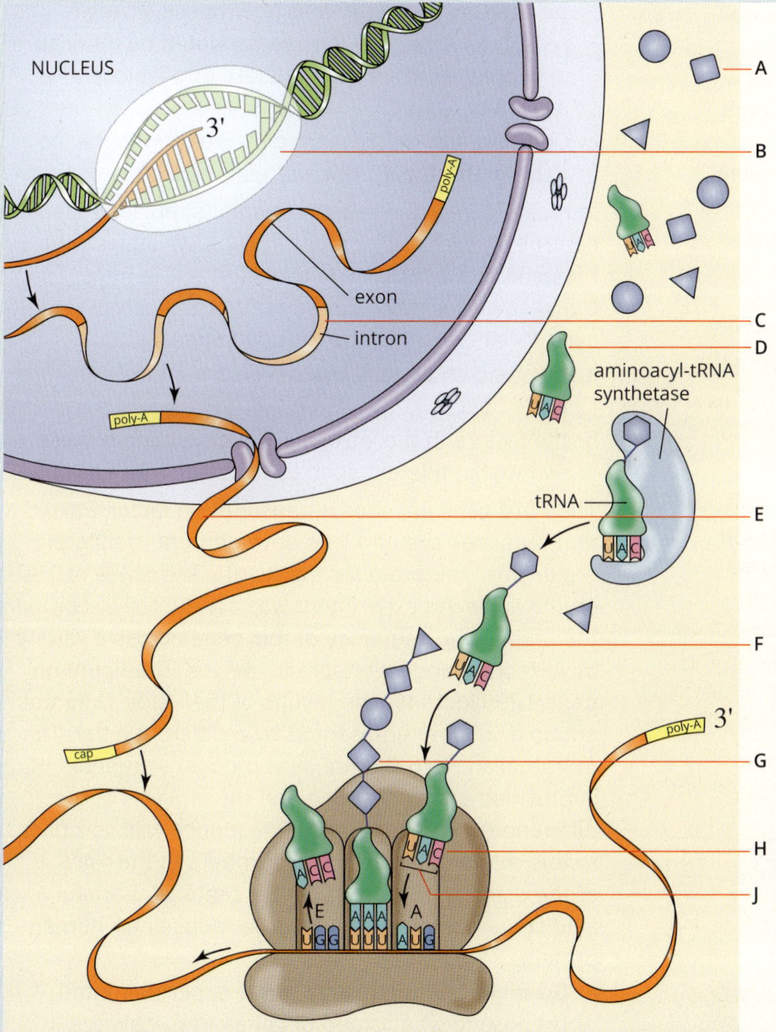

14 A nucleic acid strand is under investigation. It has been found to contain 29% A, 32% G and 17% C.

a Is the fourth base uracil or thymine?

b How do you know?

c Calculate the percentage of the fourth base in the nucleic acid strand.

15 A strand of nucleic acid is shown below.

AUG AAU CCU UAU GGU GGC UUU UAA

The protein produced as a result of the information encoded in the strand of nucleic acids is shown below.

met–asn–pro–phe

a Explain whether the strand given is DNA, pre-mRNA or mRNA.

b i During translation of the strand of nucleic acid shown, a tRNA having the anticodon UUA approached the ribosome. Which amino acid would the tRNA have been carrying?

ii Draw the tRNA molecule with its amino acid and anticodon.

16 Outline why, in humans, the mature adult intestinal cell will produce digestive enzymes while endocrine gland cells produce hormones, even though both differentiated cells contain the same human genome and secrete proteins.

17 In Section 3.2 you learnt that the longest known gene is the dystrophin gene, which is 2.5 megabases long, and that this gene is 99% introns.

a What is the maximum number of bases in the exons of the dystrophin gene?

b How many amino acids (approximately) make up the protein dystrophin?

18 Genetic engineering is used to transform bacteria by inserting human genes into their genome in order to produce human polypeptides, such as those that form insulin. Before the bacterium can be transformed, a copy of the human gene is required. A common method of acquiring the gene is to extract the appropriate mRNA from human cells and to use it as a template to make a DNA copy. This cDNA (complementary DNA) is then introduced into the bacterium, which then produces the required protein.

a Why can it be expected that a bacterium is able to decode a human gene and produce the correct protein?

b A gene made of cDNA is better for use in a bacterium than a gene cut directly from a human chromosome. Why?

c Another method of obtaining an appropriate gene is to analyse the protein needed, identify the amino acid sequence and construct a suitable section of DNA. Explain whether the gene created by this method is likely to be identical to the cDNA sequence made using mRNA as a template.

d Consider the amino acid sequence leu–pro–val. How many different DNA sequences would result in this amino acid chain? Explain how you arrived at your answer.

19 Tryptophan is an amino acid which prokaryotic cells are able synthesise when it is in short supply. Tryptophan production is controlled by a series of five enzymes that are coded for by five genes (*trp A*, *B*, *C*, *D* and *E*), which form a single unit called the tryptophan operon. This operon has been extensively studied in the bacterium *Escherichia coli*.

a What is an operon?

The *trp* operon with some of its upstream region is shown below.

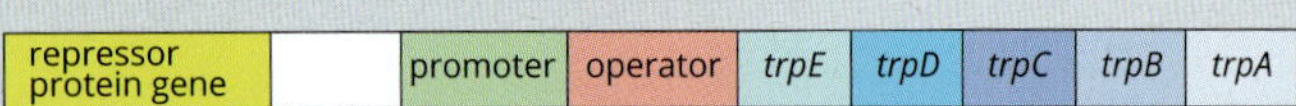

b Label the regulatory and structural genes on the diagram.

c The gene for the repressor protein is constitutively expressed. Explain the significance of this for the cell.

d The repressor protein is produced in an inactive form. It becomes activated when it binds to tryptophan. Explain how these features of the repressor protein benefit the organism.

e Once activated the repressor protein binds to the operator region. Explain how the binding of the repressor to the operator stops production of tryptophan.

20 Another well studied operon in *Escherichia coli* is the *lac* operon. It contains a cluster of genes that, when transcribed and translated, produce enzymes needed to digest the disaccharide (sugar) lactose. When glucose is available to *E. coli* but lactose is not, the operon is switched off by a repressor. When glucose is unavailable but lactose is, the lactose changes the repressor, preventing it from binding to the operator and switching the operon on.

a Explain whether or not lactose is a corepressor.

b Glucose does not require digestion before it can be used by *E. coli* for energy. Propose, giving a reason, whether the *lac* operon would be induced or repressed when both glucose and lactose are available.

c Compare the mechanism of repression in the *lac* operon and the *trp* operon.

21 In recent genetics research, scientists replaced the gene controlling eye development in *Drosophila* flies with the gene that controls eye development in mice. The transgenic *Drosophila* developed normal compound fly eyes. What does this observation suggest about:

a the gene controlling eye development in *Drosophila* and mice?

b the factors that control eye development in these two vastly different species of insect and mammal?

22 The *Pax6* gene encodes a transcription factor, Pax6, that regulates eye and lens development in different organisms. The protein consists of a sequence of 130 amino acids. An experiment was conducted to compare the amino acid sequence of this gene across a variety of vertebrate and invertebrate species. The figure on page 129 illustrates the results of the study. Different amino acids are represented by a different letter. The mammalian *Pax6* (human and mouse) sequence is provided on the first line and the analysis of the differences in the amino acid sequence was compared to this mammalian line. The numbers to the right of the diagram represent the percentage of amino acid sequence similarity with the mouse and human proteins.

a Comment on the results of the experiment and propose to what group of genes *Pax6* belongs.

b Identify any patterns in the differences between the *Pax6* amino acid sequence.

c Discuss the evolutionary implications of the similarity between the Pax6 transcription factors.

d Predict the effect of manipulating the *Drosophila* genome to replace the *Pax6* gene with that from the mouse.

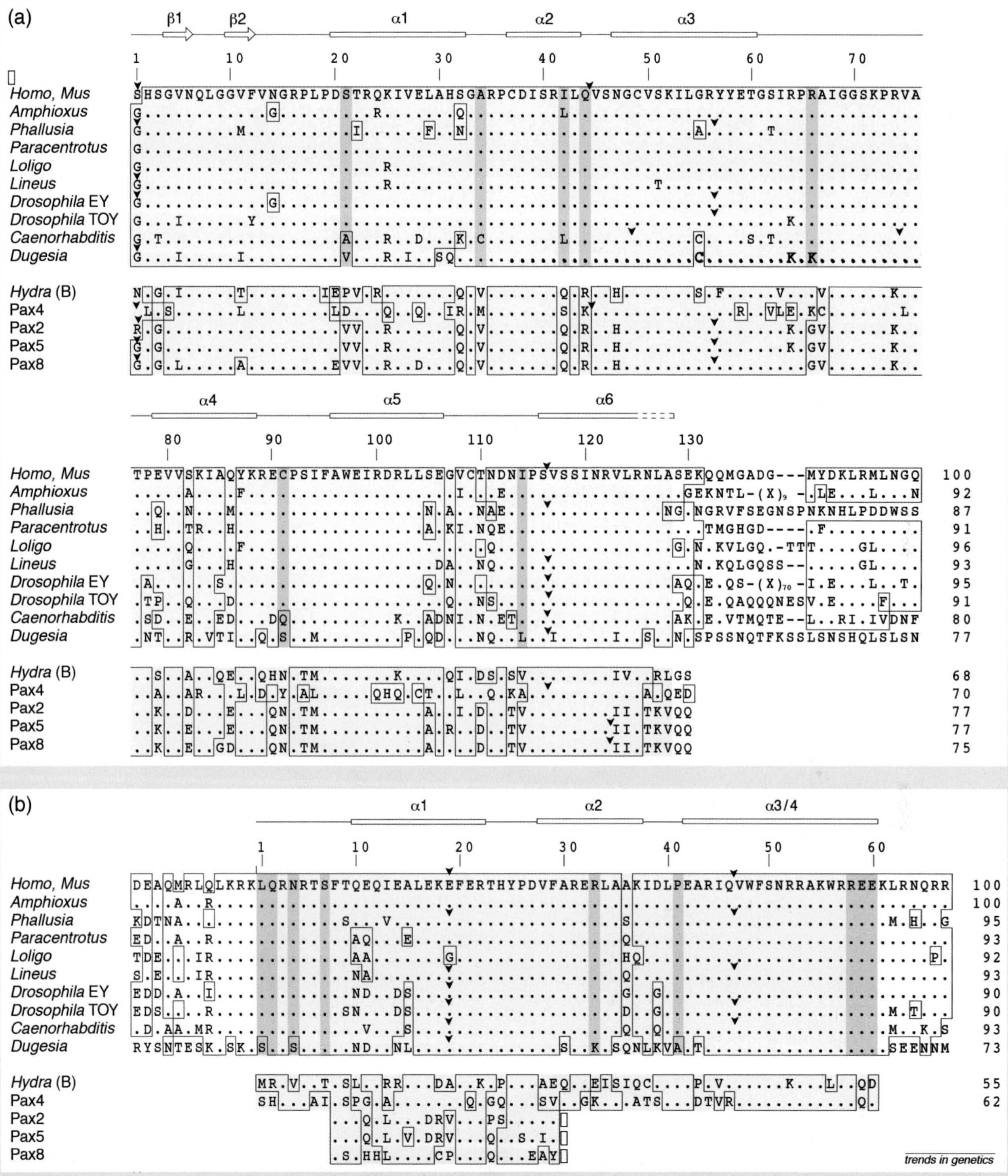

(a)

	β1 β2 α1 α2 α3	
	1 10 20 30 40 50 60 70	
Homo, Mus	SHSGVNQLGGVFVNGRPLPDSTRQKIVELAHSGARPCDISRILQVSNGCVSKILGRYYETGSIRPRAIGGSKPRVA	
Amphioxus	G............G.........R.......Q.........L..................................	
Phallusia	G.........M..........I......F..N......................A......T..............	
Paracentrotus	G...	
Loligo	G.......................R...	
Lineus	G.......................R.........................T.........................	
Drosophila EY	G............G..	
Drosophila TOY	G...I......Y...K............	
Caenorhabditis	G.T.................A...R..D...K.C.......L............C....S.T..............	
Dugesia	G...I.....I.........V...R.I..SQ.......................C........K.K..........	
Hydra (B)	N.G.I.....T.......IEPV.R.......Q.V.......Q.R..H.......S.F.....V...V......K..	
Pax4	.L.S......L........LD...Q..Q..IR.M.......S.K..............R..VLE.KC.......L.	
Pax2	R.G.................VV..R......Q.V.......Q.R..H................K.GV......K..	
Pax5	G.G.................VV..R......Q.V.......Q.R..H................K.GV......K..	
Pax8	G.G.L.....A........EVV..R..D...Q.V.......Q.R..H..................GV......K..	
	α4 α5 α6	
	80 90 100 110 120 130	
Homo, Mus	TPEVVSKIAQYKRECPSIFAWEIRDRLLSEGVCTNDNIPSVSSINRVLRNLASEKQQMGADG---MYDKLRMLNGQ	100
Amphioxus	.....A....F....................I...E.................GEKNTL-(X)$_9$ -.LE...L...N	92
Phallusia	..Q..N...M..................N.A..NAE...............NG.NGRVFSEGNSPNKNHLPDDWSS	87
Paracentrotus	..H..TR..H..................A.KI.NQE...................TMGHGD----.F.........	91
Loligo	.....Q....F.......................Q.................G.N.KVLGQ.-TTT....GL....	96
Lineus	.....G...H...................DA..NQ...................N.KQLGQSS--.....GL....	93
Drosophila EY	.A......S...................Q.N.....................AQ.E.QS-(X)$_{70}$ -I.E...L..T.	95
Drosophila TOY	.TP..Q...D....................Q..NS..................Q.E.QAQQQNESV.E....F...	91
Caenorhabditis	.SD..E..ED...DQ..........K..ADNI.N.ET...............AK.E.VTMQTE--L..RI.IVDNF	80
Dugesia	.NT..R.VTI..Q.S..M........P.QD...NQ..L..I.....I..S..N.SPSSNQTFKSSLSNSHQLSLSN	77
Hydra (B)	..S..A..QE..QHN.TM.......K....QI.DS.SV........IV..RLGS	68
Pax4	..A..AR...L.D.Y.AL.....QHQ.CT..L..Q.KA...........A.QED	70
Pax2	..K..D...E...QN.TM..........A..I.D..TV........II.TKVQQ	77
Pax5	..K..E...E...QN.TM..........A.R..D..TV........II.TKVQQ	77
Pax8	..K..E..GD...QN.TM..........A....D..TV........II.TKVQQ	75

(b)

	α1 α2 α3/4	
	1 10 20 30 40 50 60	
Homo, Mus	DEAQMRLQLKRKLQRNRTSFTQEQIEALEKEFERTHYPDVFARERLAAKIDLPEARIQVWFSNRRAKWRREEKLRNQRR	100
Amphioxus	....A..R...	100
Phallusia	KDTNA...............S...V......................S.........................M.H..G	95
Paracentrotus	ED..A..R.............AQ...E....................Q...............................	93
Loligo	TDE...IR.............AA.......G................HQ............................P.	92
Lineus	S.E...IR.............NA........................Q...............................	93
Drosophila EY	EDD.A..I.............ND..DS....................G..G............................	90
Drosophila TOY	EDS....R............SN...DS....................D..G......................M.T...	90
Caenorhabditis	.D.AA.MR..............V...S....................Q..Q......................M..K.S	93
Dugesia	RYSNTESK.SK.S..S....ND..NL..............S..K.SQNLKVA.T..................SEENNM	73
Hydra (B)	MR.V..T.SL..RR...DA..K.P...AEQ..EISIQC....P.V......K...L..QD	55
Pax4	SH...AI.SPG.A.......Q.GQ...SV..GK...ATS...DTVR............Q.	62
Pax2	...Q.L...DRV...PS.....□	
Pax5	...Q.L.V.DRV...Q..S.I.□	
Pax8	.S.HHL....CP...Q...EAY□	

trends in genetics

The amino acid sequence of the *Pax6* gene in a variety of animal species

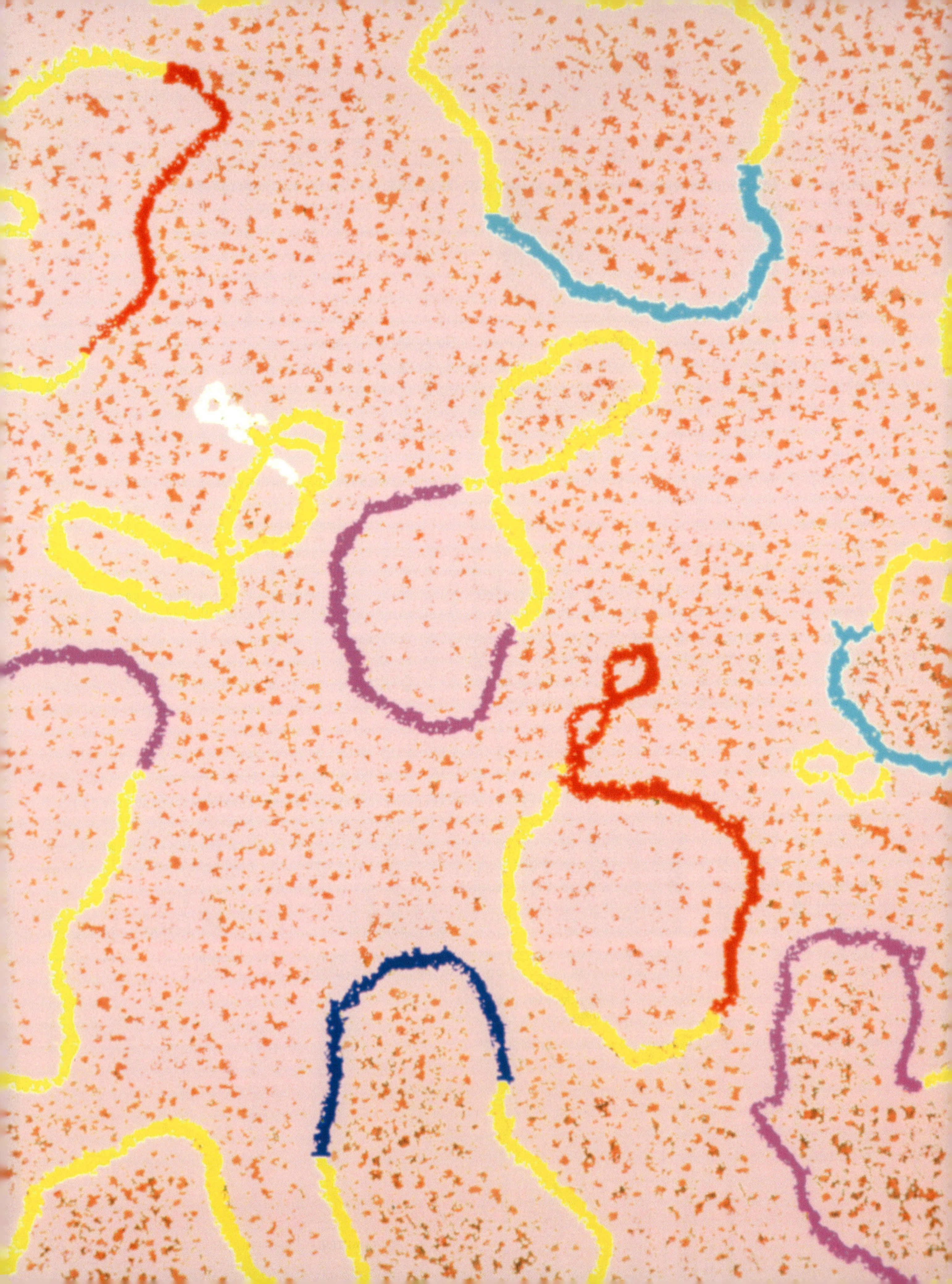

CHAPTER 04

DNA manipulation techniques and applications

Learning outcomes

By the end of this chapter, you will have developed an understanding of the molecular tools and techniques used to manipulate DNA molecules for particular purposes. You will have learnt how recombinant plasmids are created and then used as vectors in the process of bacterial transformation. You will also have an understanding of the other molecular tools and techniques explored in this chapter, including gel electrophoresis and the polymerase chain reaction (PCR).

Key knowledge

- the use of enzymes to manipulate DNA, including polymerase to synthesise DNA, ligase to join DNA and endonucleases to cut DNA **4.1, 4.2**
- the function of CRISPR-Cas9 in bacteria and the application of this function in editing an organism's genome **4.3**
- amplification of DNA using polymerase chain reaction and the use of gel electrophoresis in sorting DNA fragments, including the interpretation of gel runs for DNA profiling **4.1**
- the use of recombinant plasmids as vectors to transform bacterial cells as demonstrated by the production of human insulin **4.2**
- the use of genetically modified and transgenic organisms in agriculture to increase crop productivity and to provide resistance to disease. **4.3**

4.1 DNA manipulation

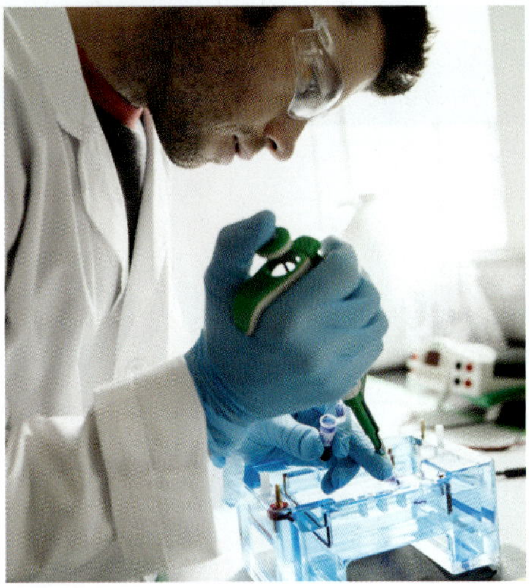

FIGURE 4.1.1 A scientist loads DNA samples into a gel electrophoresis chamber.

To work with DNA, it is necessary to have more than a few DNA molecules. The **polymerase chain reaction (PCR)** is a technique used for DNA amplification—it makes millions of identical copies of a piece of DNA. Using PCR, scientists can amplify the DNA in traces of blood left at the scene of a crime, or a particular gene from a sample of DNA.

Gel electrophoresis is a method used to separate and visualise nucleic acids and proteins according to their size (Figure 4.1.1). This is usually performed after PCR to either confirm that the correct DNA fragment was amplified, or to identify DNA fragments present in the sample (DNA profiling). It is also used to separate DNA fragments to be used in DNA manipulation techniques.

In this section, you will learn about PCR and gel electrophoresis of DNA fragments.

DNA AMPLIFICATION

Many DNA manipulation techniques require a large quantity of DNA to work with. However, sometimes only a very small sample of DNA is available for scientists. For example, only trace samples of DNA may be left at a crime scene or extracted from fossils of extinct species (Figure 4.1.2). Usually only small samples can be removed for medical tests and in embryonic or fetal DNA screening for genetic disorders. In these cases, **DNA amplification** is required to increase the amount of the target DNA sample so that it is large enough to be used or analysed in other techniques and processes.

> A microsatellite is a short repeated sequence of nucleotides found at a defined location (locus) on a chromosome.

DNA amplification uses PCR to create a large quantity of DNA that is identical to the initial trace sample. The term **target DNA** is used to describe a particular region of a DNA molecule that a scientist intends to study or manipulate (e.g. a specific gene or a **microsatellite**, which is a variable region of the genome used for DNA profiling).

FIGURE 4.1.2 A scientist extracts fossilised DNA from a Neanderthal (*Homo neanderthalensis*) bone. The DNA will be amplified for further DNA analysis.

Polymerases

As the name indicates, polymerase chain reaction (PCR) is based on the action of polymerases. **Polymerases** are enzymes that catalyse the formation of long-chain molecules (polymers), such as DNA and RNA, by linking smaller molecules (nucleotides). These enzymes are found in all living organisms and play an important role in the replication, repair and maintenance of DNA. There are two different groups of polymerases:

- DNA polymerases
- RNA polymerases.

DNA polymerases

Scientists use **DNA polymerase** enzymes in PCR and **DNA sequencing** to synthesise multiple copies of the target DNA. During PCR, the DNA polymerase attaches to the end of the target DNA sequence (called the template) and adds complementary nucleotides to create a new strand of DNA, complementary to the target DNA.

> DNA sequencing describes any process used to determine the order of the four nucleotide bases—adenine, thymine, guanine and cytosine—in DNA.

DNA polymerases used in PCR must be stable at high temperatures, have high affinity to the DNA template, and be highly specific to reduce background amplification (i.e. amplification of non-target DNA).

BIOFILE

Thermus aquaticus

A field trip to Yellowstone National Park in the 1960s radically altered the course of molecular genetics research. Thomas Brock, a bacteriologist, found a new species of bacteria in a hot spring, which he named *Thermus aquaticus* (Latin for 'hot water').

This was significant because enzymes are normally denatured if heated to temperatures of 95°C for more than a few seconds. For *T. aquaticus* to survive in the hot springs, its enzymes, including DNA polymerase, need to tolerate these high temperatures. Therefore, the DNA polymerase from *T. aquaticus* (*Taq* polymerase) has proved to be an ideal enzyme for use in PCR.

A scientist obtains a sample of *Thermus aquaticus* from a hot spring.

Taq polymerase

***Taq* polymerase** is the DNA polymerase that is most commonly used in PCR. It was originally extracted from the thermophilic bacterium *Thermus aquaticus*. The heat-resistant properties of *Taq* polymerase make it extremely useful in DNA manipulation techniques such as PCR.

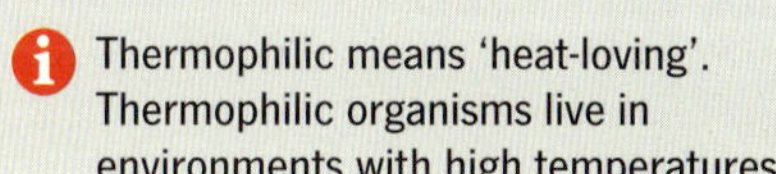

Thermophilic means 'heat-loving'. Thermophilic organisms live in environments with high temperatures.

Reverse transcriptase

Reverse transcriptase is a DNA polymerase that synthesises single-stranded DNA using single-stranded RNA as a template. This is the reverse of the usual transcription process in which DNA is transcribed into RNA. Reverse transcriptase is used in the laboratory to produce DNA molecules that can be amplified by PCR for further analysis. It is also used to make **complementary DNA (cDNA)** from modified mRNA that has already had the introns spliced out for some DNA manipulation techniques, as you will learn in Section 4.2.

RNA polymerases

RNA polymerases are enzymes that synthesise RNA from DNA during transcription. Subunits of the RNA polymerase recognise the promoter at the start of a gene. The RNA polymerase attaches to the promoter and unwinds the DNA, allowing it to add nucleotides, one at a time, in the 5' to 3' direction. RNA polymerase adds nucleotides until a 'stop' sequence is reached. A single-stranded piece of RNA is created. RNA polymerase works much more slowly than DNA polymerase. RNA polymerase is used in specialised laboratory techniques to study transcription and RNA amplification.

The polymerase chain reaction (PCR)

The polymerase chain reaction (PCR) is a method of amplifying specific target sequences of DNA. The DNA polymerase known as *Taq* polymerase is used in this process, as it is stable at the high temperatures required during PCR. Each strand of the DNA acts as a template for a new copy of itself.

PCR is carried out in cycles. Each cycle doubles the amount of DNA, resulting in exponential growth. After 30 cycles, there are over one billion copies (Table 4.1.1).

TABLE 4.1.1 Exponential growth in number of target DNA molecules during PCR

Number of PCR cycles (n)	Number of double-stranded copies of original DNA (2^n)
0	1
1	2
2	4
3	8
4	16
5	32
6	64
7	128
8	256
9	512
10	1024
20	1048576
30	1073741824

For PCR to be carried out, specialised equipment and a PCR mixture is required. The PCR mixture contains:

- DNA, including the target DNA to be amplified (the sample) (Figure 4.1.3)
- free nucleotides, to build new DNA strands
- a heat-resistant DNA polymerase (usually *Taq* polymerase), to elongate the new DNA strands by adding the free nucleotides
- two DNA **primers** complementary to the ends of the target DNA, to specify the start and finish of the DNA fragment to be amplified. The two primers are synthetic, single-stranded DNA molecules up to 30 bases in length.

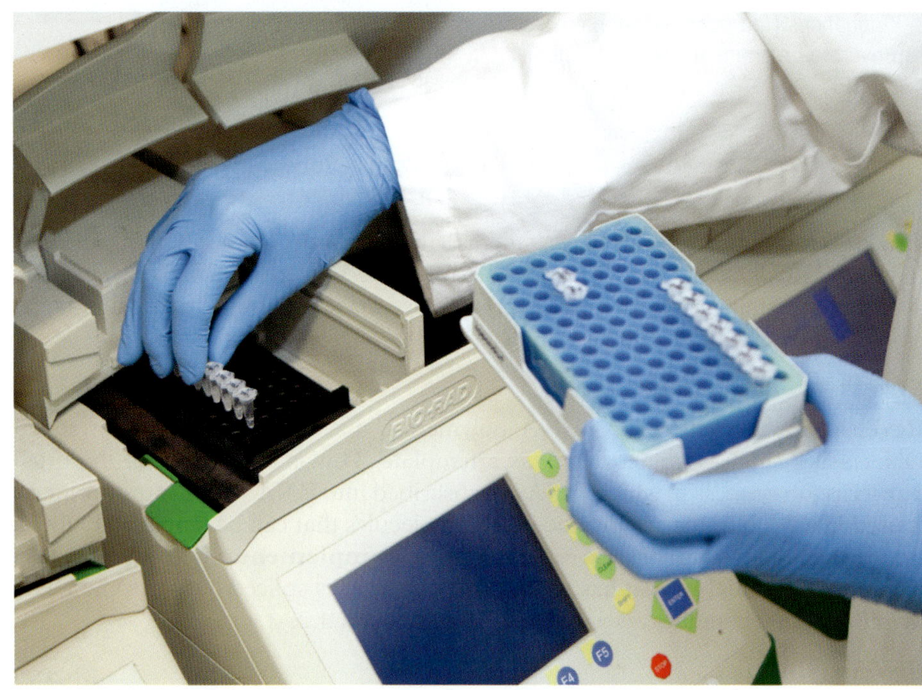

FIGURE 4.1.3 A scientist loads DNA samples into a thermocycler to be amplified by PCR.

Steps in the polymerase chain reaction

The PCR mixture is placed in a **DNA thermocycler**, which alters the temperature in pre-programmed steps. Each PCR cycle involves three steps (Figure 4.1.4):

1. Denaturation: The sample is heated to 95°C to break the hydrogen bonds between the two strands of double-stranded DNA to obtain single strands of DNA.
2. Annealing: The temperature is reduced to 50–60°C. This allows the primers to **anneal** (i.e. bind) to complementary sequences on opposite strands at each end of the target DNA sequence.

ℹ Primers determine the start and end points of a nucleotide sequence to be amplified.

3 Extension: The temperature is increased to 72°C. This allows *Taq* polymerase to attach to the primers on the DNA strands. The *Taq* polymerase moves along each strand, adding free nucleotides to form double-stranded DNA.

This three-step cycle of heating and cooling is repeated up to 50 times to ensure there is sufficient target DNA produced to work with.

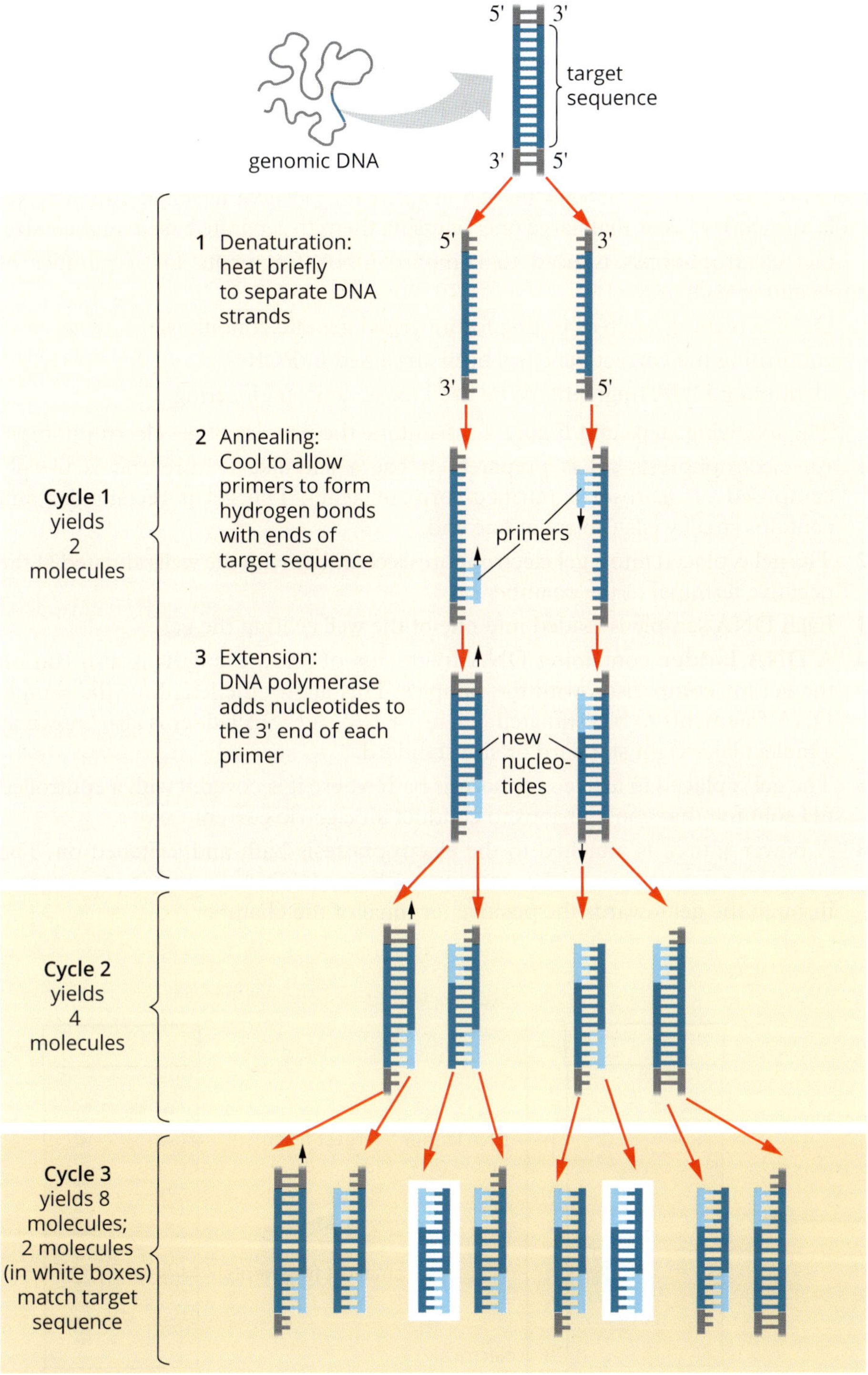

FIGURE 4.1.4 The three steps of PCR: denaturation, annealing and extension. Each PCR cycle increases the amount of target DNA exponentially.

BIOFILE

Trace samples

The process of PCR requires as few as one or two cells for DNA amplification. Scientists at the Victoria Forensic Science Centre have shown that merely touching an object deposits sufficient material for successful DNA amplification. In handling keys, opening a door or driving a car, the cellular material deposited by a criminal provides ample DNA for analysis following PCR.

DNA SEPARATION BASED ON FRAGMENT SIZE

Gel electrophoresis is commonly performed after DNA amplification using PCR. Gel electrophoresis allows scientists to separate out the different DNA fragments present in a sample based on their size. This information can be used to match fragments from different samples, as in DNA profiling, or to isolate a particular fragment for further use in another technique, such as DNA recombination and bacterial transformation. You will learn about these techniques in Section 4.2.

Gel electrophoresis

DNA is negatively charged because of the negative charge on the phosphate group in each nucleotide.

Gel electrophoresis is a technique for separating fragments of nucleic acids (DNA and RNA) or to study protein molecules. In this section, only DNA fragments will be considered. When an electric current is applied to the gel, the negatively charged pieces of DNA move through the gel towards the positive terminal. Small DNA molecules move faster than large ones, causing them to separate based on their size.

Gel electrophoresis is used to compare DNA fragments for a number of applications such as:

- DNA screening, such as testing for inherited genetic conditions
- confirming the correct gene has been amplified in PCR
- identifying DNA fragments to be used for genetic engineering.

The following steps and Figure 4.1.5 outline the process of gel electrophoresis:

1 An electrophoresis gel is prepared. It has a jelly-like texture and is usually composed of agarose (a purified form of agar). The gel is rectangular and contains small wells (holes) at one end.
2 The gel is placed into a gel electrophoresis chamber with the wells situated at the negative terminal of the chamber.
3 Each DNA sample is loaded into one of the wells within the gel.
4 A **DNA ladder** containing DNA fragments of known lengths is also run on the gel for comparison with the samples. This allows the length of the sample DNA fragments to be estimated (Figure 4.1.6). A DNA ladder is also known as a molecular weight standard or size standard.
5 The gel is placed in an electrophoresis bath where it is covered with a controlled pH solution that contains ions to conduct an electric current.
6 A power source is attached to the electrophoresis bath and switched on. The electrical current causes the negatively charged DNA fragments to migrate through the gel towards the positive terminal of the chamber.

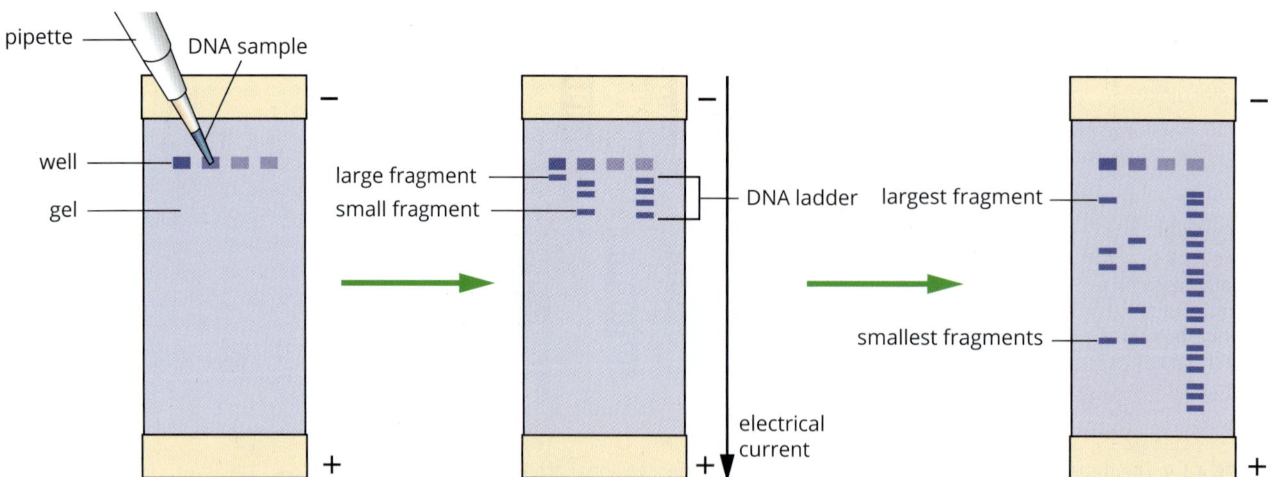

FIGURE 4.1.5 The process of gel electrophoresis, showing two DNA samples being loaded into wells and migration of the DNA fragments through the wells based on their size. The last well contains a DNA ladder for comparing and estimating the fragment length of the two DNA samples.

7 Smaller fragments move faster through the gel, so they migrate further through the gel than larger fragments in a given period of time. This sorts the fragments by length.

8 The DNA fragments are made visible by applying a stain that binds to DNA. This can be done with a fluorescent stain (which may be included in the gel or added after) or with methylene blue stain (which is added after running the gel). Fluorescent stains are viewed with ultraviolet light, while methylene blue staining can be visualised by eye. Bright-coloured bands are observed wherever there is DNA present in the gel (Figure 4.1.7).

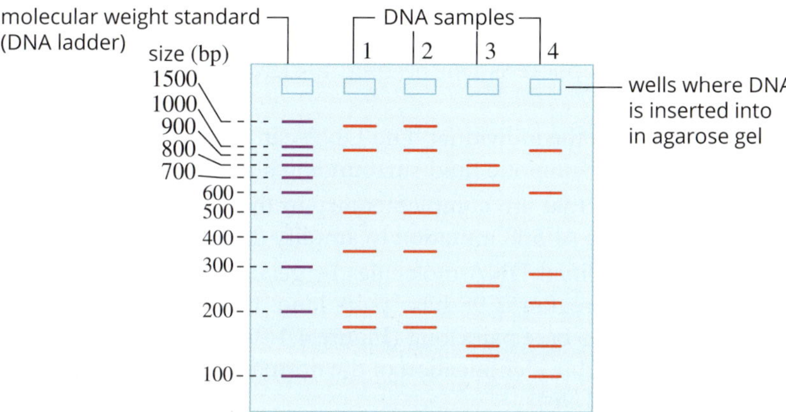

FIGURE 4.1.6 Gel electrophoresis allows a DNA fragment's length to be measured in base pairs (bp) by comparing it with the DNA ladder. Particular fragments can be identified and compared.

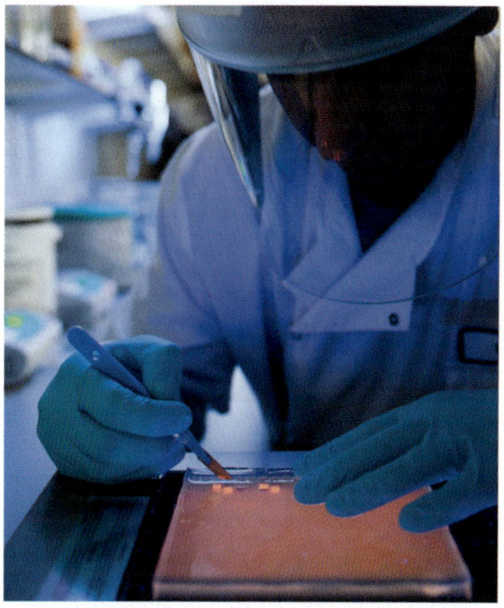

FIGURE 4.1.7 A scientist cuts a sample from an electrophoresis gel stained to show DNA fragments as pink fluorescent bands.

CASE STUDY

Surveillance of waterborne RNA viruses

Poor sanitation and faecal contamination of the water supply with viruses, bacteria and protists is a major health problem in many regions globally. The World Health Organization (WHO) provides guidelines on water safety and promotes improved methods of monitoring microbial contamination in drinking water. Outbreaks of poliovirus and hepatitis A and E virus infections may be due to contaminated water. Traditional methods of identifying viruses, by growing them in cells, take several weeks and are not successful for all viruses. Molecular methods such as PCR followed by gel electrophoresis offer faster detection and the potential to detect multiple viruses in one analysis. Viruses such as poliovirus and hepatitis viruses have an RNA genome, so a technique called RT-PCR is used. First, reverse transcriptase (RT) is used to copy the viral RNA into DNA, and then the polymerase chain reaction (PCR) amplifies the DNA.

Primers specific for different viral genes are used in the PCR. Multiple sets of primers can sometimes be used in one PCR, allowing the detection of several viruses at once (Figure 4.1.8). If a virus is present in the water sample, its RNA will be copied to DNA and amplified. The DNA resulting from the reaction can be visualised by the technique of gel electrophoresis.

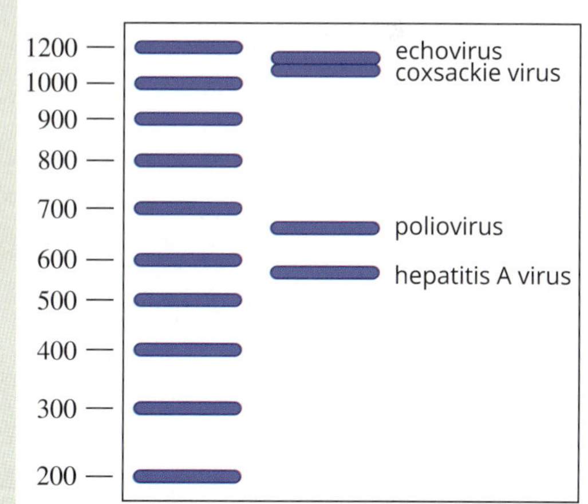

FIGURE 4.1.8 Gel electrophoresis results obtained in water sampling studies. Amplified viral genes were detected as bands of the following size: hepatitis A virus, 589 bp; poliovirus, 671 bp; coxsackie virus, 1084 bp; echovirus, 1128 bp.

COMBINING MOLECULAR TOOLS TO DETECT MUTATIONS

> **BIOFILE**
>
> **Cystic fibrosis**
>
> Cystic fibrosis (CF) is an inherited disorder that affects the respiratory and digestive systems. It can significantly shorten the lifespan of people with the condition. In a person with CF, the mucous glands secrete thick, sticky mucus, which clogs the airways, leading to breathing difficulties, respiratory infections and lung damage. The mucus also affects the pancreas, inhibiting the release of important digestive enzymes, which causes a range of nutritional problems. There is currently no cure for CF, but the first treatment that targets the defective CFTR protein, rather than simply treating the symptoms, is now available for some individuals affected with CF in Australia. The medication improves the flow of chloride ions across plasma membranes, which decreases the thickness and stickiness of mucus.

Mutations are usually discovered because of the effect they have on the individual carrying them. If an individual has the symptoms of cystic fibrosis (CF) then the sequence of the gene for CF can be analysed to detect mutations. The gene for CF is the cystic fibrosis transmembrane conductance regulator (*CFTR*) gene. The *CFTR* gene codes for a large membrane protein of the same name, which regulates chloride ion movement across plasma membranes. The *CFTR* gene is very large and many different mutations in this gene can cause disease. The most common mutation, called the ΔF508 mutation, is a deletion of three base pairs, leading to deletion of the amino acid phenylalanine (phe) from position 508 of the protein. Families with a history of CF may wish to undergo screening for this mutation and other common mutations known to cause CF.

PCR and gel electrophoresis can be used to detect the ΔF508 mutant **allele** (gene variant) by:

1 isolating DNA from the individual. The DNA can come from a mouth swab of an adult or from the amniotic fluid surrounding an unborn child.

2 using PCR primers that are complementary to the DNA sequences on either side of the site of the ΔF508 mutation to amplify the DNA.

3 comparing the amplified DNA molecules by gel electrophoresis. For a normal allele the amplified region is 98 base pairs long. In a ΔF508 mutant allele the amplified region is 95 base pairs long (Figure 4.1.9a). A DNA ladder is run next to the samples to enable identification of the normal and mutant alleles based on their size (Figure 4.1.9b).

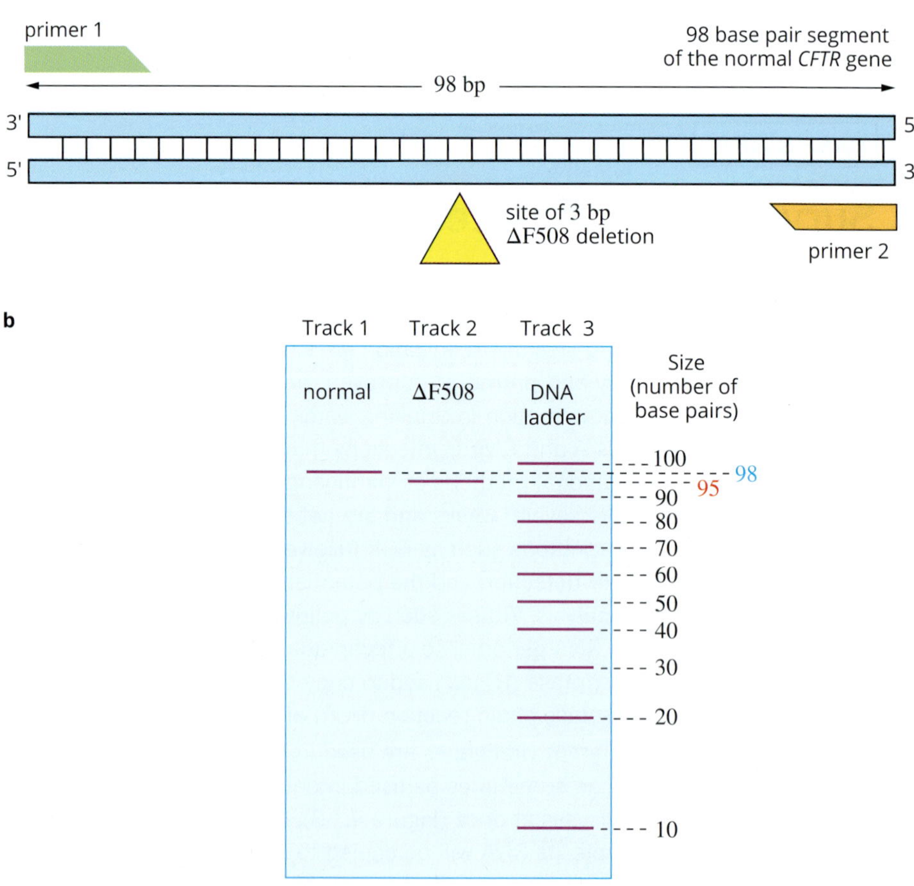

FIGURE 4.1.9 (a) PCR primers that span the site of the ΔF508 mutation are chosen to test for the presence of the CF mutation. (b) Diagram of an electrophoresis gel showing PCR products of an individual carrying two copies of the normal allele of 98 bp (track 1) and an individual carrying two copies of the ΔF508 mutation of 95 bp (track 2). The individual in Track 2 is affected with CF. The DNA ladder in track 3 allows the size of the bands in tracks 1 and 2 to be determined.

CASE STUDY ANALYSIS

Identification of cystic fibrosis carriers

A couple had a healthy daughter but their next child, a boy, failed to grow as expected, had bowel problems, lung congestion and salty-tasting skin. These symptoms led to a diagnosis of cystic fibrosis (CF) (Figure 4.1.10a). The couple were not aware of the incidence of CF in their families, yet a genetic counsellor suggested that they might be carriers of an allele for the disease. To develop CF, you need two copies of the mutant allele (that is, you need to be homozygous for the mutant allele). If you have one copy of the normal allele and one copy of the mutant allele (a heterozygous carrier) you are healthy but able to pass the mutant allele to your children. There is a 25% chance of two carriers having an affected child.

After their son's diagnosis, genetic testing was performed to determine his specific mutant alleles. Testing showed that he carried two ΔF508 mutations (he was homozygous for this mutant allele). Following this the parents underwent carrier testing for the ΔF508 mutation to confirm their carrier status. When the couple fell pregnant with a third child they chose to have genetic tests for the unborn child to determine whether this child could develop CF. A DNA sample from the unborn child was obtained from the amniotic fluid in the uterus and analysed by PCR and gel electrophoresis. The results of the unborn baby's testing, along with the results of the parents, the affected son and the unaffected daughter, are shown in Figure 4.1.10b.

Analysis

1 Using the gel electrophoresis results in Figure 4.1.10b, determine:
   a the size of the mutant allele
   b the size of the normal allele
2 The parents were confirmed as carriers of cystic fibrosis (CF) through genetic testing. Redraw the family pedigree chart, updating the symbols for the parents.
3 Study the gel electrophoresis results in Figure 4.1.10b and determine the CF status of the couple's unborn child.
4 Give your opinion on the social and ethical issues that may arise because of the unborn child's genetic result.

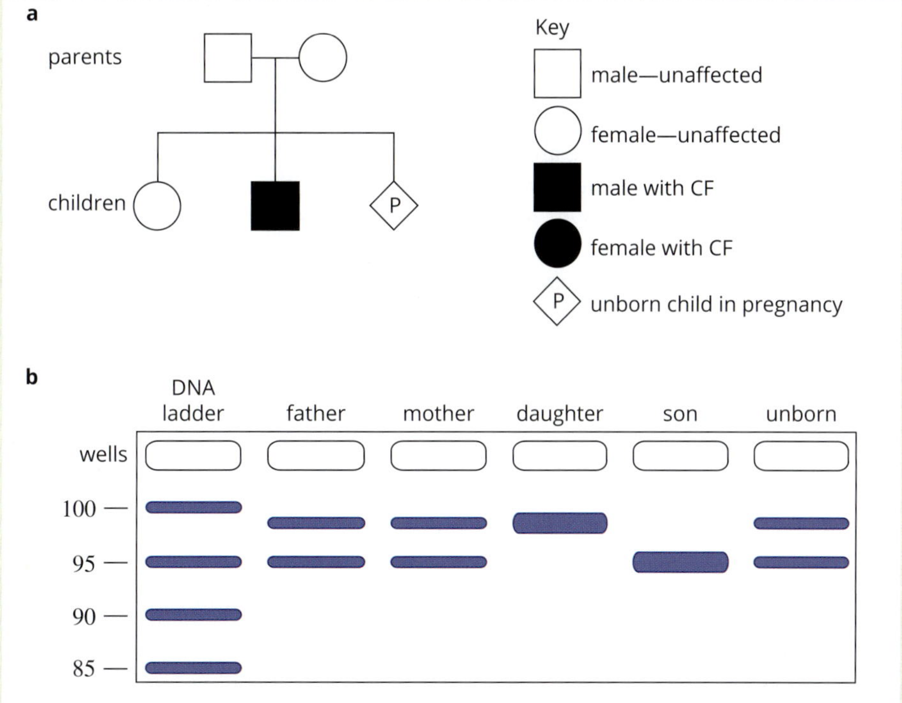

FIGURE 4.1.10 (a) Pedigree chart showing the appearance of CF in the son of two unaffected parents. The first-born daughter is unaffected. The parents want to know whether the unborn child (sex unknown) carries the CFTR mutation. (b) Gel electrophoresis results for the DNA test for the ΔF508 mutation

DNA PROFILING

DNA profiling is a technique that compares and identifies individuals based on their unique DNA sequence. DNA profiling can be used to identify one individual from any other individual. It is often used in forensics to identify the perpetrator of a crime. DNA profiling can also be used to identify bodies after disasters or to confirm if a child is genetically related to a parent (Figure 4.1.11).

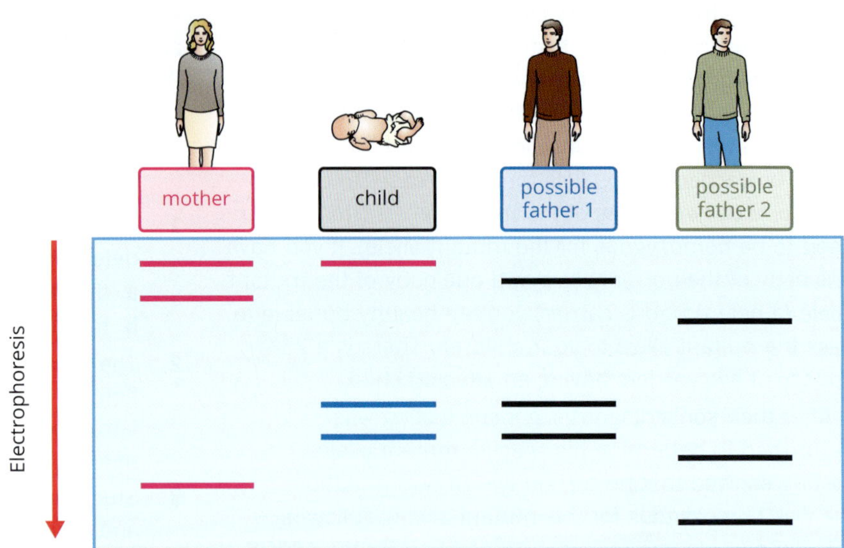

FIGURE 4.1.11 Family members share common bands in DNA profiles, although the combination of bands in each individual is unique.

> Short tandem repeats (STRs) are stretches of DNA sequences, usually 2–6 base pairs that are repeated many times. The number of repetitions varies between individuals.

DNA profiling relies on an individual's unique DNA. The non-coding sections of the DNA (introns) can vary widely between individuals. These inherited variations are called **polymorphisms**. DNA profiling uses the differences between a number of polymorphic sections to identify individuals. In particular, short, repeated sections of between two and six bases, called **short tandem repeats (STRs)**, are examined. STRs are also referred to as microsatellites.

There are thousands of STR loci throughout the human genome and 20 STRs are generally used in DNA profiling. The same STR sequence occurs on each member of a homologous pair of chromosomes, and, by chance, the number of repeats in each STR may be the same or different (Figure 4.1.12). For example, at one location on a chromosome a person might have 25 repeats, while another person might have 45 repeats at the same site.

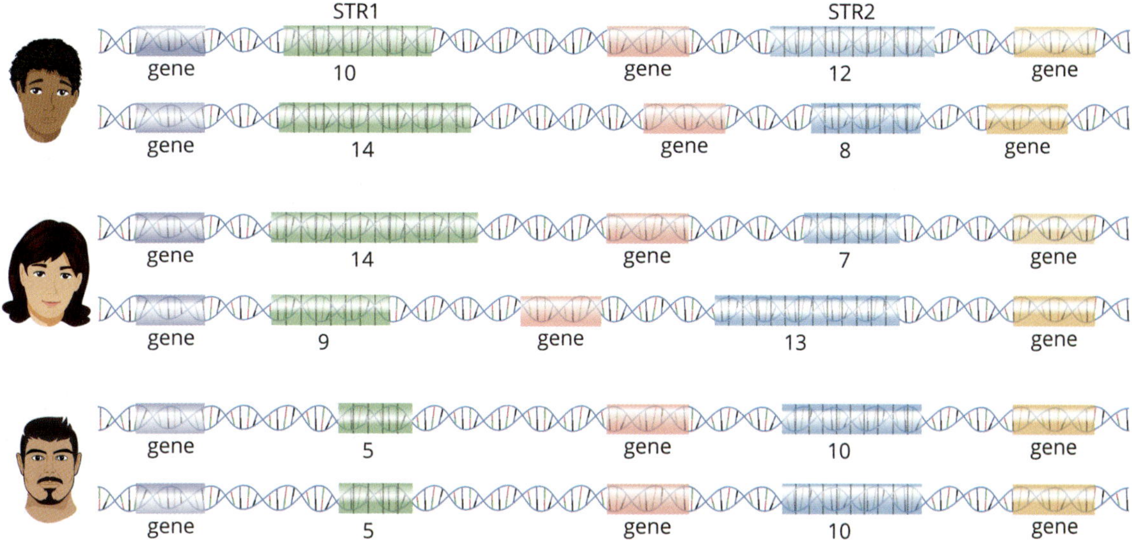

FIGURE 4.1.12 Example of the variation that can be seen in the STRs in three individuals for two STR sites. As chromosomes occur as homologous pairs, each person has two copies of each STR, which may vary in length. The numbers refer to the number of repeats.

BIOFILE

The FBI's CODIS

The 20 STR sites used in DNA profiling are part of CODIS, a United States DNA database developed by the Federal Bureau of Investigation (FBI) in 1997. CODIS stands for Combined DNA Index System. Using 20 STRs gives an extremely high probability that the DNA profile is unique and that the only perfect match of all 20 DNA sites will be with DNA from the same person. DNA profiling also includes regions on the X and Y chromosome for sex determination.

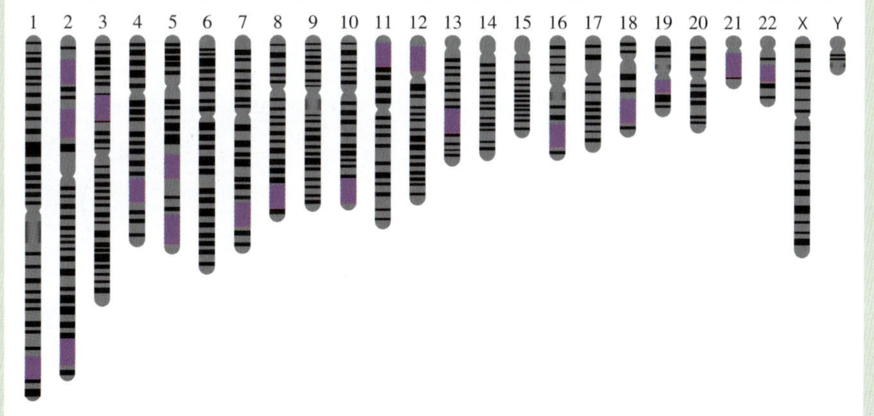

Chromosomal location of the 20 STR CODIS sites used for forensic DNA profiling. Note that there are two STR sites next to each other on chromosome 12.

Techniques involved in DNA profiling

DNA profiling is a sensitive technique that requires only a small or degraded amount of sample from blood, semen, saliva or hair.

- DNA is extracted from the sample.
- The DNA is digested into smaller fragments by specific restriction enzymes. (You will learn more about restriction enzymes in Section 4.2.)
- The STRs are amplified using PCR, with specific primers for each STR. This produces a much larger sample to test.
- Differences in the size of the STRs can be detected by standard gel electrophoresis or by capillary electrophoresis. Capillary electrophoresis is a rapid, automated system whereby the DNA fragments move in a thin tube under the influence of an electric field. The smaller the size of the fragment, the faster it moves through the capillary tube. As each fragment moves through the tube, a laser detector registers a peak on a graph. As there are two copies of each STR—one on each homologous (paired) chromosome—most STRs appear as a pair of peaks on the graph. If the same number of repeats is on each chromosome, only a single peak is seen (Figure 4.1.13).

BIOFILE

Applications of DNA profiling

DNA profiling is used for other applications such as genealogy, biogeographical population comparisons, historical population migration patterns and evolutionary relationships. For these purposes the DNA sequences used for comparisons include STRs (different sites from those used for crime scene analysis), mitochondrial DNA, Y chromosome genes and single nucleotide polymorphisms (SNPs).

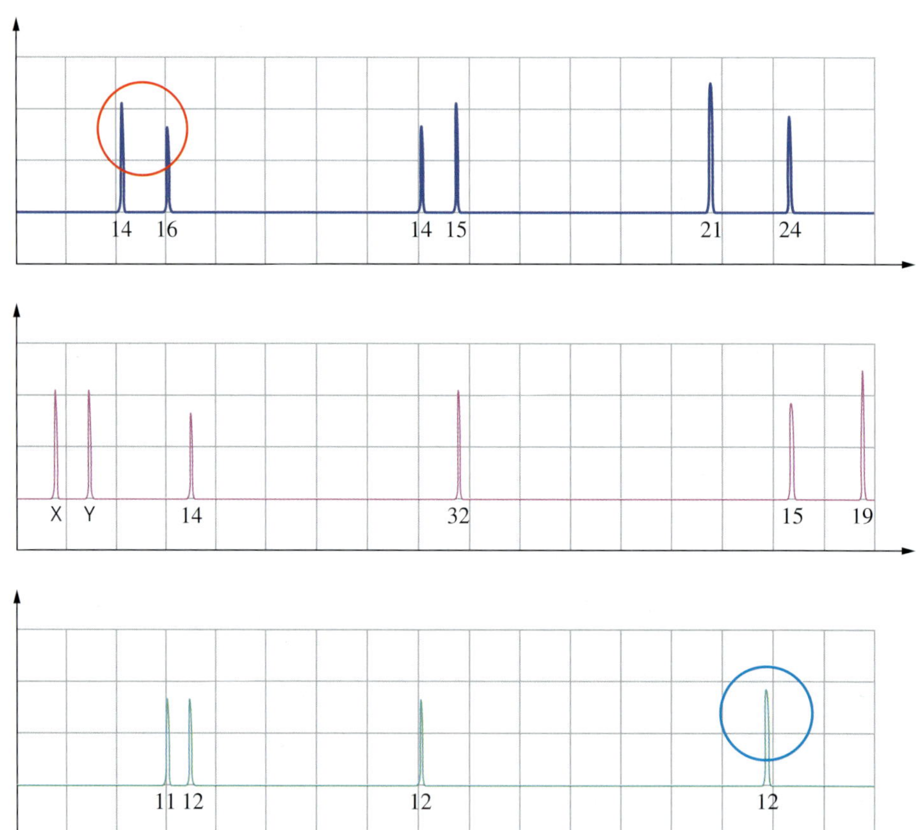

FIGURE 4.1.13 DNA profile of a male. Ten different regions (nine different STRs and the sex chromosome markers) have been analysed. The first two peaks (circled in red) show this person has 14 repeats and 16 repeats of that STR on that homologous pair of chromosomes, while the last peak (circled in blue) indicates 12 repeats of that STR on each homologous chromosome.

- A DNA profile is produced from the STR analysis. DNA fragments from samples collected from the crime scene are compared with the DNA of a suspect or suspects.
- The DNA sample is matched to that of a suspect if the lengths of the particular STRs at all sites are the same.

- If the lengths of 20 STRs from two DNA samples match perfectly, the chance that the two samples are from different people is hundreds of billions to one.

Most recently, DNA profiles have been used to identify key features of an individual's appearance. The presence of alleles for eye, skin and hair colour allows investigators to narrow down the list of suspects. Occasionally the suspect's ancestry can also be determined. This is a developing field of science and the analysis may not yet be reliable.

CASE STUDY

Tsar rediscovered using DNA

In July 1918, Tsar Nicholas II of Russia, the Tsarina Alexandra, their five children—Olga, Tatyana, Maria, Anastasia and Alexis (Figure 4.1.14)—three female servants and the royal physician were executed by a Bolshevik firing squad in the town of Ekaterinburg. Historical accounts indicate that two of the children's bodies were burned, although others claim that Anastasia escaped execution. The remaining bodies were thrown into a shallow grave and sulfuric acid was poured over them.

FIGURE 4.1.14 Tsar Nicholas II of Russia with his wife, Tsarina Alexandra, and their five children in 1913

In 1991, two amateur historians, Gely Ryabov and Alexander Avdonin, discovered nine skeletons in a grave near Ekaterinburg. The remains were tested to find out whether they came from the Tsar and his family. DNA extracted from bone tissue samples was amplified by PCR. The first step in the analysis was to identify the sex of the skeletons. This was done using PCR of a gene that is found on the Y chromosome. This indicated that there were two males and seven females.

Using DNA profiling of the bone samples, it was possible to conclude with certainty that five of the skeletons were those of two parents and their three daughters. But these could have been the remains of any family. To establish the identity of the bones, comparison to the DNA from a related person was needed. The evidence came when the DNA profiles of the skeletons were compared with those generated from known relatives of the Tsar (George, brother of Nicholas II, whose remains were exhumed from a crypt in St Petersburg) and Tsarina (Prince Philip, husband of Queen Elizabeth II).

The presence of common bands in the DNA profiles of the five bodies, Prince Philip and George indicated that all of the individuals were relatives. Indeed, it was estimated that the probability that the bones found in Ekaterinburg are the remains of the Tsar and his family is approximately 99 999 out of 100 000.

BIOFILE

Implications and issues related to DNA profiling

Privacy is a contentious issue related to DNA profiling. In Victoria, DNA samples cannot be obtained from a person unless they give their permission. They can, however, be ordered to do so if there is strong evidence that they may have committed the crime under investigation and if the DNA profile could help to confirm or deny their guilt. These DNA samples must be destroyed if the person is not guilty or is not charged. However, in some countries, the DNA may be kept for up to 10 years. This has enabled the identification of criminals who have committed crimes in unsolved cases that occurred before DNA profiling technology was developed. It has also resulted in the exoneration of people who have been wrongly accused.

Some people are in favour of the creation of a 'bank' of DNA samples, provided by everyone in the community, which could be used to solve crimes and perhaps trace the remains of unidentified missing persons. Opponents of a DNA 'bank' argue that there would be potential for these samples to be stolen or used unethically.

CASE STUDY

Freeing the innocent

DNA profiling has been used to incriminate perpetrators, but has also been used to prove the innocence of suspects or the wrongly accused. In 1987, DNA profiling was used in a case for the first time to help free an innocent man and identify the perpetrator.

Alec Jeffreys, who pioneered the technique, was asked to test semen samples found on two 15-year-old victims who were raped and murdered three years apart and compare it to a 17-year-old male, Richard Buckland, who they believed was responsible for both murders. Buckland, who had learning difficulties, had some knowledge of the murder of the second victim, and had later confessed to the second murder. The police had wanted to convict him of the first murder, though he refused to plead guilty to this. Through DNA profiling, Alec Jeffreys confirmed that the two girls were murdered by the same man but it was not Buckland. Buckland was released from custody and another man was convicted of the crimes.

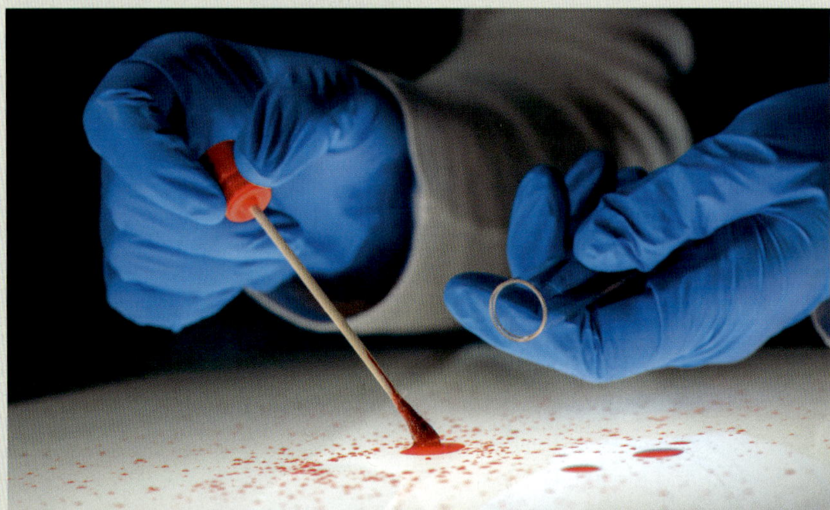

FIGURE 4.1.15 The Idaho Innocence Project uses DNA technology to free people wrongly convicted of crimes.

Greg Hampikian, a biology and criminal justice expert at Boise State University, founded the Idaho Innocence Project. Hampikian works with police and defence lawyers around the world to free innocent people wrongly convicted of crimes by using new DNA profiling technology or exposing errors with existing technology. One such case Hampikian worked on involved an error in police DNA sampling (Figure 4.1.15) that caused two innocent people to be jailed. Hampikian reviewed the DNA sampling procedures and found many errors, including one where an item of clothing containing a DNA sample was not collected until 46 days after the crime, and was passed around to several police investigators before being placed in a different position at the crime scene for photographing, allowing ample time for contamination of the sample. Hampikian also found samples from the same crime scene that were not tested by forensics teams because the quantity of the sample was smaller than what the FBI deemed 'valid for testing'. In fact, as Hampikian discovered, the samples were sufficient in quantity and quality to provide valid DNA evidence.

CASE STUDY

DNA barcodes

A DNA barcode is a short sequence of nucleotides that uniquely identifies a species. It is obtained by using the polymerase chain reaction (PCR) and DNA sequencing (Figure 4.1.16). Sequences are submitted to online databases such as BOLD, the Barcode of Life Database, or other similar databases. There are global DNA barcoding projects underway to catalogue all of life, including bees, butterflies, mosquitoes, fungi, mammals and plants. Scientists and non-scientists alike can access these sequences for research.

To compare and identify species, you need a gene sequence that is present in all organisms but differs slightly between different groups or species. For eukaryotes, the mitochondrial gene *CO1* (cytochrome oxidase subunit 1) is often used. Genes in plant plastids help to further identify plant species. For bacteria, a gene for ribosomal RNA can be used.

Examples of barcoding projects include:

- identifying species and measuring diversity; for example, barcoding organisms of the Great Barrier Reef, CSIRO scientists barcoding Australian fish species
- tracking pathogenic and non-pathogenic bacterial populations
- authenticating food—Is it shark and chips? Is the beef burger really a horse burger?
- monitoring wildlife crime, such as illegal trade in protected and endangered species
- investigating ecology and evolution; for example, seed identification and seed banking of Australian *Acacia* (wattle) genus for conservation and restoration of biodiversity.

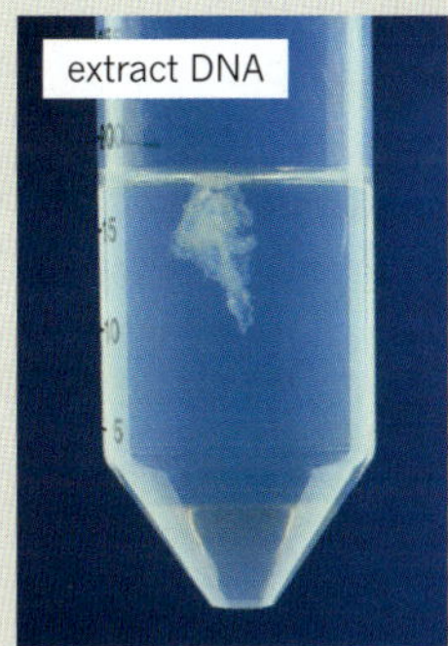

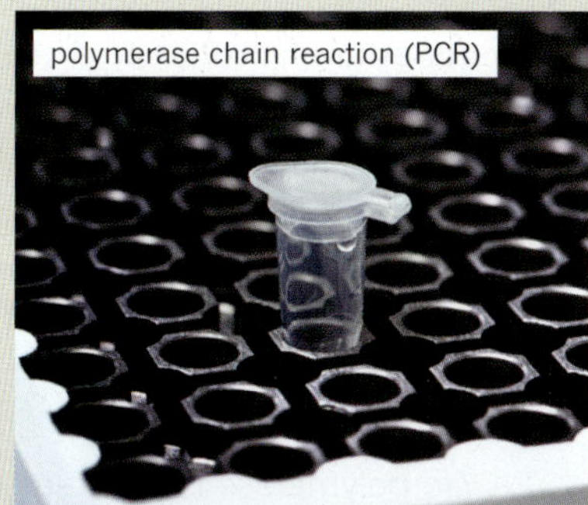

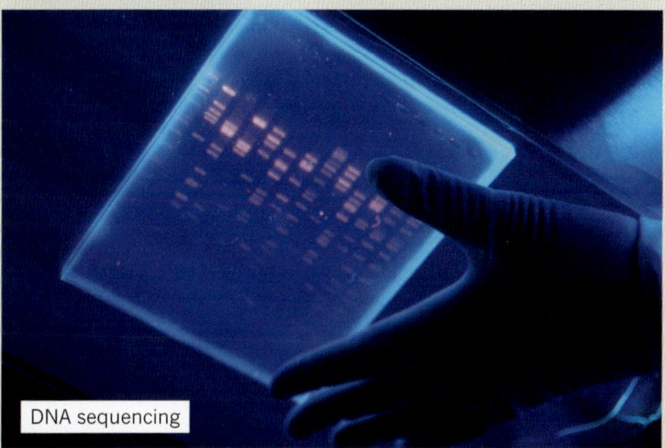

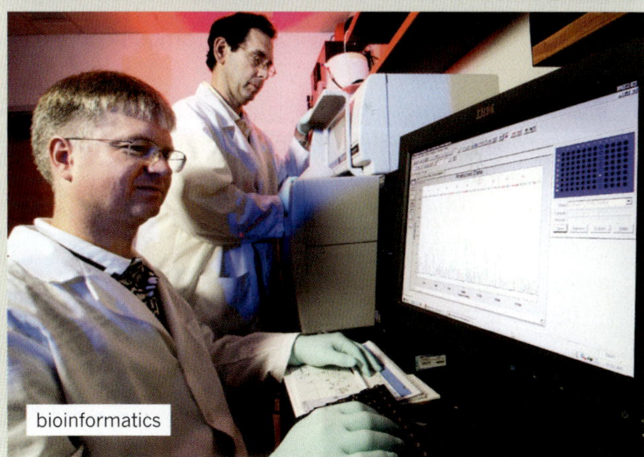

FIGURE 4.1.16 Process of obtaining a DNA barcode

4.1 Review

OA ✓✓

SUMMARY

- DNA is often found or extracted in trace amounts and requires amplification to produce a sample that is large enough for scientists to work with.
- DNA amplification uses the polymerase chain reaction (PCR) to rapidly increase identical copies of the target DNA.
- Polymerases are enzymes that catalyse the formation of long-chain molecules (polymers), such as DNA and RNA, by linking smaller molecules (nucleotides).
  - *Taq* polymerase is a heat-resistant DNA polymerase often used in PCR to synthesise multiple copies of the target DNA.
  - RNA polymerases synthesise RNA from DNA during transcription.
- The PCR mixture is added to a test tube and the test tube is placed in a thermocycler. The thermocycler alters the temperature in pre-programmed stages to enable a three-step process to be carried out:
  - denaturation
  - annealing
  - extension.
- Gel electrophoresis is a technique that separates fragments of negatively charged DNA by length.
- DNA profiling compares variable short tandem repeat (STR) regions of the genome for identification of individuals.

KEY QUESTIONS

Knowledge and understanding

1 Which one or more of the following would be a suitable target DNA for PCR?
- **A** a single gene
- **B** a genome
- **C** a short variable region within a chromosome
- **D** a microsatellite
- **E** a whole chromosome

2 **a** Describe the function of a polymerase enzyme in cells and state their function.
b Name two types of polymerase enzyme used in cells and state their function.

3 **a** Identify the function of the enzyme called reverse transcriptase.
b When is reverse transcriptase used in biomolecular techniques?

4 **a** Explain what is meant by amplifying a piece of DNA.
b What do the letters PCR stand for?
c Draw a simple, labelled flow diagram to summarise the key steps in the process of PCR.
d **i** Describe the role of the enzyme *Taq* polymerase.
ii Why does *Taq* polymerase have to be heat resistant?
e If you started a PCR reaction with one DNA molecule, determine how many molecules you would have after:
i 3 cycles of amplification
ii 32 cycles of amplification

5 **a** What is gel electrophoresis?
b Outline how this technique reveals data about the DNA fragments being tested.

6 A DNA ladder is used in gel electrophoresis. Explain the purpose of a DNA ladder.

7 **a** Explain how DNA profiling can be used to help determine the guilt or innocence of a murder suspect found with bloodstains on their clothes.
b Explain why a sample may need to be tested in more than one laboratory.

8 Explain if DNA profiling can discriminate between:
a a twin brother and sister
b identical twins.

Analysis

9 You have a single-stranded RNA molecule with the sequence 3'-AAUUGCGCA-5'. If you place it in a test tube with nucleotides and reverse transcriptase, identify the sequence that will be made on the complementary strand.

10 Gel electrophoresis was conducted with a DNA ladder (standard) and two DNA samples (A and B). The diagram illustrates the resulting gel.

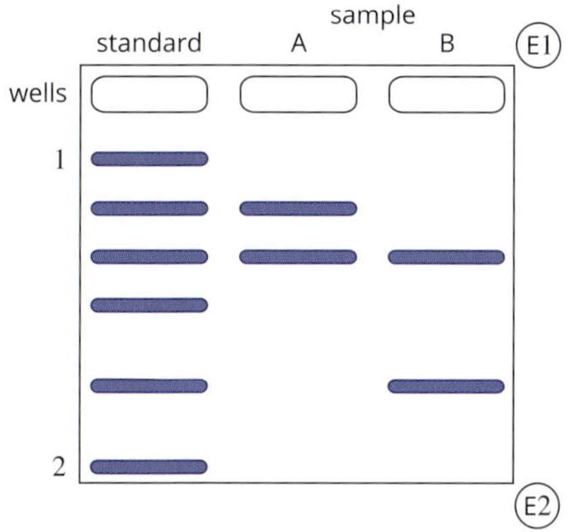

a Explain the difference between the DNA fragments labelled 1 and 2.

b E1 and E2 represent the electrodes. Which one is the positive electrode?

c Explain why the DNA migrates through the gel from E1 to E2 when an electric current is applied.

d How many DNA fragments are in:

i the standard (DNA ladder)

ii sample A

iii sample B

e You have been given information about the DNA ladder. The sizes of the fragments are 600, 500, 400, 300, 200 and 100 bp. What are the sizes of the fragments in samples A and B?

11 The following diagram represents STR regions of DNA used to identify a person who died in a natural disaster. Three people who were looking for a missing sibling submitted DNA for comparison.

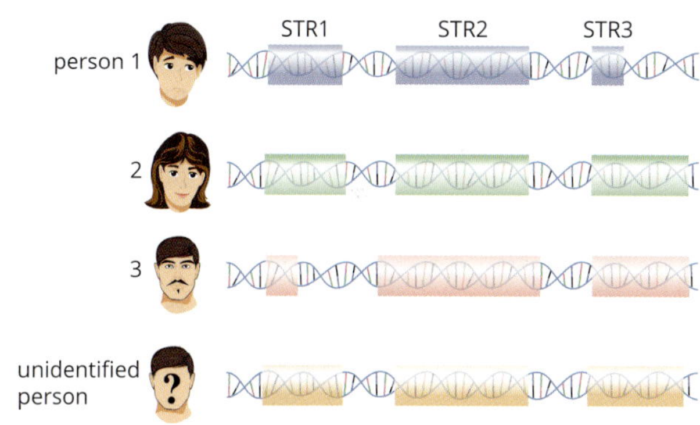

a What is an STR?

b List the steps involved in creating a DNA profile using STR analysis.

c The following illustration represents the DNA profile from the STR analysis. Which person is most likely to be the sibling of the unidentified person? Explain your answer.

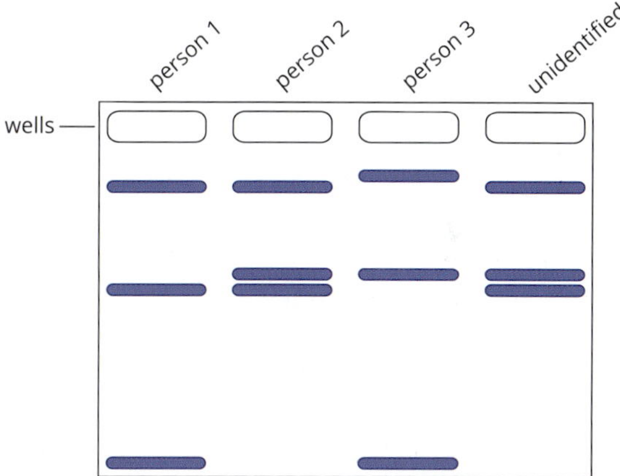

4.2 Bacterial transformation

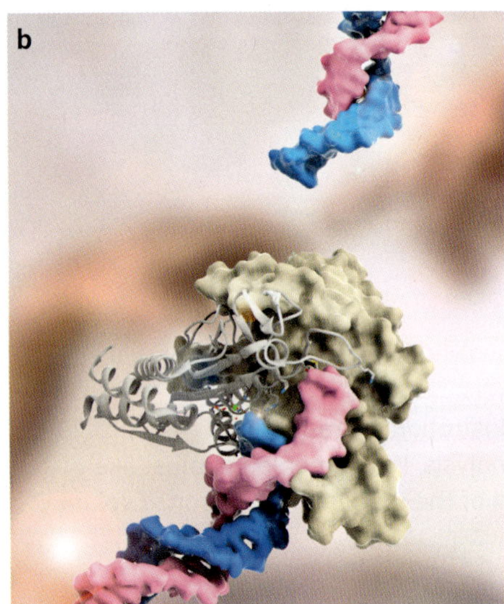

FIGURE 4.2.1 (a) Illustration representing a recombinant plasmid, showing the gene of interest (orange) combined with the plasmid DNA (purple). (b) Computer model showing the restriction enzyme EcoRI (grey) cutting a DNA strand (blue and pink)

Plasmids are small, circular DNA molecules found in bacterial cells. They are often used as **vectors** (carriers) when scientists move target DNA from one organism to another. Genes can be inserted into plasmids, which can then be incorporated into bacterial cells in a process known as **bacterial transformation**. The inserted gene can then be replicated through the self-replicating properties of the plasmid inside the bacterial cells, with the gene expressing the protein for which it codes. In this way target proteins can be mass-produced, such as the human insulin protein for use by diabetics.

In this section, you will learn how restriction enzymes (endonucleases) and ligases are used to create recombinant DNA (Figure 4.2.1), using plasmids as vectors. You will also learn how these plasmids are then incorporated into bacterial cells for replication.

RESTRICTION ENZYMES (ENDONUCLEASES)

DNA molecules are far too long for biologists to work with in their entirety. The discovery and isolation of **restriction enzymes**, also known as **endonucleases**, has enabled scientists to cut DNA into smaller, more usable fragments and isolate particular regions of interest, such as a single gene.

Restriction enzymes are a large group of enzymes that occur naturally in bacteria. They form part of a bacterial cell's defence system, targeting foreign DNA that may enter the cell, such as the DNA of **bacteriophages**. The restriction enzymes cut up foreign DNA into smaller fragments, destroying it and preventing it from replicating. Bacterial cells use a blocking process (methylation) to prevent the restriction enzymes from binding and cutting their own DNA.

Each restriction enzyme targets a specific sequence of nucleotides, usually four to six base pairs in length. This sequence is called a **recognition site**. Every time a restriction enzyme passes its recognition site, it breaks the phosphodiester backbone once on each DNA strand. As a result, the DNA molecule is cut up into fragments of different lengths.

Bacteriophages are viruses that infect bacteria.

The base-pairing ability of sticky ends allows DNA from very different species to ligate (join), forming recombinant DNA molecules.

DNA is universal to life on Earth, with the same structured molecule found in all cells. This means DNA from different species can be combined.

Types of restriction enzymes

There are two types of restriction enzyme, which cut DNA differently:

- sticky-end restriction enzymes
- blunt-end restriction enzymes.

Sticky-end restriction enzymes

Sticky-end restriction enzymes leave DNA fragments with overhanging ends. They cut the DNA backbone at a different location on each strand within the recognition site (Figure 4.2.2). This results in a staggered cut, leaving two fragments with exposed bases or 'sticky ends'. The exposed bases are then able to form complementary base pairs through hydrogen bonding with nucleotides of other DNA molecules that have complementary sticky ends.

EcoRI is an example of a sticky-end restriction enzyme extracted from *E. coli*. It cuts the recognition site GAATTC between the G and A nucleotides on each strand. The sequence on one strand is GAATTC and on the complementary strand it is CTTAAG, the same sequence when read backwards. This is called a **palindromic sequence**. EcoRI cuts at the different G and A location on each strand, so sticky end fragments are produced.

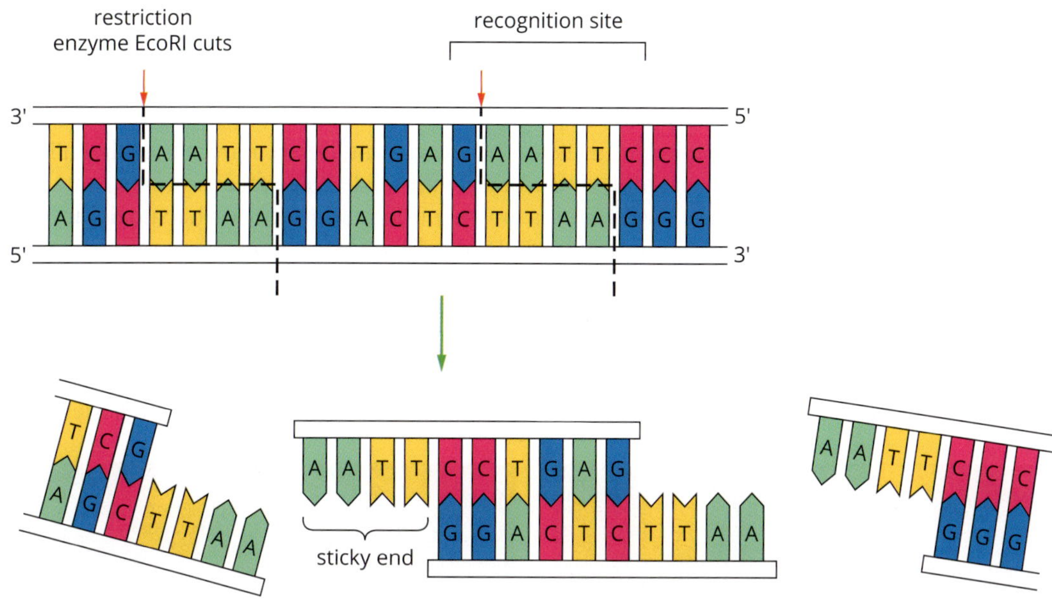

FIGURE 4.2.2 The sticky-end restriction enzyme EcoRI cuts the DNA between the G and the A of its specific recognition site, GAATTC, creating fragments of DNA with sticky ends.

Blunt-end restriction enzymes

Blunt-end restriction enzymes leave clean-cut ends by cutting the sugar–phosphate backbone on both strands of the DNA molecule at the same location within the recognition site (Figure 4.2.3).

HaeIII is an example of a blunt-end restriction enzyme, which is extracted from the bacterium *Haemophilus aegyptius*. It cuts the recognition site GGCC between the G and C nucleotides. These two bases are in the exact same location on either strand of the DNA molecule, resulting in a straight cut through both stands that leaves two fragments with blunt ends.

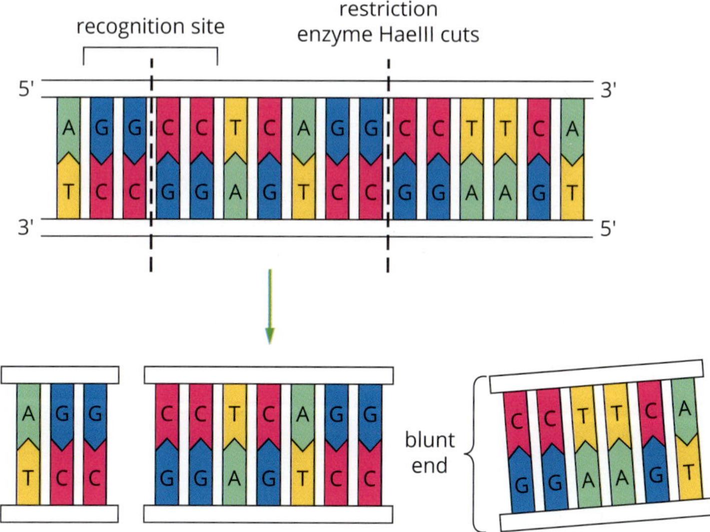

FIGURE 4.2.3 The blunt-end restriction enzyme HaeIII cuts the DNA between the G and the C of its specific recognition site, GGCC, creating fragments of DNA with blunt ends.

BIOFILE

Naming restriction enzymes

The first three letters of the name of a restriction enzyme identify the bacterial species from which they were isolated. The fourth letter refers to the particular strain of bacteria. A roman numeral is also included if more than one restriction enzyme has been isolated from this bacterial strain. For example, the restriction enzyme EcoRI comes from *E. coli* strain RY13, and was the first restriction enzyme isolated from that strain of *E. coli*.

Identifying polymorphisms and mutations

Small variations in DNA sequences (called polymorphisms) occur within a population. To be considered a polymorphism rather than a mutation, the least common allele has to have a frequency of 1% or more in a population.

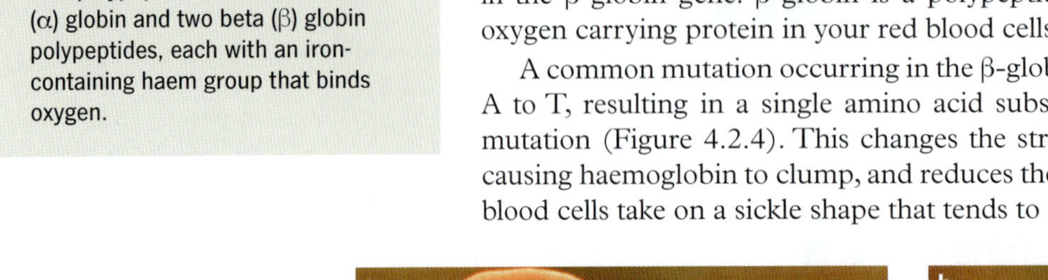

Haemoglobin is a protein made from four polypeptide chains, two alpha (α) globin and two beta (β) globin polypeptides, each with an iron-containing haem group that binds oxygen.

A mutation may change an allele to a new variant, different from the other alleles in a population. For example, individuals with sickle-cell anaemia have a mutation in the β-globin gene. β-globin is a polypeptide component of haemoglobin, the oxygen carrying protein in your red blood cells.

A common mutation occurring in the β-globin gene is a single base change from A to T, resulting in a single amino acid substitution in the protein—a missense mutation (Figure 4.2.4). This changes the structure of the β-globin polypeptide, causing haemoglobin to clump, and reduces the amount of oxygen carried. The red blood cells take on a sickle shape that tends to clog and rupture the capillaries.

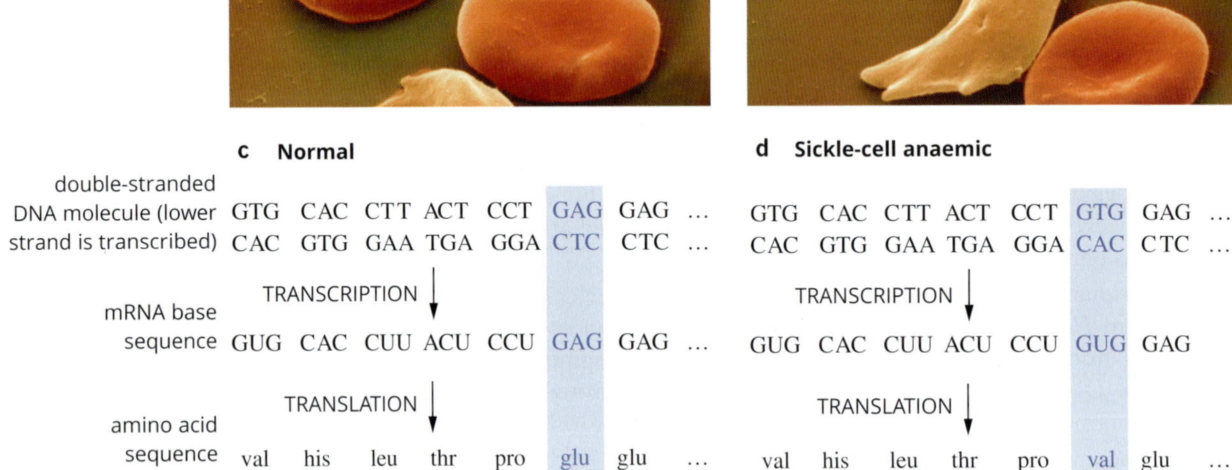

FIGURE 4.2.4 Viewed under the electron microscope there are obvious differences between the (a) normal and (b) sickle-shaped red blood cell. Under each image, expression of the (c) normal and (d) mutated genetic code is compared.

Molecular tools are used to identify individuals carrying the mutation that causes sickle-shaped red blood cells. By chance, the mutation occurs at a restriction enzyme site. The base change eliminates the recognition site for MstII (Figure 4.2.5a). To detect the mutation, DNA is extracted from individuals, the region of DNA containing the recognition site for MstII is amplified by PCR, and then the PCR products are incubated with the MstII restriction enzyme. The normal allele will be cut at the MstII recognition site. The mutant allele will not be cut. The difference in the size of the DNA fragments is identified by gel electrophoresis (Figure 4.2.5b).

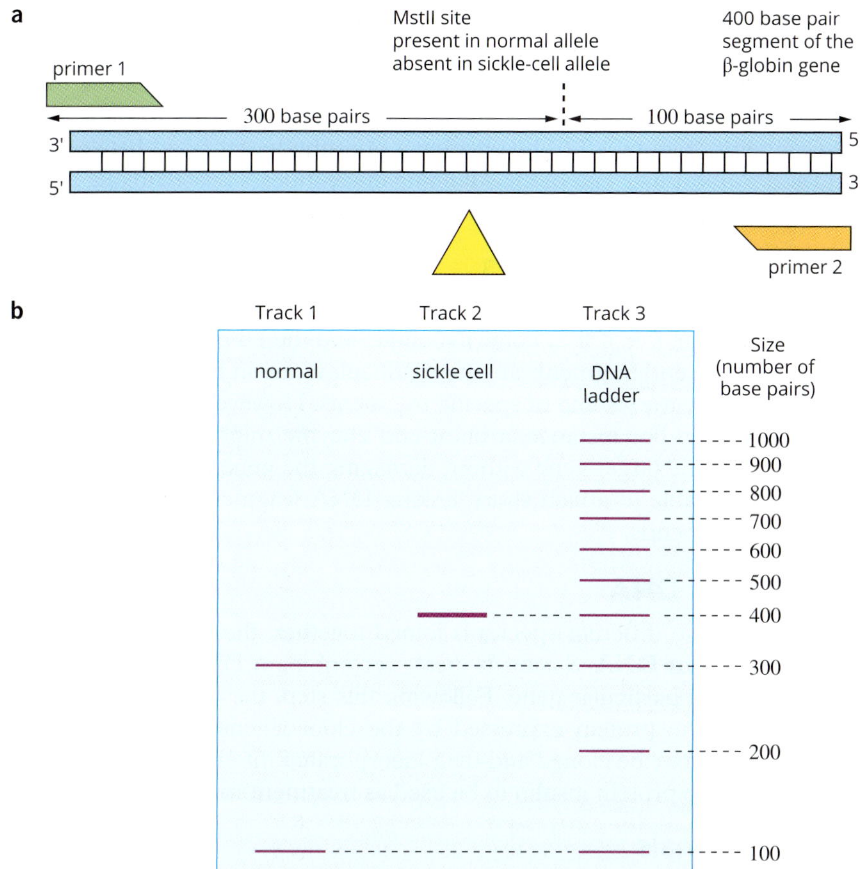

FIGURE 4.2.5 (a) A portion of the β-globin gene with primers either side of a 400 bp region. An MstII recognition site is present in the normal allele, resulting in fragments of 300 and 100 bp when the PCR product is cut with the MstII restriction enzyme. The MstII recognition site is missing in the sickle-cell allele so the PCR product is not cut. (b) Diagram of an electrophoresis gel showing DNA fragments from two individuals. The DNA ladder in Track 3 is used to determine the sizes of the Track 1 and 2 bands and identify which individual has the normal allele (Track 1) and which individual has the sickle-cell allele (Track 2).

LIGASES

Ligases are a group of enzymes that join fragments of DNA or RNA in a process called **ligation**. There are two different groups of ligases:

- **DNA ligases**
- **RNA ligases**

The role of DNA ligase in a cell is to join segments of newly replicated DNA and repair breaks in DNA molecules.

Assuming they were cut with the same restriction enzymes, DNA ligase can join DNA fragments extracted from different organisms or different species, as DNA has a universally consistent molecular structure. Under laboratory conditions, depending on the characteristics of the DNA ends to be joined (sticky or blunt ends), the conditions of the ligation reaction (e.g. incubation time and temperature) need to be adjusted to ensure efficient DNA ligation is achieved.

Ligases join two DNA fragments and create a phosphodiester bond between them.

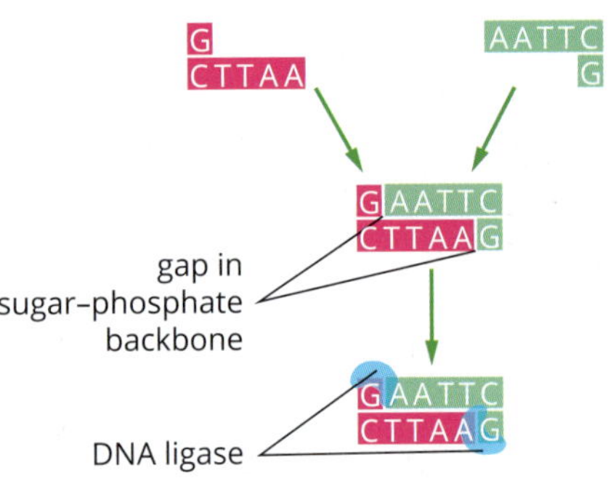

FIGURE 4.2.6 Two sticky-end DNA fragments come together by complementary base pairing, and then DNA ligase (shown as blue circles) permanently links the sugar–phosphate backbone.

Ligation of sticky-end fragments

Ligation to join sticky-end fragments is specific because the exposed bases of sticky-end fragments first bind by complementary base pairing. Complementary bases are attracted by weak hydrogen bonds that hold them together. After this, the ligase joins the fragments (Figure 4.2.6) by creating a phosphodiester bond between the 3' OH end and 5' phosphate end of the adjoining nucleotides. This technique makes recombinant DNA and is used in processes such as **gene cloning**.

Ligation of blunt-end fragments

Unlike ligation of sticky-end fragments, ligation of blunt-end fragments is random. Any two fragments can join if they come in contact and the DNA ligase joins them. For this reason, blunt end fragments are more difficult to use in DNA manipulation processes that require the joining of specific fragments. However, sometimes blunt ends are unavoidable. For instance, a blunt-end enzyme might be the only type available to cut out the target gene without damaging the gene itself. Using DNA ligase, scientists are able to attach short, linking DNA fragments onto blunt-end DNA to create sticky ends.

A clone is a genetically identical copy of a gene, cell or organism.

RECOMBINANT DNA

When DNA from two different species is joined together, the resulting molecule is called **recombinant DNA**. Scientists create recombinant DNA to clone (make multiple copies of) a particular gene. Following this step, they may also produce large quantities of the protein expressed by the cloned gene. For example, an insulin-coding gene may be cloned and then incorporated into bacteria to produce large quantities of the protein insulin to be used as treatment for diabetes.

Insulin from animals

Insulin was previously extracted from the pancreas of other animals, such as pigs and cattle, for the treatment of type 1 diabetes. This was an expensive and time-consuming method that also involved the risk of an allergic reaction to the foreign molecule and potential for contracting diseases. Porcine (pig) and bovine (cattle) insulin are similar, but not identical, to human insulin. Their biological activity is not as effective as human insulin, so it is preferable to use the human hormone. Recombinant human insulin became available for treatment in the 1980s (Figure 4.2.7).

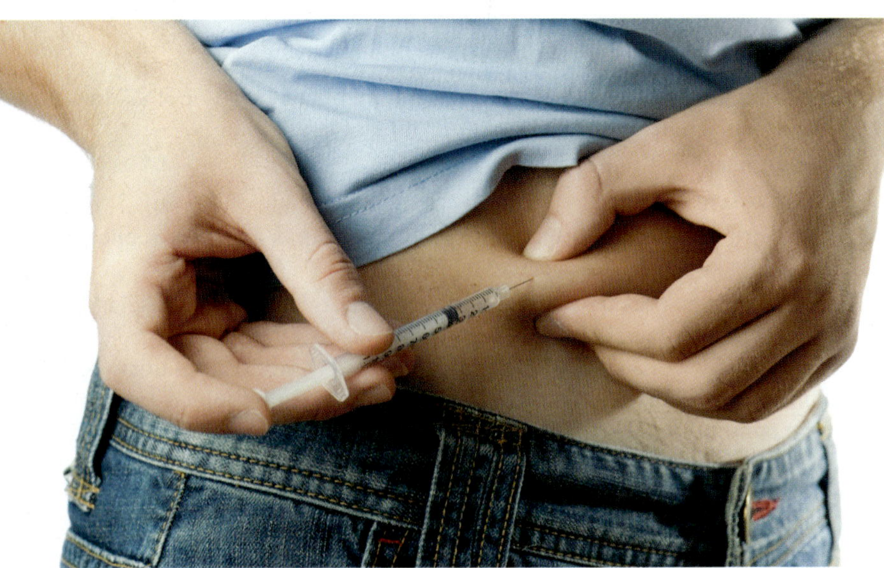

FIGURE 4.2.7 Man giving himself an injection of insulin

BIOFILE

Plasmids

In Chapter 3 you learnt about the structure of DNA and chromosomes. In eukaryotic cells, DNA forms long, linear chromosomal strands containing thousands of genes. In most prokaryotic cells, DNA forms a single double-stranded circular chromosome.

In addition to the circular chromosome, bacterial cells also contain small circular pieces of double-stranded DNA called plasmids. Plasmids replicate independently of the main chromosome. As bacterial cells do not have a nucleus, both chromosomal and plasmid DNA are located in the cytosol.

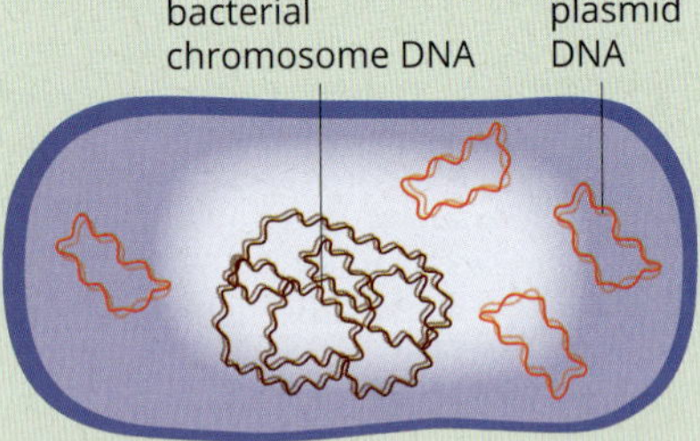

Bacterial cells contain a single, circular DNA chromosome (brown) and small rings of double-stranded DNA called plasmids (red).

Using plasmids as vectors

When scientists create recombinant human insulin, they often use a bacterial plasmid as the vector. They insert target DNA, in this case the gene required to produce insulin, the *INS* gene, into the plasmid, producing a **recombinant plasmid**. The plasmid is then placed in a bacterial cell, where the self-replicating system of the plasmid and cell replicates the plasmid genes. Each bacterial cell containing the plasmid will express the protein products of the plasmid genes, including those of the introduced target DNA (in this case, insulin).

Plasmids are used as vectors when creating recombinant DNA for the following reasons:

- Their small size makes them easy to manipulate in a laboratory.
- Plasmids carry a range of restriction enzyme sites. A plasmid containing the appropriate recognition sites can be chosen to suit your needs. For example, if the gene of interest was cut from a chromosome with EcoRI, you would use a plasmid with a single EcoRI site.
- Recombinant plasmids self-replicate independently once they are placed inside a host bacterial cell and at a faster rate than their bacterial host's chromosomal DNA. This is of vital importance in the efficient manufacture of large quantities of insulin and other proteins.

To enable the identification of cells that have incorporated the recombinant plasmid, the plasmids used as vectors must have particular characteristics including:

- an antibiotic-resistance gene
- a **reporter gene** that can be easily identified—such as a gene that produces coloured or fluorescent proteins.

One example is a plasmid containing the ***lacZ* gene** (Figure 4.2.8). Restriction enzyme sites for inserting the gene of interest are located within the *lacZ* gene. If the gene insertion is successful it will disrupt the *lacZ* gene. The use of this process in bacterial selection will be discussed later in this section. The plasmid also contains the antibiotic resistance gene amp^R, which encodes resistance to ampicillin.

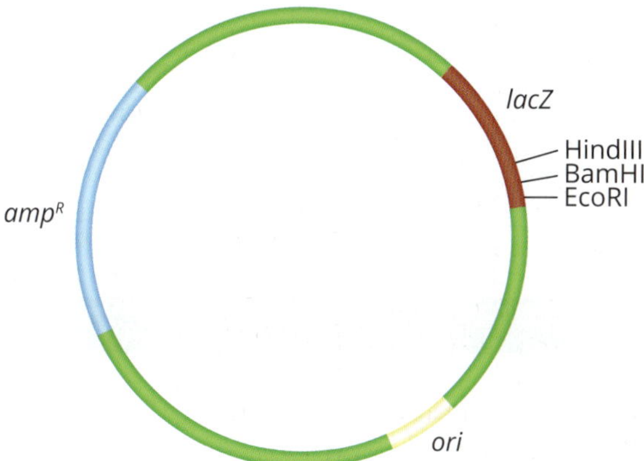

FIGURE 4.2.8 An example of a plasmid used for recombinant DNA and bacterial transformation showing the gene for ampicillin resistance (amp^R), the origin of replication (*ori*) and the *lacZ* gene. Sites for three restriction enzymes lie within the *lacZ* gene.

A recombinant plasmid is a plasmid containing a foreign gene that has been inserted using restriction enzymes and DNA ligase.

BIOFILE

Green fluorescent protein (GFP)

Green fluorescent protein (GFP) is a protein from jellyfish. The *GFP* gene can be inserted into plasmids to be used in bacterial transformation. The transformed bacterial cells express the green fluorescent protein and fluoresce green under UV light.

Reporter genes are attached to the gene under investigation and provide visual evidence when the gene of interest is expressed, since they are expressed alongside it. The *GFP* gene is a reporter gene. Other reporter genes express luciferase, an enzyme from fireflies that produces a yellow fluorescent protein, and a red fluorescent protein from a coral.

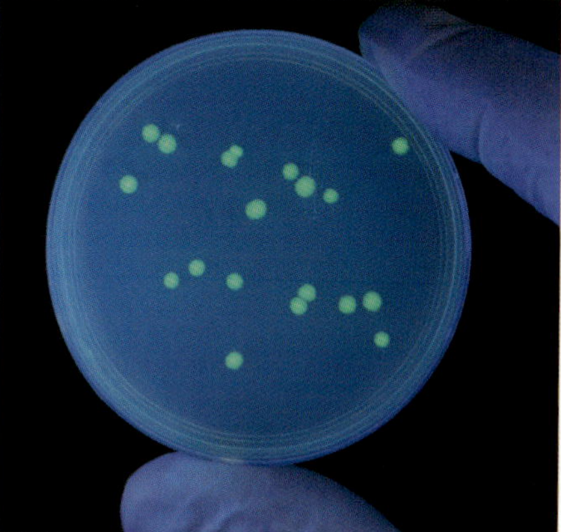

Transformed bacterial colonies can be identified as they contain the *GFP* gene and fluoresce under UV light.

> A reporter gene is a gene that allows detection of gene expression in genetic engineering, such as genes for *lacZ* and fluorescent proteins.

Creating recombinant DNA

The process of creating a recombinant plasmid is outlined below and in Figure 4.2.9:

1. The target DNA, in this case the gene responsible for producing insulin (the *INS* gene), is cut out using a sticky-end restriction enzyme and then isolated.
2. The bacterial plasmid is cut by the same restriction enzyme. The plasmid and the target DNA now have the same sticky ends with exposed bases that are complementary to each other.
3. The *INS* gene and plasmids are placed together. Some plasmids will simply close back up (known as non-recombinant plasmids), while other plasmids will incorporate the *INS* gene by complementary base pairing (known as recombinant plasmids). Reporter genes are necessary to distinguish recombinant plasmids from non-recombinant plasmids.
4. DNA ligase is added to rejoin the sugar–phosphate backbone of the DNA.

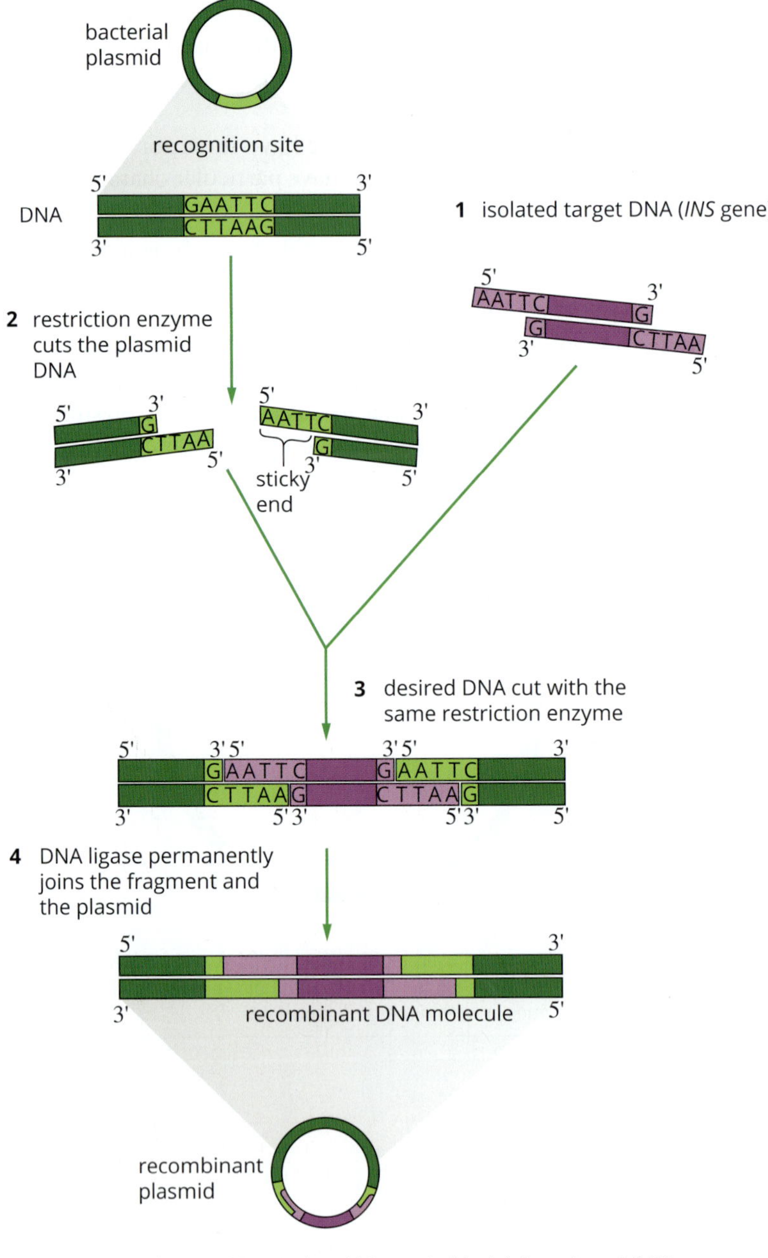

FIGURE 4.2.9 A recombinant plasmid is created by joining a target DNA fragment (the *INS* gene) and a plasmid that have both been cut with the same sticky-end restriction enzyme. They are joined using DNA ligase.

Complementary DNA

To produce a eukaryotic protein, such as insulin, in a prokaryotic cell (i.e. bacterial cell), complementary DNA (cDNA) is used as the target DNA. cDNA is DNA that has been copied from mature mRNA and contains only exons. cDNA is synthesised using the reverse transcriptase enzyme.

Reverse transcriptase

In Section 4.1, you learnt that reverse transcriptase is an enzyme with the ability to make cDNA from mRNA (Figure 4.2.10). This is useful because mature mRNA has already had the introns spliced out. Prokaryotic cells are unable to splice out introns. Reverse transcriptase allows the synthesis of DNA from mature mRNA in a test tube (*in vitro*).

> Exons are the coding regions of DNA, while introns are non-coding regions.

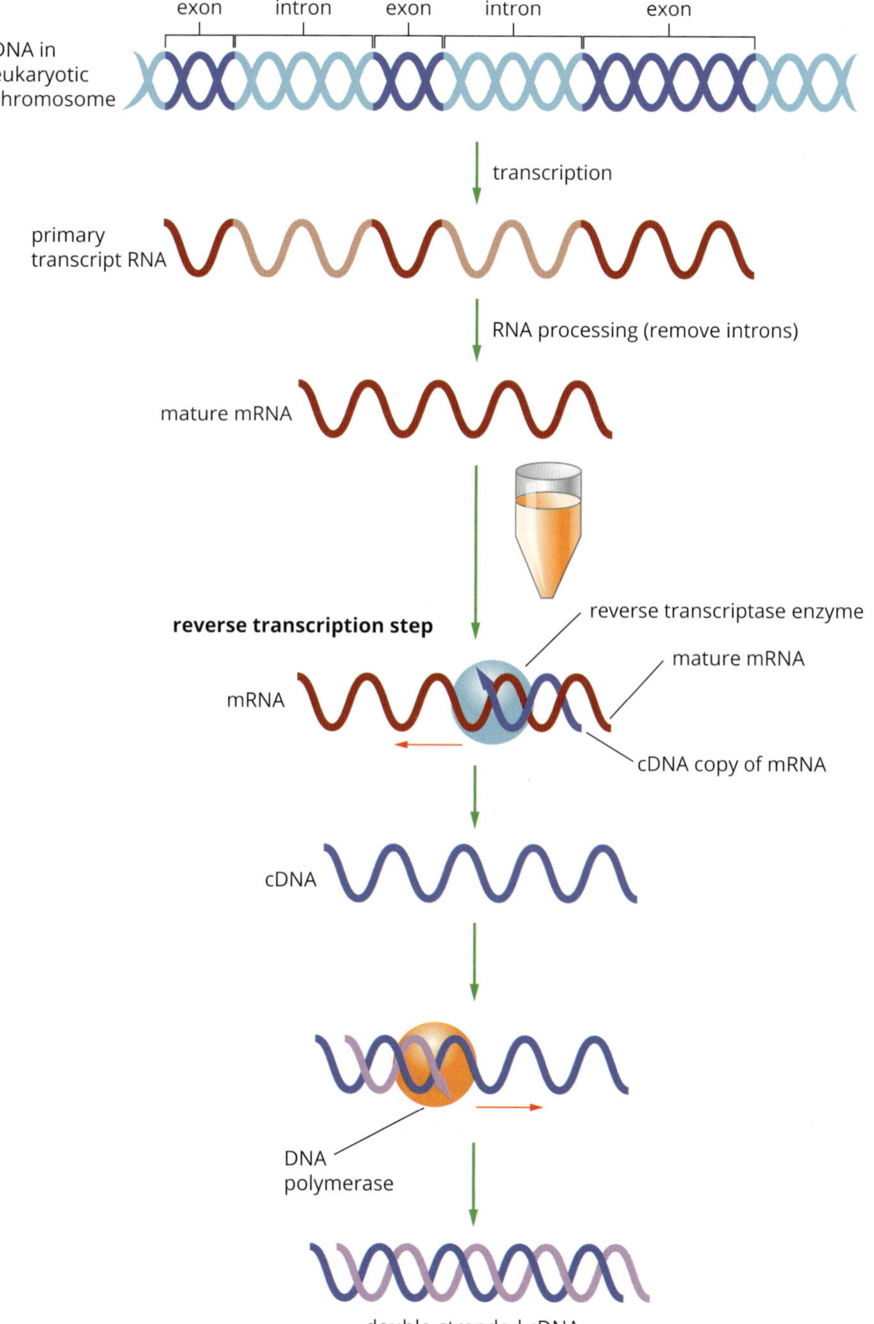

FIGURE 4.2.10 Process of creating cDNA from mRNA using a reverse transcriptase enzyme. Once the strand of cDNA is produced, DNA polymerase is used to make the cDNA double stranded.

When cDNA is inserted into a plasmid, which in turn is then incorporated into a bacterial cell, the protein encoded by this cDNA will be expressed (Figure 4.2.11). This method is now commonly used to produce vast quantities of therapeutic proteins including human insulin, growth hormone and cytokines.

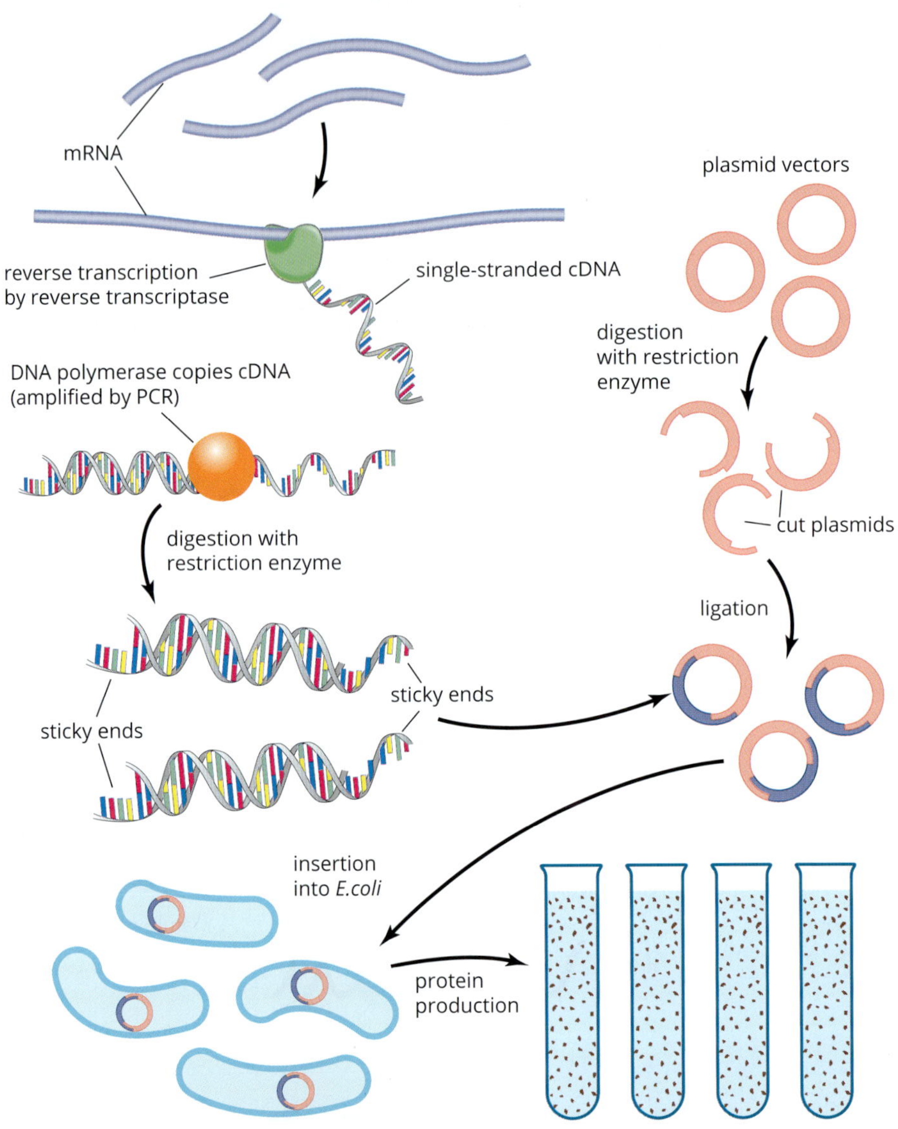

FIGURE 4.2.11 cDNA is used as target DNA to create a eukaryotic protein product within a bacterial cell. The cDNA is treated to create sticky ends so that it can be inserted into a plasmid, creating a recombinant plasmid. This plasmid is then incorporated into a bacterial cell, which expresses the protein product.

Regulatory genes in recombinant DNA

Regulatory genes may be included in plasmids for the purpose of controlling the expression of the target gene that is inserted into the plasmid. The regulatory gene is turned on by an **inducer** molecule, for example a sugar such as lactose or arabinose, or by metal ions such as iron, copper or zinc. Once the regulator is transcribed, the target gene can be transcribed and translated. Inducers are important in regulating gene expression, particularly when the aim of the recombinant DNA technology is protein production, and when a gene is being expressed and studied in plant and animal models.

TRANSFORMING BACTERIAL CELLS

Cells that have had foreign DNA incorporated into them have undergone bacterial transformation and are said to be 'transformed'. For example, when a foreign plasmid is incorporated into a bacterial cell, the bacterial cell is then 'transformed' because it can express a new gene and therefore has a new characteristic. This process is also known as **genetic transformation** and it can be both natural and artificial.

Natural transformation of bacterial cells

Bacterial cells exchange genetic material naturally through several methods. This is how an organism that reproduces asexually (e.g. by binary fission) can evolve and spread antibiotic resistance. **Bacterial competence** refers to the ability of a bacterial cell to alter its genome by taking in DNA from other cells or the environment. Antibiotic resistance genes can be taken in from other bacterial cells. Since antibiotic resistance in disease-causing bacteria is of urgent interest to scientists and the general public, bacterial competence is an important and growing area of research.

Artificial transformation of bacterial cells

Two methods of artificial bacterial transformation are used: heat shock and electroporation.

- Heat shock involves placing bacterial cells and a mixture of recombinant and non-recombinant plasmids in an ice-cold solution containing calcium ions and then rapidly increasing the temperature to disrupt the plasma membrane of the bacterial cells. The plasmids can then penetrate the plasma membrane and enter the bacteria.
- In electroporation, the bacterial cells and a mixture of recombinant and non-recombinant plasmids are subjected to an electrical current that alters the plasma membrane. Again, the plasmids are then able to enter the bacteria.

Very few of the bacterial cells will be transformed with recombinant plasmids. Some will take up the non-recombinant plasmids (plasmids without the target DNA) and others will not be transformed at all or will die from the heat shock or electroporation treatment (Figure 4.2.12). Even a small number of successful transformations is enough to grow a culture or population from.

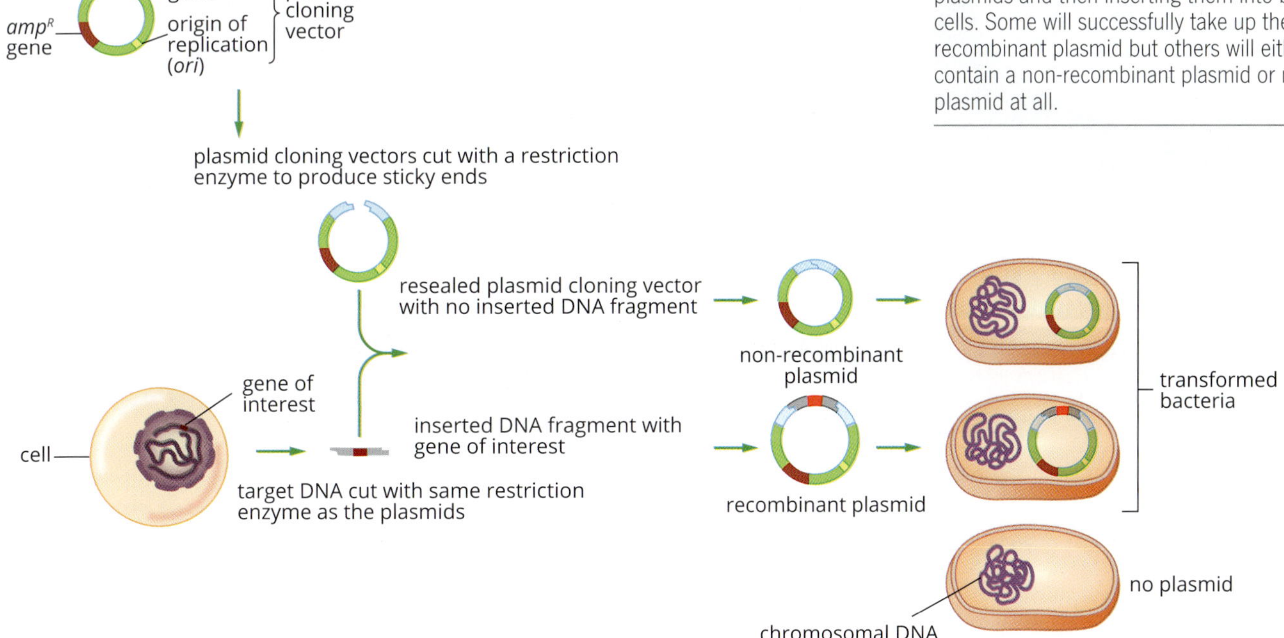

FIGURE 4.2.12 The process of bacterial transformation involves creating recombinant plasmids and then inserting them into bacterial cells. Some will successfully take up the recombinant plasmid but others will either contain a non-recombinant plasmid or no plasmid at all.

Bacteria require nutrients, such as nitrogen, carbohydrates and salt, to grow. Agar is the solid medium on which bacteria are grown (or cultured). The combination of agar and nutrients is called nutrient agar.

BIOFILE

The *lacZ* gene and X-gal

The *lacZ* gene is one of the genes in the bacterial *lac* operon. It codes for the enzyme β-galactosidase, which breaks down lactose into glucose and galactose. The *lacZ* gene is removed from the *lac* operon and inserted into plasmids to act as a reporter gene in recombinant DNA technology.

X-gal is a colourless synthetic compound with a very similar structure to lactose, so it fits in the active site of β-galactosidase and is broken down, releasing a blue reaction product. Using agar plates containing X-gal is an easy way to see whether cells produce β-galactosidase.

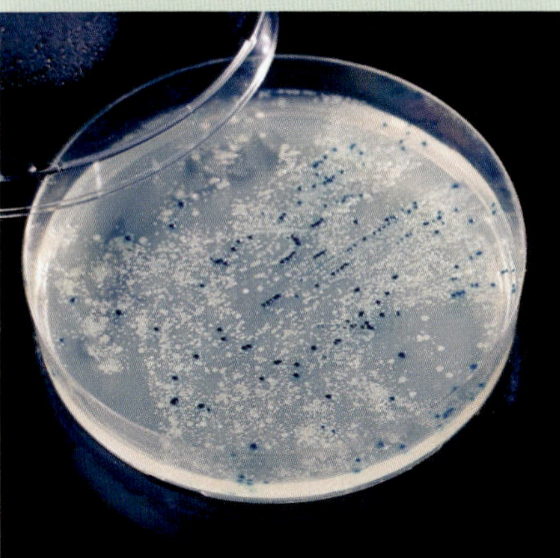

Growth of transformed bacteria on agar plates containing ampicillin and X-gal. Bacteria with the non-recombinant plasmid appear blue because the *lacZ* gene is expressed. Bacteria with the recombinant plasmid appear white.

Selection and screening of transformed bacteria

When determining which bacterial cells have been transformed with recombinant plasmids containing target DNA, the characteristics of the plasmid vectors described on page 153 become important. Recall from the earlier example (Figure 4.2.8 on page 153) that the plasmid vector contains other genes, including a gene for antibiotic resistance (in this example, ampicillin resistance) and a reporter gene that results in a particular phenotype, such as a coloured product.

Selection of transformed bacteria

To determine which of the bacterial cells have been transformed with the gene for antibiotic resistance, the bacteria are grown on nutrient agar plates that contain an antibiotic (in this case, ampicillin) and are incubated at 37°C, the optimum temperature for the bacteria to reproduce and form colonies. The only bacteria to survive will be those that have taken up the plasmid, whether it is a recombinant or non-recombinant plasmid. These bacteria have the ampicillin resistance (amp^R) gene. All other bacteria will be killed.

Screening for bacteria transformed with recombinant plasmids

In this example, the plasmid also carries the *lacZ* gene, which codes for an enzyme that breaks down an indicator called X-gal, resulting in a blue product. Bacteria carrying the non-recombinant plasmid with an intact and functioning *lacZ* gene produce blue colonies on agar plates. If the target DNA has been successfully inserted within the *lacZ* gene, expression is disrupted and the enzyme coded by this reporter gene is not produced. Therefore, bacteria transformed with recombinant plasmids appear as white colonies (Figure 4.2.13).

Bacteria transformed with the recombinant plasmids are then taken from the agar plate and cultured with nutrients in order for them to replicate and produce the protein (e.g. insulin) encoded by the target DNA.

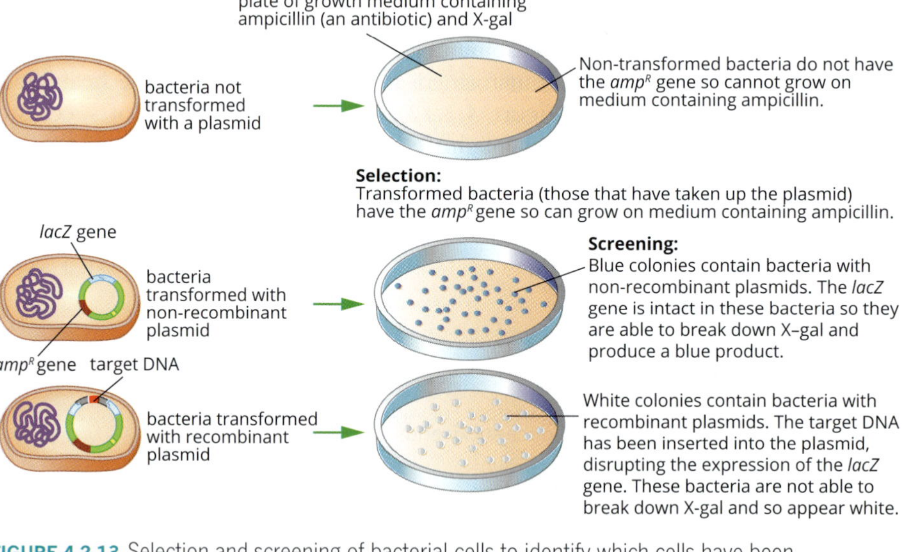

FIGURE 4.2.13 Selection and screening of bacterial cells to identify which cells have been transformed (contain a plasmid), and then which colonies contain recombinant plasmids with the target DNA and which contain non-recombinant plasmids

CASE STUDY **ANALYSIS**

Recombinant human erythropoietin

Red blood cell production is essential for maintaining oxygen homeostasis. A drop in oxygen supply to tissues (hypoxia) normally triggers the release of the protein erythropoietin (also known as EPO) from the kidneys. EPO promotes red blood cell production in the bone marrow to restore the oxygen-carrying capacity of blood and the delivery of oxygen to tissues (Figure 4.2.14). In chronic kidney disease, not enough EPO is made by the kidneys, resulting in low red blood cell counts and anaemia.

Recombinant human EPO for the medical treatment of this disease is produced in cultured mammalian cells. A copy of the human *EPO* gene is inserted into a plasmid which is introduced into mammalian host cells. EPO is a glycoprotein and must have the correct carbohydrates attached to the protein chain to function properly. Bacteria cannot do this, so mammalian cells must be used for making the recombinant protein (Figure 4.2.15). Recombinant EPO has also been developed for veterinary use, such as recombinant feline EPO for cats with chronic kidney disease.

Because EPO promotes red blood cell production and oxygen-carrying capacity, it has been used by athletes seeking an advantage. EPO has been at the centre of sports doping scandals in recent years, particularly in endurance sports such as cycling, long-distance running and triathlon. It has also been used in horse racing. The World Anti-Doping Agency (WADA) works with drug testing laboratories to develop and validate tests that can distinguish between EPO produced naturally in the athlete and pharmaceutical EPO.

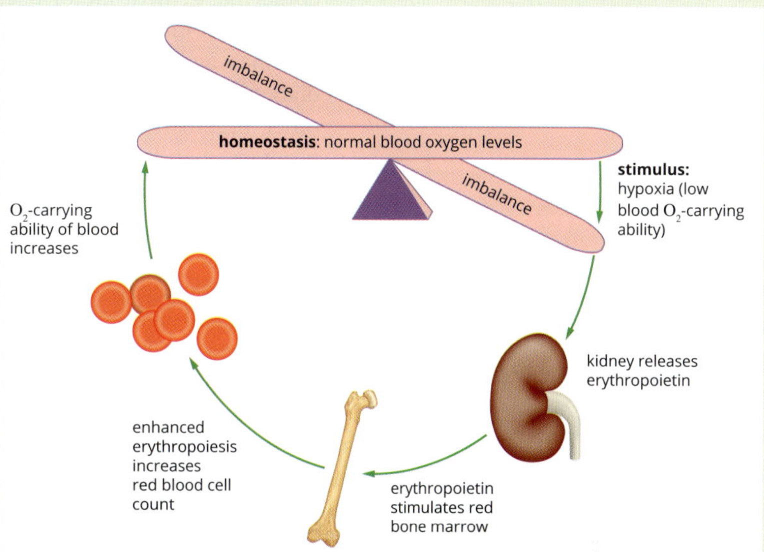

FIGURE 4.2.14 Erythropoietin is a protein released mainly by the kidney to maintain oxygen homeostasis. It promotes red blood cell production (erythropoiesis) in the bone marrow.

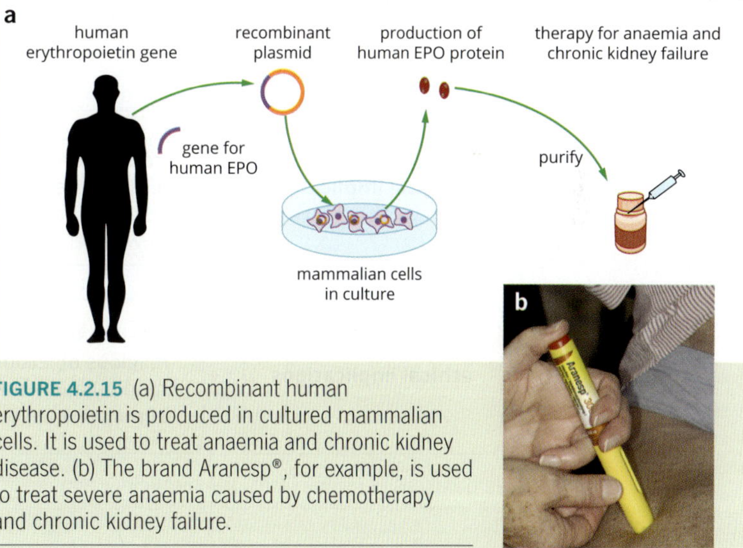

FIGURE 4.2.15 (a) Recombinant human erythropoietin is produced in cultured mammalian cells. It is used to treat anaemia and chronic kidney disease. (b) The brand Aranesp®, for example, is used to treat severe anaemia caused by chemotherapy and chronic kidney failure.

Analysis

1 Referring to Figure 4.2.14, how does normal kidney function compare with kidney function in chronic kidney disease, with respect to erythropoietin and oxygen levels?

2 Examine the text and identify the advantage of using cultured mammalian cells as opposed to cultured bacterial cells in the production of recombinant human erythropoietin.

3 Using Figure 4.2.14 as a reference, deduce what the effect of kidney disease might be on the body.

PROTEIN PRODUCTS OF RECOMBINANT DNA

Recombinant proteins are produced by introducing recombinant DNA into bacteria or eukaryotic cells and allowing them to synthesise the protein. The main types of proteins produced by this technology are hormones, cytokines, enzymes and vaccines for human therapeutic purposes. Therapeutic examples include epidermal growth factor used in the treatment of burns to improve the survival of skin grafts, interleukin-2 used in cancer treatment, antibodies for immunotherapy, vaccines against a number of viruses and to make insulin for the treatment of diabetes. This is much safer and more effective than using proteins purified from other organisms, such as insulin from pigs and growth hormone from human pituitary glands, as was done in the past. Industrial examples include enzymes such as amylase, lipase, protease and cellulase used in food processing, the textile industry and as detergent additives.

BIOFILE

Issues and implications surrounding recombinant DNA technologies

All technologies impact on people's lives and social structures. Social implications may affect people's financial position, lifestyle and reproductive decisions. Biological implications relate to the organisms used in and affected by the technology, its safety and short-term or lasting changes to human biology. Philosophical, moral and religious issues are also part of the impacts of biotechnologies.

Some of the implications, both negative and positive, of gene technologies for making therapeutic recombinant proteins are listed in the figure below.

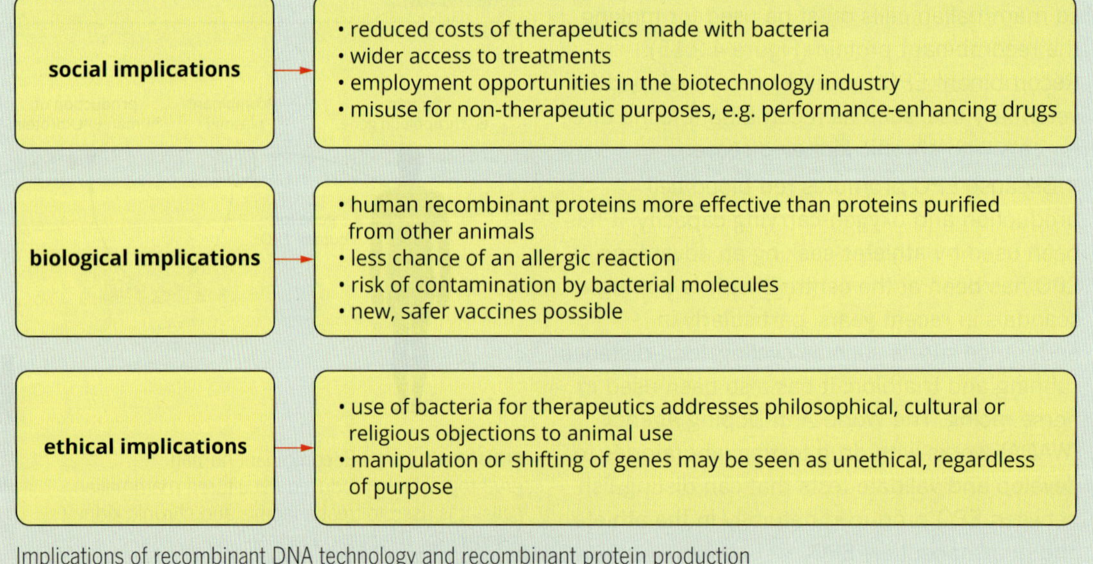

Implications of recombinant DNA technology and recombinant protein production

CASE STUDY

Natural transformers

Bacterial transformation is used in crop biotechnology to transfer genes into plants and thus introduce a desirable trait. Plants transformed in this way are called transgenic. The bacterium *Agrobacterium tumefaciens* is commonly used as it has the ability to infect plant tissue and incorporate specific parts of its plasmid into the host plant's DNA.

The study and use of *A. tumefaciens* has a long history. This common soil bacterium was identified as the cause of the tumours of crown gall disease over 100 years ago. A plasmid that can move from the bacterium into plant cells was later identified as the key agent. This is a natural transformation of plant cells by a bacterial plasmid. The plasmid in *A. tumefaciens* is called the Ti or tumour-inducing plasmid (Figure 4.2.16). It has genes that direct the movement of the plasmid from bacterium into plant cells and for insertion into the chromosomes of the plant. It also has genes for growth-promoting plant hormones that cause the tumours.

This plasmid is used for genetic engineering. The genes that cause tumour growth are removed, while the genes that ensure transfer from bacterium to plant cell remain. A foreign gene inserted into this plasmid can then be transferred into plant cells. (Figure 4.2.17a). *A. tumefaciens* inserts a section of its plasmid (containing the desired gene) into the plant cell, which is then incorporated into the plant cell DNA. This plant tissue is cultured (Figure 4.2.17b) to create a transgenic plant that expresses the foreign gene. Genes for traits such as insect resistance, herbicide resistance, and tolerance to salinity and frost are all being used in crop biotechnology.

In Australia, transgenic cotton is created using bacterial transformation to confer insect resistance. You will learn more about the production of transgenic plants such as wheat in Section 4.3.

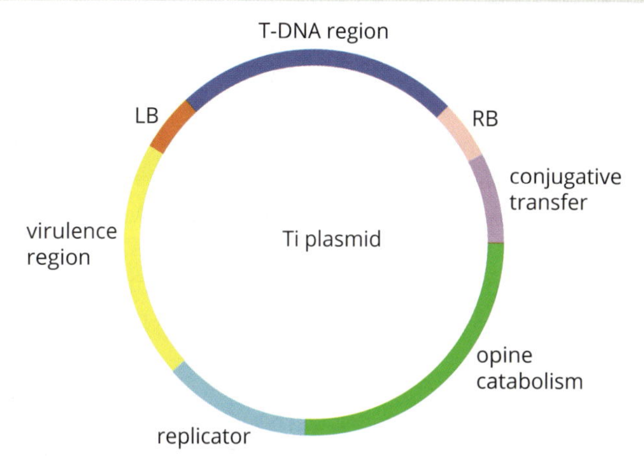

FIGURE 4.2.16 Ti plasmid from *Agrobacterium tumefaciens*. The T-DNA region carries the genes that cause tumours in crown gall disease. This region is removed and replaced with a foreign gene, making a recombinant plasmid.

FIGURE 4.2.17 (a) Plant material is exposed to *Agrobacterium* to allow transfer of recombinant plasmid from bacteria to plant cells. (b) The plants carrying the new gene are cultured and selected in the laboratory before release for field testing of the new characteristics acquired by gene transfer.

4.2 Review

SUMMARY

- Restriction enzymes (endonucleases) are enzymes that cut DNA at particular recognition sites.
  - Sticky-end restriction enzymes leave fragments with overhanging ends that have exposed bases.
  - Blunt-end restriction enzymes cut DNA to leave flat-ended fragments.
- DNA and RNA ligase enzymes permanently join fragments of DNA or RNA together in a process called ligation.
- Plasmids are small, circular pieces of double-stranded DNA found in bacterial cells. They replicate independently of the bacteria's chromosomal DNA.
- Recombinant plasmids have had target DNA inserted into them. The same sticky-end or blunt-end restriction enzyme is used to cut both the targeted gene and the plasmid, and then DNA ligase is used to permanently join the two together.
- Plasmids with antibiotic resistance are often used to enable identification of bacterial transformation, as only bacterial cells containing these plasmids will survive when grown in cultures containing the antibiotic.
- A reporter gene produces an identifiable phenotype, such as a coloured product or fluorescence, to identify transformed bacterial cells.
- There are three steps in bacterial transformation:
  - Gene uptake: Bacterial cells are induced to take up the recombinant plasmids either by heat shock or electroporation methods.
  - Selection of transformed bacteria: The bacteria are grown in the presence of an antibiotic. Bacterial cells that have been transformed will survive, as the gene for antibiotic resistance is located in the plasmid.
  - Identification of transformed colonies: Transformed bacteria containing recombinant plasmids are identified by the reporter gene in the recombinant plasmid.

KEY QUESTIONS

Knowledge and understanding

1 a Restriction enzymes (endonucleases) are a basic molecular tool in gene technology. What is a restriction enzyme and what can it do?
  b Describe the difference between sticky ends and blunt ends produced by restriction enzymes.

2 Define the following terms and give an example of each:
  a gene cloning
  b recombinant DNA

3 What is a plasmid? Describe the role played by plasmids in gene cloning.

4 Outline the purpose of DNA ligase in recombinant DNA technology.

5 a Explain what is meant by genetic transformation.
  b List three protein products manufactured using genetic transformation and outline their importance in medicine or agriculture.

6 Give two features of a plasmid that are important for identifying cells containing a recombinant plasmid.

7 Describe how antibiotics are used to select transformed bacteria.

8 Draw a table listing a positive and a negative aspect of recombinant DNA technology for making human proteins. Include social, ethical and biological issues.

Analysis

9 Three restriction enzymes and their recognition sites are illustrated below.

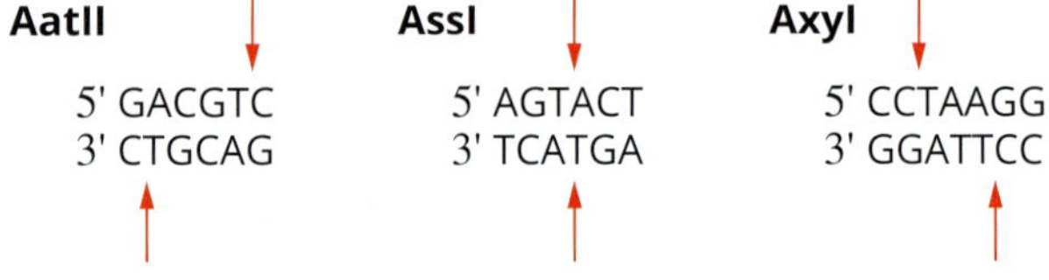

Consider the DNA sequence below.

5' AGT AGT ACT GAC GCC TAA GGA GTA CTG 3'
TCA TCA TGA CTG CGG ATT CCT CAT GAC

How many pieces would it be cut into if it was mixed with:
  a AatII only
  b AssI only
  c all three restriction enzymes

10 Draw a flow diagram outlining the different techniques and processes involved in insulin production. Use diagrams where possible to assist your explanations.

4.3 Genetically modified and transgenic organisms

Humans have used selective breeding to produce animals and plants with more useful or more attractive characteristics for tens of thousands of years. Animals or plants that expressed the desired characteristics were chosen and selectively bred, in the hope that their offspring would inherit these characteristics. In the past, selective breeding could only utilise characteristics that already existed in the gene pool of a species. We now have the knowledge and skills to use DNA manipulation techniques to alter the genetic material of an organism (Figure 4.3.1). This technology could provide many benefits, but it may also lead to questions about whether its impacts are biologically, socially or ethically acceptable.

FIGURE 4.3.1 Genetically modified cotton on the left shows its insect resistance compared to the non-modified cotton on the right, which has been ravaged by insects.

GENE EDITING WITH CRISPR-Cas9

Organisms may have their genome modified by directed mutation (mutagenesis) or by newer technologies called **gene editing**. An example of gene editing technology is a new technique called **CRISPR-Cas9**, which can cut DNA at specific locations. CRISPR stands for 'clustered regularly interspaced short palindromic repeats', which means segments of DNA with short repetitive sequences that are interspersed with unique DNA sequences.

Function of CRISPR-Cas9 in bacteria

CRISPR arrays consist of fragments of viral DNA that the bacterium has captured from invading viruses. If the same virus attacks again, the bacterium uses the CRISPR arrays to transcribe RNA sequences that are complementary to the virus's DNA sequences. The RNA sequences target the virus's DNA and cut it using endonuclease enzymes, disabling the virus. The enzyme that cuts the viral DNA is a CRISPR-associated protein called Cas9.

Application of CRISPR-Cas9 in editing genomes

The CRISPR-Cas9 system can be used to edit eukaryotic genes by combining the Cas9 enzyme with **guide RNA (gRNA)**. Scientists artificially create the gRNA, which consists of a short length of genetic code that is complementary to the target DNA sequence. The gRNA guides the Cas9 enzyme to the target site and the enzyme cuts the DNA (Figure 4.3.2). Researchers can then use the eukaryotic cell's own DNA repair mechanisms to alter the DNA, either by repairing base pair deletions or insertions that are known mutations, or by inserting new DNA sequences.

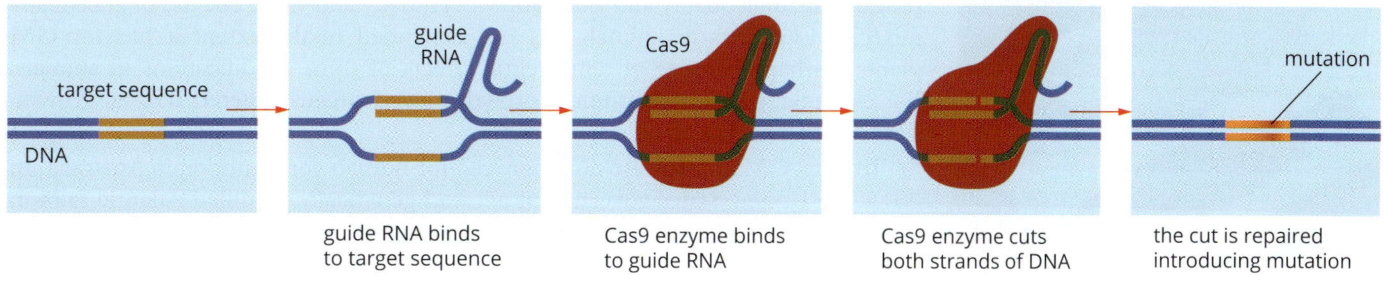

FIGURE 4.3.2 The CRISPR–Cas9 editing tool

Research with CRISPR-Cas9 in humans has already started with cancer genes and editing genes in embryos.

The current understanding of the Cas9–gRNA system is limited. The success rate for the number of genes or cells edited is not well known. Also, recent developments have shown that not every position in the gRNA needs to match the target DNA, resulting in off-target (unintended) sites being edited. Currently, research regarding CRISPR-Cas9 efficiency is inconclusive. There is a wide range of success rates for three different genes (13% to 43%), and success rates in producing transgenic mice embryos range from 2% to 88%.

The use of CRISPR-Cas9 to edit the genome in germline cells (eggs or sperm) raises bioethical concerns. If genetic modification occurs in embryos, not all the cells may carry the edited copy of the gene, but any changes made to eggs or sperm will affect all the cells in the embryo that develops after fertilisation. Any changes made will be passed on from generation to generation. Currently, the use of reproductive cells for CRISPR-Cas9 studies is illegal in most countries but patent rights to the technique are being contested by several companies.

Most of the gRNA sequences currently being tested are only around 20 nucleotides long. Such a short sequence is likely to exist somewhere else in the genome that is unrelated to the target site. This means crucial areas of other genes may also be unintentionally edited. The risks currently associated with the technology mean that it will be many years before CRISPR-Cas9 will be routinely used in humans.

GENETICALLY MODIFIED AND TRANSGENIC ORGANISMS IN AGRICULTURE

Genetically modified organisms (GMOs) have had their genetic material (DNA) altered in some way.

Transgenic organisms have had genes from another species inserted into their genetic material (DNA).

Over the last few decades, techniques were developed that allowed for the alteration of an organism's genome and for the transfer of genes from one organism to another. Organisms that have had their genetic material altered are known as **genetically modified organisms (GMOs)**. Because the DNA code is universal, almost any gene transferred from one organism to another will express the protein that it expressed in the original organism. This means that a desirable characteristic seen in one animal or plant could be transferred to another organism lacking this characteristic. **Transgenic organisms** are GMOs that have had a gene from another species inserted into their genome. The gene that came from another species is called a **transgene**. Transgenic organisms can be used in agriculture to increase crop productivity and provide resistance to disease.

Genetically modified animals

FIGURE 4.3.3 Transgenic salmon have been cleared for human consumption in the USA.

In agriculture, transgenic cows and sheep are used for improved fertility, meat production, milk quality and yield, and wool quality and yield. The use of genetically modified (GM) farm animals has not expanded to the extent it has for GM plants, perhaps because of detrimental effects of some modifications in animals. For example, genes that promote growth may also cause altered skeletal growth, arthritis, and heart and kidney problems.

To date, GM animals are not approved for human consumption in Australia. In 2015, the United States government allowed genetically modified Atlantic salmon (Figure 4.3.3) to be used for human consumption. A gene from another salmon species, along with a promoter sequence from a fish called a pout, means that the transgenic salmon eat all through the year, not just when the water temperature is warm. This increases the growth rates of these fish dramatically and means they are ready for harvest much sooner than non-modified Atlantic salmon. The eggs of the GM salmon are treated to create infertile adult fish (99% of the adults are reported to be sterile), reducing the chances of interbreeding with wild salmon if they escape from their pens. This is the first genetically modified animal of any type to be cleared for human consumption in the USA.

CASE STUDY ANALYSIS

GM mosquitoes

Genetically modified mosquitoes are being used for disease control. Some mosquito species are vectors of disease-causing viruses or protozoa. Mosquitoes of the *Aedes* genus are vectors for several disease-causing viruses, including the yellow fever virus, the dengue virus and the Zika virus. A biotechnology company has developed genetically modified mosquitoes that carry a dominant lethal gene for the purpose of reducing the population of mosquitoes carrying the viruses.

Males carrying the lethal gene are released into the wild, where they mate with normal 'wild type' females and pass the lethal gene on to their offspring. The offspring die as larvae. The DNA used to make the genetically modified mosquitoes also has a reporter gene for red fluorescent protein, enabling scientists to easily identify the adults and larvae carrying the lethal gene (Figure 4.3.4).

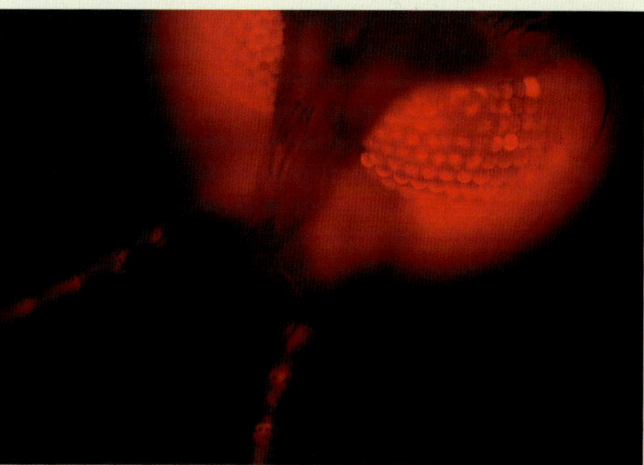

FIGURE 4.3.4 The gene for red fluorescent protein has been linked to a lethal gene to enable researchers to identify genetically modified mosquitoes.

Field trials using genetically modified *Aedes aegypti*, the vector of dengue virus, have been conducted in the Cayman Islands, Brazil and Panama, where dengue fever is a widespread and serious health problem. The recent outbreaks and rapid spread of Zika virus, which is also transmitted by *Aedes* mosquitoes, has prompted trials of these genetically modified mosquitoes in areas of Brazil affected by outbreaks of Zika virus.

Another approach to mosquito control is the release of sterile insects. Mosquitoes of the *Anopheles* genus transmit the malaria parasite *Plasmodium*. Male *Anopheles* mosquitoes have been genetically modified with genes expressed in the testes that cause the males to be unable to make sperm, so they are sterile. Female *Anopheles* mosquitoes mate only once, so mating with a sterile male limits population growth. The genes causing the sterility are linked to a reporter gene for green fluorescent protein for easy identification of the genetically modified insects (Figure 4.3.5). The aim of the research is to reduce the populations of mosquitoes that carry and transmit *Plasmodium* sp. and thus reduce the incidence of malaria, a serious health problem in many developing countries.

FIGURE 4.3.5 Internal reproductive organs of a genetically modified male *Anopheles gambiae* mosquito. The testes (T), where sperm cells develop, are fluorescent green due to the expression of a green fluorescent protein (GFP) that is linked to the genes causing sterility. The male accessory glands (M) that produce seminal secretions are not expressing GFP.

Analysis

1. What are the advantages and disadvantages of using this genetic approach to controlling insect vector populations?
2. Suggest alternative strategies to the use of transgenic mosquitoes to combat the spread of malaria. Do you think that these would be as effective and safe as transgenic organisms?

Genetically modified plants

GM crops are used in agriculture to increase crop productivity, provide resistance to insect predation and prevent disease. Several GM crops have been developed or are grown in Australia. For example, insect-resistant GM cotton has been grown since 1996, and herbicide-tolerant GM canola was approved for commercial production in Victoria in 2008. In Australia, the Office of the Gene Technology Regulator (OGTR) assesses all GM animals or plants before research, agricultural or commercial use.

Techniques for producing transgenic plants

Transferring a gene into plant cells can be limited by the presence of the cell wall. The introduction of foreign genes into plants is usually done by using a biological vector. One method utilises *Agrobacterium tumefaciens*, a soil bacterium that is able to naturally transfer a plasmid into plant cells (Figure 4.3.6). *Agrobacterium* normally causes crown gall disease because it carries a plasmid with genes that cause the growth of a tumour. A recombinant plasmid (the vector), carrying a desired gene from a different species but lacking the tumour-inducing genes, is introduced into *Agrobacterium* cells. When the transformed *Agrobacterium* is cultured with plant cells, the recombinant plasmid is transferred into the plant cells. These transformed plant cells are then grown in tissue culture into new plants for transplanting into the field as a transgenic crop (Figure 4.3.7).

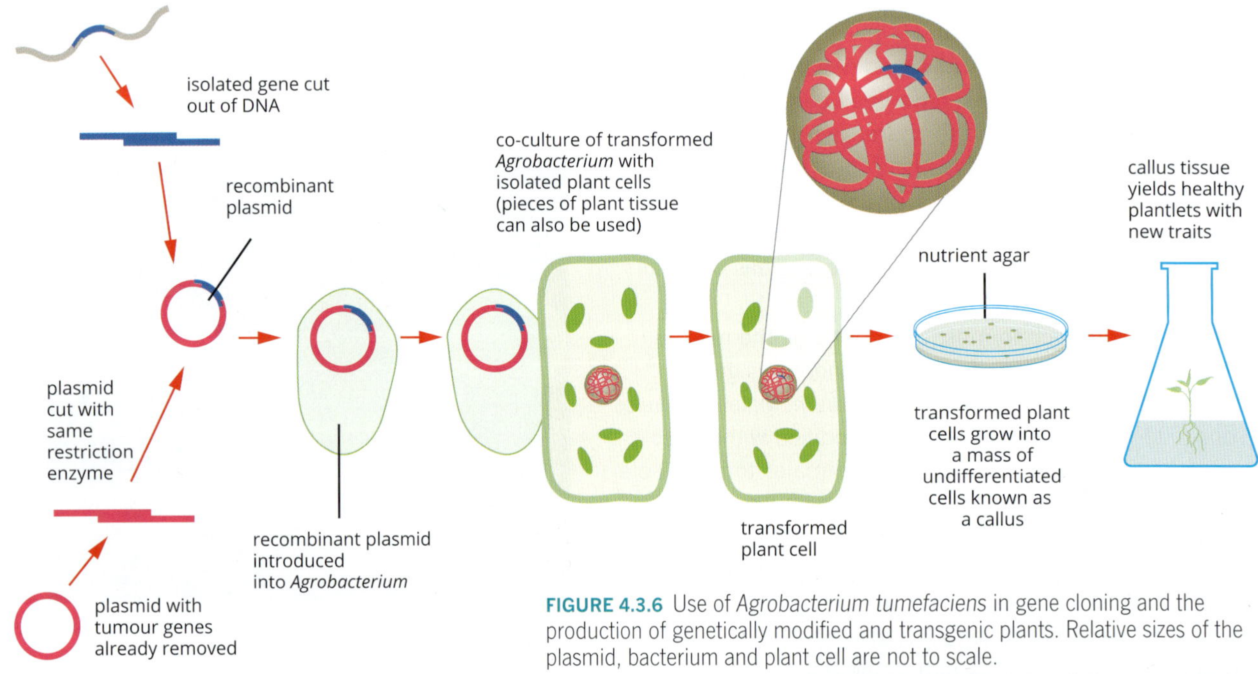

FIGURE 4.3.6 Use of *Agrobacterium tumefaciens* in gene cloning and the production of genetically modified and transgenic plants. Relative sizes of the plasmid, bacterium and plant cell are not to scale.

FIGURE 4.3.7 Plant cells carrying the new gene are grown in tissue culture and selected in the laboratory for field testing.

Increased crop productivity—salt tolerant wheat

Soil salinity is a major problem for Australian agriculture. A high level of sodium salts in the soil leads to osmotic water loss from roots and other tissues in which salt accumulates. Cells are stressed due to the altered ratios of sodium and potassium ions in cells. Salt-tolerant plants protect themselves from the effects of salinity by preventing sodium entry into cells, storing the salt in the vacuole, or pumping the sodium out of the cells. Molecular biologists have found the genes that control these features of salt-tolerant plants.

To increase crop productivity, Australian scientists from the University of Adelaide introduced a gene from a salt-tolerant Australian native plant into wheat plants. This greatly improved the grain yield of wheat grown on salty soils without affecting grain yield in normal soil (Figure 4.3.8). The salt-tolerant gene codes for a protein that removes sodium from the leaves, allowing water to move normally from the roots to the leaves. This increases the geographical range that can be used for wheat production in Australia and other countries facing salinity problems, which is becoming increasingly important as the global population grows.

FIGURE 4.3.8 Australian scientists have produced wheat plants that can grow in salty soil.

Disease resistance

In Ireland in the 1840s the introduction of a plant pathogen, *Phytophthora infestans*, resulted in the Irish potato famine, in which 1 million people died and 1.5 million people emigrated. The pathogen causes late blight in potato and tomato crops and is still a major problem, despite first being documented almost 180 years ago. It has been controlled with some success through the use of fungicides, but still causes significant financial loss. Researchers in the United Kingdom identified a gene for resistance to the disease in American black nightshade (a wild relative of the potato). The researchers were able to successfully insert the gene into potatoes, creating resistance to late blight in the genetically modified potato crop without the need for fungicides.

BIOFILE

Pharming for spider silk

Some farm animals are being used for the production of therapeutic proteins, such as antibodies that are difficult to make in bacteria and cultured cells. This process has been referred to as 'pharming' (combining the words 'farming' and 'pharmaceutical'). The products are released into blood or milk and they can be readily extracted from there.

Spider silk is strong and flexible and has many potential applications. Transgenic goats have been bred to produce spider silk protein in their milk.

Spider silk protein (figure at right) is an example of a potentially useful product made in transgenic goats. The gene for spider 'dragline' silk has been put into the genome of goats, along with regulatory genes, so that it is expressed in the milk. Spider silk is of great interest for its extraordinary strength and flexibility. Potential applications include a biopolymer for artificial ligaments and tendons, bandages, biodegradable bottles and tough bulletproof clothing.

CASE STUDY

Bt cotton

Cotton is a plant that attracts many insect pests. To protect the cotton crops, they are sprayed with insecticides up to four times before the crop is harvested. This high use of insecticides impacts the populations of both harmful and beneficial insects, and of the animals that feed on them. Insecticides may also have an impact on human health. In addition, insecticides are expensive.

Bt cotton is a transgenic crop that has been modified to contain two genes from the soil bacterium *Bacillus thuringiensis*. Expression of these genes produces proteins in the cotton plant that kill the main caterpillar pest of cotton by disrupting its digestive system. In Australia, almost all cotton grown is Bt cotton and this has reduced the use of pesticides dramatically. This decreases the environmental impacts of pesticides and reduces costs for farmers. Australian regulators have reported no adverse effects over 15 years of Bt cotton use in Australia. Cotton seed oil extracted from Bt cotton can be sold without GM labelling as the extraction processes separate the oil from the plant's proteins and nucleic acids, therefore the oil does not have any GM components.

BIOFILE

Issues and implications surrounding GMOs

GMOs are controversial for various reasons. Debates surrounding their biological, social and ethical implications are common in the scientific and general media. While the technology has many potential benefits, there is also the uncertainty and risk that comes with any developing technology.

The figure below summarises some common social, biological and ethical implications arising out of the use of genetically modified and transgenic organisms.

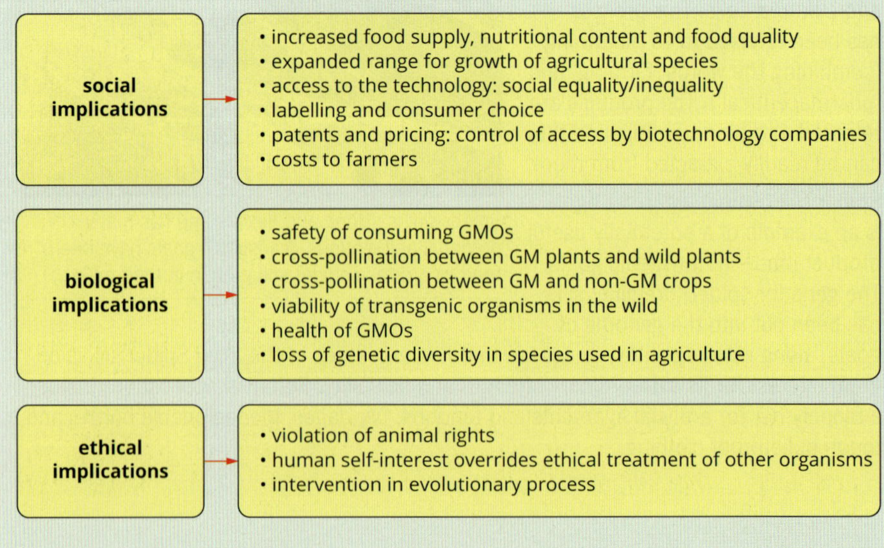

Summary of issues arising from the use of genetically modified and transgenic organisms

4.3 Review

SUMMARY

- CRISPR stands for 'clustered regularly interspaced short palindromic repeats'. In bacteria, CRISPR arrays consist of fragments of viral DNA that the bacterium has captured from invading viruses.
- CRISPR-Cas9 is used by bacteria to target an invading virus's DNA and cut it using endonuclease enzymes, disabling the virus. The enzyme that cuts the viral DNA is a CRISPR-associated protein called Cas9.
- The CRISPR-Cas9 system can be used to edit eukaryotic genes by combining the Cas9 enzyme with guide RNA (gRNA) to cut DNA at specific locations. The eukaryotic cell's own DNA repair mechanisms can then be used to alter the DNA, either by repairing mutations, or by inserting new DNA sequences.
- Organisms that have had their genetic material altered are known as genetically modified organisms (GMOs).
- Transgenic organisms are GMOs that have had a gene from a different species inserted into their genome.
- Genetically modified and transgenic animals are used in research, disease control, medicine and biomolecule production.
- *Agrobacterium tumefaciens* and plasmid transfer is an established method of transferring genes into plant cells.
- Genetically modified and transgenic plants are used in agriculture to provide varieties that resist insect attack, are herbicide resistant or have improved yield or nutritional content.
- A range of biological, social and ethical issues arise from the application of GMOs.

KEY QUESTIONS

Knowledge and understanding

1 Which Australian regulatory body oversees the development, use and commercial or medical introduction of genetically modified organisms?

2 a What does genetic modification of an organism mean?
  b Describe a successful application of genetic modification in agriculture in recent years.
  c Describe a genetically modified organism and compare it to a transgenic organism.

3 Describe how genetic modification can be useful as a tool to fight vector-borne disease, such as a disease carried and transmitted by an insect vector.

4 The following diagram illustrates two model organisms used in research and the molecular procedures being used to alter a genetic characteristic. State whether the resulting organism is genetically modified, transgenic or both.

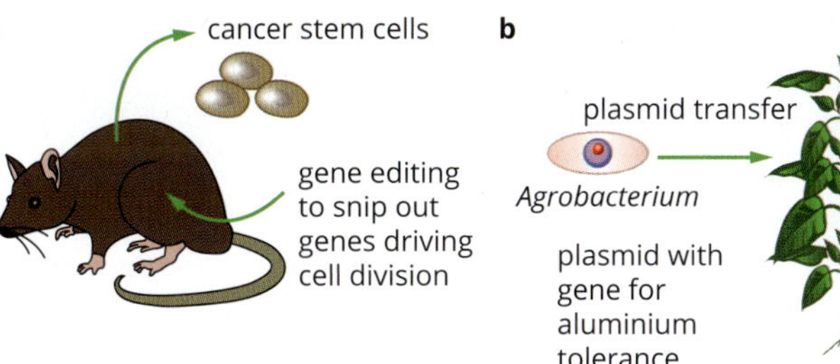

5 In your notebook, draw up a table like the one below to identify what you consider to be the key issues surrounding the use of genetically modified organisms. List the points in categories that you consider to be positive or negative aspects of the technology.

	Pros/Positives	Cons/Negatives
social		
biological		
ethical		

6 From the cases described in this section, identify one example of:
  a a genetic modification that leads to a reduced environmental impact
  b a modification that has potential commercial opportunity for increased production or a new product
  c a modification that impacts on both the viability of an animal and the benefit to human health

Analysis

7 Explain why using guide RNA sequences (gRNA) that are 20 bases long for the CRISPR-Cas9 technique is problematic. Consider a solution.

8 Consider the implications of using transgenic animals and suggest why this technology may be opposed by some religious or cultural groups.

Chapter review

OA ✓✓

04

KEY TERMS

allele
anneal
bacterial competence
bacterial transformation
bacteriophage
blunt-end restriction enzyme
complementary DNA (cDNA)
CRISPR-Cas9
DNA amplification
DNA ladder
DNA ligase
DNA polymerase
DNA profiling
DNA sequencing
DNA thermocycler
endonuclease
gel electrophoresis
gene cloning
gene editing
genetic transformation
genetically modified organism (GMO)
guide RNA (gRNA)
inducer
lacZ gene
ligase
ligation
microsatellite
palindromic sequence
plasmid
polymerase
polymerase chain reaction (PCR)
polymorphism
primer
recognition site
recombinant DNA
recombinant plasmid
regulatory gene
reporter gene
restriction enzyme
reverse transcriptase
RNA ligase
RNA polymerase
short tandem repeat (STR)
sticky-end restriction enzyme
Taq polymerase
target DNA
transgene
transgenic organism
vector

REVIEW QUESTIONS

Knowledge and understanding

1 Indicate which step in the process of PCR best describes annealing.

A binding the primers
B adding the polymerase
C separating the DNA strands
D building the complementary DNA strands

2 What is reverse transcriptase used to make?

3 A cloned organism:

A increases biodiversity
B is an identical copy of its parent
C can only be a plant
D all of the above

4 The fruit fly, *Drosophila melanogaster*, is commonly used for genetic research. One particular mutation results in the deletion of a section of DNA 200 bp long from one particular gene. The gene was extracted from a fly that is homozygous for the mutant gene and the same gene was extracted from a fly that is homozygous for the wild type (normal) version of the gene. Both versions of the gene were amplified using PCR and then run through gel electrophoresis.

Determine which gel (below) most accurately shows the PCR products.

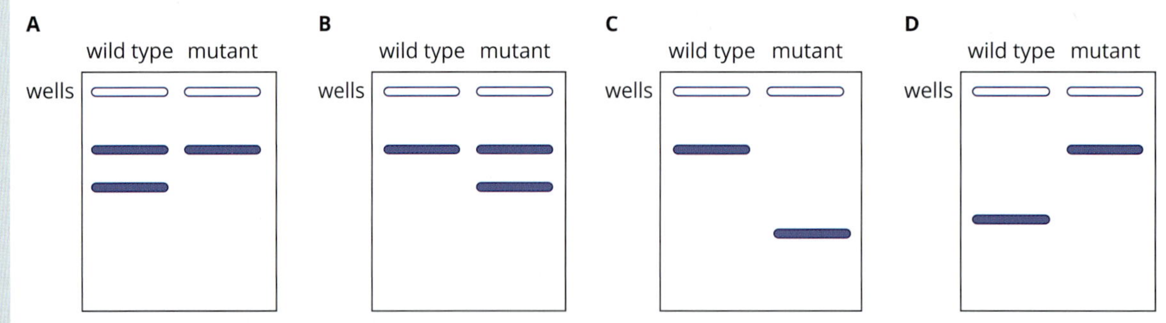

5 Genes such as the *lacZ* gene can be used as reporter genes. What are reporter genes used to determine?

6 Some students are doing an experiment involving bacterial transformation. Bacteria were incubated with plasmids containing resistance to the antibiotic ampicillin and then grown on agar plates. A plate that will have only transformed bacteria growing will have which of the following?

A nutrient agar only
B nutrient agar and ampicillin
C plain agar with ampicillin
D nutrient agar, ampicillin and penicillin

7 Define 'transgenic organism'.

8 One result of a genetic application was the Flavr Savr tomato. Tomatoes have a short shelf life due the effects of an enzyme called polygalacturonase. This enzyme catalyses the breakdown of the cell walls of the tomato, causing the tomatoes to become soft and unappetising. To slow down this process, the sequence of the polygalacturonase gene was determined and an antisense gene was produced. The antisense gene has a complementary nucleotide sequence to the polygalacturonase gene. The antisense gene was inserted into the tomatoes. When the antisense gene is transcribed, the mRNA produced is complementary to the mRNA for the polygalacturonase gene, so the two mRNAs join to form double-stranded mRNA. Double-stranded mRNA cannot be translated, so the enzyme is not formed and the cell walls are not broken down. The Flavr Savr tomatoes can be considered to be:

A only transgenic
B only genetically modified
C both genetically modified and transgenic
D none of the above

9 Before any gene can be inserted into bacteria to make proteins for human use, the number of copies of the gene must be increased. In order to do this a process called PCR is used.

a **i** What do the letters PCR stand for?
ii State the role of the 'P' in the process.
iii Identify the source of the 'P' used in this process. Why is that particular source used?

b One particular PCR machine uses the following sequence: heat to 94°C for 1 minute, cool to 56°C for 1.5 minutes, then heat to 72°C for 1.5 minutes.
i Describe what is happening at each stage.
ii How long would it take to obtain 8000 copies of the target DNA?

10 What is the purpose of a DNA ladder in gel electrophoresis?

11 How might DNA profiling be used in the legal system?

12 One of the purposes of recombinant DNA technology is to produce large quantities of proteins for therapeutic use. To do this, the gene that codes for protein production is inserted into a plasmid, prior to being placed in a bacterial cell.

a State three reasons why plasmids make good vectors for protein production.
b State three reasons why bacteria make good host organisms for the plasmids.

13 Arsenic contamination of soil is a serious problem in some countries. The arsenic contaminates groundwater and drinking wells. The flow chart below illustrates a process used to insert a gene that enables plants to absorb arsenic from the soil. Name the objects that are labelled A–E.

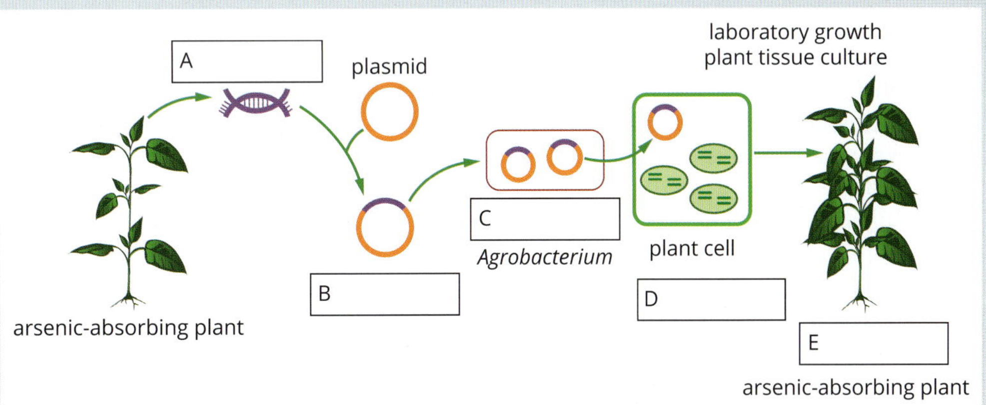

14 **a** What types of human proteins are commonly produced by recombinant DNA technology?
b Suggest an advantage of this method of production compared to a traditional approach.

15 Briefly describe the technical and ethical issues related to DNA profiling.

16 **a** Where do CRISPR arrays come from?
b Outline the steps involved in using CRISPR-Cas9 to edit genes.
c Describe two bioethical issues that arise from the use of CRISPR-Cas9

Application and analysis

17 Members of a family with a history of cystic fibrosis (CF) underwent genetic testing to determine whether they carried the common ΔF508 mutation. DNA samples obtained from cheek cells were analysed by PCR using primers specific for the ΔF508 region, followed by gel electrophoresis. The normal allele yields a 98 bp DNA fragment. The mutant allele yields a 95 bp DNA fragment.

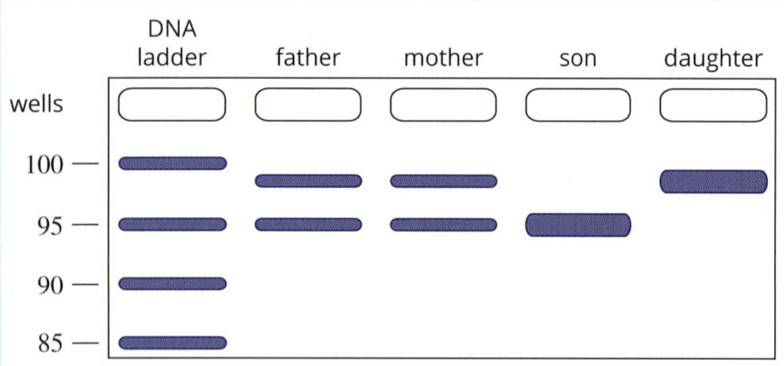

a How is PCR able to identify the allele responsible for cystic fibrosis?
b Describe the purpose of gel electrophoresis in this type of genetic testing.
c The parents are carriers of CF. Explain how the PCR and gel electrophoresis results show this.
d What does the genetic test show about the son?
e The daughter gets a cold and chest infection every winter. Is this likely to be related to the lung congestion seen in CF?

18 A couple wishes to find out if their unborn child has sickle cell anaemia. The figure top right shows the results from the gel electrophoresis of the restriction fragments of the sickle cell gene (located on chromosome 11) for the family. The mother carries the mutation, which results in sickle cell anaemia, while the father does not carry the mutation.
a How does the use of restriction enzymes in the analysis of the alleles of a gene result in different banding patterns?
b Does the child carry the mutation for sickle cell anaemia? Explain your answer.

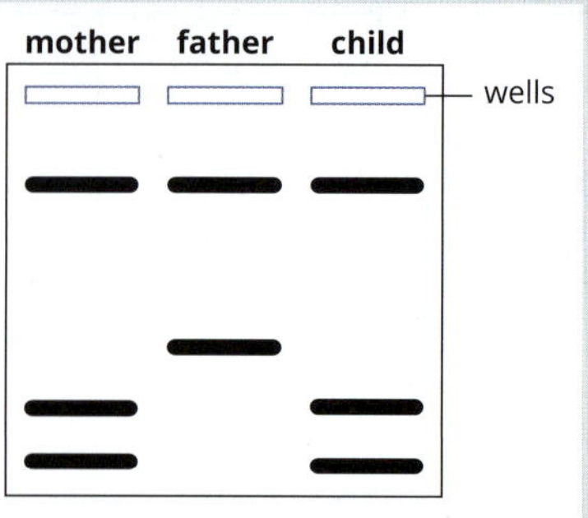

19 Some genetic disorders are sex-linked, meaning that they present differently depending on whether individuals are male or female. Males have an X and a Y chromosome, while females have two X chromosomes. Duchenne muscular dystrophy (DMD) is a sex-linked genetic disease caused by the deletion of part of the sequence of the dystrophin gene, which is located on the X chromosome. The dystrophin protein is very large. The normal gene is 2 220 390 bp long and contains many exons and introns. The total length of the coding sequences (the exons) is 11 055 nucleotides.
a How many amino acids are in the normal protein?
A young woman who has a family history of DMD is about to start trying to have a family. Her partner's family has no history of the condition.
b The first step the young woman takes is to be tested for the condition. How could she have the genetic change and not know?
c The relevant sections of the X chromosomes of the young woman were cut using an appropriate restriction enzyme and then run on an electrophoresis gel. A set of DNA standards was also run. The gel is shown below.

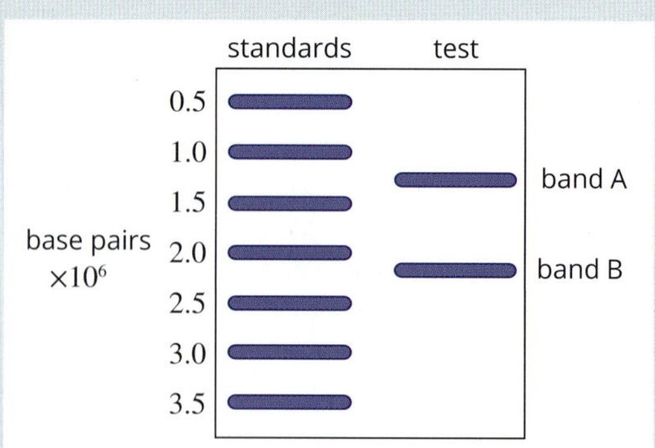

i Explain whether the woman has the DMD mutation.
ii What are the approximate lengths of the pieces of DNA represented by bands B and A?

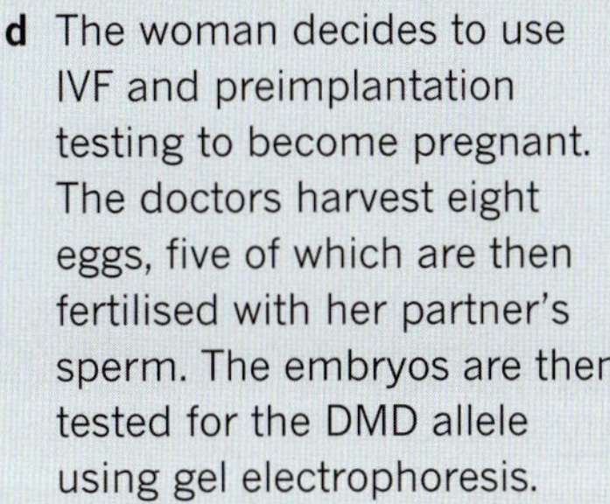

d The woman decides to use IVF and preimplantation testing to become pregnant. The doctors harvest eight eggs, five of which are then fertilised with her partner's sperm. The embryos are then tested for the DMD allele using gel electrophoresis.

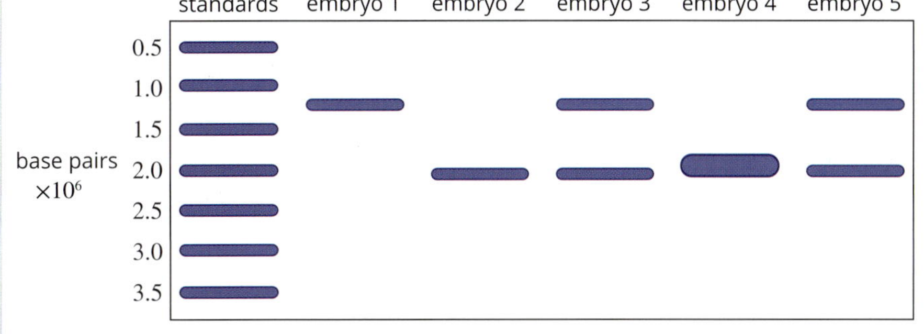

   i Explain which embryos are male.

   ii Explain which embryos are most suitable for implantation.

20 The diagram below represents a linear DNA molecule and shows the position of the recognition sites for a number of restriction enzymes.

a Examine the diagram above and determine how many fragments will be produced when the DNA molecule is cut with EcoRI.

b Examine the diagram above and determine how many fragments will be produced when the DNA molecule is cut with AatII.

c Examine the diagram above and determine how many fragments will be produced when the DNA molecule is cut with NotI.

21 Individuals with haemophilia carry a gene mutation in the *F8* gene, which codes for factor VIII, a clotting factor critical in the formation of blood clots that form after an injury. Blood clots temporarily close off injured blood vessels to stop bleeding, until the vessels are healed by the body. Individuals with haemophilia are at a significantly increased risk of bleeding after injury because they are missing this important clotting factor.

Before the availability of recombinant DNA technology, factor VIII for the management of haemophilia was obtained from blood serum provided by blood donors. Large volumes of blood serum were required in order to obtain sufficient factor VIII for treatment. In addition, the risks of transmitting other blood proteins and viruses (including HIV) was high. Many men with haemophilia contracted HIV as a result of this treatment.

Today, haemophilia is a well-managed disease thanks to the advances in recombinant DNA technology.

Discuss three advantages of using recombinant DNA technology to artificially produce factor VIII.

22 Bacteria are commonly genetically engineered to produce human proteins. The DNA sequence for the gene of one of these proteins is shown below.

5' – GCT TCT TCC CGT GCA TAT AGA TAC TCT GAA ACA CTG TGC GGC GGT GAA CTG
3' – CGA AGA AGG GCA CGT ATA TCT ATG AGA CTT TGT GAC ACG CCG CCA CTT GAC

... many base pairs ... CTG TGC ACC TAT TGT GCT ACT CCC GCA AAG TCC GAA TAG TAG GCT TCT
GAC ACG TGA ATA ACA CGA TGA GGG CGT TTC AGG CTT ATC ATC CGA AGA

CGC TGC TCC CGT GCT TCT CGC GTA TGT CCG – 3'
GCG ACG AGG GCA CGA AGA GCG CAT ACA GGC – 5'

A restriction enzyme is used to cut the gene from the human genome. Four possible enzymes have recognition sequences and cutting sites as shown.

enzyme 1	enzyme 2	enzyme 3	enzyme 4
GC TTCT CGAA GA	C TGTGC GACAC G	TCC CGT AGG GCA	CTG TGC GAC ACG

The DNA sequence is quite long, so only the beginning and end are shown along with a section before and after the gene. The start and stop triplets are underlined.

a Explain which of the restriction enzymes would be most suitable to cut out the gene so that it can be inserted into the bacterium that will produce the protein.

b A mutation can occur that changes the base indicated with the arrow from a T to a C. One way to identify individuals who have this mutation is to cut the DNA with a restriction enzyme and run the DNA on an electrophoresis gel. Explain why enzyme 4 is the most appropriate to use for this purpose.

c This mutation runs in one particular family. Ahmed and Sofia are members of the family and decide to be tested. Ahmed turns out to have two normal alleles and Sofia is heterozygous (one normal and one mutant allele).

i If enzyme 4 is used, how many DNA bands will result from the cutting of Ahmed's DNA?

ii If enzyme 4 is used, how many DNA bands will result from the cutting of Sofia's DNA?

iii Redraw and complete the picture of the electrophoresis of the DNA of Ahmed and Sofia after cutting with enzyme 4.

iv Show the positions of the positive and negative terminals on the electrophoresis set below, and explain why you placed them in those positions.

23 A plasmid of total length 1000 bp and a segment of a linear chromosome are being used to make recombinant DNA (diagram to the right). The DNA was cut with the restriction enzyme Tat1, which leaves sticky ends. The cutting sites are indicated by arrows. The resulting fragments were run on a gel (diagram to the right). The purpose of the process is to insert a segment of the chromosome into the plasmid for gene cloning.

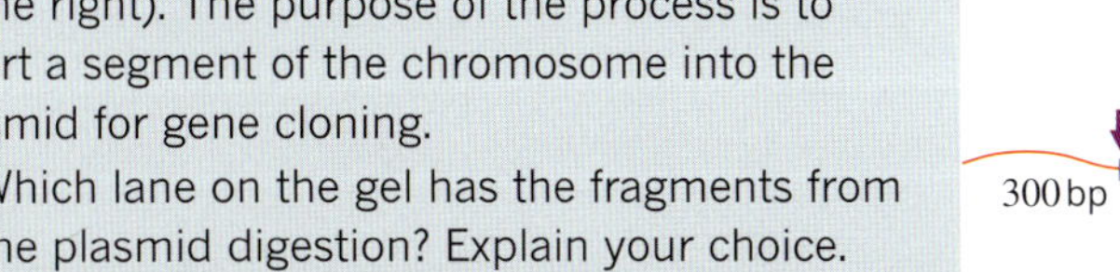

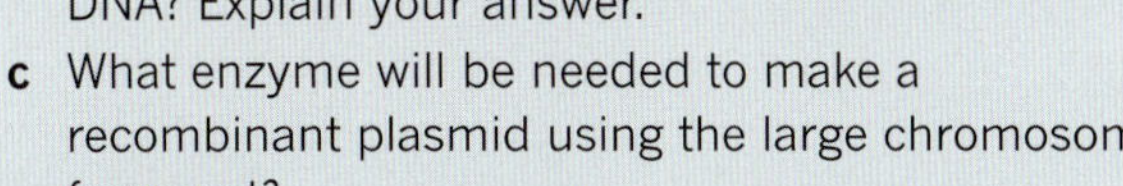

a Which lane on the gel has the fragments from the plasmid digestion? Explain your choice.

b What length is the starting chromosomal DNA? Explain your answer.

c What enzyme will be needed to make a recombinant plasmid using the large chromosome fragment?

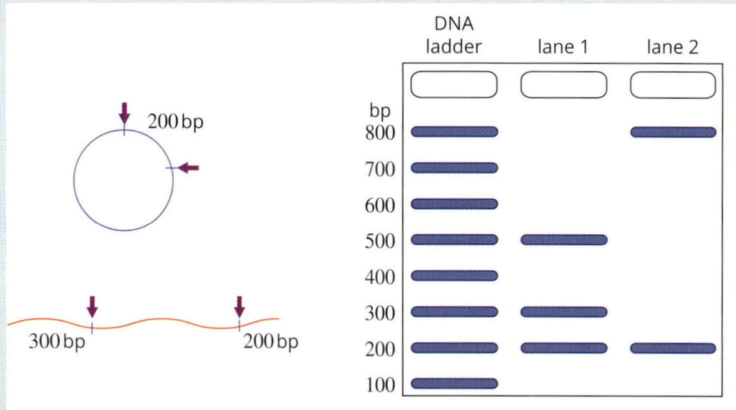

24 *E. coli* cells were transformed with a plasmid containing a gene for ampicillin resistance, the *lacZ* gene for blue/white screening and, depending on the success of gene uptake, a gene for a desired protein product. When the production of a recombinant plasmid was successful, the protein-encoding gene was inserted into the *lacZ* gene using the recognition site for the BamHI restriction enzyme. The plasmid is shown below.

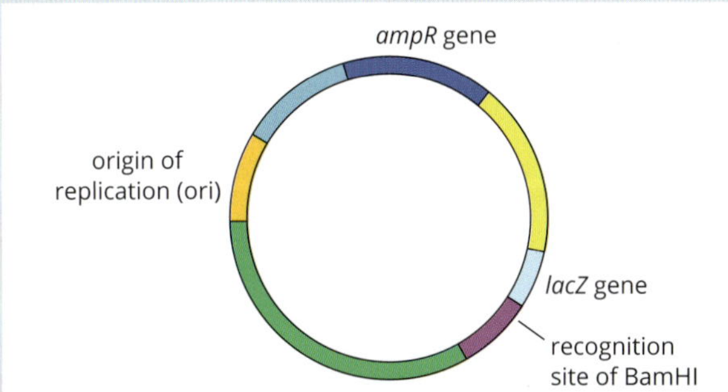

The following results were obtained:

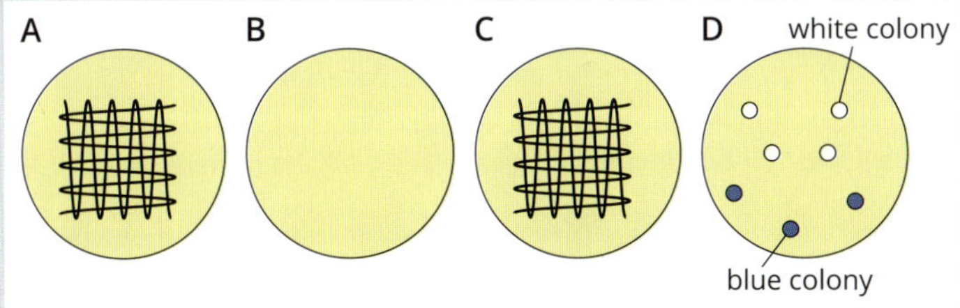

	A	B	C	D
added to plate	nutrient agar only	nutrient agar with ampicillin	nutrient agar only	nutrient agar, with ampicillin and X-gal
cultured on plate	untransformed bacteria only	untransformed bacteria only	transformed bacteria	transformed bacteria
description	lawn of bacteria	no growth	lawn of bacteria	blue and white colonies of bacteria

a Identify why plate B is significant.

b Specify the purpose of X-gal in plate D.

c Distinguish between the white and blue colonies on plate D. What evidence can you present to explain the differences?

d Consider the location of the restriction enzyme recognition site within the *lacZ* gene. Do you agree that this location is critical in allowing researchers to determine the success of the bacterial transformation? Explain.

25 At a small country hospital three babies were born on one night. This stretched the resources of the hospital to such an extent that normal procedures failed and the babies were not labelled with their mother's name. In order to ensure the correct babies were taken home by the correct parents, DNA testing was performed. A STR on chromosome 6 that has between 7 and 20 ATTG repeats was investigated in order to match the parents with their babies. The results for the couples and the babies are shown below.

Couple one		Couple two		Couple three	
Mother	Father	Mother	Father	Mother	Father
11, 14	7, 12	14, 20	12, 18	18, 20	11, 18

Baby one	Baby two	Baby three
12, 20	11, 20	12, 14

a Match each baby with its correct parents.

b Explain how you matched the couples with their children.

c Figure 4.1.13 on page 142 shows one way of analysing a series of STRs. It shows the analysis of 10 sites. Some sites have two peaks and others only one. Explain why this is the case.

26 What are some of the effects (positive and negative) of genetically modified and transgenic organisms on biodiversity?

27 Create a table listing the advantages and disadvantages of biotechnology for animal welfare.

28 Does artificial manipulation of DNA have the potential to change populations forever? Consider examples of GMOs that are already in use in Australia. Evaluate the impact these GMOs have had on the populations of these organisms in the short and long term.

OA ✓✓

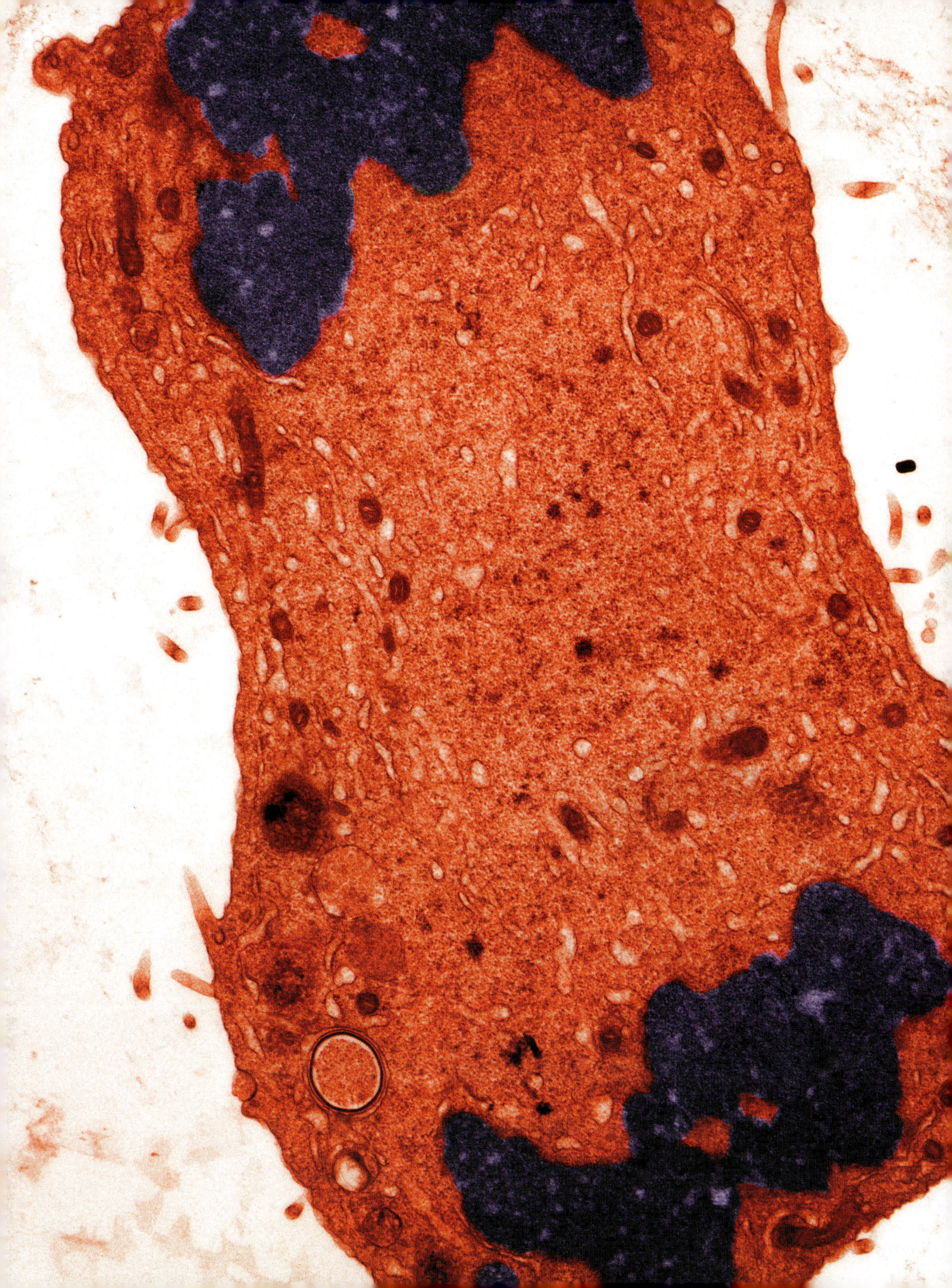

REVIEW QUESTIONS

What is the role of nucleic acids and proteins in maintaining life?

Multiple-choice questions

1 Complete this statement: The DNA molecule consists of two strands in which:
  A the percentage of adenine is the same in each strand
  B the percentage of adenine is the same as that of thymine in each strand
  C the percentage of adenine is the same as that of uracil in the whole molecule
  D the percentage of adenine is the same as that of thymine in the whole molecule

2 Consider the structures shown below.

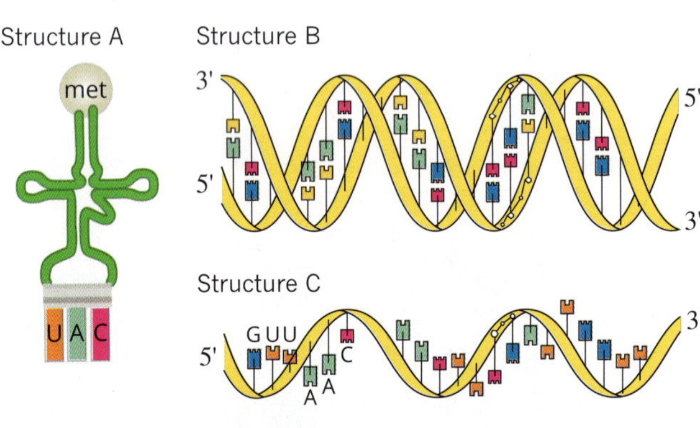

Select the answer that correctly names the type of nucleic acid represented by each structure.

	Structure A	Structure B	Structure C	Structure D
A	ribosomal RNA	messenger RNA	DNA	transfer RNA
B	transfer RNA	DNA	messenger RNA	ribosomal RNA
C	transport RNA	DNA	messenger RNA	ribosomal RNA
D	messenger RNA	DNA	ribosomal RNA	messenger RNA

3 Consider this strand of DNA:
TAACCTAAG
Conclude which one of the following sequences is the corresponding strand of mRNA.
  A UTTGGUTTC
  B ATTGGATTC
  C UAACCUAAG
  D AUUGGAUUC

4 Complete this statement: Amino acid molecules are:
  A the monomers used by cells to form polypeptide chains
  B the monomers used by cells to form DNA and RNA
  C the polymers used to form functional proteins
  D the catalysts that control biochemical reactions

5 Complete this statement: An enzyme is:
  A a type of ribonucleic acid found in some eukaryotic cells
  B a protein molecule in the secondary structural stage of folding
  C a protein molecule that acts as a catalyst in biochemical pathways
  D a protein molecule that is used as a reactant in biochemical pathways

6 Identify which part of the diagram shown below is a tertiary protein structure.

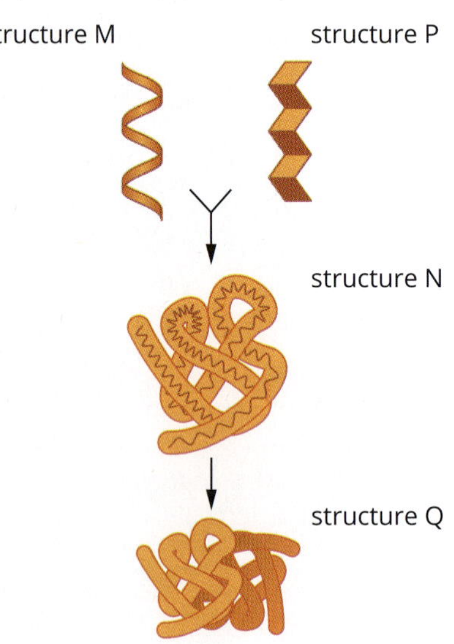

  A structure M
  B structure N
  C structure P
  D structure Q

7 What are transcription factors?

A promoters
B exons
C introns
D proteins that attach to DNA to regulate gene expression

8 Upstream areas of the gene which regulate transcription are:

A promoters
B exons
C introns
D operator regions

9 Select the statement that explains the use of genetically modified organisms most accurately.

A Genetic modification produces better species.
B Genetic modification will help humans to survive better.
C Genetically modified organisms can improve agricultural production.
D Genetically modified organisms cannot be used in medicine.

10 A plasmid is a:

A structure in prokaryotic cells used in asexual reproduction by binary fission
B short synthetic segment of DNA used in the polymerase chain reaction (PCR) to replicate DNA in a laboratory
C circular section of DNA in bacteria, which is separate from and smaller than the chromosomal DNA
D virus vector used in genetic engineering

11 Identify the most accurate sequence for the processes used to produce a transgenic organism.

A induction of ovulation, artificial insemination, normal intrauterine development
B isolation of target gene from DNA of one species, gene replication, insertion of the target gene into the cell of another species
C isolation of target gene, gene replication, insertion of the target gene into the cell of the same species
D extracting nucleus from a parental somatic cell, transferring nucleus to enucleated ovum, implanting ovum in utero

12 The figure below is a DNA profile obtained from one individual. Ten regions have been analysed: nine STRS and the sex chromosome markers. In the centre of the figure is a peak labelled 32.

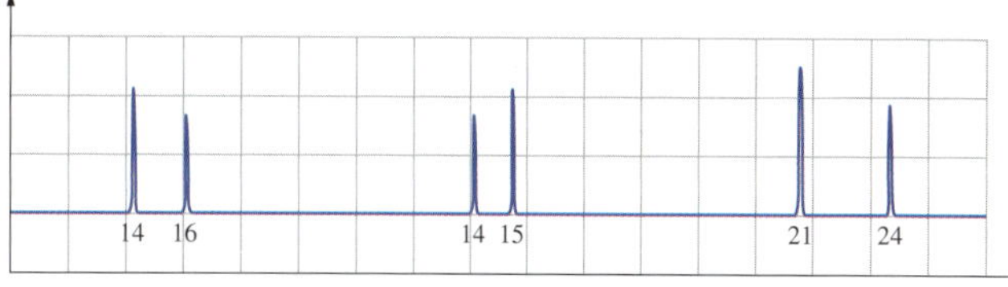

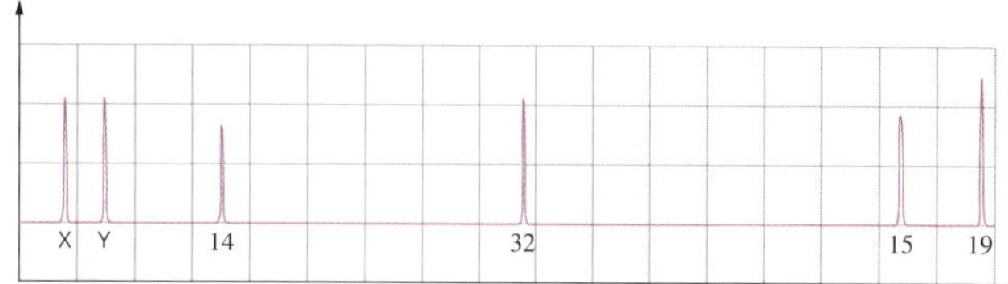

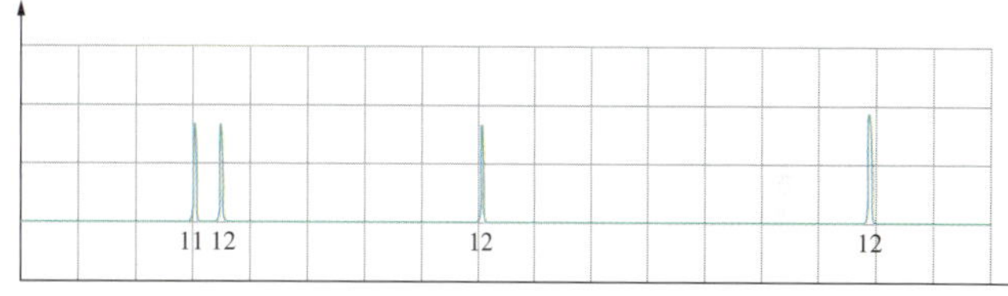

Which of the following accurately describes why there is only one peak?

A The individual has one chromosome with that STR and the STR has 32 repeats.
B The individual has two chromosomes with that STR and the STR has 32 repeats.
C The individual has two chromosomes with that STR and the STR has 16 repeats.
D The peak represents an STR on the Y chromosome and the person tested was male.

Short-answer questions

13 The following diagram shows a short section of one strand of a nucleic acid molecule.

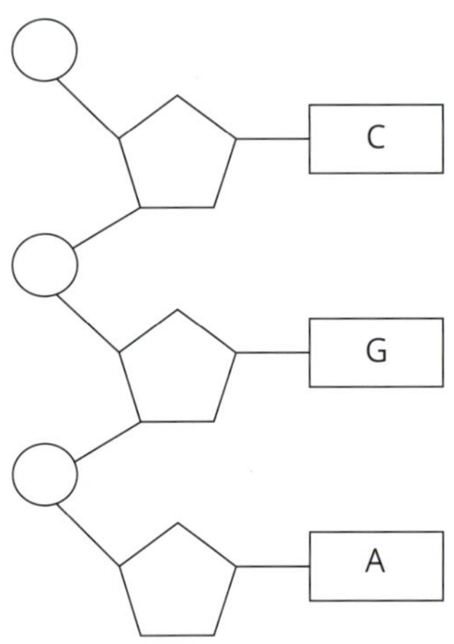

a On the diagram, circle one nucleotide and add labels for a phosphate group, a deoxyribose sugar and a nitrogenous base.

b On the diagram, draw the complementary strand of the molecule.

c Write a sequence that identifies the main steps in expression of polypeptides and functional proteins, starting from a DNA triplet code like this.

---

14 The electron micrograph below shows part of a cell that produces digestive enzymes.

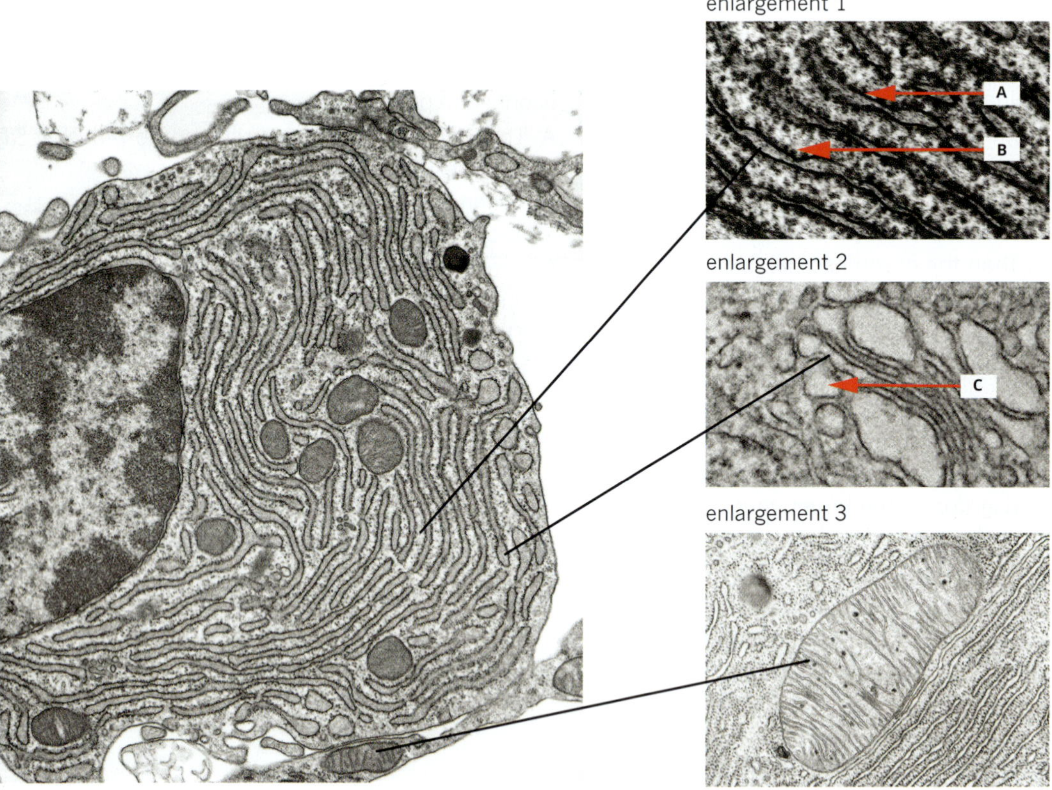

a i Identify the organelles illustrated in enlargements 2 and 3.

ii Identify the structures labelled A, B and C.

b Describe the role of the organelle shown in enlargement 3 in the production and secretion of the digestive enzymes.

c Draw a flow chart of the production and secretion of the enzymes. Refer to the relevant organelles and structures from the diagram above in your chart. Ensure you name the process by which the enzymes are secreted from the cell.

15 It is known that proteins have to proceed through a strictly defined series of structural folding to become functional.

- **a** List the hierarchical levels in the order required to form a linear molecule into a fully functional three-dimensional molecule.
- **b** Summarise the structural changes that occur during the folding process.
- **c** Describe three different examples to represent the wide functional diversity of this type of biomolecule.

16 A student intended to investigate the processing of RNA in a cell. In order to do this, the student decided to introduce a radioactively labelled RNA monomer to a culture of human skin cells.

- **a** State which type of RNA monomer would be most appropriate to radioactively label for this experiment, and give an explanation for your choice.
- **b** The radioactively labelled RNA was collected from the cell and X-ray diffraction was used to find the shapes of the various molecules. Two shapes that were identified are shown below.

  - **i** Identify the two types of RNA shown.
  - **ii** Explain where in a cell each type of RNA would be discovered.
  - **iii** Describe the role of each type of RNA in the cell.
- **c** A third type of RNA was extracted from the nucleus of the cell.
  - **i** What is this RNA?
  - **ii** Where else in the cell would this type of RNA be found?
  - **iii** What is its function?

17
- **a** Distinguish between genes, the genome and the proteome of an organism.
- **b** Distinguish functional genes from regulatory genes.
- **c** Outline the structure of eukaryotic genes.
- **d** Using the *trp* operon in *Escherichia coli* (*E. coli*) as an example, explain how gene regulation by transcriptional factors expressed by regulatory genes can occur.

18 The Ever-Open Convenience Store had experienced a number of robberies. The police were keen to catch the offender, who brandished a gun during each robbery. The police had five suspects, but were unable to gather sufficient evidence to clearly identify the perpetrator.

The robber wore rubber gloves, a mask, concealing clothing and a balaclava. After the fourth robbery the police found the little finger ripped from a pair of rubber gloves. This piece of glove was carefully collected and sent to the forensic science laboratory to be tested for DNA. Such material will contain a very small amount (if any) of DNA.

- **a** What is the potential source of any DNA found inside the glove?
- **b** Such small amounts of DNA are not suitable for preparing a DNA profile. How will the forensic scientists acquire enough DNA to create a DNA profile? Draw a flow chart describing the process.
- **c** A DNA sample from the glove was detected then amplified. A DNA profile was made using the amplified DNA from the crime scene and DNA from each of the five suspects. The profile is shown below.
  - **i** The DNA fragments in the size standard (DNA ladder) are 1000 bp, 2000 bp, 4000 bp, 5000 bp, 7000 bp and 10 000 bp. What is the size of the band indicated by the red arrow?
  - **ii** Why are size standards needed?
  - **iii** Deduce which suspect best matches the crime scene sample.
  - **iv** Does a match mean that this suspect committed the crime? Explain the reason for your decision.

19 Myotonic dystrophy is a serious disease that causes wastage of muscles. It can affect cardiac muscle, resulting in heart problems. The most severe form of the disease is caused by a mutation in the *DMPK* gene, which is found on the long arm of chromosome 19. It is caused by a CTG trinucleotide repeat. In most people, there are between 5 and 37 repeats but in individuals with myotonic dystrophy the number of repeats exceeds 50. It is often an adult-onset disease and has an autosomal dominant pattern of inheritance. This means that if the allele is inherited, it is certain that the disease will develop, but possibly not until later in life.

- **a** Explain how electrophoresis could be used to identify whether an individual has the mutated allele.
- **b** Before a person can undergo genetic testing, they must spend some time discussing associated issues with a counsellor. Outline some bioethical issues that could be associated with genetic testing for myotonic dystrophy.

20 Antithrombin is a plasma protein. Its function is to stop blood clots forming where they are not needed. People with a mutation in the gene for antithrombin production will easily develop a thrombosis (blood clot) and will generally require hospitalisation. Blood clots in the brain and heart can cause death. Like many blood-clotting diseases, antithrombin deficiency is treated by injecting required amounts of the protein. The challenge for medical researchers is the supply of the protein. In order to create a steady supply of antithrombin, goats have been genetically engineered to produce the protein in their milk. The goats are both genetically modified and transgenic.

The process is summarised at right.

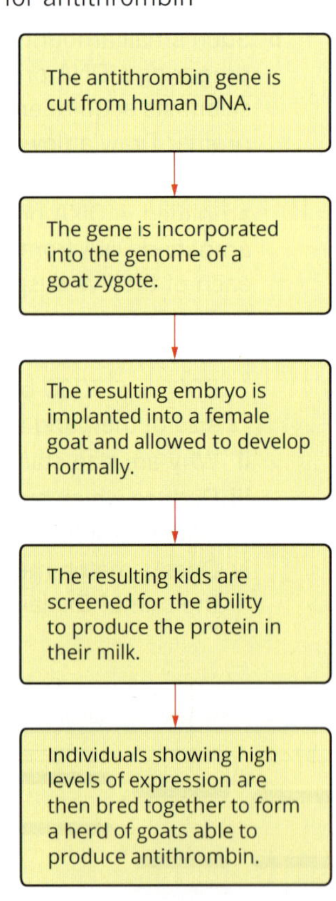

- **a** Clarify the difference between organisms that are genetically modified and organisms that are transgenic. Give examples to illustrate your understanding.
- **b** Predict any possible drawbacks to using animals to make human proteins.
- **c** Examine an ethical issue associated with the use of animals for the production of human pharmaceuticals.

21 Scientific understanding has developed rapidly since the structure and function of the DNA molecule became known. There is now a variety of genetic technologies available to artificially manipulate DNA, such as the use of specific enzyme groups, recombinant plasmids and gene editing.

- **a** Enzymes are an important tool used for manipulation of DNA. Outline how two of these types of enzymes are used to manipulate DNA in the laboratory.
- **b** **i** Another tool is the use of recombinant plasmids as vectors. Use a diagram to explain how recombinant plasmids are created.
  - **ii** Describe a successful example for the use of recombinant technology in agriculture.
  - **iii** The use of recombinant bacteria to produce human insulin has become widely accepted. Discuss why this is ethically acceptable while other proposed bacterial transformations remain controversial.
- **c** A new gene editing technology, known as CRISPR-Cas9, uses DNA sequences that are part of a bacterial defence system. Summarise how CRISPR-Cas9 technology can be used to edit an organism's genome.

22 Kuru is a disease that was once common in the highlands of New Guinea. It has been established that it is caused by a prion. Like mad cow disease in cattle and Creutzfeldt–Jakob disease (vCJD) in humans, this prion builds up in neurons, causing plaques that eventually destroy the cells, resulting in compromised neurological function.

Researchers studying the problem in New Guinea have discovered that there are some individuals who are highly resistant to the misfolding of their proteins into the prion form. Study of these individuals has established that they possess a mutated protein.

Further study of this protein is needed as it may lead to a treatment or cure for both Kuru and vCJD. In order to study this protein, a large and readily available pure supply is needed. The scientists wish to introduce the gene for the protein into bacteria, which will then produce a constant supply for research.

a Assume that the amino acid sequence of only part of the protective protein has been identified. Using this information, mRNA for the protein has been extracted from human cells. This will be used to make the gene to insert into the bacterium.

i How will the gene be produced from the mRNA?

ii Why is it better to use mRNA in this case rather than DNA?

b Once a functional copy of the gene has been created, many copies will be required. To do this a plasmid that can be inserted into a cell is needed. A plasmid containing an operon called *lacZ* and an ampicillin (an antibiotic) resistance gene (*amp*^{R}) is obtained. (These plasmids are made commercially and can be bought from a biological supplier.)

i *lacZ* has the role of being a reporter gene. Define the term 'reporter gene'.

ii Why is the ampicillin resistance gene included?

c Once the plasmid is obtained, the gene for the Kuru-protective protein must be inserted into the plasmid. The sequences shown in the diagram below are for the coding strand.

Many plasmids are cut open with a restriction enzyme and are incubated with the gene of interest. You have four restriction enzymes you could use. The enzymes have cutting sites as shown. The slash (/) indicates the cutting site. Enzymes 1 and 2 create sticky ends and enzymes 3 and 4 create blunt ends.

Enzyme	2	1	3	4
Cutting site	CTT/CCT	GGG/CCC	GA/TACT	GAA/AGC

i What is the difference between sticky ends and blunt ends?

ii Explain which enzyme should be used to cut the plasmid.

d Once the plasmid has been cut, it should be incubated with the gene to allow the gene to be incorporated into it. The bacteria will then be mixed with the plasmids and a proportion of the bacteria will be transformed.

i Name the enzyme needed to incorporate the gene for the protein of interest into the plasmid.

ii How will the proportion of bacteria that is transformed be increased?

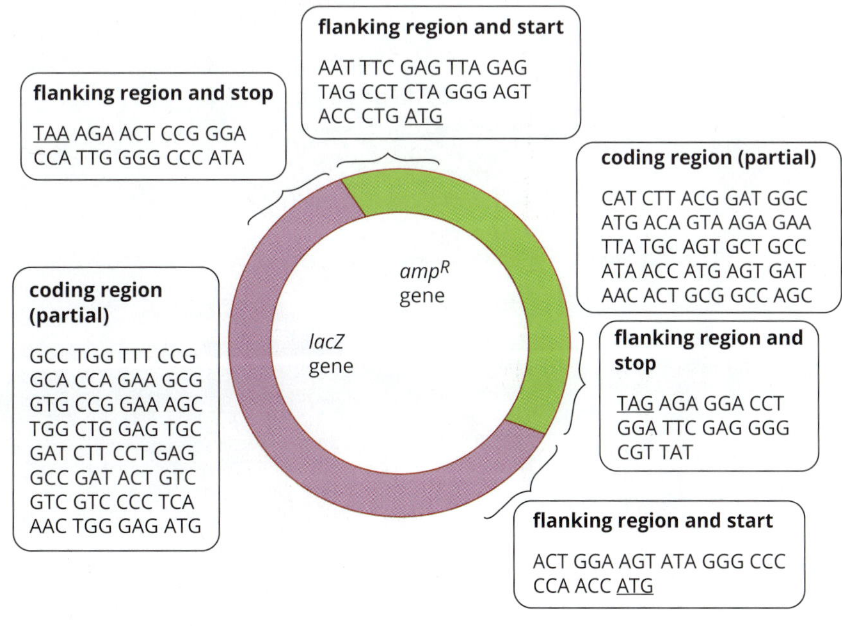

CHAPTER

05 Regulation of biochemical pathways

Learning outcomes

By the end of this chapter, you will understand the structure of the biochemical pathways in photosynthesis and cellular respiration and the role of enzymes and coenzymes in facilitating steps in these biochemical pathways. You will also understand how factors such as temperature, pH and enzyme inhibitors impact enzyme function.

Key knowledge

- the general structure of the biochemical pathways in photosynthesis and cellular respiration from initial reactant to final product **5.1**
- the general role of enzymes and coenzymes in facilitating steps in photosynthesis and cellular respiration **5.1**
- the general factors that impact on enzyme function in relation to photosynthesis and cellular respiration: changes in temperature, pH, concentration, competitive and non-competitive enzyme inhibitors. **5.2**

WS 11

OA ✓✓

5.1 Enzymes and biochemical pathways

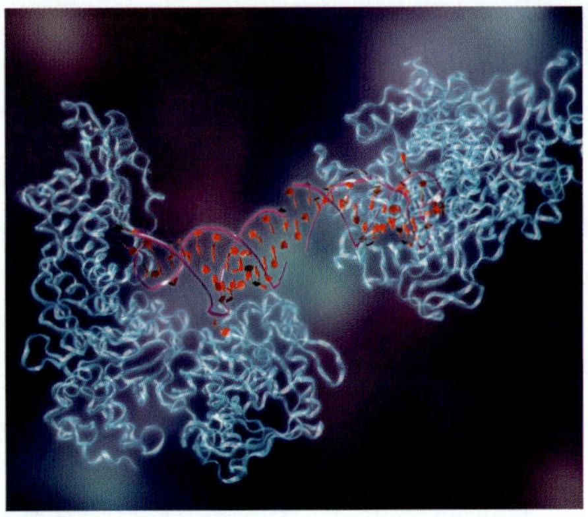

FIGURE 5.1.1 The enzyme aconitase (digitally coloured blue in this model) catalyses one step in the biochemical pathways in cellular respiration. It can also change shape to bind to ferritin mRNA (shown in red), regulating iron levels.

In this section, you will learn about the features of enzymes, including their specificity for particular substrates, and how they interact with substrates to catalyse biochemical reactions. You will also learn about the importance of enzymes in the biochemical pathways used in photosynthesis and cellular respiration.

ENZYME FEATURES

Most **enzymes** are globular proteins that have a tertiary or quaternary structure. Enzymes regulate biochemical pathways, acting on substrate molecules (**reactants**) to form a final product. During this process, enzymes interact with substrate molecules in a series of intermediate steps that involve the formation of **enzyme–substrate complexes**. The main features of enzymes are their specificity for a substrate and their catalytic power:

- **Specificity**—Different enzymes act as catalysts for different biochemical reactions by binding to a specific type of molecule called a **substrate**. Although many enzymes have evolved to be highly specific, and to act on a single substrate and **catalyse** one specific reaction, some enzymes are able to act on multiple substrates (Figure 5.1.1) and catalyse multiple reactions.
- **Catalytic power**—Enzymes do not make reactions occur that would not occur on their own; they only make reactions occur more quickly (sometimes over a million times more quickly).

To catalyse a reaction means to increase the rate of the reaction.

Enzymes are not consumed when they catalyse reactions.

Enzyme specificity

A key structure of enzymes is their **active site**. This is a pocket or groove-like part of the enzyme formed by the tertiary folding of the protein. Each enzyme's active site is a complex three-dimensional shape that interacts with a specific substrate to catalyse a specific reaction. When the active site binds to the substrate, it forms an enzyme–substrate complex (Figure 5.1.2).

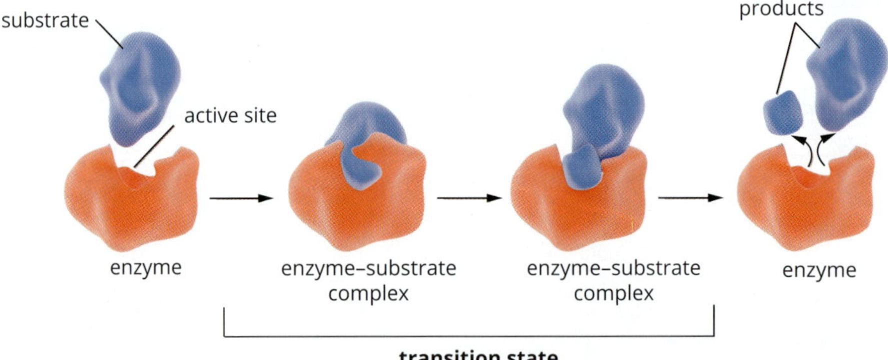

FIGURE 5.1.2 Stages of an enzyme–substrate interaction in a reaction. From left, the substrate molecule (blue) attaches to the active site on the enzyme (red), forming an enzyme–substrate complex. This complex goes through a transition state where the substrate molecule is strained, resulting in it being broken into two. The final stage is the release of the product molecules.

Enzyme–substrate interaction models

Multiple hydrogen bonds and hydrophobic interactions form between the substrate and the active site within the enzyme to stabilise the substrate in the active site. There are two models that describe how enzymes and their substrates interact:

- the lock-and-key model
- the induced-fit model.

'Hydrophobic' describes a non-polar molecule, or part of a molecule, that is unable to form energetically favourable reactions with water molecules, making it unable to dissolve in water.

The **lock-and-key model** describes the active site and the specific substrate as fitting together like a key into a lock (Figure 5.1.3a). If the 'key' (the substrate), does not fit into the 'lock' (the active site), then the reaction is not catalysed.

The **induced-fit model** states that when a substrate binds to the active site of an enzyme, a change in shape (or **conformational change**) of the active site occurs. This model is a more accurate representation of enzyme–substrate interactions, because we know that the active site is flexible and capable of changing its shape in order to conform to the shape of the substrate and achieve a tighter fit (Figure 5.1.3b).

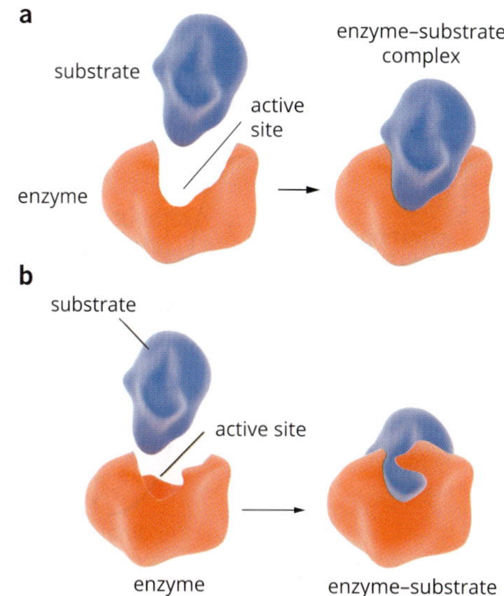

FIGURE 5.1.3 Two major models of enzyme–substrate interaction: (a) lock-and-key and (b) induced-fit

Enzyme catalytic power

Reactions are often reversible, meaning that they can be catalysed in both directions (substrate → product, and product → substrate). However, this is not always the case. For example, some of the reactions in glycolysis are reversible but others are not.

Usually different enzymes catalyse a reaction in each direction. For example, DNA polymerase builds DNA, and deoxyribonuclease (DNase) breaks it down. The direction of the reaction will depend on the concentration of substrates and products, as well as the energy requirements.

The specific enzyme itself is not changed during the reaction, so it may be used over and over again as long as that reaction is required by the cell and there is substrate present.

CASE STUDY ANALYSIS

Ribozymes

Although they are composed of only four nucleotides that are chemically similar, some ribonucleic acid (RNA) molecules can fold into three-dimensional structures that often serve as binding sites for proteins that function with the RNA molecules. In the 1980s, however, it was discovered that some RNA molecules catalyse biochemical reactions on their own, without the assistance of proteins.

The discovery of these RNA enzymes won Sidney Altman and Thomas Cech the Nobel Prize for Chemistry in 1989. It signalled the end of the long-held belief that all enzymes are proteins, and lent support to the 'RNA world' hypothesis, which suggests that RNA appeared before DNA and proteins, and was crucial to the evolution of self-replicating systems.

Catalytic RNA molecules (or ribozymes) are considered enzymes because they speed up specific chemical reactions without permanently changing themselves.

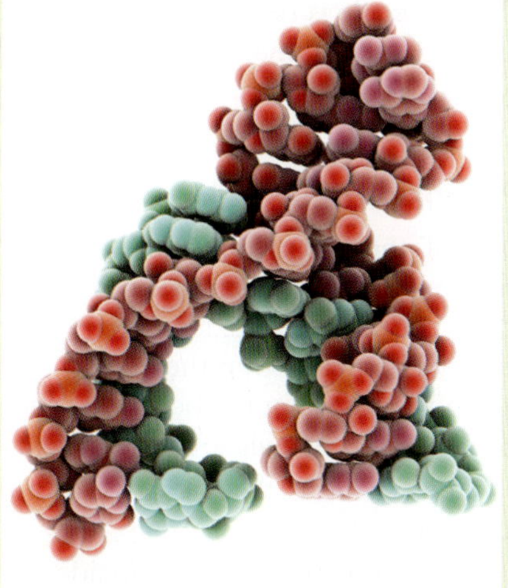

FIGURE 5.1.4 Molecular model of a hammerhead ribozyme

Since their discovery, several ribozymes have been identified. They occur within the ribosome, the cell organelle where the synthesis of proteins begins. Most ribozymes have been found to catalyse the cleavage (splitting) of themselves or other RNAs. As it is for the active sites of protein enzymes, shape is critical. The hammerhead ribozyme (Figure 5.1.4) is one of the characteristic structures that catalyses RNA cleavage.

The ability to self-cleave can make some ribozymes very rapid reproducers. One group, the viroids, are pathogenic to plants for this reason—the viroid quickly multiplies and creates physical blockages to a plant's internal structures, while also depleting the host cells of nucleotides.

The ability of ribozymes to cleave other RNA strands, however, could have highly beneficial therapeutical applications. With some laboratory modifications, natural ribozymes can be altered to disrupt RNA viruses, including HIV-1. They may also be used in the future to target the RNA involved in translating genetic disorders.

Analysis

1 What features of ribozymes allow them to be classified as enzymes?

2 Genetic disorders are coded in a person's DNA. How would a ribozyme prevent a malfunctioning protein from being synthesised?

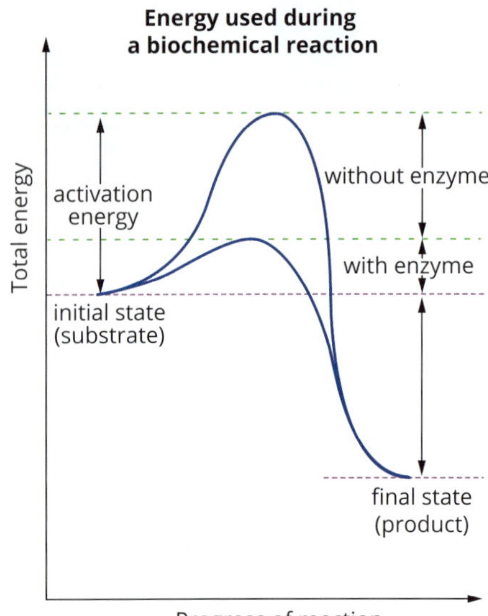

FIGURE 5.1.5 The addition of a catalyst reduces the amount of energy needed to initiate a reaction.

Enzymes reduce activation energy

All reactions need an input of energy to start. This is called the **activation energy**. Whether a reaction releases or consumes energy, activation energy is needed for the reaction to start. The catalytic power of enzymes comes from their ability to reduce the level of this activation energy, so that less energy is required for the reaction to occur (Figure 5.1.5).

Enzymes reduce the activation energy required for a reaction by influencing:

- proximity and orientation. Enzymes bring the parts of the molecules involved in the reaction closer to each other in the active site and position them where a reaction is more likely to occur.
- the micro-environment. Most active sites are hydrophobic. The absence of water results in a non-polar environment, allowing stabilising interactions such as hydrogen bonds, hydrophobic interactions and dispersion forces to occur.
- ion exchange. The amino acids in the active site can often take H^+ ions from, or donate them to, the substrate, to facilitate steps in certain reactions.

ENZYMES REGULATE BIOCHEMICAL PATHWAYS

The term **metabolism** is used to describe the set of all biochemical reactions that sustain life. Most metabolic processes occur as a sequence of reactions, in which each reaction is catalysed by a specific enzyme and the product of one reaction becomes the substrate in the next reaction (Figure 5.1.6). Such sequences of biochemical reactions form **biochemical pathways**. Some biochemical pathways are linear, some are branched (leading to many final products), and others are cycles. In a cyclic pathway, the final product is also one of the substrates of the first step. The molecule is regenerated in the cycle so that the pathway can continue without running out of materials. Photosynthesis and cellular respiration both contain cyclic biochemical pathways in their overall reactions, which are summarised in Figure 5.1.7.

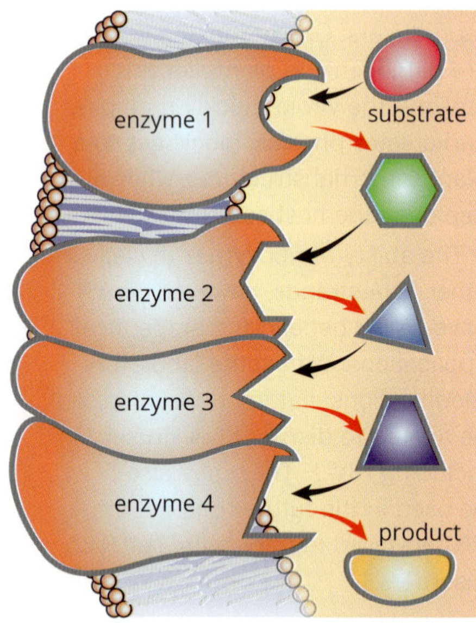

FIGURE 5.1.6 A biochemical pathway is a sequence of biochemical reactions catalysed by different enzymes. The product (indicated with the red arrow) of each reaction becomes the substrate (indicated by the black arrow) in the next reaction.

Photosynthesis

One of the most important biochemical pathways is **photosynthesis** (photo, meaning 'light', and synthesis, meaning 'putting together'). When plants have light, water and carbon dioxide (CO_2), they make glucose in their green parts, such as leaves. The plants trap the energy of sunlight and convert it into chemical energy, which they store in the bonds of glucose molecules. All photosynthetic organisms, from single-celled algae to the largest trees, produce glucose in the same way.

The reactions that occur during photosynthesis can be summarised by these equations:

Word equation: carbon dioxide + water $\xrightarrow{\text{light energy}}$ glucose + oxygen + water

Chemical equation: $6CO_2 + 12H_2O \xrightarrow{\text{light energy}} C_6H_{12}O_6 + 6O_2 + 6H_2O$

Photosynthesis involves two stages:

- a light-dependent stage (also called light reactions) that converts light energy into chemical energy (ATP)
- a light-independent stage (also called dark reactions) that uses the chemical energy (ATP) to synthesise organic compounds (glucose).

Each stage involves a series of biochemical reactions and each reaction in the pathway is catalysed by a particular enzyme (Figure 5.1.7). You will learn more about the pathways involved in photosynthesis in Chapter 6.

Cellular respiration

All cells need energy to function. They obtain this by releasing energy from organic compounds through a series of biochemical pathways. **Cellular respiration** is the name given to the combination of biochemical pathways that release energy from glucose. The energy released from glucose through cellular respiration is used to generate chemical energy in the form of ATP.

The reactions that occur during cellular respiration can be summarised by these equations:

Word equation: glucose + oxygen → carbon dioxide + water + energy (ATP)

Chemical equation: $C_6H_{12}O_6 + 6O_2 \rightarrow 6CO_2 + 6H_2O + ATP$

Cellular respiration consists of the three interconnected biochemical pathways known as glycolysis, the Krebs cycle and the electron transport chain. Each reaction in each pathway is catalysed by a particular enzyme (Figure 5.1.7). You will learn more about the pathways involved in cellular respiration in Chapter 7.

Enzymes are critical to the functioning of photosynthesis and cellular respiration. In the absence of enzymes, the reactions in these biochemical pathways would be much slower and cells could not carry out essential functions. Some important enzymes involved in photosynthesis and cellular respiration are shown in Table 5.1.1 on page 191.

Each reaction in the biochemical pathways of photosynthesis and cellular respiration is catalysed by enzymes. Without these enzymes, the pathways would not function.

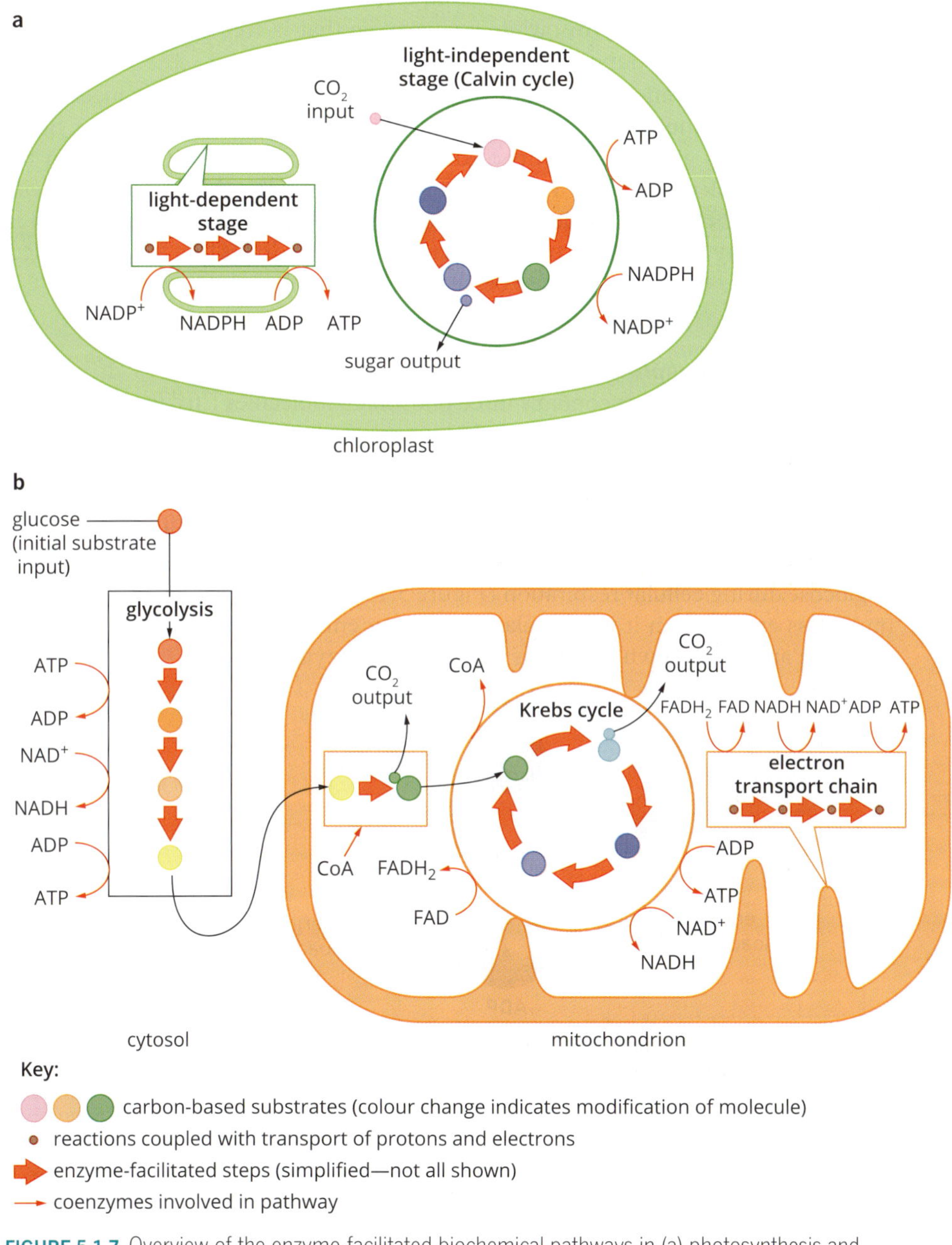

FIGURE 5.1.7 Overview of the enzyme-facilitated biochemical pathways in (a) photosynthesis and (b) cellular respiration

Coenzymes

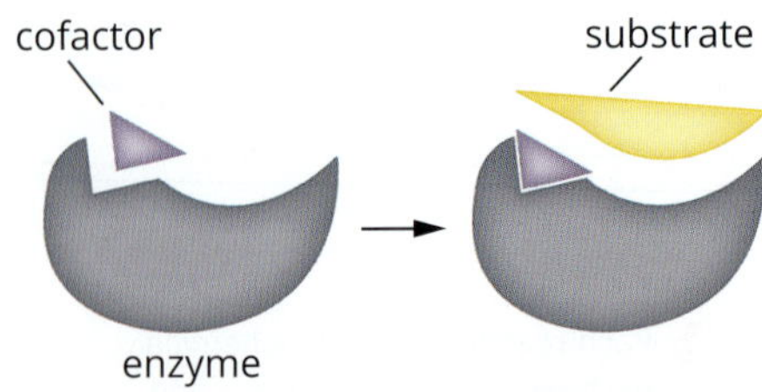

FIGURE 5.1.8 A cofactor (or coenzyme if organic) enables an enzyme to catalyse a reaction.

Some enzymes need additional non-protein components to enable them to catalyse a reaction. These components, called **cofactors**, bind to the enzyme before the substrate does (Figure 5.1.8). Cofactors can be inorganic ions such as iron (Fe^{2+}), magnesium (Mg^{2+}) and zinc (Zn^{2+}). If they are organic molecules, they are classed as **coenzymes**. Many coenzymes are vitamins, such as vitamin C, or molecules derived from vitamins, such as coenzyme A (CoA), which is modified from vitamin B_5. Some particularly important coenzymes used in metabolic processes include adenosine triphosphate (ATP), nicotinamide adenine dinucleotide (NADH), nicotinamide adenine dinucleotide phosphate (NADPH) and flavin adenine dinucleotide ($FADH_2$).

When a coenzyme is used, it is often structurally altered during the reaction, but can revert back to its original form and thus be reused. A coenzyme may also transfer **protons**, **electrons** and/or **chemical groups** from one molecule to another. A coenzyme that has a proton, electron or chemical group to donate is called a **loaded coenzyme**, while a coenzyme that is free to accept a proton, electron or chemical group is called an **unloaded coenzyme**.

> Enzymes are proteins that catalyse reactions in biochemical pathways. Coenzymes are non-protein molecules that assist enzyme activity.

In cellular respiration, for example, CoA is part of the reaction mechanism that takes the product of the first biochemical pathway into the Krebs cycle. Some of these transfers are also accompanied by an input or output of energy. In the first biochemical pathway of photosynthesis, for example, ADP is loaded with phosphate and $NADP^+$ is loaded with electrons. These coenzymes are unloaded in the next pathway and energy is transferred to the carbohydrates that are synthesised. The unloaded coenzymes can then be reused back in the first pathway.

ADP and ATP

Adenosine triphosphate (ATP) provides the energy required to drive most processes in living cells. It contains three phosphate molecules ('tri' in triphosphate means 'three'), the third of which is held by a relatively weak, unstable bond. When this third phosphate breaks free, the more stable products of **adenosine diphosphate (ADP)** and inorganic phosphate (P_i) are formed, releasing energy.

> Energy is required to break bonds, and is released when new bonds are formed.

This process is reversible, because the ADP can combine with a phosphate molecule to form an ATP molecule again, using energy derived from the breakdown of glucose during cellular respiration (Figure 5.1.9). This recycling process requires much less energy than it would take to make an entirely new ATP molecule. The synthesis and breakdown of ATP is regulated by enzymes.

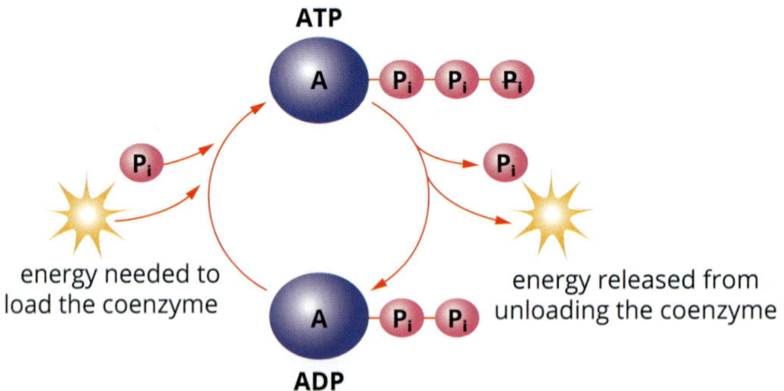

FIGURE 5.1.9 ATP transfers energy to cells when it unloads inorganic phosphate (P_i) and becomes ADP. An energy input is needed to reload it.

Hydrogen and electron carriers

NAD^+/NADH and FAD^+/$FADH_2$ are involved in the transfer of protons and electrons from hydrogen. They are used in many different processes in living cells, including cellular respiration. $NADP^+$/NADPH is another carrier found in all cells as one of its roles is in nucleic acid synthesis. It is also the coenzyme, along with ATP, which is used in photosynthesis (Table 5.1.1).

All three of these coenzymes are said to be loaded when they accept electrons from other organic molecules. In the process, they also move hydrogen atoms, sometimes separating one of them into its electron and proton (Figure 5.1.10). The electrons and associated protons are used to drive high-energy steps in biochemical pathways. For example, NADH and $FADH_2$ are unloaded in the final pathway of cellular respiration where large amounts of ATP are generated. NADPH is unloaded in the cyclic pathway of photosynthesis to give energy and hydrogen to carbon dioxide so that carbohydrates can be made.

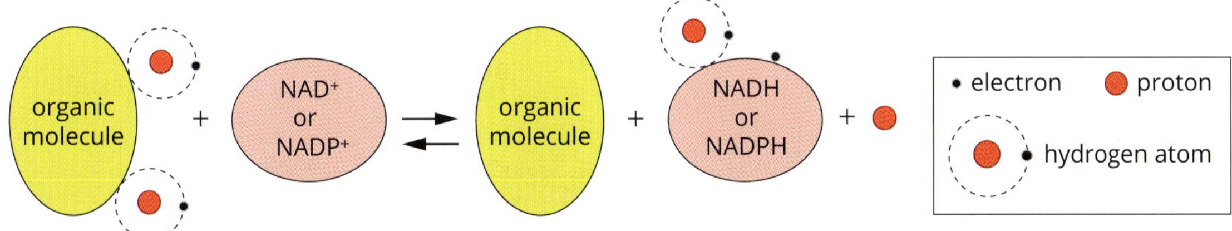

FIGURE 5.1.10 Coenzymes NAD^+ and $NADP^+$ are loaded with high-energy electrons. They also free protons from organic molecules and transport hydrogen.

TABLE 5.1.1 Roles of some enzymes and coenzymes in photosynthesis and cellular respiration

Enzyme	Cofactor/coenzyme	Function	Biochemical pathway
ATP synthase	magnesium (Mg^{2+})	forms ATP from ADP and inorganic phosphate (P_i)	photosynthesis (light-dependent stage) and cellular respiration (electron transport chain)
NADP+ reductase	FAD	transfers electrons to NADP+ to make NADPH	photosynthesis (light-dependent stage)
Rubisco	magnesium (Mg^{2+})	fixes atmospheric CO_2 into organic sugar molecules that the plant can use	photosynthesis (light-independent stage/Calvin cycle)
hexokinase	ATP	phosphorylates glucose to produce glucose-6-phosphate	cellular respiration (glycolysis)
pyruvate dehydrogenase	coenzyme A (CoA)	converts pyruvate into acetyl-CoA	cellular respiration (Krebs cycle)

CASE STUDY **ANALYSIS**

New enzyme may hold the key to beating obesity

Glucose is the main energy source used in cellular respiration, and we get it from the digestion of carbohydrates in our diet. The smaller the carbohydrate molecule, the more quickly it is broken down and can enter the bloodstream. What we call sugar is usually sucrose, which requires very little digestion and can start to raise blood glucose levels in a matter of minutes. The World Health Organization recommends a maximum daily intake of 25 g of sugar per day, yet the average Australian consumes 95.6 g per day. Rising levels of sugar consumption are closely linked to epidemics of obesity, heart disease, tooth decay and type 2 diabetes.

When blood glucose rises, the hormone insulin is released. Insulin increases uptake of glucose by muscle cells and triggers liver cells to store excess glucose as glycogen (Figure 5.1.11). Once the liver's storage capacity has been reached, the excess sugar is converted to fat. This occurs readily in people whose lifestyle includes a high-sugar diet with little energy expenditure (exercise).

Levels of another molecule, glycerol-3-phosphate (Gro3P), are also known to rise with blood glucose levels. Gro3P is formed as glucose is used in cells and is important for ATP production and lipogenesis (fat production). However, if there is excess glucose in the body, Gro3P levels get too high and damage a variety of tissues, including insulin-producing cells, leading to type 2 diabetes.

A recently discovered enzyme, Gro3P phosphatase (G3PP), has been shown to convert Gro3P to glycerol, preventing the excess formation of fat and reducing the production of glucose in the liver, which is a major problem in diabetes. The amount of G3PP enzyme produced throughout the body varies depending on the health of the tissue. In experimental trials, rats with overexpressed levels of G3PP showed reduced weight gain and increased levels of fat-removing molecules in their blood plasma. This research opens up exciting possibilities for how G3PP could be used to tackle obesity and some of its associated diseases.

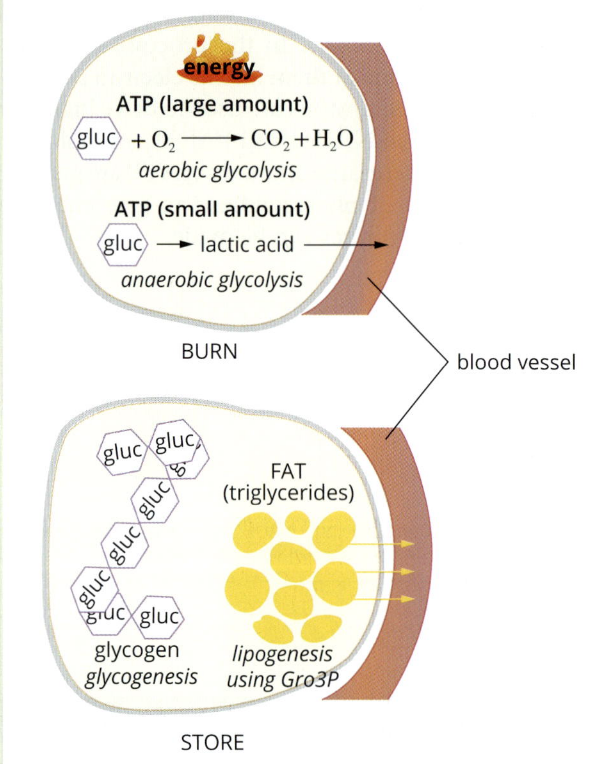

FIGURE 5.1.11 Pathways of glucose metabolism when blood glucose rises

Analysis

1. Infer a relationship between the G3PP enzyme and Gro3P using evidence from the experimental trial with rats.
2. As type 2 diabetes develops, a person's cells become less responsive to insulin. Explain how this leads to increased fat storage.

5.1 Review

OA ✓✓

SUMMARY

- Enzymes are biological catalysts that increase the rate of reactions by lowering the activation energy required for them to occur.
- Most enzymes are globular proteins, but some are made of RNA (ribozymes).
- Enzymes are not consumed in a reaction, and can be used over and over again.
- Enzymes are usually specific to particular reactions, since their active site is a three-dimensional pocket-like structure that is shaped to interact with its substrate, forming an enzyme-substrate complex. There are two models for how this functions:
  - lock-and-key: a perfect match for the two shapes
  - induced-fit: the active site changes shape when the substrate binds.
- A biochemical pathway is a sequence of biochemical reactions catalysed by different enzymes, in which the product of each reaction becomes the substrate in the next reaction.
- Biochemical pathways may be linear or branched, or they may be cyclic where the initial molecule is regenerated.
- Enzymes facilitate each reaction in the biochemical pathways of cellular respiration and photosynthesis.
- Some important enzymes in cellular respiration are ATP synthase, hexokinase and pyruvate dehydrogenase. Some important enzymes in photosynthesis are ATP synthase, $NADP^+$ reductase and Rubisco.
- Cofactors are additional components required by some enzymes to catalyse a reaction. A subset of these are coenzymes, which:
  - are small, non-protein organic molecules
  - carry chemical groups, energy, protons and/or electrons
  - cycle between loaded and unloaded forms.
- Some important coenzymes in cellular respiration include ADP/ATP, NAD^+/NADH, FAD^+/$FADH_2$ and CoA. Important coenzymes in photosynthesis include ADP/ATP and $NADP^+$/NADPH.

KEY QUESTIONS

Knowledge and understanding

1 State the two main features of enzymes and explain their importance.

2 Describe the difference between the lock-and-key and induced-fit models of an enzyme–substrate interaction.

3 Recall the word equation for:
   a photosynthesis
   b cellular respiration

4 Explain what coenzymes are, and describe the difference between an unloaded and a loaded coenzyme.

5 Explain the importance of enzymes and coenzymes in photosynthesis and cellular respiration, providing an example of each.

continued over page

5.1 Review *continued*

Analysis

6 Consider the biochemical pathway shown.

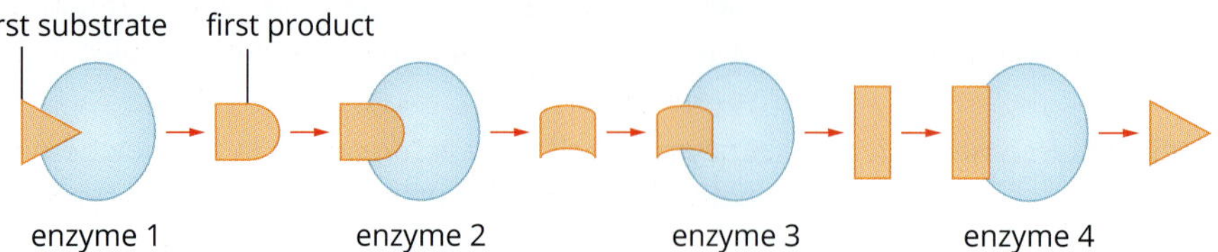

a Propose, with a reason, if the pathway is linear, branched or cyclic.

b ATP is needed to provide energy for the reaction occurring in Enzyme 2. Describe what would happen to the product of Enzyme 4 if ATP was not recycled in the cell and available for this biochemical pathway.

7 Scurvy is a disease caused by a deficiency of vitamin C. Vitamin C has many functions in the human body, including its role as a coenzyme in the production of collagen, the primary structural protein used to support cells in tissues such as skin, muscle, bone, cartilage and blood vessels. Suggest what symptoms may arise in a person suffering from scurvy.

5.2 Regulation of enzymes

There can be a negative effect on the whole organism when cells produce too much or not enough of particular substances, or are unable to properly break some substances down (Figure 5.2.1). Cells that produce excess substances are also wasting energy and resources. To account for this, cells have mechanisms that regulate biochemical reactions to ensure the final product is not over- or under-produced.

As you learnt in the previous section, enzymes control metabolism through the regulation of biochemical reactions at every step of a biochemical pathway. Because enzymes play a vital role in biochemical reactions, photosynthesis and cellular respiration can be controlled through enzyme regulation.

In this section, you will learn how enzymes are regulated, including the different ways in which enzymes are inhibited, and other factors that affect the rate of the biochemical reactions catalysed by enzymes.

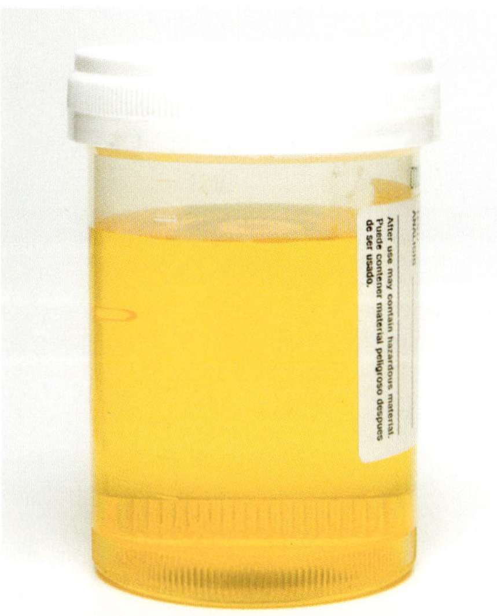

FIGURE 5.2.1 Maple syrup urine disease is caused by a defective group of enzymes that work together. It results in a build-up of branched-chain amino acids that causes the urine to smell like maple syrup and the blood to become acidic, which can result in death.

FACTORS THAT REGULATE ENZYME ACTIVITY

The amounts of final products and the speed at which they are produced in a biochemical pathway can be controlled through the regulation of individual reactions that make up that pathway. As each reaction in a biochemical pathway uses the product from the previous reaction as a substrate, slowing down one reaction will have an effect on all subsequent reactions.

All enzymes have specific conditions in which they perform at their best. Factors such as temperature, pH and the concentration of the substrate and enzyme all affect the rate of enzymatic reactions. When these conditions are optimal, enzyme activity is at its highest, and the rate of reaction is at its fastest.

Enzymes can also have their function reduced or stopped altogether by the presence of an inhibitor. Some inhibitors physically block the enzyme's active site, while others bind elsewhere but cause a change in the enzyme's shape. Because the reactions in photosynthesis and cellular respiration are catalysed by enzymes, the rate of these reactions changes as enzyme activity changes. If conditions are sub-optimal, enzyme activity will be reduced, resulting in reduced energy production by the organism's cells.

> An enzyme is denatured when it undergoes an irreversible change in its structure. Denaturation of enzymes often occurs at high termperatures.

Temperature

As with most chemical reactions, the rate of enzyme-catalysed reactions will generally increase as the temperature increases. This is because the warmer particles become during a reaction, the more rapidly they move, which makes successful collisions between them more likely to occur.

However, proteins, including enzymes, can **denature** at high temperatures. When this occurs, the hydrogen bonds and hydrophobic interactions that create the tertiary and quaternary structures of the enzyme are broken, and the shape of the enzyme's active site is changed in such a way that the substrate cannot bind to it, and the reaction cannot occur.

Most human enzymes have an optimum temperature of 36–38°C, which matches our normal body temperature (approximately 37°C). Indeed, many mammalian enzymes will begin to denature at temperatures above 40°C (Figure 5.2.2) and processes such as cellular respiration cease. However, there are some enzymes that have optimum temperatures much higher than this. For example, *Taq* polymerase is an enzyme that was originally found in bacteria living in volcanic hot springs, and it has an optimum temperature of 70°C.

If enzymes are cooled below their optimum temperature, the rate of reaction will slow down. Particles will move more slowly, making successful collisions less likely, and the bonds are not as flexible at lower temperatures, so conformational changes do not occur. Cooling an enzyme does not denature it, however, so reheating it will cause its activity to increase again.

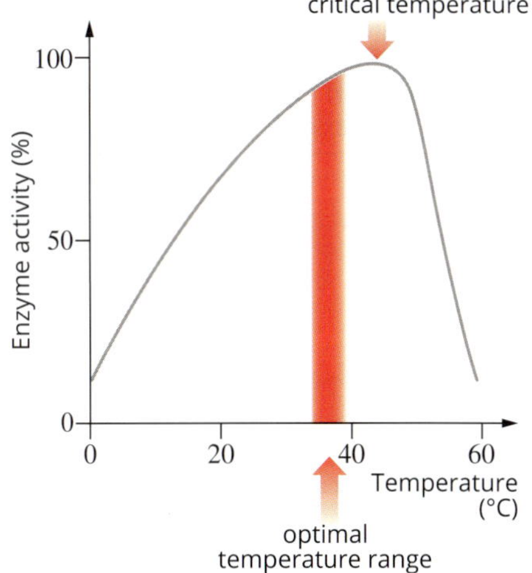

FIGURE 5.2.2 As the temperature increases, enzyme activity increases until a critical temperature is reached. The protein then denatures and the shape of the active site changes so that the substrate cannot bind. The rate of enzyme activity then decreases rapidly.

BIOFILE

Psychrophiles

Bacteria and fungi that are adapted to the extreme cold (psychrophiles) of Antarctica and the Arctic tend to have enzymes with a broader optimum temperature range, or two optimum temperature zones, including an optimum range as low as 0–15°C. This allows their cells to grow and metabolise at the constant low temperatures of their cold environments.

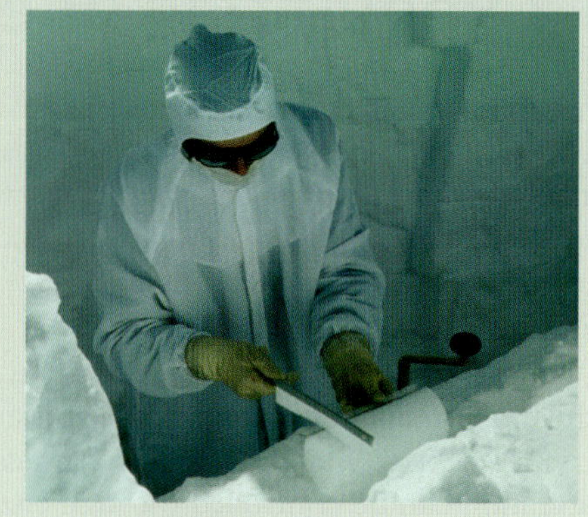

A glaciologist in Antarctica saws an ice core containing frozen psychrophiles.

pH

The pH scale is used to measure acidity (Figure 5.2.3). Enzymes have a specific pH range at which they function best. If enzymes are taken too far above or below their optimum pH, then the tertiary structure is affected, the enzymes may become denatured and the substrate may not be able to bind.

If the reaction occurs in an environment in which the pH is not ideal, the micro-environment of the active site may still provide a different pH suitable for the enzyme to function.

The optimum pH range of enzymes can be quite different, and varies depending on the function of the enzyme and where it is located in the body. Examples include the following digestive enzymes:

- Amylase starts the digestion of starch in the mouth and has an optimum pH of about 7.
- Pepsin is found in the stomach and has an optimum pH of about 2.
- Trypsin is found in the small intestine and has an optimum pH of about 8 (Figure 5.2.4).

Enzymes used within cells do not usually exhibit such a wide range of optimum pH values, although compartmentalisation of cellular components does make it possible. In photosynthesis, for example, acidic conditions are generated within the thylakoid space of chloroplasts. Outside the thylakoids, the enzyme Rubisco fixes carbon under alkaline conditions.

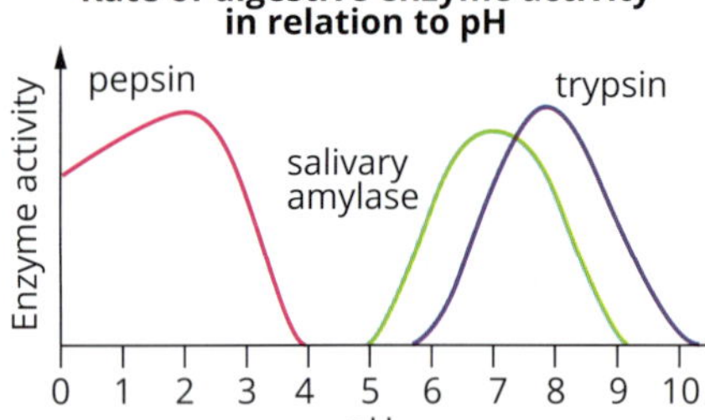

FIGURE 5.2.4 Rate of enzyme activity for three digestive enzymes in relation to pH values. At the optimum pH for an enzyme, its rate of reaction is at a maximum.

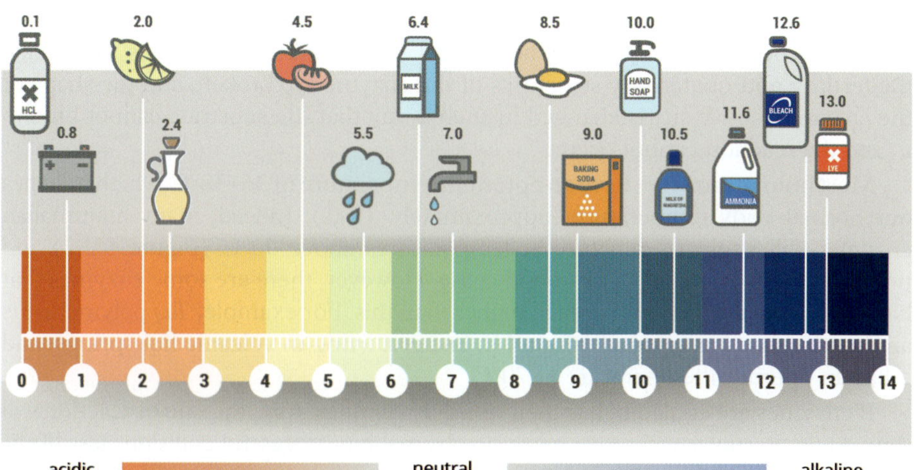

FIGURE 5.2.3 The pH scale

CASE STUDY **ANALYSIS**

Pepsin and ulcers

Chief cells in the stomach produce pepsinogen, which is a proenzyme (inactive form) of pepsin, a protein-digesting enzyme (Figure 5.2.5). The functional form of pepsin is active at pH values of around 1–4. Other cells in the stomach lining secrete acids, typically maintaining the pH level of the lumen (inside space) at around 2.

The conditions in the stomach are therefore hostile to most cells. This is an important part of both the digestive process and the body's defence against pathogens, but it also has the potential to damage the tissues that the stomach is made of. Fortunately, stomach lining cells also produce mucus that acts as a protective barrier.

Occasionally, the protection is breached and the stomach lining is painfully damaged in a localised area called a gastric ulcer. Antacid tablets and drinks can be taken to counter the stomach acid and minimise the effects of pepsin, which can digest proteins in the plasma membranes and cells lining the digestive tract.

Historically, poor diet and stress were thought to be the cause of gastric ulcers. However, as the Nobel Prize-winning research by Australians Dr Barry Marshall and Dr Robin Warren showed, the most common cause of damage leading to ulcer formation is by the bacterium *Helicobacter pylori*. Interestingly *H. pylori* does not favour the extremely acidic conditions of the stomach and will burrow into the mucus lining. It will also release the enzyme urease, which makes the product ammonia, an acid-neutralising substance. However, when the immune system responds to *H. pylori* by causing inflammation, the stomach lining is stimulated to secrete even more acid, and ulcers are formed. Antibiotics are usually required to clear the infection, although the effectiveness of this treatment is diminishing as antibiotic resistance is on the rise.

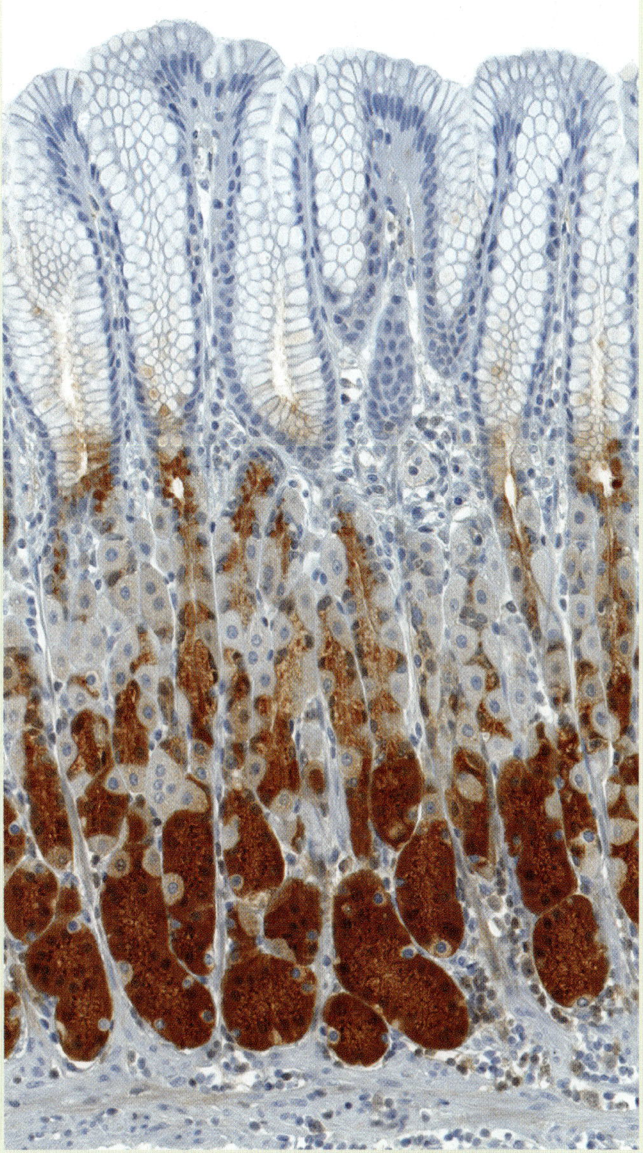

FIGURE 5.2.5 Inactive pepsin enzymes (or pepsinogens) appear brown in this stained section of stomach tissue.

Analysis

1 Why would it be necessary for chief cells to release an inactive form of the enzyme (the proenzyme pepsinogen)?

2 How would consuming an antacid affect both stomach acid and pepsin?

3 Would you expect an antacid to clear an infection of *H. pylori*?

4 The ammonia-producing enzyme urease stops functioning below pH 4. How does *H. pylori* ensure that ammonia can be secreted into the lumen of the stomach?

As a substrate is converted to product, its concentration decreases. Take care to distinguish graphs showing reaction rate over time from those plotted against different environmental conditions.

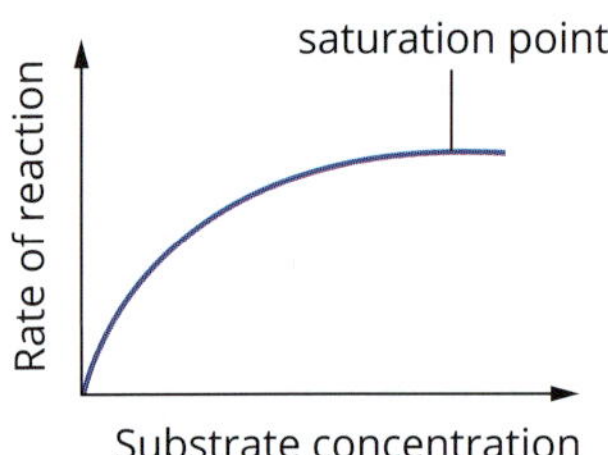

FIGURE 5.2.6 Increasing substrate concentration will increase the rate of reaction until all the active sites of the enzyme are occupied (the saturation point).

Enzyme and substrate concentration

A single enzyme molecule can be used over and over, with its active site occupied for just milliseconds at a time. This means that a reaction can be catalysed effectively with relatively small concentrations of enzyme compared to substrate. The more enzyme molecules that are present, the shorter the wait time is for substrates to reach an available active site. Increasing enzyme concentration will therefore increase the rate of reaction, unless the substrate runs out.

Raising the concentration of substrate can also increase the reaction rate. In dilute solutions, water molecules get between the enzyme and substrate molecules and reduce the chance of them connecting. By increasing substrate concentration and decreasing the number of water molecules, there is a greater chance of a substrate molecule meeting with an enzyme's active site. This means that the rate of reaction will increase to a maximum when every possible active site is filled. This point is known as the **saturation point** (Figure 5.2.6).

Inhibition of enzyme activity

The inhibition of an enzyme by an inhibiting molecule can be categorised in different ways. First, whether the inhibiting molecule binds reversibly or irreversibly. Second, whether the inhibiting molecule binds to the enzyme's active site or an allosteric site (elsewhere).

Enzyme inhibition is also classified as being competitive inhibition or non-competitive inhibition, depending on where the inhibiting molecule binds to the enzyme.

Reversible inhibition

In **reversible inhibition**, the bonds formed between the inhibitor and the enzyme are weak, such as hydrogen bonds and hydrophobic interactions, so they are easily broken. This means that diluting the inhibitor lessens its effect. Also, if the inhibitor works by blocking the active site, its effects can be reduced by increasing substrate concentration, since the two molecules are in competition for the same space and binding is only temporary. An example of this can be seen with the enzyme alcohol dehydrogenase. It usually breaks down ethanol, but methanol can compete reversibly for its active site. Since methanol breaks down to toxic byproducts, a person with methanol poisoning can be given ethanol to slow the rate of methanol digestion.

Irreversible inhibition

In **irreversible inhibition**, the bonds formed between the inhibitor and the enzyme are strong, such as covalent bonds, so they cannot be broken without also breaking apart the enzyme itself. This means the enzyme is permanently disabled and that increasing substrate concentration will have no effect on reaction rate. An example is the herbicide glyphosate, a small molecule that binds irreversibly to part of the active site of an enzyme needed by plants to synthesise three amino acids.

Competitive inhibition

Competitive inhibition occurs when the shape of the inhibitor is similar to the shape of the substrate that normally binds to the active sites of a particular enzyme. Due to their similar shapes, such inhibitors are able to bind to the active site of the enzyme, and block the substrate from binding to the site (Figure 5.2.7). Unlike a substrate, when an inhibitor binds to an enzyme's active site it does not trigger a catalytic reaction.

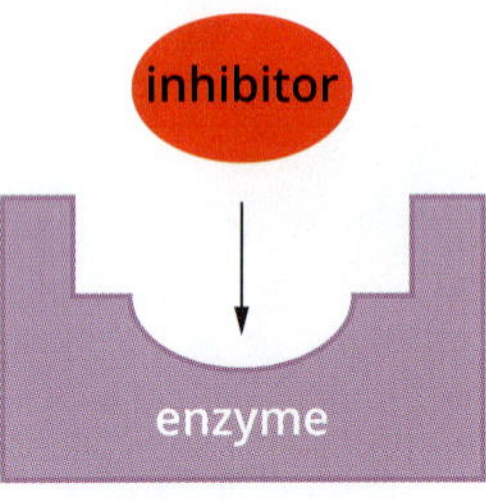

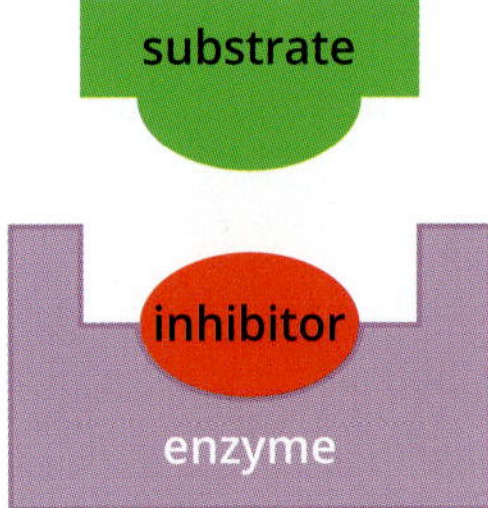

FIGURE 5.2.7 Competitive inhibition involves an inhibitor binding directly to the active site of the enzyme. The substrate is then unable to bind to the enzyme.

Non-competitive inhibition

Non-competitive inhibition (or allosteric inhibition) occurs when an inhibitor binds to an enzyme site other than the active site (known as an **allosteric site**). Binding to the allosteric site either changes the shape (or conformation) of the enzyme such that the substrate cannot bind to its active site (Figure 5.2.8), or it prevents a catalytic reaction from proceeding even if the substrate is bound.

Binding of molecules to an allosteric site does not always result in allosteric inhibition. Some molecules that bind to allosteric sites can cause conformational changes that allow reactions to occur (or allosteric activation). Whether binding to an allosteric site causes inhibition or activation, both are examples of allosteric regulation.

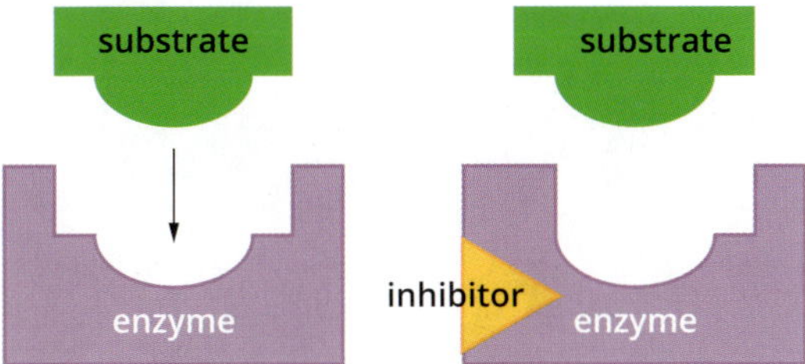

FIGURE 5.2.8 Non-competitive inhibition involves an inhibitor binding to the enzyme and causing a change in the shape of its active site, so that the substrate no longer fits.

Feedback inhibition

Feedback inhibition occurs when a product produced late in a pathway is also an inhibitor of an enzyme earlier in the pathway. As the amount of the inhibiting product increases, the number of enzyme molecules being inhibited also increases (Figure 5.2.9). This in turn reduces the amount of the inhibiting product. As the level of inhibiting product reduces, less of it will bind to the enzyme, allowing the enzyme to function again. Feedback inhibition is an important mechanism in controlling enzyme activity. In cellular respiration, the loaded coenzyme ATP can also be thought of as a product. High levels of ATP cause feedback inhibition at several stages of the biochemical pathway so that an excess of ATP (which is unstable, so not useful unless needed straightaway) is not made.

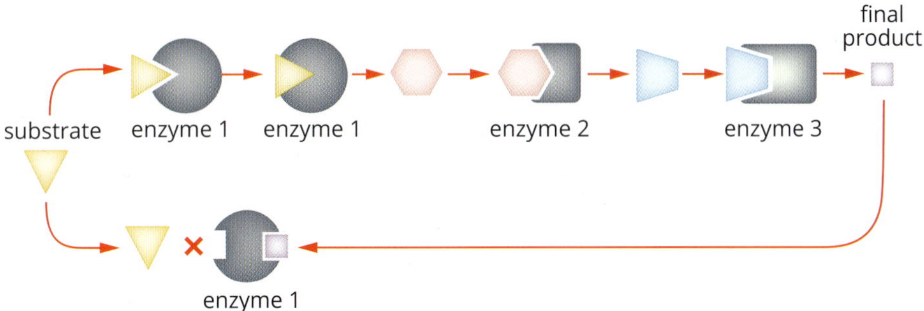

FIGURE 5.2.9 In feedback inhibition, when the amount of inhibiting product is high, the pathway slows down. When the amount of inhibiting product is low, the pathway speeds up.

CASE STUDY

Enzyme inhibition to treat Alzheimer's disease

Alzheimer's disease is characterised by a progressive degeneration of the brain that affects memory and cognitive function. A drug currently used in Australia to treat Alzheimer's disease inhibits the activity of the enzyme acetylcholinesterase in the central nervous system. The role of acetylcholinesterase is to break down the substrate acetylcholine (ACh), a neurotransmitter that is important in memory processes. Neurotransmitters are signalling molecules of the nervous system. When ACh is broken down, nerve cells (neurons) are able to return to their resting state.

In people who suffer from Alzheimer's disease, the level of ACh is low, so less memory-related signalling occurs. To compensate for the low levels of ACh, drugs that inhibit the activity of acetylcholinesterase can be used (Figure 5.2.10). These drugs can act by either competitive or non-competitive inhibition. Figure 5.2.11 shows an example of an acetylcholinesterase inhibitor that is used to treat Alzheimer's disease and acts through competitive inhibition.

FIGURE 5.2.10 Molecular model of acetylcholinesterase enzyme (purple) with a competitive inhibitor (green) bound to the active site preventing acetylcholine from binding

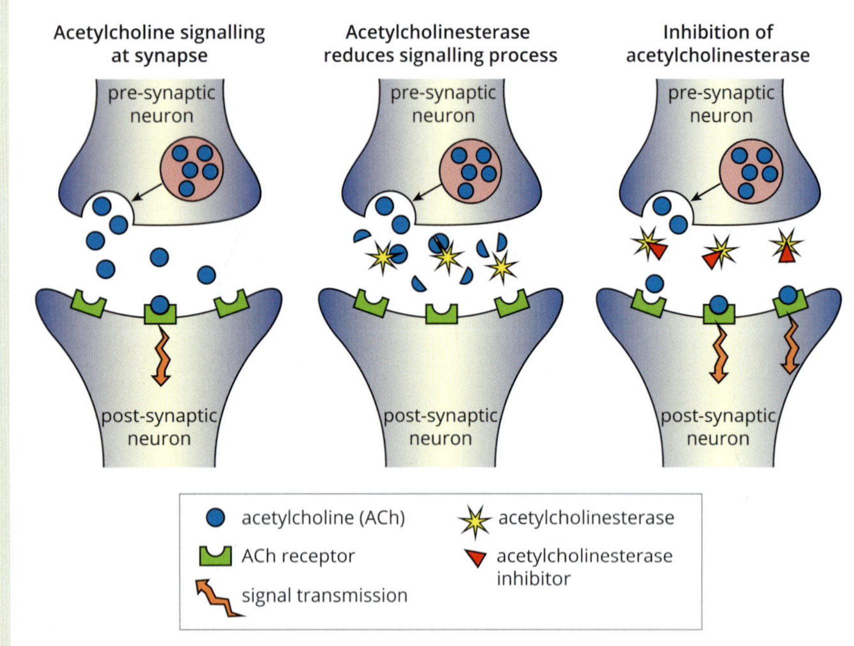

FIGURE 5.2.11 Inhibition of acetylcholinesterase blocks the enzyme's activity and allows increased nerve stimulation.

5.2 Review

OA ✓✓

SUMMARY

- Enzymes have optimum conditions in which they perform their best (fastest rate of reaction). Factors include:
  - temperature—higher temperatures increase molecular collisions up to the point where the enzyme is denatured and function rapidly ceases, while lower temperatures slow enzyme activity
  - pH—function is decreased either side of the optimum and the enzyme may be denatured outside a range of tolerances
  - substrate concentration—rate increases with more molecular collisions until all active sites are occupied. The saturation point is reached and the rate plateaus.
- Inhibition can slow or completely stop enzyme activity. Types of inhibition include:
  - reversible—weak, temporary bonds form between enzyme and inhibitor
  - irreversible—strong, covalent bonds form between enzyme and inhibitor
  - competitive—the inhibitor binds to the active site
  - non-competitive—the inhibitor binds to an allosteric site.

KEY QUESTIONS

Knowledge and understanding

1 In non-competitive inhibition, the inhibitor binds to:
   A a substrate
   B an enzyme's active site
   C an allosteric site on the enzyme
   D another inhibitor

2 Explain the difference between reversible and irreversible inhibition. Include reference in your answer to how changing substrate concentration affects each.

3 The following diagram shows an enzyme being inhibited.

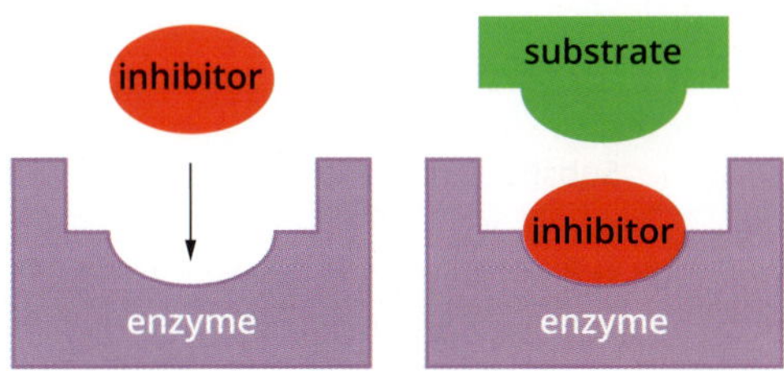

   a Is the diagram an example of competitive or non-competitive inhibition?
   b The inhibitor binds to the enzyme with covalent bonds. What effect would adding more substrate have on the enzyme's activity?

Analysis

4 The following graph shows the activity of human isocitrate dehydrogenase, which is an enzyme used in cellular respiration.

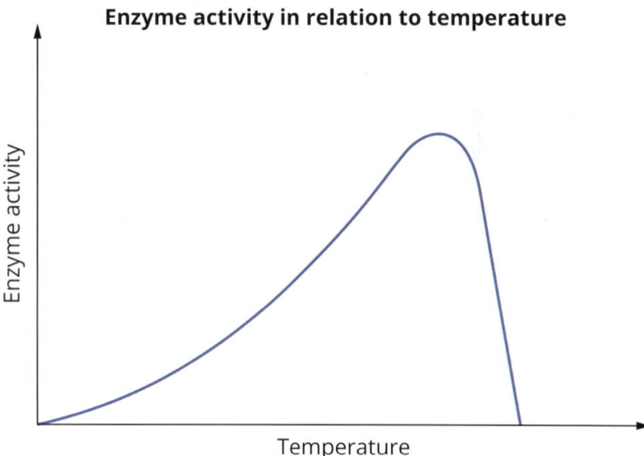

   a What is the optimum temperature of isocitrate dehydrogenase likely to be? Give a reason for your answer.
   b Explain why the shape of the curve is not symmetrical.

5 Pepsin is an enzyme that is released into the stomach of humans (pH 1.5–3.5, 37°C), where it breaks down proteins into polypeptides. Explain how you would expect the activity of pepsin to change as:
   a the temperature is increased from 37°C to 45°C
   b the pH is increased above 5.

Chapter review

KEY TERMS

activation energy
active site
adenosine diphosphate (ADP)
adenosine triphosphate (ATP)
allosteric site
biochemical pathway (metabolic pathway)
catalyse
catalytic power
cellular respiration
chemical group
coenzyme
cofactor
competitive inhibition
conformational change
denature (denaturation)
electron
enzyme
enzyme–substrate complex
feedback inhibition
induced-fit model
irreversible inhibition
loaded coenzyme
lock-and-key model
metabolism
non-competitive inhibition
photosynthesis
proton
reactant
reversible inhibition
saturation point
specificity
substrate
unloaded coenzyme

REVIEW QUESTIONS

Knowledge and understanding

1 Enzymes reduce the activation energy of a reaction by:
 A bringing the reactants close together so that a reaction is more likely to occur
 B orientating the reactants in the most favourable position for the reaction
 C providing a micro-environment favourable to the chemical reaction
 D all of the above

2 An organic molecule required by an enzyme, in order for it to function, is best described as:
 A a cofactor
 B a coenzyme
 C a chemical group
 D an enzyme activator

3 Which of the statements about NAD^+ is not true?
 A NAD^+ can accept a hydrogen ion.
 B When NADH gives up a proton, NAD^+ results.
 C Only a small amount of NAD^+ is needed in a cell.
 D NAD^+ can only be used once before it has to be resynthesised.

4 Which of the statements about photosynthesis is accurate?
 A It releases energy from organic molecules.
 B It includes a cyclic biochemical pathway.
 C It breaks down ATP to release light energy.
 D It does not require regulation by enzymes, unlike cellular respiration.

Application and analysis

5 A student investigating the activity of the enzyme pepsin, which is found in the stomach of humans, observed the change in enzyme activity as the concentration of the substrate (protein) increased. The experiment was conducted at pH 3 and 37°C. The student's data was presented in the graph shown.

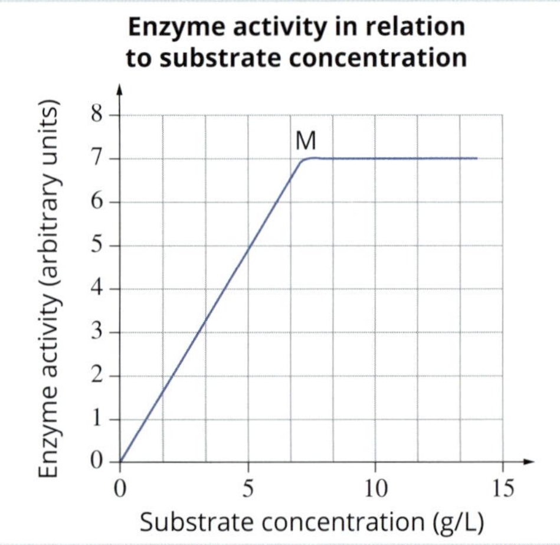

What could the student do to change the experiment so that the rate of reaction at point M was higher than that shown?

6 One of the enzymes used in cellular respiration, phosphofructokinase, functions optimally at 50°C. Despite this, the rate of cellular respiration rapidly decreases at temperatures over 40°C. Explain why this is the case.

7 The following diagram illustrates one model of enzyme activity.

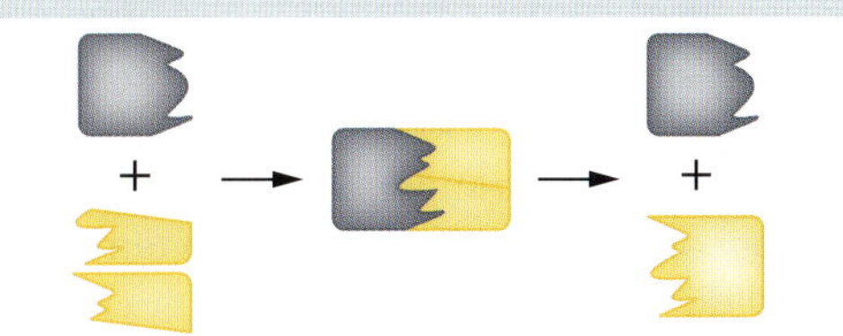

a Explain which of the two models of enzyme activity is being illustrated.

b Explain how this model of enzyme activity could be used to explain how some enzymes can act on multiple substrates.

8 The breaking and formation of bonds between atoms during biochemical reactions results in changes in the energy content of the molecules.

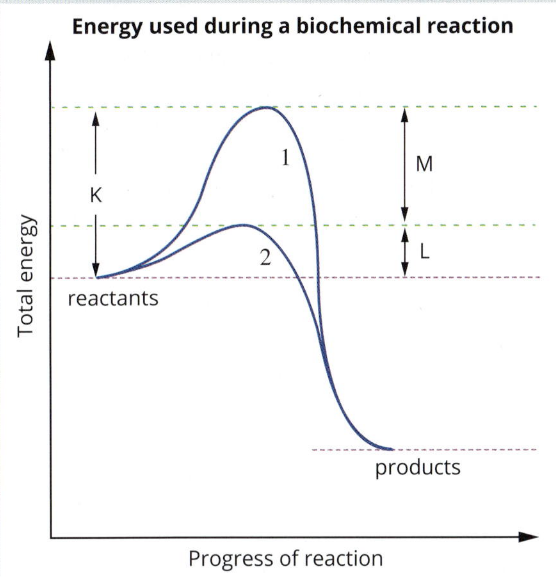

The graph above shows the energy changes during one particular chemical reaction with and without an enzyme.

a Identify the energy change represented by
   i K
   ii L
   iii M

b Overall, would you expect energy to be absorbed or released in this reaction? Explain.

c Explain which line shows the enzyme-catalysed reaction.

9 Trypsin is a digestive enzyme secreted by the pancreas of vertebrate animals. A biologist studying enzyme activity isolated trypsin from a mammal and two different species of fish and tested the enzyme activity at different temperatures. The experimental results are illustrated in the following graph.

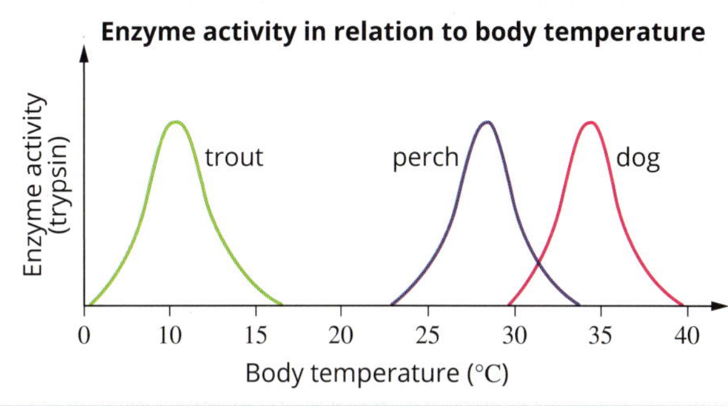

a Define 'optimum temperature of an enzyme'.

b Determine the optimum temperature for trypsin in:
   i trout
   ii perch
   iii dog

c Describe what happens to the activity of trypsin above the optimum temperature in each animal. Explain why this occurs.

d Fish are ectotherms—they depend on external heat sources to regulate their body temperature. Suggest a reason for the difference in the optimal temperatures of the enzyme in trout and perch.

10 An experiment was performed to investigate enzyme activity in three different species: the two-toed sloth (a mammal), an Arctic trout, and a bacterium from a thermal spring. The activities of the enzymes from each organism are plotted on the graph below.

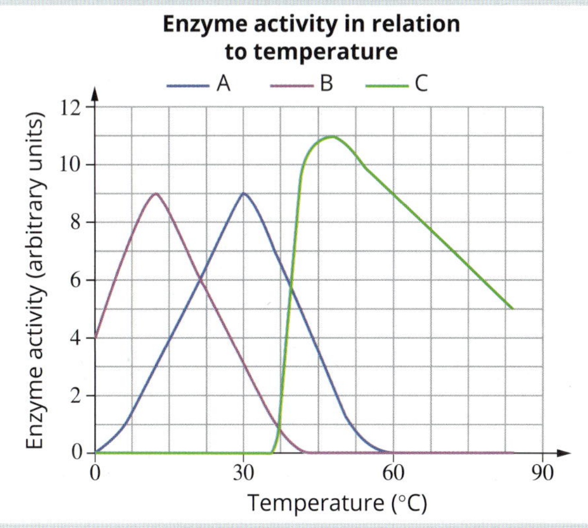

a Suggest which graph belongs to each animal, giving reasons for your answer.

b Why was no activity observed at 60°C for enzyme A?

11 The diagram below shows the enzymes involved in a metabolic pathway.

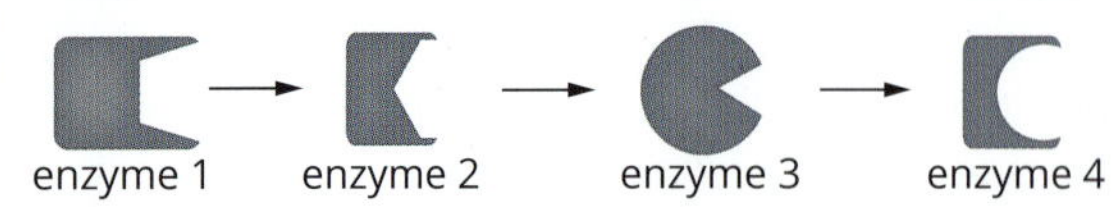

The enzyme substrates are shown below.

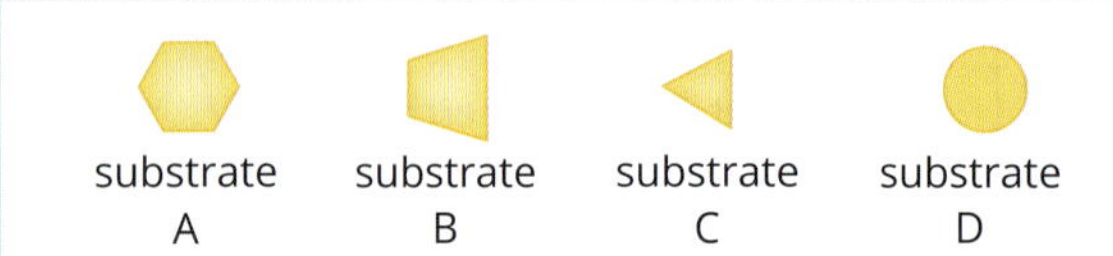

a **i** Match each substrate with its enzyme.

ii What was the basis of your decision?

b The following molecule is the final product of the reaction.

i If the concentration of the final product builds up in the cell, the reaction will stop. Explain how the increase in the concentration of the final product stops the reaction proceeding.

ii Is the product acting as a competitive inhibitor or an allosteric inhibitor? Explain your answer.

12 The rate of reaction was investigated for an enzyme at different concentrations of substrate, marked 'original' in the graph below. The experiment was repeated with a small amount of a reversible competitive inhibitor added to the mixture. Which curve (A, B, C or D) best represents the new conditions? Explain your choice.

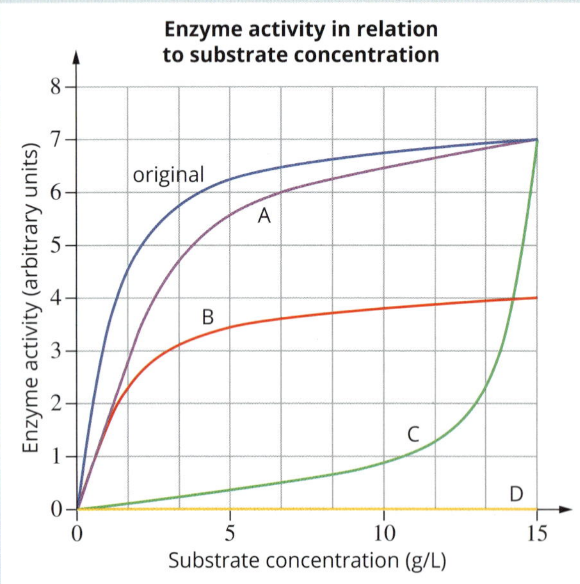

13 A cardiopulmonary bypass, or heart–lung machine, is used to redirect blood flow from a patient undergoing complex heart surgery, as seen in the image below. The blood is oxygenated and pumped by the machine. It is also cooled, which keeps the patient's body 5–9°C below normal body temperature. Why would cooling be advantageous if blood flow had to be temporarily stopped during surgery?

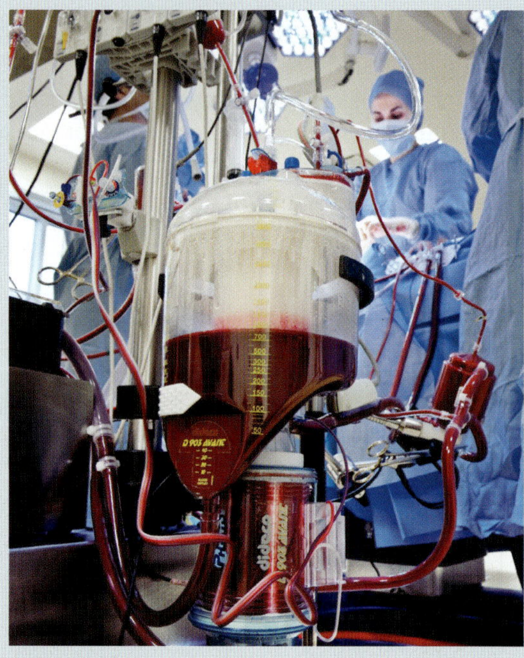

14 Conditions can vary greatly in Antarctica, from the cold of the ice sheets to the superheated water surrounding undersea volcanoes. A psychrophilic (cold-loving) bacterium, *Psychrobacter*, thrives at temperatures between –10°C and 42°C, while *Mathanopyrus*, a hyperthermopile, is known to reproduce in temperatures between 85°C and 110°C.

In a laboratory experiment, cultures of *Psychrobacter* and *Methanopyrus* were incubated at a temperature of 60°C for three hours. The bacterial cultures were then returned to their optimal temperature and the growth of the bacteria in each culture was monitored.

a What is meant by the optimal temperature for a protein?

b Which of the cultures, if any, would you expect to show growth? Explain your reasoning.

15 Robert O'Hara Burke and William Wills were famous Australian explorers. They were the first Europeans to cross Australia from south to north and nearly back again. It is thought that they perished from a lack of vitamin B_1 (thiamine), a molecule that undergoes modification to become a coenzyme.

Near the end of their lives, when they were stranded at Coopers Creek, they subsisted almost entirely on the ground sporocarps (spore cases) of the aquatic fern *Marsilea drummondii*, or nardoo (shown below). They mixed the flour from the nardoo sporocarps with water to make a type of thin uncooked porridge. Nardoo contains the enzyme thiaminase. The local Yandruwandha women also used nardoo in their flour to bake damper.

Determine why the thiaminase in the nardoo may have caused the death of Burke and Wills, but not the local Yandruwandha people.

16 Consider the following incomplete graph for an enzyme-controlled reaction in which the enzyme is present at concentration *x*. Assume there is a fixed amount of substrate present.

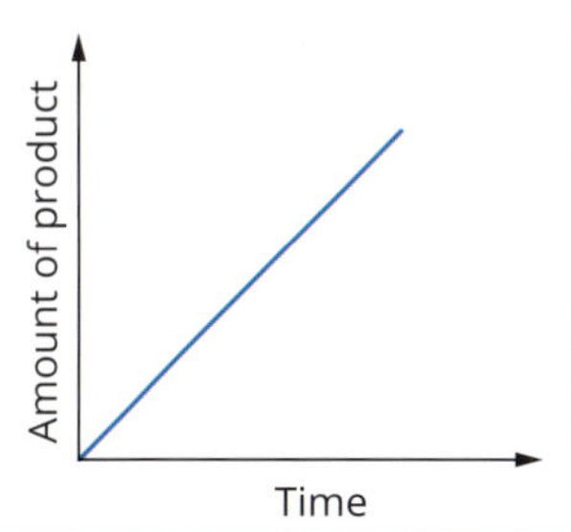

a Eventually the shape of the graph will change. Continue the line graph according to your expectations and explain what happens.

b Redraw the graph for an enzyme concentration of 2*x*.

17 The following graph illustrates the relationship between the concentration of an enzyme substrate and enzyme activity.

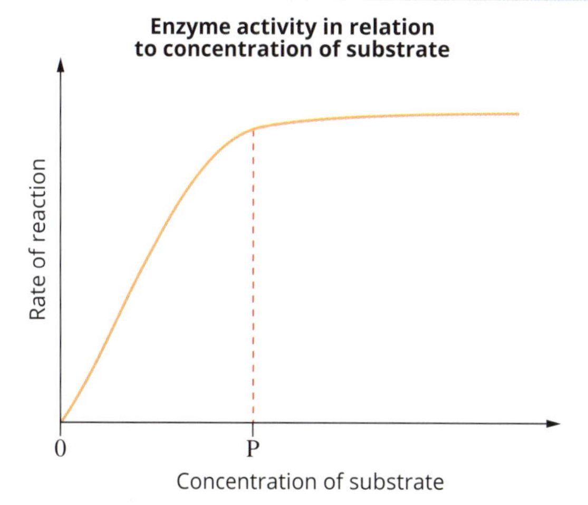

a Describe the relationship between the substrate concentration and the rate of reaction from 0 to P units of substrate concentration.

b What happens at and after point P?

c The concentration of the enzyme is described as being a limiting factor. Explain what this means.

18 During chemical reactions, changes in free energy occur. The change of energy is given as ΔG (delta G). A positive energy change means that there had to be an input of energy for the reaction to occur and a negative energy change means the reaction released energy. Energy released by one step can be used in subsequent steps.

A particular metabolic pathway involves four enzymes (K, L, M and N) in a four-step pathway. Molecule A is the initial substrate of the pathway and molecule E is the final product.

$$\mathbf{A} \xrightarrow{K} \mathbf{B} \xrightarrow{L} \mathbf{C} \xrightarrow{M} \mathbf{D} \xrightarrow{N} \mathbf{E}$$

The energy requirements of each step (in joules) are shown in the table below.

Step	ΔG	Enzyme
A → B	+5	K
B → C	+2	L
C → D	–6	M
D → E	–4	N

a Identify which steps release energy.

b Is the overall metabolic pathway likely to be building or taking apart large organic molecules?

c The shape of enzyme K is shown below.

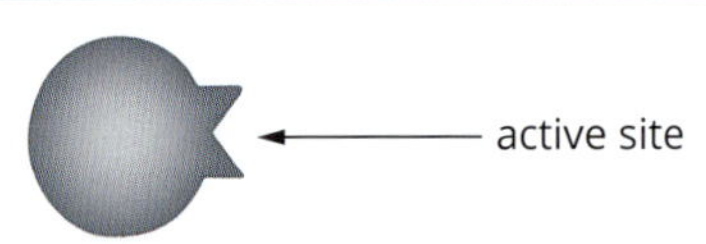

High concentrations of molecule E inhibit enzyme K.

i What is the substrate for enzyme K? Draw a possible shape for this molecule.

ii Draw a possible shape for molecule E.

iii Using annotated diagrams explain how molecule E inhibits enzyme K catalysing the formation of molecule B, thereby regulating the pathway.

19 Bile acids are processed in the liver to form bile salts. Bile salts are used in the small intestine to mechanically break down fats in the diet. The metabolic pathway for the formation of two bile acids is shown in the flow chart. The chart is simplified; some steps in the pathway between 3β, 7α-dihydroxy-5-cholestenoic acid and chenodeoxycholic acid are not known, nor are some of the enzymes in the pathway.

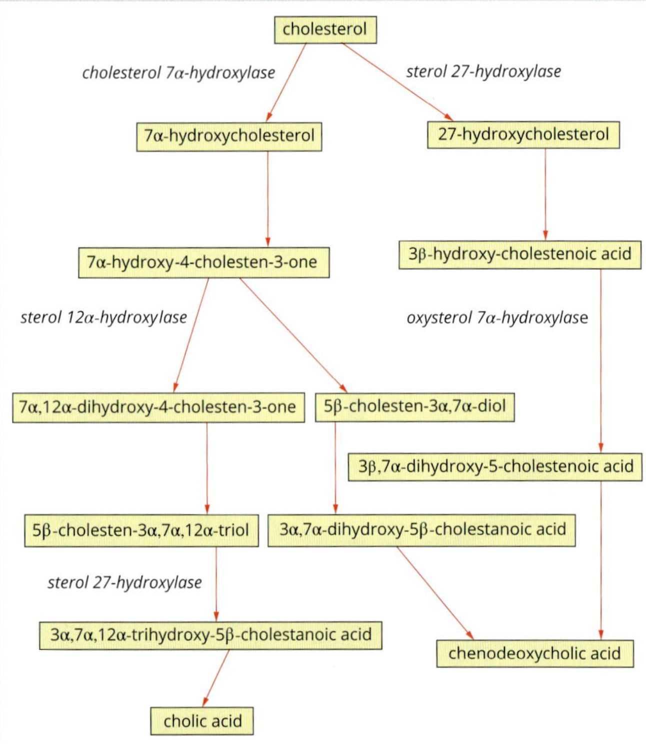

a One enzyme used in the pathway is sterol 27-hydroxylase. If an individual was unable to make functional sterol 27-hydroxylase, would they be able to make bile acids? Explain your answer.

b Infer whether it is likely that chenodeoxycholic acid acts as an inhibitor of sterol 27-hydroxylase.

c Suggest how sterol 27-hydroxylase can catalyse two different steps in the pathway.

20 One toxic product of cellular respiration is hydrogen peroxide (H_2O_2). All living cells contain the enzyme catalase to speed up the breakdown of hydrogen peroxide to water and oxygen. The chemical breakdown occurs according to the following equation:

$$2H_2O_2 \rightarrow 2H_2O + O_2$$

A group of students performed an experiment to investigate the effects of varying substrate concentration on the activity of catalase. Potatoes were used as a source of catalase. The potatoes were pureed and then the puree was strained to collect the potato juice, which contains the enzyme catalase.

The equipment was set up to collect the oxygen gas as it was produced, as shown in the experimental setup below. The measuring cylinder was filled with water before being inverted into the tank, which was also filled with water. 20 mL of potato juice was placed in the flask and the flask was sealed as shown.

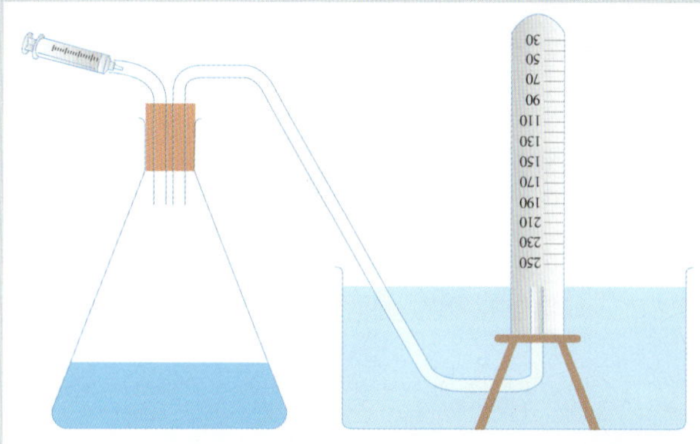

2 mL of hydrogen peroxide was drawn into the syringe and attached to the tube as shown. A 250 mL measuring cylinder and a 10 mL syringe with 1 mL graduations were used. The classroom clock was used to time the reaction. Timing began as soon as the syringe was depressed to mix the hydrogen peroxide and the enzyme mixture.

In the first test, 2 mL of a 5% solution of H_2O_2 was used. Subsequent tests used 2 mL of 10%, 15%, 20% and 25% solutions of H_2O_2. The amount of oxygen produced in 5 minutes was measured. The results obtained by the students are shown in the table below.

Volume of oxygen produced by increasing concentrations of H_2O_2 when acted on by catalase from potato.

Concentration of H_2O_2	Volume of oxygen collected
5%	9.5 cm^3
10%	20 cm^3
15%	31 cm^3
20%	40.5 cm^3
25%	52 cm^3

a Why can the volume of oxygen produced be used as a measure of the activity of catalase?

b i Plot a graph of the students' results.

ii Consider the graph. What relationship appears to exist between the concentration of H_2O_2 and the volume of oxygen produced?

c i Identify possible sources of error in the experimental design and equipment.

ii For each source of error identified, suggest a way of reducing the error or its impact on the results.

d How could the reliability and accuracy of the results be increased?

e Why must the H_2O_2 and the potato mixture be kept apart until timing starts?

f The breakdown of hydrogen peroxide releases heat.

i Explain how this might impact on the results of the experiment.

ii How might you investigate the level of influence that the heat-producing nature of the reaction is having?

g Enzymes from plants often have a much greater range of temperatures over which they maintain their activity than mammalian enzymes. Why might this be the case?

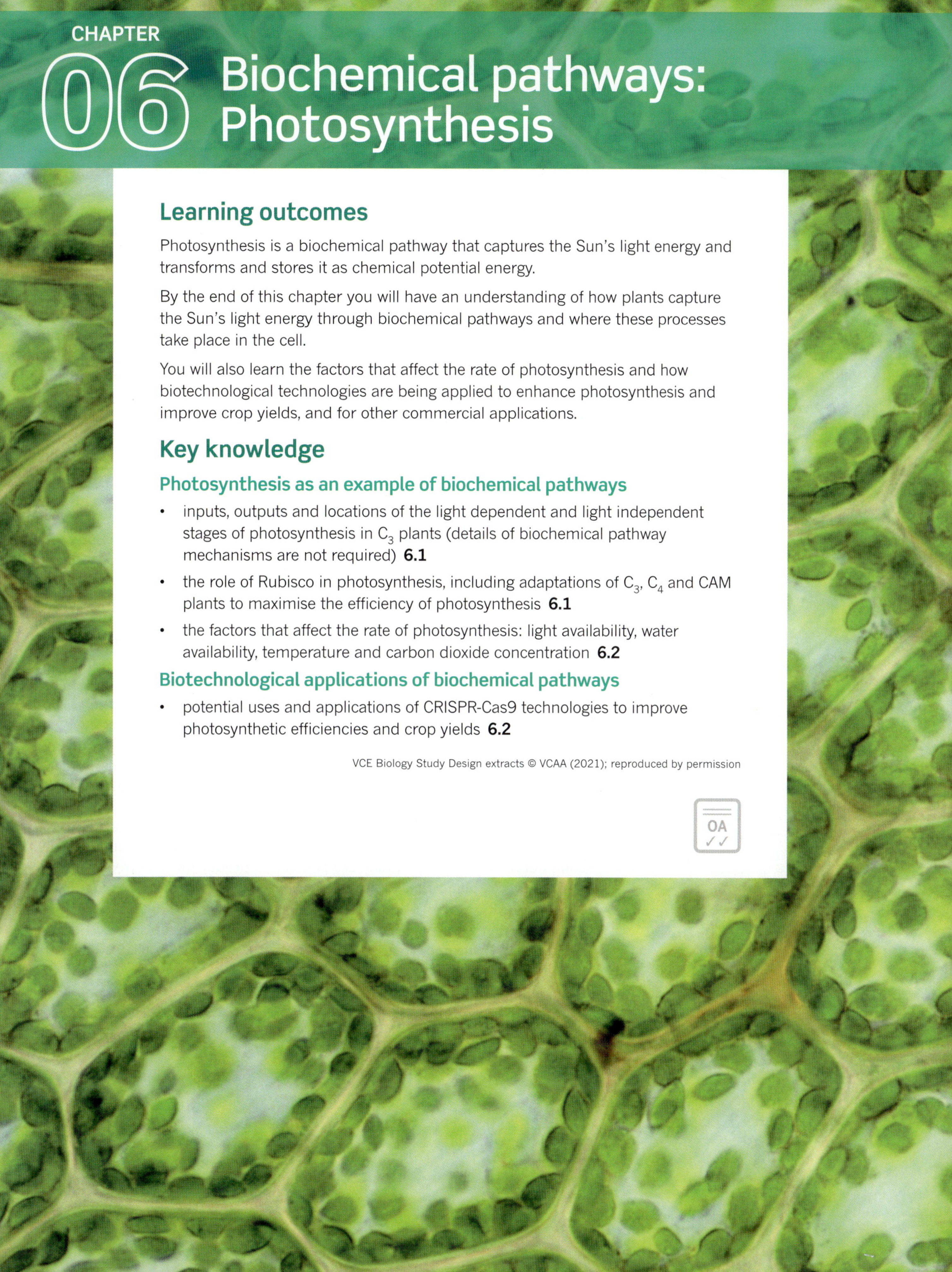

CHAPTER

06 Biochemical pathways: Photosynthesis

Learning outcomes

Photosynthesis is a biochemical pathway that captures the Sun's light energy and transforms and stores it as chemical potential energy.

By the end of this chapter you will have an understanding of how plants capture the Sun's light energy through biochemical pathways and where these processes take place in the cell.

You will also learn the factors that affect the rate of photosynthesis and how biotechnological technologies are being applied to enhance photosynthesis and improve crop yields, and for other commercial applications.

Key knowledge

Photosynthesis as an example of biochemical pathways

- inputs, outputs and locations of the light dependent and light independent stages of photosynthesis in C_3 plants (details of biochemical pathway mechanisms are not required) **6.1**
- the role of Rubisco in photosynthesis, including adaptations of C_3, C_4 and CAM plants to maximise the efficiency of photosynthesis **6.1**
- the factors that affect the rate of photosynthesis: light availability, water availability, temperature and carbon dioxide concentration **6.2**

Biotechnological applications of biochemical pathways

- potential uses and applications of CRISPR-Cas9 technologies to improve photosynthetic efficiencies and crop yields **6.2**

OA
✓✓

6.1 Characteristics of photosynthesis

Most plant species are C_3 plants, including all trees, wheat, rice and soybeans.

A biochemical pathway is a series of chemical reactions that can occur inside a cell.

Living organisms need a constant supply of energy for cellular processes, growth and reproduction. Photosynthesis is the biochemical pathway that plants and other photosynthetic organisms use to convert energy from sunlight into chemical potential energy. In this section, you will learn about the importance of photosynthesis for all life and the inputs, outputs and locations of the different stages of photosynthesis in C_3 plants. You will also learn about the role of the enzyme Rubisco in photosynthesis and the adaptations that C_3, C_4 and CAM plants have to maximise the efficiency of photosynthesis.

CELL REQUIREMENTS—ENERGY

Energy exists in many forms. It is transformed from one form of energy to another but is not destroyed or created. The study of energy is called **thermodynamics**. The energy in sunlight is **solar energy**. The heat generated by your body is thermal energy. When you turn a page, the movement involves kinetic energy. **Chemical energy** is the potential energy that can be released by a chemical reaction. Chemical reactions that require the input of energy are called **endergonic reactions**, while chemical reactions that involve the release of energy are called **exergonic reactions**.

Chemical energy is stored in the bonds or connections that join atoms together: for example, between atoms of carbon and hydrogen in organic compounds such as glucose, fats and proteins. The cells of all organisms use this chemical energy by breaking complex compounds down into simpler compounds. The chemical energy released during these biochemical reactions is then used to make a universal energy-carrying molecule called **adenosine triphosphate (ATP)**. You learnt in Chapter 5 that ATP is formed when **adenonsine diphosphate (ADP)** and inorganic phosphate (P_i) combine during photosynthesis and cellular respiration. When ATP loses a phosphate molecule, ADP is formed and energy is released for use by the cell. You will learn more about ATP in Chapter 7.

Organisms can be divided into two groups depending on the strategies they use to obtain organic compounds, which are in turn their source of energy.

Photosynthesis is sometimes called carbon fixation. This is because carbon atoms from the carbon dioxide gas in the atmosphere are incorporated (fixed) into organic molecules, such as glucose.

- **Autotrophs** (self-feeders) make their own organic compounds from inorganic compounds found in the soil and atmosphere. The conversion of inorganic compounds into organic compounds is called **carbon fixation** because the autotroph fixes inorganic carbon into organic molecules, such as **glucose**. Because autotrophs produce their own nutrients and all of the organic compounds in ecosystems, they are also called producers. Autotrophs include all of the green plants that carry out photosynthesis.
- **Heterotrophs** (other-feeders) obtain organic compounds by consuming (eating) other organisms (autotrophs or other heterotrophs). Because heterotrophs consume organic compounds, they are also called consumers. Heterotrophs include all animals and fungi.

Both autotrophs and heterotrophs use matter (organic and inorganic compounds) to produce the energy required for all biological processes. **Photosynthesis** and cellular respiration are the biochemical processes that cells use to transform matter into energy (Figure 6.1.1).

Cyanobacteria, algae, phytoplankton and terrestrial plants produce glucose through the process of photosynthesis (Figure 6.1.2). This glucose is then used in the biochemical pathway cellular respiration. You will learn more about cellular respiration in Chapter 7.

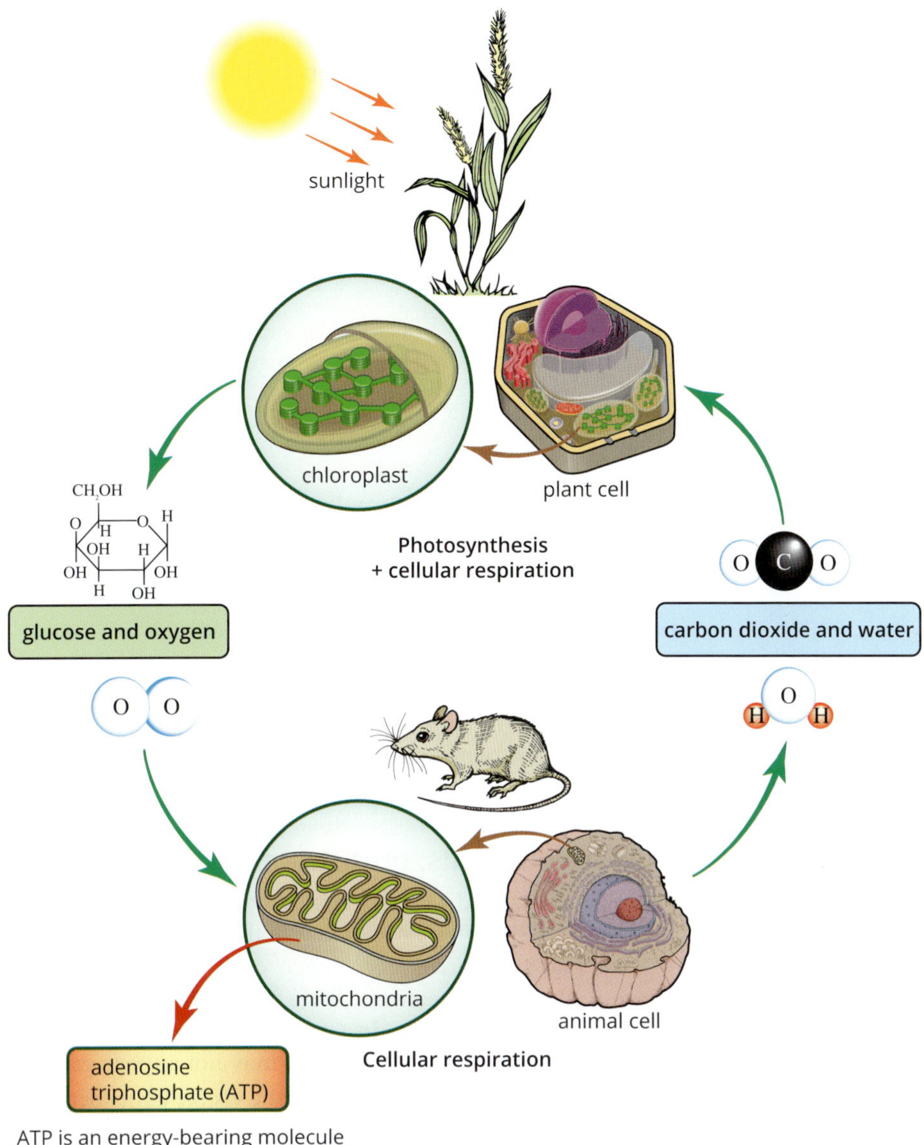

FIGURE 6.1.1 This cycle shows how plants obtain their energy using photosynthesis and cellular respiration. Animals obtain their energy via cellular respiration.

FIGURE 6.1.2 Photoautotrophs use light energy to synthesise organic molecules from carbon dioxide and water. (a) On land, plants are the main photoautotrophs. In aquatic environments, the main photoautotrophs are (b) algae, like this kelp, and (c) prokaryotes called cyanobacteria.

Photosynthesis is an enzyme-regulated biochemical process.

PHOTOSYNTHESIS—CONVERTING SOLAR ENERGY

Photosynthesis ('photo' meaning 'light', 'synthesis' meaning 'putting together') is the process in which plants and other photoautotrophic organisms obtain energy from sunlight to make their own organic compounds (Figure 6.1.3). Photosynthesis is critically important for life on Earth. Carbon fixation and the production of plant matter is essential for providing energy and biomass in both aquatic and terrestrial ecosystems. Plants can produce all the organic compounds they need from the glucose they produce during photosynthesis, as long as they obtain necessary minerals, such as nitrates, from the soil (Figure 6.1.4).

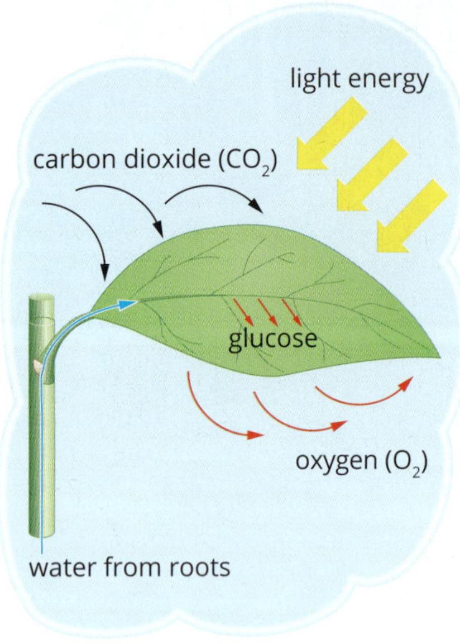

$$6CO_2 + 6H_2O \xrightarrow[\text{chlorophyll}]{\text{light energy}} C_6H_{12}O_6 + 6O_2$$

FIGURE 6.1.3 Simplified representation of photosynthesis

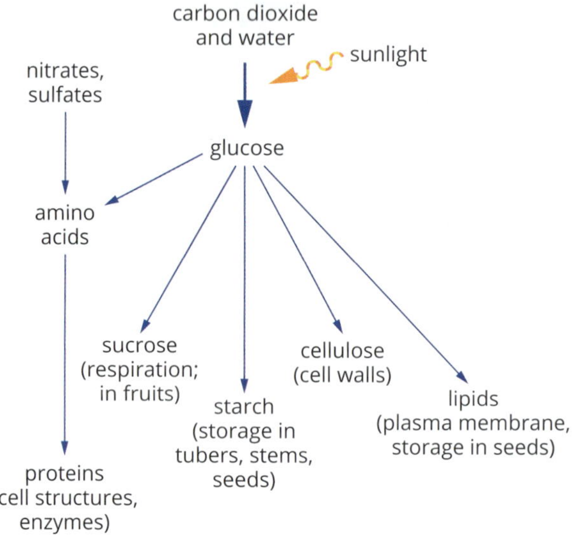

FIGURE 6.1.4 Plants can make all the organic compounds they need from the glucose they produce through photosynthesis.

Photosynthesis in non-green plants, protists and cyanobacteria

All plants (as well as many protists and cyanobacteria) contain **chlorophyll**, a green **pigment** that is able to absorb different light wavelengths (colours), except for green. However, not all pigments are green. Since each pigment captures only a narrow range of the visible light spectrum, there are usually several kinds of pigments present, each of a different colour. In some plants that contain other pigments, these mask the green chlorophyll and the plant may not appear green.

These other pigments absorb light at various ranges of the visible light spectrum, giving the leaves different colours, including red (phycobiliprotein), orange and yellow (carotenoids) and purple (anthocyanin) (Figure 6.1.5). These pigments broaden the range of light that can be absorbed by the plant. They are known as **accessory pigments** because they cannot pass their absorbed energy directly to the photosynthesis biochemical pathway. Instead, they transfer it to chlorophyll to then be used in photosynthesis. Accessory pigments also have functions other than a role in photosynthesis. For example, carotenoids are involved in protecting plants from overexposure to excess sunlight by dissipating excess sunlight as heat. Other accessory pigments act as antioxidants or are used to attract pollinators.

FIGURE 6.1.5 (a) Deciduous plants such as the red maple decrease chlorophyll production in autumn before losing their leaves, revealing the colours of carotenoids. (b) This purple-leaved oxalis plant (*Oxalis triangularis*) contains the pigment anthocyanin.

CASE STUDY **ANALYSIS**

Photosynthetic pigments

Photosynthetic pigments are coloured substances that collect the light energy that is used in photosynthesis. Chlorophyll is the most abundant and visible photosynthetic pigment found in plants.

The rate of photosynthesis is highest under red light (between 650 and 700 nm). The rate of photosynthesis is also high under blue and violet light (between 400 and 450 nm). The lowest rate of photosynthesis occurs under green light (between 500 and 600 m) (Figure 6.1.6a).

Chlorophyll absorbs many of the colours that make up the spectrum of white light, but it reflects green light. There are several types of chlorophyll found in photosynthetic organisms.

- Chlorophyll *a* is found in all photosynthetic organisms.
- Chlorophyll *b* is found in some plants.
- Chlorophyll c is found in algae.
- Chlorophyll *d* and *f* are found in cyanobacteria.

The different types of chlorophyll have slightly different molecular shapes and therefore absorb different wavelengths of light (Figure 6.1.6b). The peaks in the spectrum show that red, blue and violet light are all strongly absorbed, while yellow is absorbed to a lesser extent and green is not absorbed at all.

By comparing Figures 6.1.6a and 6.1.16b, you can see that the pattern of photosynthetic activity is very similar to the absorption spectrum of chlorophyll.

Accessory pigments absorb and capture light at different wavelengths to chlorophyll (Figure 6.1.7).

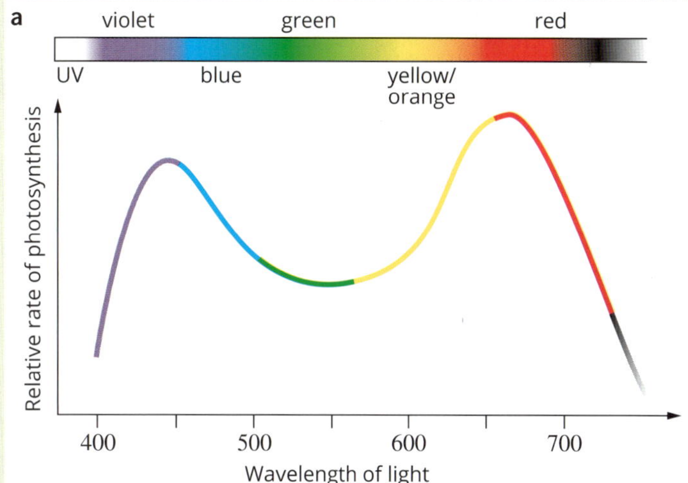

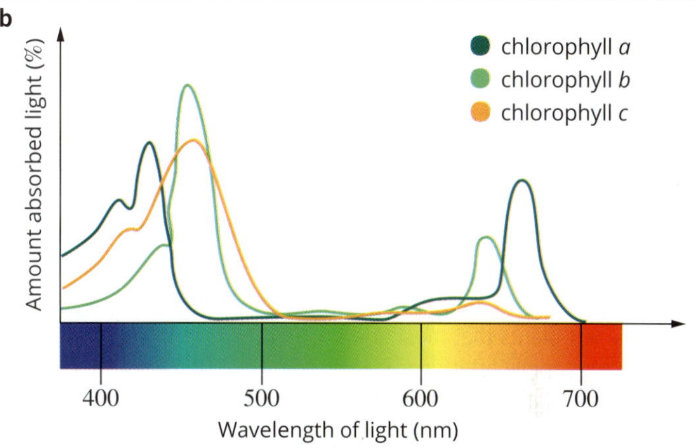

FIGURE 6.1.6 (a) The rate of photosynthesis occurs at different rates in light of different wavelengths. (b) Different types of chlorophyll absorb different wavelengths of light.

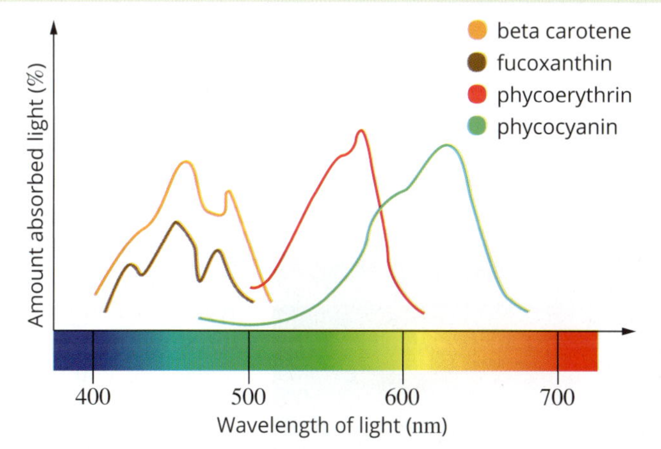

FIGURE 6.1.7 Absorption spectra of different photosynthetic accessory pigments

Analysis

1 How can a range of accessory pigments assist the survival of a photosynthetic organism?

2 Describe how the wavelength ranges for the absorption of light of chlorophyll *a* and beta carotene differ.

3 Phycocyanin is an accessory pigment found in cyanobacteria. It is commercially valued, harvested and sold for nutritional supplements and as a natural dye for food and cosmetics.
   a Using the absorption spectrum in Figure 6.1.7, determine what colour phycocyanin is. Give reasons for your answer.
   b Samples of cyanobacteria were grown under either orange, yellow or white light. Determine which wavelength would produce the greatest rate of photosynthesis and therefore more biomass of cyanobacteria for harvesting.

PHOTOSYNTHETIC REACTIONS

Photosynthesis is a complex process in which solar energy is converted into chemical energy and stored in the form of glucose. The glucose formed in photosynthesis may be:

- used as an immediate source of energy by the plant
- stored by the plant as starch for later conversion back to glucose and use as a source of energy
- used as a chemical starting point for the synthesis of complex compounds, such as cellulose and proteins.

The oxygen (O_2) formed in photosynthesis may be used by the plant for cellular respiration or released into the atmosphere.

Photosynthesis is a multistage process, but it is usually summarised in the following way:

INPUTS				OUTPUTS				
carbon dioxide	+	water	$\xrightarrow[\text{chlorophyll}]{\text{light energy}}$	glucose	+	water	+	oxygen
$6CO_2$	+	$12H_2O$	$\xrightarrow[\text{chlorophyll}]{\text{light energy}}$	$C_6H_{12}O_6$	+	$6H_2O$	+	$6O_2$

Each stage involves a series of biochemical reactions, often referred to as a biochemical pathway (or metabolic pathway). Each reaction in the pathway is catalysed (accelerated) by a particular enzyme and spread over two stages: light-dependent reactions and light-independent reactions.

> ℹ During photosynthesis 12 H_2O are consumed and 6 H_2O are produced. A simpler equation for photosynthesis only shows the net water consumed:
> $6CO_2 + 6H_2O \rightarrow C_6H_{12}O_6 + 6O_2$

> ℹ Light energy is written above the arrow in the equation for photosynthesis, as it is not a reactant of photosynthesis, but it is required for the reaction to take place.

Location of photosynthetic reactions—chloroplasts

In the cells of eukaryotic autotrophs, photosynthesis occurs in organelles called **chloroplasts**. Chloroplasts are found in the cells of the leaves and green stems of plants, which are green because of the pigment chlorophyll that is inside the chloroplasts. The chlorophyll molecules absorb energy from sunlight, the essential first stage of photosynthesis.

In vascular plants, chloroplasts are found in the **mesophyll cells**, which make up the upper surface of the leaves. The location of mesophyll cells near the upper surface of the leaf means that the chloroplasts have maximum exposure to sunlight, increasing the rate of photosynthesis.

Each chloroplast has an outer and an inner membrane, which together regulate the movement of materials into and out of the organelle. Inside these membranes is a fluid matrix called **stroma** and a highly complex inner **thylakoid membrane** system (Figure 6.1.8).

Light-dependent reactions

The first stage of photosynthesis is the **light-dependent reactions** (or light reactions), which can only take place in the presence of light. The thylakoid membrane is the site of the light-dependent reactions of photosynthesis. Molecules of chlorophyll are embedded in the thylakoid membrane where they absorb light and convert the light energy to chemical energy. The thylakoid membrane is extensively folded with a high surface-area-to-volume ratio. The folded structure of the thylakoid membrane increases the efficiency of the light-dependent photosynthetic reactions. Figure 6.1.8 shows that the thylakoid membranes fold to form flat hollow discs, which form stacks called **grana** (singular granum). Each granum looks like a stack of coins. Between the grana are flat membrane sheets called **thylakoid lamellae**.

Chlorophyll captures solar energy and uses it to produce the energy-carrying ATP molecules and **nicotinamide adenine dinucleotide phosphate (NADPH)**. During this process, photolysis occurs, where water is split into hydrogen ions and oxygen gas (Figure 6.1.9).

BIOFILE

Glucose reserves in flowering plants

Plants often produce more glucose than they need and there is an obvious survival advantage in being able to store energy reserves. Energy reserves enable plants that were dormant over winter to grow rapidly in spring. Energy stores are also important for plants such as desert plants that must flower and fruit very rapidly after rain. Plants store carbohydrates mainly in the form of starch, a large molecule made up of glucose subunits. The starch molecules cluster into dense granules in cells. The seeds of many plants contain a store of oils and carbohydrates to provide energy for the growth of the new seedling (see figure).

Young plant growing from a seed

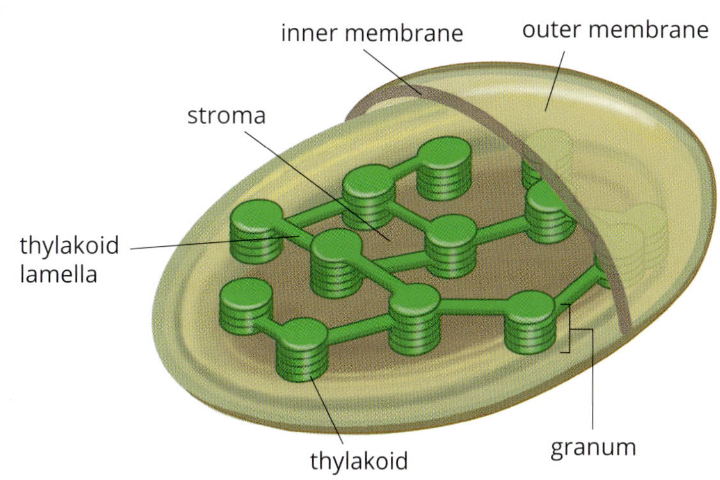

FIGURE 6.1.8 (a) Transmission electron microscopy image of two chloroplasts in a cell in the leaf of a pea plant (*Pisum sativum*). The chloroplasts are seen here in side view, and the grana are coloured yellow. (b) Diagram showing the main structures of a chloroplast

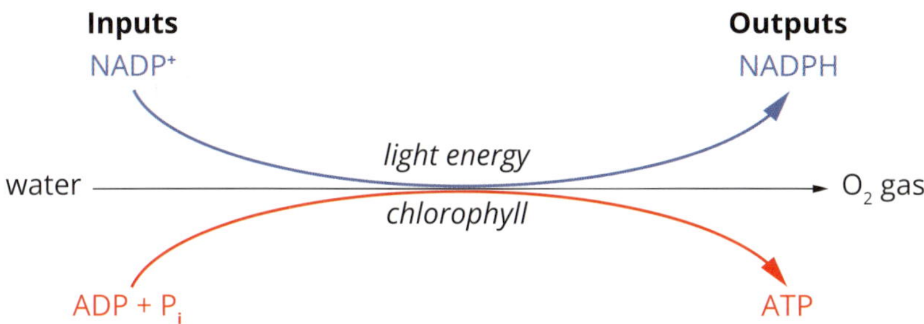

FIGURE 6.1.9 In the first stage of photosynthesis (the light-dependent reactions), water is split into O_2 and hydrogen. The O_2 is released as a gas. The NADPH and ATP are used in the second stage of photosynthesis.

> $NADP^+$ is a carrier molecule. It can receive hydrogen ions to become NADPH.

> ADP can create a temporary bond with inorganic phosphate (P_i) to become ATP. ATP is an energy-carrying molecule.

Light-independent reactions

The second stage of photosynthesis is the **light-independent reactions** (or dark reactions), which do not require solar energy. They do require the NADPH and ATP produced in the light-dependent reactions. ATP provides the energy for the light-independent reactions. This energy is needed to combine carbon dioxide (CO_2) with hydrogen ions (also from the light-dependent reactions) to form glucose and water (Figure 6.1.10). The light-independent reactions take place in the stroma (fluid part) of the chloroplasts.

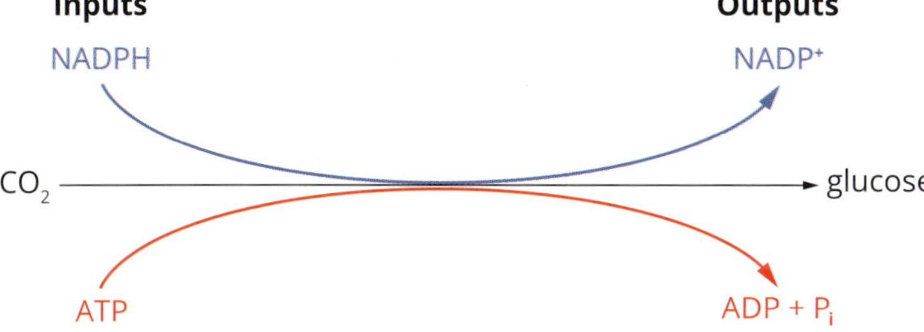

FIGURE 6.1.10 In the second stage of photosynthesis (the light-independent reactions), CO_2 is reduced to form the sugar glucose.

The main stage of the light-independent reactions is the **Calvin cycle** (Figure 6.1.11). The identifiable product of the Calvin cycle is a three-carbon carbohydrate called glyceraldehyde-3-phosphate (GAP). In the cytoplasm, two GAP molecules combine to produce the glucose molecule that is normally identified as the product of photosynthesis. The remaining GAP molecules are recycled into ribulose 1,5-bisphosphate (RuBP), which is the starting molecule of the light-independent reaction.

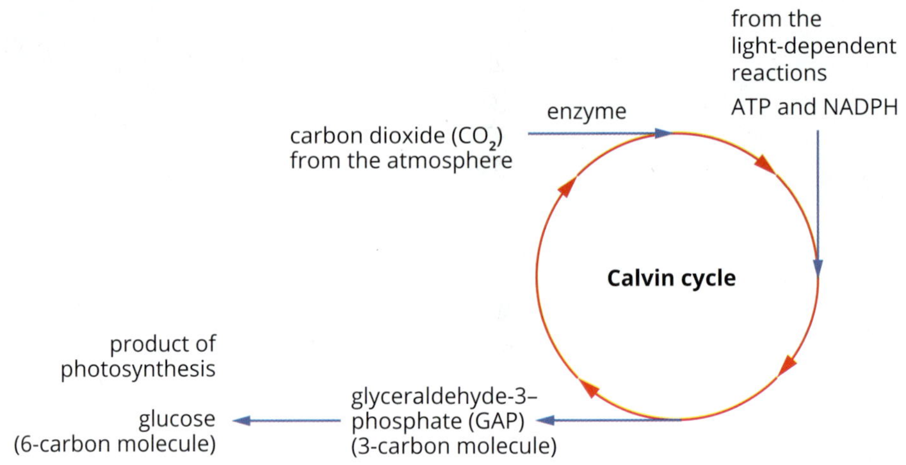

FIGURE 6.1.11 Carbon dioxide from the atmosphere feeds into the Calvin cycle, which uses energy carried by ATP and NADPH from the light-dependent reactions.

For a plant to produce energy, the two stages of photosynthesis are tightly interlinked. The first stage captures light and stores the energy in the bonds of the molecules ATP and NADPH. In the second stage, the ATP is broken down to ADP, releasing energy for the plant to use and the electrons from NADPH to convert the carbon dioxide to glucose (Figure 6.1.12).

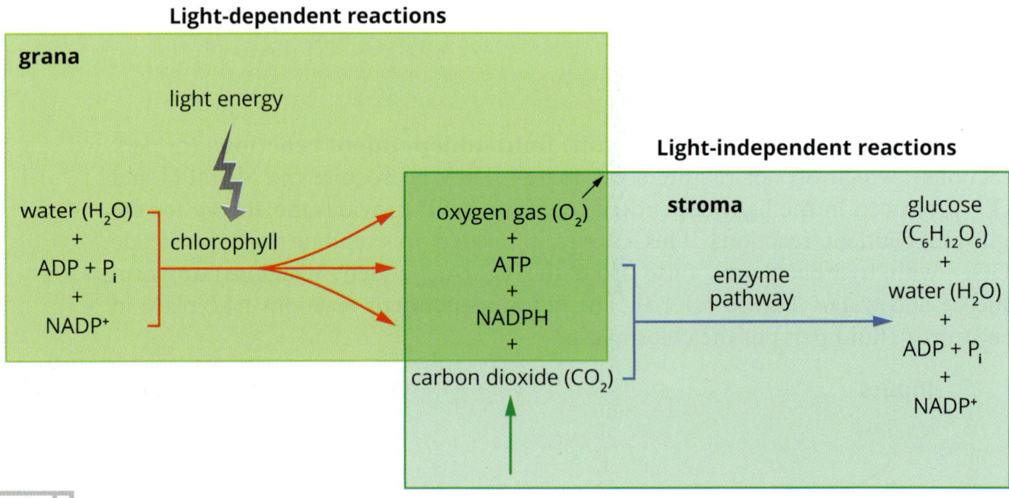

WS 14 PA 7

FIGURE 6.1.12 Photosynthesis occurs in two phases. The light-independent reactions require the energy-carrying molecules, ATP and NADPH, generated by the light-dependent reactions.

Photosynthetic pathways—C_3, C_4 and CAM

Plants living in different environments have different metabolic needs. For example, plants living in hot, dry environments need to conserve more water than those living in cool, wet environments. In response to different environmental conditions, plants have evolved three photosynthetic pathways, known as C_3, C_4 and **crassulacean acid metabolism (CAM)** photosynthesis.

C_3 plants

C_3 plants always use the Calvin cycle for carbon fixation. These plants can fix carbon directly from carbon dioxide. The enzyme responsible for this is **Rubisco** (ribulose-1,5-bisphosphate carboxylase/oxygenase). Rubisco removes the carbon from carbon dioxide, adds it to a 5-carbon molecule (ribulose bisphosphate) and then splits this into two 3-carbon molecules (phophogycerate) that can be used by the cell (Figure 6.1.13).

The enzyme Rubisco is an abundant protein in chloroplasts that fixes inorganic carbon (atmospheric CO_2) into organic sugar molecules that the plant can use.

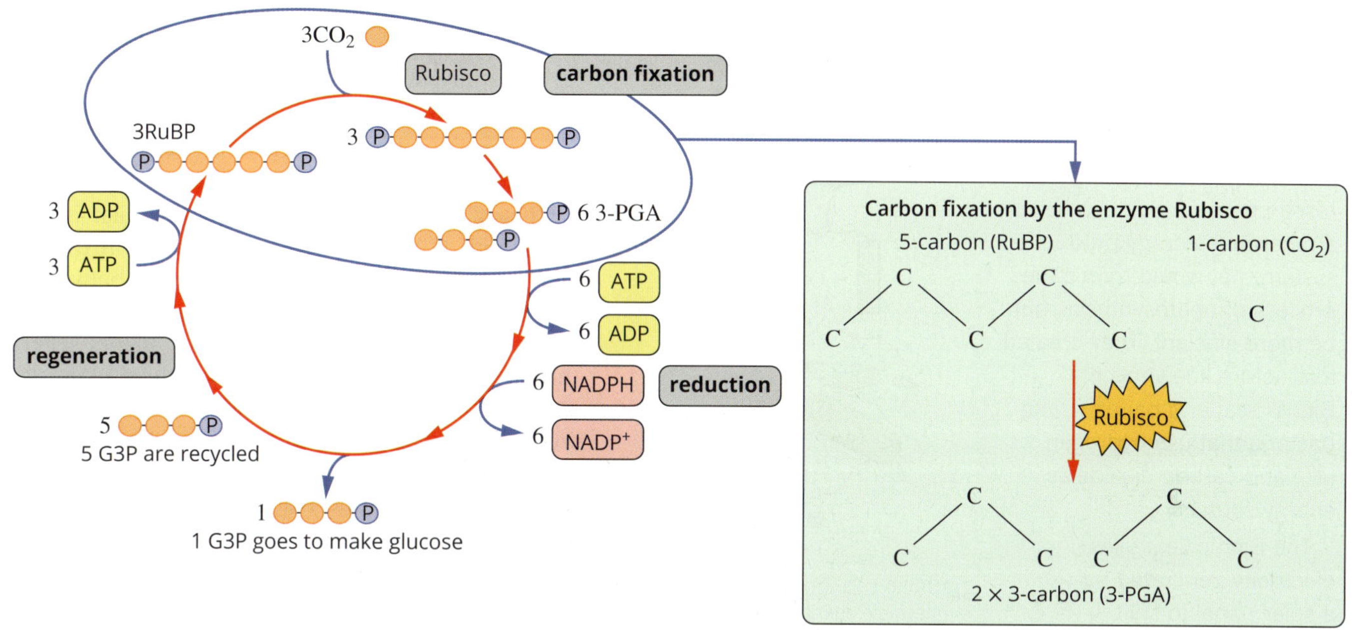

FIGURE 6.1.13 Role of the enzyme Rubisco in the Calvin cycle

Although Rubisco plays an essential role in photosynthesis, it is relatively slow compared to other enzymes. Plants have adapted by producing high concentrations of Rubisco within each chloroplast. Rubisco is believed to be the most abundant protein on Earth.

The other factor that makes Rubisco inefficient in creating sugar molecules is its lack of specificity for carbon dioxide. Oxygen competes with carbon dioxide for the binding site, leading to the pathway **photorespiration**. This results in the loss of between one-quarter and one-half of the fixed carbon, which can no longer be synthesised in the Calvin cycle. At higher temperatures, the loss is even greater, posing a distinct problem for the C_3 plant. This happens because, under higher temperatures, the plant closes its **stoma** (pores in the leaf surface) to reduce water loss by evaporation. Oxygen builds up inside the leaf, leading to a low $CO_2:O_2$ ratio and increasing the chances of oxygen binding to Rubisco. This has led to the evolution of physical and biochemical adaptations in C_4 and CAM plants.

C_4 and CAM photosynthesis are adaptations to conserve water and to maximise the efficiency of photosynthesis.

FIGURE 6.1.14 (a) C_4 (sugar cane) and (b) CAM (pineapple) plants are important commercial crops in tropical Queensland.

C_4 plants

C_4 plants include many grasses and crops, such as sugar cane (Figure 6.1.14a) and sorghum, and operate differently from C_3 plants in the following two ways:

- Instead of Rubisco, an enzyme called phosphoenolpyruvate carboxylase (PEPCase) is used to capture carbon dioxide. PEPCase is less likely to bind to oxygen and therefore photorespiration is less likely to occur. This is an advantage for plants living in hot, dry environments.

CASE STUDY

Bionic Leaf and bacteria make liquid fuel

Scientists from Harvard University have created a system that uses bacteria and solar energy to manufacture a liquid fuel from water (H_2O) and carbon dioxide. The researchers set out to develop a renewable energy production system that would mimic the process of photosynthesis but be more efficient. They created a structure known as the Bionic Leaf and paired it with bacteria that use hydrogen (H_2) and carbon dioxide as energy sources.

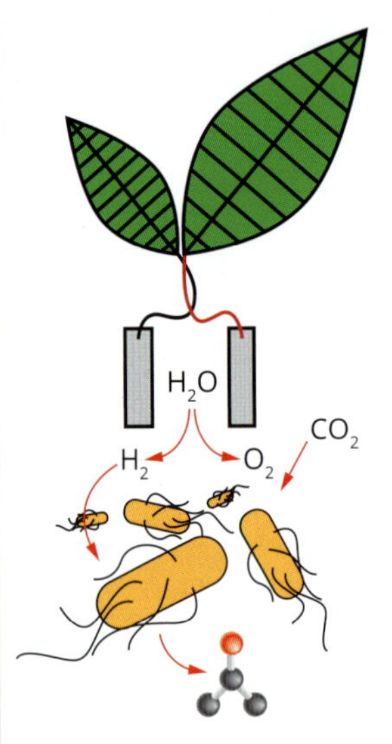

FIGURE 6.1.15 The Bionic Leaf is a renewable energy production system that mimics the natural process of photosynthesis.

The Bionic Leaf uses electricity generated by a solar panel to split water into its component elements (hydrogen and oxygen) by photolysis, just as photosynthesis does. The electrodes of the Bionic Leaf are submerged in a vial containing water and the soil bacterium *Ralstonia eutropha* (Figure 6.1.15). The water-splitting reaction occurs when an electric voltage from the solar panels is applied to the electrodes of the artificial leaf. The bacteria feed on the hydrogen generated from the reaction, along with carbon dioxide bubbles that are added to the system. The bacteria use this food source and produce isopropanol as a by-product.

This system can now convert water and carbon dioxide to fuel at an efficiency of 3.2%—triple the efficiency of photosynthesis. This is due to the solar panels, which have a greater capacity to harvest sunlight than most plants. The researchers' findings were published in 2015 and have great potential for use in many powerful applications. Efficient renewable energy production and storage is one of the important areas where this technology could be applied.

Genetic engineering of bacteria also creates many possibilities for the synthesis and metabolism of a wide variety of chemicals. This might create countless applications for the technology, in both the production of compounds and the removal of chemical pollutants from the environment.

- Photosynthetic reactions occur in different cells in the plant. Carbon is taken up by PEPCase in the mesophyll cells, where it is added to the other carbons, forming a four-carbon molecule, rather than the three-carbon molecule made in the C_3 pathway. The four-carbon molecule (called oxaloacetate) is converted to pyruvate and transported to the adjacent **bundle sheath cells**, where it releases carbon dioxide for carbon fixation by Rubisco in the Calvin cycle. Carbon is transported from many mesophyll cells to fewer bundle sheath cells, resulting in a high ratio of carbon dioxide to oxygen in the bundle sheath cells, which further inhibits photorespiration.

The C_4 and CAM pathways are similar in many ways, with CAM plants using Rubisco in the same way as C_4 plants. The main difference between the two pathways is in how and where the plants take up and concentrate carbon dioxide. C_4 plants separate the uptake of carbon dioxide and the Calvin cycle spatially (in mesophyll cells and bundle sheath cells), while CAM plants separate these reactions over time (day and night). Both types of plants concentrate carbon dioxide around Rubisco in the chloroplasts to increase the efficiency of photosynthesis.

CAM plants

CAM photosynthesis is common in plants that live in hot, dry environments, such as deserts. In **CAM plants**, the stomata only open at night to collect carbon dioxide. Rather than using the carbon dioxide immediately, as non-CAM plants do, the plant stores the carbon dioxide in cell vacuoles as an organic compound called malic acid. During the day, the malic acid is transported to the chloroplasts, where it is used to produce the carbon dioxide needed for photosynthesis. By storing the carbon dioxide required for photosynthesis at night, the plant can close its stomata during the heat of the day to reduce water loss. While this pathway allows CAM plants to survive in extreme heat and aridity, it limits the amount of sugars the plant can produce and is the reason that many desert plants are slow growing. Examples of CAM plants are cacti, orchids and pineapples (Figure 6.1.14b on page 217).

CASE STUDY

Plants and microbes provide electricity in the Amazon rainforest

When floodwater damaged cables supplying electricity to an Indigenous community in the eastern Amazonian region of Ucayali, 65% of the community was left without power.

A research team at the University of Engineering and Technology (UTEC) in Lima, Peru, wanted to provide a clean, renewable source of electrical light to the community. Using findings from research on plant–microbial fuel cells, a research team of seven professors and eight students developed a lamp powered by plants and microbes (Figure 6.1.16).

The 'plant lamp' technology takes advantage of a natural energy exchange between photosynthesis, plant roots and soil bacteria. During photosynthesis, excess energy that is generated by the plant is excreted through the root system into the soil in the form of organic matter. Microbes, known as *Geobacter*, inhabit the soil around the plant roots and break down the plant's organic matter, releasing electrons as a waste product. The 'plant lamp' uses a grid of electrodes to capture the free electrons in the soil and stores the energy in a battery, which is used to power a low energy-consumption, high-illumination LED light bulb.

This innovation has very real economic and educational benefits for people in the remote Ucayali community. The 'plant lamps' can generate a clean and safe source of light for two hours per day and have the potential to be used in many renewable energy applications.

FIGURE 6.1.16 The plant lamp generates a clean and safe source of light.

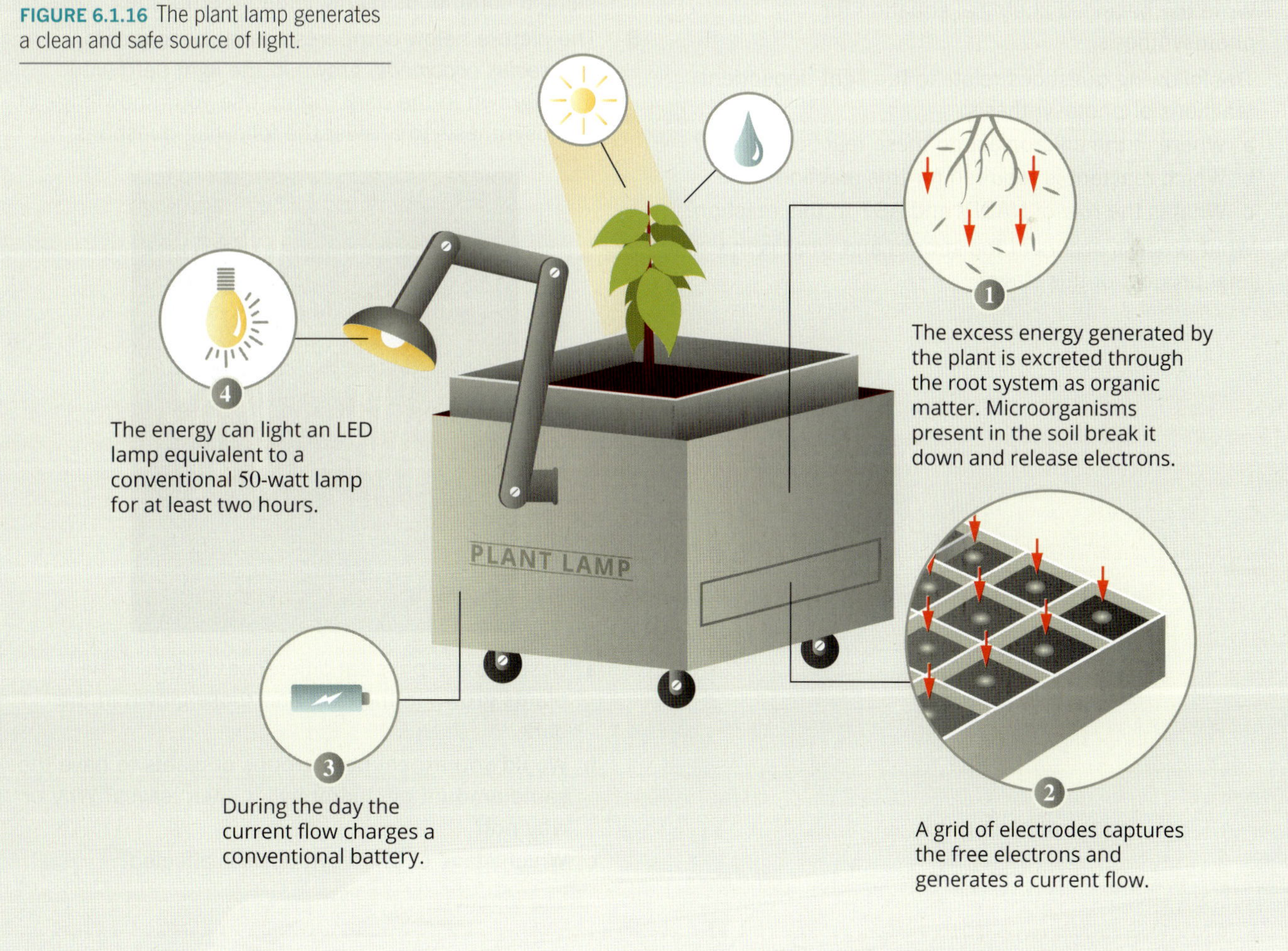

6.1 Review

OA ✓✓

SUMMARY

- Photosynthesis is a complex process in which solar energy is converted into chemical energy and stored in the form of glucose.
- Photosynthesis is carried out in two stages: light-dependent reactions and light-independent reactions.
- Light-dependent reactions occur in the grana in the chloroplasts. Water is split and oxygen is released as a product.
- Light-independent reactions occur in the stroma in the chloroplasts. Carbon dioxide is reduced to form glucose.
- Rubisco (ribulose-1,5-bisphosphate carboxylase/ oxygenase) is an essential enzyme involved in carbon dioxide fixation in the Calvin cycle.
- Plants have evolved three photosynthetic pathways in response to different environmental conditions: C_3, C_4 and CAM.

KEY QUESTIONS

Knowledge and understanding

1. Name the organisms that make their own organic compounds from surrounding inorganic compounds.
2. Identify three possible uses by a plant for glucose formed by photosynthesis in the plant.
3. Write the balanced chemical equation for photosynthesis.
4. The following questions relate to the light-dependent reactions of photosynthesis.
   a Where in the chloroplast do these reactions occur?
   b Which reactant is required for this reaction?
   c What is the role of $NADP^+$ and ADP in this reaction?
5. What process occurs in the stroma and what is the final product of this stage?

Analysis

6. Name and differentiate between the photosynthetic pathways plants have developed in response to different environmental conditions.
7. Explain the significance of the thylakoid membrane being a continuous highly folded membrane.
8. The picture below compares runner bean plants (*Phaseolus coccineus*) grown in the light (left) with those grown in the dark (right). Use your knowledge of photosynthesis to answer the following questions.

   a Describe the physical differences between the plants grown in the light and those grown in the dark.
   b Would you expect both groups of plants to have the same amount of chlorophyll in their leaves? Why or why not?
   c Which stage of photosynthesis is affected?

6.2 Factors that affect the rate of photosynthesis

As for all biochemical processes, the rate of photosynthesis varies according to the internal conditions of the cell, which in turn are often affected by the external environmental conditions.

The main factors that control the rate of photosynthesis are:

- light availability
- water availability
- temperature
- carbon dioxide concentration.

The availability or concentrations of these factors may limit the rate of the reactions in the photosynthesis pathway.

For each factor, there is an optimum amount at which photosynthetic reactions proceed at the fastest rate. Below the optimum level, reactions slow down. Above the optimum, reactions do not proceed any faster—in fact, they sometimes occur more slowly.

The factor that is present in the smallest amount is the limiting factor. Only one factor will be limiting at a particular time. For example, if a plant is in the process of photosynthesis and there is a large amount of carbon dioxide available but not enough light, then light is a limiting factor. If there is enough light but not enough carbon dioxide, then the concentration of carbon dioxide is the limiting factor (Figure 6.2.1).

In this section, you will learn about the factors that affect the rate of photosynthesis.

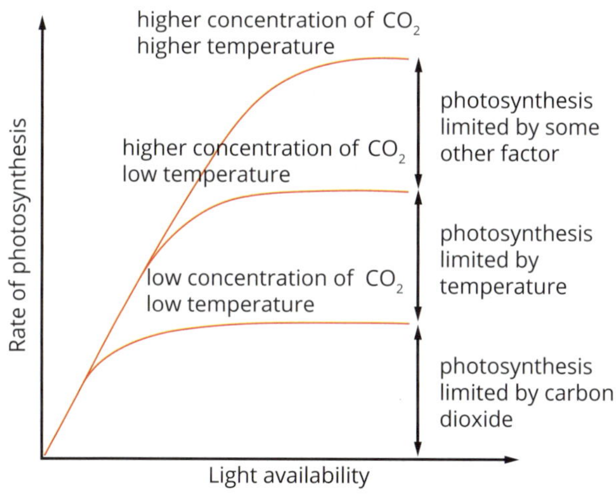

FIGURE 6.2.1 Carbon dioxide concentration, light availability and temperature are limiting factors for photosynthesis.

BIOFILE

The unexpected result of rising carbon dioxide levels

The rising atmospheric CO_2 levels and consequent global warming will lead to increased rates of photosynthesis in plants. This means plants will produce more glucose, and it has long been predicted that crops will therefore grow more quickly. Scientists studying the effects of elevated CO_2 levels on some important crop plants have, however, made some unexpected and alarming findings. In experiments in which crops were grown in an atmosphere with artificially elevated CO_2 levels, it was shown that although overall plant growth increased, levels of zinc, iron and protein in the plants decreased. People rely on food from crops for these important nutrients, and decreased levels may lead to serious malnutrition for people throughout the world.

CARBON DIOXIDE

The level of carbon dioxide in the air remains relatively constant. The factors that affect the amount available for photosynthesis in most terrestrial plants are the number of stomata in the leaves and whether these stomata are open or closed. If the stomata are closed, photosynthesis will use up the carbon dioxide inside the leaf, therefore lowering the concentration of carbon dioxide in the leaf. With less carbon dioxide available, the rate of photosynthesis, even in the presence of light, will be limited.

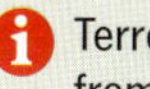
Terrestrial plants receive their CO_2 from the air. Aquatic plants can utilise the CO_2 dissolved in the water.

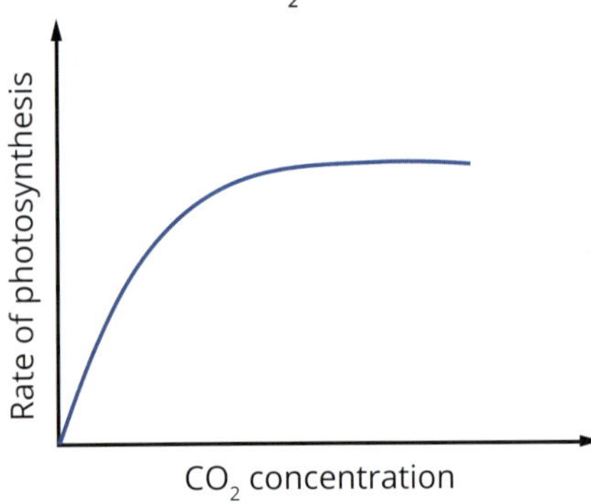

FIGURE 6.2.2 Effect of CO_2 levels on the rate of photosynthesis

In the laboratory it is possible to control the concentration of carbon dioxide to which plants are exposed without changing other factors. Figure 6.2.2 shows a comparison of the rate of photosynthesis for a particular species of plant exposed to different concentrations of carbon dioxide.

WATER AVAILABILITY

Photosynthesis requires water, carbon dioxide and light energy. Because the amount of water used in photosynthesis is small compared with the amount needed to keep the cells alive, a living plant cell normally has sufficient water for photosynthesis to occur. Therefore, water does not have a direct effect on the rate of photosynthesis in nature. However, it does have an indirect effect. If a plant is suffering from water stress, the stomata in the leaf close and reduce the availability of carbon dioxide (Figure 6.2.3). When water is not available, the plant cells lose water, reducing the turgor (pressure) inside the cell. Therefore, the cells cannot maintain a rigid shape and the plant collapses (Figure 6.2.4). This will have a further impact in regions where the climate is changing.

FIGURE 6.2.4 This plant has wilted due to water loss reducing the turgor inside the cells.

FIGURE 6.2.3 This coloured scanning electron micrograph image of a tobacco leaf shows one open stoma and one closed stoma. When a stoma is open, gas exchange can occur and water molecules can escape.

LIGHT AVAILABILITY

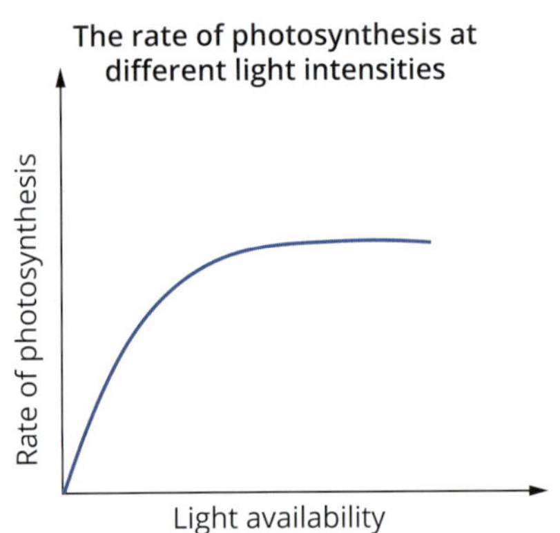

FIGURE 6.2.5 Effect of light availability on the rate of photosynthesis

In the laboratory, chloroplasts can be extracted from plant cells and tested in isolation. By varying the amount of light shining on these isolated chloroplasts, while keeping the carbon dioxide levels constant, it is possible to measure the rate at which photosynthesis occurs at different light levels. The results of this experiment are shown in Figure 6.2.5 and present what we refer to as a **light saturation curve**. The curve shows a steady increase in the photosynthesis rate with an increase in light intensity until a plateau is reached. The plateau indicates that there is a maximum rate at which photosynthesis can occur. Assuming unlimited amounts of carbon dioxide (and water), the limit will be the point at which all of the photosynthesis systems and enzymes in the chloroplasts are working at their optimum rate.

In the natural environment, the amount of light available to a plant for photosynthesis will be determined by the amount of sunlight in its environment. Sunlight will vary during the cycle of a day and will change with the seasons and the weather. Trees and taller plants will shade plants on the forest floor, and the amount of light available to aquatic plants will be dependent on how far underwater they grow (Figure 6.2.6).

FIGURE 6.2.6 (a) Plants growing on the forest floor will receive less light than taller trees. (b) Aquatic photosynthetic organisms, such as this sea kelp, can only grow near the water's surface. As you move deeper underwater, there is less light available.

Figure 6.2.7 shows that tomato plants grown in unshaded conditions have greater rates of photosynthesis than plants grown in shaded conditions.

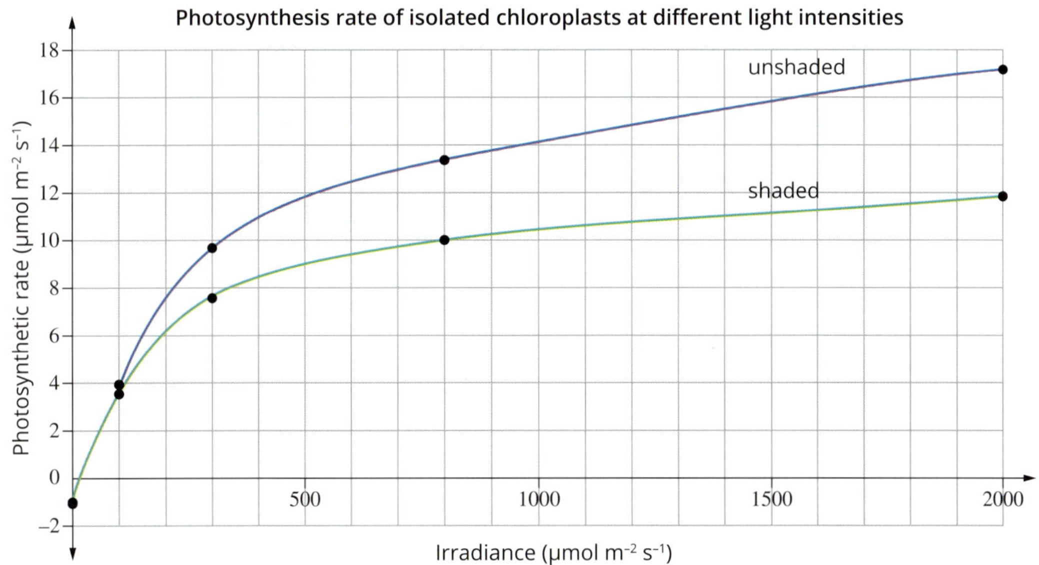

FIGURE 6.2.7 Light saturation curves of chloroplasts extracted from tomato plants grown under shaded and unshaded conditions

TEMPERATURE

All light-dependent and light-independent reactions are catalysed by enzymes. In Chapter 5 you learnt that the rate of an enzyme reaction is affected by temperature. Enzyme activity initially increases as temperature increases, but enzyme molecules become denatured above optimum temperatures and can no longer function. For this reason, the rate of photosynthesis will approximately double for every increase in temperature until the optimum temperature is reached. Above the optimum temperature, the rate of photosynthesis will not stay level. It will decline steeply as the enzymes become denatured and the chemical processes involved in the light-independent reactions cannot be catalysed (Figure 6.2.8). Different plants are adapted to different environments so there is no fixed optimum temperature for plant enzymes. The enzymes in a cactus will have a higher optimum temperature than those in an aquatic plant in cool water.

When the temperature is below the optimum range of an enzyme, there is low kinetic energy (movement) in the molecules. Therefore the rate of photosynthesis is low, as the water and CO_2 molecules are moving slower.

Plants from Arctic regions will have a lower optimum temperature range than plants found in tropical climates.

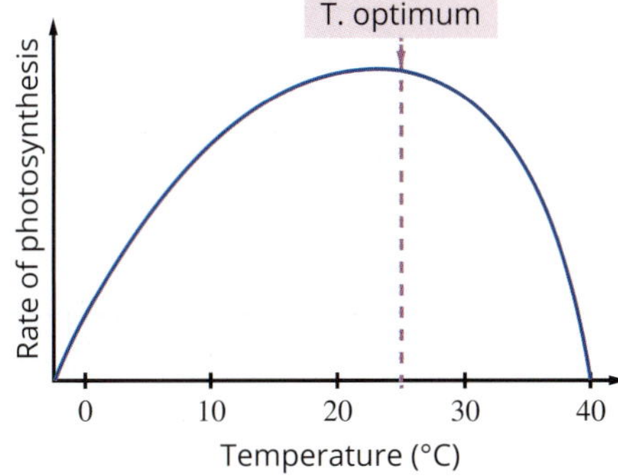

FIGURE 6.2.8 Rate of photosynthesis at different temperatures. The shape of the curve is typical of those shown for enzyme reactions. The rate rises with temperature to the optimum temperature (T. optimum) and then falls to zero at high temperatures as the photosynthetic enzymes denature.

CRISPR-Cas9 TECHNOLOGY—IMPROVING CROP YIELD AND PHOTOSYNTHETIC EFFICIENCY

The CRISPR-Cas9 system is a gene editing technique that has been used to edit plant genomes since 2013. You learnt about the mechanisms of the CRISPR-Cas9 system in Chapter 4. CRISPR-Cas9 technology is expanding rapidly, due to its simplicity, efficiency and precision. It is now being used to study the genomes of agricultural crops and target genes of interest to ensure our increasing demand for food, feed and fuel is met. As well as being a faster way to produce beneficial characteristics than traditional crossbreeding techniques, CRISPR-Cas9 allows genetic diversity to be maintained within a species.

The steps used to edit an organism's genomes using CRISPR-Cas9 technologies and an example of a recent alternative approach based on the CRISPR-Cas9 system are illustrated in Figure 6.2.9.

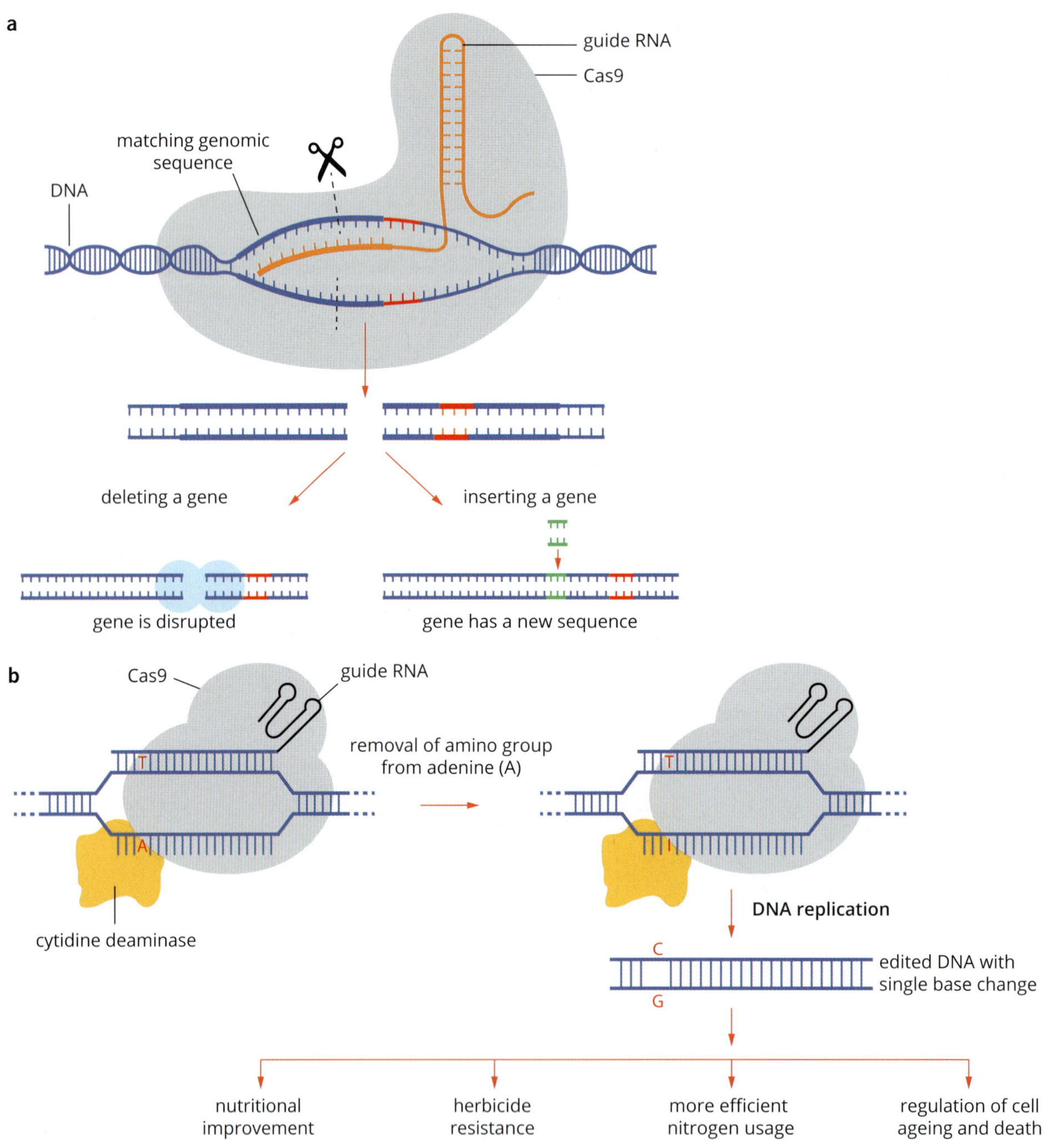

FIGURE 6.2.9 (a) The CRISPR–Cas9 editing tool. (b) CRISPR-Cas9 technology introducing a point mutation into a rice genome to disrupt the gene function, resulting in beneficial characteristics for the organism

The application of CRISPR-Cas9 technologies is vast. It includes improving resistance to abiotic and biotic stresses, increasing photosynthesis efficiency, genomic functional studies, improving fruit quality (texture, colour and flavour), enhancing shelf life and increasing levels of nutrition and bioactive compounds. In rice crops, CRISPR-Cas9 technology has been used to increase the levels of amylase (an enzyme that breaks down starch) and amylose (resistant starch) to improve human health. This technology may allow developing countries to grow crops that are nutrient-fortified or better adapted to environmental challenges. Figure 6.2.10 and Table 6.2.1 outline some of the uses and applications of the CRISPR-Cas9 technology on a variety of crops.

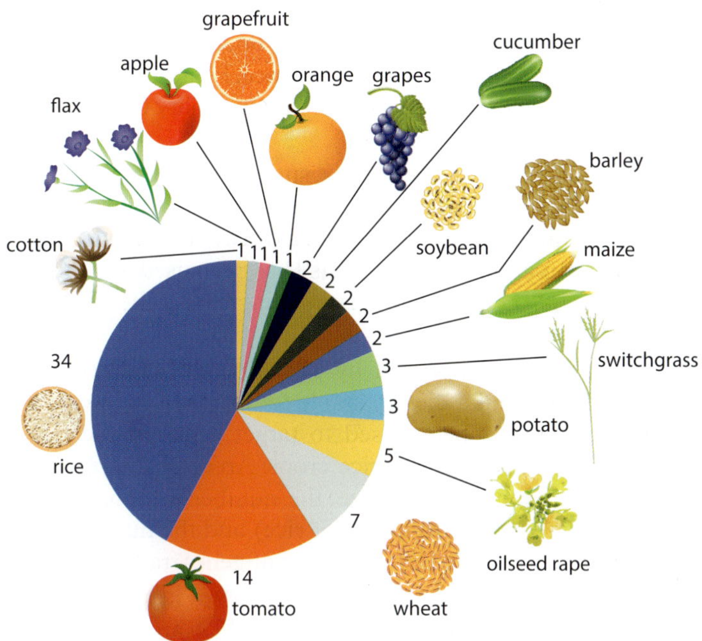

FIGURE 6.2.10 Number of genes modified in crop plants using CRISPR-Cas9 technology

TABLE 6.2.1 Examples of genome editing using CRISPR-Cas9 technology for crop improvement

Application	Crop	Trait
resistance to abiotic stresses	tomato	heat tolerance, frost tolerance
	watermelon, corn, rice, soybean	herbicide tolerance
resistance to biotic stresses	tomato, grape, wheat	resistance to powdery mildew
	cacao	resistance to *Phytophthora palmivora*
	cucumber	resistance to cucumber vein yellowing virus (ipomovirus)
	apple	resistance to fire blight disease
improved fruit quality	tomato	purple colour, yellow colour, increase in lycopene (red pigment with antioxidant properties)
	tomato, rice, potatoes	enhanced shelf life
	rice	increase in amylose (resistant starch) content
	wheat	increase in grain weight and protein content

Crop yield

CRISPR-Cas9 technology has been used to modify the genomes of many crop plants, allowing them to resist stresses and increasing crop yield. CRISPR-Cas9 technology has been used to make crops resilient to abiotic stresses such as flooding, drought, extreme temperatures and soil salinity, which can lead to the loss of crops and income (Table 6.2.1 on page 225). Conventional plant breeding methods have not been very successful in producing plants with increased resilience to these stresses. One reason may be that these mechanisms involve the coordination of a number of genes.

CRISPR-Cas9 technology can precisely target one or several genes at the same time to determine which genes are involved and target them appropriately. For example, plants grown in areas that are prone to flooding are being produced with tall, strong stems and mechanisms for water balance. Plants grown in arid areas with poor soil can have their genes altered to improve the ability of their root systems to capture water and nutrients.

Biotic stresses (for example, pathogens such as viruses, bacteria and fungal infections) can also have an enormous impact on crop viability. Citrus canker, caused by the bacterium *Xanthomonas citri*, is a disease that can lead to devastating economic losses in citrus crops. CRISPR-Cas9 has been used to target the promoter region of the citrus gene (*LOB1*) and disrupt it, increasing the plant's resistance to infection from *X. citri*.

Crop yields can also be improved by changing the architecture of a plant. Rice is a major food crop and a dietary staple for much of the world's population. CRISPR-Cas9 technology is being used to target genes in the rice genome to increase both the yield and quality of rice crops. Areas of the plant's architecture that have been targeted include grain size, the number of grains per panicle (the cluster of branches that give rise to grains of rice) and the number of panicles per plant. These features are governed by a number of genes, rather than a single gene.

One of the advantages of using the CRISPR-Cas9 system is the ability to target multiple genes at once. Quantitative trait loci (QTL) are chromosomal regions that are associated with a particular trait. Many QTLs have been targeted using CRISPR-Cas9 to produce novel rice lines. Studies have shown that introducing mutations in three QTL regions that are associated with grain weight in rice has led to significant increases in grain weight.

Photosynthetic efficiency

CRISPR-Cas9 technology can also be used to improve photosynthetic efficiency by increasing the photosynthetic protein abundance and minimising photorespiration. This has applications in several areas including the production of biofuels (discussed in Chapter 7), food, pharmaceuticals and nutraceuticals.

One promising application is the use of CRISPR-Cas9 technology to edit the genome of green algae (*Chlamydomonas* sp.) used in biofuel production. By increasing the photosynthetic efficiency of green algae, biofuel production is more economically feasible at a commercial scale. Using CRISPR-Cas9, scientists have been able to truncate the antenna molecules that absorb light in green algae. When the antenna molecules are long they absorb more light than the system can process and therefore release the excess energy as heat. This decreases the viability of algal cells and becomes a constraint in high-density cultures of algae. When the antenna molecules are shorter, the algae are photosynthetically more productive and are able to replicate in mass cultures, making photosynthesis and biofuel production more efficient.

6.2 Review

OA ✓✓

SUMMARY

- The four main factors that control the rate of photosynthesis are:
  - light availability
  - water availability
  - temperature
  - carbon dioxide concentration.
- An increase in carbon dioxide levels can increase the rate of photosynthesis.
- An increase in light availability can increase the rate of photosynthesis.
- The availability of water has an impact on the rate of photosynthesis.
- Plants have an optimum temperature range for photosynthesis. Too cold and the rate of reaction will be slow. Too hot and the enzymes in chloroplasts can denature.
- CRISPR-Cas9 technology is a gene editing technique that can been used to edit plant genomes to:
  - increase crop yields
  - increase photosynthetic efficiency.

KEY QUESTIONS

Knowledge and understanding

1. If a plant is grown at optimal temperature with unlimited levels of carbon dioxide and access to light, will the rate of photosynthesis keep increasing? Explain your answer.
2. When a plant closes its stomata it can no longer exchange oxygen and carbon dioxide, therefore the rate of photosynthesis decreases. What is the benefit to the plant of closing its stomata?
3. List factors that directly affect the rate of photosynthesis. Give brief explanations for each factor.
4. List the improvements to crop production achieved using CRISPR-Cas9 technology.

Analysis

5. The graph below shows the rate of photosynthesis at different temperatures.

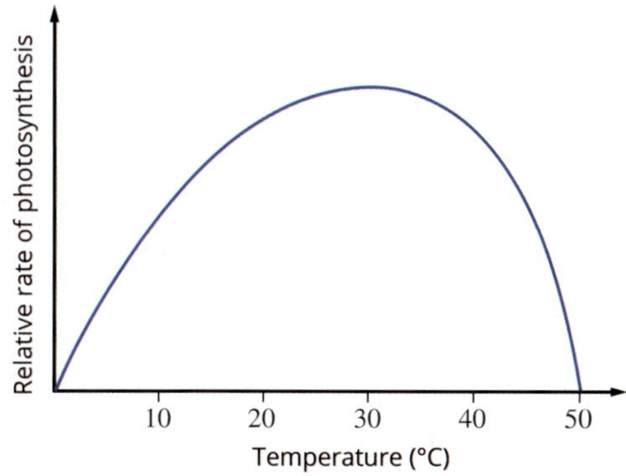

   a. Describe the rate of photosynthesis between 0°C and 20°C and provide a reason for the trend.
   b. At what temperature does the rate of photosynthesis begin to decrease? Explain why this occurs.

6. The following graphs represent the changes in rate of photosynthesis when temperature, light availability or distance from a light source are increased.
   a. Label each graph with the factor that is best represented by the data presented.

i

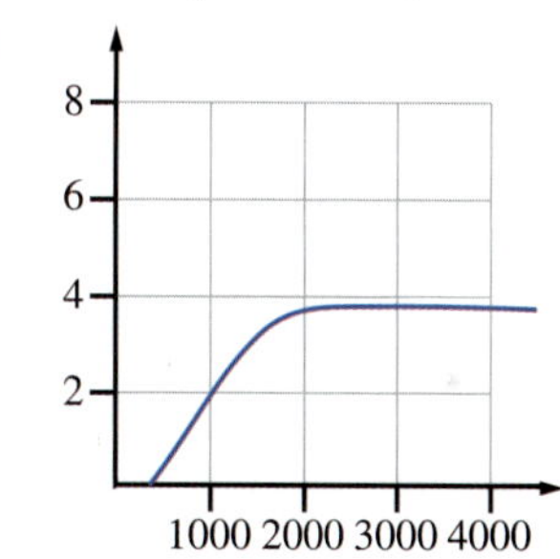

ii

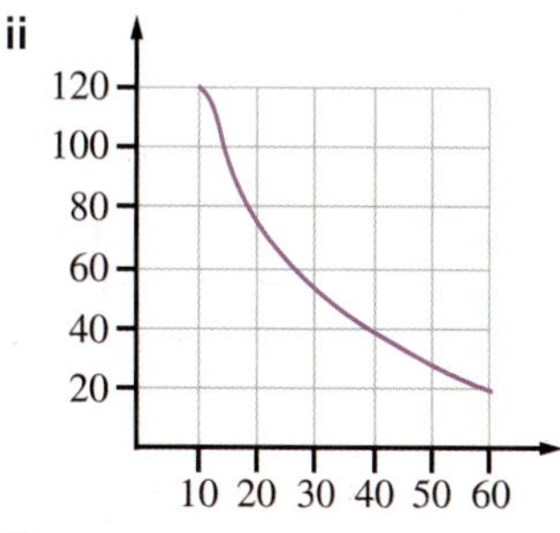

iii

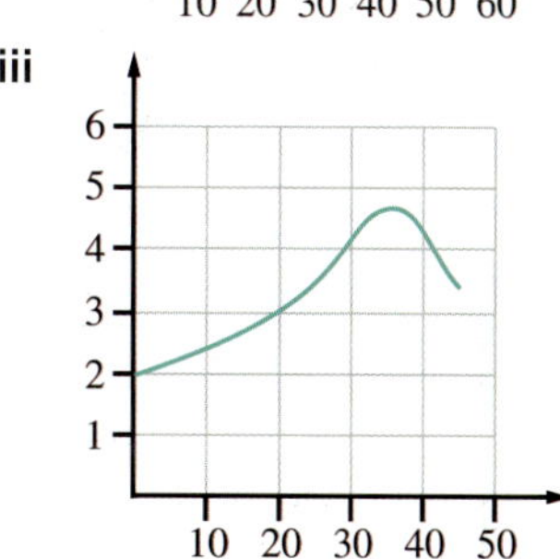

   b. Describe what is occurring in graph ii.

7. Discuss the benefits of and concerns about using CRISPR-Cas9 technology.

Chapter review

KEY TERMS

accessory pigment
adenosine diphosphate (ADP)
adenosine triphosphate (ATP)
autotroph
bundle sheath cell
C_3 plant
C_4 plant
Calvin cycle
CAM plant
carbon fixation
chemical energy
chlorophyll
chloroplast
crassulacean acid metabolism (CAM)
endergonic reaction
exergonic reaction
glucose
grana (singular granum)
heterotroph
light-dependent reaction
light-independent reaction
light saturation curve
mesophyll cell
nicotinamide adenine dinucleotide phosphate (NADPH)
photorespiration
photosynthesis
pigment
Rubisco
solar energy
stoma (plural stomata)
stroma (plural stromata)
thermodynamics
thylakoid lamella (plural thylakoid lamellae)
thylakoid membrane

REVIEW QUESTIONS

Knowledge and understanding

1 Recall the difference between autotrophs and heterotrophs.

2 Select the biochemical pathway/s that plants use to generate energy.
- **A** photosynthesis
- **B** photorespiration
- **C** cellular respiration
- **D** photosynthesis and cellular respiration

3 Which molecule is broken down by solar energy during photosynthesis?

4 Identify whether the following statements are true or false.
- **a** Chlorophyll is a pigment that causes green colouring of leaves.
- **b** Chloroplasts are pigments that absorb light.
- **c** Chlorophyll is located in the thylakoid membranes.
- **d** Chlorophyll absorbs light energy, which is then transformed into chemical energy.

5 Which of the following is true of light-independent reactions of photosynthesis in C_3 plants?
- **A** They occur on the inner membranes of chloroplasts.
- **B** They involve the production of organic compounds containing three-carbon atoms in the Calvin cycle.
- **C** They are adapted to maximise glucose production in hot, dry environments.
- **D** They represent a special kind of photosynthesis that can produce organic compounds completely in the absence of light.

6 How do C_4 plants concentrate carbon dioxide around Rubisco?

7 Where do the following reactions take place?
- **a** light-dependent reaction
- **b** light-independent reaction

8 What occurs in the light-dependent reaction?

9 Answer the following questions about the enzyme Rubisco.
- **a** In C_3 plants, where is Rubisco located in the cell?
- **b** What is the main purpose of Rubisco?

10 What is the adaptation of plants that have evolved using the CAM pathway?

11 What are the advantages of using CRISPR-Cas9 technology for editing the genomes of plants?

12 Explain why the availability of water for plants has an indirect effect on photosynthesis rather than a direct effect, like carbon dioxide, light and temperature do.

13
- **a** What leads to photorespiration?
- **b** Photorespiration is a wasteful pathway. Give reasons to support this statement.

Application and analysis

14 To meet increasing food demands scientists are attempting to increase crop yields, exploring the possibility of enhancing photosynthesis by engineering plants to use near-infrared light. This would be aimed at the lower leaves of the canopy. Chlorophyll *d* is found in a genus of bacteria that absorbs light in the far-red range of the spectrum. In recent years chlorophyll *f* was discovered in cyanobacteria; it can absorb light in the near-infrared range (706 nm).

a How does light absorption differ with chlorophyll *a* and chlorophyll *f*?

b Explain why scientists are targeting the lower leaves of the canopy.

15 How can CRISPR-Cas9 technology be used to study the effects of abiotic stressors on gene expression in a plant?

16 Cellulose is a complex carbohydrate that is made up of many individual units of glucose. The molecule shown below, composed of carbon, hydrogen and oxygen, was found in the gut of an animal. Determine how a photosynthetic autotroph would have obtained this molecule.

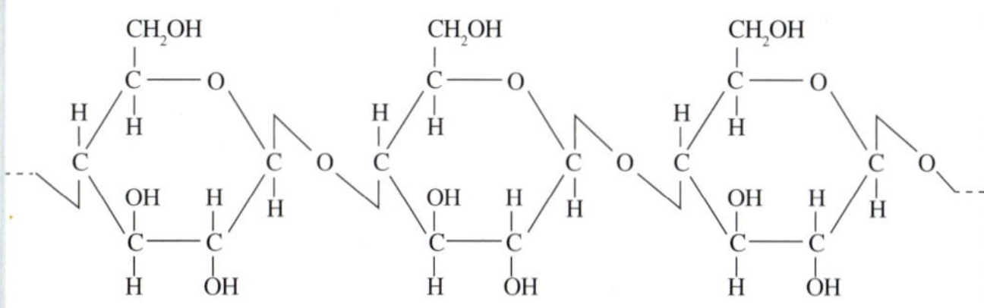

17 The following graph shows the relationship between net oxygen output and uptake and light availability for a green plant. Explain what is happening when light availability is at:

a 2 units

b 16 units

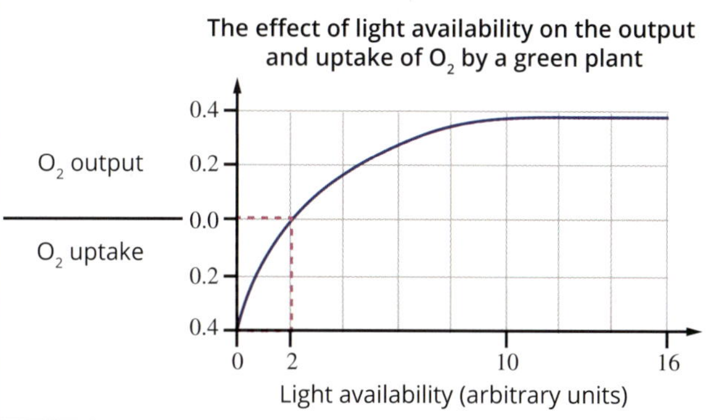

18 Photosynthetic organisms living in deep water have especially difficult challenges to survival. As shown on the graph below, light has limited ability to penetrate water.

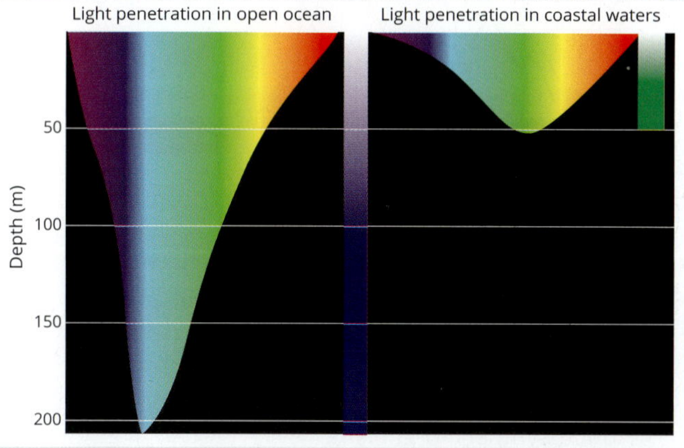

a Light availability is a significant limiting factor influencing the rate at which photosynthesis can occur. Why is light significant?

b What is the depth beyond which photosynthetic organisms are unable to survive in the ocean?

c All photosynthetic organisms must contain chlorophyll. Why is this?

Question continued overleaf

As well as chlorophyll, photosynthetic organisms may also contain a range of pigments called accessory pigments. Together, accessory pigments are able to absorb most wavelengths of light. The absorption spectrum for various photosynthetic pigments is shown in the graphs below.

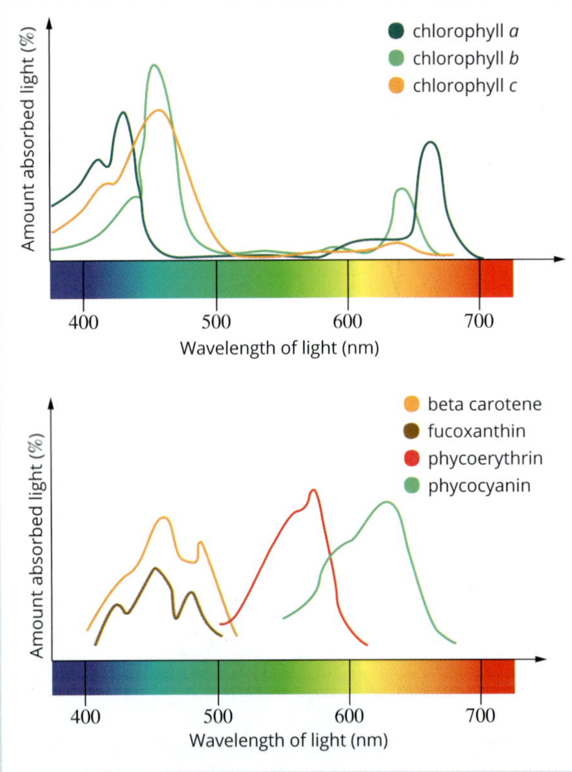

d Using the data presented in the graphs above and your knowledge of the pigments used in photosynthesis, explain how having a range of accessory pigments could assist the survival of a photosynthetic organism.

Plants are one of the major groups of photosynthetic organisms. Another is algae. Seaweeds are a type of alga. The diagram below shows the results of a chromatography experiment in which the pigments of various types of seaweed and an oak tree are compared.

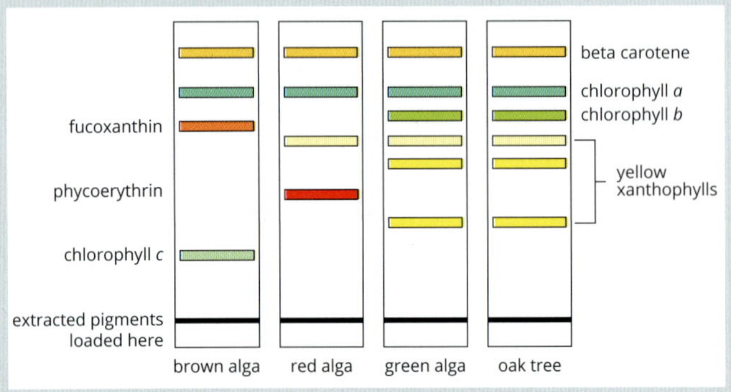

Chromatography is a method used to separate chemicals according to their solubility. The material to be separated is placed at one end of a special type of paper and that end is dipped into a solvent that is allowed to diffuse up the paper. The chemicals to be separated dissolve in the solvent and are carried up the paper. The chemicals separate according to their solubility. More soluble chemicals move further than less soluble chemicals.

e i Explain which type of seaweed is most likely to be found in the deepest waters in coastal regions.

ii The brown algae tested lack chlorophyll *b*, yet they are very successful, with some growing to 60 m in length. Explain how, despite the lack of chlorophyll *b*, the brown algae can grow so large.

19 An experiment was set up to investigate photosynthesis. A plant was placed in a sealed container at 20°C. The air in the container was 0.09% carbon dioxide at the beginning of the experiment. (Carbon dioxide concentration in room air is between 0.03 and 0.04%.) The experiment was undertaken at three different light intensities: dim, moderate and bright. The carbon dioxide concentration in the container was monitored over a 4-hour period. The results for each light availability were collected and are shown in the table below.

Time (min)	Percentage CO_2		
	Dim light	Moderate light	Bright light
0	0.090	0.090	0.090
30	0.084	0.080	0.080
60	0.076	0.070	0.070
90	0.069	0.060	0.054
120	0.060	0.053	0.045
150	0.054	0.045	0.038
180	0.048	0.038	0.030
210	0.041	0.030	0.027
240	0.039	0.027	0.021

a Draw graphs of the information in the table. Use the same set of axes for the three light intensities. Ensure that the graphs comply with all of the graphing conventions.

b Describe the trends evident in the data.

c i Why can carbon dioxide uptake be used as a measure of the rate of photosynthesis?

ii Explain why the carbon dioxide decreased fastest in bright light.

d Why were all of the experiments carried out at 20°C?

e What hypothesis could this experiment have been testing?

20 In an experiment, photosynthetic algae filaments were placed in a solution that was compartmentalised so that each compartment was exposed to different spectral light colours. Aerobic bacteria were added to the solution and were able to move through all compartments. The following diagram indicates the distribution of these bacteria after one hour.

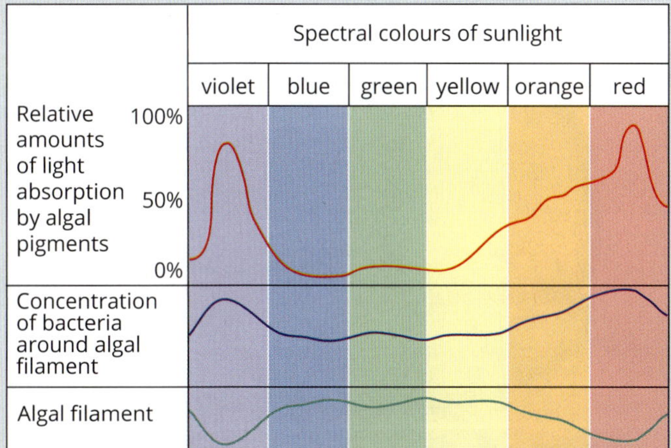

a Explain why aerobic bacteria were used in this experiment.

b Draw conclusions about what the distribution of bacteria implies about the relationship between the spectral light colour and the rate of photosynthesis.

21 Atrazine is a chemical commonly used as a weed killer. It is absorbed by roots from the soil. In the leaves, it attaches to a protein called D1, which is part of the electron transport chain used to generate ATP during photosynthesis. Atrazine blocks the movement of electrons along this chain.

a Which phase of photosynthesis is disrupted by atrazine?

b ATP is not a final product of photosynthesis. What happens to the ATP normally produced in the electron transport chain?

c The advertising blurb on one particular brand of weed killer which has atrazine as its active ingredient says, ‘This product works by starving the plant’. Is this an accurate description of the action of atrazine? Explain your answer.

22 The rate of photosynthesis for a particular species of plant was monitored under different wavelengths of light and varying temperature conditions.

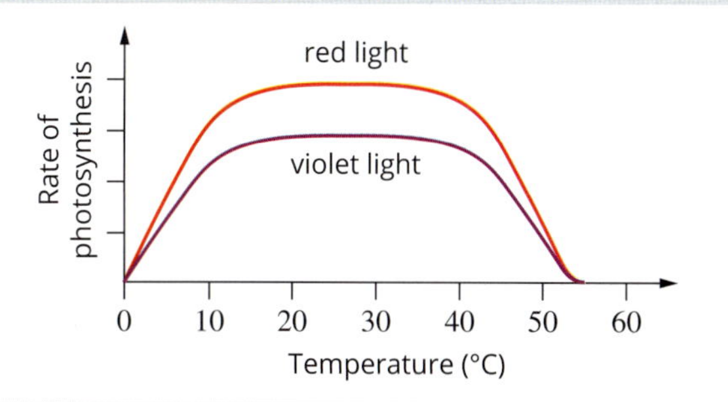

a Use information from the graph to name two factors that affect the rate of photosynthesis.

b Name two other environmental factors that can affect the rate of photosynthesis.

c Under which wavelength of light—red or violet—is the greater amount of oxygen gas produced? Explain your answer.

d Plants grown under both red and violet light showed a sharp decline in the rate of photosynthesis at temperatures above approximately 40°C. Photosynthesis stopped completely at approximately 55°C.

i Explain this observation.

ii What does this suggest about the process of photosynthesis?

e Is photosynthesis an endergonic or an exergonic reaction? Explain.

OA ✓✓

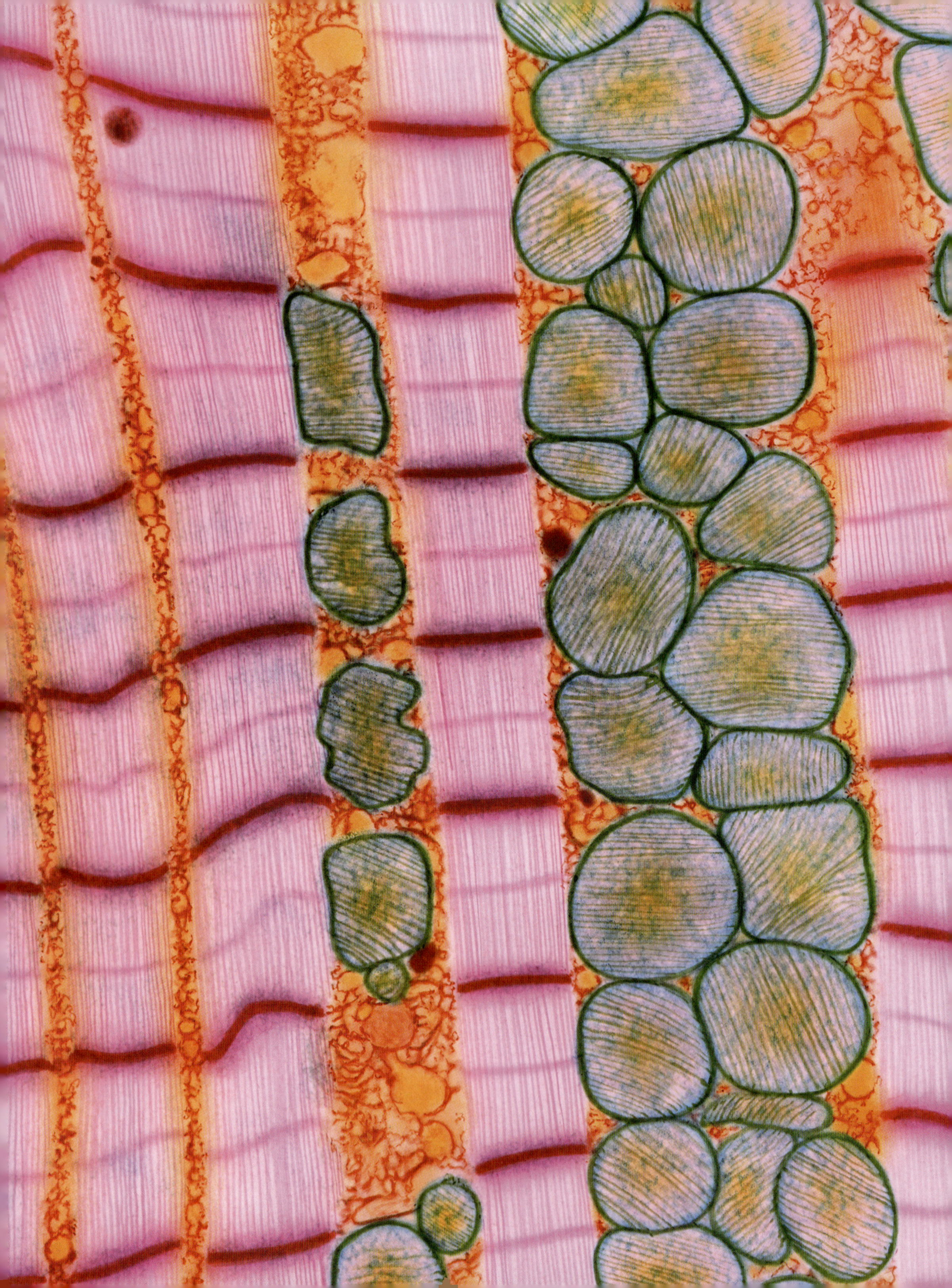

CHAPTER 07

Biochemical pathways: Cellular respiration

Learning outcomes

Energy is vital for life. Whether growing, moving, reproducing, responding or excreting, living organisms are using energy. 'Using energy' involves transforming energy from one form to another, and transferring it from one place to another. Energy is produced via several biochemical pathways that have evolved over time.

By the end of this chapter, you will have an understanding of biochemical pathways, glycolysis, cellular respiration and anaerobic fermentation, and how these provide living cells with the energy they need to survive.

You will also learn how cells adjust their metabolism to account for changes in environmental conditions and how biotechnological applications of biochemical pathways are being explored.

Key knowledge

Cellular respiration as an example of biochemical pathways

- the main inputs, outputs and locations of glycolysis, Krebs Cycle and electron transport chain including ATP yield (details of biochemical pathway mechanisms are not required) **7.1**
- the location, inputs and the difference in outputs of anaerobic fermentation in animals and yeasts **7.2**
- the factors that affect the rate of cellular respiration: temperature, glucose availability and oxygen concentration **7.3**

Biotechnological applications of biochemical pathways

- uses and applications of anaerobic fermentation of biomass for biofuel production. **7.2**

OA ✓✓

7.1 Cellular energy production

Cells need energy to do work. The energy needed by organisms and their cells is stored in the chemical bonds of organic compounds. Energy cannot be created or destroyed, but it can be changed from one form into another.

Cells obtain energy from organic compounds, such as glucose. They transform the chemical energy stored in these organic compounds into a more usable form of chemical energy, which is stored in the bonds of **adenosine triphosphate (ATP)**. This transformation of energy occurs through a combination of biochemical processes which varies according to the availability of free oxygen (O_2) (Figure 7.1.1).

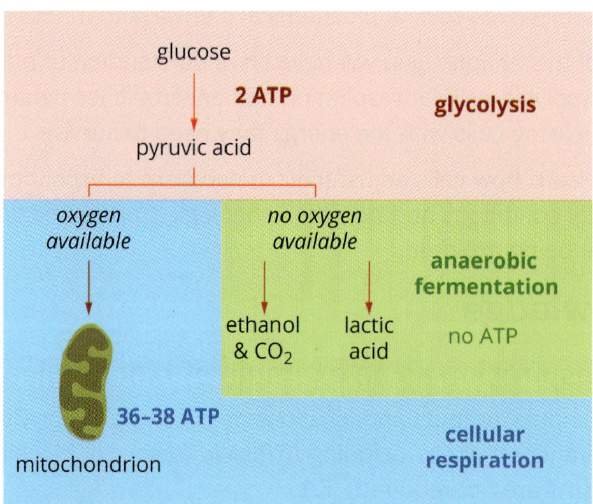

FIGURE 7.1.1 Simplified pathway of the process of cellular energy production in the presence of oxygen (cellular respiration) and absence of oxygen (anaerobic fermentation)

Cellular respiration consists of the three interconnected biochemical pathways known as:

- glycolysis
- the Krebs cycle
- the electron transport chain.

Anaerobic fermentation also involves glycolysis. You will learn more about anaerobic fermentation in Section 7.2.

In this section, you will learn about the first biochemical pathway in aerobic and anaerobic cellular energy production, glycolysis. Glycolysis is the first biochemical pathway; it then branches into different pathways, depending on oxygen availability. You will learn how the energy produced in glycolysis is transferred from glucose to coenzymes to be used in the Krebs cycle and electron transport chain when oxygen is available.

ENERGY FROM GLUCOSE

The biochemical pathway for cellular energy production not only depends on the availability of oxygen but also on the availability of glucose. Animals (heterotrophs) obtain their glucose by eating other organisms and breaking down the organic molecules (Figure 7.1.2). Plants (autotrophs) make their own glucose from inorganic substances (light, water and carbon dioxide [CO_2]). Cellular respiration forms part of a larger cycle of energy transformation (Figure 7.1.3).

Cells can obtain energy from high-energy fats and other organic compounds, including proteins. However, most cells use glucose as their immediate source of energy. For cells to extract energy from other organic molecules, the molecules must either be converted into glucose or broken down to small molecules that can enter the cellular respiration pathways.

FIGURE 7.1.2 Complex carbohydrates are broken down by our digestive system into simple sugars, such as glucose. Glucose is the primary source of energy for most of our cells.

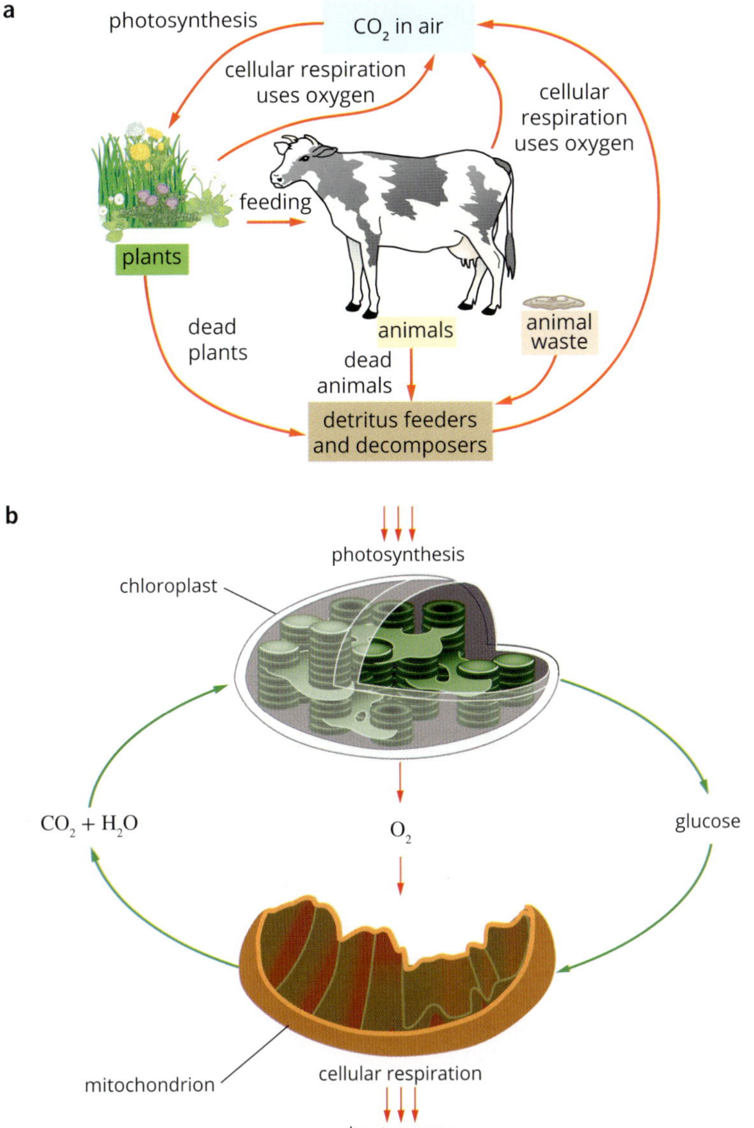

FIGURE 7.1.3 (a) Cellular respiration and photosynthesis operate together as part of the carbon and oxygen cycles in an ecosystem. (b) The cycling of materials is driven by biochemical processes in two cell organelles, chloroplasts and mitochondria, using energy derived from sunlight.

ATP can be used for vital cellular processes, including protein synthesis and active transport.

Cellular respiration requires oxygen. When oxygen supplies are low, a cell can carry out anaerobic fermentation.

Cellular respiration is the name given to the combination of biochemical pathways that occur together within a cell to release energy from glucose. The energy released from glucose through cellular respiration is used to generate the coenzyme ATP. ATP is the universal carrier of energy in living organisms. Molecules of ATP are the cell's store of immediately usable chemical energy that is required for cell processes. The energy is transferred when ATP is formed from **adenosine diphosphate (ADP)** and inorganic phosphate (P_i), and it is stored in the bond between ADP and phosphate. This can be represented as:

$ADP + P_i \rightarrow ADP{\sim}P_i$ (ATP)

The '~' represents the high-energy bond in which the energy is stored.

When the high-energy bond in ATP is broken, energy is released for use in the many energy-demanding processes that occur in the cell. The transfer of energy is summarised in Figure 7.1.4. The recycling process of ADP requires much less energy than it would take to make an entirely new ATP molecule.

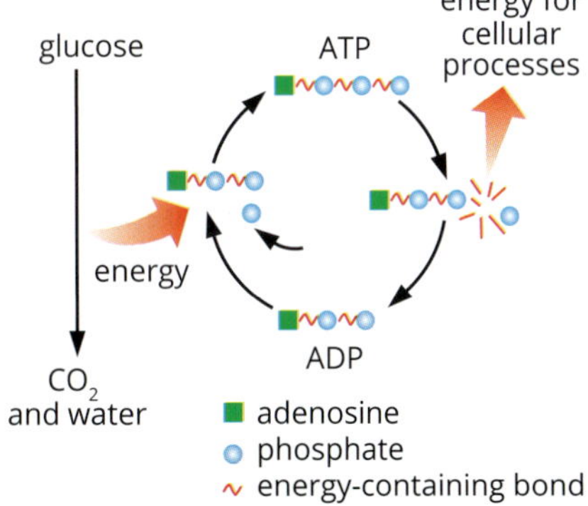

FIGURE 7.1.4 Energy from glucose is transferred in the synthesis of ATP from ADP and phosphate. The energy is stored in the phosphate bond. When the bond is broken, the energy is released to drive cellular processes. The ADP and phosphate are recycled.

CASE STUDY ANALYSIS

Glycolysis appeared early in evolution

Cells require a source of energy to perform basic functions. Initially, they were limited to energy that they could obtain from their immediate surroundings. Cells required a process to both develop their own energy and convert stored energy into usable energy. During evolution, three metabolic systems developed: glycolysis, photosynthesis and cellular respiration (Figure 7.1.5).

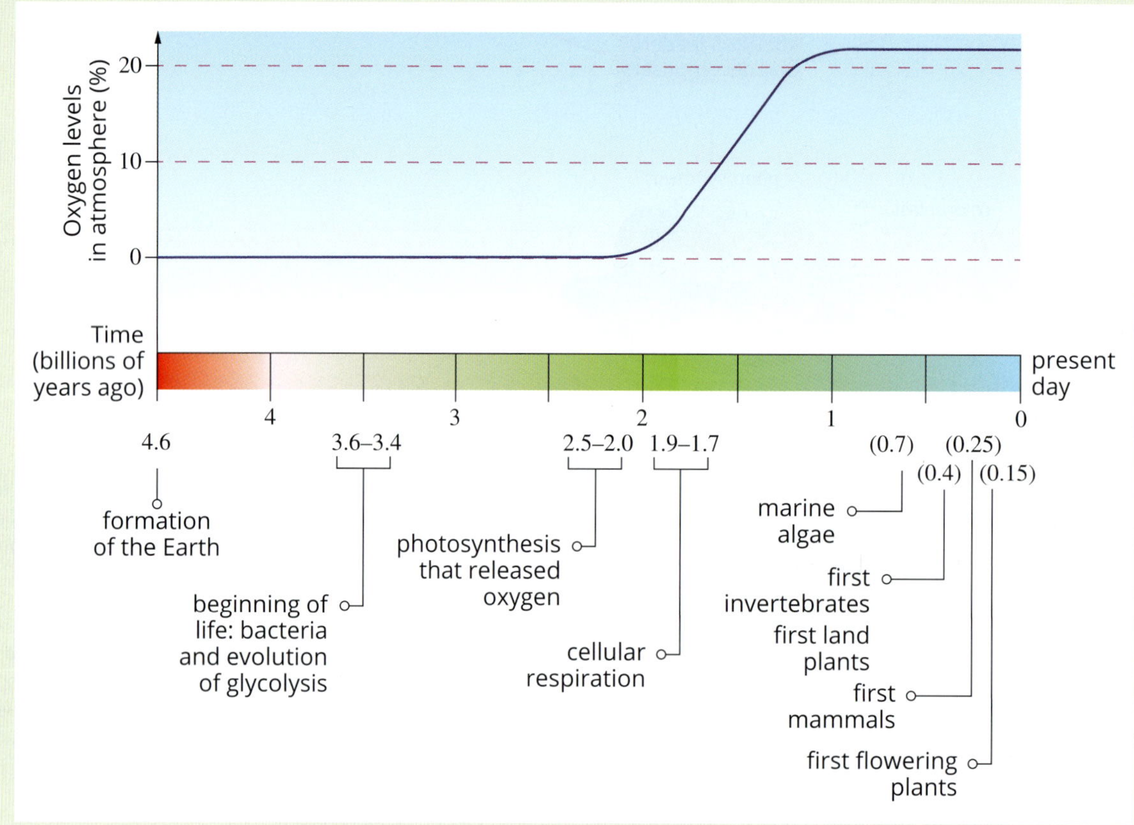

FIGURE 7.1.5 Glycolysis evolved in the low-oxygen atmosphere that existed before the evolution of photosynthesis.

Analysis

Use Figure 7.1.5 to answer the following questions.

1 Explain why glycolysis was the predominant process for generating ATP in living organisms before the appearance of photosynthetic prokaryotes.

2 Nearly all present-day cells carry out glycolysis. Does this statement support the notion that glycolysis is one of the oldest metabolic pathways? Explain your reasoning.

3 Approximately 2 billion years ago, the process of photosynthesis developed in cells. Describe what occurred following the evolution of photosynthesis and the release of oxygen into the atmosphere in terms of the diversification of life.

GLYCOLYSIS

Glycolysis is the first stage of cellular respiration and occurs in the **cytoplasm** of the cell. This involves the splitting (lysis) of glucose (a six-carbon molecule) into two three-carbon pyruvate molecules (also known as pyruvic acid). In the initial stage of glycolysis, two ATP molecules are required to break down one glucose molecule, but four ATP molecules are produced. There is a net gain of two ATP molecules. A small amount of energy is released from the glucose molecule; however, as you will learn in Section 7.2, the steps following glycolysis produce more energy for the system.

The energy released from glycolysis is transferred to the coenzymes ATP and NADH, which act as energy carriers. The overall reaction of glycolysis is:

INPUTS		OUTPUTS
glucose + 2ADP + $2P_i$ + $2NAD^+$	→	2 pyruvate + 2ATP + 2NADH

NADH and $FADH_2$

An important coenzyme in cellular respiration is **nicotinamide adenine dinucleotide** (**NAD^+**). It acts as an energy carrier. There are various steps in cellular respiration in which energy is transferred in the making of NADH from NAD^+. During the final stage of cellular respiration (the electron transport chain), NADH is converted back to NAD^+ and the energy released is used in the formation of ATP.

Another important coenzyme is **flavin adenine dinucleotide** (**FAD**). It is also an energy carrier. FAD functions in the second stage of cellular respiration, cycling between FAD and $FADH_2$ and releasing energy. This will be examined more closely in Section 7.2.

CELLULAR RESPIRATION

Following glycolysis, the substrate formed (pyruvate) can be diverted into one of two biochemical pathways, depending on the availability of oxygen—cellular respiration (in the presence of oxygen) or anaerobic fermentation (in the absence of oxygen). You will learn about anaerobic fermentation in Section 7.2.

These additional biochemical pathways yield a greater number of ATP than using glycolysis alone (Figure 7.1.6).

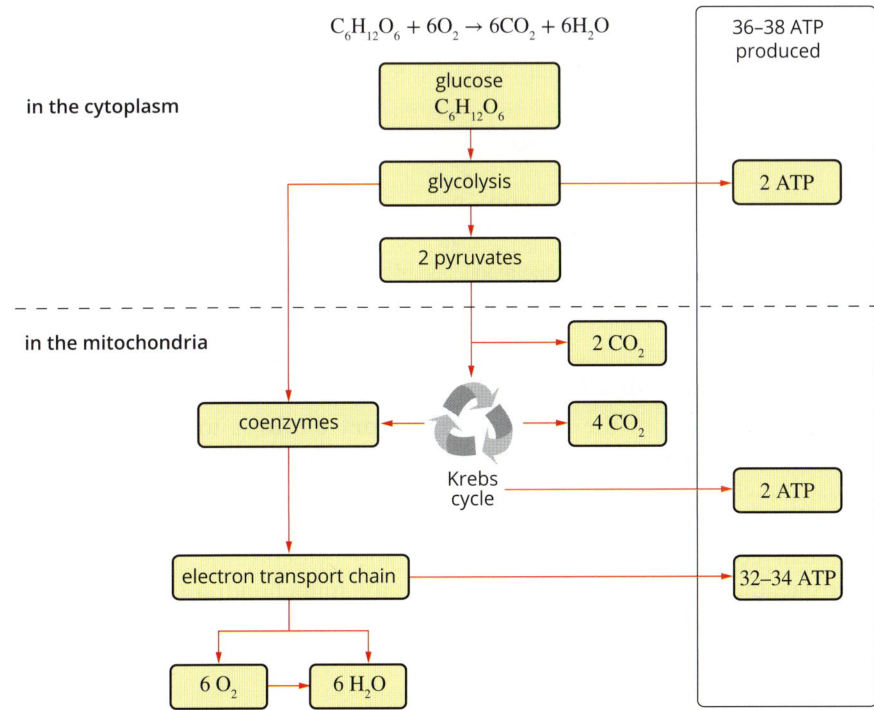

FIGURE 7.1.6 Overview of the stages of cellular respiration

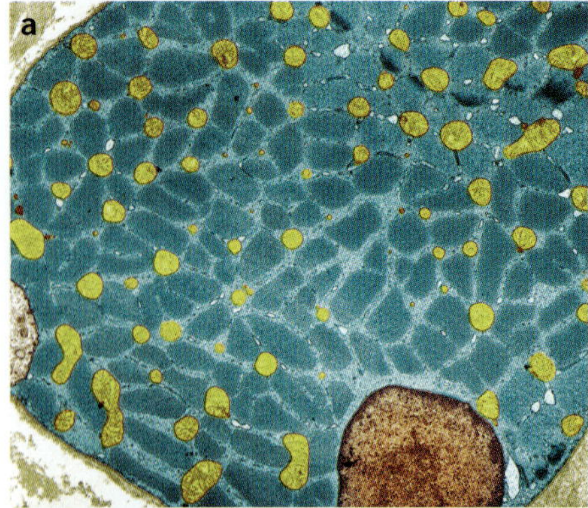

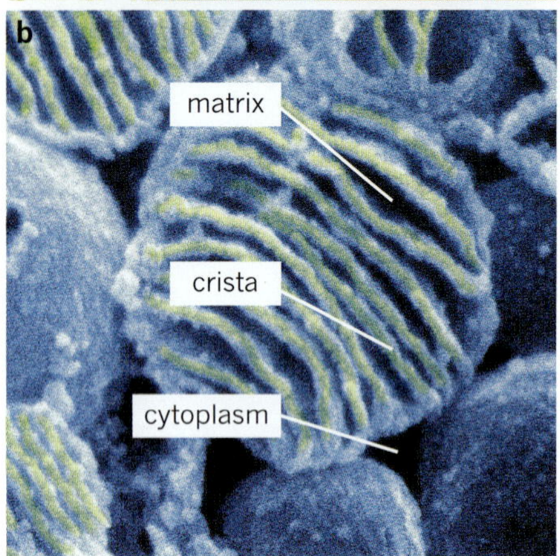

FIGURE 7.1.7 Glycolysis occurs in the cytoplasm of the cell and cellular respiration occurs in the mitochondria of the cell. (a) Coloured transmission electron microscopy image of a skeletal muscle cell. The mitochondria are coloured yellow. (b) The Krebs cycle occurs in the matrix of the mitochondria and the electron transport chain occurs on the cristae of the mitochondria, shown in this coloured scanning electron microscopy image.

When the cell has a supply of oxygen, cellular respiration will occur in the **mitochondria**. The pyruvate formed from glycolysis passes into the mitochondria (Figure 7.1.7a), where it is further broken down into carbon dioxide and water through a series of biochemical steps. The Krebs cycle and the electron transport chain occur in the mitochondria (Figure 7.1.7b).

The Krebs cycle

The second important stage of cellular respiration occurs after glycolysis and is called the **Krebs cycle** (or the citric acid cycle). This takes place in the mitochondrial matrix. The Krebs cycle is a series of eight reactions, each catalysed by a different enzyme. The pyruvate formed from glycolysis diffuses from the cytoplasm through the outer membrane of the mitochondria and is then moved by active transport through the inner membrane.

When the pyruvate, a three-carbon molecule, is in the mitochondrial matrix, it is converted into acetyl coenzyme A (acetyl CoA), a two-carbon molecule, which is the substrate for the first of a series of reactions that make up the Krebs cycle. In the formation of acetyl CoA from pyruvate, one carbon dioxide molecule is formed. In addition, in one turn of the Krebs cycle two carbon dioxide molecules are formed. That is a total of three molecules of carbon dioxide formed for every pyruvate molecule and six molecules of carbon dioxide for every glucose molecule metabolised.

During the reactions of the Krebs cycle, energy is transferred to energy-carrying coenzymes such as NADH, $FADH_2$ and ATP. From each turn of the Krebs cycle, one molecule of acetyl CoA is metabolised into two molecules of carbon dioxide, three molecules of NADH, one molecule of $FADH_2$ and one molecule of ATP (Figure 7.1.8). The Krebs cycle does not use oxygen but it will stop if oxygen is not available. This is because oxygen is the electron acceptor that is required to regenerate NAD^+ and FAD.

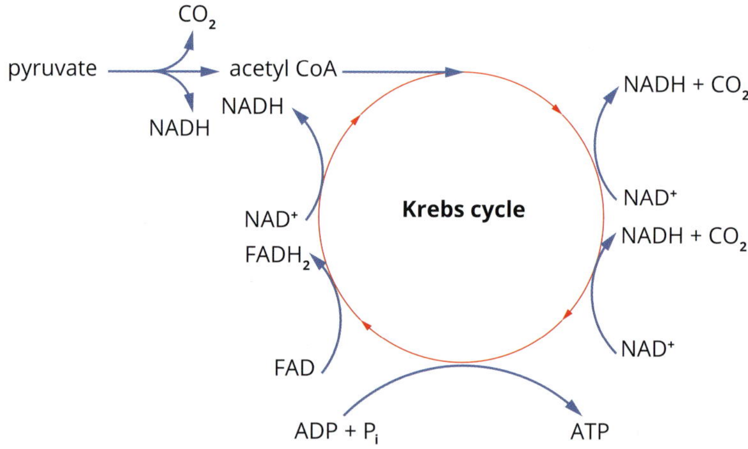

FIGURE 7.1.8 The three-carbon molecule pyruvate produced in glycolysis is converted to a two-carbon molecule acetyl CoA with the production of a CO_2 molecule. The acetyl CoA enters the Krebs cycle, where it results in CO_2, ATP, NADH and $FADH_2$ being formed.

Moving pyruvate into the mitochondria uses the ATP generated by glycolysis.

FAD (flavin adenine dinucleotide) is capable of carrying two hydrogen ions, becoming $FADH_2$.

NAD^+ is capable of carrying one hydrogen ion, becoming NADH.

The electron transport chain

The overall purpose of the **electron transport chain** is to move protons and electrons across a membrane as a system for generating ATP. Oxygen is essential because it picks up electrons at the end of the chain. If oxygen is not available, the electron transport chain stops.

Protein complexes, including enzymes and cytochromes, are embedded in the cristae, the inner membrane structure of the mitochondria (Figures 7.1.9 and 7.1.10). The cristae is convoluted to increase the surface area for these complexes, which form an interconnected series that together make up the electron transport chain, the third stage of cellular respiration.

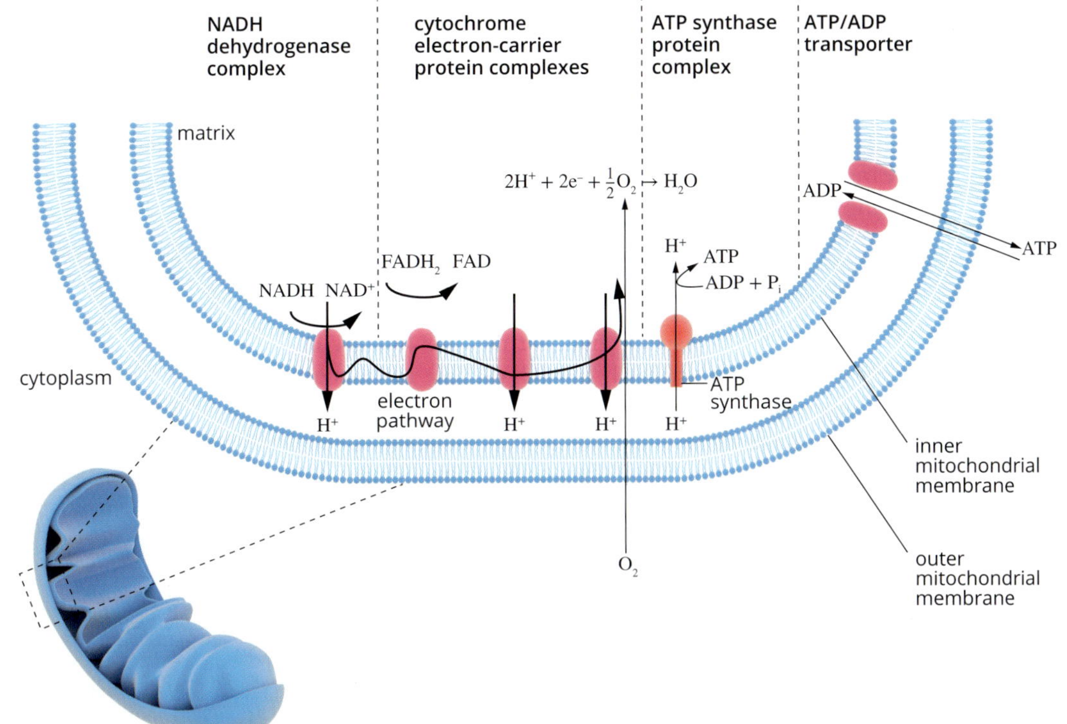

FIGURE 7.1.9 The inner mitochondrial membrane has several embedded proteins that pass along the energy and H^+ ions from carriers (NADH and $FADH_2$). The final enzyme in the pathway, ATP synthase, generates large amounts of ATP.

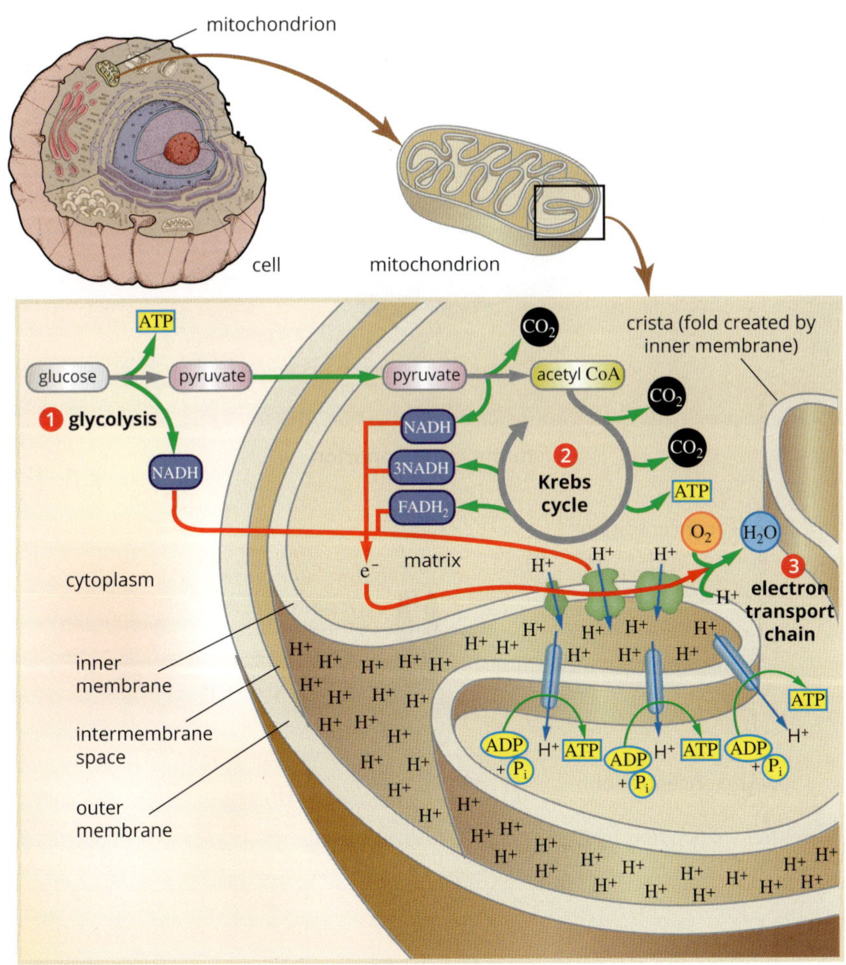

FIGURE 7.1.10 Summary of cellular respiration

Energy-carrying molecules from the Krebs cycle feed into the electron transport chain. NADH is converted back to NAD^+ by interacting with the first complex at the beginning of the electron transport chain, and $FADH_2$ is converted back to FAD by interacting with the second complex.

The hydrogen ions (H^+) originating from the conversion of NADH and $FADH_2$ are moved into the intermembrane space and the electrons are transferred along the chain. The energy obtained in this process is used to make ATP. At the end of the electron transport chain, hydrogen ions and electrons combine with oxygen to form water:

$$2H^+ + 2e^- + \tfrac{1}{2}O_2 \rightarrow H_2O$$

The electron transport chain forms 32–34 molecules of ATP using the energy that was contained originally in each glucose molecule.

Some tissues and cells are more efficient at carrying out cellular respiration than others, but in general if we add up the ATP molecules formed at each stage of cellular respiration we find that for every molecule of glucose metabolised, 36–38 ATP molecules can be formed:

- 2 ATP molecules from glycolysis
- 2 ATP molecules from the Krebs cycle
- 32–34 ATP molecules from the electron transport chain.

Figure 7.1.10 on page 239 summarises cellular respiration.

BIOFILE

ATP synthase—the cell's 'hydro' scheme

Hydroelectricity is generated using the potential energy in water stored in a dam. The potential energy is converted to kinetic energy and then to electrical energy by running the water downhill through water turbines that spin electricity generators.

Similarly, the electron transport chain builds up potential energy by pumping protons (hydrogen ions, H^+) released from NADH into the intermembrane space. This forms a concentration gradient across the inner membrane. The potential energy accumulated in the protons in the intermembrane space is transferred to ATP as the protons run through an ATP synthase enzyme complex embedded in the membrane and into the matrix.

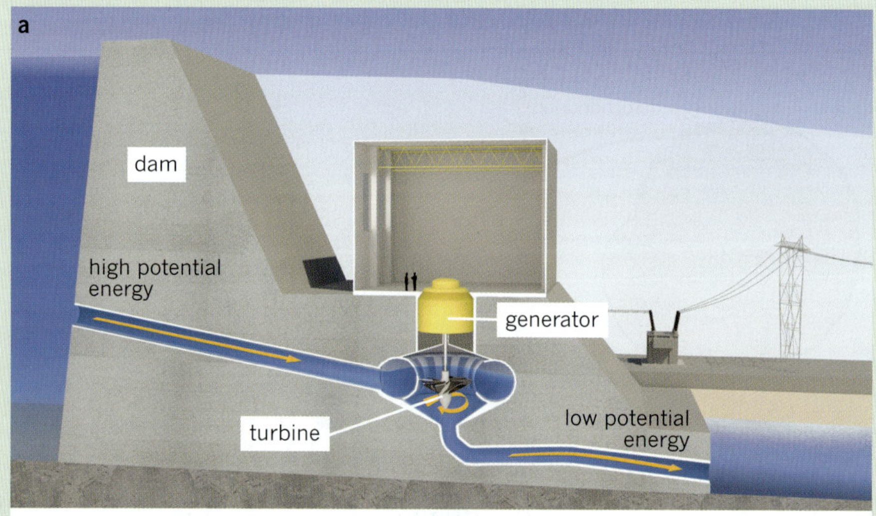

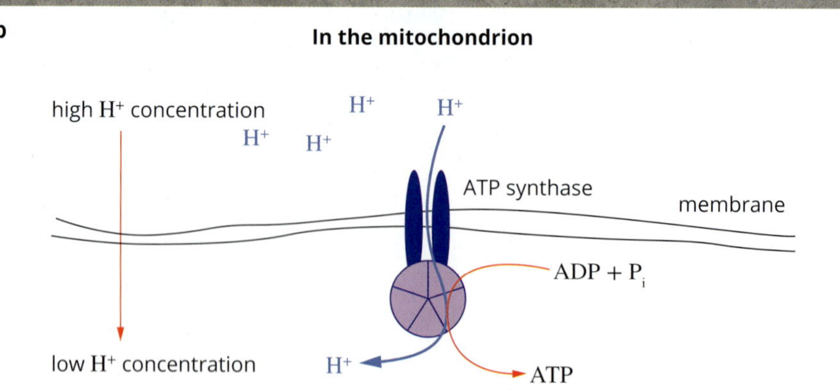

(a) Water from a dam runs through a hydroelectricity generator. (b) A similar process occurs in the mitochondrion.

7.1 Review

SUMMARY

- Cells use chemical energy in the form of ATP (adenosine triphosphate) to carry out cell functions.
- Cells produce ATP by the process of cellular respiration.
- Cellular respiration has three stages: glycolysis, the Krebs cycle and the electron transport chain.
- Glycolysis occurs in the cytoplasm of the cell and generates two ATP (net) molecules and two molecules of pyruvate per molecule of glucose.
- The Krebs cycle occurs in the matrix of the mitochondria and yields two ATP molecules. Pyruvate is broken down into carbon dioxide.
- The electron transport chain occurs in the cristae of the mitochondria and yields 32–34 ATP molecules. Oxygen accepts the hydrogen ions that are used to generate a large amount of energy.
- Cellular respiration is more efficient than anaerobic fermentation. Cellular respiration produces 36–38 ATP molecules per glucose molecule; anaerobic fermentation only yields two ATP molecules per glucose molecule.

KEY QUESTIONS

Knowledge and understanding

1. Which molecule carries a usable form of energy for cellular processes?
2. Does glycolysis require an input of energy to proceed?
3. Name each biochemical pathway that takes place under aerobic conditions. Identify the location of each pathway.
4. What is the main purpose of the electron transport chain?
5. What is the role of oxygen in cellular respiration?

Analysis

6. A cell is undergoing glycolysis and is beginning to accumulate an excess of ATP. Would this increase or decrease the rate of glycolysis? Justify your answer.

7.2 Anaerobic fermentation

If there is little or an absence of oxygen, or an organism does not have genes for the enzymes used in the Krebs cycle or electron transport chain, the organism can still release a small amount of energy using the **anaerobic fermentation** pathway, following glycolysis. Animals, bacteria and yeasts use anaerobic fermentation to produce various final products (Table 7.2.1) along with the recycling of NADH to NAD^+ to maintain glycolysis for energy production. There are many commercial products made through the anaerobic fermentation process (Figure 7.2.1).

TABLE 7.2.1 Comparison of anaerobic fermentation outputs in animals, bacteria and yeasts

Organism	Output
animals	lactic acid
bacteria	• lactic acid • ethanol • butanol • acetic acid
yeasts	ethanol

FIGURE 7.2.1 Beer and wine are commercial products of anaerobic fermentation by yeast. Yoghurt and cheese are made through anaerobic fermentation by bacteria.

Anaerobic fermentation is an essential biochemical pathway to recycle NADH to NAD^+, to be used as an electron carrier in glycolysis.

There are anaerobic organisms that undergo anaerobic respiration. Rather than using oxygen as an electron acceptor, these organisms use nitrogen or sulfur. This process is different from anaerobic fermentation.

In animals, glycolysis and the final reaction that converts pyruvate to lactic acid are together called lactic acid fermentation (or lactate fermentation).

IMPORTANCE OF ANAEROBIC FERMENTATION

Biochemical pathways are interlinked and highly regulated within a system. For a biochemical pathway to continue, the products of each reaction must be processed by the next reaction in the chain. If the product is not removed, it slows down the reaction. This causes the reaction before it to be slowed down and so on back through the pathway until the whole pathway is slowed down or even stopped. Think of it like people moving in a queue. If the person at the head of the queue stops moving, the person behind must stop and so on, until the whole queue stops.

The final reaction in the electron transport chain requires oxygen. Lack of oxygen will limit the rate of this final reaction, which in turn will limit the reaction before it and so on back through the pathways. NADH will not be converted back to NAD^+. The Krebs cycle will also be slowed down. When the Krebs cycle slows down, pyruvate will begin to accumulate. An accumulation of pyruvate will cause glycolysis to slow down.

In addition, for glycolysis to occur, a constant supply of NAD^+ is required. If NAD^+ is not being regenerated, glycolysis will slow down or even stop.

Fortunately, to allow glycolysis to continue at low oxygen levels, some protists, fungi and some animal cells, such as muscle cells, contain an enzyme that can catalyse the conversion of pyruvate to lactic acid and NADH to NAD^+. This reaction solves both problems and enables glycolysis to continue. It first removes pyruvate so that the pathway is no longer blocked and then it provides a source of NAD^+ that is essential for glycolysis.

ANAEROBIC FERMENTATION IN ANIMALS

The lactic acid produced through anaerobic fermentation can leave the cell by diffusion through the plasma membrane and via a special membrane transport protein. This means the lactate level in the cell can remain low so that the pyruvate-to-lactate reaction can continue to occur, glycolysis can occur and ATP can continue to be produced to satisfy the cell's needs. In humans, when lactic acid produced by muscle cells is diffused, it can be circulated in the blood to other tissues in the body, including liver and heart muscle, where it is converted back to pyruvate and enters aerobic cellular respiration pathways to produce ATP (Figure 7.2.2).

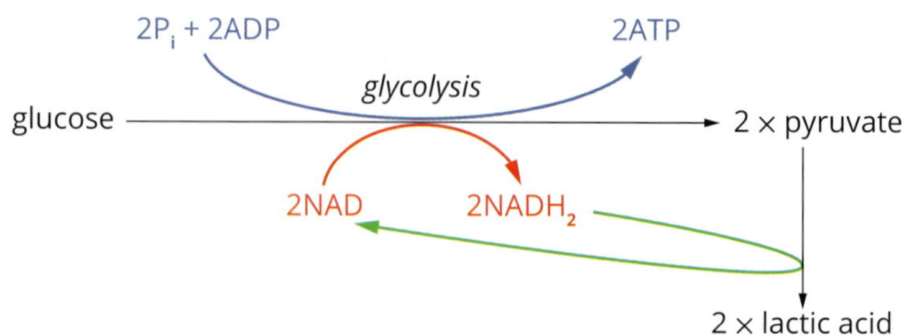

FIGURE 7.2.2 In animals, pyruvate is converted into lactic acid during anaerobic fermentation to prevent a build-up of pyruvate. Later the lactic acid can be converted back into pyruvate for aerobic cellular respiration.

ANAEROBIC FERMENTATION IN YEAST

Yeast cells (Figure 7.2.3) carry out a different type of anaerobic fermentation called ethanol fermentation (Figure 7.2.4). As in animal cells, one glucose molecule is broken down to two pyruvate molecules and NADH is formed from NAD^+. In yeast anaerobic fermentation, the pyruvate is then broken down to ethanol (alcohol) and carbon dioxide, and NADH is converted back to NAD^+ through two reactions. In the first step, pyruvate is broken down to acetaldehyde and carbon dioxide. In the second step, the acetaldehyde is broken down to alcohol and NADH is converted to NAD^+.

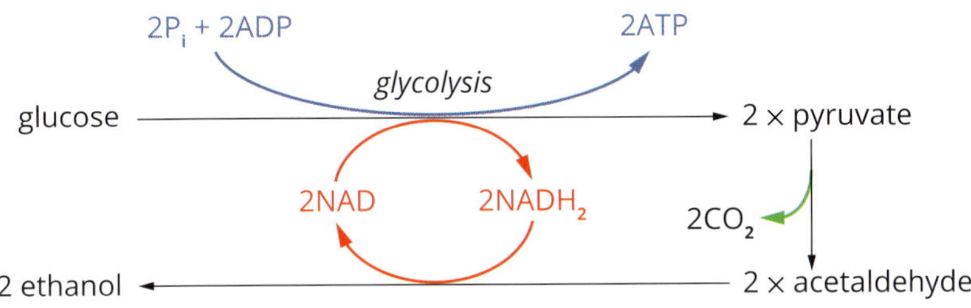

FIGURE 7.2.4 In yeast, pyruvate is converted into ethanol and CO_2 during anaerobic fermentation.

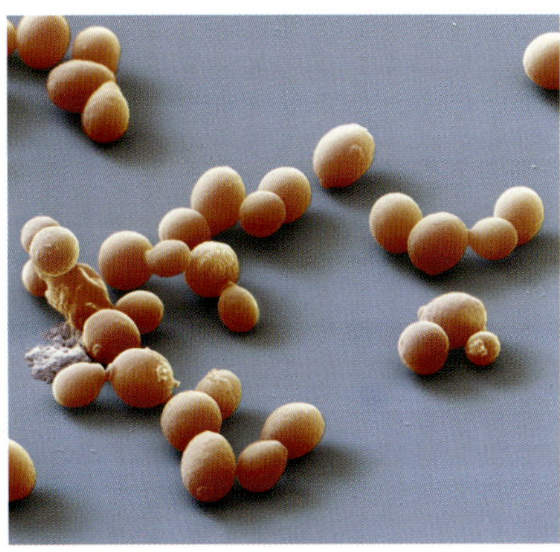

FIGURE 7.2.3 Coloured scanning electron micrograph of baker's yeast (*Saccharomyces cerevisiae*) cells

The alcohol readily diffuses out of the cell by passive diffusion into the surrounding environment. The loss of alcohol from the cytoplasm means that the reactions of anaerobic fermentation can continue until the alcohol builds up to a level in the external environment where alcohol no longer diffuses from the cell; that is, there is no longer a concentration gradient. Alcohol builds up in the cell to a point at which the anaerobic fermentation reactions are blocked and they stop.

> In yeast, glycolysis and the final reaction that converts pyruvate to ethanol and CO_2 are together called ethanol fermentation.

APPLICATIONS OF FERMENTATION FOR BIOFUEL PRODUCTION

Concerns about the limited supply of non-renewable fossil fuels and of greenhouse gas emissions have led scientists, entrepreneurs and companies to invest in development and refinement of renewable fuels. One of these strategies involves using the biochemical process of anaerobic fermentation on **biomass** by microorganisms such as yeast, bacteria, cyanobacteria and algae for production of biofuels. Biomass waste contains a high sugar content, which slowly releases stored chemical energy over time. Before microorganisms are used to ferment the sugars for biofuel production, the biomass needs to be liquefied using heat treatment, mechanical disruption or enzymatic digestion.

> Anaerobic fermentation can be conducted commercially on a large scale to produce fuels (such as bioethanol) and alcohol.

Biofuels are a diverse range of technologies that generate fuel with at least one component based on a biological system. Biofuels include bioethanol, biodiesel and biogas and are produced as first-, second- and third-generation fuels.

- First-generation biofuels are produced using biomass from food crops grown for the sole purpose of biofuel production. To date, biofuel production has mainly focused on first-generation fuels, converting crops such as corn and sugar to ethanol. One major drawback to crop-based biofuels is the enormous amount of agricultural land required to replace a fraction of petroleum with biofuels.
- Second-generation biofuels are produced using non-food crops or organic waste. Second-generation biofuels are more sustainable than first-generation biofuels because the biomass is sourced from agricultural waste and industrial food waste.
- Third-generation biofuels are produced from algae. Research has shown that algae biofuels could relieve the current reliance on fossil fuels since they avoid some of the previous drawbacks associated with crop-based biofuels.

Biofuels from algae

Microalgae are single-celled organisms that have the potential to offer a variety of solutions for our fuel requirements. Because microalgae only need light, water, carbon dioxide and a small amount of minerals to grow, they offer a potentially more stable energy supply. Additionally, their cells divide quickly, meaning that they can be harvested faster than land-based biomasses and can be harvested all year round.

Microalgae have several advantages over terrestrial plants.

- They are single-celled organisms that reproduce by division.
- High-throughput technologies can be used to rapidly evolve strains, reducing the time required to generate sufficient yields to a few months.
- They impact the environment less than terrestrial sources of biomass used for biofuels.
- They can be grown on land not typically used for traditional agriculture.
- They can be grown throughout the year in various water conditions.
- They do not require herbicide or pesticide use.
- They use nutrients from water more effectively, meaning that not only would production of algae biofuels minimise land use compared with biofuels produced from terrestrial plants, it would also reduce the amount of waste generated.
- They have the potential to be bioengineered, allowing improvement of specific traits and production of valuable co-products, enabling algal biofuels to compete economically with petroleum.

The production of microalgal biofuels may be further enhanced through more efficient algae cultivation methods and genetic engineering. Researchers have shown that algae cultivation has the potential to produce cost-competitive biofuels; however, significant challenges need to be overcome for algal fuel commercialisation. These challenges include:

- how and where to grow algae
- improved oil extraction and fuel processing
- strain identification and isolation
- improvement of crop protection
- nutrient and resource allocation and use
- the production of co-products to improve the economics of the entire system.

Although there is optimism about their potential, much work is still required to realise the economic viability of algal biofuels.

Bioethanol

Bioethanol is a widely used biofuel that has been used in the transportation sector since the early 1900s. Yeast plays an essential role in the production of bioethanol. Yeast produces a high ethanol yield, has high productivity, a tolerance to ethanol exposure and is inexpensive and easily grown. There are several yeast species that are being employed. The strains *Saccharomyces* and *Kluyveromyces* are commonly used as they are able to break down a variety of sugars in addition to glucose, making them more viable.

However, there are some challenges for large-scale production. At an industrial scale, yeast needs to be cultured under strict environmental conditions for optimal growth and ethanol production. Although yeast can tolerate a high ethanol concentration, if the concentration is too high the yeast's plasma membrane is damaged, leading to inhibition of growth and viability. Temperature and pH are other factors that need to be monitored for optimal growth, as well as avoiding bacterial contamination of the culture. The starting material is also important. Yeast can easily break down simple sugars, such as glucose found in sugar cane juices. The process becomes more difficult when the starting material consists of complex carbohydrates, such as starch or cellulose (Figure 7.2.5).

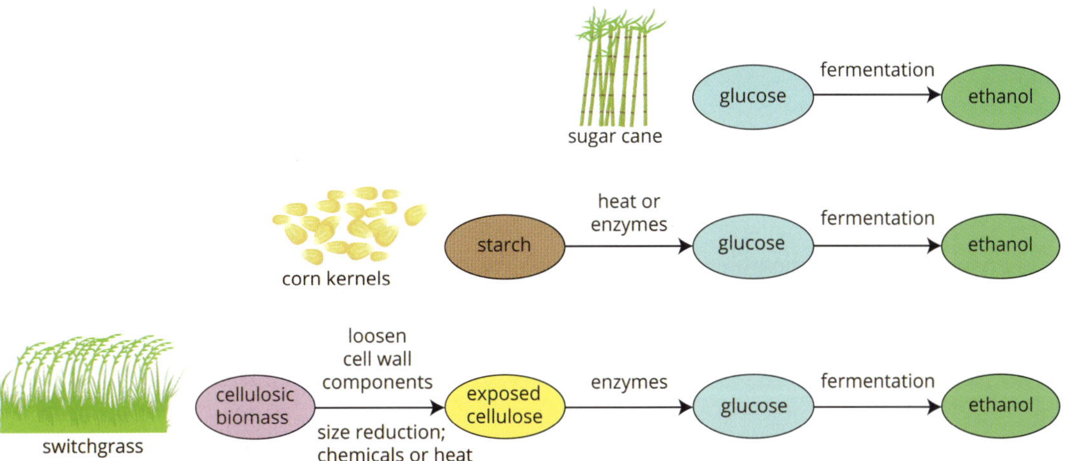

FIGURE 7.2.5 Comparison of the steps necessary to convert sugar cane, corn and switchgrass (plant with high composition of cellulose) to ethanol

There are many possible avenues to improve and make biofuels more viable. These include genetically modifying yeast to increase its tolerance to ethanol exposure, and to enable it to ferment a range of sugars, such as 5-carbon (e.g. xylose) as well as 6-carbon (glucose) sugars, to broaden its range of use to an industrial scale.

Biotechnology and engineering strategies are being employed to make economical clean biofuels a viable option. This includes the challenges of upscaling for industrial volumes, improving yield by using the appropriate microorganisms, including possibly developing genetically engineered microorganisms, increasing efficiency of production and ensuring that the generation of waste is minimised.

COMPARING AEROBIC AND ANAEROBIC PATHWAYS

The anaerobic fermentation pathway in eukaryotes produce only two ATP molecules per molecule of glucose during glycolysis, whereas aerobic pathways (cellular respiration) produce 36–38 ATP molecules per molecule of glucose during glycolysis, the Krebs cycle and the electron transport chain. Therefore, cellular respiration is much more efficient in supplying the cell with energy. The organic products of anaerobic pathways still contain much energy and both can be further metabolised to release energy. Table 7.2.2 compares cellular respiration and anaerobic fermentation.

TABLE 7.2.2 Summary of cellular respiration and anaerobic fermentation pathways for organisms to generate ATP molecules

	Cellular respiration	Anaerobic fermentation
Pathway	glycolysis Krebs cycle electron transport chain	glycolysis fermentation
Location	cytoplasm and mitochondria	cytoplasm
Reactants (inputs)	glucose and O_2	glucose
Products (outputs)	CO_2 and water	lactic acid and water (animals/yeasts/bacteria) ethanol and CO_2 (plants/fungi/bacteria) acetic acid and CO_2 (bacteria) Note: There are several end products in anaerobic fermentation in bacteria
Energy output (per glucose molecule)	36–38 ATP molecules	2 ATP molecules

BIOFILE

The biochemistry of a hangover

A headache and nausea are the common symptoms of a 'hangover' experienced after drinking too much alcohol. Cells in the liver convert alcohol to acetyl CoA in three enzyme-catalysed reactions. The Krebs cycle metabolises the acetyl CoA to CO_2 and water. However, there is a build-up of intermediate products that accumulate in the blood and circulate to the brain, where they cause the toxic effect experienced as a hangover.

There are additional effects due to the diuretic effect of alcohol, including the depletion of electrolytes from the body.

It is not the alcohol that causes a hangover but the intermediate substances formed in its breakdown by the body.

7.2 Review

SUMMARY

- In animals, the product of anaerobic fermentation is lactic acid.
- In yeast, the products of anaerobic fermentation are ethanol and carbon dioxide.
- Anaerobic fermentation of biomass to produce biofuels is being investigated to provide clean and renewable sources of fuel.
- Anaerobic fermentation is less efficient than cellular respiration. Cellular respiration produces 36–38 ATP molecules per glucose molecule; anaerobic fermentation only yields two ATP molecules per glucose molecule.

OA ✓✓

KEY QUESTIONS

Knowledge and understanding

1 What is the purpose of anaerobic fermentation?

2 Which molecule serves as a final electron acceptor in lactic acid fermentation?

3 Describe the efficiency of cellular respiration and anaerobic fermentation.

Analysis

4 Compare lactic acid fermentation (in animals) and ethanol fermentation.

5 Students wanted to observe anaerobic fermentation in yeast and set up the experiment shown at right. They set up flasks (three replicates for each condition) with different materials and under different anaerobic conditions by sealing the flask with a stopper and a syringe in the top of the flask.

  a Suggest a hypothesis for this experiment.

  b How did the students measure gas production in this set-up?

  c Looking at the average results in the table on the right, is this the outcome you would expect? Explain why there was a difference between flasks 3 and 4.

  d Flask 5 contained the same materials as flask 3. Suggest a reason for the different outcome in carbon dioxide production.

  e Identify the flask(s) used as the controls, the independent and dependent variables.

  f What would you expect to observe if pyruvate was used instead of glucose?

  g One student plotted the data in the graph at right. Identify any elements missing from the graph.

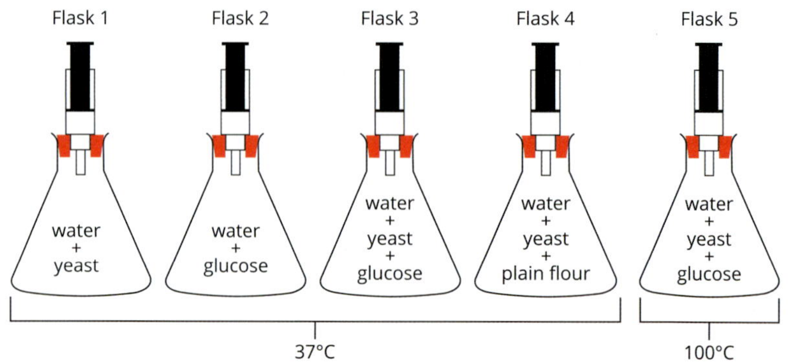

	Average carbon dioxide production (cm^3)				
Time (h)	**Flask 1**	**Flask 2**	**Flask 3**	**Flask 4**	**Flask 5**
0	0	0	0	0	0
3	0	0	1.75	0.5	0
6	0	0	3.8	1	0
9	0	0	5.4	1.9	0
12	0	0	6.4	2.9	0

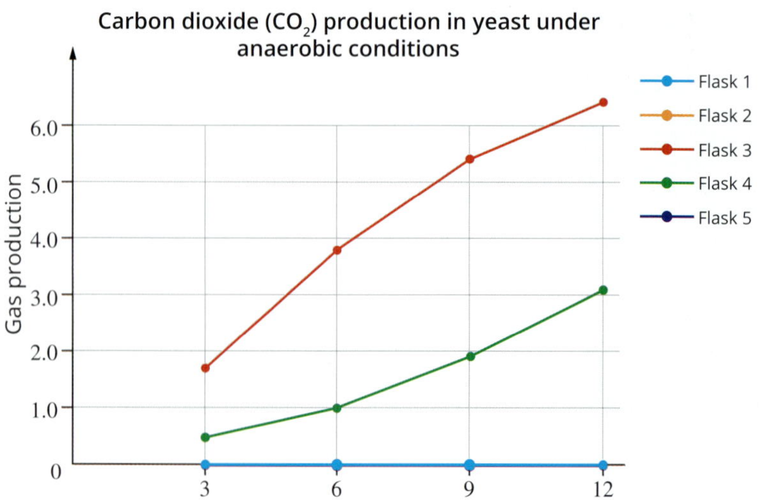

7.3 Factors that affect the rate of cellular respiration

Cellular respiration is affected by a number of factors, including temperature, glucose availability and oxygen concentration.

In this section, you will learn how each of these factors affects cellular respiration.

TEMPERATURE

You have learnt that cellular respiration is comprised of a number of interconnected biochemical pathways and that each pathway is a series of chemical reactions catalysed by specific enzymes. You have also learnt that the rate of an enzyme reaction is affected by temperature and that each enzyme has an optimum temperature.

As the temperature either rises above or falls below the optimum temperature, the rate of the enzyme reactions, and therefore the rate at which cellular respiration occurs, will slow down (Figure 7.3.1).

- When the temperature drops, the reactant molecules contain less kinetic energy and so do not react as quickly.
- When the temperature rises above the optimum level, the increased heat energy can disrupt the hydrogen bonds in the enzyme, causing the enzyme to denature. This means that the active site of the enzyme has lost its three-dimensional functional shape. This distortion in the shape means that the enzyme cannot bind to the substrate effectively, slowing down the rate of reaction.

Living organisms have particular temperature tolerance limits within which they will survive. More complex organisms, such as birds and mammals, control their body temperature at levels that are optimal for the functioning of their enzymes. Other organisms do not have the capacity to control the temperature of their cells, and the cellular respiration for these organisms is affected by the temperature of the external environment.

When an enzyme is denatured it cannot be repaired. The loss of shape is permanent and it will no longer catalyse the reaction.

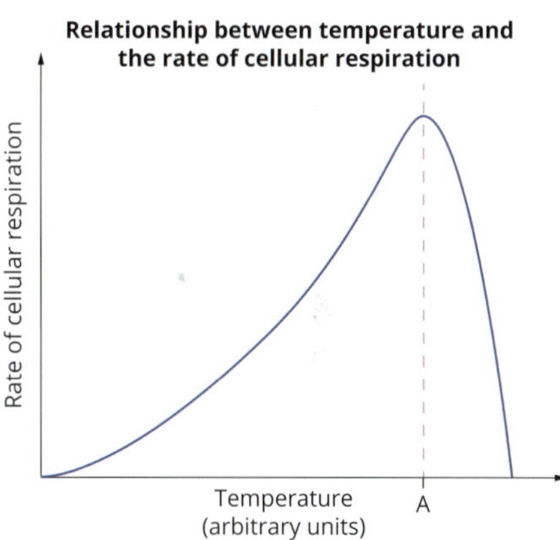

FIGURE 7.3.1 Graph showing the relationship between temperature and the rate of cellular respiration. At the optimum temperature (point A), cellular respiration will occur at the maximum rate.

BIOFILE

Birds use their beaks

Toucans and many other birds, including Australian parrots, use their beaks as heat exchangers to lose excess heat to cool their bodies. When it is hot, the bird pumps more blood to its beak, where heat can be lost to the air. The ability to lose excess heat is beneficial to the bird because the enzymes that catalyse the reactions in the biochemical pathways of animals have an optimum temperature. The optimum temperature for enzyme reactions in birds is about 40°C—slightly higher than in humans. At temperatures above the optimum, the bird's metabolism will begin to function poorly. The rate of cellular respiration will slow and the supply of ATP to the cells will drop.

The large beak of a toucan is used as a heat exchanger to regulate body temperature.

GLUCOSE AVAILABILITY

All chemical reactions are limited by the concentration of reactants. An enzyme's reaction is limited by the availability of its substrate(s). Glucose is the substrate for glycolysis, and therefore it is the substrate for the first reaction in cellular respiration. The availability of glucose will affect the rate at which this first reaction occurs. The products of the first reaction become the substrates for the next and so on along the pathway. Hence the availability of glucose will affect the first and subsequent reactions in the cellular respiration biochemical pathways.

If the temperature remains constant, increasing the availability of glucose will increase the rate of cellular respiration up to a maximum level. This maximum level will be determined by the concentration of the enzymes and other cofactors required for cellular respiration.

A reactant is a molecule or substance that is involved in a chemical reaction. A substrate is the term given to a molecule that an enzyme acts upon.

The concentration of enzymes for cellular respiration is linked to the availability of mitochondria in the cell.

BIOFILE

Energy storage in animals

In winter, food generally becomes scarce for animals, and their ability to store energy is necessary for survival. Animals store carbohydrates as **glycogen**, a large molecule made from glucose subunits. In humans, about 100 g of glucose is stored as glycogen in the liver and a further 200 g is stored in the muscles. This provides enough energy for about half a day at a moderate level of activity. The remainder of our energy reserves are stored as fats.

Animals use fats rather than carbohydrates as their main form of energy storage because:

- almost 25% more ATP is produced (per carbon atom) from fats than from carbohydrates
- fat is almost 50% lighter than carbohydrate
- stored carbohydrates attract and bind water molecules, increasing their weight between two and five times, whereas fats do not
- one gram of carbohydrate or protein can provide up to 17 kJ of energy, whereas one gram of fat provides 39 kJ of energy.

An average 70 kg male human stores about 11 kg of fat, which provides enough energy to last about one month without eating food. If the 70 kg man stored this amount of energy as carbohydrates, he would weigh more than 100 kg.

CASE STUDY

Other energy molecules

Glucose is the molecule most commonly used as the source of energy in cells. However, cells can also release chemical energy from other organic compounds, such as fats and proteins, to make ATP. In animals, if most of the available glucose stores are gone (such as during times of food shortage or extreme prolonged activity), fat stores are used to provide the ATP needed for cells to continue functioning (Figure 7.3.2).

In extreme cases, such as during long periods of starvation, even the proteins in muscles and other body tissues will be broken down to provide the energy necessary to survive. Fats provide more energy per gram (39 kJ) than carbohydrates or proteins (17 kJ), but the conversion process is slower and more complex.

FIGURE 7.3.2 (a) The cells of this underweight horse have converted stores of fat into ATP to help the horse survive. (b) Rebecca Clarke of New Zealand shows the exhaustion of completing a triathlon at the World Championship in Stockholm. During the race, her glucose and oxygen levels would have been severely depleted.

OXYGEN CONCENTRATION

In cellular respiration, a constant supply of oxygen is necessary. Oxygen is the final reactant of the electron transport chain, so oxygen concentration affects the rate of cellular respiration. When the concentration of oxygen is low, the rate at which the electron transport chain can occur will be reduced. As you learnt in the previous section, when oxygen is in very short supply or absent, some cells will use fermentation reactions so that pyruvate does not accumulate and the glycolysis reactions can continue. This means ATP will continue to be produced, although at much lower rates, and the cell can remain alive.

BIOFILE

Myoglobin

Myoglobin is a protein that binds oxygen. It is present in muscle tissue of mammals and some other vertebrates. Diving mammals such as whales have large amounts of myoglobin in their muscles to enable them to store the oxygen needed for cellular respiration during a long dive.

Whales have large amounts of myoglobin in their muscles.

7.3 Review

OA ✓✓

SUMMARY

- The rate of cellular respiration is affected by temperature, glucose availability and oxygen concentration.
- When the temperature is above or below the optimum range, the rate of cellular respiration is reduced.
- Glucose is a reactant in glycolysis, therefore an increase in glucose availability will increase the rate of cellular respiration.
- Oxygen is a reactant in the electron transport chain, therefore an increase in oxygen concentration will increase the rate of cellular respiration.

KEY QUESTIONS

Knowledge and understanding

1 What factors limit the rate of cellular respiration?

2 What would happen to the rate of cellular respiration at temperatures above the optimum range? Explain your answer.

3 List three types of organic compounds that can be used as a source of energy for cellular respiration.

4 What will limit cellular respiration if there is an excess supply of glucose and oxygen, and the reactions are taking place under optimal temperature?

Analysis

5 Mitochondria were extracted from some cells and isolated from the other cell contents. The mitochondria were suspended in a nutrient solution containing pyruvate in order to investigate cellular respiration.

  a Explain why the nutrient solution contained pyruvate rather than glucose.

  b Assuming that the mitochondria were in a sealed container, would carbon dioxide or oxygen be the limiting factor in reducing the rate of the electron transport chain?

Chapter review

KEY TERMS

adenosine diphosphate (ADP)
adenosine triphosphate (ATP)
anaerobic fermentation
biofuel
biomass
cellular respiration
cytoplasm
electron transport chain
flavin adenine dinucleotide (FAD)
glycolysis
Krebs cycle
mitochondrion (plural mitochondria)
nicotinamide adenine dinucleotide (NAD^+)

REVIEW QUESTIONS

Knowledge and understanding

1 Identify the reactants and products of glycolysis.

2 What molecule acts as an energy carrier in glycolysis?

3 Identify whether the following statements about glycolysis are true or false.

- **a** Glycolysis takes place in the mitochondria.
- **b** Glycolysis produces 30–32 molecules of ATP.
- **c** Glycolysis can occur without light.
- **d** Glycolysis is the second step of cellular respiration.

4 During cellular respiration, free oxygen:

- **A** combines with water and helps produce ATP
- **B** is produced as pyruvate is broken down
- **C** is the final electron acceptor in the electron transport chain
- **D** combines with carbon to produce carbon dioxide

5 What is the role of the two energy carriers NAD^+ and FAD in glycolysis and the Krebs cycle?

6 Study the diagram below that summarises the processes of energy release in cells.

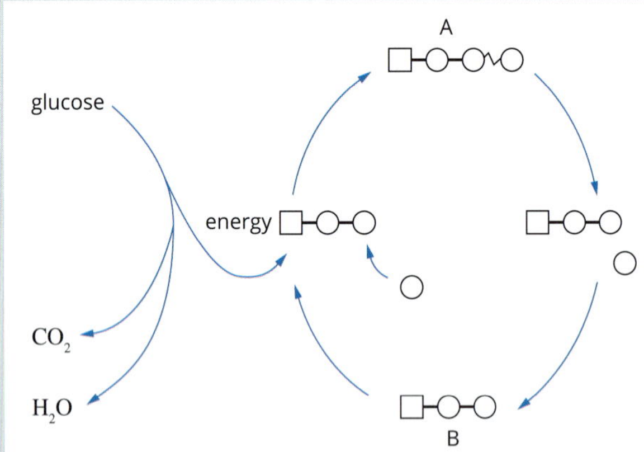

- **a** Name the molecule represented by each letter.
- **b** Describe how molecules A and B are related.
- **c** Glucose is the energy-rich molecule that enters glycolysis. Complete this sentence:
  In glycolysis, a single glucose molecule is split to produce ______________________.

7 The function of cellular respiration is to produce ATP. It does so by the following chemical reaction.

ADP + molecule X → ATP

Which of the following statements about this reaction in cellular respiration is false?

- **A** Molecule X is P_i.
- **B** The electron transport chain is the major site of this reaction.
- **C** The energy needed for the reaction to proceed can come from $FADH_2$.
- **D** The energy needed for this reaction to proceed can come from $NADP^+$.

8 Which stage of cellular respiration requires oxygen as a reactant? What happens at low oxygen concentrations?

9 $FADH_2$ is produced during:

- **A** glycolysis in cellular respiration
- **B** the Krebs cycle in cellular respiration
- **C** the electron transport chain of cellular respiration
- **D** the electron transport chain of the light-dependent reactions of photosynthesis

10 What are the products of anaerobic fermentation in yeast?

11 **a** Where does the Krebs cycle of cellular respiration take place?

- **b** What are the final products of this process?

12 What is the major benefit to cells in using cellular respiration rather than anaerobic fermentation?

13 Briefly describe what happens to pyruvate in the two biochemical pathways that can occur after glycolysis when oxygen is:

- **a** present (in aerobic organisms)
- **b** absent (in anaerobic organisms)

14 Why would a prokaryotic organism use anaerobic fermentation rather than cellular respiration?

15 What strategies could a scientist use to engineer yeast to increase the efficiency of biofuels?

Application and analysis

16 Discuss the benefits and the current limitations and disadvantages of biofuels.

17 A study was conducted to investigate the best protocol for increasing the efficiency of biofuel production and to compare the theoretical versus practical yield of ethanol. The study used the microorganism *Saccharomyces cerevisiae* (yeast) strain and compared its ability to produce ethanol using two growing methods: batch and fed-batch cultures. The batch method (graphs A) involved providing the yeast with all the glucose (food) it required at the start of the experiment (26%), whereas in the fed-batch culture (graphs B), the addition of glucose was staggered over time (18% initially + 4% glucose supplemented at 24 and 48 h, totalling 26%). Below are the results of the yeast grown under the two conditions. Ethanol yield (%) is the ratio of the concentration of ethanol produced compared to the concentration of glucose consumed.

a Interpret the two graphs in regards to glucose and ethanol levels over time.

b Discuss the relationship between dissolved oxygen levels (%) and ethanol concentration (g/L).

c Which protocol would you recommend be followed for increased production and why?

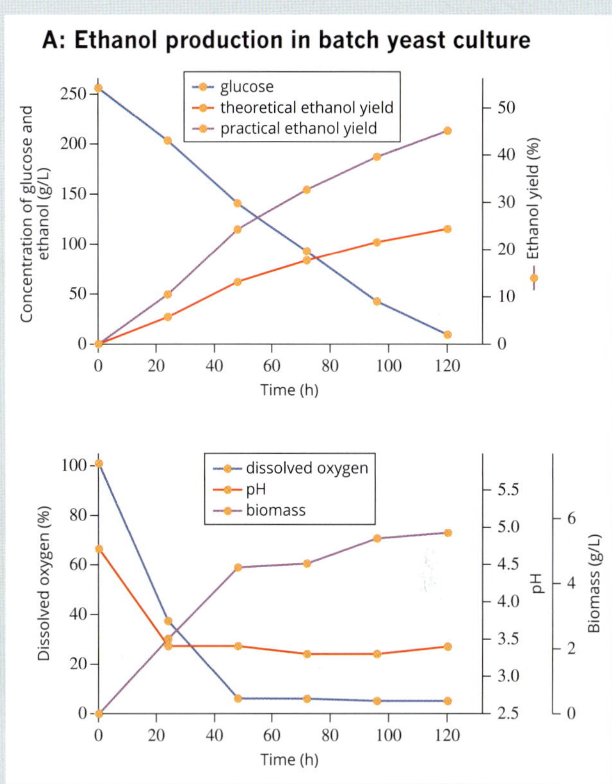

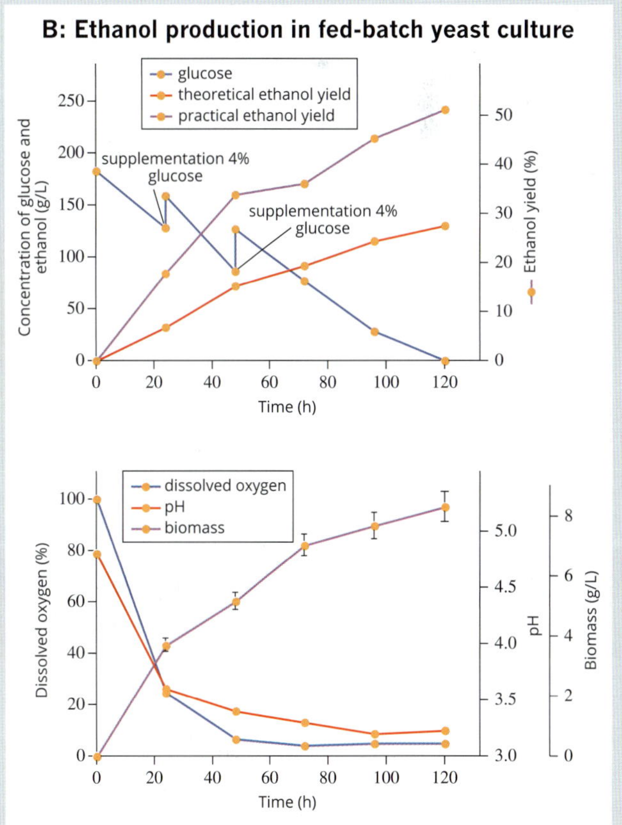

18 The Krebs cycle is an important stage of cellular respiration. The diagram below shows the chemical changes that occur during the cycle for one molecule of pyruvate.

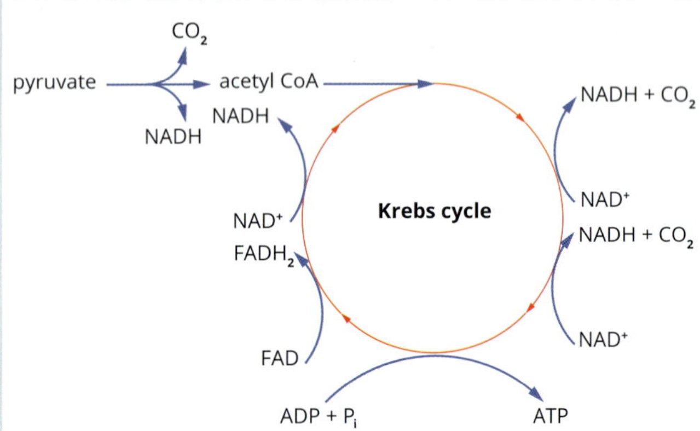

Use the information in the diagram above and your knowledge of cellular respiration to answer the following questions.

a How many molecules of ATP are formed in the Krebs cycle from one molecule of glucose?

b How many H^+ ions are loaded onto carriers during one turn of the Krebs cycle?

c NAD and FAD are sometimes described as proton carriers. Why?

d **i** How many molecules of carbon dioxide are produced during one turn of the Krebs cycle?

ii Which stage produces the remainder of the carbon dioxide?

e Most of the ATP produced during the aerobic phase of respiration comes from the electron transport chain, yet this could not occur without the Krebs cycle. Explain.

19 Describe and compare the pros and cons of first-, second- and third-generation biofuels generated by anaerobic fermenting organisms.

20 The enzyme cytochrome c oxidase is found embedded in the cristae of mitochondria. It is the last enzyme in the electron transport chain. It transfers electrons to oxygen and also binds protons (H^+) to oxygen to form water.

a What is the significance of the electron transport chain to living cells?

b Sodium azide is a toxic chemical that binds irreversibly to cytochrome c oxidase. Explain why sodium azide is sometimes used in agriculture to control pathogens in the soil.

c An experiment was performed using human skin cells. Skin cells were grown in culture so that the cells were all separated. The culture was divided in half and one half of the culture was placed in a test tube. The other half of the culture was treated to separate the mitochondria from the rest of the cell contents. The mitochondria were placed in one tube and the residue was placed in a third test tube. Samples from each of the tubes were grown in the following solutions:

S glucose + sodium azide
T pyruvate + sodium azide
U glucose only
V pyruvate only

The test tubes were supplied with oxygen. All other variables were kept the same. After 30 minutes, the tubes were tested for the presence of carbon dioxide and lactic acid. The results are shown in the table below.

	S glucose + sodium azide	**T pyruvate + sodium azide**	**U glucose only**	**V pyruvate only**
Whole cells	lactic acid	lactic acid	CO_2	CO_2
Mitochondria	neither	neither	neither	CO_2
Cell residue without mitochondria	lactic acid	neither	lactic acid	neither

i What is/are the control(s) in this experiment?

ii Explain why carbon dioxide was produced in the 'pyruvate only' tube but not in the 'glucose only' tube, of the tubes that contained only mitochondria.

iii The experiment could not be extended beyond 30 minutes because after that time the whole cells in series T died. The whole cells in series S continued to live for some time. Explain why the cells with pyruvate + sodium azide died but those with glucose + sodium azide did not.

iv Do the results of this experiment support the claim that sodium azide disrupts the electron transport chain?

v How would the results have differed if plant cells had been used for the experiment?

OA ✓✓

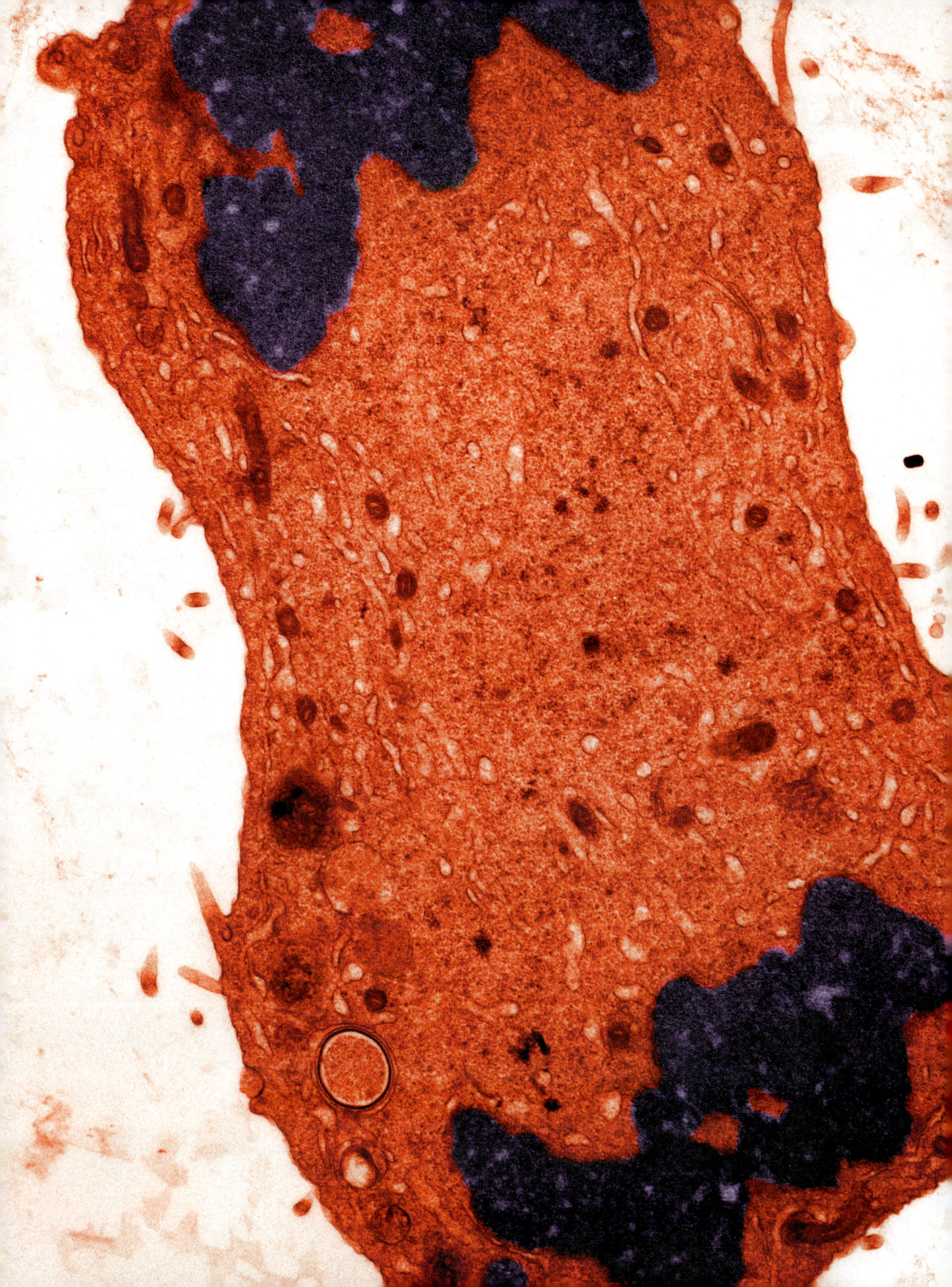

REVIEW QUESTIONS

How are biochemical pathways regulated?

Multiple-choice questions

Use the following graph to answer questions 1 and 2. The graph shows the energy levels of a reaction in the presence and absence of an enzyme.

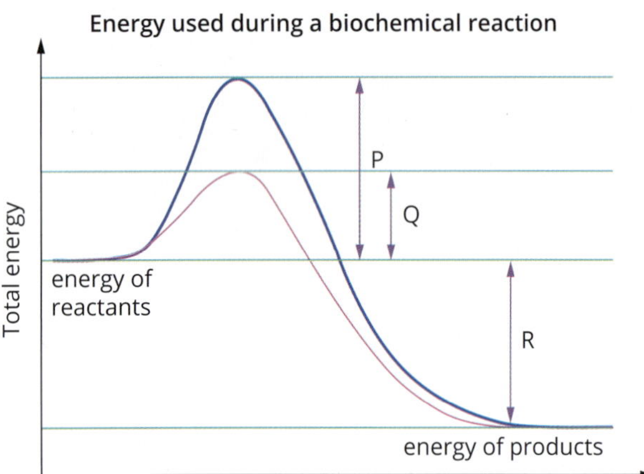

1 Conclude the best explanation for the different energy amounts labelled P, Q and R.

	P	Q	R
A	absence of an enzyme	presence of an enzyme	endergonic reaction
B	presence of an enzyme	absence of an enzyme	exergonic reaction
C	absence of an enzyme	presence of an enzyme	exergonic reaction
D	presence of an enzyme	absence of an enzyme	endergonic reaction

2 Which of the following also represents the activation energy of the enzyme-catalysed reaction?

A P
B Q
C P + R
D Q + R

3 The following graph shows the effect of changing substrate concentration on the amount of product formed.

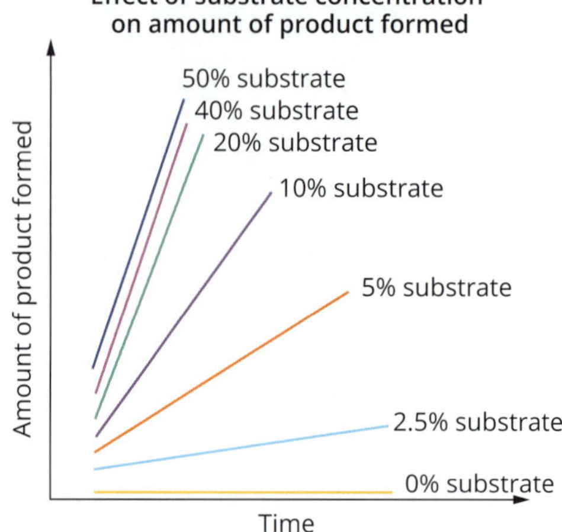

Identify the conclusion that can be drawn.

A The rate of reaction increases exponentially with an increase in substrate concentration.
B The rate of reaction decreases exponentially with an increase in substrate concentration.
C The rate of reaction increases greatly up to a point as the substrate concentration increases, and then the rate of increase starts to decrease.
D The rate of reaction is not affected by any change in the substrate concentration.

4 The graph below shows the rate of photosynthesis of two different plant species when the plants are experiencing the same environmental conditions.

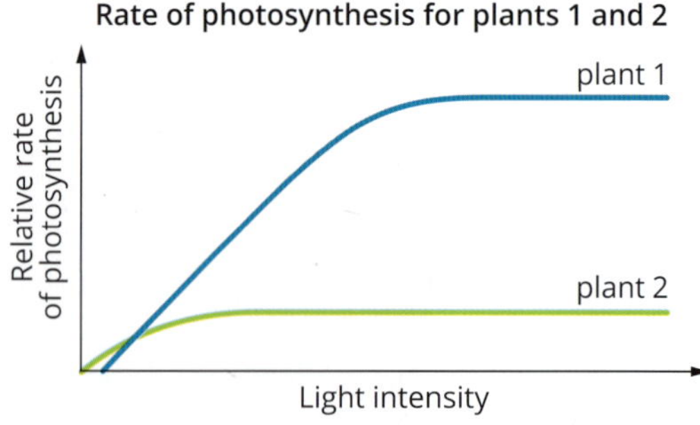

Which limiting factor is most likely to be causing the difference in photosynthetic rates between the two plants?

A oxygen availability
B chloroplast availability
C carbon dioxide availability
D water availability

5 Identify which of the following statements outlining the key differences between C_3, C_4 and CAM plants is correct.

A C_3 and C_4 plants both use Rubisco to fix carbon. CAM plants do not.

B C_3 plants are able to use Rubisco to fix a three-carbon molecule. C_4 and CAM plants do not.

C C_3 and C_4 plants are both able to use Rubisco to fix a three-carbon molecule; however, CAM plants must first fix a four-carbon molecule.

D C_3, C_4 and CAM plants are all able to use Rubisco to fix a three-carbon molecule; however, C_4 and CAM plants will first fix a four-carbon molecule.

6 Identify which of the following is not a product of the Krebs cycle.

A CO_2

B NADH + H^+

C NADPH + H^+

D ATP

7 Identify the most accurate description of ATP.

A a competitive inhibitor

B an energy-carrying molecule

C the main product of photosynthesis

D aerobic transport pathway

Short-answer questions

8 Study the following graph of enzyme activity.

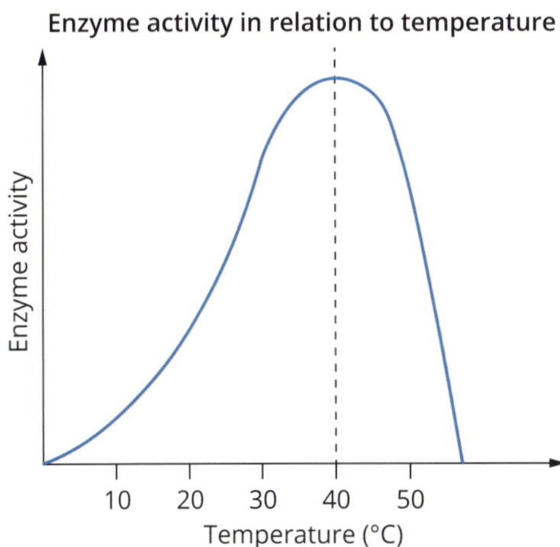

a Outline what happens to enzyme activity as the temperature increases from 0°C to 40°C.

b Identify the optimum temperature for the enzyme.

c Explain what happens to the enzyme above 40°C.

d Name the other factors that affect enzyme activity.

9 A transmission electron micrograph of structure Z is shown in the following figure.

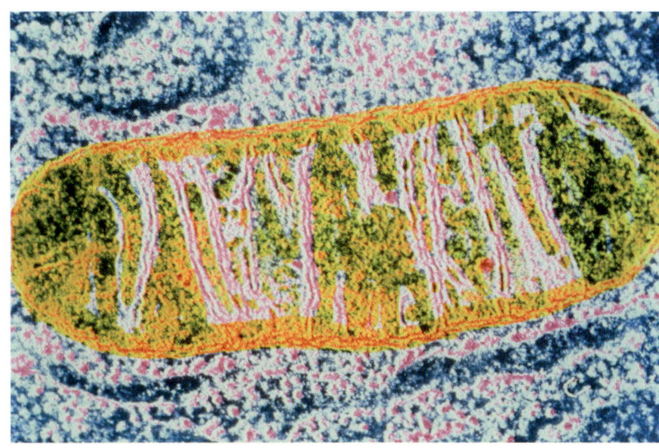

a Identify structure Z.

b State the overall chemical equation for the reaction that involves this structure, providing total inputs and outputs.

10 Scientists investigated the effect of temperature on the rate of photosynthesis on two different plants: plant A and plant B. The graph shows the results of the experiment.

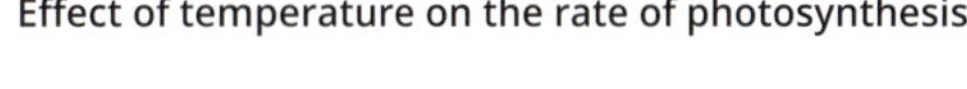

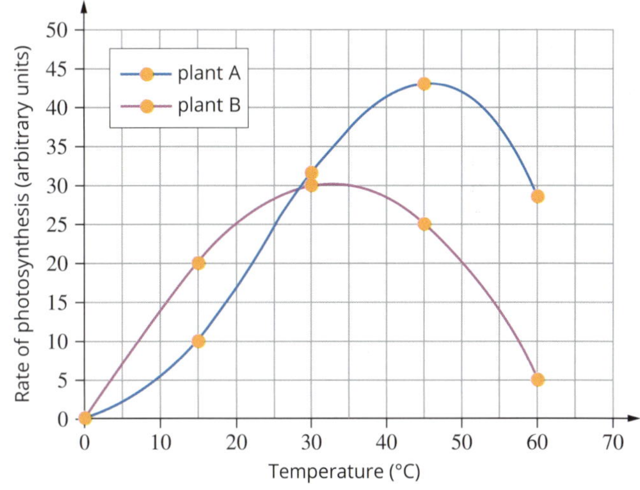

a Compare the general effect of temperature on the rate of photosynthesis in plant A and plant B.

b Explain why the rate of photosynthesis falls at temperatures higher than 50°C.

c Which plant is most likely to be a CAM plant? Justify your answer.

11 The following graph shows the rates of photosynthesis and cellular respiration in an indoor plant that receives plenty of natural light through a large window.

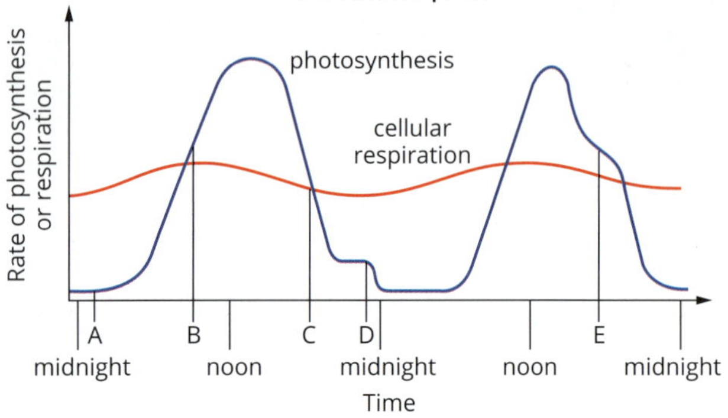

a Deduce whether there is a net oxygen uptake or output at each of these times: A, B, C, D, E.

b Propose what change in the plant's environment might have caused the rate of photosynthesis shown at times D and E.

12 The graph below shows the contributions of the two energy-producing pathways (aerobic cellular respiration and anaerobic fermentation) to physical activity.

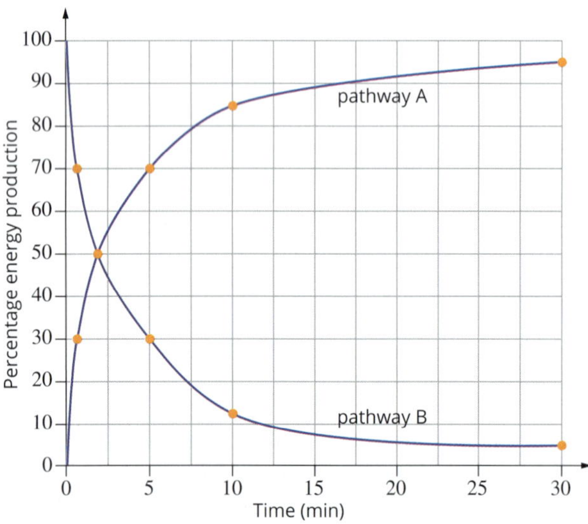

a Athletes competing in sports that require short-term power output, such as sprinting, obtain most of their ATP from the anaerobic pathway, but athletes requiring sustained energy use aerobic cellular respiration to meet most of their energy needs.

i Why do cells need ATP?

ii What is the name of the process that produces ATP during anaerobic fermentation?

iii Explain which pathway (A or B) is most likely to represent ATP production by a sprinter.

iv Why can't anaerobic fermentation supply the energy needs of athletes in events requiring energy over a sustained period of time?

b Animals make lactic acid during anaerobic fermentation, but yeasts and plants produce ethanol and carbon dioxide. Suggest a reason why the products are different in animals and plants.

13 Enzymes are used to catalyse cellular reactions during both photosynthesis and cellular respiration.

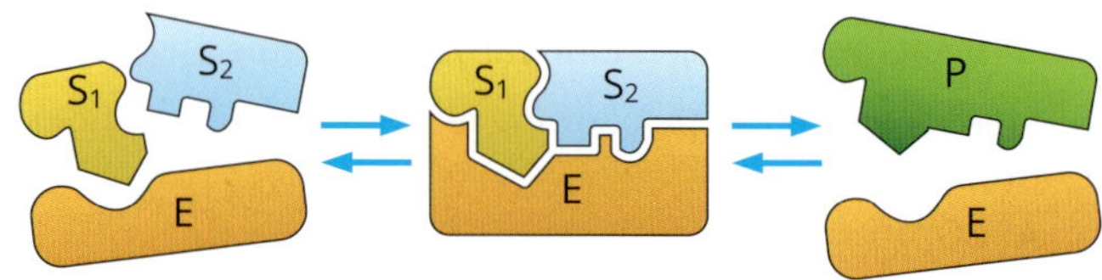

a What enzyme activation model does the diagram above represent?

b If heated above its critical temperature, an enzyme denatures. Describe what happens to an enzyme's structure when it denatures.

c Why does denaturation affect the enzyme's activity during these processes?

14 Blood and tissue fluid in a human are generally maintained in a narrow pH range, between 7.35 and 7.45. This means blood is slightly basic. A family of enzymes called carbonic anhydrases is responsible for the conversion of CO_2 to HCO_3^- in the tissues, and HCO_3^- back to CO_2 in the lungs. Imidazole is a chemical that inhibits the action of carbonic anhydrase. An experiment was run at 37°C, with and without imidazole, at varying CO_2 concentrations. The amount of imidazole added was kept constant throughout the experiment.

The results are shown in the table.

CO_2 concentration (mmol/L)	Rate of conversion	
	Without imidazole (mmol/min)	With imidazole (mmol/min)
0.1	3.2	1.6
0.2	4.5	2.7
0.3	6.7	4.1
0.5	7.1	6.3
1.0	8.8	7.8
2.0	10.3	10.1

a **i** Name the independent variable(s) in this experiment.

ii Imidazole is a competitive inhibitor of CO_2 conversion. Explain how imidazole slows the rate of reaction. Ensure you describe the structures involved in your answer.

iii Explain the difference in the rate of inhibition as the concentration of CO_2 changes.

b In a further experiment on the activity of carbonic anhydrase, its activity during changes to pH was investigated. The results were graphed. Four graphs are shown.

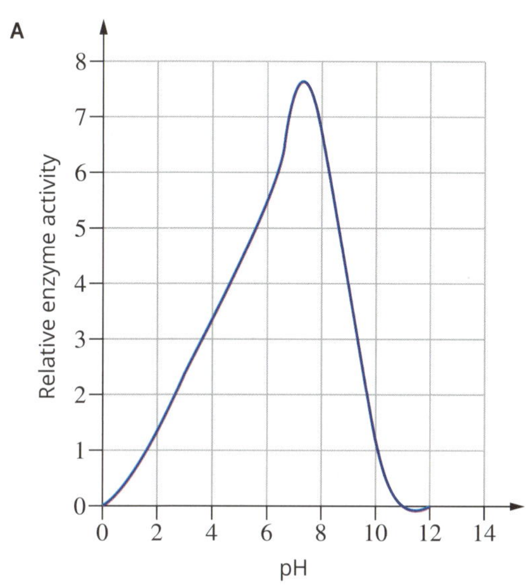

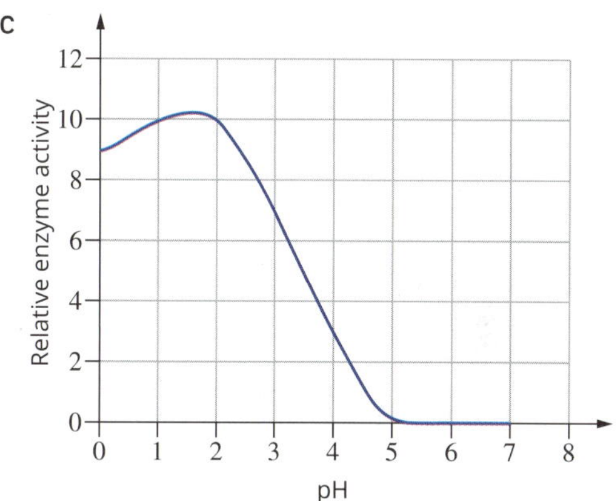

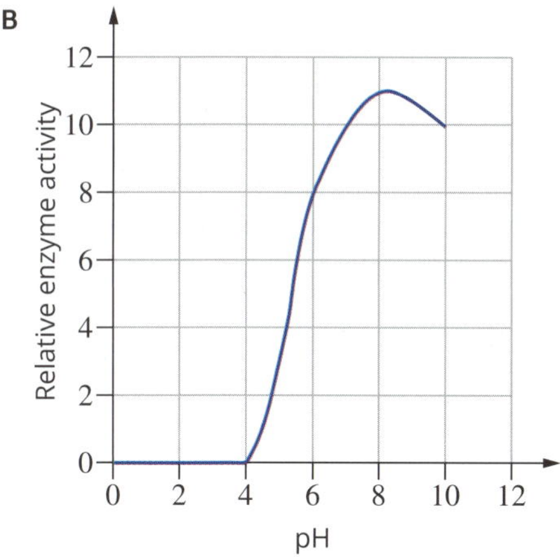

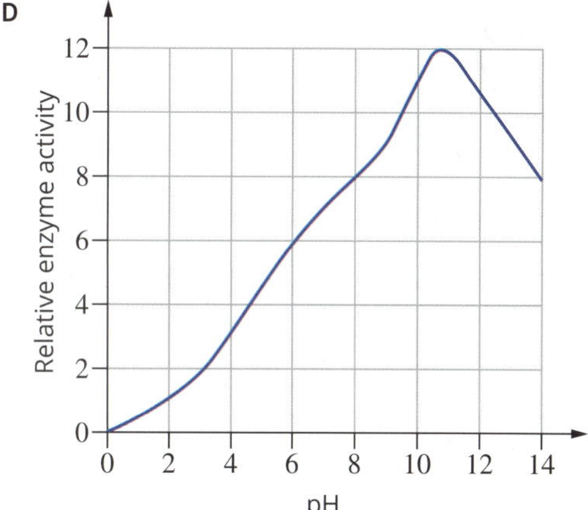

i Propose which of the graphs is most likely to show the activity of this enzyme at various pH values.

ii Justify your choice and explain what is occurring to the enzyme.

iii State why the other graphs are not correct.

c Imidazole could be used as a poison. Explain how it would affect the metabolism of an organism.

15 A section of the glycolysis pathway showing various substrates, products and enzymes is illustrated in the diagram below.

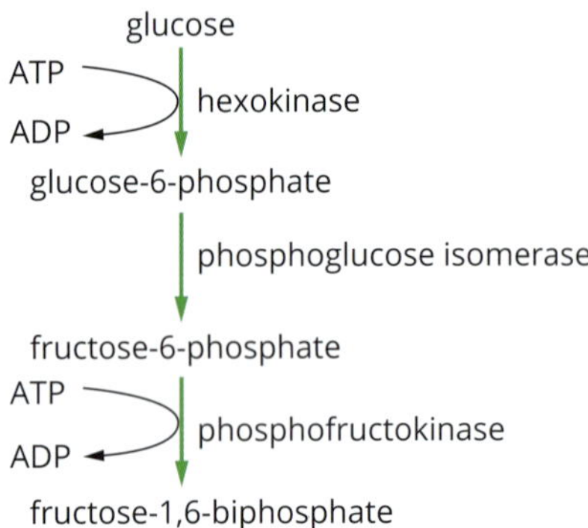

Scientists measured the effect of different concentrations of fructose-6-phosphate on phosphofructokinase activity. Phosphofructokinase activity was also measured with low and high concentrations of ATP in the reaction mixture. The graph below shows the results.

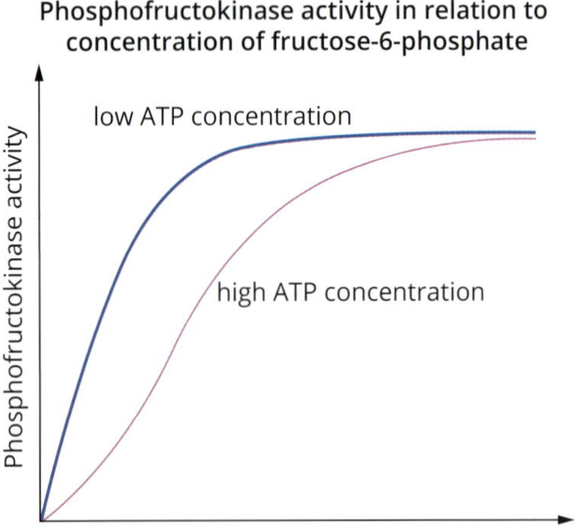

a Outline the effect of increasing fructose-6-phosphate concentration.

b Explain how increasing the concentration of fructose-6-phosphate affects the activity of phosphofructokinase.

c Outline the effect of increasing ATP concentration on the activity of phosphofructokinase.

d Given that one of the functions of glycolysis is to produce ATP, explain how the concentration of ATP affects phosphofructokinase.

16 Cellular respiration is a process that is essential for the survival of most life forms. The diagram below shows the stages of cellular respiration and the inputs and outputs of each stage.

a Complete the diagram by filling in the empty boxes.

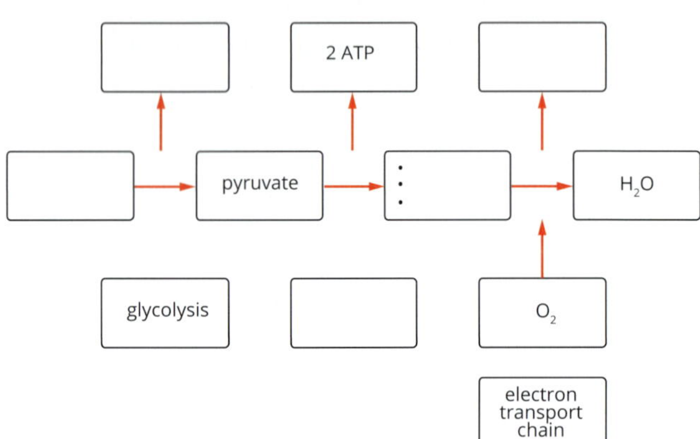

b DNP (2,4-dinitrophenol) is a chemical that was used during World War I to make explosives. Workers handling the DNP were found to have extremely high body temperatures (up to 44°C) and suffered from severe weight loss. Some workers died as a result of absorbing DNP through their skin, ingesting it or inhaling it. Research into DNP has shown that one of its actions is to block the movement of phosphate ions into the mitochondria.

i Normally, during cellular respiration much of the energy released from the breakdown of glucose is used to build ATP. What happens to the rest of the energy?

ii Suggest how the action of DNP could lead to the very high body temperatures observed.

iii Very high body temperatures have been associated with a number of fatalities. Explain how high body temperatures can lead to cell death.

c A student was investigating the effects of DNP on cellular respiration. A cell culture was supplied with glucose and monitored for the products of cellular respiration. During the experiment, DNP was added to the culture. The graph to the right shows the results of the experiment.

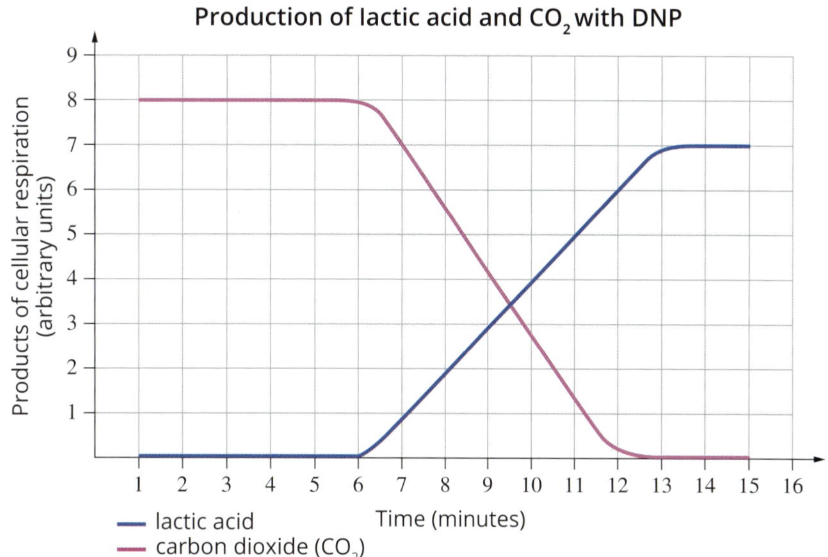

i At what time was DNP added to the culture? Justify your response.

ii Why is lactic acid being produced by the cells?

iii DNP is highly toxic. Identify two safety precautions that the student should implement while performing this experiment.

17 The Calvin cycle is a series of chemical reactions. Each reaction is catalysed by its own enzyme. The first step in the Calvin cycle adds a carbon dioxide molecule to a ribulose 1,5-bisphosphate molecule. The new molecule (called 3-phosphoglycerate) is then further modified, ultimately forming glucose. The enzyme that catalyses the first step is called Rubisco.

ribulose 1,5-bisphosphate + CO_2 $\xrightarrow{\text{Rubisco}}$ 3-phosphoglycerate

a Which stage of photosynthesis is the Calvin cycle part of?

Examine the diagram of a chloroplast and the visual representation of photosynthesis in the chloroplast.

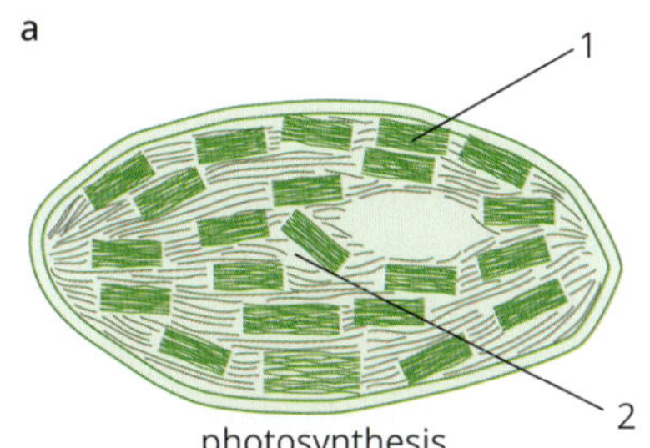

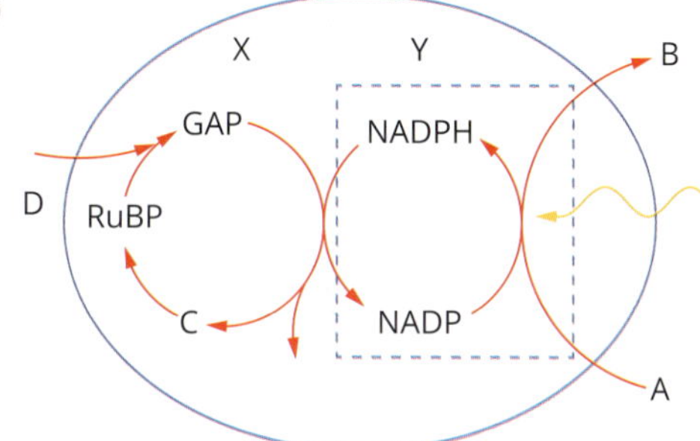

Different compartments of the chloroplast are represented by 1 and 2 in figure a, and X and Y in figure b.

b Match and name the structures identified in figure a with X and Y in figure b.

c In figure b, identify substances A to D from the following list: CO_2, O_2, H_2O, glucose.

d On an overcast day, the inputs and outputs of each stage in photosynthesis can be affected. Determine which substance would be directly affected by an overcast day and propose how this would affect the production of glucose.

18 **a** An experiment was performed in which muscle cells were incubated in a low-oxygen environment at 20°C. The cumulative uptake of glucose was measured in grams. The results for the first 10 minutes are shown in the following table.

Glucose use in the absence of oxygen	
Time (minutes)	**Glucose uptake (g)**
2	5
4	10
6	15
8	20
10	25

After 10 minutes, oxygen was infused into the culture and measurement of the uptake of glucose continued. The results for the next 10 minutes are tabulated below. Temperature was maintained at 20°C.

Glucose use in the presence of oxygen	
Time (minutes)	**Glucose uptake (g)**
12	26
14	27
16	28
18	29
20	30

i Graph the uptake of glucose versus time for the 20 minutes of the experiment. Clearly mark the point at which oxygen was introduced into the culture.

ii Explain why glucose uptake declined so significantly after oxygen was added to the culture.

iii What is the independent variable in the experiment?

iv Why was it necessary to ensure that the temperature remained at 20°C throughout the experiment?

b Some mitochondrial diseases are caused by mutations in the genes needed for cellular respiration.

i One mitochondrial disease is caused by a mutation in the gene that encodes the protein cytochrome *c* oxidase. Cytochrome *c* oxidase is the last of the cytochrome proteins forming the electron transport chain. Where in the mitochondria will this protein be found?

ii An experimenter investigating mitochondrial mutations performed the experiment from part a using cells with mitochondria possessing mutations. The scientist noted that when oxygen was added after 10 minutes, no change in glucose use was observed. Explain this observation.

19 Increasing global temperatures have had an impact on the crop yield of many edible plants and their fruits. Scientists are currently researching the use of CRISPR-Cas9 technology to reduce the impact of these changes and improve the crop yield to address food security at a global level. One possible area of research is the modification of the enzyme Rubisco activase to increase the rate and efficiency of photosynthesis in food crops. In some fruiting plants, such as tomato, Rubisco is inhibited by the presence of sugar–phosphate bonds at the active site. Rubisco activase is used to modify these bonds of the Rubisco enzyme to allow it to perform its function. It has been suggested that Rubisco activase is temperature-sensitive and can reduce function at temperatures above 45°C, which could be a limiting factor in rates of photosynthetic activity.

a In which stage of photosynthesis is Rubisco active?

b What impact on photosynthesis would inactive Rubisco have in tomato plants?

c Outline an experiment that uses hypothetical CRISPR-Cas9 enhanced tomato plants with modified Rubisco activase to improve photosynthetic activity.

WS 18

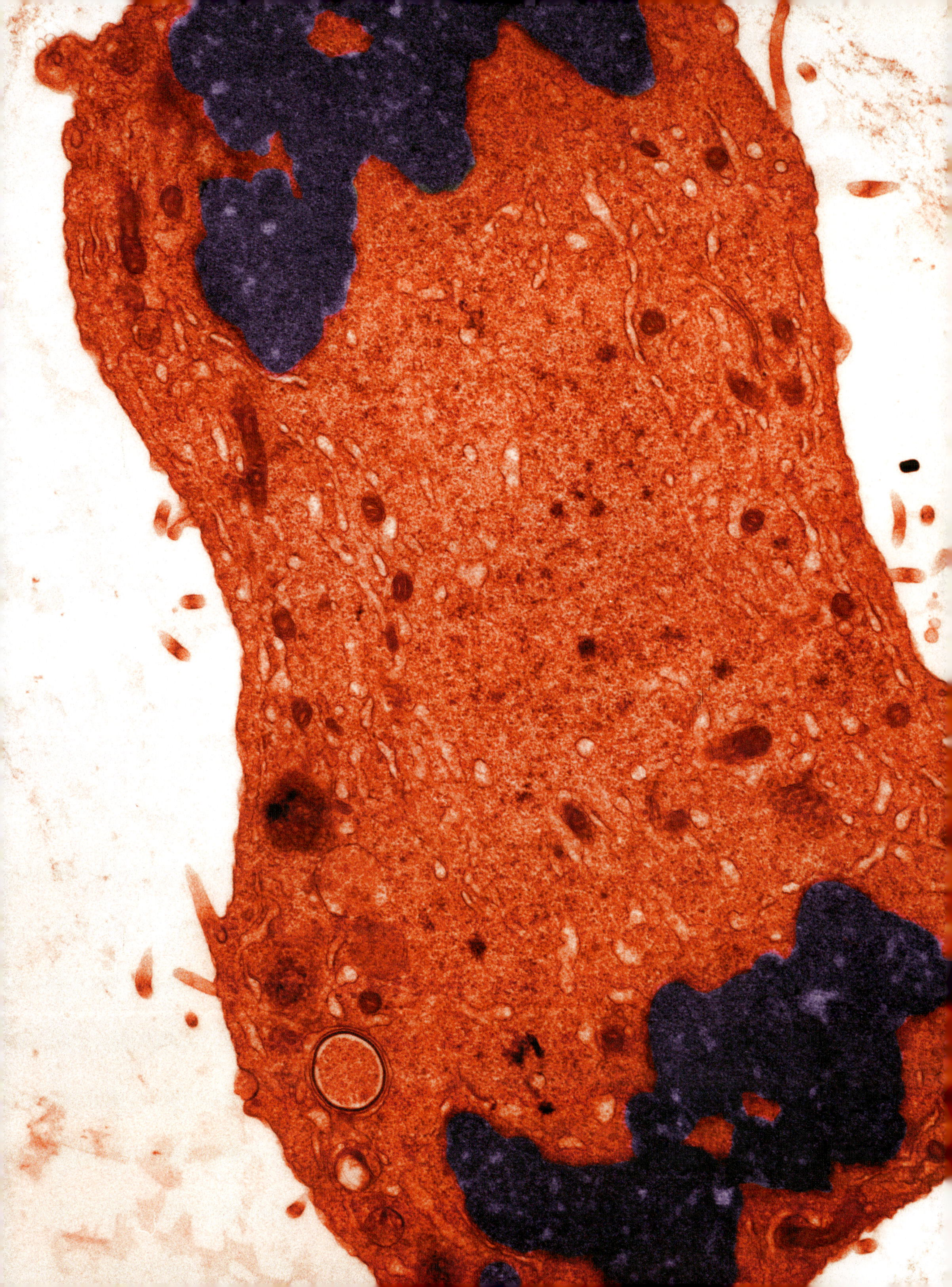

UNIT 4 How does life change and respond to challenges?

To achieve the outcomes in Unit 4, you will draw on key knowledge outlined in each area of study and the related key science skills on pages 7–9 of the study design. The key science skills are discussed in Chapter 1 of this book.

AREA OF STUDY 1

How do organisms respond to pathogens?

Outcome 1: On completion of this unit the student should be able to analyse the immune response to specific antigens, compare the different ways that immunity may be acquired and evaluate challenges and strategies in the treatment of disease.

AREA OF STUDY 2

How are species related over time?

Outcome 2: On completion of this unit the student should be able to analyse the evidence for genetic changes in populations and changes in species over time, analyse the evidence for relatedness between species, and evaluate the evidence for human change over time.

AREA OF STUDY 3

How is scientific inquiry used to investigate cellular processes and/or biological change?

Outcome 3: On the completion of this unit the student should be able to design and conduct a scientific investigation related to cellular processes and/or how life changes and responds to challenges, and present an aim, methodology and methods, results, discussion and a conclusion in a scientific poster.

CHAPTER

08 Responding to antigens

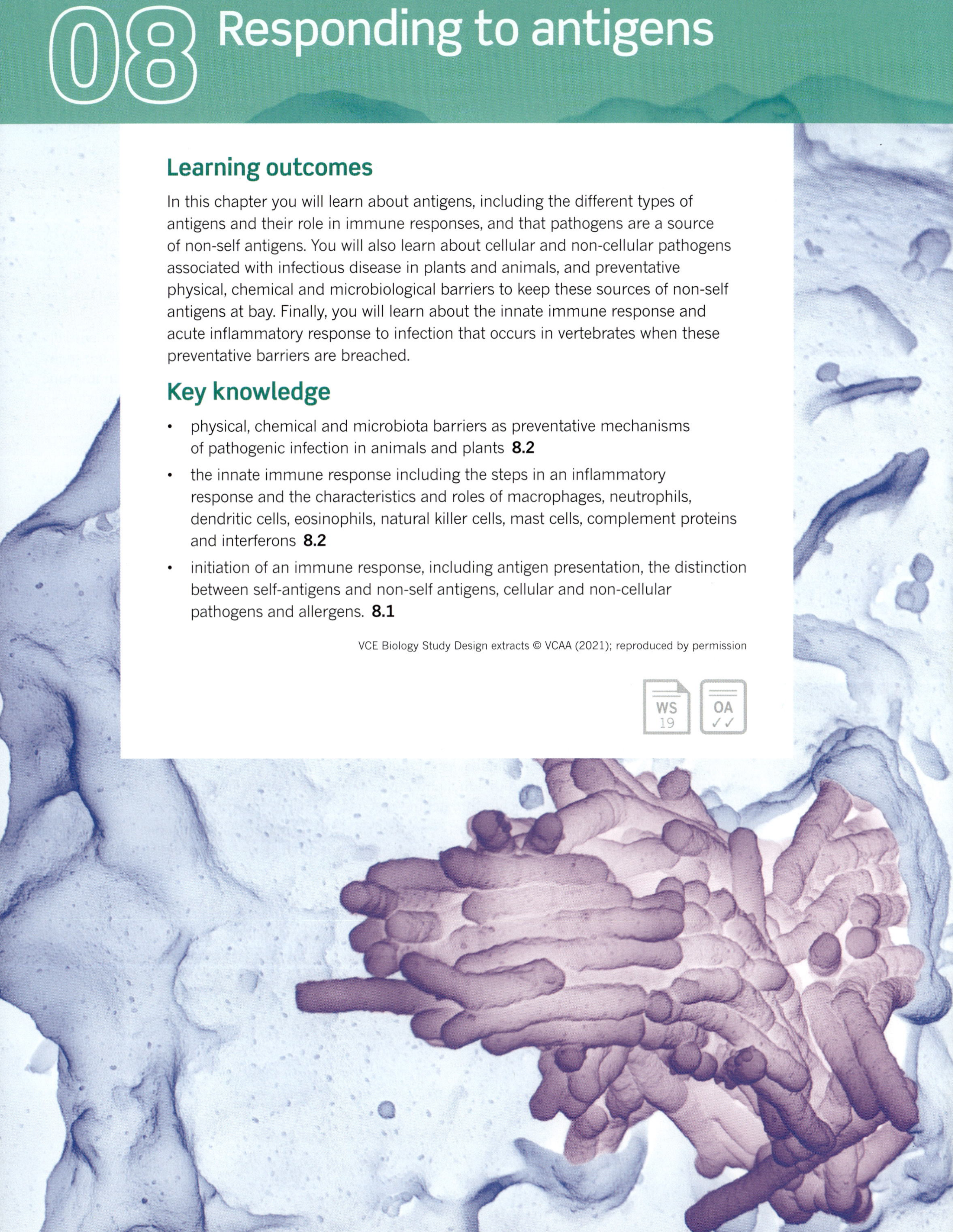

Learning outcomes

In this chapter you will learn about antigens, including the different types of antigens and their role in immune responses, and that pathogens are a source of non-self antigens. You will also learn about cellular and non-cellular pathogens associated with infectious disease in plants and animals, and preventative physical, chemical and microbiological barriers to keep these sources of non-self antigens at bay. Finally, you will learn about the innate immune response and acute inflammatory response to infection that occurs in vertebrates when these preventative barriers are breached.

Key knowledge

- physical, chemical and microbiota barriers as preventative mechanisms of pathogenic infection in animals and plants **8.2**
- the innate immune response including the steps in an inflammatory response and the characteristics and roles of macrophages, neutrophils, dendritic cells, eosinophils, natural killer cells, mast cells, complement proteins and interferons **8.2**
- initiation of an immune response, including antigen presentation, the distinction between self-antigens and non-self antigens, cellular and non-cellular pathogens and allergens. **8.1**

WS 19

OA ✓✓

8.1 Antigens and pathogens

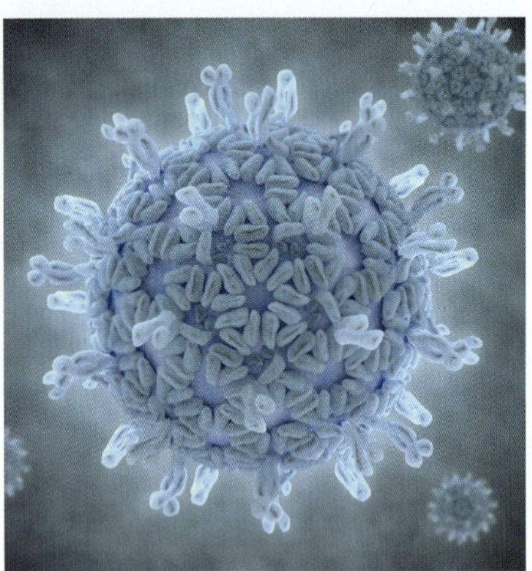

FIGURE 8.1.1 Artist's impression of rotavirus, a virus that is a common cause of gastroenteritis and diarrhoea in infants. Proteins on the surface of the virus act as antigens, which are recognised by the body's immune cells.

Immune cells (also called leukocytes or white blood cells) are cells produced by the immune system to protect your body from foreign substances, such as pathogens. Immune cells include eosinophils, neutrophils and lymphocytes.

Antibodies (also known as immunoglobulins) are proteins produced by B lymphocytes that bind to specific antigens.

Antigens are unique molecules, or parts of molecules, that can often elicit an immune response, and so play a crucial role in immunity. Antigens can be classified as self-antigens or non-self antigens, and an organism's immune cells can usually differentiate between self-antigens and non-self antigens and respond accordingly.

This section will focus on the nature of antigens, the distinction between different types of antigens, and on pathogens as sources of non-self antigens (Figure 8.1.1).

THE NATURE OF ANTIGENS

Antigens are unique molecules, or parts of molecules, that can be recognised by receptors on **T lymphocytes** (also known as T cells), or by **antibodies** produced by **B lymphocytes** (also known as B cells). You will learn more about T and B lymphocytes in Chapter 9. Antibodies, also known as **immunoglobulins (Ig)**, can be bound to the surface of, or secreted by, B lymphocytes.

Antigens are important because they allow the body to recognise potentially harmful pathogens and mount an immune response against them. Although many antigens trigger an immune response, some do not. Antigens that elicit an immune response are more properly known as **immunogens**; however, in the context of an immune response it is still common to simply refer to them as antigens.

Structure of antigens

Most antigens are protein-based and can be composed of one or more polypeptide chains. However, antigens can also be composed of carbohydrates, lipids and even nucleic acids. For example, the complex carbohydrates of the human ABO blood group are antigens.

Types of antigens

Antigens are expressed or presented on the surface of the plasma membrane of cells, where they act as recognition sites for the immune system. However, not all antigens are attached to a cell; for example, some antigens, such as toxins released by bacteria, circulate freely in body fluids.

Antigens that result in immediate hypersensitivity reactions are called **allergens**. Immediate hypersensitivity reactions (or **allergic responses**) are due to a rapid and vigorous overreaction of the immune system to antigens that would otherwise be harmless. Typical allergenic substances include pollen, fur, house dust, latex and foods such as peanuts, lobster and monosodium glutamate (MSG). Depending on the particular individual and antigen, the hypersensitivity reaction can range from mild to being a life-threatening reaction known as **anaphylaxis**.

The immune system is normally able to distinguish antigens that are expressed by its own cells (**self-antigens**) from those that are not (**non-self antigens**), and respond accordingly.

CASE STUDY ANALYSIS

Determining blood groups for successful transfusion

The first successful human-to-human blood transfusion is reported to have occurred in the 1800s. At that time, a blood transfusion was a risky procedure. It might help a patient, but it could also make them much worse. This is because human blood groups (ABO) were not discovered until 1901 and the idea that transfusions should be matched to the recipient's blood group was not suggested until 1907.

The A and B blood type antigens are carbohydrate molecules attached to proteins and lipids in the red blood cell membrane. The structure of the carbohydrate makes the A antigen different from the B antigen. If the blood type transfused into a patient is different from the patient's own blood type, an immune response will be elicited by the patient's immune system. Antibodies will recognise the transfused blood cells as foreign and will bind to their antigens. This causes clumping (or agglutination) of red blood cells (Figure 8.1.2). Agglutination destroys the red blood cells, which normally transport oxygen throughout the body, and so can result in severe anaemia and even death.

The presence or absence of A and B antigens on the surface of red blood cells determines whether the blood group is A, B or AB. Group O blood has neither A nor B antigens on the surface of red blood cells (Table 8.1.1). If the blood type transfused into a patient is different from the patient's own blood type, an immune response may be elicited by the patient's immune system, which can lead to death.

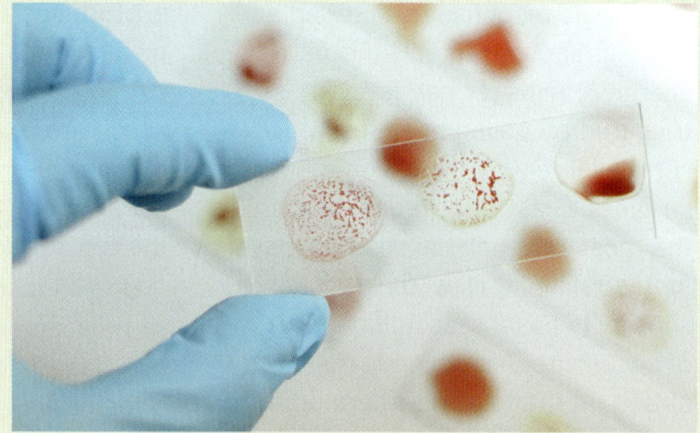

FIGURE 8.1.2 Agglutination test: red blood cells have clumped together (or agglutinated) in the left and middle drops of blood on the microscope slide. The drop of blood on the right of the microscope slide has not agglutinated.

TABLE 8.1.1 The four different blood groups, as determined by the presence or absence of A and B antigens

Blood type	Red blood cells	Antibodies present in plasma	Antigens present on cells
A		anti-B	A
B		anti-A	B
AB		none	A and B
O		anti-A and anti-B	none

An individual who presents at a hospital today with serious bleeding will receive a safe, antigen-matched transfusion. Blood group matching is a fairly simple and quick procedure in which antibodies to the blood proteins are mixed with samples of the patient's blood in order to identify the correct blood group for transfusion. The antibodies for antigen A are called 'anti-A' and the antibodies for antigen B are called 'anti-B'. The matching antibody and antigen are never found in the same individual and when they are mixed, the blood will agglutinate (clump together).

Analysis

A patient has presented to the emergency department of the local hospital needing a blood transfusion. The plates below show the results of the test carried out to determine the patient's blood group.

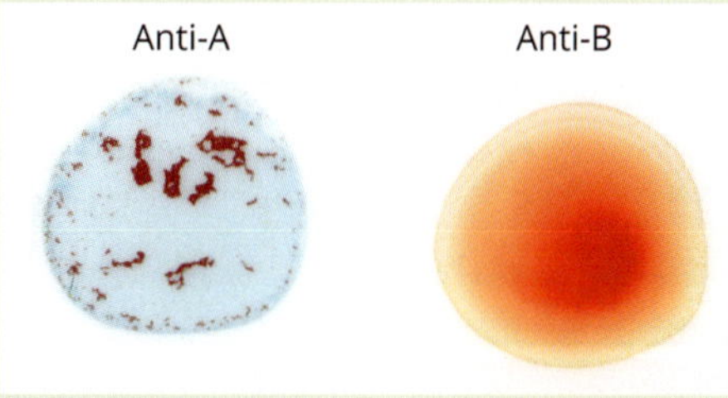

1. What is the patient's blood group?
2. How do you know?
3. What blood groups could be used for the patient's transfusion?

CASE STUDY

Allergic responses

An allergic response (also called an allergic reaction) to pollen is called allergic rhinitis (or hay fever). It is triggered by pollen particles, which carry allergenic antigens on their surfaces. Grass and tree pollens are the most common cause of hay fever in Australia and New Zealand. Pollen sensitivity has a seasonal pattern of occurrence, as pollen is most abundant in the atmosphere during spring and early summer.

Whatever the allergen, mast cell release of histamine is central to immediate hypersensitivity reactions. Allergic responses are mediated by a specific type of antibody called immunoglobulin E (IgE). IgE is produced by plasma cells and travels in the bloodstream. When IgE comes into contact with mast cells, which are common in epithelial and mucosal tissues, the tail end of the IgE antibody binds to receptors on the cell surface.

Upon subsequent exposure to the same allergen, the allergen binds to a pair of adjacent IgE molecules, bridging the gap between (or crosslinking) the two IgE molecules. This binding triggers a cascade of signalling molecules that causes the mast cells to release histamine (and other mediators of inflammation) from their intracellular vesicles by exocytosis (Figure 8.1.3).

Histamine is an organic nitrogenous compound that binds to specific receptors on various cell types. Histamine causes:

- blood vessel dilation
- a decrease in blood pressure
- an increase in the permeability of blood vessels to immune cells and fluids for a better immune response at the site of antigen contact
- contraction of smooth muscles lining the airways, which can make it more difficult to breathe
- activation of fluid-secreting cells that results in a runny nose, teary eyes and sneezing, which expels foreign antigens.

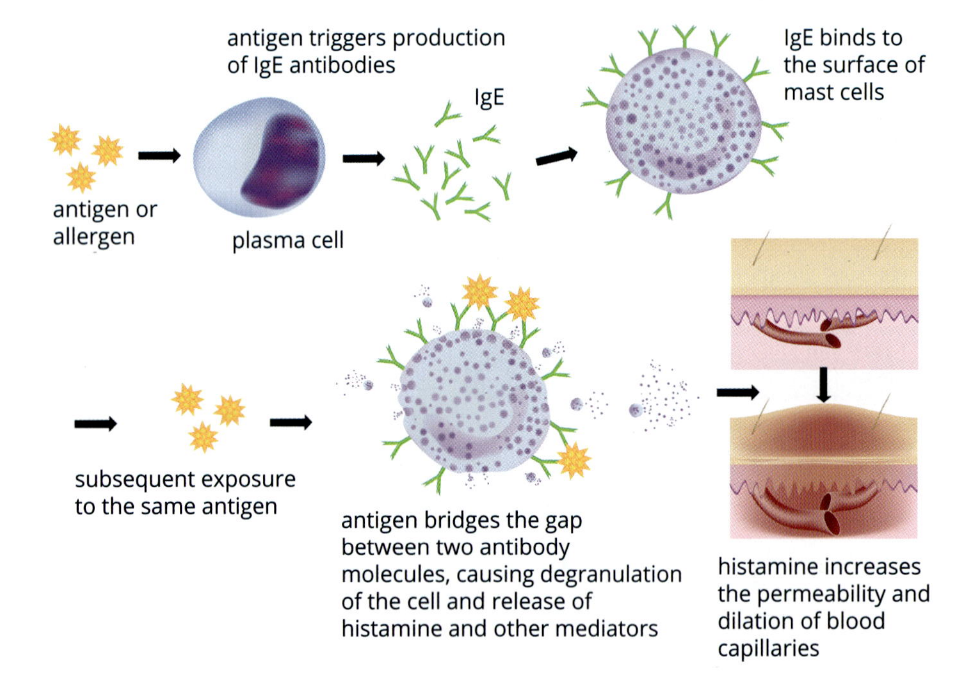

FIGURE 8.1.3 Step 1: Initial exposure to allergens (e.g. pollen) triggers plasma cells to produce IgE molecules specific to the antigen. Step 2: The tail end of the IgE binds to receptors on mast cells. Step 3: Subsequent exposure to the same allergen causes the antigen to bind to two IgE molecules on a mast cell. Step 4: This binding triggers a cascade of signalling molecules that ultimately result in the release of histamine. Step 5: Histamine binds to receptors on various cells in the body, which produces the classic features of an allergic response.

Responding to antigens

Antigen recognition is dependent on the detection of antigens by receptors:

- The receptors on B lymphocytes are membrane-bound antibodies that recognise free antigens or antigens that are on the surface of a pathogen (Figure 8.1.4a). Antibodies can also be secreted by the B lymphocytes (Figure 8.1.4b).
- The receptors on T lymphocytes are different from the membrane-bound antibodies of B lymphocytes, and recognise antigens presented by the organism's own **antigen-presenting cells (APCs)** (Figure 8.1.4c).

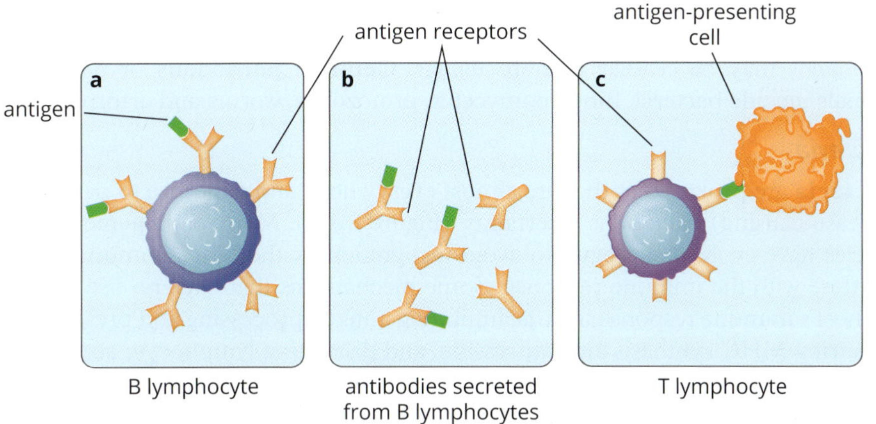

FIGURE 8.1.4 Antibodies are antigen receptors that can be (a) bound to the plasma membrane of B lymphocytes or (b) secreted from B lymphocytes. (c) T lymphocytes have their own type of receptors that recognise antigens presented by specialised antigen-presenting cells.

There are many different receptors and they are specific to particular antigens.

The **major histocompatibility complex (MHC)** proteins, also called **human leukocyte antigens (HLA)**, are proteins on the surface of your body's cells that present self-antigens or non-self antigens to T lymphocytes. There are different classes of MHC proteins, which you will learn more about in Sections 8.2 and 9.2.

In the thymus, T lymphocytes undergo a maturation stage called positive selection, in which the T lymphocytes that do not interact with MHC proteins are destroyed by programmed cell death (apoptosis). They then undergo a second stage of maturation, called negative selection, in which T lymphocytes that react with self-antigens in the thymus bind tightly to the cells in the thymus and eventually die. This two-stage process of selecting T lymphocytes that can recognise MHC proteins, and eliminating T lymphocytes that react to self-antigens, is called clonal deletion.

The inability to respond to self-antigens is called tolerance, or **self-tolerance**. If self-tolerance breaks down and the immune system responds to self-antigens, it results in autoimmune diseases. You will learn more about autoimmune diseases in Chapter 10.

As mentioned earlier, not all antigens (including not all non-self antigens) elicit an immune response, and antigens that do elicit an immune response are called immunogens.

BIOFILE

Antigens and antibodies

The term 'antigen' stands for 'antibody generator'. Antibodies are proteins made when immune cells called B lymphocytes become activated. Antibodies bind to specific antigens (see figure below) and play an essential role in removing pathogens and preventing disease.

Digital illustration showing antibodies (pink) binding to specific antigens on the surface of influenza virus (yellow)

MHC proteins are sometimes also called human leukocyte antigens because they were first discovered through antigenic differences between leukocytes of different individuals.

PATHOGENS AS SOURCES OF NON-SELF ANTIGENS

Pathogens are agents that cause **disease**. Depending on their ability to cause disease, pathogens are divided into two groups:

- Primary pathogens—cause disease any time they are present.
- Opportunistic pathogens—only cause disease when the host's defences have been weakened, for example, by poor nutrition or stress.

Most pathogens contain unique antigens that can be recognised by the immune system. For example, the tuberculosis bacterium, the fungus that causes tinea, and the virus that causes influenza each have antigens that are unique to them. Toxins secreted by pathogens can also act as antigens.

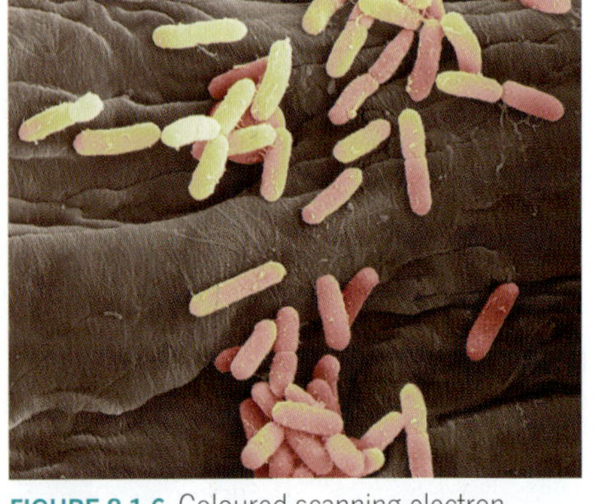

FIGURE 8.1.6 Coloured scanning electron micrograph (SEM) of *E. coli* bacteria (rod-shaped) found in a urine sample from a patient with a urinary tract infection

Cellular pathogens

Pathogens may be cellular or non-cellular. **Cellular pathogens** of plants and animals include bacteria, fungi, oomycetes, protozoans, worms and arthropods.

Bacteria

Bacteria are prokaryotes that are almost everywhere, and exposure to pathogenic (disease-causing) bacteria is a certainty (Figure 8.1.5). Many pathogenic bacteria species have evolved strategies to avoid recognition by the host's immune cells or interfere with the immune response. Some mechanisms that bacteria use to avoid the host's immune response include inhibiting antigen processing and presentation, impairing MHC synthesis and expression, and disrupting lymphocyte activation.

Not all bacteria species are pathogenic and therefore an immune response is not always required. The human body supports and relies on a range of bacteria that reside on and inside it. For example, humans benefit from the metabolic products of non-pathogenic *Escherichia coli*, an inhabitant of the intestine. However, it is possible for the same strain of *E. coli* that is beneficial in the intestine to cause infection if it enters the urinary tract (Figure 8.1.6).

FIGURE 8.1.7 Although mushrooms sold at the supermarket are safe to eat, many species found in the wild are poisonous.

FIGURE 8.1.8 Coloured SEM of the yeast *Candida albicans*, which causes thrush (candidiasis). Depending on environmental conditions, *C. albicans* takes a unicellular yeast-like form or a multicellular filamentous form.

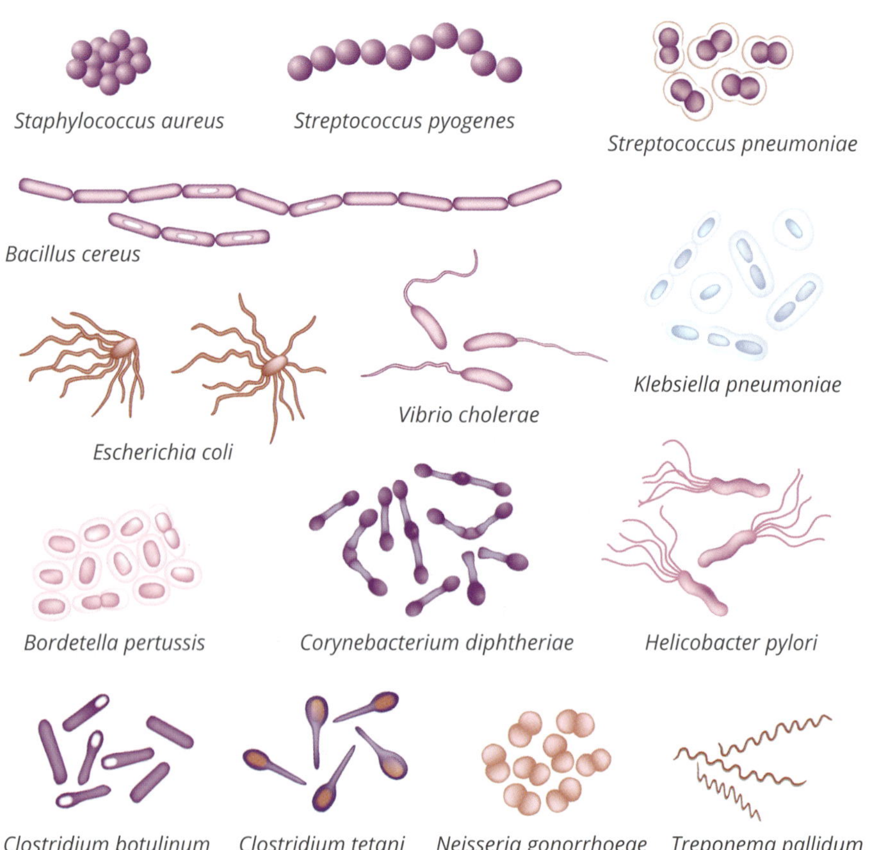

FIGURE 8.1.5 Common pathogenic bacteria

Fungi

Fungi are a diverse family ranging from the macroscopic (mushrooms, Figure 8.1.7) to the microscopic (moulds, unicellular yeasts and yeast-like fungi, Figure 8.1.8). Fungi can secrete digestive enzymes into their environment to break down organic matter, which can then be absorbed into the fungus. It is these secreted substances that are usually responsible for causing disease in animals and plants. Fungal cells produce surface glycoproteins and polysaccharides that act as antigens, allowing them to be identified by cells of the immune system.

Oomycetes

The **oomycetes** (phylum Oomycota) include organisms that cause blight and downy mildew on plants and life-threatening infections in animals. Originally thought of as fungi, the oomycetes, including *Phytophthora* (which means 'plant destroyer'), are now classified in the kingdom Protista.

Oomycetes have motile cells (with flagella), walls of cellulose, and many cellular processes that are not found in fungi. When spores of oomycetes are released on a leaf, they may be carried in water droplets to other leaves, swim to a germination site, or germinate directly, sending out a hypha (fungal thread) that branches and invades plant tissue. These branching hyphae (haustoria) penetrate living cells and absorb nutrients or release enzymes that digest cytoplasm into molecules that can be absorbed. In plants, it has been shown that ooymcetes release molecules that suppress their host's immune response and inhibit apoptosis.

There are about 35 species of *Phytophthora*, and they infect many crops including potato, tomato, apple, tobacco plants and citrus trees. In Australia, *P. cinnamomi* has destroyed tens of thousands of hectares of valuable eucalypt timberland (Figure 8.1.9). The spores of *P. cinnamomi*, some of which can survive for years in moist soil, are attracted to the roots of the plants they infect by a chemical released from the roots.

> *Pythium insidiosum* is the only oomycete known to infect mammals, including humans.

> Motile cells are capable of motion.

FIGURE 8.1.9 Eucalypt forest in the Brisbane Ranges National Park infected with cinnamon 'fungus', *Phytophthora cinnamomi*, which causes dieback disease. Species of *Xanthorrhoea* (grass tree) are also very susceptible to this disease and rapidly turn brown and die.

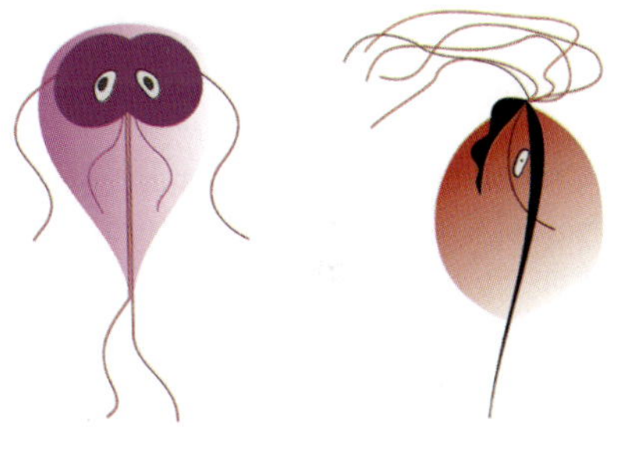

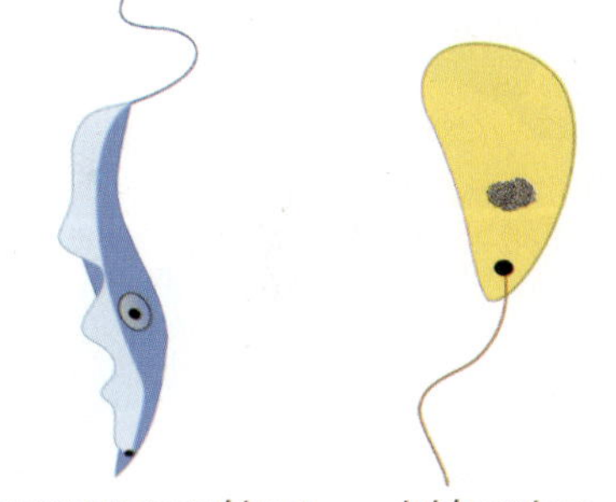

FIGURE 8.1.10 Common pathogenic protozoa

Protozoans

Protozoans are unicellular eukaryotes (Figure 8.1.10). Some reproduce within their host's cells, while others, like *Giardia lamblia* (Figure 8.1.11), demonstrate extracellular reproduction. The life cycles of some protozoans include multiple stages in different hosts. Many of these protozoans express different proteins on the surface of their plasma membranes at different stages of their life cycle. These proteins act as antigens. The mechanism by which their surface antigens change is known as **antigenic variation**, and it assists protozoans in evading detection by the host.

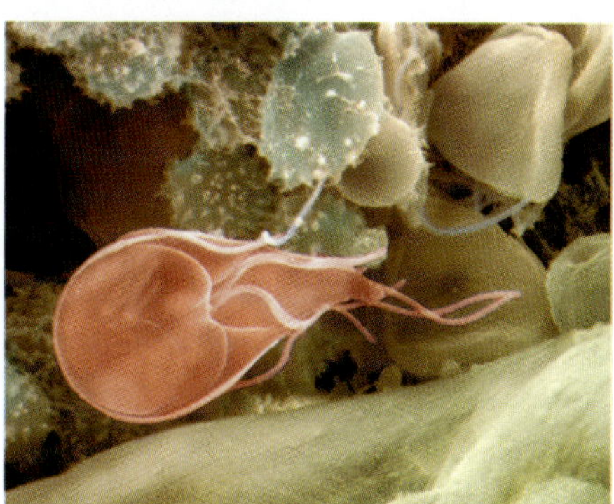

FIGURE 8.1.11 Coloured SEM of a *Giardia lamblia* protozoan (pink), which undergoes asexual reproduction in the small intestine and causes diarrhoea

BIOFILE

The changing face of malaria

The malarial protozoan *Plasmodium* infects red blood cells, where it resides to evade recognition by circulating immune cells (see figure below). *Plasmodium* produces adhesion proteins, which it presents on the surface of the red blood cell. These proteins interfere with the cell's activities within capillaries.

The immune system recognises the adhesion proteins as non-self antigens, but before it can mount an effective immune response, the parasite replaces the adhesion protein with a different adhesion protein. *Plasmodium* has approximately 60 different adhesion proteins that it can continually interchange to remain a step ahead of the immune system. Scientists have recently discovered that an enzyme known as ribonuclease (RNase) causes this process of antigenic variation.

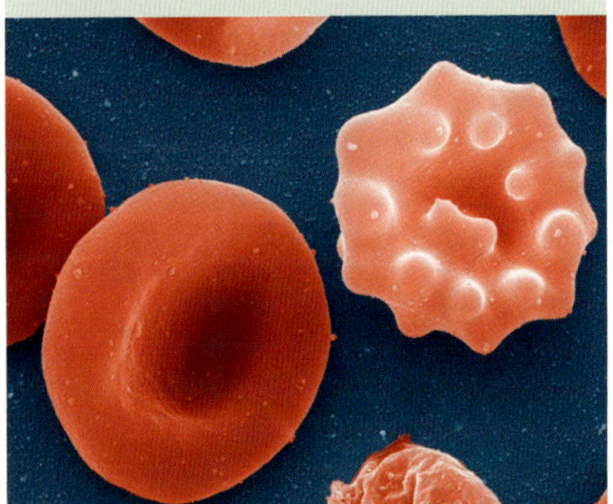

Coloured SEM of red blood cells infected with *Plasmodium* parasite. Changes to the normal red blood cell membrane (left) are clearly seen in the infected red blood cell (top right).

Worms

Parasitic worms can infect plants and animals. They include flatworms such as tapeworms, and roundworms such as hookworms, pinworms and threadworms. In plants, roundworms infect roots and are major pests of orchard trees and crops. In animals, parasitic worms can regulate the immune system in a number of ways so that the immune response against them is suppressed. For example, some roundworms (Figure 8.1.12), such as *Nippostrongylus brasiliensis*, secrete inhibitors that block the action of enzymes needed for antigen presentation, which is an important part of the immune response you will learn more about in Section 8.2.

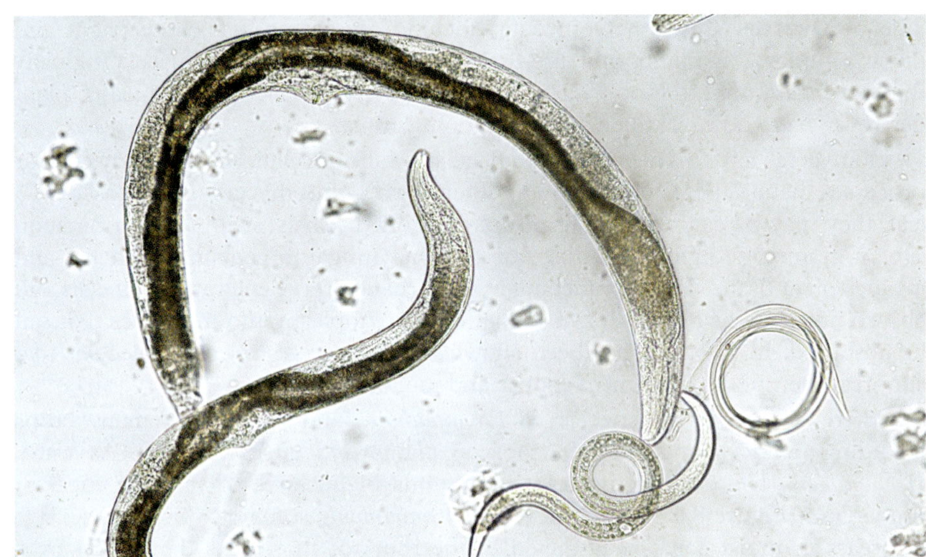

FIGURE 8.1.12 Light micrograph (LM) of female and juvenile roundworms (or nematodes)

Arthropods

Arthropods are invertebrates with external skeletons (or exoskeletons). Arthropods able to transmit or cause disease in humans include insects such as mosquitoes, ticks, lice and mites (Figure 8.1.13). Arthropod saliva contains molecules that modulate the host immune response and inhibit inflammation. These molecules create a favourable environment for pathogen transmission. Arthropod saliva also contains antigens that can trigger an immune response and their immunogenic properties are being used to help develop vaccines against some vector-borne diseases.

Some arthropods can also damage plants. For example, psyllids (lerp insects) are small insects that in their larval stages induce the formation of swollen areas of leaf tissue known as galls. The larval stages of many psyllid species construct a covering (a lerp) under which they feed on the leaf surface. The saliva from feeding psyllids kills leaf tissue, causing extensive discolouration of the leaf.

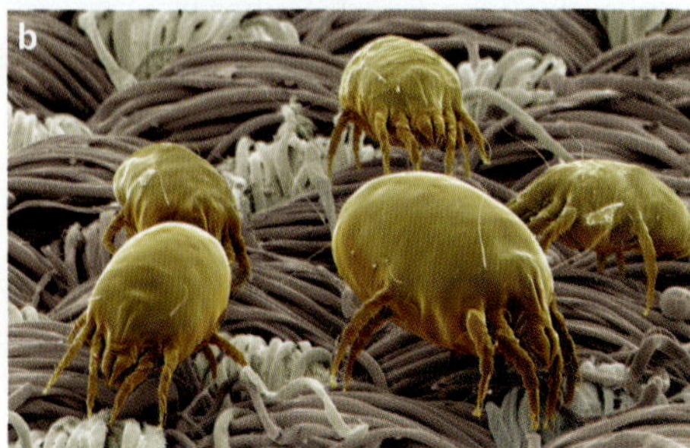

FIGURE 8.1.13 (a) Ticks feed on blood, so they can act as vectors, able to transmit a number of bacterial and viral pathogens when they bite. (b) Common household dust mites feed on flakes of dead skin in dust, and their waste products can cause skin and respiratory allergies.

Non-cellular pathogens

Viruses, viroids and prions are **non-cellular pathogens** (not living), but they have the ability to cause disease.

Viruses

A **virus** is an infectious agent that is composed of genetic material, either DNA or RNA, enclosed in a protein coat. Some viruses also have a lipoprotein envelope (Figure 8.1.14).

During replication, viruses gradually accumulate genetic mutations and some of these mutations cause changes to the viral antigens. This process is known as antigenic drift and results in viruses that are very similar but not identical. They will usually be recognised by the immune system if a similar virus has infected the host on a previous occasion. Over time, the accumulation of antigenic changes can result in a new virus.

Antigenic shift, by contrast, is a much more abrupt change in the genetic code of a virus due to re-assortment of genes from different viral strains, resulting in significantly different antigens on the coat of the virus. The abrupt genetic changes can result in a virus with different characteristics, enabling it to infect new hosts.

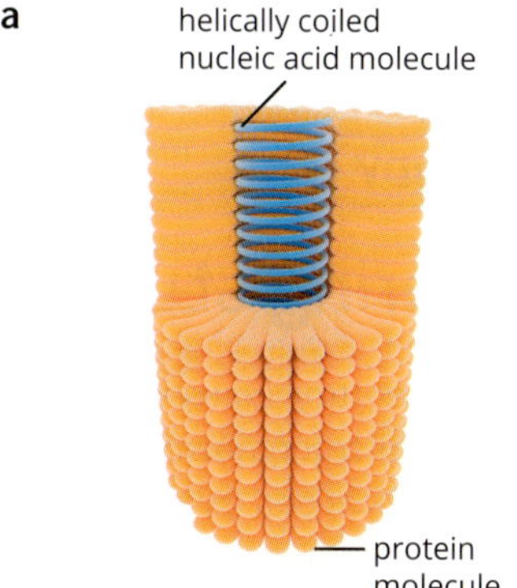

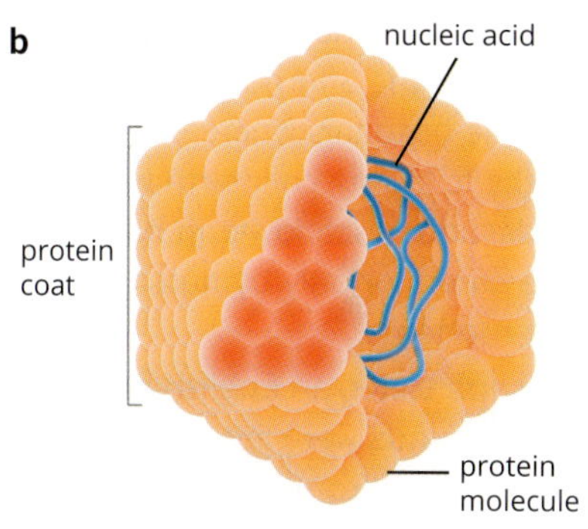

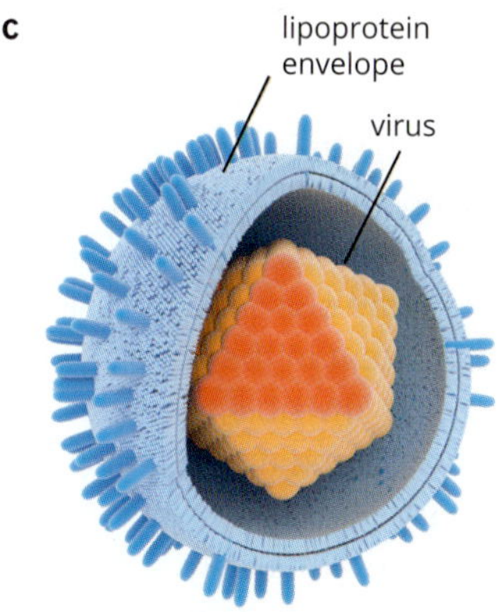

FIGURE 8.1.14 Different structures of virions: (a) a rod-shaped virion with proteins (orange) surrounding a helically coiled nucleic acid molecule (blue); (b) an isometric virion with an icosahedral protein coat surrounding a nucleic acid core (blue); (c) an icosahedral virion (orange) enclosed by a lipoprotein envelope (blue)

Viroids

A **viroid** is a type of self-cleaving RNA enzyme (or ribozyme) that is composed of short, circular strands of RNA that lack a protein coat. Viroids are only known to be pathogens of plants. They damage plants by competing for nucleotides and forming viroid bundles, which mechanically interfere with the internal structures of plants, much like a tumour. Viroids are known to have an unusually high mutation rate, which creates antigenic variation and allows them to avoid host resistance mechanisms.

Prions

The only pathogens smaller than viroids are prions. **Prions** are the only known infectious agents that do not contain genetic material. Prions are proteins that are similar to normal cellular prion proteins (PrP), which are located mainly in the central nervous system. However, unlike PrP, prions have an abnormal shape.

Prions stimulate the organism's normal PrP to misfold into the infectious prion form. Prions are resistant to being denatured or broken down by proteases (enzymes that break down proteins). Prions cause neurodegenerative diseases in mammals. The first of these to be discovered was scrapie in sheep (Figure 8.1.15).

In humans, prions cause Creutzfeldt–Jakob disease (CJD). CJD prions cause vacuoles and misfolded proteins (or plaques) to form in the brain, which kills neurons and makes the brain appear 'spongy' under a microscope. Symptoms include dementia and sudden muscle contractions, leading to death. The equivalent disease in cattle, bovine spongiform encephalopathy (BSE), commonly known as mad cow disease, has been linked to human variant CJD through human consumption of BSE-contaminated beef.

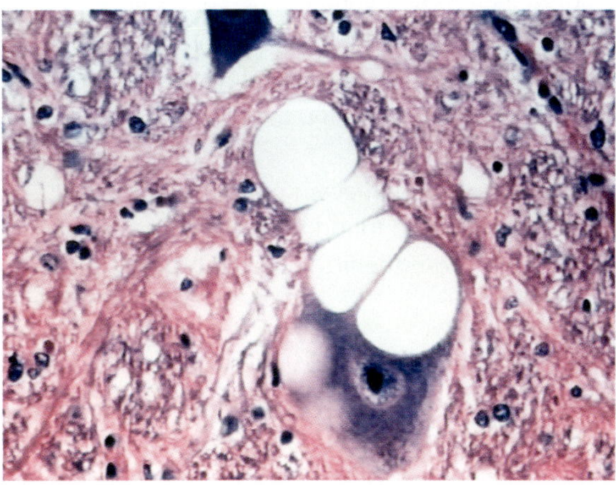

FIGURE 8.1.15 LM showing a section through the brain of a sheep infected with scrapie. The large empty vacuoles (white) in the centre show the effects of the disease. As scrapie progresses, an increase in the number of empty vacuoles makes the brain tissue appear spongy and destroys neurons.

Prions do not trigger an immune response because they are very similar to normal PrP, and any T lymphocytes that would have responded to normal PrP would have been destroyed to prevent an autoimmune reaction. Another reason could be that prions cannot be broken down and presented by antigen-presenting cells. You will learn more about antigen presentation in Section 8.2.

CASE STUDY

Pioneering studies of disease

In the 19th century, Louis Pasteur established the existence of microorganisms and showed that infectious diseases were caused by microbes. Prior to this, deaths in hospitals due to post-operative infection were commonplace. This was not helped by the fact that many doctors would do post-mortems in the mornings and surgery in the afternoons, without changing their clothes. In some hospitals where this practice was followed, deaths following childbirth were very high in doctors' wards, and much lower in nurses' wards where mothers were cared for only by nurses who did not participate in post-mortems.

Joseph Lister, an English surgeon, had observed that when wounds were left open to the air almost half the patients died from infection. But with other wounds that were closed, infection was not nearly so severe. He concluded that infection was due to 'something in the air'. When he heard of Pasteur's experiments showing microbes to be the cause of putrefaction (rotting) of food, he considered they might also be the cause of the infections in his patients. It was known that carbolic acid was highly poisonous to living organisms, so Lister decided to use it in the hospital wards in the hope that it would kill the 'invisible microbes'. He used it on patients, his own hands and around hospital rooms, and required nurses to use it also. The incidence of infection in his patients was dramatically reduced. This was the first practice of antiseptic surgery.

One of the earliest experimental studies by scientists who investigated the cause of disease was the work of a German doctor Robert Koch (Figure 8.1.16). Towards the end of the 19th century, with the improvement of the microscope, scientists were able to identify different species of bacteria and protozoa. In Koch's early work, he studied anthrax.

Koch's experimental method involved examining blood samples taken from patients with different diseases, then growing microbes from the blood on nutrient plates. When he injected specific microbes into mice he found that they developed diseases similar to those of the original patient. As a result of these studies, specific microbes became recognised as the cause of particular diseases.

FIGURE 8.1.16 Dr Robert Koch (1843–1910), German bacteriologist, in his laboratory in South Africa

Koch formulated a set of criteria, known as Koch's postulates, which were used to establish whether a specific microorganism was the cause of a particular disease. Koch's postulates are:

1 The microorganism must be present in the tissues of the infected organism and not in a healthy organism.
2 The microorganism must be able to be cultivated in isolation from the infected organism.
3 When an uninfected organism is then inoculated with the culture, it should develop symptoms of the disease.
4 Samples from the second infected organism should be able to be isolated and found to be the same as the microorganism from the first infected organism.

8.1 Review

OA ✓✓

SUMMARY

- Antigens are molecules, or parts of molecules, that interact with the receptors of T lymphocytes, B lymphocytes and with antibodies.
- Antigens:
  - have a unique molecular structure
  - are composed of one or more polypeptide chains but can also be composed of nucleic acids, carbohydrates or lipids
  - can identify cells as self or non-self
  - can be found on the surface of the plasma membrane of cells or circulating freely in body fluids (e.g. bacterial toxins).
- Some, but not all, non-self antigens elicit an immune response.
- Antigens that elicit an immune response are called immunogens.
- Antigens that elicit an allergic response in susceptible individuals are called allergens.
- Self-antigens do not normally elicit an immune response. This is known as self-tolerance.
- Under normal conditions, any foreign molecule is recognised by the immune system as a non-self antigen.
- Pathogens are sources of non-self antigens and agents that cause disease.
- Cellular pathogens include bacteria, fungi, oomycetes, protozoans and some worms and arthropods.
- Non-cellular pathogens include viruses, viroids and prions.

KEY QUESTIONS

Knowledge and understanding

1 Which of the following is true?

A All immunogens are antigens.

B All antigens are immunogens.

C Antigens are produced by T lymphocytes to defend against non-self antigens.

D Antigens are produced by B lymphocytes to defend against immunogens.

2 Define antigen.

3 What is the difference between an immunogen and an allergen?

4 Are pathogens sources of self-antigens or non-self antigens? Explain the difference between self-antigens and non-self antigens in your answer.

5 What happens if an immune response is directed against a self-antigen?

Analysis

6 Lymphocytes recognise cells as non-self because the foreign cells have antigens that are complementary in shape to their receptors for detecting foreign antigens.

a Explain what is meant by the word 'complementary' in this context.

b A lymphocyte (cell Z) with its receptor is shown below. Explain which of the cells (A, B, C or D) will be identified as non-self by the lymphocyte.

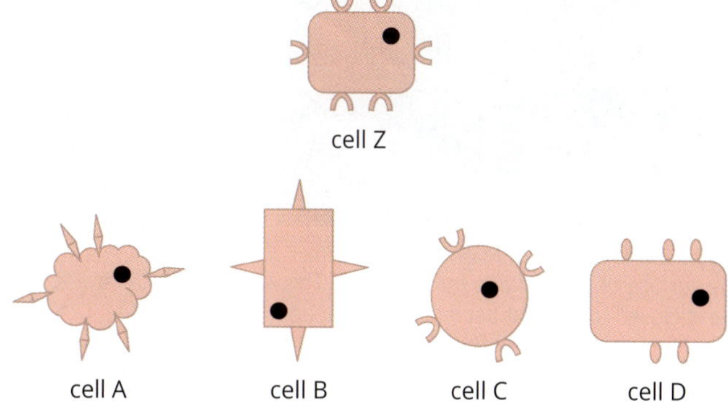

8.2 Innate immunity

There are several mechanisms by which the immune systems of organisms respond to non-self antigens and defend against pathogens. These defence mechanisms include barriers that help prevent infection, and immune responses to pathogens that breach these barriers. In vertebrates, immune responses are divided into innate (or non-specific) and adaptive (or specific) immune responses. You will learn about the adaptive immune response in Chapter 9.

In this section you will learn about **innate immunity**, which consists of physical, chemical and microbiological barriers that provide innate resistance to infection. You will also learn about the innate immune response to infection that occurs when these barriers are breached, which includes the response of innate immune cells such as macrophages (Figure 8.2.1).

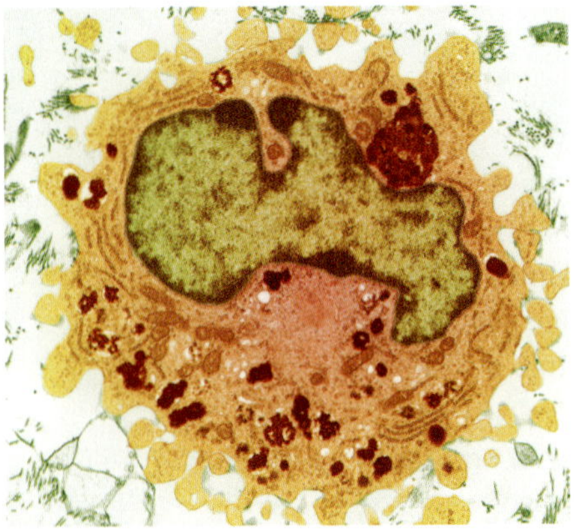

FIGURE 8.2.1 Coloured transmission electron micrograph (TEM) of a macrophage

Macrophages are cells of the innate immune response in vertebrates that recognise and engulf foreign material.

BARRIERS TO INFECTION

Organisms have a number of first-line defences (or barriers) that provide innate resistance against pathogens, including:

- **physical barriers**, such as skin or bark
- **chemical barriers**, such as the **lysozyme** enzymes in saliva and other body secretions
- **microbiological barriers**, formed by an organism's **microbiota**, namely **microflora**.

Physical barriers in plants

Physical barriers in plants largely involve cell walls that provide strength and flexibility. Cutin and waxes are fatty substances that make up the cuticle, which is found on the outer cell wall. A thicker cuticle generally prevents more pathogens from infecting the plant than a thinner cuticle. In trees, a thicker layer of bark is better able to prevent pathogens from entering the plant. Stomata (singular, stoma) create openings in the physical barriers of plants, providing an entry point for pathogens, but these openings can be closed when signalled (Figure 8.2.2).

Stomata are tiny epidermal pores bound by two specialised guard cells; they are the main way in which gas exchange occurs in plants.

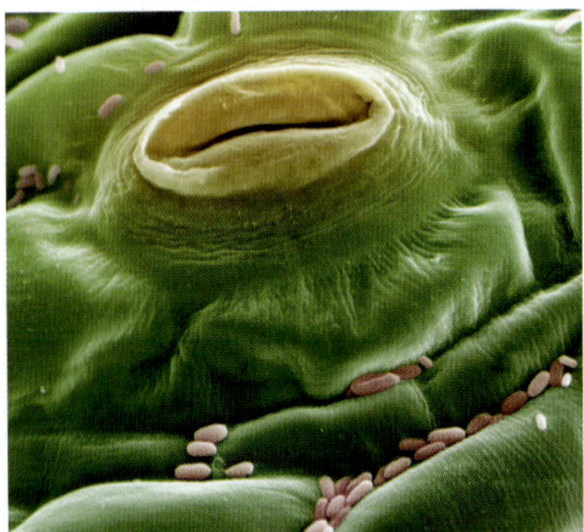

FIGURE 8.2.2 Coloured SEM of a single stoma on the surface of the leaf of a tomato plant. This stoma is closed to prevent bacteria (rod-shaped, pink) entering and infecting the plant.

The orientation of leaves can also play a role in defence. By positioning leaves vertically, water is unable to collect on the surface of leaves. This prevents infection by pathogens that are reliant on water for motility.

Physical barriers in animals

In vertebrates, epithelial cells create a physical barrier that prevents pathogens from entering the organism. Epithelial cells line the skin, as well as the respiratory, gastrointestinal and urogenital tracts. They are joined tightly by specialised membrane proteins to form a continuous barrier against pathogens.

In addition to toughened (keratinised) skin, adaptations that provide physical barriers to pathogens in animals include mucus-secreting membranes that trap invading organisms in mucus, and membranes lined with cilia that sweep foreign bodies away (Figure 8.2.3).

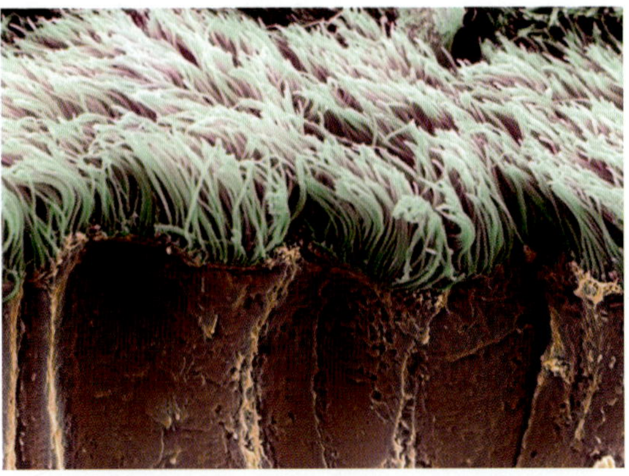

FIGURE 8.2.3 Coloured SEM of the mucous membrane (or bronchial epithelium) that lines the major airways of the lung. Mucus traps potential pathogens and foreign particles, and the rhythmic movement of hair-like cilia moves bacteria and other particles away from the lung and towards the throat.

Chemical barriers in plants

Plants produce a range of chemicals that help defend against infection when physical barriers are breached. Many of these chemicals are usually present in the plant tissues, although their levels may increase when the plant is attacked by a pathogen.

Chemical barriers in plants include the following.

- **Alkaloids** are toxic to many organisms, from fungi, bacteria and insects to humans. Caffeine, nicotine, morphine, capsaicin and atropine are all alkaloids. Their toxicity is usually dose dependent. Some alkaloids are used in low doses as drugs. Atropine is used as a cardiac stimulant in minute doses but is lethal in large doses.
- **Cyanogenic glycosides** are compounds that break down to form hydrogen cyanide. Hydrogen cyanide is extremely toxic to all eukaryotic cells because it disrupts ATP production in the mitochondria by blocking the electron transfer chain, leading to cell death.
- **Phenolics** include phytoalexins, flavonoids and tannins. Phytoalexins and many flavonoids have antibiotic properties. They disrupt cellular metabolism in the pathogen. Tannins are water-soluble chemicals that are stored in vacuoles until required and are highly toxic to plant pathogens. Tannins bind to salivary proteins and digestive enzymes such as trypsin, which can result in the death of the pathogen through inadequate energy intake. Tannins are effective against antibiotic-resistant bacterial strains.
- **Saponins** have soap-like properties. They break down lipids, disrupting the plasma membranes of pathogens.
- **Terpenes** make up many of the essential oils found in plants. Pyrethrins are terpenes that are used in insecticides. Phytoectysones are terpenes that mimic the hormones involved in insect moulting and cause fatal disruption to larval development, which reduces infestation.

Chemical barriers in animals

External chemical barriers in vertebrates include lysozyme enzymes and toxic metabolites, for example lactic acid and fatty acids, which are found in secretions such as tears, sweat and saliva (Figure 8.2.4). Here, they have protective functions and provide a generalised defence, for example, by destroying bacterial cell walls.

Other chemical barriers include stomach acid and digestive enzymes, which are primarily involved in the digestion of food, but also kill many pathogens. The fluid in the lungs contains proteins that act as surfactants. Surfactants coat the pathogens, making it easier for the pathogens to be eliminated by macrophages. In mammals, male and female genital mucosa produce secretions that serve several functions, including defence against pathogens.

> Surfactants are detergent-like substances found in lung secretions that lower the surface tension of lung fluids and prevent the alveoli from collapsing on exhalation.

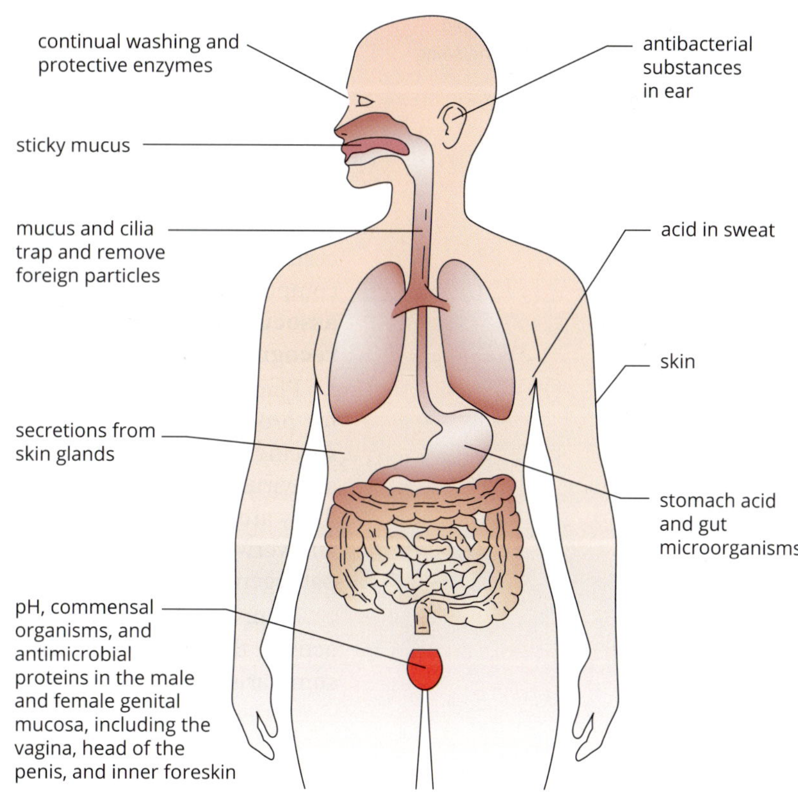

FIGURE 8.2.4 Some of the physical and chemical defence mechanisms that prevent foreign organisms from gaining access to the human body

Microbiological barriers in animals

Non-pathogenic bacteria, referred to as normal flora, are found on the skin, and in the mouth, nose, throat, lower part of the gastrointestinal tract and the urogenital tract in healthy individuals.

The presence of normal flora prevents the growth and colonisation of pathogenic bacteria because normal flora compete with other bacteria for space and resources, and produce chemicals that reduce the pH of the micro-environment.

Taking a course of antibiotics can disrupt the normal flora of the intestine, as the antibiotics do not discriminate between beneficial normal flora and harmful bacteria. This can disturb the normal gut function and predispose a person to various infections until the normal flora return to their pre-treatment levels.

Although not a problem in healthy individuals, in people with weakened immune systems normal flora can sometimes grow unchecked and cause disease.

THE INNATE IMMUNE RESPONSE

The innate immune response is non-specific and does not result in immunological memory.

The adaptive immune response is specific and results in immunological memory.

Immunological memory is the ability of lymphocytes of the adaptive immune system to 'remember' antigens after primary exposure, and to mount a larger and more rapid response when exposed to the same antigen again.

If pathogens manage to breach the barriers that act as a first line of defence, they are immediately met by attacking cells and molecules.

The **innate immune response** is found in all organisms, and its persistence over millions of years of evolution indicates its fundamental importance. Indeed, even when the innate immune response is unable to eliminate a pathogen, it remains critical for keeping infections under control until the adaptive immune response (see Section 9.2), which can take up to several days to develop, kicks in.

Innate immune responses in vertebrates:

- are non-specific—they do not target a specific antigen
- are rapid—they occur within hours
- are present in all animals
- are fixed responses—they do not adapt
- do not result in **immunological memory** of the pathogen that caused the infection.

Innate immune responses in plants

The main response of plants to infection is a chemical response. Some of these chemicals are always present in plant tissues and provide innate resistance (the chemical barrier), whereas others are only produced upon exposure to pathogens as part of the plant's innate immune response.

The innate immune response in plants is triggered when plant cells recognise certain molecules, such as certain lipopolysaccharides, or other common cell wall components, which form the cell walls of pathogens. These are called **pathogen-associated molecular patterns (PAMPs)**, and they are recognised by **pattern recognition receptors (PRRs)**.

Plants possess specific genes called resistance genes. Resistance genes code for proteins (R proteins), which switch on a plant's defences when it recognises specific PAMPs. The specific pathogen molecules are generally proteins, known as avirulence proteins (AVr), and are coded for by AVr genes in the pathogen. They are avirulence genes because the presence of their protein product stimulates an overwhelming response by the infected plant tissues that rapidly destroys the pathogen, and so disease does not eventuate.

Plants use hormone-like chemicals, such as jasmonic acid and salicylic acid, to activate their responses in the recognition pathways. The plant response pathway is summarised in Figure 8.2.5.

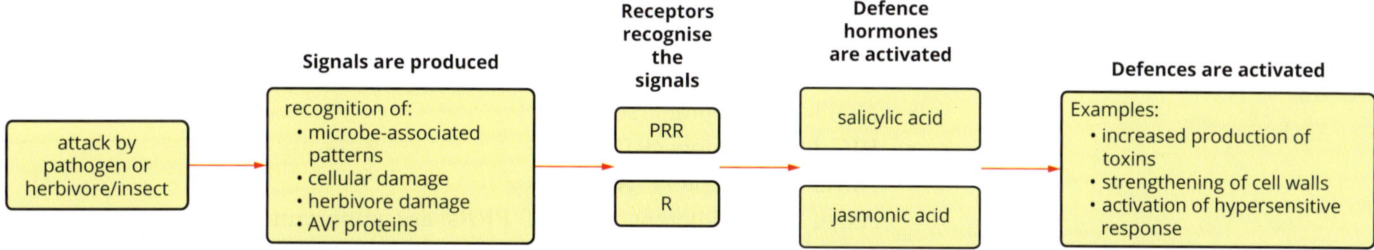

FIGURE 8.2.5 Plants recognise attack by pathogens or plant-eating organisms and mount defences that strengthen natural barriers and increase production of toxic and signalling chemicals.

Plants produce a range of proteins and enzymes that provide defence against pathogens. Because it requires a lot of energy to produce these substances, most plants only make them when they are under attack by a pathogen. Proteins produced by plant tissues include the following:

- **Defensins** are small proteins that act against digestive enzymes and are also thought to act against microbes by disrupting the plasma membrane. Many of these proteins are rich in the amino acid cysteine. Bread wheat can produce defensins (Figure 8.2.6).
- **Protease inhibitors** inhibit enzymes such as trypsin, an important digestive enzyme.
- **Digestive enzyme inhibitors** are proteins that block normal digestion. They include lectins, which block the digestion of starch by insects, and ricin, which is so toxic that 0.2 mg is enough to kill an adult human.
- **Hydrolytic enzymes** break down cell walls. Chitinases break down chitin, the main constituent of fungal cell walls and insect exoskeletons. Glucanases break the chemical bonds between the molecules that form glucans, which comprise the cell walls of members of the oomycetes. Lysozymes are enzymes that can digest bacterial cell walls.

In many plants, recognition of a pathogen may also activate enzymes that strengthen the cell wall as a barrier to infection and cause surrounding cells to thicken their cell walls. Many plants also undergo cell-mediated responses when all other pathways have failed. These responses can result in the self-destruction of the infected tissues, the hypersensitive response, in an attempt to wall off the pathogen and protect the rest of the plant.

FIGURE 8.2.6 Bread wheat, *Triticum aestivum*, contains small cysteine-rich proteins that act as plant defensins to inhibit the growth of bacteria and fungi.

Innate immune responses in animals

Like plants, animals can recognise and respond to pathogens through the identification of PAMPs. PAMPs include:

- lipopolysaccharide (a major component of the outer layer of Gram-negative bacteria)
- peptidoglycan (the main component of the cell wall of Gram-positive bacteria)
- flagellin
- microbial nucleic acids.

In vertebrates, the immune response to PAMPs involves many specialised cells, namely white blood cells (or **leukocytes**). Leukocytes have pattern recognition receptors (PRRs) on their surface, which are able to recognise PAMPs.

PAMP means 'pathogen-associated molecular pattern'.

PRR means 'pattern recognition receptor'.

There are five main classes of PRRs (Table 8.2.1):

- Toll-like receptors (TLRs)
- C-type lectin receptors (CLRs)
- NOD-like receptors (NLRs)
- RIG-like receptors (RLRs)
- AIM2-like receptors (ALRs).

All of these different types of PRRs are important in the innate immune response. The locations of PRRs determine whether they respond to **intracellular pathogens** or **extracellular pathogens**.

TABLE 8.2.1 Pattern recognition receptor (PRR) families and locations

Family	Receptor location	Responsive to extracellular or intracellular pathogens
TLR	cell surface, endosomal compartments	extracellular or intracellular
CLR	cell surface	extracellular
NLR	cytoplasm, endosomal membrane associated	intracellular
RLR	cytoplasm	intracellular
ALR	cytoplasm, nucleus	intracellular

PAMPs are common to a range of pathogens, meaning that the innate immune response to them is not specific to a particular pathogen. In contrast, the adaptive immune response is able to target a specific pathogen by the specific antigens it expresses.

Granulocytes

Granulocytes are leukocytes that contain cytoplasmic granules.

Granulocytes are leukocytes that contain many cytoplasmic granules, which are released during an immune response. **Neutrophils**, **basophils**, **eosinophils** and **mast cells** are all granulocytes, and all secrete a range of defensive molecules during an innate immune response. Basophils and mast cells also release histamine and are involved in allergic responses. High numbers of eosinophils are associated with parasitic infections.

Mast cells reside in connective tissue. Granulocytes in blood can be identified by the colour their granules stain using compound dyes such as hematoxylin and eosin. Basophils are attracted to the basic dye hematoxylin, and stain blue. Eosinophils are attracted to the acid dye eosin and stain red. Neutrophils are neutral and stain a neutral pink, as shown in Table 8.2.2 on page 280.

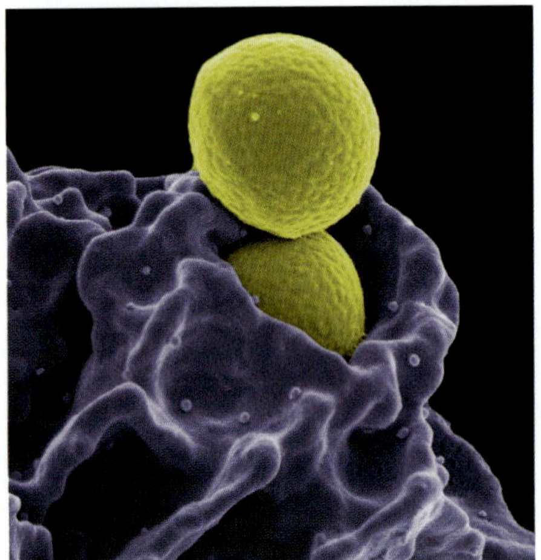

FIGURE 8.2.7 This neutrophil (purple) is engulfing *Staphylococcus aureus* bacteria (yellow), which it will then phagocytose.

Phagocytes

Phagocytes are leukocytes that are able to engulf and break down pathogens in a process known as **phagocytosis** (Figure 8.2.7). Phagocytes include neutrophils, **macrophages**, monocytes and **dendritic cells**.

For all TLRs, interaction with PAMPs triggers cascades of signalling molecules within the cell that lead to phagocyte activation, inflammation, and enhanced phagocytosis and pathogen destruction.

When phagocytes are activated, their enzyme NADPH oxidase releases reactive oxygen species that damage pathogens.

Some phagocytes, namely macrophages and dendritic cells, also act as antigen-presenting cells (APCs). When APCs phagocytose a pathogen, fragments of digested antigen are linked to MHC-II proteins and displayed (or presented) on the surface of the plasma membrane (Figure 8.2.8).

Antigen presentation

As you learnt in Section 8.1, there are major histocompatibility (MHC) proteins on the surface of your body's cells, which present self-antigens or non-self antigens to T lymphocytes. There are different classes of MHC proteins, including MHC class I and class II, which are both involved in **antigen presentation**.

MHC class I (MHC-I) proteins are normally found on all nucleated cells, and present peptide antigens derived from the proteins of pathogens in the cytoplasm of non-phagocytic cells to **cytotoxic T cells** in the adaptive immune response. You will learn more about T lymphocytes and their role in the adaptive immune response in Section 9.2. MHC-I proteins are also important in the innate immune response. Some intracellular pathogens down-regulate MHC-I expression to avoid recognition by cytotoxic T cells, and **natural killer cells** appear to have evolved in response to this. Natural killer cells recognise the absence of MHC-I in infected and damaged cells, and respond by releasing molecules that destroy the cell (Figure 8.2.9).

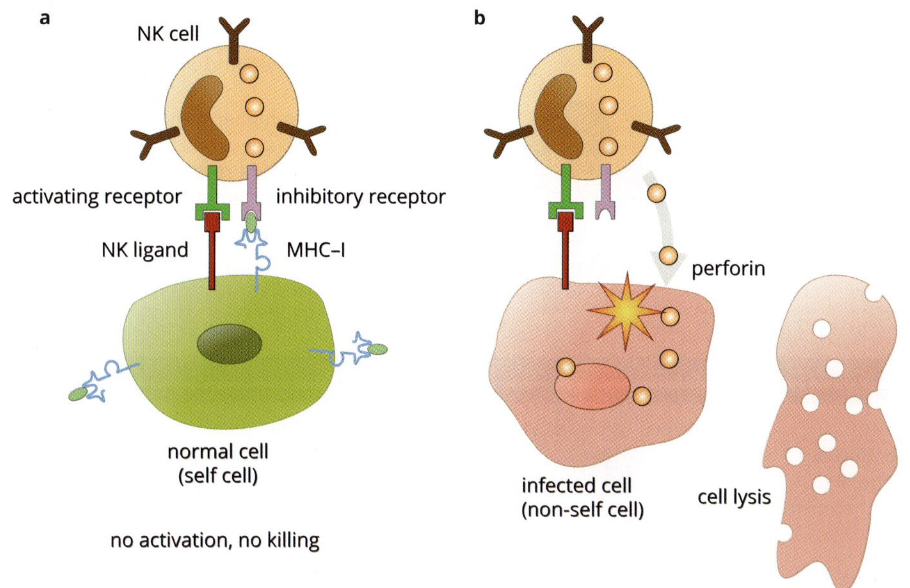

FIGURE 8.2.9 Action of natural killer (NK) cells. (a) The NK cell recognises a normal host cell by the presence of MHC-I and does not elicit an attack. (b) MHC-I is absent from the host cell's surface and the NK cell recognises that the cell is infected or damaged. The NK cell elicits a response to destroy the infected cell.

MHC class II (MHC-II) proteins can be conditionally expressed on all cells, but are most commonly found on the surface of APCs such as dendritic cells, macrophages, monocytes and B lymphocytes. This presentation of antigens activates the **helper T cells** of the adaptive immune response, linking the innate and adaptive immune responses (Figure 8.2.10).

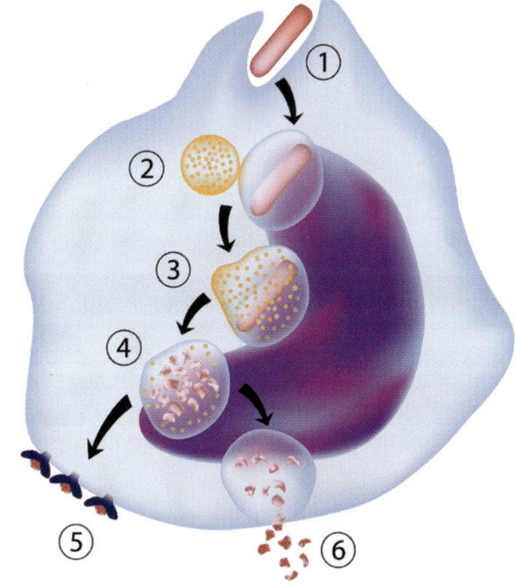

1 engulfing of foreign material
2 fusion of lysosome and phagosome
3 enzymes start to degrade foreign material
4 foreign material broken into small fragments
5 antigen fragments are bound to MHC-II and presented on the APC surface
6 leftover fragments released by exocytosis

FIGURE 8.2.8 Phagocytosis and antigen presentation in an antigen-presenting cell

Exocytosis is the release of substances enclosed within a vesicle to the outside of a cell. It occurs by fusion of the vesicle with the plasma membrane.

Antigen presentation links the innate and adaptive immune responses.

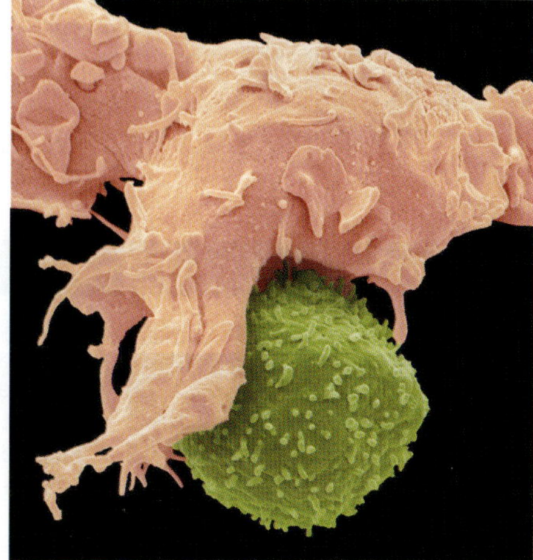

FIGURE 8.2.10 Coloured SEM showing the interaction between a macrophage (pink) and a helper T cell (yellow)

SUMMARY OF INNATE IMMUNE CELLS

Table 8.2.2 shows some of the leukocytes involved in innate immune responses and indicates whether they are involved in phagocytosis, antigen presentation, or the release of cytokines that promote inflammation. You will learn more about cytokines later in this section.

TABLE 8.2.2 Some of the leukocytes involved in innate immune responses and their function

Cell type	Function	Cell type	Function
neutrophil (granulocyte)	• phagocyte • secretes antimicrobial peptides such as defensins and reactive oxygen species that disrupt pathogen cell membranes • secretes a range of cytokines and chemokines • antigen presentation under certain conditions	basophil (granulocyte)	• secretes histamine, which dilates blood vessels and promotes inflammation • involved in allergic responses • secretes a range of cytokines, chemokines and antimicrobial peptides • has a limited role in phagocytosis • antigen presentation under certain conditions
macrophage	• phagocyte • antigen-presenting cell • secretes a range of cytokines, chemokines and antimicrobial peptides	eosinophil (granulocyte)	• secretes a range of cytokines, chemokines and antimicrobial peptides • found in higher numbers in parasitic infections • has a limited role in phagocytosis • antigen presentation under certain conditions
monocyte	• phagocyte • antigen-presenting cell • secretes a range of cytokines and chemokines	mast cell (granulocyte)	• phagocyte • secretes histamine, which dilates blood vessels and promotes inflammation • involved in allergic responses • secretes a range of cytokines, chemokines and antimicrobial peptides • antigen presentation under certain conditions
dendritic cell	• phagocyte • antigen-presenting cell • has many grooves that increase its surface area and permit contact with a large number of nearby cells • secretes a range of cytokines and chemokines	natural killer cell Immunofluorescent LM of natural killer cells: cytotoxic granules (green), nuclei (blue), cytoplasm (red)	• recognises virus-infected and cancerous cells • secretes cytotoxic chemicals from granules, such as perforin, which punches holes in plasma membranes, triggering apoptosis and cell death of abnormal and virus-infected cells • not considered a granulocyte as its granules are far less numerous than true granulocytes • secretes a range of cytokines and chemokines

DEFENSIVE MOLECULES

Complement proteins and cytokines are defensive molecules involved in both the innate and adaptive immune responses. You will learn more about adaptive immune responses in Chapter 9.

Complement proteins

The **complement proteins** are an array of more than 30 proteins that circulate in the blood and are able to help kill foreign cells. They are found in body fluids in an inactive form, and are activated as part of the non-specific (innate) immune response to certain antigens and carbohydrates on the surfaces of some bacteria and parasites.

> Complement proteins and cytokines are involved in both the innate and the adaptive immune responses.

Activation of complement proteins results in an enzyme-triggered reaction that leads to the **lysis** of the invading pathogens. For example, complement proteins destroy bacteria directly by punching holes in their plasma membranes, causing them to lyse. The release of the bacterial contents attracts phagocytes to the site of infection. Complement proteins activated by **antigen–antibody complexes** are also involved in specific (adaptive) immune responses.

Cytokines

Cytokines are small signalling molecules of the immune system that coordinate many aspects of our immune responses. Cytokines can be peptides, proteins or glycoproteins, and are released by body cells in response to cell damage or the presence of pathogens.

There are many different cytokines and they trigger a variety of responses, both non-specific and specific. For example, cytokines can promote the proliferation of lymphocytes, induce inflammation and fever, promote antibody responses and activate macrophages. Interferons and chemokines are two different types of cytokines, and they each have different functions.

Interferons

Interferons are a class of cytokines that are produced by, and act on, a host cell infected by a virus. Interferons act in an autocrine manner, activating the infected cells to produce enzymes that break down viral RNA and proteins that block translation. This limits viral replication and release from the cell. Interferons also attract NK cells, which release cytotoxic peptides to kill the virus-infected cell.

> Autocrine refers to a substance secreted by a cell that also has an effect on that cell.

Interferons are non-specific and will act against any virus. However, viruses vary widely in their susceptibility to interferons. Many viruses can evade interferon-induced defences and the more virulent viruses may be able to inhibit the production of interferon. Interferons also play a smaller role in combating bacterial and parasitic infections.

> While an important defence against viruses, interferons also regulate the immune system in a number of ways, such as enhancing T lymphocyte activity.

Chemokines

Chemokines are a type of cytokine and act as chemical attractants (or chemo-attractants). Chemokines are important for attracting leukocytes to sites of infection and inflammation.

THE INFLAMMATORY RESPONSE

Inflammation is the accumulation of fluid, plasma proteins and leukocytes that occurs when tissue is damaged or infected, and results in heat, pain, swelling, redness and loss of function.

The interaction between leukocytes (especially phagocytes) and pathogens triggers the inflammatory response that results from the production, activation or release of peptides and proteins such as complement proteins and cytokines.

Acute inflammation involves phagocytes, and occurs soon after infection as part of the innate immune response, but inflammation can also involve lymphocytes and occur later as part of the adaptive immune response.

A number of steps are involved in the initiation of an acute inflammatory response to infection (Figure 8.2.11):

1 Bacteria or other pathogens breach the barriers that provide a first line of defence, such as through an open cut or wound in the skin.
2 Injured cells release cytokines (chemokines) that attract neutrophils, and mast cells release **histamine**, which increases blood vessel dilation and permeability. The dilated, more permeable blood vessels allow leukocytes and fluid containing peptides and proteins such as complement proteins to enter the infected tissue. Platelets release clotting factors at the site of the wound.
3 Neutrophils migrate towards the cytokines and are activated, causing the neutrophils to recruit macrophages and secrete factors, such as defensins and hydrogen peroxide, that degrade and kill pathogens.
4 Macrophages in turn become activated and secrete cytokines and, along with neutrophils, phagocytose pathogens and debris at the site of infection. This may lead to the production of pus, which is fluid containing leukocytes, dead pathogens and cell debris.
5 The inflammatory response continues until the pathogen is eliminated and the wound has healed.

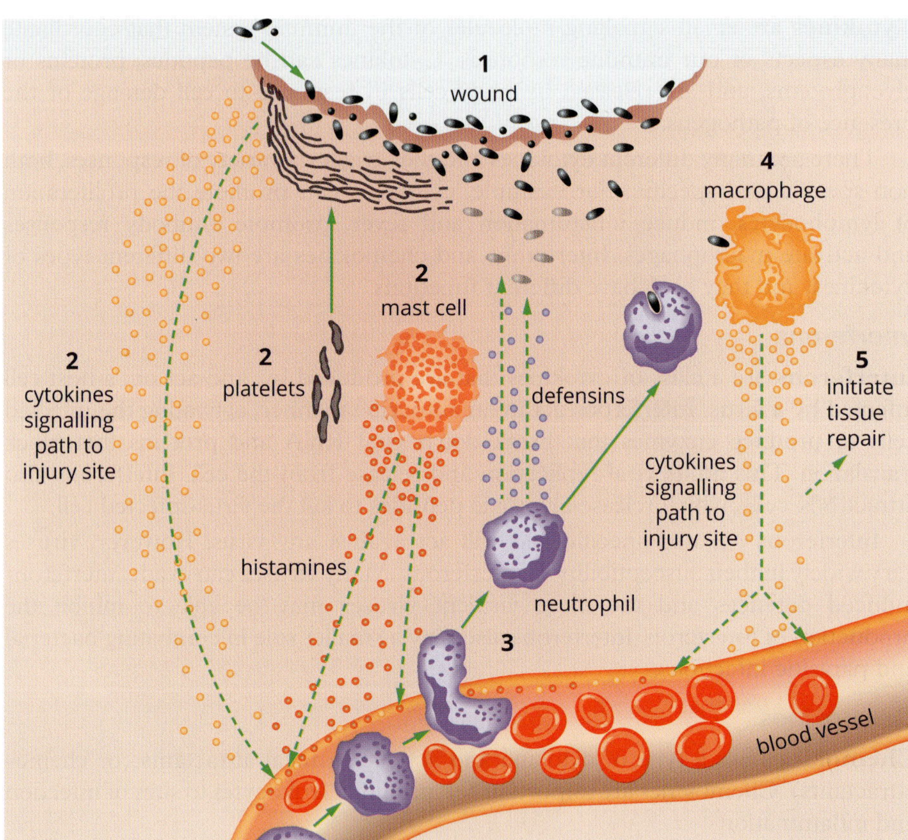

FIGURE 8.2.11 Process of acute inflammation (cell colours for illustrative purposes only)

Fever

A **fever** is an increase in body temperature that results when the regulated body temperature set point in the hypothalamus of the brain is set to a higher level by inflammatory cytokines. In humans, normal body temperature is around 37°C. Fever occurs when body temperature is above normal.

Fever slows the replication of bacteria and viruses by shifting the temperature away from their optimal range, and so allows more time for other defences to intervene. Additionally, moderate increases in temperature increase the activity and proliferation of leukocytes, so fever also improves the immune response.

8.2 Review

OA
✓✓

SUMMARY

- Barriers that provide innate resistance to infection include physical, chemical and microbiological barriers.
- Innate immune responses occur when these barriers are breached.
- Innate immune responses are:
  - non-specific—they do not target a specific antigen
  - rapid—they occur within hours
  - present in all animals
  - fixed responses—they do not adapt.
- Innate immune responses do not result in immunological memory.
- Leukocytes have pattern recognition molecules called toll-like receptors (TLRs) on their surface, which can recognise pathogen-associated molecular patterns (PAMPs).
- Phagocytes are leukocytes that can engulf and break down pathogens in a process known as phagocytosis.
- Some phagocytes also act as antigen-presenting cells.
- Defensive molecules include complement proteins and cytokines:
  - Activation of complement proteins results in an enzyme-triggered reaction that leads to the lysis of the invading pathogens.
  - Cytokines are small signalling molecules of the immune system that coordinate many aspects of our immune responses.
- Cytokines include interferon and chemokines:
  - Interferons are produced by virus-infected cells and inhibit viral replication.
  - Chemokines attract white blood cells to the site of infection.
- Inflammation is the accumulation of fluid, plasma proteins and leukocytes that occurs when tissue is damaged or infected.
- Fever slows the replication of bacteria and viruses and improves the body's immune response.

continued over page

8.2 Review *continued*

KEY QUESTIONS

Knowledge and understanding

1 Innate immune responses are:
  A specific and delayed
  B non-specific and rapid
  C non-specific and delayed
  D specific and rapid

2 **a** Define fever.
  b Explain how fever occurs.

3 How do macrophages and dendritic cells link the innate and adaptive immune responses?

4 Explain how the increased permeability of capillaries, which occurs during inflammation, helps to defend against pathogens.

5 The diagram shows a natural killer (NK) cell attacking an infected cell. What is the missing inhibitory receptor that indicates to the NK cell that this cell is an infected non-self cell?

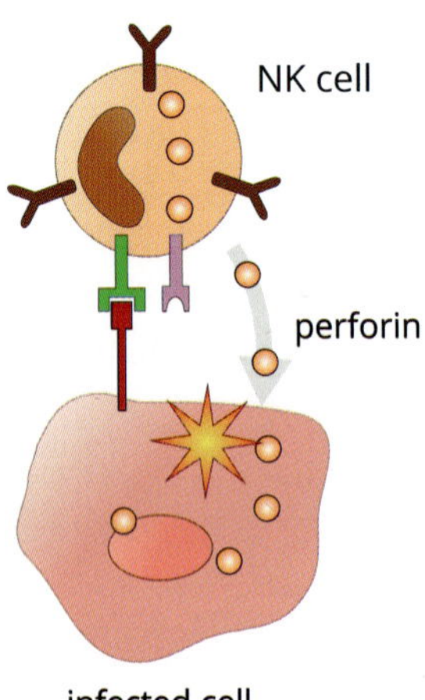

Analysis

6 Indole glucosinolate and camalexin are phenols involved in protecting plants from pathogens. The graph below shows the resistance to infection in plants with mutations that result in reduced production of camalexin, indole glucosinolate or both. Using only this data, draw conclusions about the contributions of these two chemicals to plant resistance to infection.

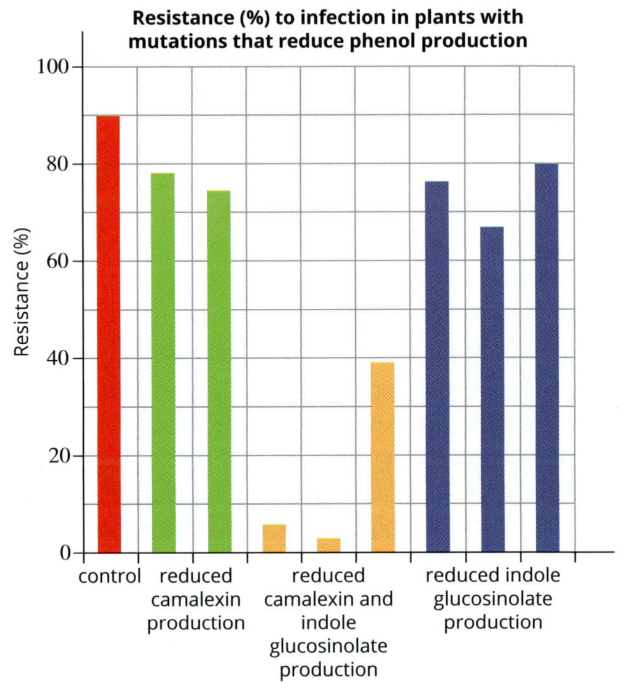

Chapter review

KEY TERMS

alkaloid
allergen
allergic response
anaphylaxis
antibody
antigen
antigen–antibody complex
antigen presentation
antigen-presenting cell (APC)
antigenic variation
B lymphocyte
bacterium (plural bacteria)
basophil
cellular pathogen
chemical barrier
chemokine
complement protein
cyanogenic glycoside
cytokine
cytotoxic T cell
defensin
dendritic cell
digestive enzyme inhibitor
disease
eosinophil
extracellular pathogen
fever
fungus (plural fungi)
granulocyte
helper T cell
histamine
human leukocyte antigen (HLA)
hydrolytic enzyme
immunogen
immunoglobulin (Ig)
immunological memory
inflammation
innate immune response
innate immunity
interferon
intracellular pathogen
leukocyte
lysis
lysozyme
macrophage
major histocompatibility complex (MHC)
mast cell
microbiological barrier
microbiota
microflora
natural killer cell (or NK cell)
neutrophil
non-cellular pathogen
non-self antigen
oomycete
pathogen
pathogen-associated molecular pattern (PAMP)
pattern recognition receptor (PRR)
phagocyte
phagocytosis
phenolic
physical barrier
prion
protease inhibitor
protozoan
saponin
self-antigen
self-tolerance
T lymphocyte
terpene
viroid
virus

REVIEW QUESTIONS

Knowledge and understanding

1 Which of the following is not one of the initial barriers to infection in animals?
  A intact skin
  B the action of complement proteins
  C lysozymes in saliva
  D competition from microflora

2 What is the role of histamine in the inflammatory response?
  A Histamine is primarily released by mast cells and basophils and increases blood vessel dilation and permeability, allowing immune cells and complement proteins to enter infected tissue.
  B Histamine is primarily released by mast cells and basophils and reduces blood vessel dilation and permeability, making it more difficult for bacteria to enter the bloodstream.
  C Histamine is primarily released by eosinophils and increases blood vessel dilation and permeability, allowing immune cells and complement proteins to enter infected tissue.
  D Histamine is primarily released by eosinophils and reduces blood vessel dilation and permeability, making it more difficult for bacteria to enter the bloodstream.

3 Which of the following innate immune cells are phagocytes ?
  A neutrophils, mast cells, natural killer cells, dendritic cells
  B neutrophils, macrophages, natural killer cells, dendritic cells
  C neutrophils, macrophages, monocytes, dendritic cells
  D neutrophils, macrophages, natural killer cells, dendritic cells

4 Which major histocompatibility complex proteins are involved in antigen presentation?
  A MHC-I
  B MHC-II
  C both MHC-I and MHC-II
  D neither MHC-I nor MHC-II

5 A natural killer cell will release perforins and destroy a cell when:
  A MHC-I presents self-antigens produced by a host cell
  B MHC-I presents pathogenic non-self antigens on a host cell
  C MHC-II is absent from the host cell surface
  D MHC-I is absent from the host cell surface

6 Recall examples of physical barriers in plants.

7 Recall examples of chemical barriers in plants.

8 What is the role of mucus and cilia?

9 a Define cytokines.
  b Describe the role of interferons.
  c Describe the role of chemokines.

10 Innate immunity consists of innate resistance to infection and innate immune responses. Describe the difference.

11 a Are bacteria prokaryotes or eukaryotes?
  b Are all bacteria pathogens? Explain your response.

12 What genetic material does a single virus contain?

13 a What are viroids?
  b Do viroids infect plants, animals or both?

14 a What are prions?
  b What type of disease do prions cause?

15 Define inflammation.

16 What are the symptoms of inflammation?

17 Describe the process of the acute inflammatory response that occurs when bacteria enter an open wound.

Application and analysis

18 An important function of phagocytes is to destroy bacteria. This is done through the endocytosis of the bacterium, followed by its digestion by lysosomal enzymes. As with all cellular functions, this process is regulated by proteins. Chediak–Higashi syndrome is a rare inherited disorder of the immune system in which the proteins that regulate the joining of lysosomes with endosomes are defective.
The failure of lysosomal breakdown of engulfed bacteria will seriously undermine not only the innate immune response, but also the adaptive immune response. Explain.

19 It has been suggested that salicylic acid is involved in stimulating the pathway leading to the production of phenols—chemicals that defend plants from attack by pathogens and herbivores. The graph shows the results after a group of plants were sprayed with salicylic acid (SA) at various concentrations.

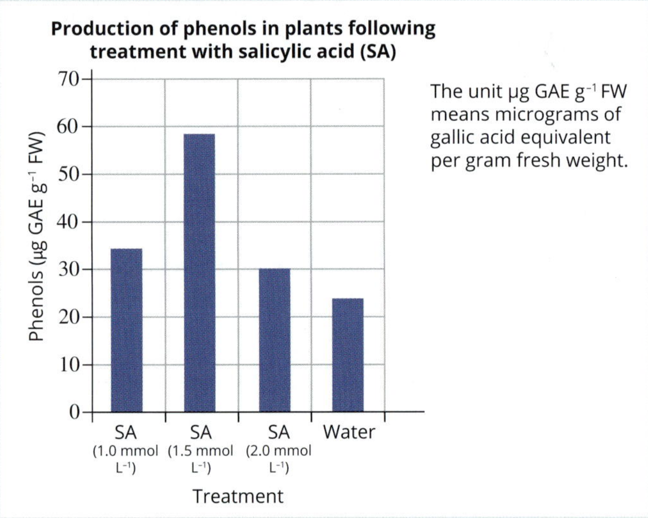

  a Describe the features of the graph that support the contention that salicylic acid promotes the production of defensive chemicals in plants.
  b Identify the concentration of salicylic acid that is most effective and state how you know this.

20 Saponins are a group of chemicals that defend plants against attack by pathogens. A study using one type of saponin, avenacin, was undertaken to investigate the research question, 'Will the absence of avenacin result in an increase in plant disease?' A group of plants from the species *Avena strigosa* was mutated so they could not produce avenacin. They were allowed to mature and were then exposed to spores of the pathogenic fungus *Gaeumannomyces graminis*. The results of one study with the mutant plants are shown in the following table. Plants were examined 21 days after infection with the fungal spores.

Note: Wildtype plants do not have reduced avenacin production. Plants A–I are mutants: those marked by * have reduced avenacin levels; all other mutants lack avenacin.

Draw conclusions about the effectiveness of avenacin in combating fungal infections in plants.

Plant group	Percentage of seedlings			
	No disease	Low disease	Moderate disease	Severe disease
Wild type (not mutant)	100	0	0	0
A	0	25	12	63
B	28	28	28	16
C*	62	38	0	0
D	31	44	6	19
E	12	71	17	0
F	27	40	33	0
G	6	44	50	0
H*	56	38	6	0
I*	74	21	5	0

21 Mild infections such as the common cold are a regular experience for children attending childcare. One symptom of these infections is usually a mild fever (up to 39°C). The usual treatment given to children is a medication such as paracetamol, which reduces their temperature to normal.

a Explain why reducing the body temperature of a patient with a mild fever may prolong the infection.

b Explain why very high temperatures associated with a severe fever can reduce the body's ability to fight off an infection.

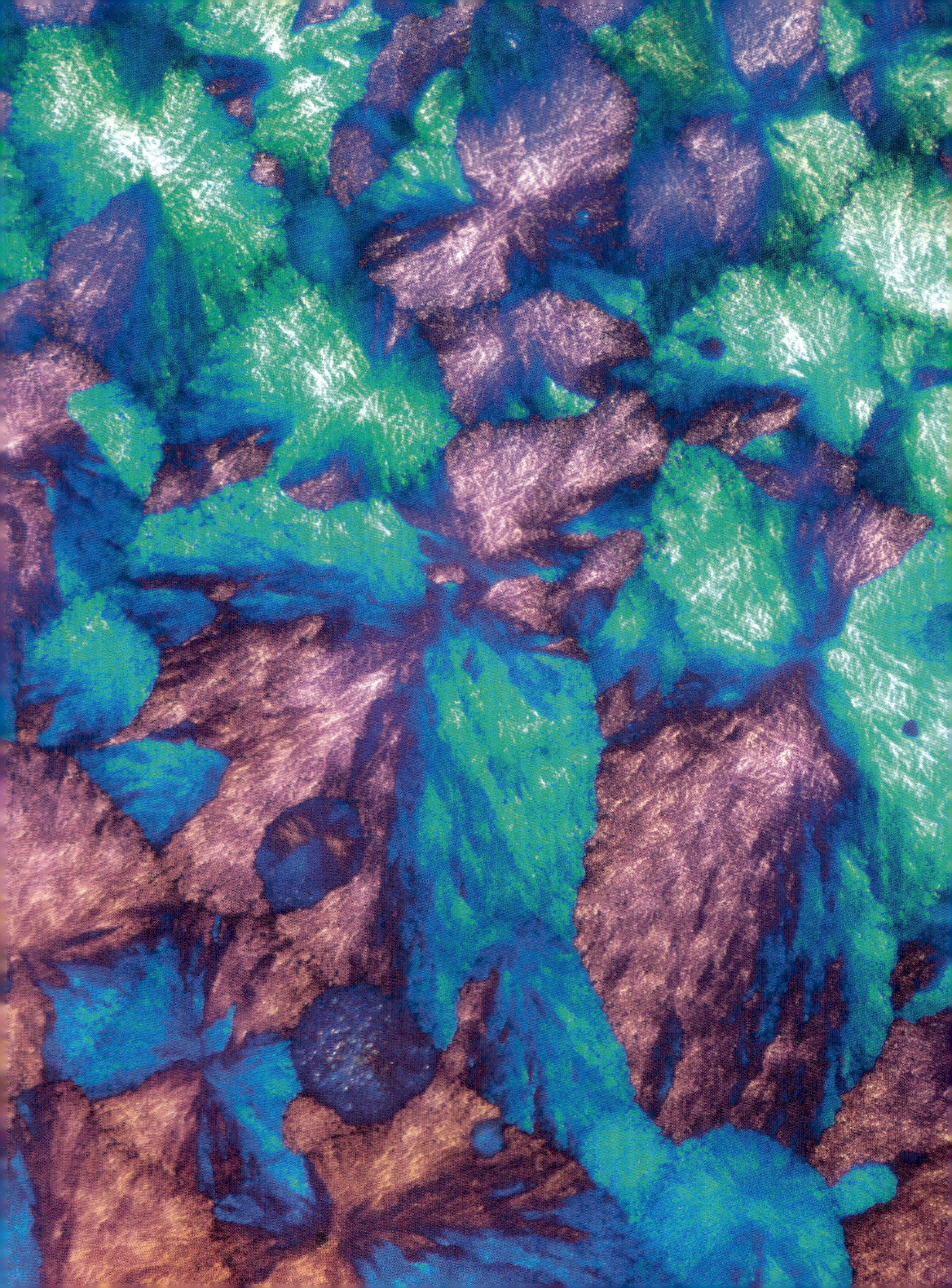

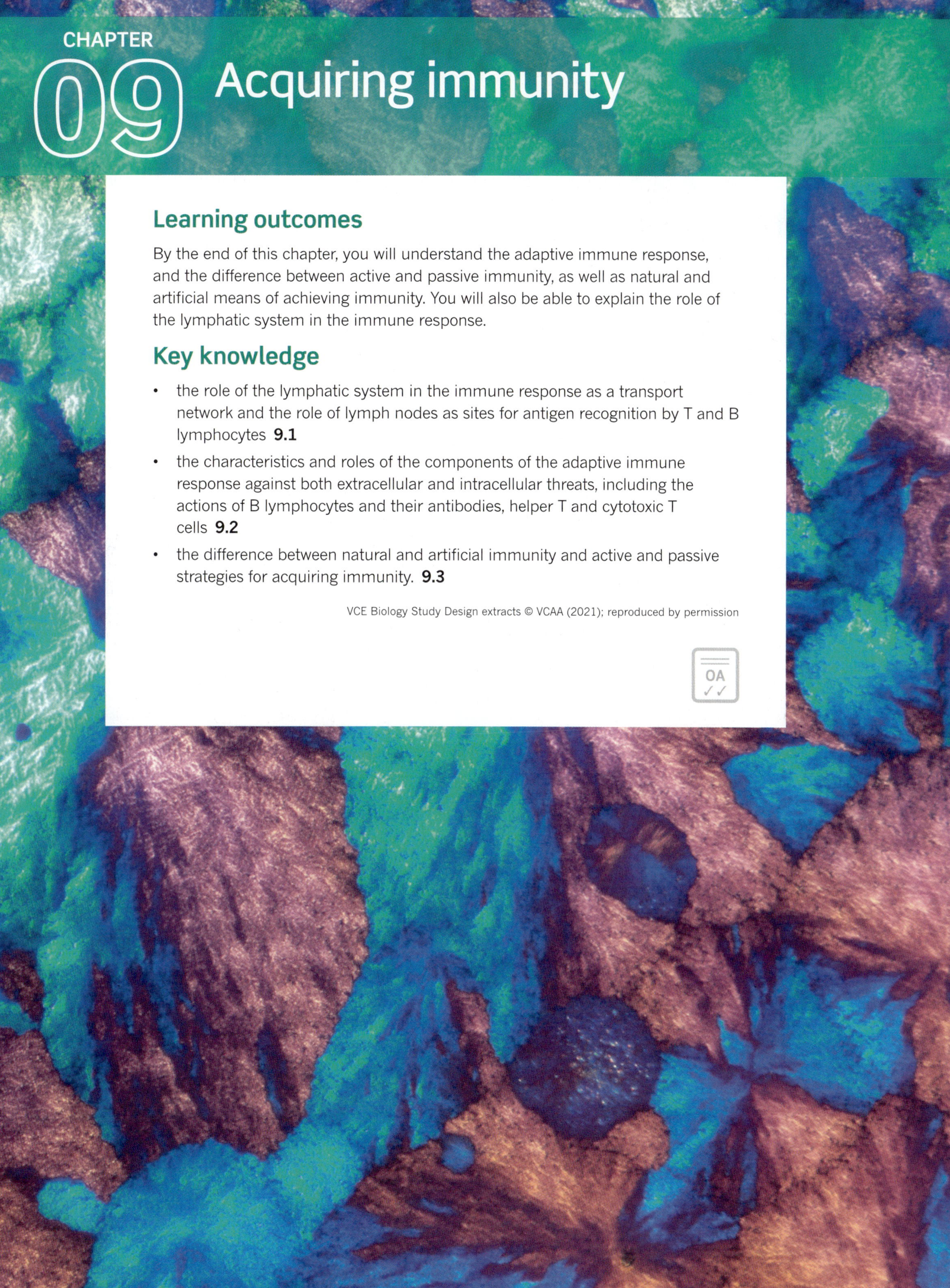

CHAPTER 09

Acquiring immunity

Learning outcomes

By the end of this chapter, you will understand the adaptive immune response, and the difference between active and passive immunity, as well as natural and artificial means of achieving immunity. You will also be able to explain the role of the lymphatic system in the immune response.

Key knowledge

- the role of the lymphatic system in the immune response as a transport network and the role of lymph nodes as sites for antigen recognition by T and B lymphocytes **9.1**
- the characteristics and roles of the components of the adaptive immune response against both extracellular and intracellular threats, including the actions of B lymphocytes and their antibodies, helper T and cytotoxic T cells **9.2**
- the difference between natural and artificial immunity and active and passive strategies for acquiring immunity. **9.3**

OA ✓✓

9.1 The lymphatic system

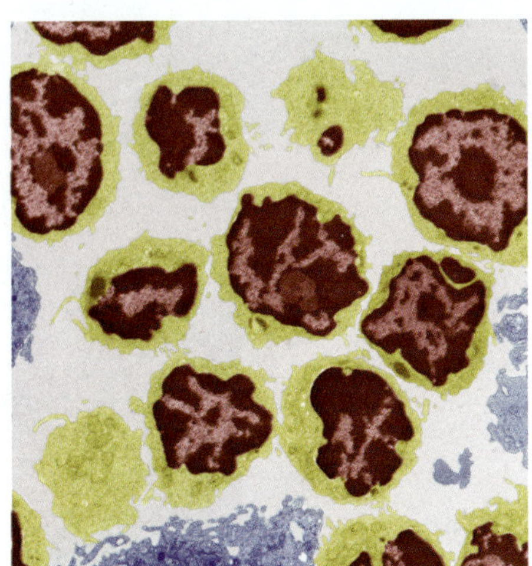

FIGURE 9.1.1 Coloured transmission electron micrograph of a section through a lymph node, showing a variety of lymphocytes (yellow)

The mammalian lymphatic system plays a key role in defending the body against infection. It transports immune cells, including antigen-presenting cells, throughout the body, and is where antigen recognition by lymphocytes occurs. In this section, you will learn about the lymphatic system, and how its structures (Figure 9.1.1) are involved in adaptive immune responses.

THE ROLE OF THE LYMPHATIC SYSTEM

The mammalian **lymphatic system** has several roles, including:

- returning fluid that seeps out of the blood vessels into tissues back to the circulatory system
- absorbing and transporting fatty acids and fats from the digestive system
- providing a place for lymphocytes to mature
- transporting lymphocytes and antigen-presenting cells to the lymph nodes, stimulating the adaptive immune response.

The lymphatic system is vital to the immune response. Invading pathogens are transported in the lymph to the lymph nodes, where bacteria, viruses and cancer cells are trapped and destroyed by phagocytes and lymphocytes. This is why your lymph nodes swell up when you have an infection.

THE STRUCTURE OF THE LYMPHATIC SYSTEM

The lymphatic system is made up of lymph, lymphatic vessels and primary and secondary lymphoid organs and tissues (Figure 9.1.2).

When the fluid that surrounds the tissues (or interstitial fluid) is drained into the lymphatic vessels, it is considered **lymph**. Lymph contains immune cells such as lymphocytes and phagocytes.

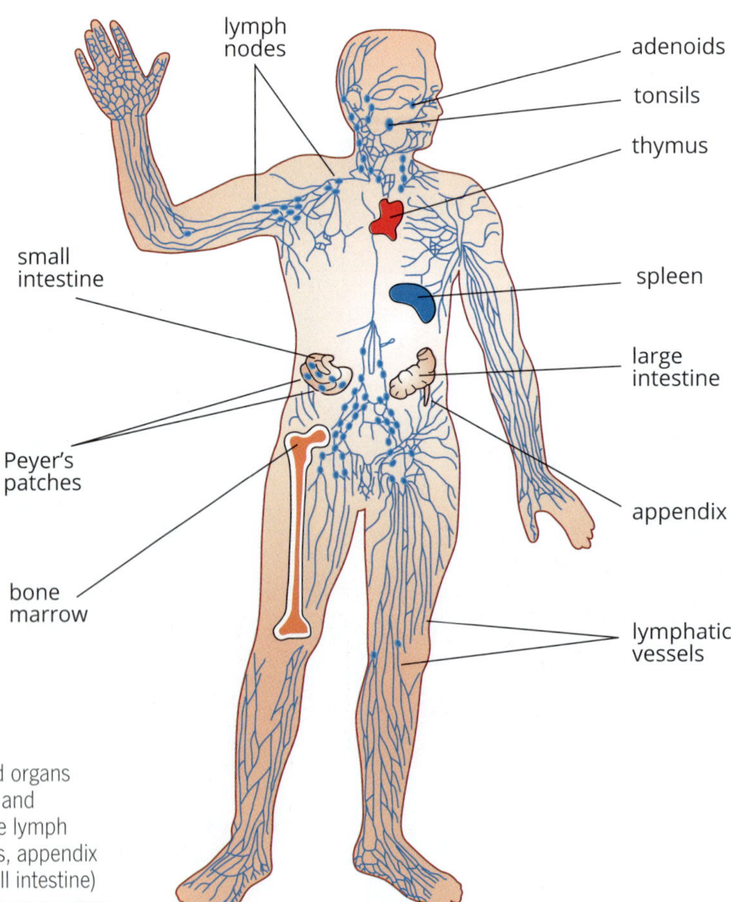

FIGURE 9.1.2 Primary lymphoid organs (the thymus and bone marrow) and secondary lymphoid organs (the lymph nodes, spleen, tonsils, adenoids, appendix and Peyer's patches of the small intestine)

CASE STUDY

Where lymph and blood capillaries meet

The structure of the lymphatic system is similar to the venous part of the circulatory system, where blood flows from the veins to the heart, then to the lungs to become oxygenated, and then through the arteries to tissues. In the lymphatic system, fine lymphatic capillaries join to increasingly larger vessels. The lymph drains through a lymphatic duct called the thoracic duct, and into a vein called the left subclavian vein, as well as through the right lymphatic duct into the right subclavian vein and the right internal jugular vein (Figure 9.1.3a).

Lymphatic capillaries are widespread, but they are absent from bones and the central nervous system (where excess tissue fluid drains into cerebrospinal fluid). Although blood and lymph capillaries are closed to each other, cells and fluid are able to pass between them through a process called extravasation (Figure 9.1.3b).

Some of the larger lymph vessels can contract, but most lymph flow results from the external compression of lymph vessels by muscular activity, such as during movement and breathing.

When vessels are compressed, the lymph fluid is forced in one direction because of numerous one-way valves, like those in veins, located along the vessels (Figure 9.1.3c).

When a person is inactive (such as standing still or sitting) for a long time, the fluid drainage from tissues decreases and causes swelling. This is especially so in the legs, because fluid drainage must work against gravity.

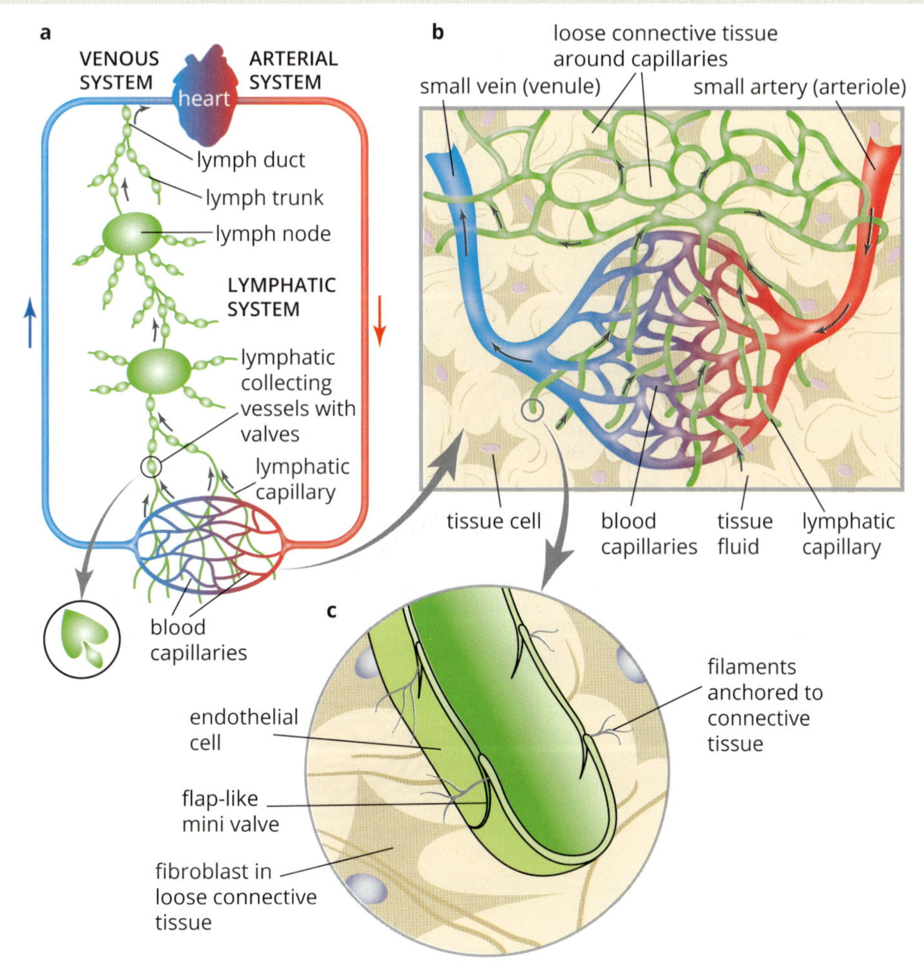

FIGURE 9.1.3 Lymphatic vessels weave through tissue cells and blood capillaries in loose connective tissues of the body. (a) Lymph drains through the thoracic duct and into the left subclavian vein, as well as through the right lymphatic duct into the right subclavian vein and the right internal jugular vein.
(b) Process of extravasation
(c) Lymphatic capillaries are closed-ended tubes in which adjacent endothelial cells overlap each other, forming flap-like mini valves.

Primary lymphoid organs and tissues

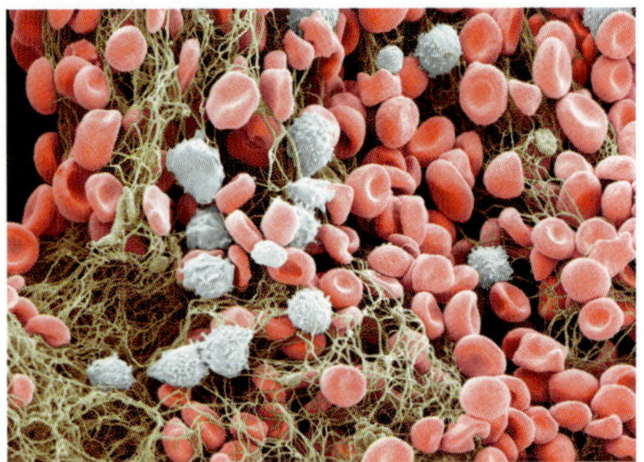

FIGURE 9.1.4 Coloured scanning electron micrograph (SEM) of a fractured rib. Bone marrow lies between the spongy bone and contains stem cells that give rise to red blood cells (red) and white blood cells such as B and T lymphocytes (grey).

The **primary lymphoid organs and tissues** are bone marrow and the thymus.

Bone marrow contains stem cells from which B and T lymphocytes originate (Figure 9.1.4). B lymphocytes undergo several stages of development in the bone marrow then enter the bloodstream and travel to the spleen and other secondary lymphoid tissues, where they complete their maturation, and also where they become activated after being exposed to antigen.

Immature T lymphocytes travel from the bone marrow to the thymus, where they mature. The thymus is considered a primary lymphoid organ because of its role in the maturation of T lymphocytes. The size of the thymus peaks at puberty, and then gradually shrinks each year, as it becomes replaced by fat (or adipose) tissue. The shrinking of the thymus does not immediately reduce the immune response, because of the already established pool of peripheral T lymphocytes, but it does contribute to the higher risk of infection and cancer that comes with age.

Secondary lymphoid organs and tissues

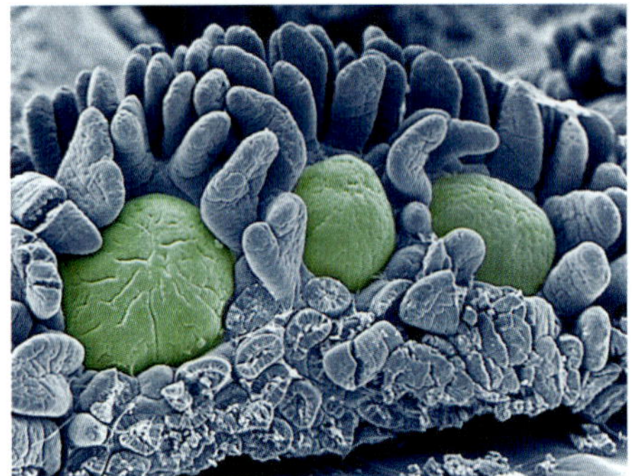

FIGURE 9.1.5 Coloured SEM of Peyer's patches (green) of the small intestine. Peyer's patches defend against infection by supplying lymphocytes to the local intestinal tissue, and are named after the Swiss anatomist Johann Conrad Peyer, who first described them in 1677.

The **secondary lymphoid organs and tissues** are the lymph nodes, spleen, tonsils, adenoids, appendix and Peyer's patches (Figure 9.1.5). It is in these organs and tissues that adaptive immune responses begin.

Lymphocytes are activated in secondary lymphoid tissues, where they recognise and respond to non-self antigens that are specific to their receptors.

Lymph nodes

Lymph nodes are composed of lymphoid tissue, and are located at regular intervals along the lymphatic system. Lymph passes through lymph nodes on its way back to the bloodstream (Figure 9.1.6). Lymph nodes act as filters, trapping foreign particles, cellular waste, toxins and pathogens.

The structure of lymph nodes maximises the chance of encounters between antigens and immune cells. Some dendritic cells and macrophages are stationed in the lymph nodes, where they phagocytose pathogens, and present the foreign antigens to helper T cells. Antigen-presenting cells in body tissues also migrate to the lymph nodes after phagocytising pathogens, to present foreign antigens to helper T cells.

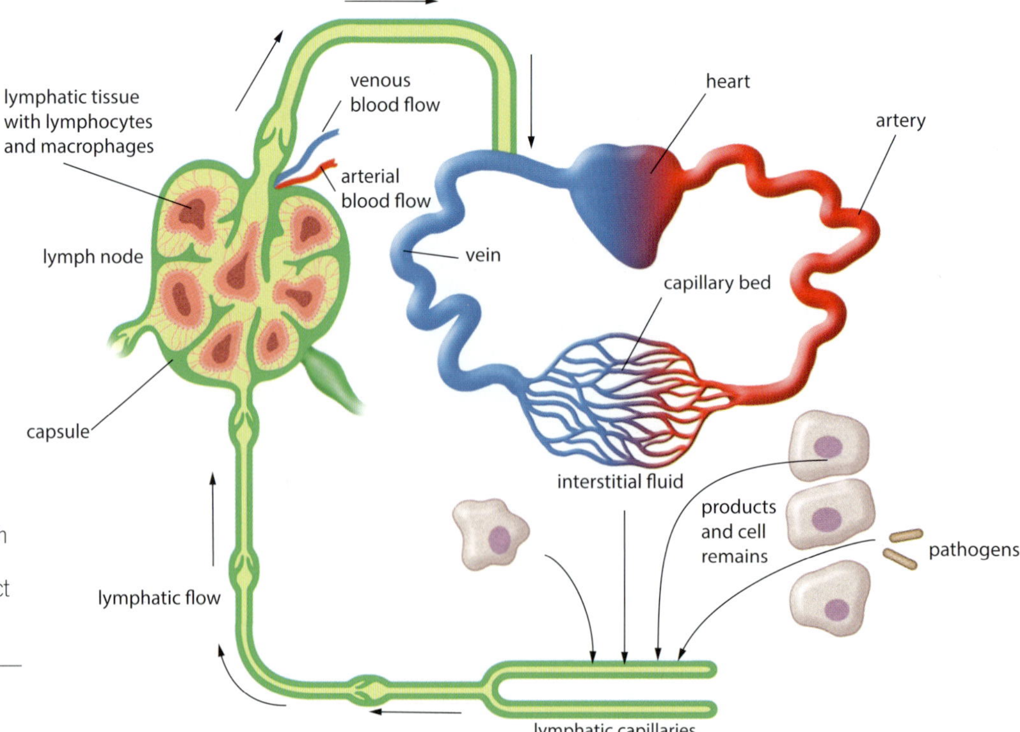

FIGURE 9.1.6 The flow of lymph through the lymphatic system is one-way due to the presence of valves. Lymph nodes act as filters and are important centres of immune cell activity.

B and T lymphocytes interact in the follicles of the lymph nodes. B lymphocytes that identify an antigen undergo clonal expansion and differentiation to plasma cells. Antibodies are released into the bloodstream to travel throughout the body. Cytotoxic T cells are activated, proliferate, and travel through the bloodstream to sites where they are needed.

The size of lymph nodes can expand markedly when cell proliferation is occurring in response to an infection. For example, during a respiratory tract infection, it is common for swollen lymph nodes to occur on the side of the neck (Figure 9.1.7).

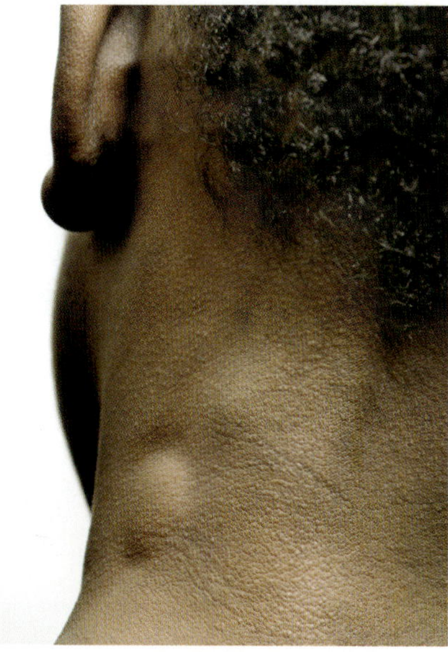

FIGURE 9.1.7 Swollen lymph nodes in a boy's neck

Spleen

The spleen's primary function is to control the number of red blood cells in the body by destroying old and defective red blood cells. The spleen also stores up to a quarter of the body's lymphocytes and is a site of B lymphocyte maturation and activation.

If for some reason the spleen needs to be removed, this does not have a disastrous effect on B lymphocyte maturation, because B lymphocytes can mature in other secondary lymphoid tissues.

BIOFILE

Sentinel lymph nodes

Lymph nodes are filters for antigens and invading microbes, but they can also trap abnormal cells, such as cancer cells that have separated from a primary tumour and travelled in the lymph until reaching a lymph node.

A sentinel lymph node, the first node to which cancer cells are most likely to spread, may be removed and examined under the microscope. The presence of cancer cells in the lymph node indicates that a tumour is malignant.

Tattoos do not cause cancer, but tattoo ink migrates through to the lymph nodes and can mimic the appearance of cancer, making proper diagnosis of cancer more difficult.

Whether or not tattoos look good on the outside is open to interpretation, but on the inside, tattoo ink migrates to your lymph nodes and can make them look cancerous.

9.1 Review

OA ✓✓

SUMMARY

- The lymphatic system produces lymphocytes and transports them, along with antigen-presenting cells, to the lymph nodes to stimulate adaptive immune responses.
- The primary lymphoid organs and tissues are the bone marrow and thymus.
- T lymphocytes develop in the bone marrow and mature in the thymus.
- Secondary lymphoid organs and tissues include lymph nodes, spleen, tonsils, adenoids, appendix, and Peyer's patches of the small intestine.
- B lymphocytes develop in the bone marrow and mature in the secondary lymphoid organs and tissues.
- Secondary lymphoid organs and tissues are the sites where lymphocytes identify and interact with antigen-presenting cells and are then activated to divide and differentiate.

KEY QUESTIONS

Knowledge and understanding

1 Which of the following are the primary lymphoid organs and tissues?
  - **A** bone marrow and lymph nodes
  - **B** bone marrow and spleen
  - **C** lymph nodes and spleen
  - **D** bone marrow and thymus

2 B lymphocytes complete their maturation in:
  - **A** bone marrow
  - **B** the thymus
  - **C** the gall bladder
  - **D** the spleen

3 Explain why the lymphatic system is an important part of immune responses in general, and of adaptive immune responses in particular.

4 Where do B lymphocytes originate and develop, and then mature?

5 Identify the missing labels (A–E) for the diagram (at right) of the lymphatic system.

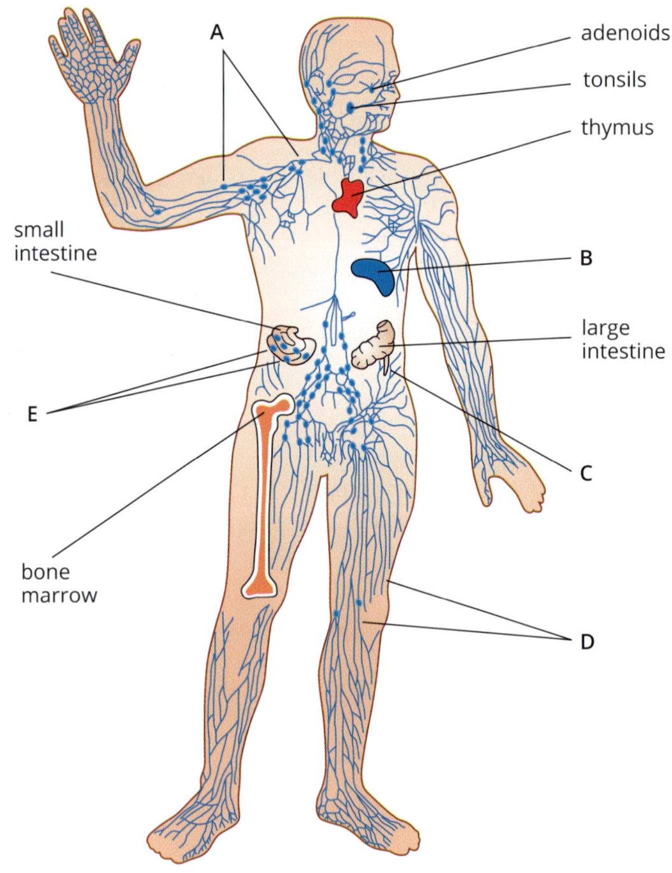

Analysis

6 One way in which the structure of lymph nodes enhances their efficiency is that the nodes have more vessels carrying fluid into the nodes than carrying fluid out.
  - **a** **i** What is the fluid travelling in lymphatic vessels called?
    - **ii** What effect would more vessels leading into the node than out of the node have on the rate of flow?
  - **b** The lymphatic system is responsible for draining the fluid that leaks from blood vessels. How might the flow of fluid in the lymphatic system be affected by inflammation?
  - **c** Explain why it is beneficial to the immune process that lymphocytes accumulate in the lymph nodes.

9.2 Adaptive immune responses

In Chapter 8, you learnt that if a vertebrate's first line defences are breached by a pathogen, that pathogen is met with a non-specific innate immune response. However, this innate immune response may or may not be successful in eliminating the invader. Fortunately, vertebrates have evolved an additional immune response to pathogens, known as the adaptive immune response. In this section, you will learn about the adaptive immune response.

The innate immune response is non-specific and does not result in immunological memory.

The adaptive immune response is specific and results in immunological memory.

B and T lymphocytes are specialised for adaptive immune responses.

Plants do not have an adaptive immune response and they lack mobile immune cells that can travel to the site of infection. Every plant cell has to respond to pathogens independently.

THE NATURE OF THE ADAPTIVE IMMUNE RESPONSE

There are two distinguishing features of the **adaptive immune response**:

- **specificity**—the ability to recognise and respond exclusively to specific antigens (Figure 9.2.1). On recognising a specific foreign antigen on a pathogen, cells of the adaptive immune system trigger an array of defensive mechanisms that destroy the pathogen.
- **immunological memory**—the ability of cells of the adaptive immune system to 'remember' antigens after primary exposure, and to mount a larger and more rapid response when exposed to the same antigen again.

Lymphocytes: cells of the adaptive immune response

The cells that are crucial to the adaptive immune response are **B lymphocytes** (or B cells) and **T lymphocytes** (or T cells). Each **lymphocyte** has a different receptor for a particular antigen, and is able to proliferate, creating clones of the initial lymphocyte with the specific receptor for the antigen. This is called **clonal selection**.

B and T lymphocytes interact with and respond to antigens differently. The B and T lymphocyte sub-populations have distinct roles but are both key to the adaptive immune response. Lymphocytes travel through the lymphatic system and become activated when they encounter antigens specific to their receptors. You learnt about the lymphatic system in Section 9.1.

FIGURE 9.2.1 B lymphocytes produce antibodies for a specific antigen.

Mechanisms of adaptive immune responses

There are two mechanisms of immunity in the adaptive immune response (Figure 9.2.2):

- **humoral immunity**, in which macromolecules, such as complement proteins, and antibodies produced by B lymphocytes, are secreted into the extracellular fluid
- **cell-mediated immunity**, which involves the action of antigen-presenting cells and T lymphocytes.

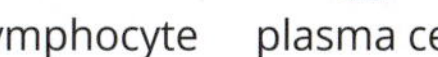

FIGURE 9.2.2 (a) B lymphocytes are involved in humoral or antibody-mediated immunity. (b) T lymphocytes are involved in cell-mediated immunity. Except for plasma cells, the different types of lymphocytes look very similar under a microscope. The only way to know which is which is to identify their different surface proteins.

HUMORAL IMMUNITY

In medieval times, the term 'humor' referred to body fluids.

Humoral immunity involves B lymphocytes, which produce specific antibodies against non-self antigens and release them into the blood and lymph (Figure 9.2.3).

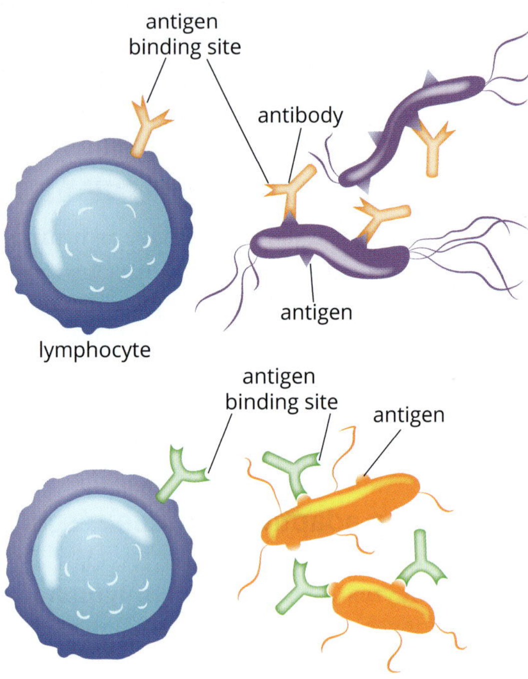

FIGURE 9.2.3 Antibodies specific to a foreign antigen will bind to it, helping to eliminate the invading pathogen.

CASE STUDY

Clonal selection theory

An almost infinite number of different antigens exist, and the immune system is able to produce lymphocytes specific to each antigen upon exposure. The clonal selection theory is a scientific theory that explains how lymphocytes are able to produce a large number of antibodies specific to an antigen.

When B and T lymphocytes form, each has a receptor that will react to a single antigen. The clonal selection theory states that a specific antigen will only activate a lymphocyte with a receptor that specifically recognises that antigen. Once activated, this lymphocyte will proliferate into a clone of millions of effector cells dedicated to eliminating the specific antigen that stimulated the immune response (Figure 9.2.4).

Clonal selection theory was developed in 1957 by Sir Frank Macfarlane Burnet, one of Australia's most celebrated scientists.

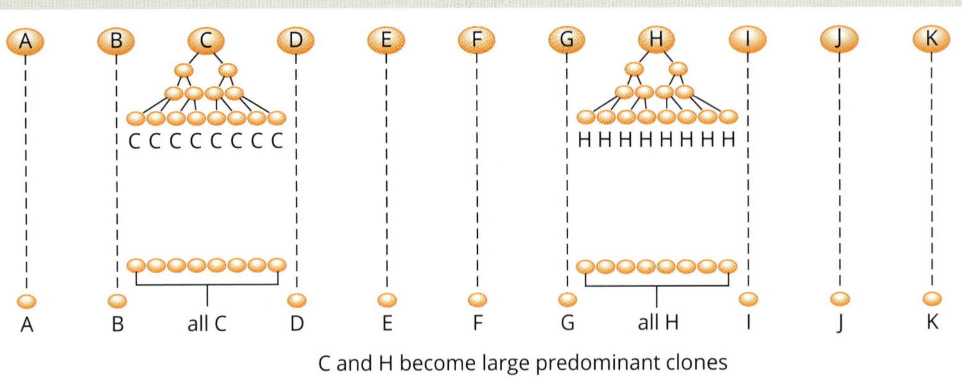

FIGURE 9.2.4 Lymphocytes that encounter or interact with an antigen, such as lymphocytes C and H, begin to proliferate. This increases the number of lymphocytes with identical receptors, or clones, for the specific antigen that was first encountered.

B lymphocytes

B lymphocytes (also known as B cells) originate and commence **differentiation** in the bone marrow and complete their maturation in the secondary lymphoid organs and tissues. At any time there are billions of B lymphocytes circulating in the blood.

> B lymphocytes develop in the bone marrow and complete their maturation in the secondary lymphoid organs and tissues.

B cell receptors

Each B lymphocyte has thousands of **B cell receptors (BCRs)**, which are membrane-bound antibodies. BCRs on any individual B lymphocyte are the same, but different B lymphocytes have different BCRs that detect different antigens.

> Mature lymphocytes that have not been activated by an antigen are said to be 'naive'.

When BCRs bind to antigens, B lymphocytes engulf and process the antigens, and function as **antigen-presenting cells (APCs)** by displaying processed antigens to helper T cells.

The binding of BCRs to antigens also results in the activation and proliferation of B cells with the same specific BCR variants. Cytokines released by helper T cells are also important for helping to activate B lymphocytes. When B lymphocytes are activated, they divide and further differentiate into two types of daughter cells:

> Activated B lymphocytes divide to form antibody-secreting plasma cells or memory B cells.

- plasma cells
- memory B cells.

Plasma cells

Activation of B lymphocytes leads to the production of **plasma cells**, which are essentially 'factories' specialising in antibody production (Figure 9.2.5). The antibodies produced are specific to the antigen that activated the B lymphocyte. Plasma cells can produce thousands of antibodies per second.

Memory B cells

Memory B cells can remain in lymphoid tissues for long periods (even for the lifetime of the animal) and are responsible for the immunity that often follows infection or vaccination. These cells can divide and give rise to plasma cells if secondary exposure to the antigen occurs (Figure 9.2.5).

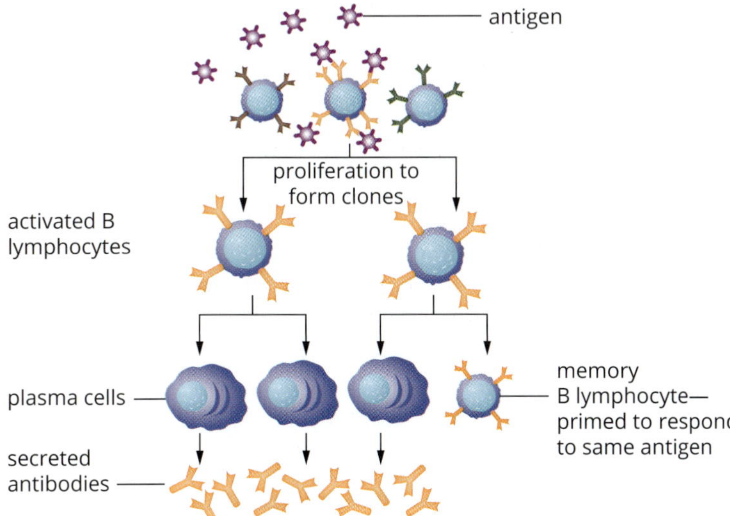

FIGURE 9.2.5 Many B lymphocytes differentiate into plasma cells, which produce and secrete antibodies for immune protection. Others become memory B cells and are retained in lymph nodes.

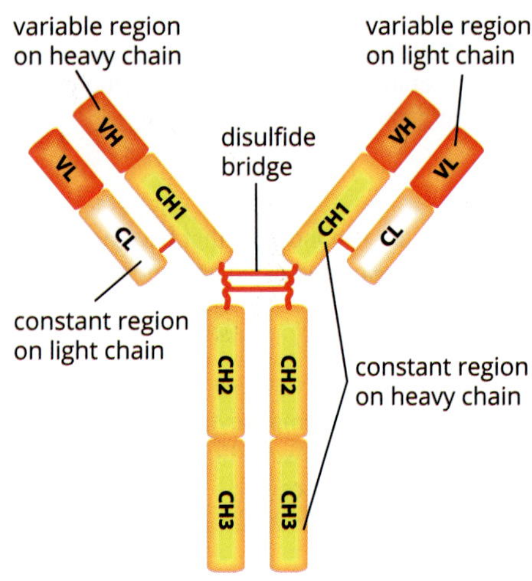

FIGURE 9.2.6 Antibodies have two long heavy (H) chains and two short light (L) chains. Both heavy and light chains have a variable (V) and a constant (C) region. Naturally-produced antibodies consist of two identical variable regions that are specific for a particular antigen. The constant region is capable of binding to and initiating other immune components, such as the complement proteins.

Antibodies are produced by B lymphocytes and bind to specific antigens.

Antibodies

Antibodies, which are also known as **immunoglobulins (Ig)**, are produced by B lymphocytes and released into the blood and lymph. Antibodies are proteins that bind to specific, intact antigen molecules.

The basic unit of an antibody molecule is a Y-shaped protein, formed by four polypeptide chains: two long **heavy chains**, and two short **light chains** (Figure 9.2.6). The amino acid sequences that form the top of the 'arms' of the Y-shaped antibody are known as the **variable regions**. It is the variation of these variable regions that allows antibodies to bind to different antigens. The two variable regions are identical antigen-binding sites and attach to identical antigens. The single 'stem' of the Y-shaped antibody is a conserved sequence in all antibodies and is called the **constant region**. The constant region binds to and recruits other components of the immune system.

Antibodies may act singly (monomers), in pairs (dimers) or in groups of five (pentamers). Mammals have five main classes of antibody molecules with different structures and functions (Table 9.2.1).

TABLE 9.2.1 Structure and function of mammalian immunoglobulins

Class	Half-life in serum	Presence	Functions	Structure
IgG	21 days	blood, lymph and extracellular fluid; most circulating antibodies (>80%); crosses placenta	agglutination, complement activation	
IgM	10 days	blood and lymph; produced early in infection response	agglutination, complement activation	disulfide bond; joining chain
IgA	6 days	found in secretions such as tears, saliva and milk	mucosal immunity	joining chain; secretory protein
IgD	3 days	blood and lymph; mostly present on B lymphocyte surfaces; small amount in circulation; binds to basophils and mast cells	functions not well understood; possible role in regulating innate immune responses	
IgE	2 days	blood and lymph; attaches to mast cells	involved in allergic reactions	

Antibody function

Antibodies do not directly destroy pathogens, but carry out several important mechanisms to interfere with the function of the pathogen (Figure 9.2.7):

- **neutralisation** of bacterial toxins: Antibodies bind to bacterial toxins, blocking the action of the toxin.
- neutralisation of pathogens: Antibodies bind to antigens on the surface of the pathogen, which are required for entry into host cells, thereby preventing pathogen invasion of host cells.
- **agglutination**: Antibodies bind to antigens and form **antigen–antibody complexes** that clump together, and which activate phagocytes and the complement cascade, leading to antigen or pathogen destruction.
- **precipitation**: Antibodies bind to soluble antigens, causing them to become insoluble and precipitate out of solution.

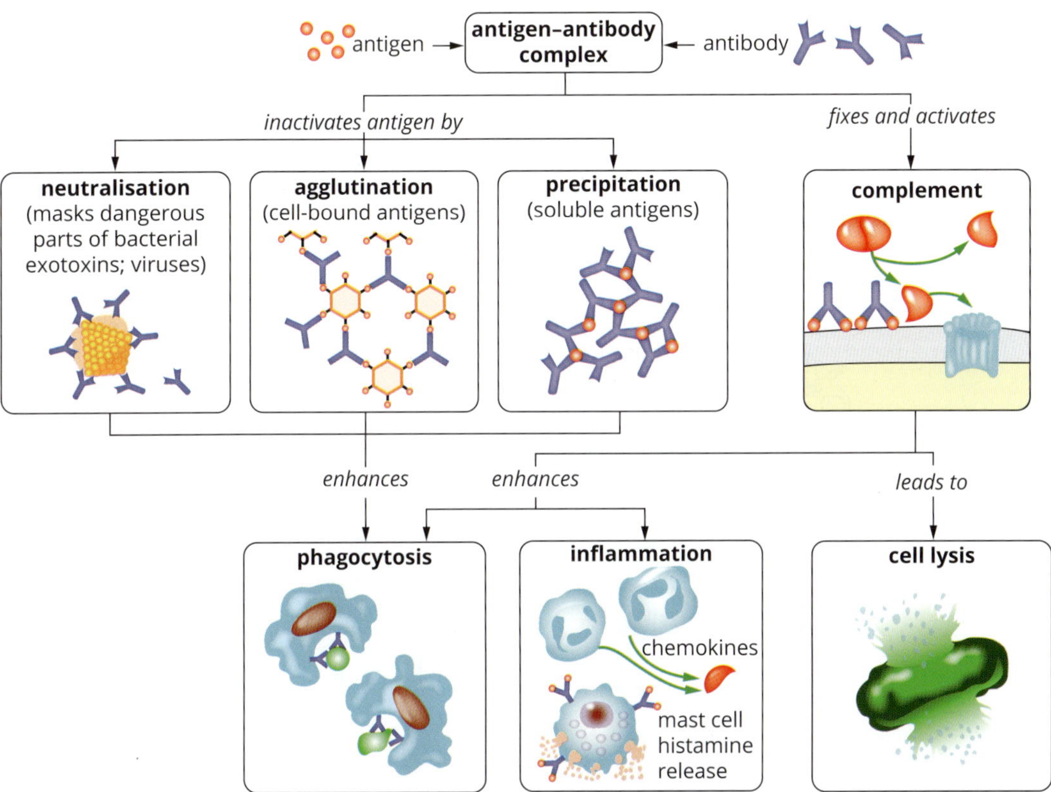

FIGURE 9.2.7 Antibodies function in a number of different ways to help eliminate pathogens.

CELL-MEDIATED IMMUNITY

Unlike humoral immunity, which involves B lymphocytes, cell-mediated immunity is regulated by T lymphocytes (Figure 9.2.8). The response is mediated by the **T cell receptors (TCRs)**.

T lymphocytes develop in the bone marrow and mature in the thymus.

T lymphocytes

Depending on their function, T lymphocytes (also known as T cells) are classified as helper, cytotoxic or memory T cells.

Helper T cells

Helper T cells (T_H cells) do not directly kill pathogens but instead 'help' with immune responses. There are two major types of T_H cells:

- T_H1 cells secrete cytokines to activate cytotoxic T cells (Figure 9.2.8). In other words, T_H1 cells promote cell-mediated immunity.
- T_H2 cells secrete cytokines that stimulate naive B lymphocytes (mature B cells that have not been activated by antigens) to differentiate into antibody-secreting plasma cells (Figure 9.2.8). In other words, T_H2 cells promote humoral immunity.

The nature of the invading pathogen, and the cytokines secreted by innate immune cells in response to it, will determine whether a T_H1 or T_H2 response predominates. T_H1 responses are driven by intracellular pathogens, whereas T_H2 responses are driven by extracellular pathogens.

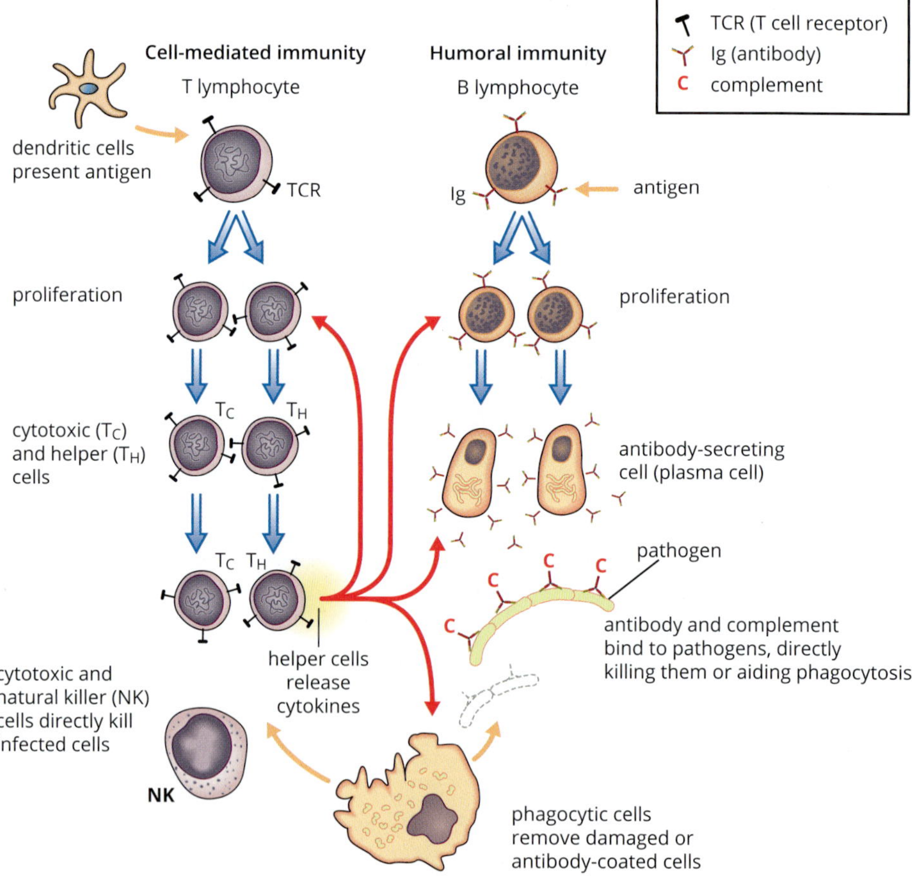

FIGURE 9.2.8 Summary of cell-mediated immunity and humoral responses

Cytotoxic T cells

Cytotoxic T cells recognise and kill foreign (non-self), infected or abnormal host cells by releasing toxic compounds.

Cytotoxic T cells are important in combatting infections with intracellular pathogens, including all viral infections, and some bacterial and parasitic infections. Intracellular pathogens typically destroy infected cells when they replicate enough that they cause the cell to lyse. Cytotoxic T cells destroy infected cells before the pathogen replicates enough to lyse the cell and escape. This limits the progression of intracellular infections. Cytotoxic T cells also secrete cytokines, such as interferons, that alter the expression of surface proteins on other infected cells, making them easier to identify and destroy.

Cytotoxic T cells support natural killer cells to destroy cancer cells, and T_H1 cells secrete cytokines that enhance the ability of cytotoxic T cells to identify and destroy cancer cells (Figure 9.2.9).

FIGURE 9.2.9 Digital illustration of cytotoxic (white) T cells attacking migrating cancer cells (yellow)

Memory T cells

Memory T cells are produced after helper and cytotoxic T cells have been activated during an infection. Activated helper T cells and cytotoxic T cells differentiate into memory T cells that are antigen-specific. The memory T cells persist after the infection is resolved, to ensure a stronger and faster response should the same pathogen reinfect the organism.

T cell receptors

T cell receptors (TCRs) are central to the function of T lymphocytes in the adaptive immune response. TCRs are made up of two polypeptide chains. Like antibodies, TCRs have a variable and constant region (Figure 9.2.10). Unlike antibodies, which have two antigen-binding sites, TCRs have only one antigen-binding site.

Each T cell produces one type of T cell receptor (TCR) and is specific to one antigen.

Recall that BCRs bind to intact antigens that have not been processed by APCs. By comparison, TCRs bind to fragments of antigens that are displayed or presented on the surface of APCs. Receptor binding triggers signal transduction in the T lymphocyte, resulting in proliferation, cytokine release and activation of cytotoxic function.

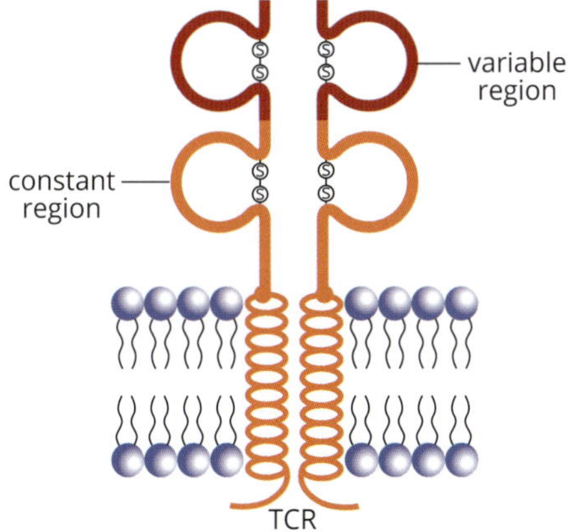

FIGURE 9.2.10 Structure of the T cell receptor (TCR), which is found on helper T and cytotoxic T cells and binds to fragments of antigen

ANTIGEN RECOGNITION BY T LYMPHOCYTES

T lymphocytes check the antigens of cells they come into contact with in the body, differentiating between cells that belong to the organism (self) and cells that are foreign (non-self). Remember that during their development, lymphocytes that react to self-antigens are normally destroyed. This inability of lymphocytes to respond to self-antigens is known as self-tolerance.

All nucleated cells have surface proteins that present peptide antigens of the proteins being synthesised in that cell. These antigens are presented to cytotoxic T cells by major histocompatibility complex I (MHC-I) molecules. Infected cells display non-self antigen fragments on MHC-I molecules, which are recognised by cytotoxic T cells that subsequently destroy the infected cells (Figure 9.2.11).

When an APC engulfs a pathogen, the antigens of the pathogen are broken into small peptides in the cell. These antigen fragments bind to MHC-II molecules inside the cell. The antigen–MHC-II complexes then move to the cell surface to present the antigens to helper T cells. The TCRs on the helper T cells recognise the antigen–MHC-II complex. Signal transduction in the T lymphocyte leads to activation of the cell, which then proliferates and releases cytokines (Figure 9.2.11).

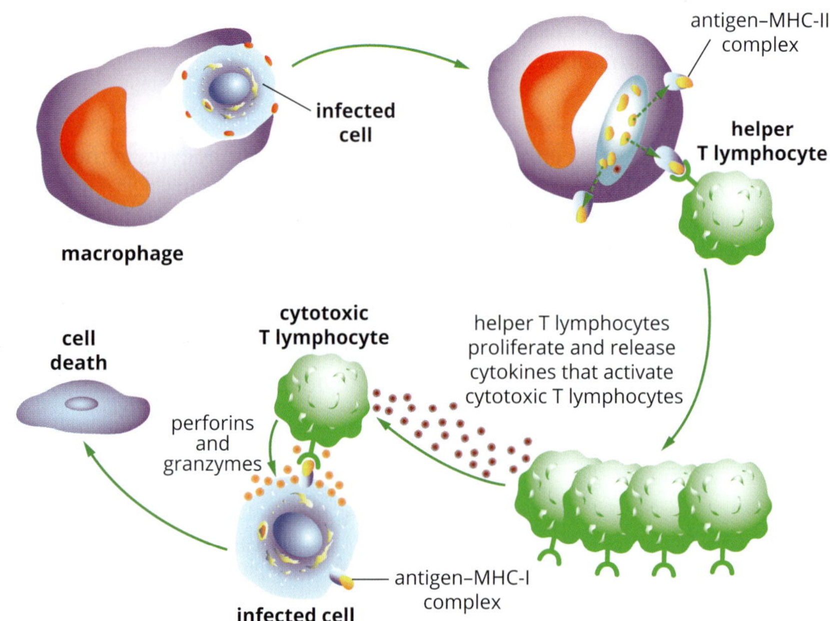

FIGURE 9.2.11 T lymphocytes are activated in cell-mediated immunity. Antigen-presenting cells present antigen fragments of phagocytosed pathogens to helper T cells using MHC-II. Infected cells present antigen fragments of pathogens to cytotoxic T cells using MHC-I.

IMMUNOLOGICAL MEMORY

The response arising from the first encounter of a T or B lymphocyte with a specific antigen is known as the **primary immune response** (Figure 9.2.12). After the initial exposure, B and T lymphocytes form B and T memory cells. IgM antibodies are the predominant antibodies produced in a primary response.

The response arising from subsequent encounters with the same antigen is known as the **secondary immune response**. Lymphocyte proliferation and production of antibodies occurs much more quickly during the secondary immune response, because the existing memory cells, which were produced during the first encounter and which remain for months or years, allow faster proliferation of the required lymphocytes (those with the receptor specific to the antigen). IgG antibodies are the predominant antibodies produced in the secondary response.

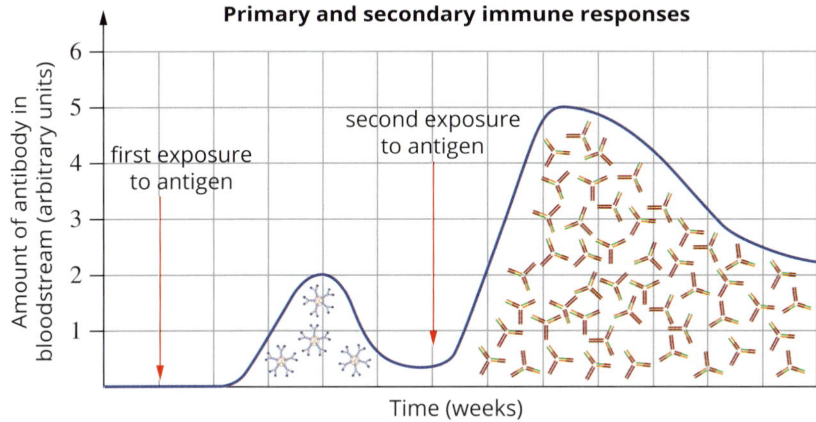

FIGURE 9.2.12 Primary and secondary immune responses after initial (first) and secondary (second) exposure to the same antigen

9.2 Review

OA ✓✓

SUMMARY

- An adaptive immune response is one that is specific to a certain antigen.
- The adaptive immune response in vertebrates is classified as humoral or cell-mediated.
- Humoral immunity involves B lymphocytes, which become activated and proliferate when stimulated by specific antigens or cytokines released by helper T cells. Activated B lymphocytes become plasma cells that produce antibodies and memory cells that remain in lymphoid tissues and provide immunological memory.
- Antibodies, also known as immunoglobulins, are proteins that bind to specific antigen molecules.
- Antibodies are Y-shaped proteins that have a constant 'tail' and variable 'arm' regions. The variable regions have antigen-binding sites and the constant region recruits components of the immune system.
- Cell-mediated immunity involves T lymphocytes:
  - Cytotoxic T cells recognise and kill foreign, infected or abnormal host cells by releasing toxic compounds.
  - Helper T cells secrete cytokines that activate leukocytes, including cytotoxic T cells and B cells.
- The major histocompatibility complex (MHC) is important in antigen presentation:
  - MHC-I is expressed on all nucleated cells and presents peptide antigens of proteins produced within cells to cytotoxic T cells.
  - MHC-II is expressed on antigen-presenting cells (APCs) and presents peptides of phagocytosed antigens to helper T lymphocytes.
- Cytotoxic T cells kill infected cells, which are identifiable by pathogen antigens on MHC-I.
- Memory B and T cells persist after an infection to enable a larger and faster response upon reinfection with the same pathogen.
- The first infection with a pathogen produces a primary immune response, while reinfection with the same pathogen produces a secondary response due to the presence of memory cells from the primary response (this is known as immunological memory).

KEY QUESTIONS

Knowledge and understanding

1 What part of an antibody interacts with antigens on a pathogen?
  A the constant region
  B the disulfide bridge
  C the constant region of the heavy chain
  D the variable region

2 Define immunological memory.

3 Describe the innate and adaptive immune responses.

4 Which MHC class interacts with helper T cells, and which interacts with cytotoxic T cells?

5 Explain how T lymphocytes recognise antigens presented by cells.

Analysis

6 Look at the graphs and determine which label on the graphs (A, B, C, D or E) represents:
  a the period when the virus has just entered the body and started to multiply (the incubation period)
  b the day the patient became infected
  c the period the patient felt most ill
  d body temperature
  e the period when the patient's antibodies destroy the virus

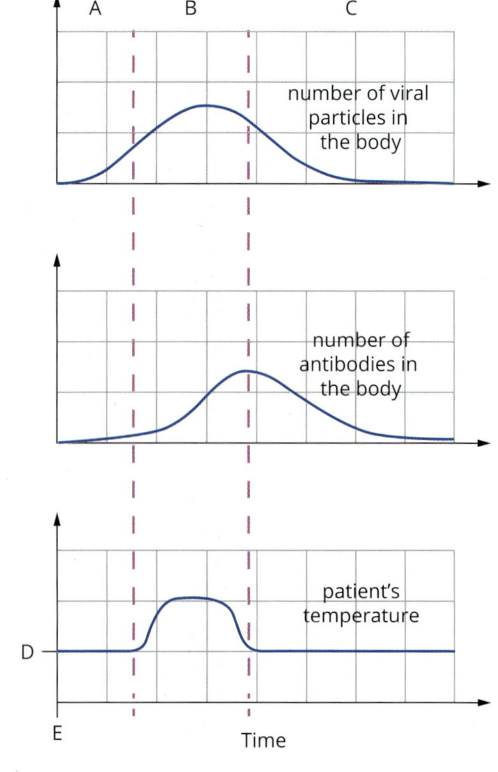

9.3 Strategies for acquiring immunity

In this section you will learn that immunity can be active or passive, natural or artificial. You will also learn that vaccination (Figure 9.3.1) is an example of artificial active immunity.

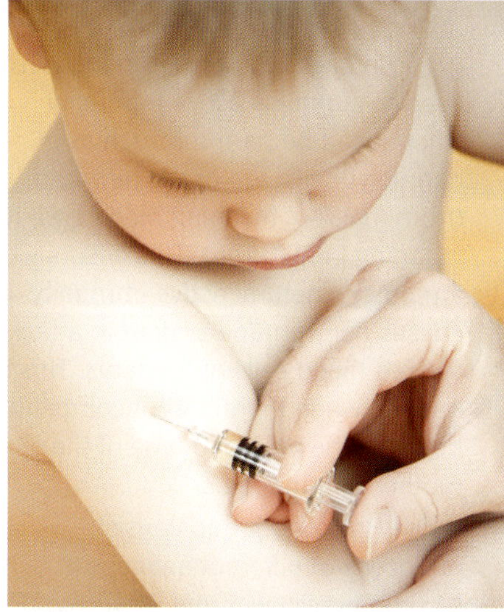

FIGURE 9.3.1 A baby being vaccinated. The vaccine stimulates an adaptive immune response that protects against infection.

TYPES OF IMMUNITY

Immunity is active or passive, depending on the origin of the immune response.

- **Active immunity** is protection provided by an individual's own adaptive immune response. This type of immunity takes time to develop, but the memory B and T cells that result can provide immunological memory that can last for many years, or even a lifetime.
- **Passive immunity** is protection provided to an individual by the transfer of antibodies produced by another organism. This type of immunity is immediate, but will only protect the recipient for a limited time because it does not result in immunological memory. The transferred antibodies degrade over time and are removed from the body.

Table 9.3.1 provides a summary of active and passive immunity.

TABLE 9.3.1 A summary of the differences between active and passive immunity

Active immunity	Passive immunity
• Adaptive immune response to antigen occurs in the individual. • The individual's immune system is activated against the antigen and achieves immunological memory. • Immunity can be maintained by stimulating memory cells, i.e. with booster vaccinations. • Immunity develops over weeks.	• Adaptive immune response occurs in another organism that is exposed to the antigen and antibodies are then transferred to a recipient. • The recipient's immune system is not activated against the antigen and does not achieve immunological memory. • Immunity cannot be maintained. • Immunity is immediate.

Immunological memory is the retention of B and T lymphocytes sensitised to specific antigens. It enables a stronger and more rapid immune response should the same antigens be encountered again.

Immunity can develop naturally through exposure to a pathogen, or be induced artificially through purposeful introduction of antigens or antibodies into the body. Both active and passive immunity can arise naturally or artificially.

Natural passive immunity

Natural passive immunity involves the passive transfer of antibodies from mother to fetus through the placenta prior to birth, and from mother to baby through breastfeeding. These maternal antibodies provide protection to the baby for weeks or months, while its own immune system is developing.

Artificial passive immunity

Artificial passive immunity involves an individual receiving antibodies produced by another organism, usually by injection of antiserum. **Serum** is the fluid portion of blood that remains after blood cells and material involved in blood clotting have been removed (Figure 9.3.2). **Antiserum** is serum that contains specific antibodies. When these transferred antibodies bind to the antigens on the pathogen or toxin, they form an antigen–antibody complex that inhibits the pathogen or toxin before it does much damage.

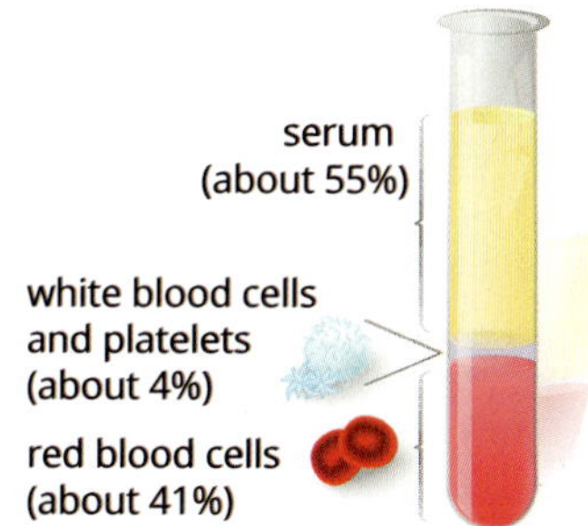

FIGURE 9.3.2 Serum is the fluid portion of blood that remains after blood cells and clotting factors (platelets) have been removed.

Antiserum is serum containing specific antibodies. It is injected to treat or protect against disease.

Antivenom is venom antiserum.

Artificial passive immunisation can be a useful means of treatment of an infection by a pathogen, or a bite or sting by a venomous animal, when death is likely to occur before the primary immune response has had time to develop. For example, the administration of tetanus antiserum protects against tetanus in at-risk patients, such as those with a deep or dirty puncture wound. The antiserum contains antibodies specific for the toxin, called antitoxins, which bind to the tetanus toxin and inhibit it.

CASE STUDY ANALYSIS

Haemolytic disease of the newborn

Artificial passive immunisation is also used to suppress active immunity when it can be harmful, such as in haemolytic disease of the newborn (from 'haemo' meaning blood and 'lysis' meaning breakdown). This occurs when a mother's natural active immunity causes her immune system to attack the red blood cells of her fetus. This can happen when there is an incompatibility between the Rh blood type of the mother and the fetus.

The Rh blood group consists of dozens of different antigens. Incompatibility with the D antigen (Rh D) is the most common cause of severe haemolytic disease of the newborn.

People who have the Rh D antigen are Rh D positive and people who lack it are Rh D negative (Figure 9.3.3). While it is still common to refer to people who are Rh D positive as being 'rhesus positive' or having the 'rhesus factor' and those without it as being 'rhesus negative', these terms only relate to Rh D antigen and are obsolete. The positive or negative suffix that follows the ABO blood group refers to the presence or absence of Rh D.

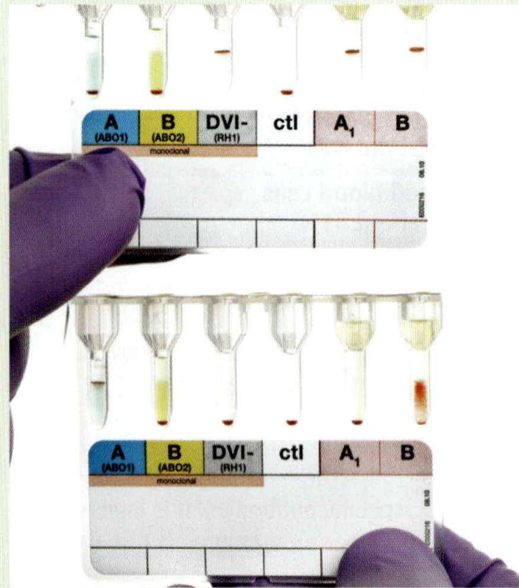

FIGURE 9.3.3 Laboratory tests showing an Rh D positive result (top), in which there is D antigen present on red blood cells (third vial from the left), and an Rh D negative result (bottom), in which there is a lack of D antigen.

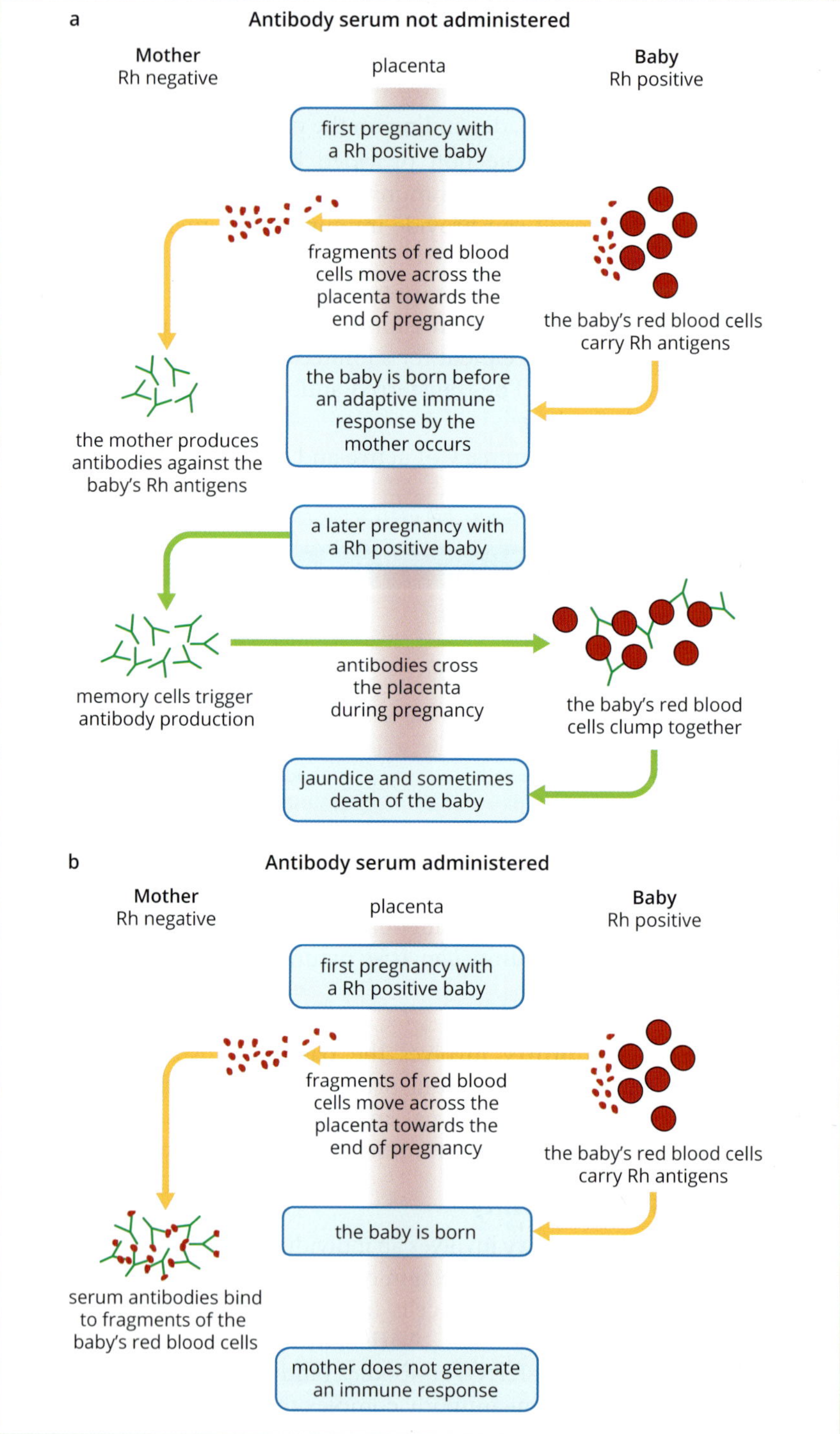

a **Antibody serum not administered**

Mother Rh negative

placenta

Baby Rh positive

first pregnancy with a Rh positive baby

fragments of red blood cells move across the placenta towards the end of pregnancy

the baby's red blood cells carry Rh antigens

the baby is born before an adaptive immune response by the mother occurs

the mother produces antibodies against the baby's Rh antigens

a later pregnancy with a Rh positive baby

memory cells trigger antibody production

antibodies cross the placenta during pregnancy

the baby's red blood cells clump together

jaundice and sometimes death of the baby

b **Antibody serum administered**

Mother Rh negative

placenta

Baby Rh positive

first pregnancy with a Rh positive baby

fragments of red blood cells move across the placenta towards the end of pregnancy

the baby's red blood cells carry Rh antigens

the baby is born

serum antibodies bind to fragments of the baby's red blood cells

mother does not generate an immune response

FIGURE 9.3.4 A mother who is Rh negative has a baby who is Rh positive. (a) When red blood cell fragments cross the placenta, the mother has an adaptive immune response that makes memory cells and antibodies against the specific Rh antigen (called anti-Rh antibodies). In a later pregnancy with another Rh positive fetus, memory cells trigger the production of antibodies that can cross the placenta and will damage any subsequent Rh positive fetuses. (b) A dose of anti-Rh antibodies can be administered to the mother to neutralise any fetal Rh antigens before an immune response occurs, protecting any future Rh positive fetus.

However, introducing antibodies to contain the threat before the person's own adaptive immune response can be mobilised means the protection provided is only temporary, as no immunological memory is formed.

Natural active immunity

Natural active immunity develops from the adaptive immune response to a natural infection, and the immunological memory that results. This means that if exposed to the same antigen again in the future, the immune system will recognise it immediately, and a secondary immune response will occur (Figure 9.3.5).

Secondary immune responses are much faster and stronger than primary immune responses, and are therefore more likely to minimise disease. For example, if you have had chickenpox, you are unlikely to get it again because your immune system has developed immunological memory specific to the antigens of *Varicella zoster* virus, the virus that causes chickenpox.

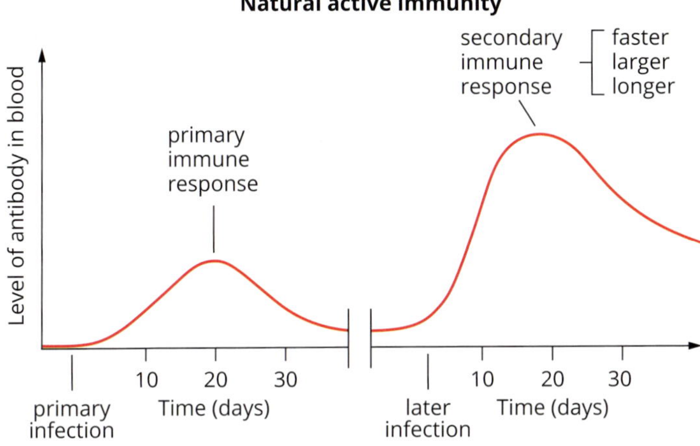

FIGURE 9.3.5 A subsequent infection with the same infectious agent will trigger a secondary immune response that is faster and stronger than the primary immune response.

When a mother is Rh D negative and has a baby who is Rh D positive, the mother is likely to develop an adaptive immune response to the Rh D antigen, as fragments of fetal red blood cells cross the placenta during birth (Figure 9.3.4a). If the same Rh D negative woman has another pregnancy with an Rh D positive baby, her memory cells will trigger the production of antibodies that cross the placenta and damage the baby's blood cells, causing them to develop haemolytic disease. Incompatibility with other Rh antigens can also cause haemolytic disease. Haemolytic disease of the newborn can be prevented by artificial passive immunisation.

To prevent the mother having an adaptive immune response and producing anti-Rh antibodies during her first pregnancy, she is given a dose of anti-Rh antibodies (Figure 9.3.4b). These administered antibodies will neutralise any fetal Rh antigens before an adaptive immune response by the mother occurs.

Analysis

Occasionally, if Rh D negative blood is in short supply, a patient who is Rh D negative can receive a transfusion of Rh D positive blood.

1 This procedure can only be done once. Why?

2 This procedure would not be recommended for a young female patient. Why not?

Vaccines are made of altered, weakened or killed microorganisms, such as bacteria or viruses, or inactivated forms of toxins or proteins.

Artificial active immunity

Artificial active immunity results from the administration of antigens to induce an adaptive immune response. This technique of inducing an adaptive immune response to produce active immunity is known as **vaccination** (or immunisation), and a substance used to induce artificial active immunity is called a **vaccine**.

As with natural active immunity, the primary response to vaccination takes time to develop, and booster vaccines are often needed to stimulate the stronger secondary immune response that provides longer-lasting immunity (Figure 9.3.6).

Vaccines need to be highly specific to initiate an adaptive immune response resulting in immunological memory. Increased understanding of microbiology and immunology has led to the development of very safe vaccines that induce the desired immune response with minimal side effects.

Table 9.3.2 provides a summary of examples of the different types of immunity.

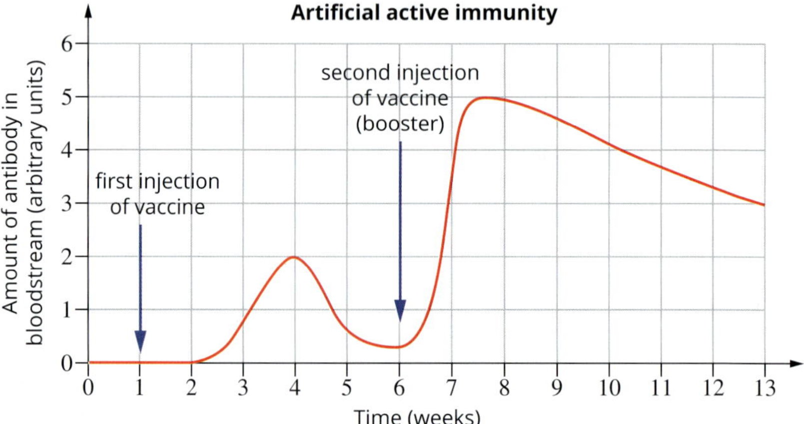

FIGURE 9.3.6 A booster vaccine is often needed to stimulate a stronger secondary immune response.

TABLE 9.3.2 Examples of types of immunity

	Active	Passive
Natural	the adaptive immune response of an individual	the passive transfer of antibodies from mother to fetus through the placenta prior to birth, and from mother to baby through breastfeeding
Artificial	vaccination of an individual to stimulate an adaptive immune response	administration of antibodies that have been produced by another organism against a pathogen or toxin

CASE STUDY **ANALYSIS**

Influenza vaccines

Because of their high rate of mutation, influenza (or flu) viruses evolve so rapidly that the immunity developed to one year's strains usually doesn't provide protection against the next year's strains. This is true whether you have caught the flu and developed natural active immunity, or have received a flu vaccine and developed artificial active immunity. The reason the immune system has trouble recognising new strains of influenza viruses is that genetic changes have caused the flu viruses to express different antigens. This change in antigens is called antigenic drift.

To keep up with antigenic drift, new flu vaccines are released every year. Although it is a single injection, each new flu vaccine is a cocktail of vaccines for different influenza strains, and the decision about which strains to vaccinate against is based on which strains are predicted to be the most common in the coming flu season. However, sometimes the most common strains are unexpected, in which case the latest flu vaccine provides little protection.

One reason a new influenza strain might be unexpected is that influenza viruses can swap genetic material in a single host. This is known as antigenic shift, because it occurs more rapidly and results in more major changes than antigenic drift. An example of antigenic shift is the H1N1 strain of swine flu virus that was first detected in 2009 (Figure 9.3.7). The reassortment of genes that resulted in H1N1 is thought to have occurred in North American and Eurasian pig herds. The eight RNA strands of H1N1 include one strand from a human influenza strain, two strands from bird (or avian) influenza strains and five strands from pig (or swine) influenza strains.

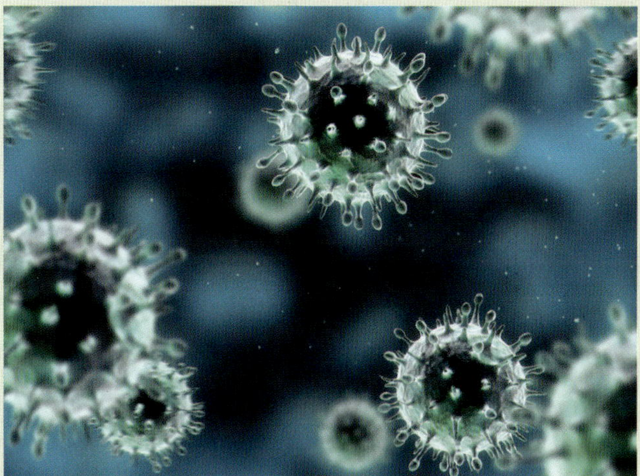

FIGURE 9.3.7 Digital illustration of the H1N1 strain of swine influenza virus

Analysis

1 In a particular community, avian influenza strain H3N2 and a human strain H1N1 have been circulating. Without any warning, there is a sudden increase in people arriving at hospital emergency departments with influenza. The strain is analysed and found to be a new strain, H3N1. This new strain is likely to have arisen as a result of:
   - **A** lack of people in the population having vaccination
   - **B** antigenic shift
   - **C** antigenic drift
   - **D** the development of antibiotic resistance by the influenza virus

2 Consider the graph below.
   - **a** Why is the 2013 to 2016 data effective for predicting the future 2017 notifications?
   - **b** What can be predicted about the number of notifications in 2017?

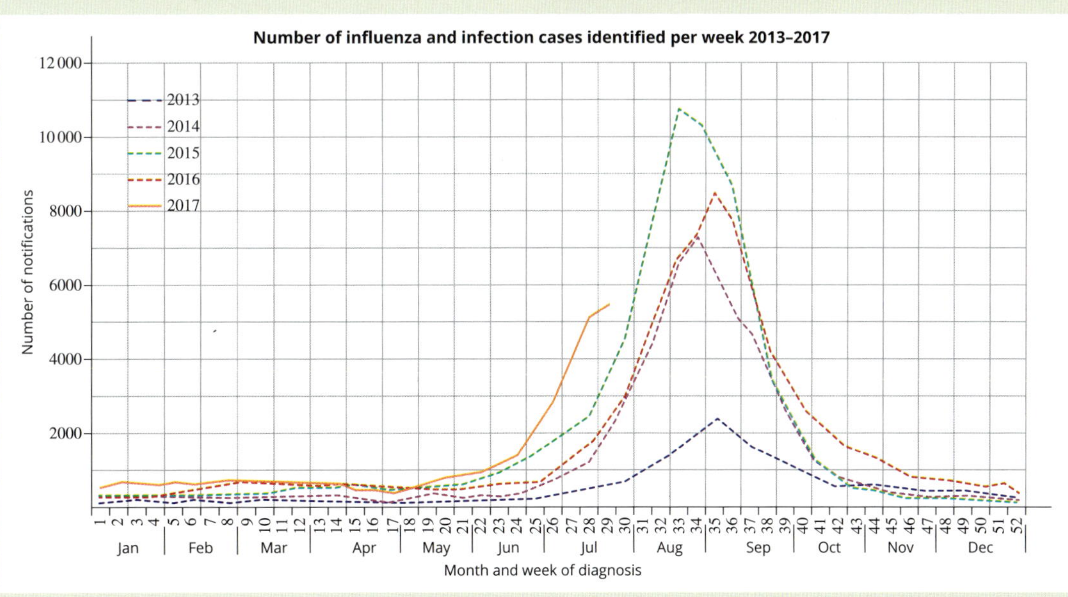

9.3 Review

SUMMARY

- Immunity can develop naturally or be induced artificially.
- Passive immunity involves the transfer of antibodies produced in another organism. It does not result in immunological memory and is temporary.
- Active immunity involves the individual's adaptive immune response. It results in immunological memory that can be long-lasting.
- Natural passive immunity is the result of antibodies naturally produced by another organism providing immunity.
- Artificial passive immunity involves an individual receiving antibodies produced by another organism.
- Natural active immunity develops from the adaptive immune response to a natural infection, and the immunological memory that results.
- Artificial active immunity results from the administration of antigens to induce an adaptive immune response, i.e. vaccination. This results in the generation of immunological memory.
- Immunological memory provides a stronger and faster secondary immune response if exposed to the same antigen again.

KEY QUESTIONS

Knowledge and understanding

1 Vaccination is an example of:
  A artificial passive immunity
  B natural active immunity
  C artificial active immunity
  D natural passive immunity

2 Define vaccination.

3 Explain the difference between active and passive immunity.

4 Explain the difference between natural passive immunity and artificial passive immunity. Give examples.

Analysis

5 A group of scientists think that they have developed a vaccine that would protect people at risk from developing allergic responses to peanuts. Design an experiment using mice that have a genetic tendency to peanut allergy that would allow the scientists to test their new drug. State the results that would show that the vaccine is a success.

Chapter review

09

KEY TERMS

active immunity
adaptive immune response
agglutination
antibody
antigen–antibody complex
antigen-presenting cell (APC)
antiserum
artificial active immunity
artificial passive immunity
B cell receptor (BCR)
B lymphocyte
cell-mediated immunity
clonal selection
constant region
cytotoxic T cell
differentiation
heavy chain
helper T cell
humoral immunity
immunoglobulin (Ig)
immunological memory
light chain
lymph
lymphatic system
lymphocyte
memory B cell
memory T cell
natural active immunity
natural passive immunity
neutralisation
passive immunity
plasma cell
precipitation
primary immune response
primary lymphoid organ and tissue
secondary immune response
secondary lymphoid organ and tissue
serum
specificity
T cell receptor (TCR)
T lymphocyte
vaccination
vaccine
variable region

REVIEW QUESTIONS

Knowledge and understanding

1 Pregnant women who are Rh D negative but who are carrying an Rh D positive fetus are injected with anti-Rh D antibodies shortly after birth. These injections make the mothers temporarily immune to Rh D antigens and prevent the formation of maternal anti-Rh D memory cells. This protects a future Rh D positive fetus from a maternal adaptive immune response. What type of immunity is this?

A artificial passive immunity
B artificial active immunity
C natural passive immunity
D natural active immunity

2 Which type of antibody is found in the highest concentration in the blood of a newborn baby?

A IgA
B IgG
C IgE
D IgD

3 Why can vertebrates respond to a secondary infection with significantly greater potency than the first infection?

A Vertebrates have innate immune responses that provide immunological memory.
B Vertebrates have adaptive immune responses that provide immunological memory.
C Vertebrates have first line defences/barriers that provide immunological memory.
D Vertebrates have innate resistance that provides immunological memory.

4 Which immune cells are involved only in adaptive immune responses?

A B and T lymphocytes
B complement proteins
C leukocytes
D antigen-presenting cells

5 Which of the following is not a role of lymphatic vessels?

A returning fluid that seeps out of the blood vessels into tissues back to the circulatory system
B absorbing and transporting fatty acids and fats from the digestive system
C providing a site for lymphocytes to mature
D transporting lymphocytes and antigen-presenting cells to the lymph nodes

6 Breastfed babies tend to be healthier than bottle-fed babies. Give a reason why.

7 Explain the specificity of the adaptive immune response.

8 Where do T lymphocytes mature?

9 Mammalian antibodies (or immunoglobulins) are generally grouped into five types. Draw a table that lists and summarises the role of each type.

10 Why doesn't the adaptive immune system respond to prions?

11 Describe the humoral mechanism of the adaptive immune response, including its relationship, if any, to immunological memory.

12 Describe the cell-mediated mechanism of the adaptive immune response, including its relationship, if any, to immunological memory.

13 *Pseudomonas aeruginosa* is an extracellular pathogen. Would *P. aeruginosa* trigger a T_H1 or T_H2 response?

14 *Legionella pneumophila* is an intracellular pathogen. Would *L. pneumophila* trigger a T_H1 or T_H2 response?

15 Describe how the recognition of antigens by B and T cell receptors differs.

Application and analysis

16 The following graph represents changes in antibody concentrations that occur during a primary and secondary adaptive immune response. Provide labels to describe A, B, C and D.

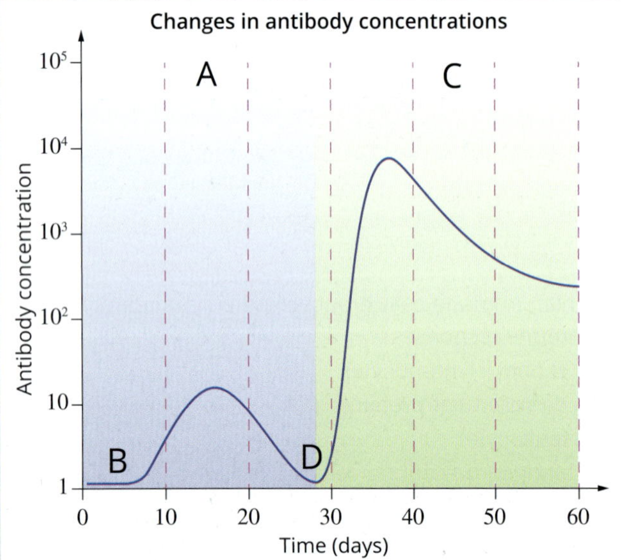

17 Sometimes the blood bank in Melbourne will advertise for people who have recently recovered from chickenpox to donate blood. The blood bank takes a blood donation from these individuals and then separates the blood using a centrifuge. The blood serum is collected. The serum is then purified and some of the proteins specific to chickenpox are extracted.

- **a** What are these proteins?
- **b** These proteins are given to patients. Which group of patients is most likely to be in need of these proteins?
- **c** Why does the extraction of these proteins not increase the likelihood of the donors developing chickenpox with possible future exposure?
- **d** Explain whether the injection of these proteins gives the recipients long-term immunity.

18 Vaccinations against measles are included in Australia's National Immunisation Program. As part of this program, it is recommended that children receive measles vaccines at 12 months, 18 months and 4 years of age.

The following graph shows the levels of measles antibodies in a child from birth to seven years of age.

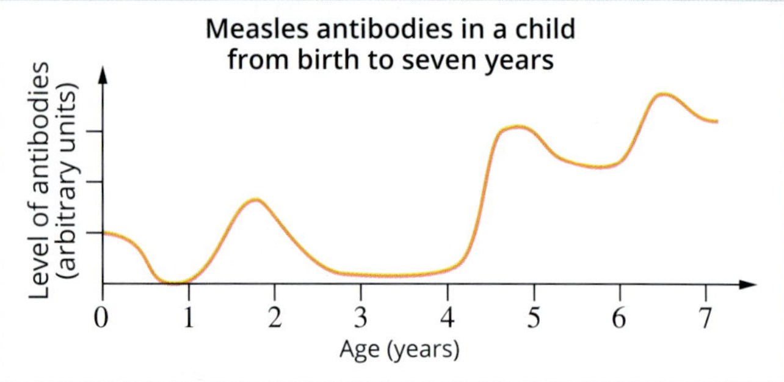

- **a** This child had antibodies against the measles virus at birth. Explain how.
- **b** Why did the antibody levels drop off to zero in the months following birth?
- **c** The child was immunised against measles at one year of age and again at four years of age. Explain why antibody production occurs more rapidly and to a higher level after the second vaccination compared with the first vaccination.
- **d** At six years of age, antibody production increases greatly again, but no vaccination has occurred. Explain.
- **e** What could explain an increase in cases of measles among children?

19 The following graph shows the level of different antibodies before and after birth.

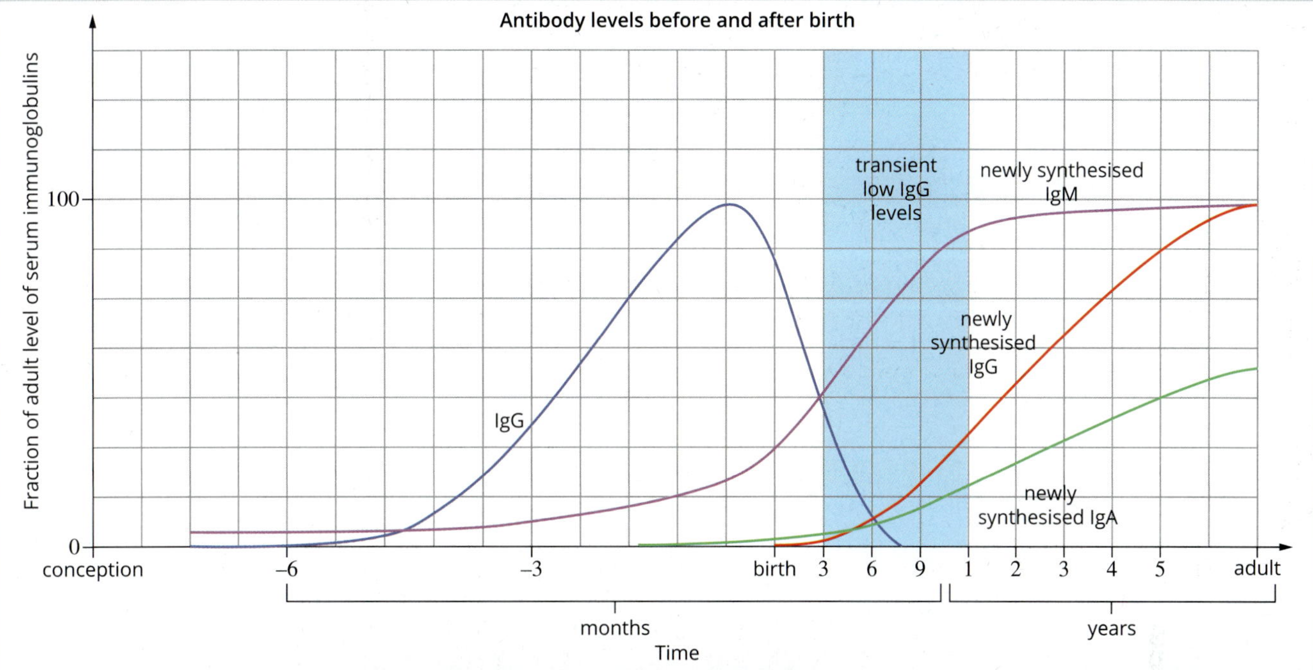

a Why is IgG present in high levels before birth?

b What are the trends from birth for the two sources of serum IgG?

20 An experiment was performed to determine the effectiveness of a vaccine against rabies in stimulating the production of antibodies in cattle. Two groups of cattle, A and B, were assessed. Both groups were first exposed to the virus on day 0, then only one of the groups was exposed again on day 120. Antibody levels were measured on days 0, 30, 210, 390, 450 and 540. The results are shown in the following graph.

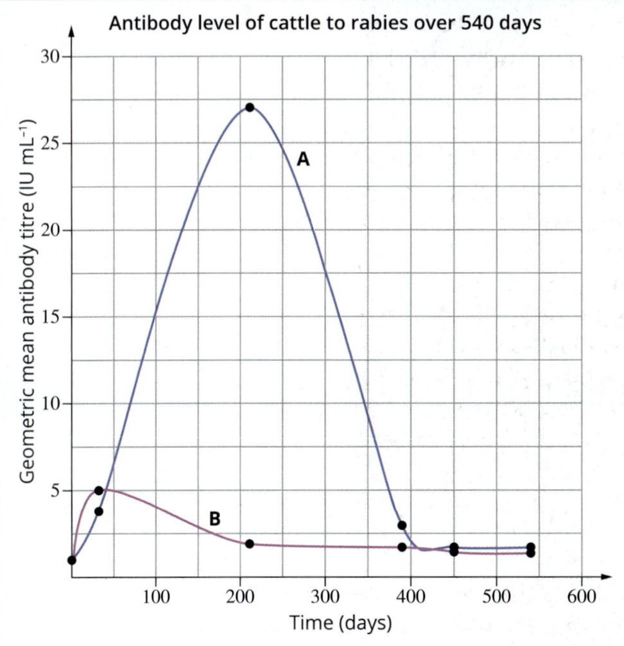

a Deduce which result, A or B, represents the group that was exposed only once. Explain your response.

b i Explain why the interpolation of the antibody levels shown on the graph for group A between days 30 and 210 is unreliable.

ii Predict what a scientist who measured the antibody titre of group A would measure if they sampled at day 100.

c Along with increased production of antibodies, T lymphocyte concentration would be expected to rise significantly, especially in the group having two exposures to the pathogen. Explain.

OA ✓✓

CHAPTER

10 Disease challenges and strategies

Learning outcomes

By the end of this chapter, you will understand how new treatments that utilise the immune system are being used to treat cancer and autoimmune diseases. You will also learn about the impact of pathogens in a globally connected world, the strategies that can be employed to control the spread of disease and the importance of vaccination programs in maintaining herd immunity.

Key knowledge

- the emergence of new pathogens and re-emergence of known pathogens in a globally connected world, including the impact of European arrival on Aboriginal and Torres Strait Islander peoples **10.2**
- scientific and social strategies employed to identify and control the spread of pathogens, including identification of the pathogen and host, modes of transmission and measures to control transmission **10.2**
- vaccination programs and their role in maintaining herd immunity for a specific disease in a human population **10.3**
- the development of immunotherapy strategies, including the use of monoclonal antibodies for the treatment of autoimmune diseases and cancer. **10.1**

OA

10.1 Immunotherapy

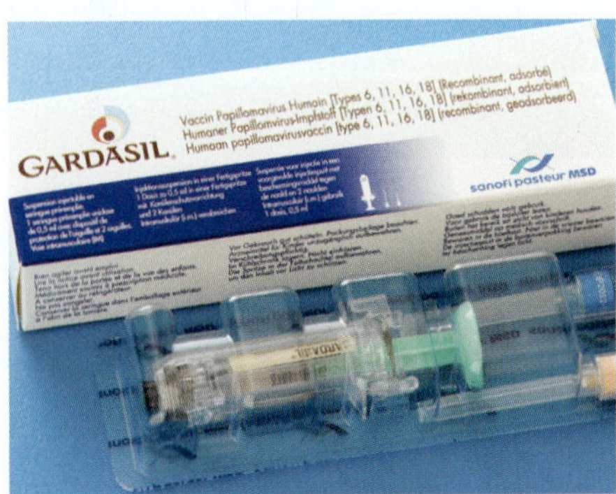

FIGURE 10.1.1 Gardasil is the brand name for a vaccine that provides immunity against certain strains of human papillomavirus (HPV). The vaccine reduces the risk of developing certain types of cancers associated with HPV infection.

> Immunotherapy activates or suppresses the immune system to treat disease.

> Carcinogens are substances that cause damage to cell DNA.

Immunotherapy is any treatment that activates or suppresses the immune system to fight disease. In this section, you will learn about different types of immunotherapies used to treat cancer, including monoclonal antibody therapy, and about immunotherapies for autoimmune conditions. You will also learn about cancer and vaccines that prevent certain types of cancer (Figure 10.1.1).

CANCER

There are many different types of cancer that affect many different types of cells, but in all cases cancer results from a single abnormal cell that multiplies uncontrollably and spreads throughout the body.

This uncontrolled growth is the result of changes to genes that control how cells grow and divide, and a resistance of these abnormal cells to apoptosis (programmed cell death). These genetic changes can be inherited, or develop as a result of damage to the cell's DNA over a lifetime.

Substances that can damage DNA are called **carcinogens**. Carcinogens can be physical (e.g. radiation), chemical (e.g. asbestos), or biological (e.g. certain viruses; approximately 15–20% of human cancers are thought to be the result of viruses).

Tumours

A **tumour** forms when the number of abnormal cells has increased significantly, forming a clump of cells. Depending on where the tumours form, they may damage or block the normal function of organs and tissues. Not all tumours are cancerous.

Benign tumours are not cancerous, because their abnormal cells do not invade nearby tissue or spread throughout the body. Some benign tumours can become cancerous.

Malignant tumours are cancerous because their cells can invade nearby tissue and spread (or metastasise) from the site at which they originate.

Metastasis occurs when cancer cells break away from the original tumour (or primary tumour), travel through the blood and lymph vessels, and form secondary tumours at other locations (Figure 10.1.2).

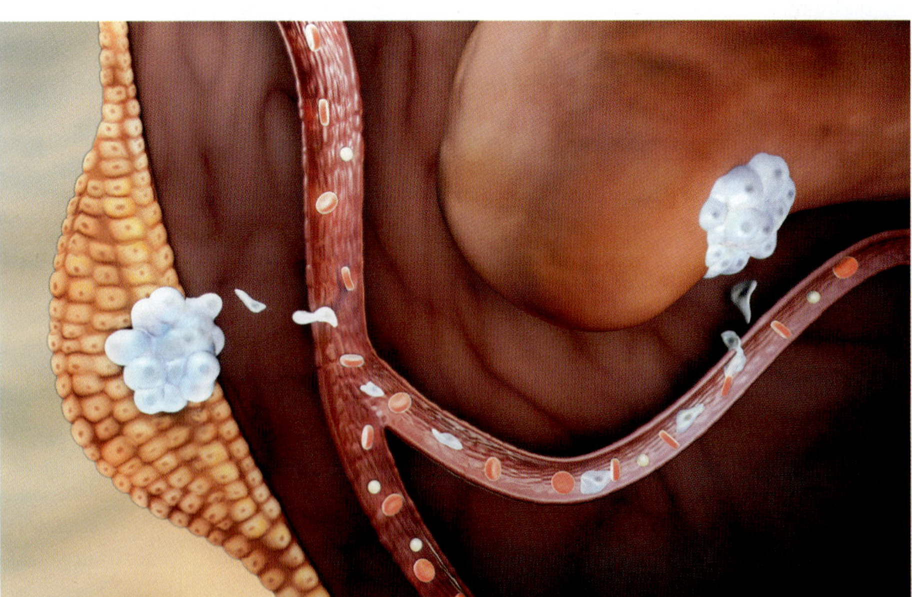

FIGURE 10.1.2 Cancer cells (white) migrating from a tumour and forming a secondary tumour in another organ

Cancer treatments

Cancer treatments over several decades have included chemotherapy, radiation therapy, and surgery to remove tumours.

- Chemotherapy involves administering drugs that are **cytotoxic** to cells that multiply rapidly. Although the chemotherapy drugs available today are more specific than those of the past, they are not specific enough to avoid damage to healthy cells that also divide rapidly, such as bone marrow and hair follicle cells. This is why chemotherapy can have negative side effects.
- Radiation therapy indiscriminately kills cells by damaging their DNA. In cancer treatment, radiotherapy is directed at the cancerous cells, but some damage to surrounding tissue is inevitable.
- Surgery is beneficial in removing solid tumours but it also has its disadvantages. Any surgery takes a toll on the body, and often it is very difficult to ensure that all malignant cells are removed from the body.

The immune response to cancer

Tumour cells evade the immune response in a number of ways. They do this by expressing defective MHC-I molecules (so that cytotoxic T cells cannot detect that the cells are defective), by producing immunosuppressive cytokines, or by releasing enzymes that suppress T lymphocyte responses. Because of this, even people with a normal immune system may be unable to control the growth of tumours. People who are older, who use immunosuppressive medications for a long period of time, or who have immunodeficiency have weakened immune systems and are at increased risk of developing cancer.

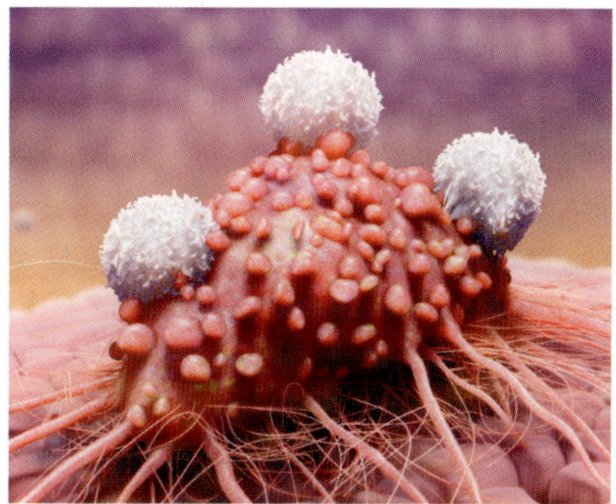

FIGURE 10.1.3 Digital illustration of cytotoxic T cells (white) attacking a cancer cell (red)

CANCER IMMUNOTHERAPY

Immunotherapy is a new frontier in the treatment of cancers. It enables more specific (or personalised) medicines than other types of treatments, and improves outcomes while simultaneously minimising side effects.

Cancer immunotherapy activates the immune system to destroy tumours, and can be non-specific or specific.

- Non-specific immunotherapies stimulate the immune system in general; for example, by the injection of cytokines. Cytokines do not directly target cancer cells, but the stimulation of the immune system can result in a better immune response against cancer cells.
- Specific immunotherapies act on cancer cells by directly stimulating the adaptive immune response against them (Figure 10.1.3). Specific immunotherapies include cancer vaccines and monoclonal antibody therapy.

Cytokines are a group of peptides and proteins released from cells that are important in cell signalling, particularly between cells of the immune system.

Cancer vaccines

Cancer vaccines stimulate the immune system to attack cancer cells. Some cancer vaccines contain peptides or whole proteins of cancer cells and adjuvants to help stimulate an immune response against them. **Adjuvants** are substances that enhance the effect of a vaccine or other medical treatment. Sometimes a patient's own immune cells are harvested, exposed to these antigens, and then injected back into the body to produce an immune response. Cancer vaccines have few or no side effects. They can be classified as preventive, therapeutic or personalised.

Preventive cancer vaccines

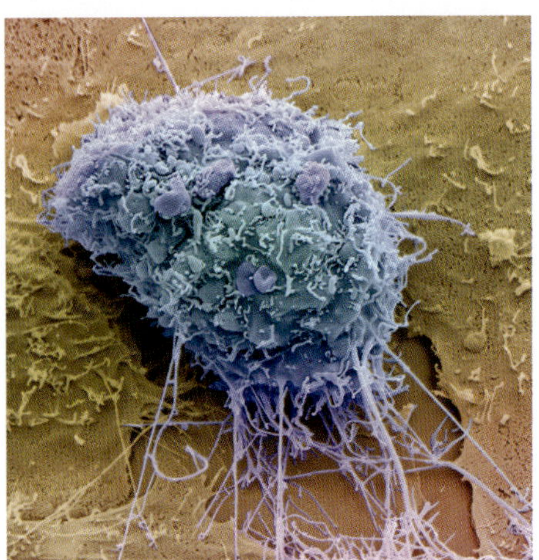

FIGURE 10.1.4 Coloured scanning electron (SEM) micrograph of a cervical cancer cell

Preventative cancer vaccines are vaccines directed against viruses that cause cancer. Examples include vaccines for human papillomavirus (HPV), which causes cervical cancer (Figure 10.1.4), and hepatitis B virus (HBV), which causes cancer of the liver. These vaccines introduce specific viral antigens into the body, creating an adaptive immune response that will lead to immunological memory and a stronger and faster response towards the virus if the body is exposed to it.

An example of a preventative cancer vaccine is the HPV vaccine called Gardasil. Professor Ian Frazer and Dr Jian Zhou from the University of Queensland developed Gardasil using a gene for the coat protein of HPV (Figure 10.1.5).

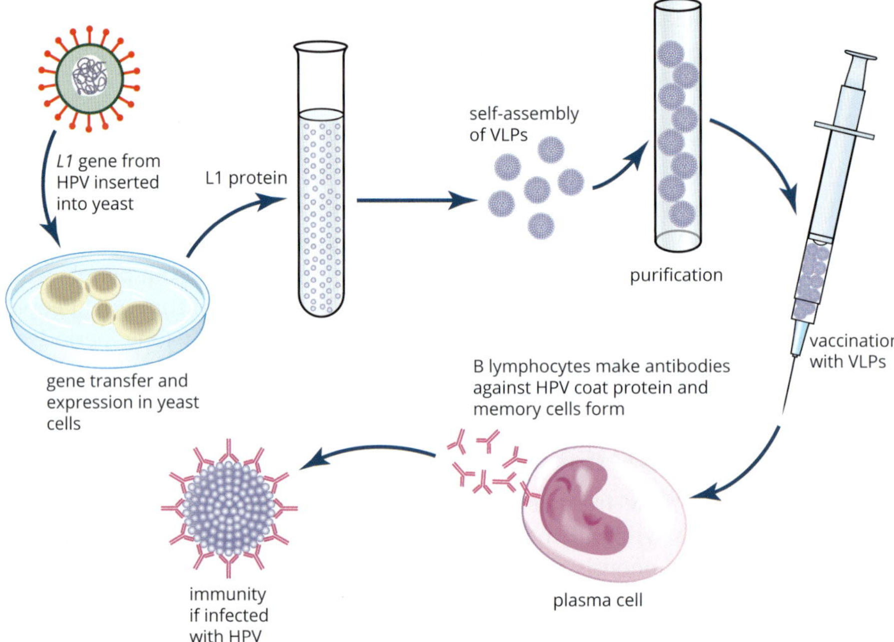

FIGURE 10.1.5 The gene for the HPV coat protein L1 was taken from the virus and inserted into yeast cells, which produce large amounts of the protein. This L1 protein 'self assembles' into particles that look like the virus, and are therefore known as virus-like particles (VLPs). These VLPs do not contain any viral DNA, so they do not cause disease and are not infectious. The VLPs are the antigen used in the Gardasil vaccine.

Therapeutic cancer vaccines

Therapeutic cancer vaccines are given to people who already have cancer. These vaccines are made up of antigens for a specific type of cancer cell (usually proteins or parts of proteins), and often adjuvants are included to help boost the immune response, increasing its ability to identify and destroy cancer cells.

Personalised cancer vaccines

Personalised cancer vaccines are therapeutic vaccines developed for an individual patient. Some involve tumour cells that have been removed from the patient, altered in the laboratory to make them more obvious to the immune system, and then injected back into the patient.

One of the most effective personalised cancer vaccines involves the patient's own tumour and **dendritic cells**, which are antigen-presenting cells of the immune system. This therapy works by presenting the antigen isolated from the patient's tumour sample to the patient's dendritic cells. The activated dendritic cells are then injected back into the patient, where they present antigens to helper T cells, eliciting an adaptive immune response.

> Antigen-presenting cells present foreign antigens attached to MHC-II molecules on their surface to helper T cells.

CASE STUDY

A cancer vaccine for Tasmanian devils

Tasmanian devils (*Sarcophilus harrisii*) have been plagued for decades by a contagious facial cancer called devil facial tumour disease (DFTD). The cancer originated from a single cancerous Schwann cell in a single Tasmanian devil many years ago, and spread from one Tasmanian devil to another by bites. DFTD has devastated the Tasmanian devil population and threatens the survival of the species.

Contagious cancer is very unusual, and at first it was thought the reason DFTD spread so easily between Tasmanian devils was that they had weakened immune systems, or low genetic diversity. Low genetic diversity would make it easy for the cancer to spread, because it would mean the immune system of an individual devil wouldn't recognise the cancer cells from another devil as foreign. However, both these hypotheses have been ruled out.

In fact, it appears the cancer cells evade the devil's immune system by destroying their major histocompatibility complex (MHC) molecules, which are vital for immune recognition. Without MHC molecules on the cancer cells, the devil's immune cells do not detect them.

Researchers from the University of Tasmania have shown that a vaccine using killed DFTD tumour cells and adjuvant is able to stimulate a protective adaptive immune response (Figure 10.1.6). The vaccine is currently being trialled, and if successful, it may help bring the Tasmanian devil back from the brink of extinction.

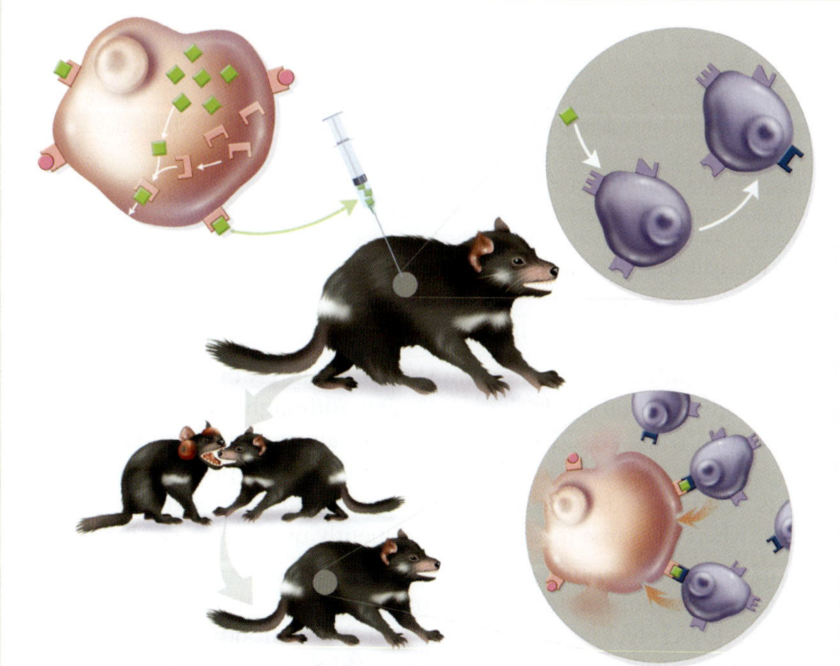

FIGURE 10.1.6 The cancer vaccine for devil facial tumour disease uses tumour-specific antigens from killed tumour cells to stimulate an adaptive immune response involving T lymphocytes.

Monoclonal antibody therapy for cancer treatment

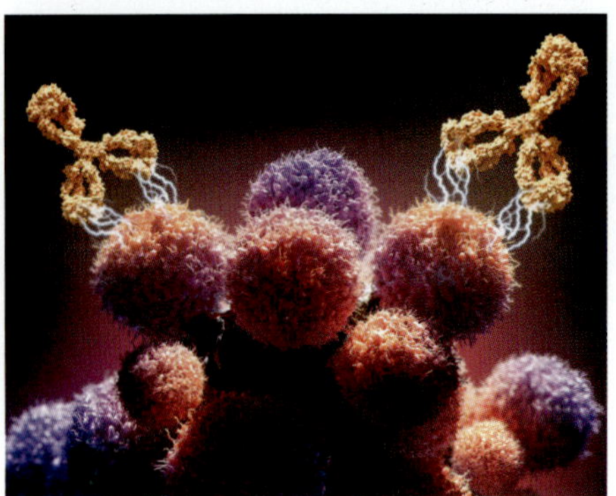

FIGURE 10.1.7 Digital illustration of monoclonal antibodies binding to antigens on cancer cells

Immortal cell lines can continually undergo division without mutation, which would normally occur as a cell ages, and can therefore be cultured for long periods.

Monoclonal antibodies (mAbs) are antibodies produced by a single clone of a B lymphocyte that is grown in cell culture to produce a large volume of the same clone. The mAbs produced by the clones are all identical and specific to the same antigen.

One of the ways mAbs are used to treat cancer is by targeting specific antigens present on tumour cells (Figure 10.1.7). But they can also be used to target cells of the immune system and direct the immune response in a way that helps destroy tumour cells.

Production of monoclonal antibodies

Figure 10.1.8 illustrates the steps required to make monoclonal antibodies.

- First, mice are injected with a particular antigen, which in the case of cancer therapy is an antigen from a cancer cell.
- This induces the mice's B lymphocytes to produce antibodies specific to the antigen, and these B lymphocytes are then isolated from the spleens of the mice.
- In isolation the B lymphocytes only have a limited lifespan, so in order to produce the large quantity of antibodies needed, the isolated B lymphocytes are fused with **myeloma cells**, which are an **immortal cell line**.
- The fusion of the two cells results in a hybridised cell called a **hybridoma**.
- The hybridoma is more stable in tissue culture conditions and the cell secretes multiple copies of the specific antibody (the mAbs), which are then harvested.

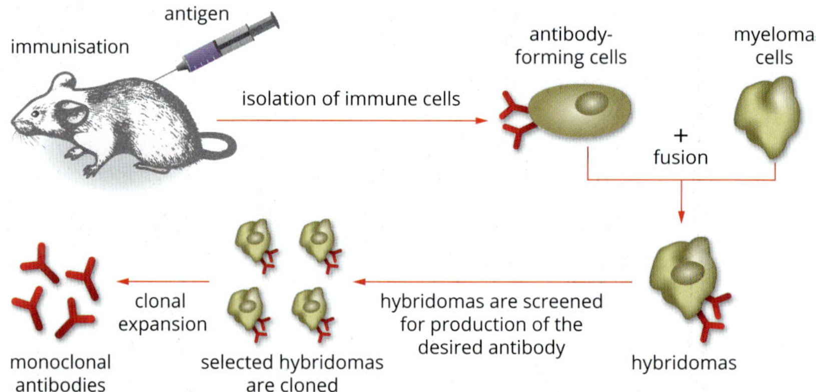

FIGURE 10.1.8 Production of monoclonal antibodies. Mice are injected with an antigen. B lymphocytes sensitised to the antigen are then taken from the mice and fused with myeloma cells. The fused cells, called hybridomas, make antibodies to the antigen, and are grown in culture dishes to produce large quantities of monoclonal antibody specific to the antigen.

Humanised monoclonal antibodies

The first mAbs produced were mouse mAbs made entirely by mouse B lymphocytes, and many mAbs are still made this way today. Although these types of mAbs are initially effective when used in human therapy, an immune response is mounted against them once they are identified as foreign (mouse) proteins. Immunological memory is formed, and the adaptive immune response recognises and destroys them faster when the same mAbs are subsequently used again.

To help prevent an immune response directed against them, researchers have replaced some components of mouse antibodies with human components using recombinant DNA techniques (Figure 10.1.9). Antibodies with a mixture of mouse and human components are known as **chimeric monoclonal antibodies (chimeric mAbs)**. As mAbs contain more and more human components, they are termed **humanised monoclonal antibodies (humanised mAbs)** (Figure 10.1.10a). Some mAbs are now fully human antibodies produced by transgenic mice. Although chimeric, humanised and human mAbs are all still produced by mice, they may be safer and potentially more effective than earlier mAbs. Antibodies that contain only human components are known as **human monoclonal antibodies (human mAbs)** (Figure 10.1.10b).

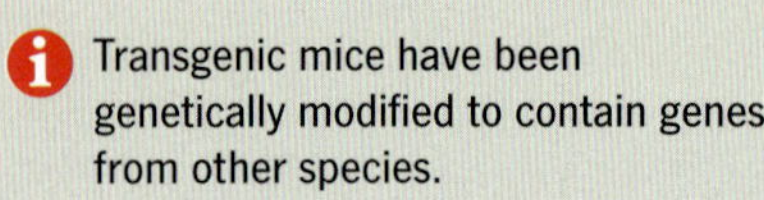
Transgenic mice have been genetically modified to contain genes from other species.

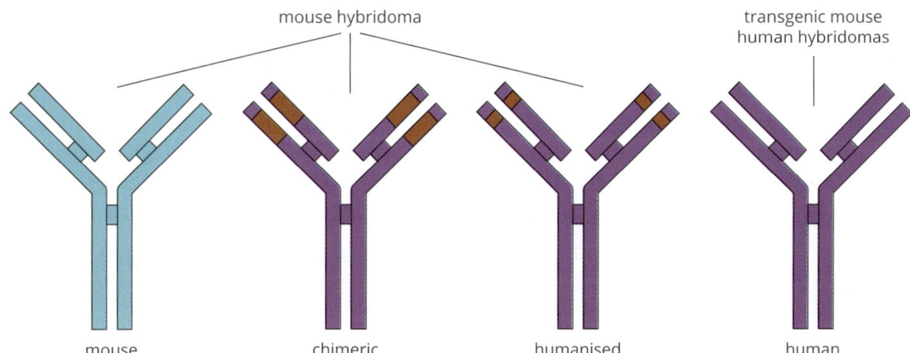

FIGURE 10.1.9 Monoclonal antibodies can be categorised into four types based on their protein composition: animal (most commonly mouse), chimeric (combination of mouse and human), humanised (mostly human) and human.

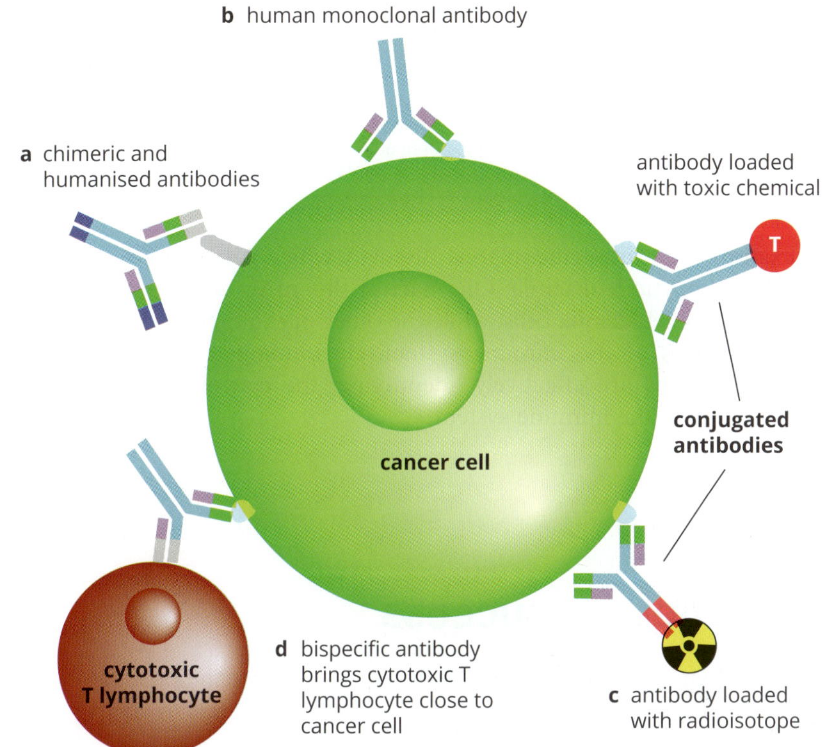

FIGURE 10.1.10 (a) Chimeric antibodies contain parts from mouse and human antibodies, and humanised antibodies contain mostly human parts, but are still produced in mice. (b) Human monoclonal antibodies contain only human components. (c) Conjugated antibodies are joined to toxins or radioisotopes to specifically target and kill cancer cells and spare normal body cells. (d) Bispecific antibodies are those engineered to bind two different antigens, in order to bring cytotoxic T cells close to tumour cells.

Conjugated monoclonal antibodies

Conjugated monoclonal antibodies (conjugated mAbs) are mAbs that have been attached to a chemotherapy drug, a toxin or a **radioactive** particle (Figure 10.1.10c). In this way, conjugated mAbs are used as carriers modified to deliver treatments directly to the specific cancer cells. For example, radioimmunotherapy can be used to treat pancreatic cancer. A radioactive isotope (lead-212) is combined with a specific antibody capable of targeting cancerous cells. This combination of antibody and lead-212 radioisotope is injected intravenously into the body. When it reaches the pancreas, it locks onto the cancerous cells' antigens and the lead-212 destroys the cells by irradiating them. This treatment limits the toxic effects on healthy cells to those located near the cancerous cells.

BIOFILE

Pembrolizumab

Pembrolizumab is a humanised monoclonal antibody approved for use in Australia to treat metastatic melanoma (see figure below). It works by binding to a receptor on T lymphocytes called programmed cell death 1 (or PD-1). PD-1 normally interacts with two ligands on antigen-presenting cells, called PD-L1 and PD-L2. This interaction between PD-1 and the PD-L1 and PD-L2 ligands inhibits T lymphocyte activation and cytokine production. (A ligand is a substance that binds specifically and reversibly to another substance, forming a complex.)

A range of tumour cells, including melanoma cells, have PD-L1 expressed on their surface, so binding of T lymphocyte PD-1 and tumour cell PD-L1 inhibits T lymphocyte responses against the tumour cells (PD-L2 is also expressed on a variety of tumour cells, but it has not been studied as extensively as PD-L1, and its impact on anti-tumour immunity is less clear). The binding of pembrolizumab to PD-1 receptors blocks the inhibition of T lymphocytes, allowing them to become activated and produce inflammatory cytokines. This improves the immune response against tumour cells, but it also results in autoimmune reactions in which healthy cells are damaged.

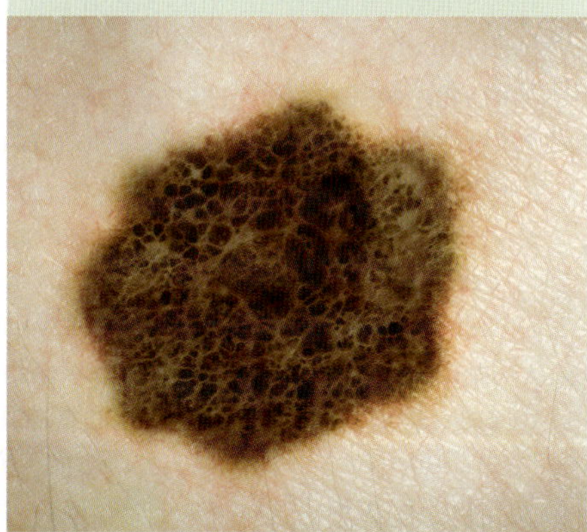

Photograph of melanoma

Bispecific monoclonal antibodies

Bispecific monoclonal antibodies (bispecific mAbs) are artificially produced using recombinant DNA technology and are used to target cancer cells and activate the immune system simultaneously (Figure 10.1.10d on page 321). This type of mAb is used to indirectly activate an adaptive immune response. Naturally produced mAbs have two binding sites, but each site binds to the same antigen. Bispecific mAbs can attach to two different antigens at the same time, because they are composed of parts from two different mAbs and have two different antigen-binding sites (Figure 10.1.11).

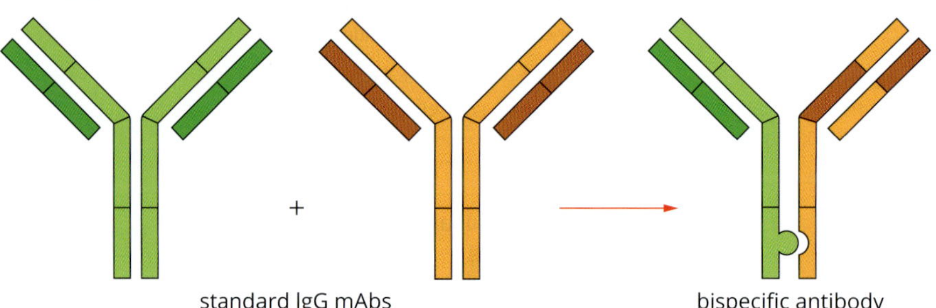

FIGURE 10.1.11 Diagram of a bispecific monoclonal antibody, in which parts of two different antibodies have been fused to form a hybrid that can bind to cancer cells and to T lymphocytes

An example of a bispecific mAb is Blincyto, which is used to treat some types of acute lymphocytic leukaemia. One part of this mAb attaches to a protein on the surface of the leukaemia cells, while the other part attaches to a protein found on T lymphocytes of the immune system. By binding to both these proteins, the bispecific mAbs are effectively 'identifying' the cancer cells as foreign and 'delivering' them to the immune system.

CASE STUDY ANALYSIS

Gut microbes boost immunotherapy success

Immunotherapy is a rapidly growing field of medicine that has shown a lot of promise as a highly effective treatment for cancer.

Immunotherapy works by using the patient's immune system to fight cancer cells. There are four main types of immunotherapy:

- monoclonal antibodies, which are synthetic antibodies designed to destroy, slow the growth of, or directly deliver medicine (chemotherapy) to cancer cells
- immune checkpoint inhibitors, which prevent immune cells from being switched off by cancer cells, enabling the immune system to recognise and destroy cancer cells (Figure 10.1.12)
- cancer vaccines, which trigger the immune system to prevent cancer cell growth or destroy existing cancer cells
- non-specific immunotherapies, which boost the immune system in a non-targeted way to slow or halt cancer cell growth.

In clinical trials, melanoma (skin cancer) patients who received immunotherapy treatments had increased survival rates, with less toxicity and fewer negative side effects than patients who received chemotherapy or radiotherapy. Immunotherapy has also been successful in treating several other types of cancer, including cancer of the lung, breast and colon. Although this form of treatment holds a lot of promise, the success rates with patients have been varied and there is more research to be done.

Researchers from the University of Chicago found that introducing the bacteria species *Bifidobacterium* to the digestive systems of mice with melanoma markedly increased their anti-tumour T lymphocyte response.

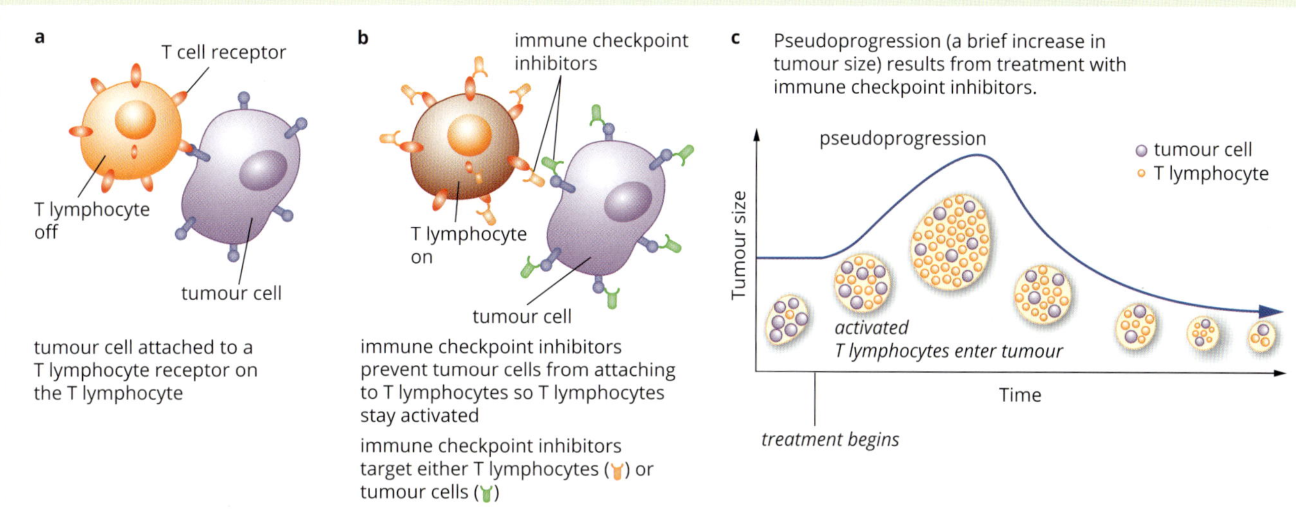

FIGURE 10.1.12 (a) Some tumour cells switch off T lymphocytes by binding to T cell receptors. (b) Immune checkpoint inhibitors, a type of immunotherapy drug, block inhibitory molecules on tumour cells from binding to T cell receptors, allowing T lymphocytes to remain active and infiltrate tumours, keeping them from growing. (c) Treatment with immune checkpoint inhibitors will briefly increase the size of a tumour (pseudoprogression) due to T lymphocyte infiltration, before its size is reduced due to tumour cell death.

When mice with the bacterial strain were compared to mice without the bacteria, but which were receiving immunotherapy via the drug anti-PD-L1 (an immune checkpoint inhibitor), the researchers found that tumour growth was slowed in both groups. By combining the bacterial treatment with immunotherapy, tumour control was dramatically improved. Another study by researchers at the Institut Gustave Roussy in Paris found that antibiotics reduced the effects of an immunotherapy drug. By replenishing gut microbes in antibiotic-treated and infection-free mice, the anti-cancer effects of the immunotherapy drug were restored.

Further investigation revealed that the *Bifidobacterium* triggered an immune response by interacting with dendritic cells in the intestinal tract. Dendritic cells are antigen-presenting cells that are responsible for detecting potential threats to the immune system and presenting them to T lymphocytes, thereby triggering an adaptive immune response. A genome-wide scan of mice with *Bifidobacterium* also showed upregulation of several genes involved in anti-tumour responses. Both of these studies have demonstrated the important role that a healthy gut microbiome plays in the immune response and the significant implications for the treatment of cancer using immunotherapy.

Analysis

Examine evidence that gut microbes can influence immunotherapy success and prepare an argument that could be presented to government to request more research funding.

AUTOIMMUNE DISEASE

As you learnt in Chapter 7, normally T and B lymphocytes that are reactive against self-antigens are destroyed. This means that when your immune response is working properly it is directed against non-self antigens, not against self-antigens, and this is known as self-tolerance. **Autoimmune diseases** result from a failure of self-tolerance, which leads to an adaptive immune response directed against specific self-antigens.

When autoimmune diseases occur, the cytotoxic T cells of the adaptive immune response attack the tissues directly. B lymphocytes act indirectly by secreting antibodies. Mast cells are also activated and release histamines, which results in inflammation around the affected tissues.

Particular autoimmune diseases and combinations of autoimmune diseases tend to be inherited and are generally more common in females. Environmental factors also seem to have an effect in the development of autoimmunity. Over 80 autoimmune diseases are currently known, including Crohn's disease, systemic lupus erythematosus, type 1 diabetes and rheumatoid arthritis (Figure 10.1.13).

Researchers are trying to understand the causes and risk factors, as well as the rise of autoimmune diseases in industrialised countries.

Autoimmune diseases result from an adaptive immune response directed against self-antigens.

BIOFILE

Cancer immunotherapy and autoimmune diseases

Cancer immunotherapy generally aims to stimulate the immune response, and autoimmune disease immunotherapy aims to suppress it, so there has been little research conducted on the use of immunotherapies in people who have both cancer and autoimmune conditions.

People with autoimmune conditions have been excluded from cancer immunotherapy trials due to concerns that cancer immunotherapies designed to stimulate the immune response might worsen autoimmune conditions, which are the result of an overactive immune response. However, with the significant benefits of cancer immunotherapy now well established, and roughly one-third of cancer patients having autoimmune conditions, there is a growing recognition of the need to study how different immunotherapies can be used to treat people who have particular cancers and autoimmune conditions.

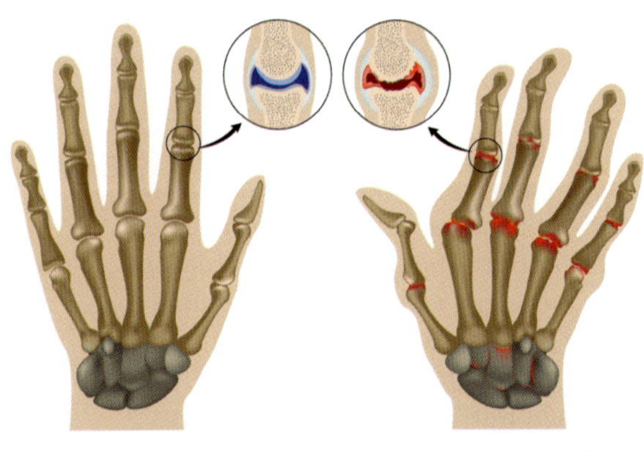

FIGURE 10.1.13 Rheumatoid arthritis causes inflammation in the joints and tends to begin by affecting the small joints of the hands and feet, which over time can lead to disfigurement.

Autoimmune disease treatments

Autoimmune conditions have traditionally been managed with anti-inflammatory and immunosuppressive drugs such as corticosteroids and methotrexate, which reduce inflammation and suppress the immune response. However, due to the chronic nature of autoimmune diseases, long-term use of such medications is needed and has undesirable side effects.

Drugs that suppress the immune system have the potential to result in cancer or infection, as the suppressed immune system is unable to control the replication of abnormal cells or pathogens. Despite these drawbacks, immunosuppressive drugs are effective in managing autoimmune symptoms, and as with all medications the benefits of their use are weighed against the risks.

Methotrexate is used to manage autoimmune conditions that do not respond to other treatments, and also as a chemotherapy agent to manage cancer.

Monoclonal antibody therapy for autoimmune disease treatment

In the same way mAb therapy complements traditional cancer therapies, mAb therapy also complements traditional immunosuppressive medications used to treat autoimmune conditions. While much remains unknown about the use of different mAbs to treat cancers and autoimmune diseases that occur at the same time, particular mAbs are known to be useful treatments for both different types of cancers and autoimmune diseases.

The growing range of mAbs used to treat autoimmune diseases includes the following.

- Tumour necrosis factor alpha (TNFα) inhibitors. Infliximab, adalimumab and golimumab bind to and inhibit TNFα, a pro-inflammatory cytokine involved in systemic inflammation (Figure 10.1.14). Anti-TNFα mAbs are used to treat autoimmune diseases including rheumatoid arthritis, psoriatic arthritis, and inflammatory bowel diseases such as Crohn's disease and ulcerative colitis.
- Interleukin (IL) inhibitors. Interleukins are a class of cytokines that are important in immune cell differentiation and activation. A number of different mAbs targeting different interleukins have been used to treat various autoimmune diseases (Table 10.1.1).
- B and T lymphocyte inhibitors. Ritixumab targets and eliminates B lymphocytes that express the CD20 protein. It is used to treat a range of different blood cancers and autoimmune diseases. Alemtuzumab targets the CD52 protein expressed by T and B lymphocytes, as well as natural killer cells and monocytes. It is used to treat a type of B lymphocyte blood cancer called chronic lymphocytic leukaemia, as well as to reduce the relapse rate in multiple sclerosis.

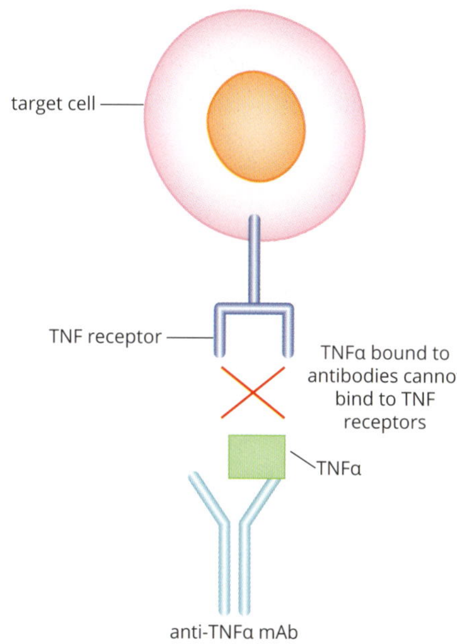

FIGURE 10.1.14 Infliximab, adalimumab and golimumab bind to soluble TNFα and prevent it from binding to TNF receptors.

TABLE 10.1.1 Examples of monoclonal antibodies used to treat autoimmune diseases

Monoclonal antibody	Target	Used to treat (lists not exhaustive)
infliximab	TNFα	Crohn's disease, ulcerative colitis, rheumatoid arthritis, psoriatic arthritis and ankylosing spondylitis
adalimumab		Crohn's disease, ulcerative colitis, rheumatoid arthritis, psoriatic arthritis, ankylosing spondylitis and uveitis
golimumab		rheumatoid arthritis, psoriatic arthritis, ankylosing spondylitis and ulcerative colitis
daclizumab	IL-2	multiple sclerosis
tocilizumab	IL-6	rheumatoid arthritis
ustekinumab	IL-12 and IL-23	Crohn's disease, psoriasis and psoriatic arthritis
rituximab	CD20	rheumatoid arthritis, systemic lupus erythematosus and graft-versus-host disease
ocrelizumab	CD20	multiple sclerosis
alemtuzumab	CD52	multiple sclerosis

BIOFILE

Immune system suppression

Pimecrolimus is a drug that blocks T lymphocyte activation and prevents the release of pro-inflammatory cytokines by mast cells. Pimecrolimus is an immunotherapy used to treat inflammation of the skin (dermatitis) when topical corticosteroids cannot be used or fail to work. Some studies have shown a correlation between pimecrolimus use and the development of skin cancers and cancer of the lymphatic system (lymphoma), but causation has not yet been proven. The benefits of medications must be weighed against their risks by informed consumers.

CASE STUDY

More than mAbs: Other treatments for cancer and autoimmune diseases

In addition to mAbs, there is a range of other types of immunotherapeutic drugs that are used to treat cancer and autoimmune diseases, and many more are currently in development. Drugs that inhibit the Janus kinase (JAK) family of enzymes are one such example.

Cytokines function by binding and activating cytokine receptors, and these receptors rely on JAK enzymes to phosphorylate their activated receptors, which in turn allows further cell signalling events to occur. JAK inhibitors are drugs that block the action of JAK enzymes, which prevents cell signalling events that would normally be triggered by the activation of cytokine receptors and result in cellular responses that promote inflammation.

There are several different JAK inhibitors approved for use, such as baricitinib and tofacitinib for rheumatoid arthritis. Clinical trials for several new JAK inhibitors, including some that could be used to treat blood cancers, are currently taking place.

Researchers are also currently working to develop an immunotherapy that would stimulate the immune system to destroy the specific immune cells that cause type 1 diabetes. Type 1 diabetes occurs when there is a failure of self-tolerance and immune cells attack the insulin-producing beta cells of the pancreas (Figure 10.1.15). Insulin regulates blood glucose levels, so people with type 1 diabetes need daily insulin injections to maintain glucose balance. If successful, a novel treatment that selectively destroys the rogue immune cells that attack the beta cells in the pancreas of type 1 diabetics would protect these insulin-producing cells, and limit disease progression in people who are diagnosed in the early stages of the disease.

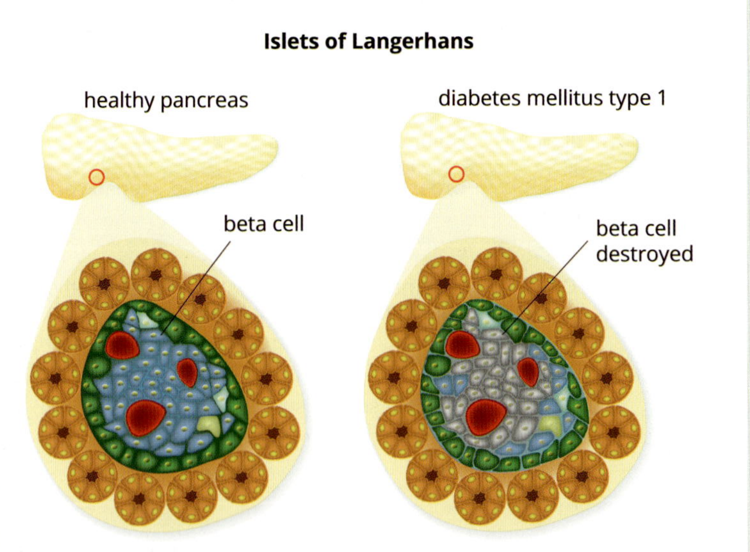

FIGURE 10.1.15 In type 1 diabetes, immune cells attack and destroy beta cells in the pancreas.

10.1 Review

OA ✓✓

SUMMARY

- Cancer occurs when a single rogue (or abnormal) cell multiplies uncontrollably and spreads throughout the body.
- Cancer treatments over the last several decades have included chemotherapy, radiation therapy and surgery. These treatments come with significant side effects.
- Immunotherapy is any treatment that activates or suppresses the immune system of the patient to fight autoimmmune diseases and other diseases such as cancer.
- Immunotherapies that stimulate the immune system are used to treat cancer.
- Immunotherapies can be non-specific, such as the injection of cytokines, or specific, such as cancer vaccines, personalised immunotherapy and monoclonal antibody therapy.
- Cancer vaccines are made using specific antigens from cancer cells or pathogens that cause cancer. Like other vaccines, they are administered to a patient to stimulate an immune response that results in the production of an immunological memory.
- Cancer vaccines can be preventative, therapeutic or personalised.
- Monoclonal antibody (mAb) therapy involves antibodies produced by a single clone of B lymphocytes that is replicated in culture. mAbs are all identical and specific to the same antigen. Targeting specific cells reduces harm to healthy cells, but identifying the specific antigen in order to create mAbs is a difficult task.
- The first monoclonal antibodies (mAbs) were mouse mAbs made entirely by mouse B lymphocytes. To avoid an immune response against mAbs, chimeric, humanised and human monoclonal antibodies can now be produced using transgenic mice:
  - Chimeric mAbs are a mix of human and mouse components.
  - Humanised mAbs are also a mix but are mostly human.
  - Human mAbs are fully human.
- mAbs can be used as carriers of treatments (drugs, toxins and radioactive particles) for delivery specifically to cancer cells. These types of mAbs are called conjugated mAbs.
- Bispecific mAbs are made up of two different mAbs and have two different binding sites: one is usually for a cancer cell and the other for an immune cell, such as a T lymphocyte. This enables an 'identify' and 'deliver' approach.
- Autoimmune diseases are caused by the body triggering an immune response against its own cells, leading to tissue and organ inflammation and damage.
- Immunotherapies that suppress the immune response are used to treat autoimmune diseases.
- Tumour necrosis factor alpha (TNFα) inhibitors, interleukin (IL) inhibitors and B and T lymphocyte inhibitors are examples of mAbs that are used to treat autoimmune diseases.

KEY QUESTIONS

Knowledge and understanding

1 If a monoclonal antibody has a toxin or radioactive substance attached to it, what kind of monoclonal antibody is it?

A bispecific
B conjugated
C chimeric
D humanised

2 Define cancer.

3 Explain the difference between a preventative cancer vaccine and a therapeutic cancer vaccine, including reference to the type of antigen in each vaccine.

4 What is the difference between chimeric, humanised and human monoclonal antibodies?

5 Bispecific antibodies are produced artificially. How are these antibodies different from those produced naturally by the immune system?

continued over page

10.1 Review *continued*

Analysis

6 The results of a preliminary trial comparing two methods of drug administration for cancer treatment are shown in the table below. The same drug was administered with (antibody, A) and without (conventional, C) an antibody attached to it. In both cases, there was one round of treatment, which consisted of administration of the drug once per week for four successive weeks.

Patient number	Antibody (A) or conventional (C) therapy	Tumour size at commencement of treatment	Tumour size after one round of treatment	Change in tumour size	% change in tumour size
1	A	12.5 mm^3	9.6 mm^3		
2	A	23.9 mm^3	15.5 mm^3		
3	A	54.2 mm^3	26.8 mm^3		
4	A	46.8 mm^3	27.9 mm^3		
5	A	53.6 mm^3	56.4 mm^3		
6	C	54.8 mm^3	48.5 mm^3		
7	C	84.1 mm^3	66.9 mm^3		
8	C	36.9 mm^3	30.8 mm^3		
9	C	56.1 mm^3	49.1 mm^3		
10	C	38.9 mm^3	31.5 mm^3		

a Complete the table by calculating the percentage change in the size of the tumour. Use the formula:

$$\frac{\text{change in tumour size}}{\text{original tumour size}} \times 100 = \%\text{ change in size}$$

b Why is it necessary to calculate percentage change in tumour size before analysing the results of the trial?

c Do the results indicate that further trials of this approach should be undertaken? Explain your reasoning by referring to the data.

d Suggest a possible explanation for the results observed in patient 5.

7 Choose an autoimmune disease to research, and write about how the disease affects the body and how it is managed. Include any traditional treatments, or monoclonal antibodies or other immunotherapies used or currently in development to treat it.

10.2 Emerging diseases and infection control strategies

As the world's population has grown and become more connected, new and rapidly spreading diseases have proved globally challenging. Urbanisation and globalisation have brought people into closer contact with each other through the development of large cities and rapid transport between cities and countries (Figure 10.2.1).

FIGURE 10.2.1 As transport and larger cities bring people in closer contact with each other, diseases can become more prevalent and spread more rapidly.

The study and surveillance of newly emerging and re-emerging diseases aims to find ways to predict, prevent and respond to outbreaks of disease. Many of the diseases of local and global importance are infectious diseases and controlling these diseases requires a coordinated global effort.

Although scientific knowledge has increased over the decades, allowing us to identify, diagnose and treat more diseases, other factors have led to new infectious diseases becoming a challenge for governments and medical personnel around the world. In this section you will learn about the impact of the emergence of new pathogens and re-emergence of known pathogens, and the scientific and social strategies that are used to identify them and control their spread.

THE EMERGENCE OF NEW PATHOGENS AND RE-EMERGENCE OF KNOWN PATHOGENS

As you learnt in Chapter 8, **infectious diseases** can be caused by a range of cellular and non-cellular **pathogens** (agents that cause disease). **Emerging infectious diseases** may be defined as:

- new or previously unrecognised pathogens and diseases
- diseases that have increased in **incidence**, **prevalence** or geographic range over the past 20 years
- diseases that may increase in the near future.

New diseases can emerge as pathogens adapt to a new host under different conditions. They may also emerge due to genetic mutations that increase the pathogen's **virulence** or ability to infect a wider range of host organisms, including humans (Table 10.2.1 on page 330). New pathogens can have serious consequences for a population or species. Without previous exposure to a pathogen's antigens, it is unlikely that any individuals in the population will have developed antibodies and immunological memory, leaving the entire population susceptible to infection.

When a disease passes from another animal to a human host it is known as a **zoonotic** disease. Many emerging and re-emerging diseases are zoonotic. For example, the severe acute respiratory syndrome coronavirus 2 (SARS-CoV-2) virus that caused the COVID-19 pandemic is believed to have been first transmitted to humans by other animals at a wildlife wholesale market in Wuhan, China, in late 2019.

Incidence is the rate of occurrence of new cases of a medical condition in a population in a specified period of time.

Prevalence is the proportion of cases of a medical condition in a population at any given time.

Virulence is the disease-producing power or severity of a pathogen.

Zoonotic diseases are infectious diseases that are transmitted from different species of animals to humans.

TABLE 10.2.1 Examples of emerging diseases of human populations

Prevalence of disease	Example
new or previously unrecognised	• HIV: other primate → human • SARS-CoV: bat → civet cats → human • SARS-CoV-2: bat → currently unknown, possibly pangolins or snakes → human • MERS: camel → human • Hendra virus: bat → horse → human • Zika virus: mosquito → human • vCJD prion BSE: cattle → human
increased in incidence, virulence or range over past 20 years	• Ebola: bat → human • dengue virus • West Nile virus • cholera • MRSA • *Clostridium difficile*
may increase in the near future	• influenza • antibiotic-resistant bacteria • cholera • dengue virus • prion diseases • non-infectious diseases: diabetes, obesity, Alzheimer's

An epidemic is the sudden increase in the number of cases of a disease above what is normally expected in that population in that area.

A pandemic is an epidemic that has spread over several countries or continents, usually affecting a large number of people.

An **epidemic** is the rapid spread of a disease to a large number of people. More specifically, an epidemic is a sudden increase in the occurrence of a particular disease among the population of a given area (Figure 10.2.2). When the spread of the disease reaches global proportions, it is known as a **pandemic**.

There are many factors that influence the emergence and spread of diseases. Some of these factors are:

- human migration and demographics (size, structure and distribution of populations)
- human behaviour
- changes in farming practices and food production
- uncontrolled or inappropriate use of antimicrobials
- lack of sanitation and poor hygiene.

Human migration

Over the centuries, there has always been movement of human populations either due to war or for socio-economic reasons. In any migration, there is the potential to introduce pathogens to a new area. Infectious diseases can have devastating consequences for populations that have never encountered them because the entire population has little or no immunity. While the disease can be serious or potentially lethal for those who have never been exposed to it, the individuals carrying the disease into the new area may have mild symptoms or no symptoms at all due to their previous exposure to the pathogen and immunological memory.

Today, with people travelling in large numbers across long distances in short periods of time, emergent diseases have the potential to spread rapidly at a global scale. This was demonstrated by the 2009 swine flu outbreak, caused by the H1N1 virus, which killed more than 18 000 people. This virus was a derivative of two strains of influenza virus. Although believed to have originated in a local geographic area in Mexico, the virus rapidly spread through the United States (US), with cases identified in most countries within months. The first case in Australia was a woman who arrived in Brisbane after a flight from Los Angeles.

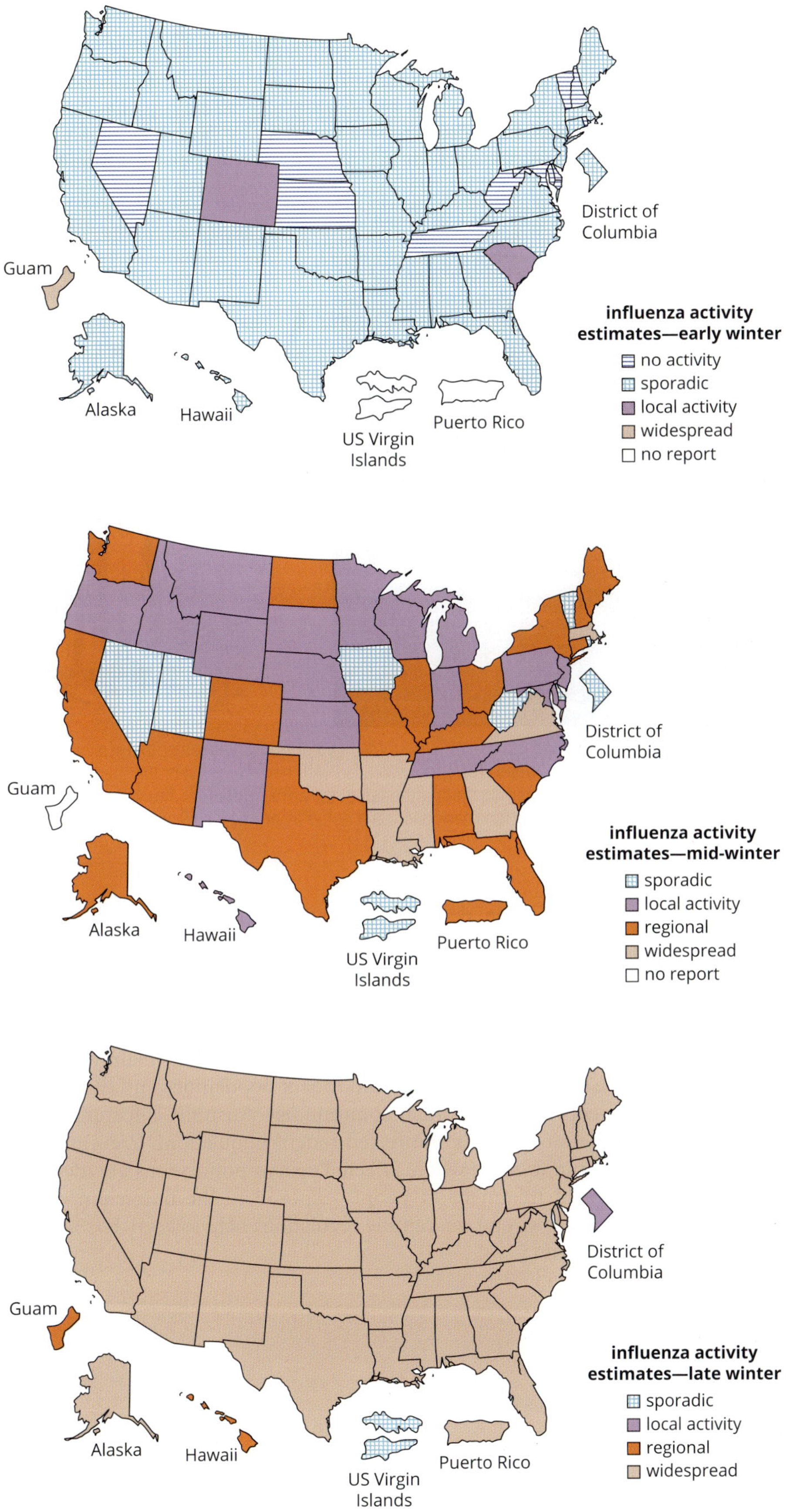

District of Columbia
Guam
Alaska
Hawaii
US Virgin Islands
Puerto Rico

influenza activity estimates—early winter

- no activity
- sporadic
- local activity
- widespread
- no report

District of Columbia
Guam
Alaska
Hawaii
US Virgin Islands
Puerto Rico

influenza activity estimates—mid-winter

- sporadic
- local activity
- regional
- widespread
- no report

District of Columbia
Guam
Alaska
Hawaii
US Virgin Islands
Puerto Rico

influenza activity estimates—late winter

- sporadic
- local activity
- regional
- widespread

FIGURE 10.2.2 The spread of influenza in the United States over the winter 2017–2018

BIOFILE

Refugees and displaced persons

Conflict, natural disasters and political shifts can cause a large, and often quick, migration of people to new regions. Refugees and displaced persons often reside in high-density, temporary accommodation that may lack appropriate sanitation, medical care and infrastructure. It is in these environments that disease can quickly spread. In 1994 almost 1 million people fled from Rwanda to Zaire where they sheltered in refugee camps. Epidemics of cholera and dysentery soon broke out, killing 50 000 people. The recent displacement of Syrian refugees (see figure below) has also seen an increase in disease, particularly leishmaniasis, a disease historically known as 'Aleppo boil'. Leishmaniasis is a parasitic infection spread by sandflies. It causes skin ulcerations that can fatally spread to internal organs.

Leishmaniasis has become prevalent among displaced Syrian refugees, many of whom reside in refugee camps such as the Atma refugee camp on the Turkish–Syrian border.

Bubonic plague

The bubonic plague historically caused devastation to human populations and remains a re-emerging disease in several parts of the world. Sometimes simply called 'the plague', it is caused by the bacterium *Yersinia pestis*. This bacterium infects rodents and is transmitted from rodent to rodent by fleas. It is transmitted to humans when a flea carrying the bacterium bites a susceptible person. Once bacteria reach the lungs, they become airborne and highly contagious (Figure 10.2.3). Symptoms appear 7–10 days after infection.

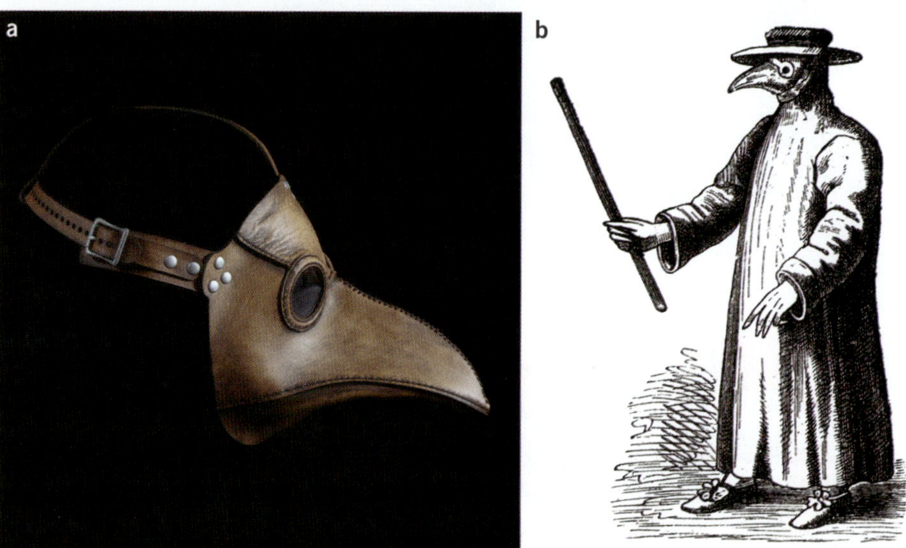

FIGURE 10.2.3 (a) A mask used by sixteenth-century Venetian doctors to protect themselves against the highly contagious bubonic plague. The beak was filled with herbs to prevent the 'bad' air from being inhaled. (b) An illustration demonstrating the full outfit worn by doctors attending to patients suffering from the bubonic plague. The mask and outfit acted as protective wear so the doctor did not become infected.

The first pandemic plague recorded occurred in the sixth century and is believed to have been brought to Europe from Africa by the fleas on rats in trade ships. There have been several epidemics and pandemics of the plague throughout the centuries, which have claimed the lives of millions of people.

The incidence of plague has dramatically declined, largely due to improved living standards. However, it continues to appear in several countries, mainly in Africa (Figure 10.2.4). The bacterium resides in wild rodent populations and can transmit the pathogen to rodents living in human communities. Australia is fortunate to be the only continent that does not have infected rodent populations. Prevention and control measures are put in place where affected rodent populations are identified, and early detection and treatment of people exposed to the bacterium enables rapid recovery. Australia's strong quarantine regulations help to keep *Yersinia pestis* infections out of the country.

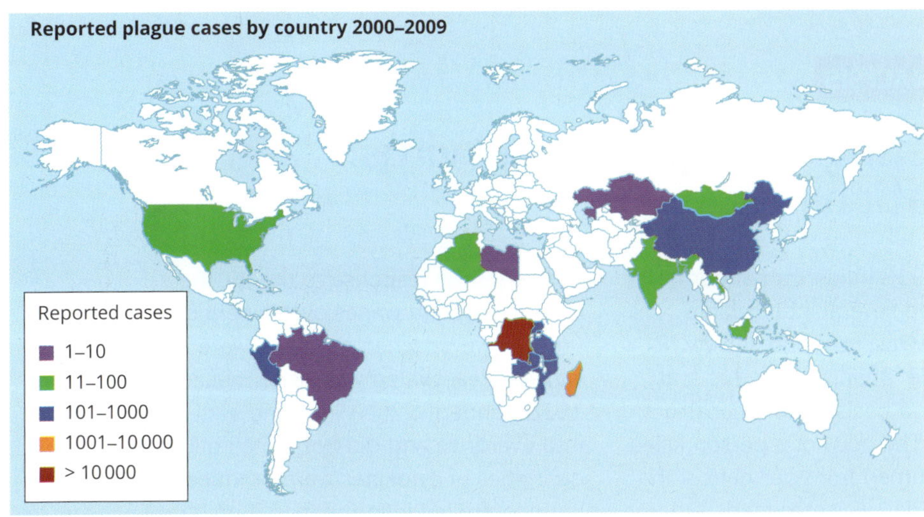

FIGURE 10.2.4 World Health Organization data on plague cases for the years 2000–2009

Influenza

From 1918 to 1920, approximately 50 million people were killed worldwide during an influenza (flu) pandemic. The influenza virus was highly contagious and deadly, even to healthy young adults. The disease was spread by air as the pathogen was breathed or coughed out of an infected person's lungs. World War I was occurring during the same period and infected soldiers spread the disease throughout Europe, the USSR and the US in three deadly waves. It was dubbed the 'Spanish flu', although it did not originate in Spain. The origins of the flu are still debated.

Influenza viruses circulate in populations of migratory birds, domestic poultry, pigs and humans. In humans, a new variant emerges almost every year and infects large numbers of people; this is called seasonal flu. New strains evolve from genetic mixing or reassortment of genes. This occurs when two or more different viruses infect the same human or other host animal. New strains resulting from genetic reassortment have new combinations of the surface antigens, haemagglutinin (H) and neuraminidase (N), and may be more virulent. New strains arising in birds or pigs that gain the ability to infect human cells are a high risk for pandemic flu.

Seasonal flu is often the H3N2 type of virus. The 2009 pandemic, an H1N1 type, was also called swine flu because it originated in pigs.

Strategies for preventing the emergence of new pandemic flu strains include regular gene sequencing of human, bird and pig viruses to identify problem variants, the culling of animals that are infected with dangerous new strains and limiting close interaction between humans and animals. Strategies to contain the virus include good hygiene, wearing face masks to prevent the spread of airborne viral particles (Figure 10.2.5), vaccination and **antiviral** medication. Vaccination before each flu season limits the number of infected people. When fewer people and other animals are infected with influenza virus, there is a reduced chance of genetic reassortment giving rise to dangerous new strains.

FIGURE 10.2.5 Influenza virus infects lung tissue. It is spread in water droplets released through coughing and sneezing.

The Victorian Infectious Disease Reference Laboratory in Melbourne is part of the World Health Organization Global Influenza Surveillance and Response System. It analyses influenza viruses circulating in the human population to help determine which viral strains should be included in the flu vaccine each year.

CASE STUDY ANALYSIS

Reconstructing the 'Spanish flu' virus: The implications

The severity of the 1918 influenza pandemic has been attributed to several inter-related factors between the virus, host and society. To better understand the 'Spanish flu' virus and why it was so deadly, a decision to reconstruct the virus was made. The body of a person killed by the virus was exhumed from permafrost in Alaska and the viral RNA was extracted. Genome sequencing of the virus was completed in 2005 and it was identified as H1N1 influenza A, originating in birds. The reconstructed virus was cultured in cells and used to infect mice and monkeys (Figure 10.2.6). The virus was highly virulent, killing the experimental animals more quickly than any other influenza virus tested.

In a normal response to the flu, the immune system wanes over a period of time. However, when monkeys were infected with the reactivated virus that caused the 1918 pandemic, their immune systems went into overdrive and did not switch off. Large quantities of cytokines were produced, resulting in uncontrolled inflammation. Similar responses were observed in the H5N1 avian flu, an epidemic that appeared in Asia in 1997, and COVID-19, the pandemic that began in China in 2019. Researchers are using this information to investigate possible treatments.

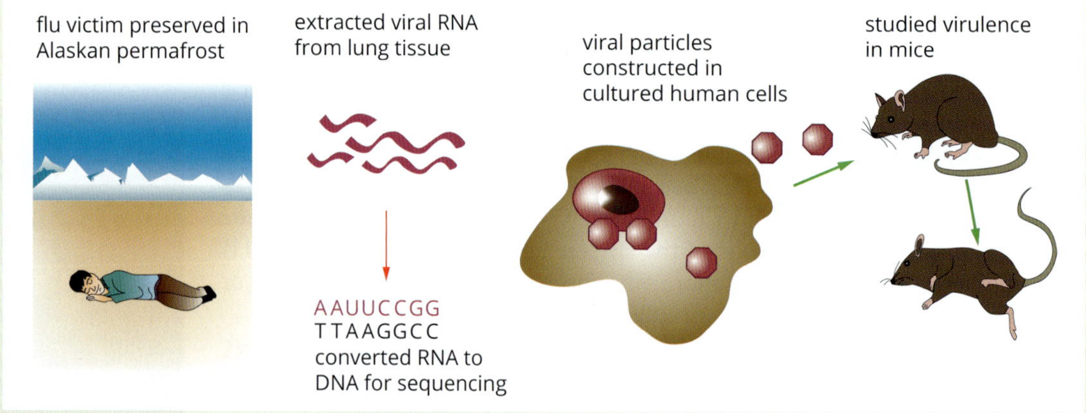

FIGURE 10.2.6 The 'Spanish flu' virus was reconstructed in a laboratory and tested in animal models.

Analysis

1. The 'Spanish flu' surprisingly affected young healthy people more than older adults. Use your understanding of immunological memory to suggest why this was so.
2. Discuss the benefits of researchers understanding the virulence and spread of the virus for future potential deadly outbreaks.
3. Do you believe that a 'better understanding of the virus' is a crucial reason for reconstructing such a virulent virus? What ethical issues need to be considered? Do the risks outweigh the benefits?

Introduced diseases among Indigenous Australians

In 1788 when Europeans arrived on the continent now known as Australia, they brought with them infectious diseases. The original inhabitants of Australia had never been exposed to the pathogens that caused these diseases, so their immune systems were vulnerable.

More than 500 different groups of Indigenous peoples, with an estimated population of 750 000, inhabited, cultivated and actively managed the resources of the continent for over 60 000 years. In the 10 years that followed the arrival of Europeans, the impact of introduced diseases, along with the dispossession of Indigenous peoples' lands and resulting conflict over land and resources, had reduced the Indigenous population by as much as 90%.

Major epidemic diseases in the early stages of European habitation of Australia were influenza, tuberculosis, measles and smallpox, as well as sexually transmitted infections such as syphilis that reached epidemic proportions due to the sexual abuse and exploitation of Indigenous women and children. As a consequence of both disease and abuse, Indigenous peoples experienced significantly higher morbidity and mortality than the Europeans. Governor Phillip reported that smallpox alone was responsible for halving the Aboriginal population in the Sydney region within 14 months of the First Fleet's arrival.

The impact of these diseases on the Europeans was less severe, as they had been previously exposed to the pathogens and so had immunological memory that allowed them to mount a rapid immune response. The Europeans also had access to medical care and treatments that were not made available to the Indigenous people. In addition, the spread of disease to Europeans was unintentionally limited by the various government policies of 'protection' that began in 1837, where Indigenous people were segregated in reserves.

Human behaviour

Human behaviour also has an impact on the emergence and re-emergence of diseases, increasing the spread of infectious disease, and may include:

- hygiene practice
- sanitation
- dietary habits
- human-to-human contact
- sexual activity
- medical procedures
- exposure to environmental agents of disease.

HIV

Human immunodeficiency virus (HIV) (Figure 10.2.7) is the virus that, if left untreated, causes acquired immunodeficiency syndrome (AIDS). HIV was first recognised as a new pathogen in 1981, and appears to have a zoonotic origin, as a genetically similar virus is found in some African monkeys. Researchers suggest that there were two independent transmissions from monkeys to humans. These transmissions might have occurred between pet monkeys and their owners or following the slaughter of a monkey by a person, perhaps for bush meat.

HIV is spread through contact with contaminated bodily fluids including semen and blood, but it is not spread through saliva. You cannot get HIV from kissing someone who is HIV-positive or sharing cutlery with them. Transmission can occur through unprotected sexual contact, or from the use of hypodermic needles, blood transfusions or organ transplants contaminated with HIV.

Some people present with flu-like symptoms within two weeks of acquiring HIV; in others, the virus incubation period lasts many years and they show no symptoms. The only way that someone can know if they have HIV is by getting tested. Frequent sexual health testing is an important responsibility for anyone who is sexually active and is no cause for shame.

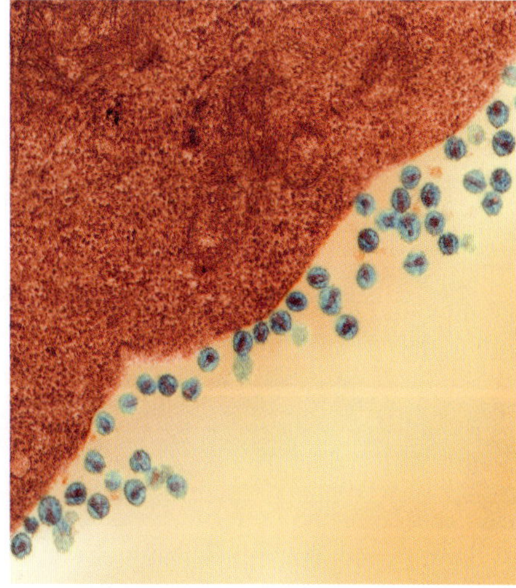

FIGURE 10.2.7 HIV infecting a cell. Transmission electron micrograph (TEM) of a section showing human immunodeficiency virus (HIV) particles (virions, round) on the surface of a cultured cell

While there is not yet an effective vaccine for HIV, thanks to significant investment in research and development, there is a suite of antiretroviral drugs that keep HIV under control and allow people living with HIV to live normal lifespans. In addition to condoms, a range of biomedical prevention strategies that greatly reduce the risk of HIV transmission has been developed, and ultimately these prevent many new cases of HIV.

Much progress has been made since the initial epidemic in the 1980s. It is now possible to imagine the end of new HIV transmissions in Australia. However, much remains to be done to ensure people everywhere can access the full suite of HIV treatment and prevention options needed to achieve the ambitious aim of ending new HIV transmissions. In particular, social, political and economic factors continue to contribute to unequal access to these treatment and prevention options both in Australia and globally.

Kuru

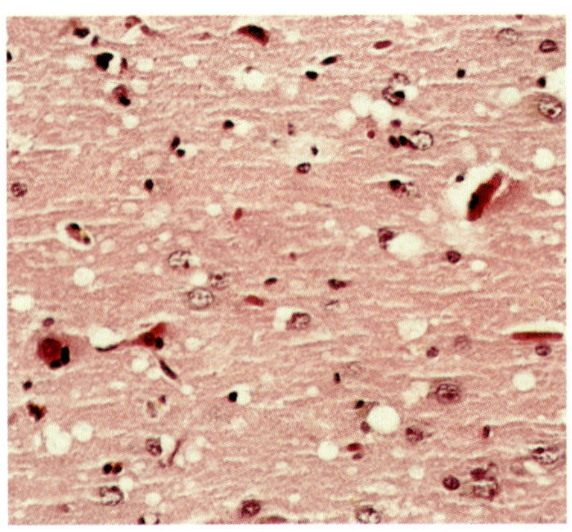

FIGURE 10.2.8 A slice of brain tissue showing the spongy appearance of spongiform encephalopathy disease such as CJD or Kuru. The white holes are gaps where neurons have died.

Kuru is a neurodegenerative disease caused by prion proteins that cause normal proteins to change shape and form abnormal protein clusters in the brain, impeding normal function and resulting in the death of neurons (Figure 10.2.8). These changes are similar to those of bovine spongiform encephalopathy (BSE) affecting cows, scrapie affecting sheep and Creutzfeldt–Jakob disease (CJD), another human prion disease. Death occurs approximately one year after the onset of symptoms, which include behavioural changes, coordination problems, tremors and pain.

An epidemic of Kuru began in the 1920s, emerging in the Fore people of Papua New Guinea. The Fore people, who had a complex belief system, believed that eating their deceased relatives was a sign of respect as part of the mourning ritual. During funeral rites, the women and the children would eat the brain of the deceased. The infectious prion protein in the brain was then transmitted to a new host. During the 1950s the ritual was banned, leading to a decline in the disease. However, the incubation period can be as long as 50 years and cases continue to be reported.

Changes in farming practices and food production

Farming practices and food production have changed over the centuries to adapt to globalisation and an increasing human population. A greater demand for meat has created pressures on the farming industry. Throughout the world, dense farming practices have developed and this has occurred in close proximity to human populations. This has allowed pathogens in farm animals to transfer to humans more easily. These new zoonotic diseases can have devastating effects because the human population has never been exposed to the pathogen and has no immunological memory. With so many vulnerable hosts within a population, the pathogen can spread quickly, resulting in severe illness and death.

Mad cow disease

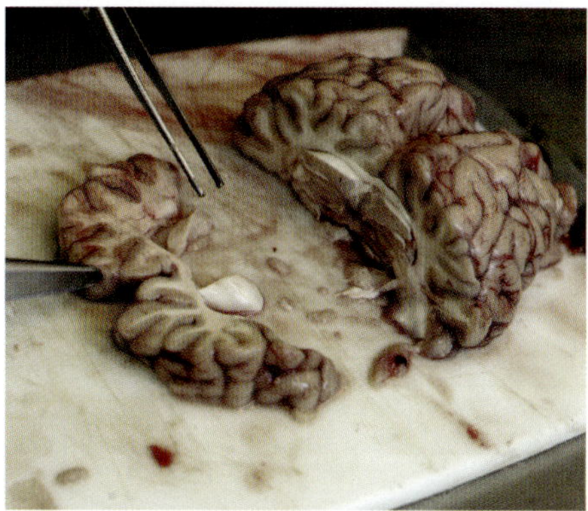

FIGURE 10.2.9 A vet slicing a cow brain to look for signs of BSE and other diseases. Tissue showing signs of damage must not enter the food chain.

Bovine spongiform encephalopathy (BSE) or mad cow disease is caused by an infectious prion protein and is a fatal neurological disease in cows. The first documented case of BSE in the UK was recorded in 1986 and the incidence grew rapidly. At the height of the epidemic in 1993, up to 1000 new cases were reported weekly. Over 180 000 cattle were affected. BSE is caused by abnormal prion proteins that misfold; these prions then cause normal prion proteins to misfold and aggregate. This damages the brain tissue.

In the 1990s, a new human disease appeared, variant Creutzfeldt–Jakob disease (vCJD). It was acquired by eating products from BSE-infected cattle. Transmission occurred because the abnormal prion proteins are not denatured or destroyed when cooked, as occurs with most proteins.

A farming practice that contributed to the rapid and extensive spread of BSE in the UK was feeding meat and bone meal to farm animals. Meat and bone meal is the ground up remains of farm animals, and was used to increase the protein content of animal feed. It is possible that BSE started when this meal contained tissue from sheep with scrapie (a prion disease). The knowledge gained from the spread of the Kuru epidemic, which was caused by humans eating infected material, led to the immediate actions of culling potentially infected animals to control the spread. The practice of feeding mammalian protein to cattle was banned and stringent slaughter practices, testing (Figure 10.2.9) and surveillance was introduced. Australia's thorough screening and tight importation regulations on animal products has kept the country free of BSE.

Uncontrolled use of antimicrobials

The overuse and mismanagement of **antimicrobial drugs** against pathogens such as bacteria, viruses and parasites has led to a global issue that has the potential to cause a major crisis. Although **antimicrobial resistance** is a natural phenomenon, it has been made worse by the excessive use of antimicrobial drugs, particularly in medicine and agriculture. This has resulted in the emergence of some drug-resistance in pathogens. Resistance in pathogens can develop in several ways and two examples are given.

Antibiotic-resistant pathogens ('super bugs')

Antibiotics act against bacteria and are not effective against viruses. Antibiotics target specific biological pathways that disrupt cellular function and kill or slow the growth of the bacteria.

There is a chance that a population of bacteria will become resistant to a particular antibiotic when treated with it. This can occur because some bacteria in a population already carry resistance genes; they are not killed by the antibiotic and pass on the resistance genes. Alternatively, new spontaneous mutations may occur in the bacterial DNA, making the bacteria resistant to the antibiotic. These bacteria are then able to multiply rapidly without competition from antibiotic-sensitive strains and produce cloned offspring that carry the resistance gene.

Bacterial cells also exchange plasmid DNA between each other in a process known as **horizontal gene transfer**. Plasmids carrying a drug resistance gene can also be acquired in this way.

Transmission of resistant bacteria can be indirect, through food and water, or through direct contact with an infected organism. Methicillin-resistant *Staphylococcus aureus* (MRSA) and vancomycin-resistant *Enterococcus* (VRE) are examples of bacteria that are now resistant to top-line antibiotics, making them almost impossible to treat. They are frequently acquired in hospitals. You will learn more about bacterial resistance to antibiotics in Chapter 11.

Malaria

Despite repeated and ongoing attempts to eliminate the malaria parasite and its vector, the mosquito, malaria continues to re-emerge and is likely to spread as climate change continues to warm the planet. Approximately half the world's population is at risk of malaria, and there were 214 million cases in 2015.

The parasite *Plasmodium falciparum* has developed an intricate relationship with two hosts, the *Anopheles* mosquito and primates, including humans. The parasite is carried by the female mosquito and transmitted to humans when an infected mosquito bites them. The disease may also be passed between humans during blood transfusions. In humans, the parasite replicates inside red blood cells, causing them to burst. Symptoms include fever, vomiting and headaches.

The antimalarial drug chloroquine, discovered in 1934 and used in the 1940s, was a safe and highly effective drug. However, in the late 1950s chloroquine-resistant parasites emerged in Asia. A mine along the border of Thailand and Cambodia attracted many workers from neighbouring regions. Environmental changes that occurred from the construction of the mine favoured the breeding of mosquitoes, and workers were given inadequate doses of chloroquine. A mutation in the DNA of the parasite allowed the parasite to survive in the presence of chloroquine and spread to infect the many workers in the region, and then further afield as the workers migrated (Figure 10.2.10).

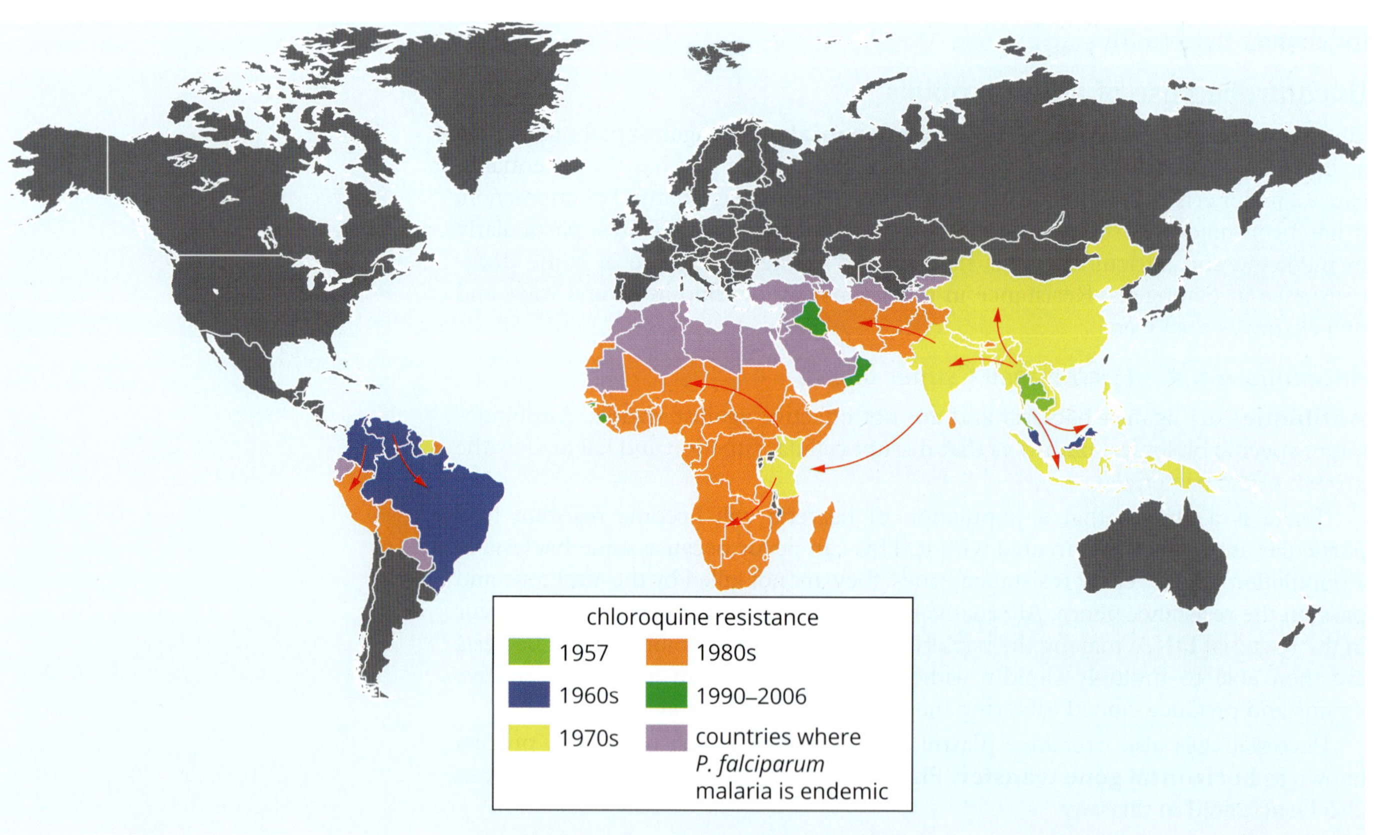

FIGURE 10.2.10 History of chloroquine-resistant *Plasmodium falciparum*

Artemisinin, a compound from the sweet wormwood plant (*Artemisia annua*), was identified by the Chinese scientist Tu Youyou in the 1970s. The discovery of this potent antimalarial compound earned her a Nobel Prize for Medicine in 2015. Sweet wormwood has been used in Chinese traditional medicine for many years. It is now a key component of combination therapy for malaria and is effective against drug-resistant malaria.

Lack of sanitation and poor hygiene

Outbreaks of various diseases are common in developing countries, and in isolated communities in developed countries. These places often lack the infrastructure for reliable sanitation, such as clean running water, effective sewerage systems and adequate health services.

Cholera

The cholera epidemic that occurred in Haiti in the wake of the devastating 2010 earthquake is the worst cholera epidemic in recent history, with more than 700 000 cases and almost 9000 deaths. Prior to this, Haiti had not experienced a cholera epidemic in more than a century. The epidemic was caused by a toxigenic strain of *Vibrio cholerae* (Figure 10.2.11). Cholera affects the gastrointestinal tract, causing watery diarrhoea, vomiting and dehydration, and can be fatal.

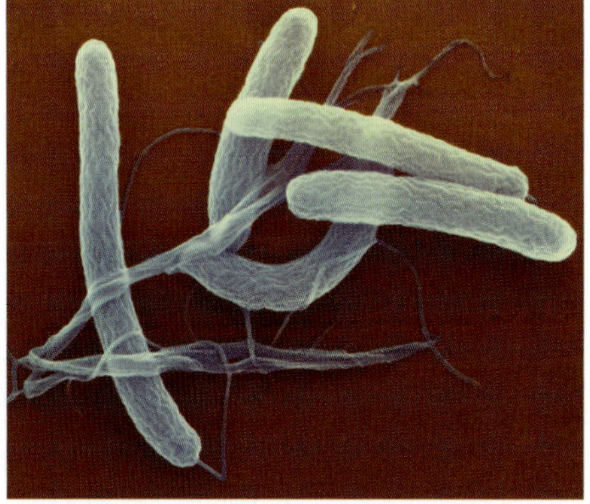

FIGURE 10.2.11 Coloured SEM of *Vibrio cholerae*

The origin of the Haitian epidemic was traced to a camp of Nepalese army personnel who came to the country to aid the relief effort. Epidemiology and scientific methods were used to trace the source and identify the pathogen. DNA sequencing matched the epidemic strain to a *V. cholerae* strain previously identified in Bangladesh, rather than strains that already existed at low levels in Latin America. The *V. cholerae* bacterium was most likely introduced by the Nepalese soldiers, who had recently been in a region experiencing a cholera outbreak. Contaminated sewage from the camp was released into the local river. The communities downstream used this water for washing and cooking, because the water supply network had been damaged and was still being repaired.

Despite improved sanitation and repairs to water infrastructure, new infections continue to occur. Reasons why cholera continued to be a problem in Haiti some time after the initial cause of the epidemic include:

- limited sanitation infrastructure
- limited access to adequate medical services
- gaps in the water quality systems and limited chlorination
- limits to the alert and coordination systems
- displacement of peoples.

A different strain of *V. cholerae*, which emerged in 1992–93, was responsible for a cholera epidemic in a large area along the Bay of Bengal. Scientific studies of this *V. cholerae* strain are providing clues about how bacteria evolve into new variants with the potential to cause epidemics.

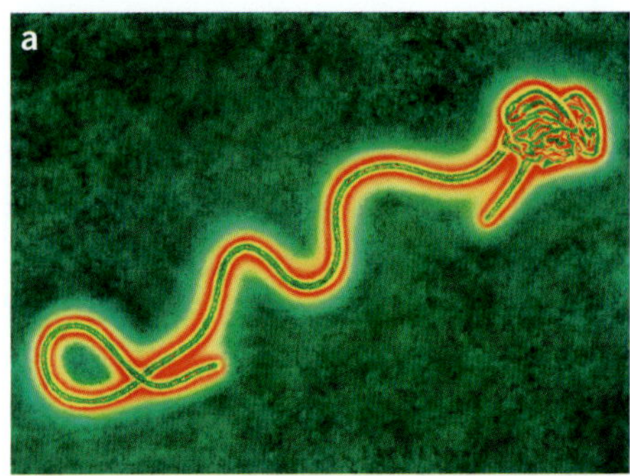

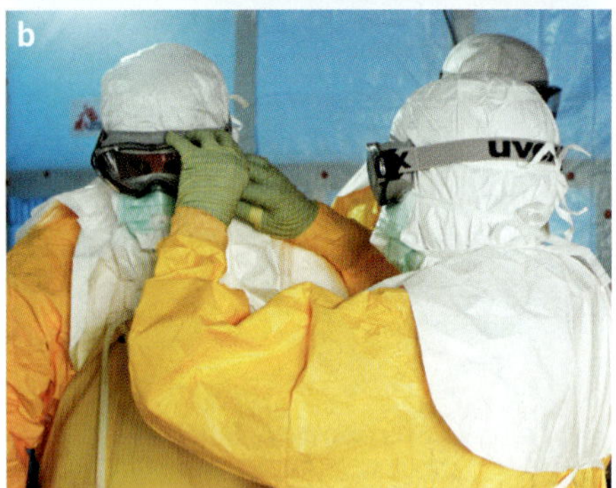

FIGURE 10.2.12 (a) Coloured TEM of the Ebola virus. (b) Personal protective equipment that healthcare workers were issued during the 2014 Ebola outbreak in Sierra Leone

Multiple factors

Multiple factors frequently contribute to an epidemic or pandemic. Existing pathogens may emerge in new locations and among populations that lack the ability to recognise and deal with them early enough to prevent their spread.

Ebola

In 2014, an epidemic of the Ebola virus (Figure 10.2.12) occurred in Western Africa, affecting over 11 000 people in multiple countries. Although this was the largest and longest outbreak, Ebola was first identified in 1976 and had affected small numbers of people previously at least 26 times, mainly in equatorial African countries.

Ebola is an often-fatal virus in humans. The virus is carried by fruit bats, which do not show symptoms of the disease. Ebola entered the human population when people ate bat meat. Ebola is spread through bodily fluids and has a relatively short incubation period—as little as two days.

The 2014 epidemic was an example of an old virus in a new context. Equatorial African countries with a history of local Ebola outbreaks had the experience to recognise it early, the laboratory facilities to identify it and the means to contain an outbreak. West African countries, however, had none of this experience.

Several other factors contributed to Ebola's spread: years of poverty and civil unrest, poor transport and road systems, a highly mobile population, and cultural practices, especially burial practices. In addition, the affected countries had a limited number of healthcare workers and health facilities, and they lacked the knowledge to deal with Ebola. Hospitals lacked the equipment and facilities to handle the numbers of infected people. A delay in the foreign aid response, mistrust by locals towards health workers and a reluctance to change practices all allowed Ebola to spread rapidly.

Today the virus is identified by an immunoassay that detects either the viral antigen or antibodies present in the blood of people who have been exposed to the virus (called an enzyme-linked immunosorbent assay or ELISA). The virus is also detected by viral RNA genome sequencing.

EMERGING DISEASES IN AGRICULTURE AND WILDLIFE

Wildlife and agricultural species are at risk of new diseases emerging from pathogens already in the environment and from introduced species. Australian biosecurity agencies and research organisations conduct intensive surveillance to identify potential pathogens, prevent their entry into the country and prevent them spreading throughout vulnerable habits and farmlands. Some examples of emerging diseases of importance to agriculture and wildlife biodiversity are listed in Table 10.2.2.

TABLE 10.2.2 Examples of emerging diseases affecting Australian wildlife and agricultural plants and animals

Prevalence of disease	Example
new or previously unrecognised	• Hendra virus in horses • bat lyssavirus • devil facial tumour disease • mad cow disease (BSE) • bee *Varoa* mite • amphibian chytrid fungus • sugar cane orange rust • Coral Sea fan fungus
increased in incidence, virulence or range over the last 20 years	• foot-and-mouth disease virus in cattle • blue tongue virus in sheep • wheat stem rust fungus • jarrah dieback (*Phytophthora cinnamomi)* • loggerhead turtle fungus

Exotic species originate in another country. Native species have not coevolved with them and so often do not have any defence against the pathogens of exotic species. Infectious fungal diseases, for example, are emerging globally as major threats to whole ecosystems and to particular plants and animals. Species extinctions are occurring because organisms have not evolved any natural defences against the new diseases (Figure 10.2.13). Human activity, such as travel within and between countries, bushwalking and mountain biking through affected areas, and the import and export of agricultural products, increases the speed at which these pathogens move around the globe.

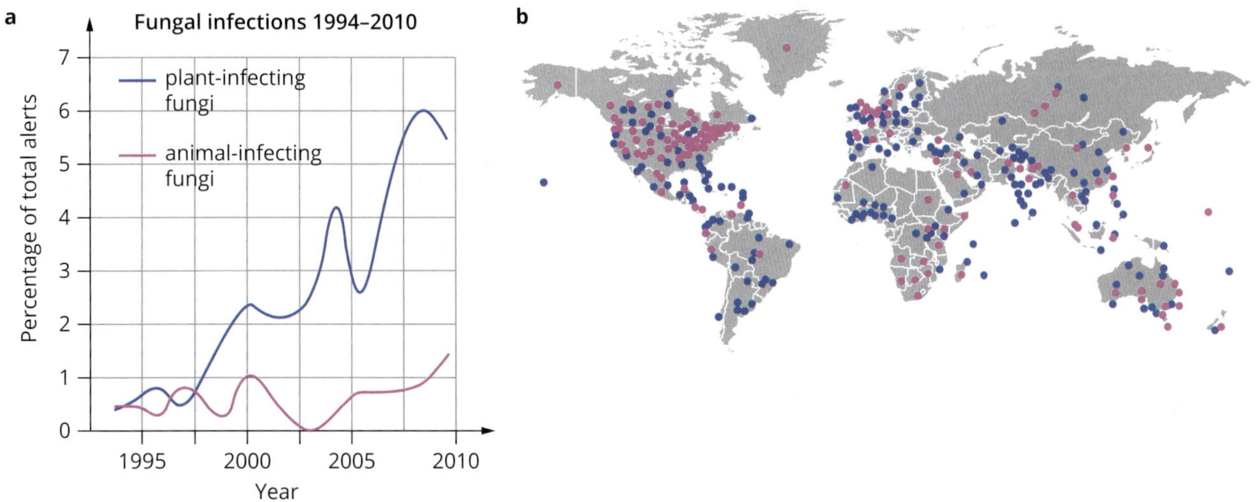

FIGURE 10.2.13 (a) Data showing the global rise in emerging fungal infections of animals and plants. (b) Location of the reported diseases

A significant fungal disease of plants in many parts of the world, and now Australia (including Victoria), is dieback, caused by *Phytophthora*, a genus of water moulds. Although the pathogen has been in Australia since the 1930s, it has recently spread to regions that were previously unaffected and poses a significant threat to Australian biodiversity. *Phytophthora cinnamomi* penetrates the roots and stem to get nutrients, clogging the vascular tissue and starving the plant (Figure 10.2.14a). Other *Phytophthora* species affect commercially important fruit crops (Figure 10.2.14b). Strategies to control dieback range from simple steps for everyone visiting an affected area, such as washing shoes and boots after bushwalking (Figure 10.2.14c), to aerial spraying of large affected areas with potassium phosphonate.

FIGURE 10.2.14 The exotic water mould *Phytophthora* affects many plant species in Australia. (a) *Phytophthora cinnamomi* causes root rot or dieback disease in jarrah forests in Western Australia. (b) *Phytophthora palmivora* causes fruit rot in papaya. (c) A forest worker washes his boots to avoid carrying *Phytophthora* outside an affected area.

PREDICTION AND SURVEILLANCE OF EMERGING DISEASES

Surveillance of infectious diseases involves detecting pathogens and notifying public health organisations at local, regional and global scales. Public health surveillance can vary between cultures, with less developed countries often not having public health bodies to monitor and control disease, or not having the resources for effective management. Further, identifying and treating diseases can vary from culture to culture.

Statistical analysis and mathematical modelling are also used to predict the risk and spread of an infectious disease at an early stage of an epidemic. Global health maps present geographic location data on infectious diseases. National and international bodies use monitoring and surveillance to collect this data. Medical health personnel, public health regulators and private citizens can then use this data, which is presented in health maps. Global health maps can show:

- incidence and prevalence of disease according to geographic location (Figure 10.2.15)
- changing patterns of disease
- mortality rates according to geographic location
- designations of medical aid.

Public health and government organisations, such as the World Health Organization, have websites with interactive global health maps. These are updated with the latest data on current epidemics and pandemics.

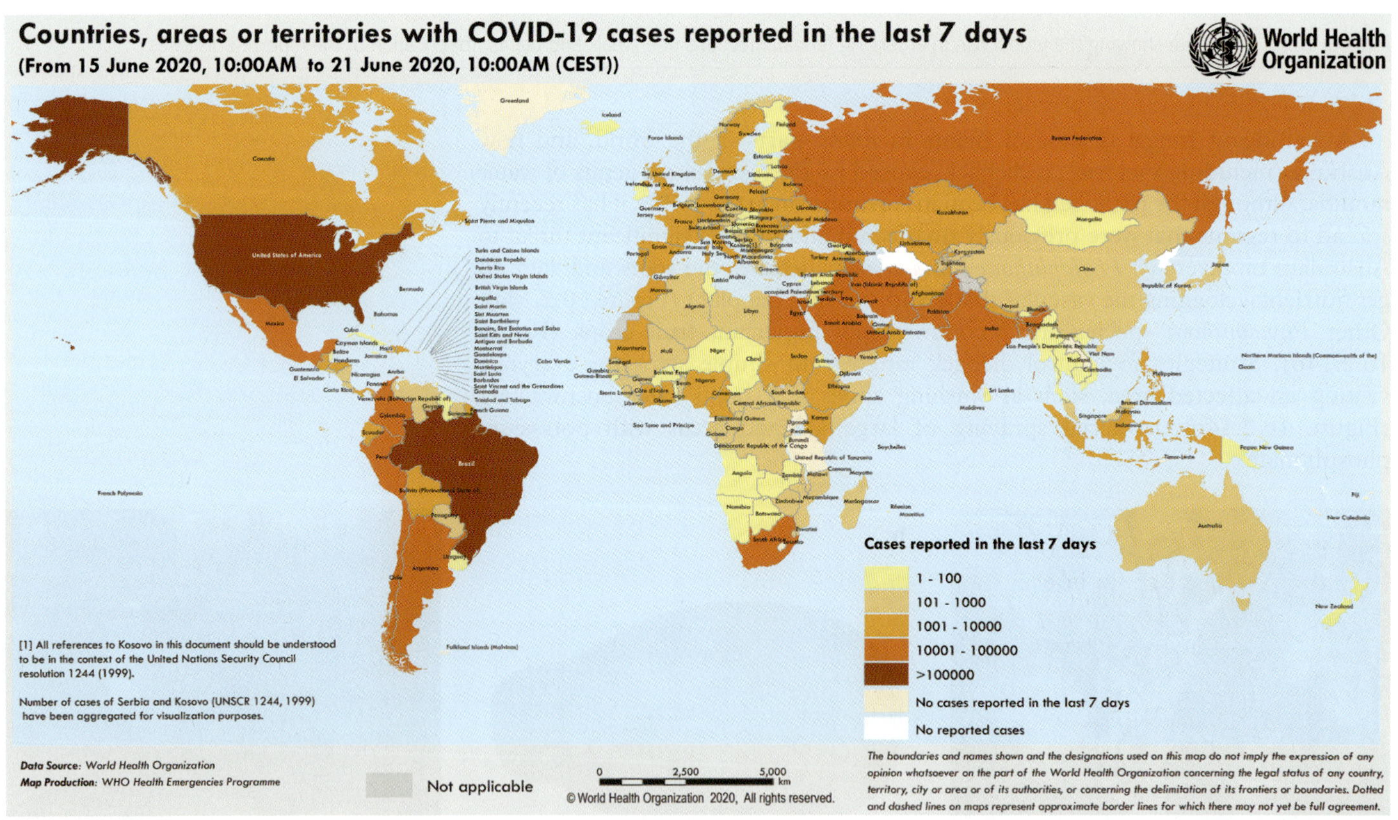

FIGURE 10.2.15 World Health Organization map showing the number of COVID-19 cases reported over 7 days in June 2020

Identification of pathogens

To control a fast-spreading infectious disease, the cause of the disease must be identified and a test that detects the pathogen developed. Testing for the presence of the pathogen is more effective than relying on diagnosing the symptoms of the disease, as many diseases have similar symptoms and some have long incubation periods in which people are infectious without displaying symptoms.

To determine if a person is infected with a particular pathogen, a sample of tissue or blood is taken from the host and tested for the presence of pathogenic proteins or genetic material (DNA or RNA). A commonly used technique is the polymerase chain reaction (PCR), which is used to amplify the genetic material of the pathogen. The amplified genetic material is then sequenced to determine if the pathogen is present in the host's sample.

Another approach to detecting pathogens is to look for the presence of antibodies against the pathogen. A test called an **enzyme-linked immunosorbent assay (ELISA)** can be performed to identify antibodies in blood samples. If a host has been infected by a particular pathogen and had an immune response, antibodies that bind to the pathogen's antigens will be present in the host's blood.

The development of rapid molecular techniques, databases of protein and gene sequences and bioinformatics has made identification of pathogens much faster, enabling a more rapid response and a greater chance of control and containment of disease outbreaks.

Biosecurity and border control

Australia is fortunate to be free of many of the diseases that have damaged agricultural industries in other countries around the globe. As a geographically isolated continent with a natural border of water, many pathogens have been prevented from reaching Australia's shores. However, as travel, trade and global connection increases, so do the chances of introducing new diseases.

Because Australia has been isolated for millions of years, its flora and fauna are highly susceptible to new disease-causing agents. Australia has a strong emphasis on **biosecurity** to prevent, respond to and recover rapidly from pests and diseases, and to help protect agriculture, biodiversity and human health. Australia's Federal Government spends hundreds of millions of dollars on biosecurity each year. Australia has strict regulations to prevent the introduction of new diseases. If you have ever travelled overseas, you will have noticed the screening and inspection of people, luggage and cargo coming into the country. X-ray machines and detection dogs are used for inspection and severe consequences apply to those who do not declare items that could import disease. Australia also implements strict quarantine conditions for animals arriving in the country.

It is also important for Australia to have interstate biosecurity measures in place, as it is such a large country with diverse environments. Some devastating agricultural diseases have entered Australia but have been contained to a particular region. Restrictions apply in each state and territory on the movement of fruit, vegetables, plants, soils, flowers, plant products, agricultural machinery, animals or animal products, and recreational equipment.

> The chain of infection of a pathogen is: reservoir → portal of exit → mode of transmission → portal of entry → susceptible host.

CHAIN OF INFECTION

Infectious disease results from the interaction between a pathogen, a host and the environment. Transmission of a pathogen occurs when it leaves its reservoir, or host, through a portal of exit, and by some mode of transmission enters a portal of entry and infects a susceptible host. This sequence is referred to as the **chain of infection**. Understanding the chain of infection is crucial to containing and controlling the spread of infectious disease.

Reservoir

The **reservoir** of a pathogen is the habitat in which it normally reproduces or replicates. Many infectious diseases have human reservoirs, but as you have learnt, pathogens also have reservoirs in other animals. For example, the natural reservoir for Hendra virus is bats. Environmental reservoirs include plants, soil and water. For example, the bacteria that causes Legionnaires disease is often traced to water in cooling towers and evaporative condensers.

Portal of exit

The **portal of exit** is the path by which a pathogen leaves a host, and it often corresponds to where the pathogen is localised. For example, respiratory infections leave by the respiratory tract, gastrointestinal infections leave by faeces, blood-borne infections can leave by cuts or blood-sucking vectors.

Mode of transmission

The **mode of transmission** is the way in which a pathogen can be transmitted from its natural reservoir to a susceptible host. Modes of transmission can be direct or indirect.

Direct transmission includes direct contact and droplet spread:

- Direct contact includes kissing, skin-to-skin contact, sexual activity and contact with soil or vegetation.
- Droplet spread occurs through sneezing, coughing and even talking. It results from contact with aerosolised droplets and it is classified as direct because it occurs at short range before droplets fall to the ground.

Indirect transmission can be airborne, vehicle-borne or vector-borne:

- Airborne transmission of pathogens occurs when dust or droplet nuclei are suspended in the air. It includes transmission via material that has settled on surfaces and become resuspended in the air. Droplet nuclei are different from droplets; they are very small, dried residue that can be suspended in the air for long periods. The virus that causes measles is spread by droplet nuclei.
- Vehicles are inanimate objects such as food, water, door handles and other inanimate objects that you might ingest or touch before touching your face or mouth.
- Vectors include mosquitoes, ticks, fleas and other organisms that can carry a pathogen (mechanical transmission), and which may also support its growth (biological transmission).

Portal of entry

A **portal of entry** is the manner in which a pathogen enters a host. Portals of entry include the respiratory tract, the gastrointestinal tract (faecal–oral route), the skin, and mucous membranes. Some pathogens can have the same portal of entry and exit. For example, influenza enters and leaves hosts through the respiratory tract. Others can have different portals of entry and exit. For example, hookworm larvae enter through the skin, mature in the gastrointestinal tract and release their eggs in faeces.

Susceptible host

A **susceptible host** is the last link in the chain of infection. Susceptibility is determined by a range of factors including genetics, and whether or not an individual has immunity to a pathogen, or is malnourished or immunocompromised.

CONTAINMENT AND CONTROL OF INFECTIOUS DISEASE

Organisations that monitor global disease, such as the US Centers for Disease Control and Prevention (CDC) and the World Health Organization (WHO), generate priority lists of emerging diseases to direct global surveillance and data collection. They work on early detection, containment (preventing the spread of pathogens to other areas) and early treatment. Local and federal governments also play a critical role in preventing the spread of infectious disease and protecting their population's health. A range of strategies is used to prevent and contain the spread of pathogens and infectious diseases.

- Prevention: Preventative measures include public education campaigns that encourage regular and thorough handwashing with soap, the use of hand sanitiser or other antiseptics when soap is not available, and coughing or sneezing into your elbow instead of hands if no tissues are available. Enhanced cleaning and/or disinfection of surfaces in public transport and other public spaces may be implemented. As cases increase, social (i.e. physical) distancing may become necessary, along with appropriate use of personal protective equipment, such as gloves and face masks, as advised by health authorities. Borders may be closed and travel limited. Contact tracing of people who have come into contact with a confirmed case is vital so they can be isolated to prevent further spread of infection, tested, and treated if necessary.
- Isolation and quarantine: People suspected of being infected, or at high risk of being infected, or at high risk of complications if they become infected, may voluntarily self-isolate. In some circumstances, people may be ordered by government to isolate for a period of time to see if they develop symptoms. The reason for this measure is that it is often possible to be infectious without displaying any symptoms. Infected people may be quarantined in separate hospital rooms or wards, or their own home if they are well enough but still pose a risk of infecting others. Airports may have medical personnel to test people who show symptoms on arrival. Employees may be encouraged to work from home where possible. Schools and other communal places may also be closed until the epidemic or pandemic has slowed.
- Control carriers: For example, cattle and sheep infected with mad cow disease (BSE) and scrapie, and ducks and chickens infected with avian flu are euthanased. Temporary bans on the consumption and commercialisation of meat and egg products may also be put in place.
- Eradication of vectors: Insect vectors such as mosquitoes and midges may be eliminated, or repelled, to control malaria and other vector-borne diseases.
- Response plans: Plans by governments or communities need to be put in place to instigate and enforce all of the above controls for both existing infectious diseases and outbreaks of new diseases. Governments often work together to share information and control the spread of diseases.
- Vaccination: Vaccines often take some time to develop but are effective in preventing future infections.

Figure 10.2.16 summarises the key steps in the containment and control of infectious disease.

Antiseptics such as rubbing alcohol and iodine are substances that stop or slow the growth of microorganisms. The antiseptics can be applied externally in line with the directions, to living tissue such as skin.

Disinfectants such as bleach are substances that inactivate or destroy pathogens on inert surfaces. They are not safe to use topically (on living tissue) or for ingestion.

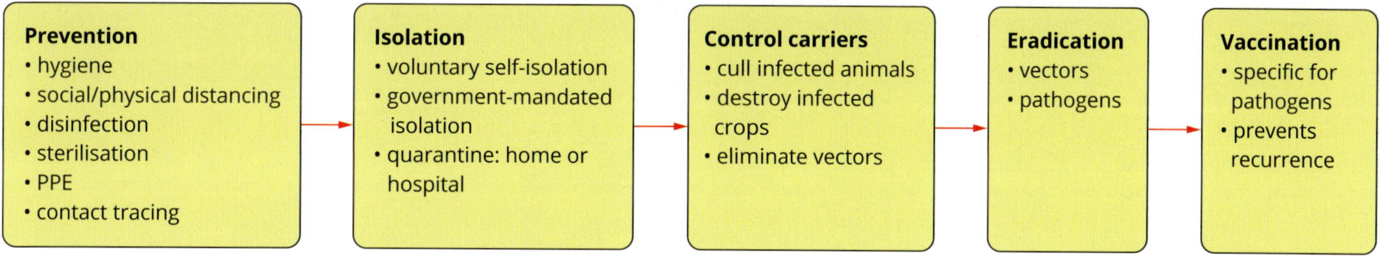

FIGURE 10.2.16 Key steps in the containment and control of infectious diseases

CASE STUDY ANALYSIS

Bats, horses and Hendra virus

A new disease struck racehorses and their handlers on a Queensland property in 1994. Within just a few days, 14 horses and their owner had died. Government departments and the CSIRO Australian Animal Health Laboratory in Geelong moved into action to identify the mystery disease. The culprit was the Hendra virus, an RNA virus in the genus *Henipavirus* (Figure 10.2.17).

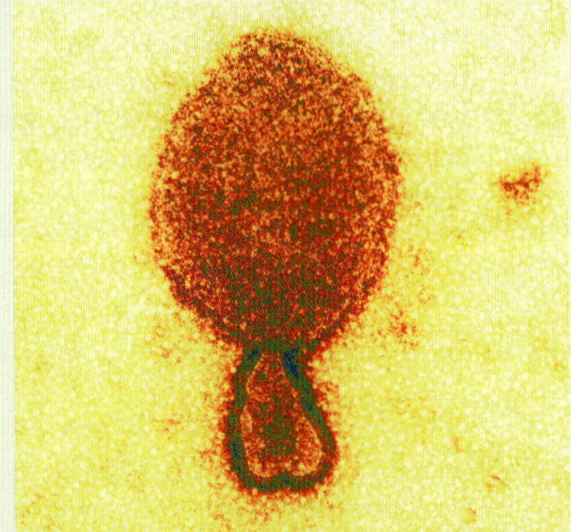

FIGURE 10.2.17 Coloured TEM of Hendra virus

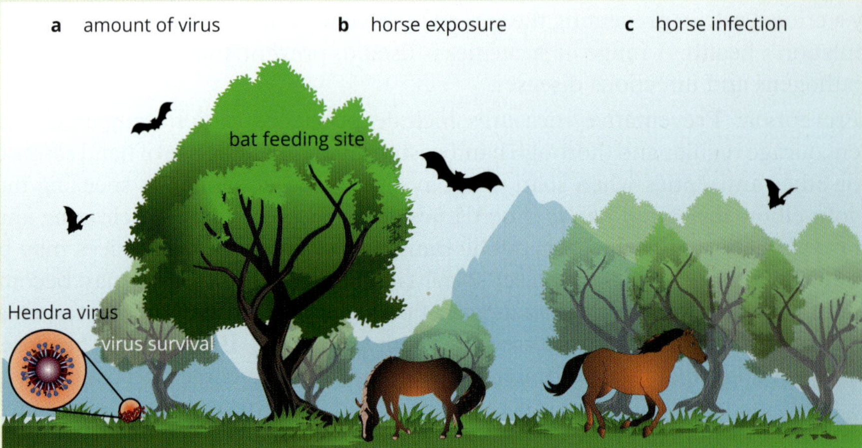

FIGURE 10.2.18 Risk factors for the development of Hendra virus infection by horses include (a) the concentration of virus, (b) the exposure of horses to the virus and (c) whether the horse contracts the disease.

The natural host of the Hendra virus is the Australian fruit bat, also called the flying fox. The fruit bat species most associated with the Hendra virus is the black flying fox (*Pteropus alecto*) that resides in Queensland. The virus has not been found in Victoria's grey-headed flying foxes (*Pteropus poliocephalus*), so the risk of this disease in Victoria is low. However, it is possible that a horse incubating the virus could be brought into Victoria and transfer the virus to other horses and to human handlers.

Exactly how the Hendra virus is transmitted from bats to horses is not fully understood. It is likely that horses are infected by ingesting or inhaling the virus in droplets of fluid secretions from bats, when horse properties overlap with bat roosts. The virus is transmitted from horse to horse through direct contact with infectious body fluids (Figure 10.2.18).

The virus can move from horse to human through contact with infected mucus and/or blood or other body fluids. Hendra virus does not appear to be transmitted from person to person. Horse owners and vets need to be aware of the signs of Hendra virus infection, which include fever, increased heart rate and breathing difficulty, and they must seek treatment early. Prevention by immunisation is best.

Equivac HeV is a vaccine for the immunisation of horses against Hendra virus. This is the most effective way to prevent the spread of Hendra virus and protect horses, horse owners and vets.

Analysis

1 Outline three risk factors that could lead to the spread of the Hendra virus.

2 Outline how communities, government organisations and private industries play a role in disease control.

10.2 Review

OA ✓✓

SUMMARY

- The emergence of new and recurring pathogens poses a challenge to global health.
- An epidemic is the sudden increase in the number of cases of a disease in a localised area.
- A pandemic is an epidemic that has spread over several countries or continents.
- Human demographics and behaviours, mobility, cultural traditions, changes in lifestyle, sanitation and general hygiene contribute to the emergence and spread of pathogens.
- Agricultural practices and procedures may contribute to the introduction and spread of pathogens in animals and crops.
- Overuse and inappropriate application of antimicrobial drugs have contributed to the evolution of resistant pathogens.
- Zoonotic diseases pose a threat to human populations that have close contact with natural host animals.
- Control of zoonotic and vector-borne diseases requires control of the natural host, the vector and the pathogen.
- Control of disease outbreaks requires detection, identification and containment of the affected population.

KEY QUESTIONS

Knowledge and understanding

1 What is an emerging infectious disease?

2 How can an organism acquire the ability to become pathogenic and cause disease in humans?

3 **a** Define epidemic.
  b Define pandemic.

4 What is a zoonotic disease? Give an example of this type of disease.

5 Give examples of emerging pathogens influenced by each of the following factors:
  a human demographics, demographic change and/or mobility
  b human behaviour
  c farming practices
  d overuse of an antimicrobial agent
  e poor sanitation
  f limited social, transport or health infrastructure
  g close association between wildlife and domestic animals

6 Why does the influenza virus repeatedly emerge each year, and sometimes cause a pandemic?

7 Describe briefly how government agencies and research organisations can help prevent the emergence of new pathogens that may impact on agriculture and wildlife biodiversity.

Analysis

8 How is scientific knowledge applied to deal with the emergence of new pathogens?

10.3 Vaccination programs

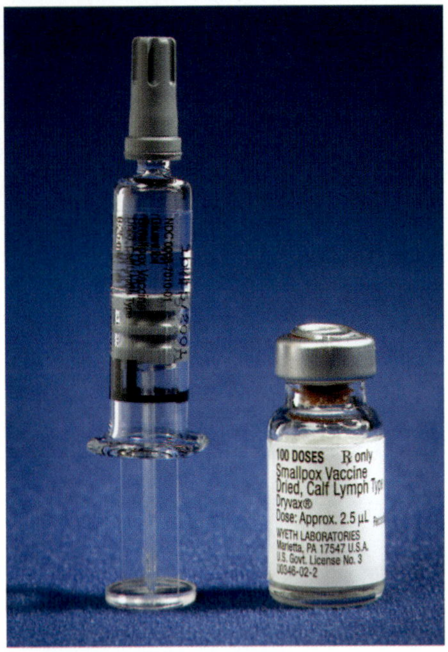

FIGURE 10.3.1 Vaccination helped eradicate smallpox.

In Section 9.3 you learnt that vaccines result in artificial active immunity. In this section you will learn about the different types of vaccines and how vaccination programs can reduce the spread of infection in a population through herd immunity.

Vaccination programs aim to decrease the incidence of many diseases, with the purpose of ultimately eradicating them (Figure 10.3.1). Through the introduction of vaccination programs, many countries have dramatically reduced new cases of diseases that were common a century ago. For example, in the early 1900s, Australia still experienced epidemics of diseases such as smallpox, polio and measles, but these diseases have either been eliminated (in the case of smallpox) or are now very rare.

Australians enjoy the benefits of vaccination not only because safe and effective vaccines were developed, but also because everyone in Australia receives many vaccinations free of charge as part of the National Immunisation Program (Table 10.3.1). Since the introduction of vaccination for children in 1932, deaths from vaccine-preventable diseases have fallen by 99%, despite a threefold increase in the Australian population over that time.

TABLE 10.3.1 Australian child and school immunisation programs

Child programs	
Age	**Vaccine**
birth	• hepatitis B (hepB)
2 months	• hepatitis B, diphtheria, tetanus, acellular pertussis (whooping cough), *Haemophilus influenzae* type b, inactivated poliomyelitis (polio), (hepB-DTP-a-Hib-IPV) • pneumococcal conjugate (13vPVC) • rotavirus
4 months	• hepatitis B, diphtheria, tetanus, acellular pertussis (whooping cough), *Haemophilus influenzae* type b, inactivated poliomyelitis (polio), (hepB-DTP-a-Hib-IPV) • pneumococcal conjugate (13vPVC) • rotavirus
6 months	• hepatitis B, diphtheria, tetanus, acellular pertussis (whooping cough), *Haemophilus influenzae* type b, inactivated poliomyelitis (polio), (hepB-DTP-a-Hib-IPV) • pneumococcal conjugate (13vPVC) • rotavirus b
12 months	• *Haemophilus influenzae* type b, meningococcal C (Hib-MenC) • measles, mumps and rubella (MMR)
18 months	• measles, mumps, rubella and varicella (chickenpox) (MMRV)
4 years	• diphtheria, tetanus, acellular pertussis (whooping cough) and inactivated poliomyelitis (polio) (DTPa-IPV) • measles, mumps and rubella (MMR) (to be given only if MMRV vaccine was not given at 18 months)
School programs	
Age	**Vaccine**
10–15 years	• varicella (chickenpox) • human papillomavirus • diphtheria, tetanus and acellular pertussis (whooping cough) (dTpa)

TYPES OF VACCINES

Live attenuated vaccines

Live attenuated vaccines involve a living microbe that has been weakened in the laboratory, usually through repeated culturing.

The advantage of live attenuated vaccines is that a single dose usually provides long-lasting immunity, because the vaccines induce a strong adaptive immune response that produces many types of antibodies directed against multiple antigens.

The disadvantages of live attenuated vaccines are that although they are safe for most people, they may cause disease in those with weakened immune systems. Also they may cross the placenta in pregnant women and cause damage to developing fetuses.

Attenuated vaccines are more commonly used for viruses than for bacteria, because bacteria have thousands of genes and thus are much harder to control. Examples of attenuated vaccines include those against measles, mumps, rubella and polio.

Inactivated vaccines

Inactivated vaccines, also known as killed vaccines, contain microbes that have been inactivated by heat, radiation or chemical means.

The advantage of inactivated vaccines is that they result in the production of many different antibodies, due to the fact they contain many different antigens. Inactivated vaccines can safely be used in people who have weakened immune systems.

The disadvantage of inactivated vaccines is that they stimulate a relatively weak immune response compared to live, attenuated vaccines, so they require booster doses to achieve and maintain long-term immunity. Adjuvants can be added to inactivated vaccines to help boost the immune response.

Most vaccines against bacteria are inactivated vaccines. Examples of inactivated vaccines include the inactivated rabies and hepatitis A vaccines.

> Adjuvants are added to vaccines to stimulate a stronger immune response against antigens administered at the same time; examples include aluminium phosphate and aluminium hydroxide.

Subunit vaccines

Like inactivated vaccines, **subunit vaccines** do not contain any live microbial components. Unlike inactivated whole-cell vaccines, subunit vaccines contain only parts of microbes selected for their ability to induce an adaptive immune response. Subunit vaccines can contain a fraction of an antigen, a single antigen or multiple antigens, and these antigens can be proteins, detoxified toxins or polysaccharides. Those that contain multiple antigens induce a broader immunity, because they will induce the production of antibodies directed against multiple antigens.

Subunit vaccines share the advantages of inactivated vaccines in that they are safer, more stable, and easier to store than live attenuated vaccines. However, subunit vaccines also share the disadvantages of inactivated vaccines in that they require multiple doses and an adjuvant to improve the strength of the immune response.

Subunit vaccines are made by growing the pathogen in the laboratory and chemically extracting the antigens, or by using **recombinant DNA technology**. An example of a recombinant subunit vaccine is one that has been genetically engineered to produce a purified component of the protein coat of the virus that causes foot-and-mouth disease. Recombinant DNA technology can also be used for live vaccines, to genetically modify microbes so that they elicit an immune response but do not cause illness.

BIOFILE

Human papillomavirus vaccine

Human papillomavirus (or HPV) is a virus that can result in the development of certain types of cancer, including cervical cancer. Professor Ian Frazer and Dr Jian Zhou from the University of Queensland developed a subunit vaccine for HPV. In 2007, Australia became the first country to roll out a national HPV vaccination program. The program originally only covered girls, but was extended to cover boys in 2013.

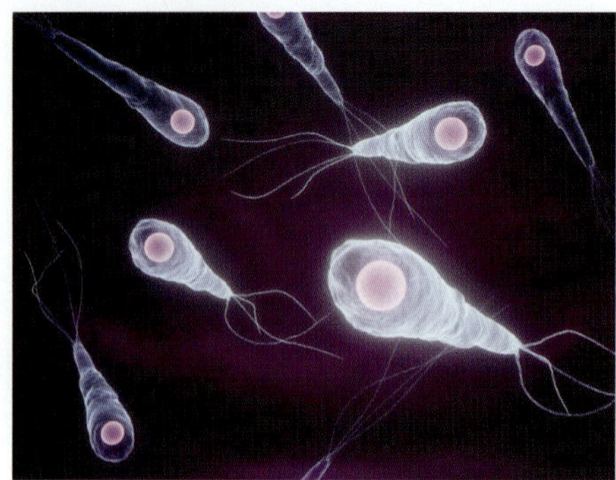

FIGURE 10.3.2 Digital illustration of *Clostridium tetani*, the bacterium that produces a toxin called tetanospasmin

Toxoid vaccines, a type of non-recombinant subunit vaccine, use toxins inactivated by formalin (called toxoids) to stimulate an adaptive immune response. Although the toxoid is inactivated, it remains similar enough to the original toxin that the immunological memory for the toxoid is also effective for the toxin. For example, *Clostridium tetani* is a bacterium that produces a neurotoxin that causes tetanus, and the inactivated toxin is used in the vaccine for tetanus (Figure 10.3.2). Another example of a toxoid vaccine is the diphtheria vaccine. Toxoid vaccines often require multiple doses to achieve immunity.

HERD IMMUNITY

Immunisation is critical, not only for the person immunised, but also for the health of the wider community. For an immunisation program to be successful, a sufficient number of people need to be vaccinated—a phenomenon called **herd immunity** (Figure 10.3.4). The more people who are vaccinated, the less chance there is of a pathogen spreading throughout a population, because there will be fewer potential carriers. Herd immunity is essential for the protection of those who cannot be vaccinated or who have suppressed immune systems. This includes newborn babies, the elderly, people suffering from an immune disease and people taking immunosuppressant medication.

CASE STUDY

Edward Jenner

Edward Jenner (Figure 10.3.3) was the English doctor who discovered a vaccine for smallpox. Jenner had heard of a milkmaid who could not catch smallpox because she had already had cowpox (a mild infection that is not fatal). To test this, in 1796, Jenner infected a young boy with cowpox in the hope of preventing infection with smallpox. He then infected the boy with smallpox, by injecting pus from a smallpox lesion under the boy's skin. The boy did not develop smallpox, providing evidence that inoculating a person with cowpox virus provided immunity against smallpox.

Despite the development of the smallpox vaccine, smallpox persisted in Asia, Africa and South America for many years. In 1966, there was worldwide action to eradicate smallpox, and the last case of epidemic smallpox was registered in 1977 in Somalia, Africa. The only confirmed cases of smallpox after this time were in 1978. They involved a medical photographer called Janet Parker from the University of Birmingham, and her mother. Parker became infected after smallpox travelled through air ducting connecting her office and a laboratory. Although her mother survived, Parker did not.

FIGURE 10.3.3 Edward Jenner (1740–1823)

On 8 May 1980, the World Health Organization announced smallpox had been eradicated. Since then, there have been no vaccinations against smallpox. Laboratories have also greatly improved infection control standards since the time of Parker, and the smallpox virus is now stored safely in only two laboratories (one in Atlanta, USA, and the other in Novosibirsk, Russia). The virus is stored in case there is a need to produce the vaccine again in the future.

Factors that contributed to the eradication of smallpox included the lasting immunity achieved by vaccination and in those who recovered from infection, as well as the fact that the smallpox virus only infected humans and there were no natural reservoirs of infection.

a

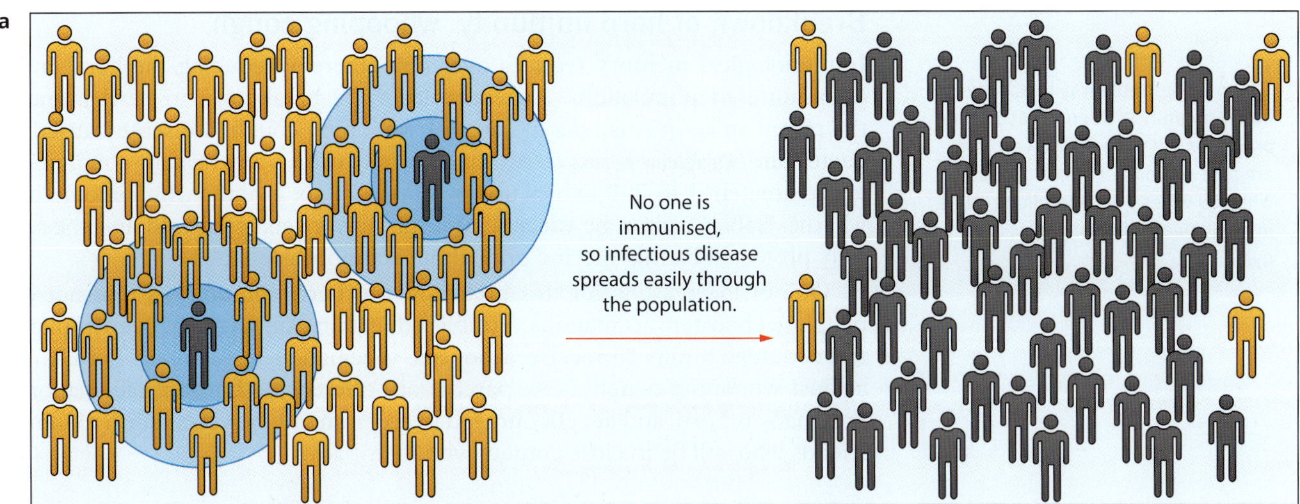

b

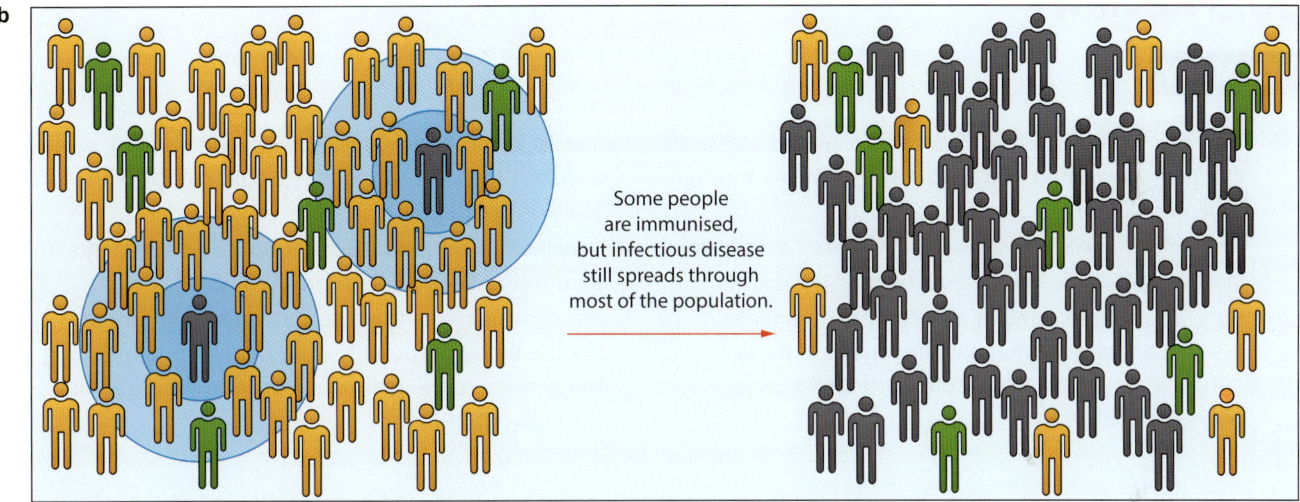

c

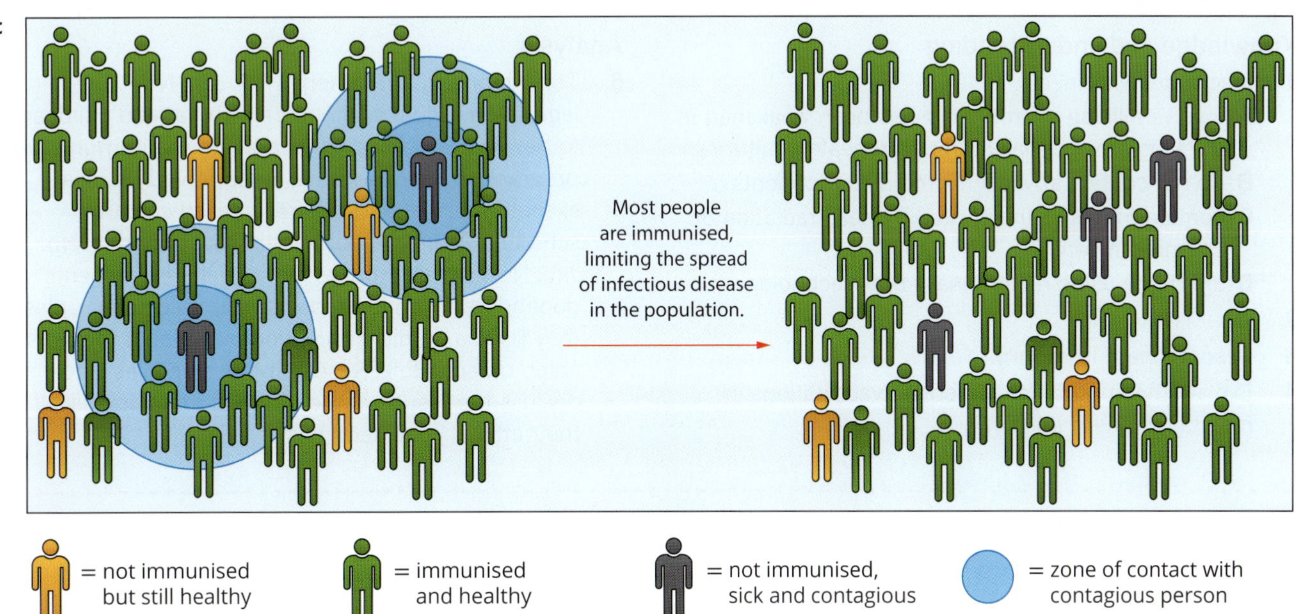

= not immunised but still healthy

= immunised and healthy

= not immunised, sick and contagious

= zone of contact with contagious person

FIGURE 10.3.4 Three diagrams illustrating the effectiveness of herd immunity. (a) With no immunisation in a community, infectious diseases spread easily. (b) With some immunisation in a community, infectious diseases spread less easily. (c) When most of the community is immunised, there are few carriers or infected people and minimal spread of infectious diseases. This is known as herd immunity.

Breakdown of herd immunity: whooping cough

> The most desirable way to achieve herd immunity is through vaccination programs. This is because vaccines are safe and do not cause disease, whereas widespread infection results in many people becoming sick and dying.

Immunological memory reduces over time, thereby reducing the herd immunity of immunised populations. An example of the breakdown in herd immunity is the recent spike in Australia in cases of whooping cough, a disease caused by the bacterium *Bordetella pertussis*. Although it only causes a persistent cough in adults, approximately 1 in 200 babies under the age of six months who become infected will die. Babies cannot be vaccinated until they are six weeks old and they are not fully protected by this vaccine until about six months of age.

One of the reasons for this breakdown in herd immunity is that not enough people get booster vaccinations. A public education campaign has been implemented to encourage adults to receive a booster vaccination to maintain herd immunity against whooping cough. New parents are offered the booster vaccination when their baby is born, and are encouraged to recommend the vaccination to family and friends who will be in close contact with their baby.

10.3 Review

OA ✓✓

SUMMARY

- Live attenuated vaccines involve a living microbe that has been weakened in the laboratory, usually through repeated culturing.
- Inactivated (or killed) vaccines are composed of microbes inactivated by heat, radiation or chemical means.
- Subunit vaccines contain a fraction of an antigen, a single antigen or multiple antigens, and these antigens can be proteins, detoxified toxins or polysaccharides.
- Toxoid vaccines are non-recombinant subunit vaccines that use toxins inactivated by formalin (called toxoids).
- Vaccination programs are set up by governments in an effort to produce herd immunity.
- Herd immunity is the result of large numbers of people being immune to a pathogen, thus reducing the chance of the pathogen successfully spreading through a population.

KEY QUESTIONS

Knowledge and understanding

1 Inactivated vaccines:
 A involve a living microbe that has been weakened in the laboratory, usually through repeated culturing
 B do not contain any live microbial components
 C contain microbes inactivated by heat, radiation or chemical means
 D are made using recombinant DNA technology

2 What is a toxoid?

3 Describe herd immunity.

4 Explain the importance of booster vaccinations in maintaining herd immunity.

Analysis

5 The Victorian Government's 'No Jab, No Play' legislation requires all children who attend childcare to be vaccinated and up-to-date on the immunisation program, unless they have an approved medical exception or are part of a recognised catch up schedule. The Commonwealth Government's 'No Jab, No Pay' also makes family welfare payments dependent on children being fully vaccinated unless they have a medical exception or are on a recognised catch up schedule. Research these policies, and discuss the results of their implementation as well as their ethical considerations.

Chapter review

10

KEY TERMS

adjuvant
antibiotic
antimicrobial drug
antimicrobial resistance
antiviral
autoimmune disease
benign tumour
biosecurity
bispecific monoclonal antibody (bispecific mAb)
cancer vaccine
carcinogen
chain of infection
chimeric monoclonal antibody (chimeric mAb)
conjugated monoclonal antibody (conjugated mAb)
cytotoxic
dendritic cell
emerging infectious disease
enzyme-linked immunosorbent assay (ELISA)
epidemic
herd immunity
horizontal gene transfer
human monoclonal antibody (human mAb)
humanised monoclonal antibody (humanised mAb)
hybridoma
immortal cell line
immunotherapy
inactivated vaccine
incidence
infectious disease
live attenuated vaccine
malignant tumour
metastasis
mode of transmission
monoclonal antibody (mAb)
myeloma cell
pandemic
pathogen
portal of entry
portal of exit
prevalence
radioactive
recombinant DNA technology
reservoir
subunit vaccine
susceptible host
toxoid vaccine
tumour
virulence
zoonotic

REVIEW QUESTIONS

Knowledge and understanding

1 The antibody shown below is:
A chimeric
B bispecific
C conjugated
D humanised

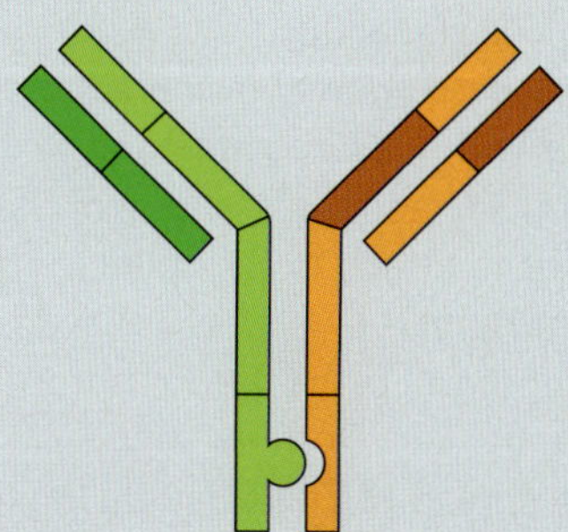

2 The use of alcohol-based hand sanitiser has become very common. This is a type of:
A antiseptic
B disinfectant
C antibiotic
D antiviral

3 Major epidemic diseases introduced when the British colonised Australia that were responsible for decimating Indigenous Australian populations included:
A coronavirus, tuberculosis, measles, smallpox, syphilis
B influenza, tuberculosis, measles, smallpox, syphilis
C SARS-CoV, tuberculosis, Kuru, smallpox, syphilis
D influenza, HIV, measles, smallpox, syphilis

4 As a result of introduced diseases and conflict, by up to how much did Indigenous Australian populations decrease in the ten years following the arrival of the First Fleet?
A 50%
B 70%
C 90%
D 99%

5 Which list only contains monoclonal antibodies?
A Adalimumab, golimumab, infliximab, pimecrolimus, ritixumab
B Adalimumab, baricitinib, golimumab, infliximab, ritixumab
C Adalimumab, golimumab, infliximab, methotrexate, ritixumab
D Adalimumab, belimumab, golimumab, infliximab, ritixumab

6 Define virulence.

7 Briefly describe what autoimmune diseases are and how they occur.

8 What is the difference between non-specific and specific immunotherapies?

9 Effective immunotherapy is one of the goals of modern research into treatment for cancer.
a Briefly describe the three traditional methods of combatting cancer.
b Explain the advantages of immunotherapy compared to traditional cancer treatments.

10 What is the difference between a benign tumour and a malignant tumour?

11 Why are malignant tumours more difficult to treat than benign tumours, and much more likely to recur?

12 Using an example, explain how an epidemic can be caused by increased exposure to disease.

13 Give an example of a disease epidemic caused by a virus and explain how it managed to reach epidemic levels.

14 Why are biosecurity and border control important for Australia's agricultural industry?

15 How does antibiotic resistance arise and spread through bacterial populations?

16 MRSA has become a serious problem in most major hospitals around the world.

- a What is MRSA?
- b Why is it an issue?
- c Explain the relationship between MRSA and the inappropriate use of antibiotics.
- d Why is it important to continuously search for new antimicrobial substances?

Application and analysis

17 The use of monoclonal antibodies to deliver cancer drugs has been shown to have fewer side effects than conventional methods of drug delivery, and much lower doses are needed to achieve the same level of response in patients. Using your understanding of how monoclonal antibody therapy works, explain these two observations.

18 Classify the following as prevention or treatment.

- a applying iodine before making a surgical incision
- b taking a course of antibiotics for a bacterial chest infection
- c having a vaccination against cholera before going overseas
- d washing hands after going to the toilet
- e isolating an exotic bird in quarantine after it has been found in luggage by customs officials

19 Scientists at a biotechnology company have developed a vaccine that they say will immunise people against peanut allergy. They wish to test their vaccine in mice. They have a number of genetically similar laboratory mice that they are going to use for their initial experiments.

- a What is one characteristic that the mice must all have in order to make the experiment valid?
- b Design an experiment to test the vaccine using the mice as subjects.
- c What result would provide evidence that this vaccine is successful?
- d Name the dependent and independent variables in your experiment.
- e One mouse in the trial has an adverse reaction to the vaccine. Should the company abandon the trial at this point? Explain your reasoning.

20 Cancer is characterised by cells that undergo unchecked reproduction, competing with body tissues for space and nutrients. Leukaemia is a cancer of the white blood cells. In this condition, the white blood cells of the bone marrow reproduce prolifically. One treatment for this cancer is radiation therapy to destroy a large percentage of the cancerous cells, followed by a bone marrow transplant.

- a Suggest why radiation therapy is an important part of the treatment regime for leukaemia.
- b Explain why it is important that the closest possible tissue match is made between a donor and the recipient for a bone marrow transplant to be successful.

21 a Research and describe the fire blight infection, including the pathogen, the type of organisms affected and how it can be controlled.

- b Discuss any biosecurity issues relating to fire blight and the Australian fruit industry.

22 Zika virus is a mosquito-transmitted virus that has been linked to an increase in birth defects in South America. It has been postulated that it has caused an increase in the number of babies born with microcephaly (a small skull and brain). There is currently no vaccine or treatment for Zika virus, but considerable research efforts are being directed towards the development of both. Preliminary research indicates that the virus uses a receptor on cell surfaces called AXL to attach to cells and gain entry. This receptor is one in a family that is very important in cell signalling.

- a i Draw a labelled diagram showing a hypothetical AXL receptor with the virus attached to it.
  - ii What can you say about the shape of the protein the virus uses to attach to the cell and the shape of the AXL receptor?
- b Design a drug that will block the virus from attaching to and entering the cell. Draw its shape.
- c How might the spread of Zika virus be limited even in the absence of a vaccine or treatment?

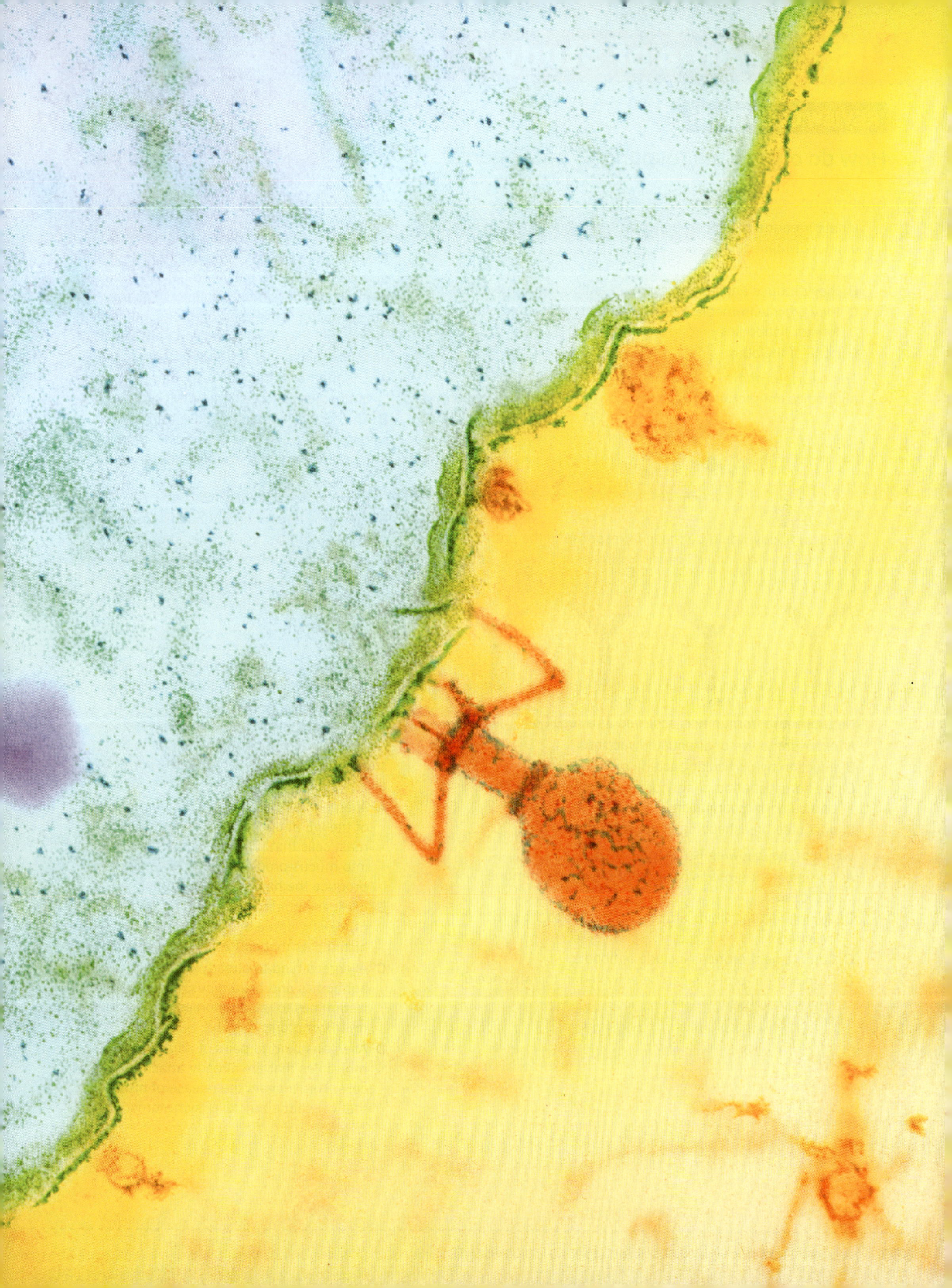

REVIEW QUESTIONS

How do organisms respond to pathogens?

Multiple-choice questions

1 Innate immune responses are critical to maintaining the health of an individual because:

A they are specific to the antigens on pathogenic organisms
B they produce antigens, which bind antibodies to mast cells
C they provide immediate and continuous protection against foreign antibodies
D none of the above

2 The pathogen shown below enters a human body and antibodies are produced against it.

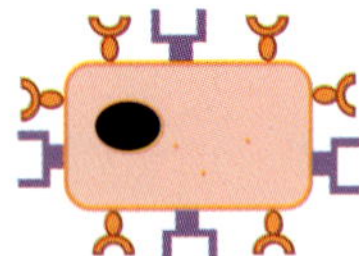

Which antibody would be made in response to the pathogen?

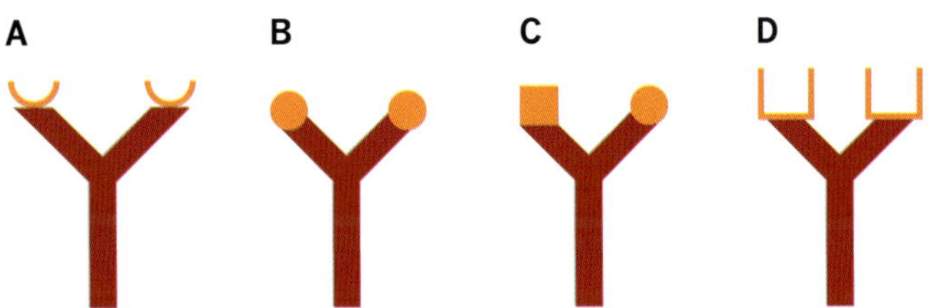

3 Natural active immunity is achieved as a result of:

A exposure to live or attenuated vaccines
B infection by particular bacteria or virus
C the administration of antibodies or antitoxin specific to a particular microorganism
D adequate breastfeeding in newborn infants

4 Which of the following is not true?

A Complement proteins are part of the innate immune response.
B Complement proteins attract phagocytes to the site of the infection.
C Complement proteins include antibodies.
D Complement proteins can be activated in the absence of antibody–antigen reactions.

5 A baby was born severely jaundiced and anaemic. The baby recovered after a blood transfusion, but concerned doctors tested the blood of both the baby and her mother. It was discovered that the baby and her older brother were positive for an antigen known as the Rhesus factor, while their mother's antigen was Rhesus-negative. The jaundiced appearance and anaemia of the newborn occurred because:

A The mother's antigens recognised the baby's antibodies as foreign and attacked them.
B The mother had an adaptive immune response to the Rhesus factor in the blood of her first child, and memory cells triggered an immune response to the Rhesus factor in the blood of her second child.
C The baby developed antibodies in response to her mother's Rhesus-negative blood and caused an immune response.
D The Rhesus antigen present on the mother's red blood cells caused an immune response to the baby's blood, which was free of the antigen.

6 Hay fever is an overreaction to previously encountered allergens such as animal fur, pollen or dust. The body produces IgE in response to the first contact with these allergens. This antibody then binds to the surface of mast cells. Which is the correct allergic response?

A Allergens bind to the specific binding sites of the IgE antibodies, which then bind to mast cells that migrate via the bloodstream into mucus-producing tissues where they produce the hay fever symptoms.
B Allergens bind to mast cells in the mucus-producing tissue and trigger them to release histamines that cause the allergic symptoms.
C Allergens bind to already-bound IgE antibodies and histamines. The binding of histamines to IgE antibodies causes the hay fever symptoms.
D Allergens bind to pairs of adjacent IgE molecules that are already attached to mast cells. This triggers the release of histamines that cause the hay fever symptoms.

7 Blincyto is a bispecific monoclonal antibody that is used to treat certain types of acute lymphocytic leukaemia. It is successful in treating cancer by binding the B cell leukaemia with a T lymphocyte. Which of the following scenarios best describes the structure of the monoclonal antibody used and the role of the T lymphocyte in reducing the cancerous cells?

A The monoclonal antibody would be loaded with a toxic chemical that is presented to the infected B cell. This induces programmed cell death (apoptosis) directly with the B cell, and T lymphocytes phagocytose the remnants of the cellular material to prevent the cancer spreading.

B The monoclonal antibody would have two different variable regions present. One variable region attaches to a protein on the leukaemia cell and the other attaches to a T lymphocyte. Once the T lymphocyte is activated, it releases cytokines that induce apoptosis of the leukaemia cell.

C The monoclonal antibody would have identical antigen binding sites to bind directly with the infected B cell. The T lymphocyte then releases cytokines, which increase inflammation in the region of the infected cancerous cells. Inflammation triggers rapid proliferation of plasma cells to secrete more antibodies.

D The monoclonal antibody would have both B cell leukaemia and T lymphocyte antigen-binding sites present. The T lymphocyte would release antibodies that are complementary to the B cell and this would result in agglutination of the infected cells for macrophages to engulf.

8 Media reports sometimes refer to large-scale diseases as epidemics or pandemics. Identify the correct distinction between the two terms.

A The bubonic plague bacterium caused epidemics and pandemics in earlier, less hygienic times.

B A pandemic is a rapid and unusual spread of an infectious disease; an epidemic is a pandemic that has become more global.

C An epidemic is a rapid and unusual spread of an infectious disease; a pandemic is an epidemic that has become more global.

D Human movements around the globe tend to cause epidemics and pandemics, such as the SARS outbreak.

Short-answer questions

9 Hay fever is an allergic response in which the immune system overreacts to the presence of a previously encountered foreign antigen. The diagram below illustrates the key steps that occur in an allergic reaction.

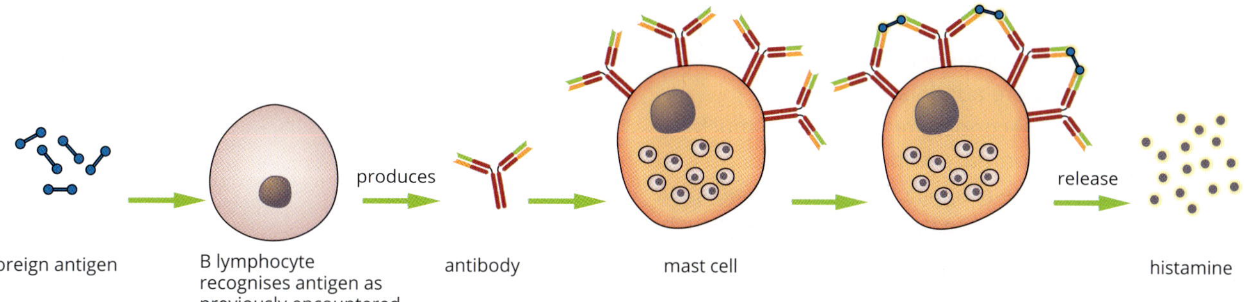

a Define the term antigen.

b Identify the type of antibody involved in allergic reactions.

c Where in the body are mast cells located?

d Describe the event that causes mast cells to release histamine.

e Describe two effects resulting from the release of histamine.

10 The lymphatic system has a number of important functions.

a List the major functions of the lymphatic system.

b How do lymph nodes assist the efficiency of the adaptive immune response?

c How do primary and secondary lymphoid tissues differ?

11 A number of cases of measles has been reported to authorities in recent times. Measles is a preventable disease caused by a virus. In 2013 there were 96 000 deaths worldwide from measles. It is the disease with the highest mortality rate of all vaccine-preventable diseases. Most people in Australia are vaccinated against it in childhood. The vaccination schedule requires children to be given one injection at 12 months and a second injection at 18 months. The antibody response to the vaccinations is shown in the graph below.

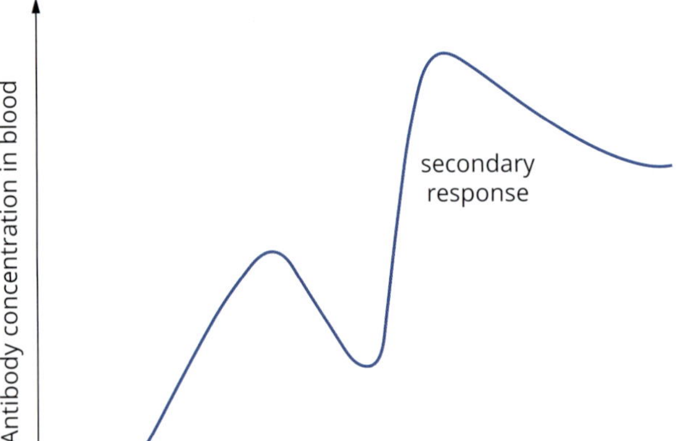

a Mark on the graph when the first and second immunisations were administered.

b Explain why the secondary response is so much greater than the primary response.

c T lymphocytes play a significant role in the adaptive immune response of the body against viruses such as the one that causes measles. Describe the role of T lymphocytes in immunity.

d In February 2016, Department of Health statistics showed that 93.58% of five-year-old children in Victoria were fully immunised against measles. In some areas, the immunisation rate is as low as 73%. Why is an immunisation rate below 90% a cause for concern?

12 Tetanus is a serious, often fatal disease caused by the bacterium *Clostridium tetani*. The most serious of the symptoms are caused by the toxin produced by the bacteria. The toxin enters neurons in the central nervous system where it blocks the release of the neurotransmitters glycine and GABA. These neurotransmitters stimulate neural pathways that inhibit the contraction of muscles.

a Explain what effect the toxin will have on muscle behaviour.

b **i** A vaccination is available for tetanus. It involves injection of extremely minute amounts of the toxin. Explain how the injection of this tiny amount of toxin gives protection from tetanus.

ii Does this produce active or passive immunity? Provide evidence to support your answer.

iii It is recommended that everyone has a booster vaccination for tetanus every 10 years. Why is this necessary?

c An individual who steps on a rusty nail is at significant risk of developing tetanus if they have never been vaccinated. Such individuals are given an injection of a preparation that has been created in horses. This preparation will protect the person against tetanus.

i What is the active constituent of the injection?

ii Is this an example of active or passive immunity?

iii Explain why this type of treatment would not give long-term protection against tetanus.

13 Over the course of two years from 1789, after the arrival of Europeans in Australia, a smallpox outbreak is estimated to have killed between 50% and 70% of the Indigenous Australian populations.

a Explain why the disease had such a devastating impact on Indigenous Australian populations.

b Discuss why Europeans did not succumb to the outbreak as badly as Indigenous Australians, with reference to their immune function.

14 A range of infectious diseases was examined for their ability to cause potential damage within the Australian population.

Disease	Probability of infection given contact between an infected person and a susceptible person (transmissibility)	Rate of contact (probability) between an infected and a susceptible person	Duration of infectiousness (days)	Mortality (%)	Pathway of spread	Type of pathogen
A	0.9	0.95	7	4	foodborne	bacterium
B	0.2	0.05	14	30	airborne droplets	virus
C	0.35	0.4	20	0.2	insect vector	protozoan
D	0.8	0.6	11	12	contaminated water	bacterium
E	0.1	0.9	6	0.2	direct contact	virus

a A student stated that disease D is less contagious than disease B because more people have died from disease B.

i Explain, using data, if the student's statement is correct or incorrect.

ii List two features of the immune system that would prevent the bacterium in disease D from infecting the body.

b Which disease(s) would antibiotics be most effective in treating?

c Outline two public health initiatives that could be implemented to reduce the infection rate of disease A.

15 **a** One important treatment for HER2-positive breast cancer is a drug called trastuzumab, marketed in Australia as Herceptin. This medication is a monoclonal antibody. It binds to the external domain of the HER2 receptor.

i What are monoclonal antibodies?

ii Herceptin is a humanised monoclonal antibody. What does this mean?

iii Why do scientists humanise monoclonal antibodies?

b A drug company has invented a new monoclonal antibody that it claims will be effective against HER2-positive breast cancers that have been shown to be resistant to current therapies. The company is ready to commence human trials of its proposed treatment.

i Describe an experiment that could be used to test the effectiveness of the new treatment.

ii What results would suggest that the drug company's claims are justified?

iii Discuss two ethical concepts that the researchers should apply when performing human trials.

iv Many human trials enlist only a few individuals. How does that affect the validity of the results?

16 There is a number of strategies that can be used to control the spread of pathogens. Discuss three examples of these strategies.

17 In 2019–20, a pandemic caused by COVID-19 virus, also referred to as coronavirus, swept through human populations around the world. The virus causes respiratory symptoms, usually starting with fever, a dry cough and breathing difficulties, sometimes becoming severe and leading to lung infection. Millions were infected and many thousands died. Starting in the city of Wuhan in China, the infection spread from country to country, initially transmitted by international travellers. Governments and health authorities in each country reacted with a range of responses to manage the pandemic, including lockdown of public movement in many places. People were advised to use physical distancing to limit spread of the virus, but this strategy was not always successful. As well as the health issue, there were severe economic impacts around the globe.

a Why was the COVID-19 infection not declared a pandemic until after it spread from its origin in China?

b COVID-19 disease is caused by SARS-CoV2, which was identified as a novel virus even though it is part of a known family of coronaviruses. A model of the virus is shown in the illustration below.

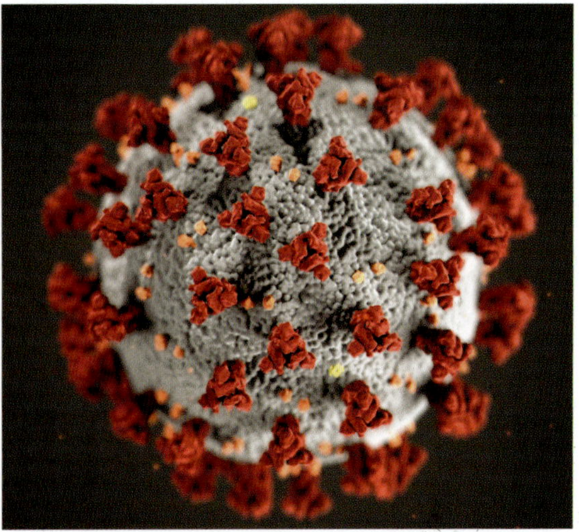

i Explain what is meant by a novel virus.

ii There have been previous pandemics caused by viruses of the coronavirus family, notably MERS-CoV, originating in the Middle East in 2012, and SARS-CoV, starting in China in 2002. Despite these previous outbreaks, there was no vaccine available when COVID-19 started to spread rapidly. Explain why health researchers and authorities were unprepared for this particular coronavirus pandemic.

c During the COVID-19 pandemic, several new terms became familiar to the public, some of which are listed below. Outline the meaning of each term and how it could reduce or control the spread of this virus.

i physical distancing

ii self-isolation and quarantine

iii lockdown

iv herd immunity

d One strategy used to manage the COVID-19 pandemic is known as 'flattening the curve'. With reference to the graphs below, clarify the meaning of this term and evaluate the effectiveness of such a strategy.

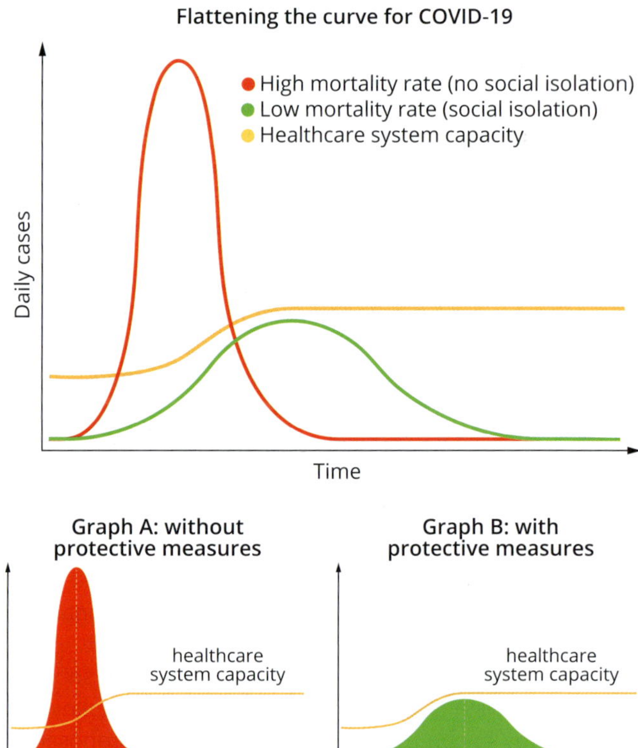

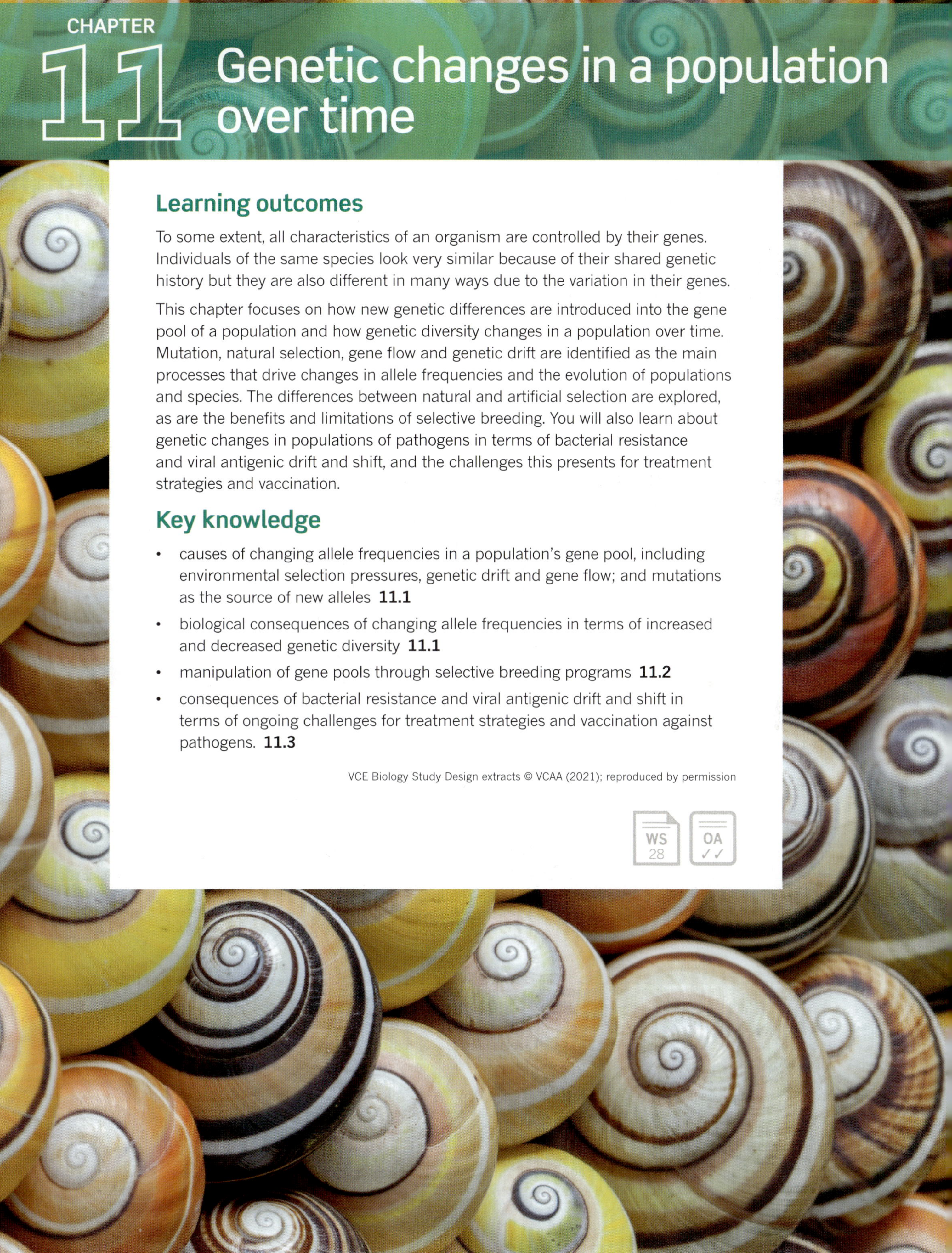

CHAPTER 11

Genetic changes in a population over time

Learning outcomes

To some extent, all characteristics of an organism are controlled by their genes. Individuals of the same species look very similar because of their shared genetic history but they are also different in many ways due to the variation in their genes.

This chapter focuses on how new genetic differences are introduced into the gene pool of a population and how genetic diversity changes in a population over time. Mutation, natural selection, gene flow and genetic drift are identified as the main processes that drive changes in allele frequencies and the evolution of populations and species. The differences between natural and artificial selection are explored, as are the benefits and limitations of selective breeding. You will also learn about genetic changes in populations of pathogens in terms of bacterial resistance and viral antigenic drift and shift, and the challenges this presents for treatment strategies and vaccination.

Key knowledge

- causes of changing allele frequencies in a population's gene pool, including environmental selection pressures, genetic drift and gene flow; and mutations as the source of new alleles **11.1**
- biological consequences of changing allele frequencies in terms of increased and decreased genetic diversity **11.1**
- manipulation of gene pools through selective breeding programs **11.2**
- consequences of bacterial resistance and viral antigenic drift and shift in terms of ongoing challenges for treatment strategies and vaccination against pathogens. **11.3**

WS 28 OA ✓✓

11.1 Changing allele frequencies

FIGURE 11.1.1 Flower colour (phenotype) in hydrangeas (*Hydrangea macrophylla*) varies depending on the acidity of the soil in which they grow.

While individuals of the same species share similar DNA sequences, there are also many differences. These genetic differences are evident in the wide variety of traits seen in individuals, populations and species. The variation in traits and genes in a population can be influenced by many factors and changes over time. The proportion of a trait or genetic variant (allele) in a population is called the allele frequency. In this section, you will learn about some of the factors that cause changes in the allele frequencies in a population and the biological consequences of these changes in terms of increased and decreased genetic diversity.

ALLELE FREQUENCIES IN A GENE POOL

You learnt in Chapter 3 that a gene is a sequence of **deoxyribonucleic acid (DNA)** that codes for a particular characteristic or **trait**. Slight variations in the DNA sequence of a gene can result in different forms of a trait. These variations of a gene are called **alleles**. The various combinations of alleles in an individual make up its **genotype**. The genotype, together with the environment, determine an organism's observable traits (or **phenotype**) (Figure 11.1.1). Although species share the same genome, the individuals within a species are not genetically identical because they have different combinations of alleles (Figure 11.1.2).

> Alleles are different forms of a gene. A genotype is the combination of alleles in an organism. A phenotype is the observable characteristics of an organism.

FIGURE 11.1.2 Height in humans is a trait controlled by different combinations of alleles.

The variation in genes or alleles within a population or species is known as **genetic diversity** (or genetic variation). The total genetic diversity in a population is known as the **gene pool**.

For a variety of reasons, allele frequencies within a gene pool change over time. The main processes that drive changes in allele frequencies and the **evolution** of populations and species are:

- mutation
- natural selection
- gene flow
- genetic drift.

You will learn more about each of these processes throughout this chapter.

Calculating allele frequencies

> Allele frequency refers to the proportion of the gene pool that carries a particular allele, relative to other alleles for that gene.

> Homozygotes have two copies of the same allele for one gene (e.g. *AA* or *aa*). Heterozygotes have two different alleles for one gene (e.g. *Aa*).

The relative proportion of a particular allele in a population is referred to as the **allele frequency**, and is often expressed as a percentage or as a decimal (e.g. 25% or 0.25). The following equation shows you how to calculate the frequency of an allele.

$$\text{allele frequency} = \frac{2(\text{number of homozygotes}) + (\text{number of heterozygotes})}{2(\text{total number of individuals})} \times 100$$

When calculating the allele frequency of a particular trait, it is important to remember that **homozygotes** with the trait will have two copies of the allele (e.g. *AA*) and **heterozygotes** will only have one copy (e.g. *Aa*).

MUTATION

All genetic diversity between species and between individuals of the same species is a result of **mutation**. Mutations are changes in DNA and they are the source of new alleles in a species or population. It is important to remember that mutation means change, not the introduction of a fault or disease. Mutations can have a beneficial or harmful effect, or no effect at all, on the organism.

Mutation means change. Changes in DNA may be harmful, beneficial or neutral.

Mutations can occur randomly as errors during cell replication and can affect a single gene, multiple genes or may involve entire chromosomes. They occur spontaneously or as a result of **mutagens**—factors that induce mutation. Common mutagens include different forms of radiation. For example, UV radiation can cause mutations in skin cells that result in skin cancers. Most mutations are detected and repaired by enzymes. Those that cannot be repaired fall into one of three categories—neutral, beneficial or harmful.

Mutations are a source of new alleles in a gene pool.

- Neutral mutations have no effect on survival.
- Beneficial mutations increase the likelihood of survival.
- Harmful mutations decrease the likelihood of survival.

Somatic mutations occur in body cells and only affect that individual. **Germline mutations** are heritable because they affect **gametes** (sperm and egg cells) and can therefore be passed on to offspring. A germline mutation may bring a new allele into a gene pool, potentially influencing the allele frequencies.

CASE STUDY

A beneficial mutation

An example of a beneficial mutation is one that involves the *APOA-1* gene. This gene codes for a protein (apolipoprotein A-1) that is normally involved in the transport of cholesterol and phospholipids to the liver, where they are then redistributed or broken down and excreted. One of the mutated forms of the protein, ApoA-1 Milano, involves a substitution of the amino acid arginine (arg) for cysteine (cys). This mutated protein acts as an antioxidant, reducing cholesterol deposition in arteries, and thereby significantly decreasing the risk of cardiovascular disease.

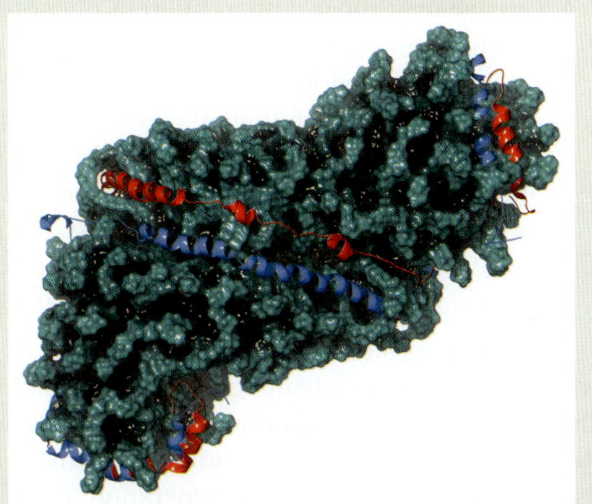

FIGURE 11.1.3 ApoA-1 Milano is a mutated form of a protein that can reduce cholesterol levels in humans.

The mutant form of the ApoA-1 protein (Figure 11.1.3) was first identified in Milan, and so the mutated gene was named after this city. Further investigation, including blood tests of an entire Italian village, traced the origin of the mutation to a single man. The 3.5% frequency of the gene in that village population can be attributed to the descendants of this one man.

NATURAL SELECTION

The genetic diversity between individuals in a population results in different phenotypes (traits) that have varying advantages for survival and reproduction. Individuals that are well-suited to their environment are more likely to survive and reproduce—this process is known as **natural selection**.

The alleles of the individuals that survive and reproduce will persist in the population and increase in frequency. This is the mechanism by which natural selection changes the allele frequencies and phenotypes in populations.

A selection pressure is an environmental factor that affects the survival and reproductive success of an individual based on their phenotype.

Environmental selection pressures

The conditions or factors that influence which phenotypes are most successful in a population—and therefore, influence allele frequencies in that population—are known as **selection pressures**. Selection pressures can be natural environmental pressures or artificial pressures brought about by humans through selective breeding. You will learn more about selective breeding in Section 11.2.

Examples of environmental selection pressures include:

- climatic conditions such as extreme temperature changes and drought
- competition for resources such as food and water, as well as competition for shelter
- mate availability
- predator abundance.

Environmental selection pressures affect individuals in a population differently for a number of reasons:

- Genetic diversity—There are genetic differences between individuals of a population that produce different phenotypes. Some phenotypes are better suited to the environmental conditions and give the individual a survival advantage over those with a different phenotype.
- Reproduction—Individuals have different levels of **reproductive success**. That is, some individuals produce more offspring than others. Those individuals that reproduce pass on their alleles to their offspring. The ability of an organism to survive and produce viable offspring is known as **biological fitness**. The offspring are genetically similar (as in sexual reproduction) or genetically identical (as in asexual reproduction) to the parents.
- Survival— Individuals have different rates of survival. Not all individuals survive long enough to reproduce and produce offspring.

BIOFILE

Adaptation in the desert

Thorny devils have a number of physical adaptations that enable them to thrive in the very arid ecosystems of central Australia. Their mottled camouflage colouring and hard spikes have high adaptive value because these features reduce the likelihood of predation. Thorny devils also have highly textured skin, which allows capillary action to collect any moisture in their environment and channel it directly into their mouths.

Thorny devil (*Moloch horridus*)

When it comes to survival, some phenotypes have a high **adaptive value** and give the individual an advantage over individuals with phenotypes of lower adaptive value. This concept is often referred to as 'the survival of the fittest'. Having an advantageous trait means the individual is more likely to survive to reproduce and pass their alleles on to the next generation. If the environment remains unchanged, the traits of organisms will continue to have high adaptive value, and there will be little change to allele frequencies over time. Traits that are well suited to an organism's environment are known as **adaptations**.

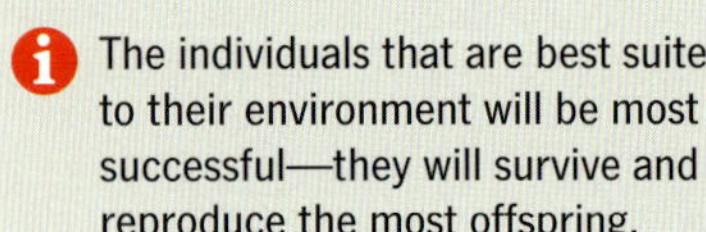

The individuals that are best suited to their environment will be most successful—they will survive and reproduce the most offspring.

Alleles of the advantageous trait tend to increase in frequency in the gene pool, while alleles of the less advantageous trait tend to decrease in frequency. Advantageous traits of high adaptive value may persist in the population until all individuals possess them (100% allele frequency).

GENE FLOW

Gene flow is the transfer of alleles between populations and can result in changes in the allele frequencies in a gene pool. This occurs when new individuals join the population from a different gene pool or when individuals leave a population (Figure 11.1.4). Movement of individuals in and out of populations can result in new alleles being introduced or alleles being lost, increasing or decreasing the genetic diversity in the population.

Gene flow occurs when individuals from different populations interbreed and when individuals enter or leave a population.

When gene flow exists between two different populations, the gene pools may remain fairly similar. When gene flow is not possible between populations, the gene pools are said to be isolated. Genetic isolation is another key factor in the process of evolution and you will learn more about it in Chapter 12.

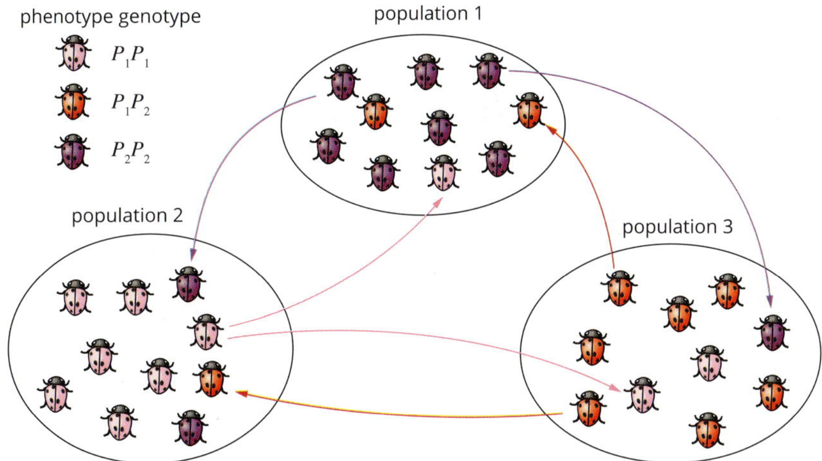

FIGURE 11.1.4 Gene flow occurs by migration between neighbouring populations of a species. Gene flow can introduce genetic diversity into populations.

GENETIC DRIFT

Allele frequencies in a gene pool may also change randomly over time as a result of chance events, such as births and deaths. This is called **genetic drift**. Genetic drift has the greatest impact on small populations with little to no gene flow, as the random death of one individual can significantly alter the allele frequencies. Generally, in small populations genetic drift results in the loss of genetic diversity over time as alleles are lost from the gene pool (Figure 11.1.5).

Genetic drift is the random change in allele frequencies of a population due to chance events, such as births and deaths.

Genetic drift can occur when populations decrease for a period of time (a bottleneck effect) or in small founding populations (the founder effect).

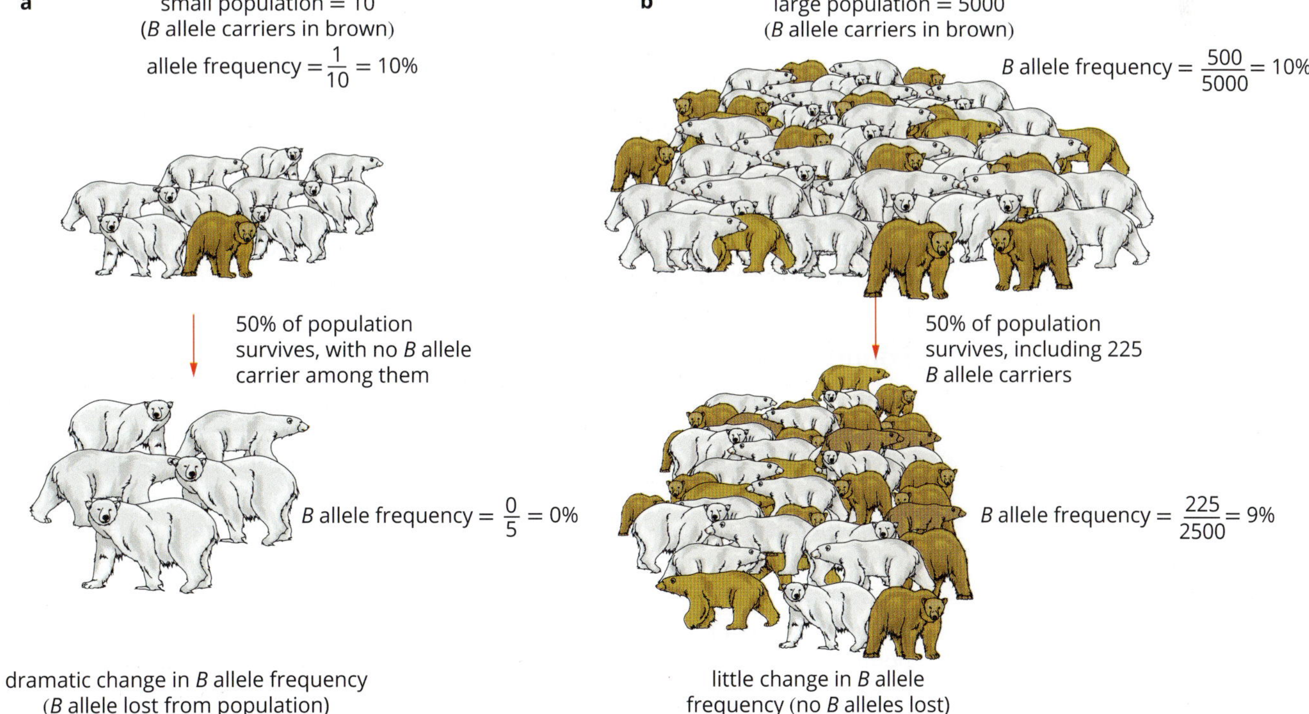

FIGURE 11.1.5 Genetic drift can result in greater changes to allele frequencies in (a) a small population than (b) a large population. In this example, individuals carrying a particular allele are shaded brown.

Bottleneck effect

The number of individuals in a population can be drastically and quickly reduced as a result of a random event, often a natural disaster. Human activities, such as hunting and land clearance, have also greatly and quickly reduced the number of individuals in wild populations of plants and animals. The phenotype of an individual is unlikely to significantly increase its chances of surviving a natural disaster, such as a volcanic eruption, tidal wave or landslide. The individuals that survive will do so by chance. The allele frequencies of the remaining population are unlikely to reflect those of the original population.

A sudden and substantial reduction in a population's size is referred to as a 'bottleneck' and the **bottleneck effect** describes the impact of the population reduction on the remaining population. Because of the reduced population size, the possible reproductive pairings are limited, which leads to high levels of inbreeding. Inbreeding results in reduced genetic diversity in the population and an increase in the numbers of homozygous individuals (Figure 11.1.6). The smaller the population, the greater the effect of genetic drift. Alleles may be lost from the gene pool immediately after the natural disaster or be 'bred out' in only a few generations. The lowered genetic diversity may make the population more vulnerable to environmental change.

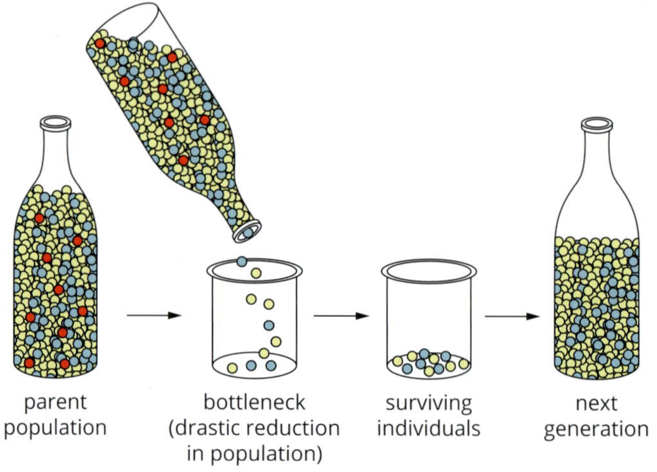

FIGURE 11.1.6 The bottleneck effect is a form of genetic drift (random sampling of alleles) that occurs when a population decreases for a period of time. The loss of individuals from the population means that there will be less genetic diversity in subsequent generations.

BIOFILE

Population bottleneck in the mountain pygmy possum

The mountain pygmy possum (*Burramys parvus*) is an example of a population that has experienced the bottleneck effect. The Mt Buller population reduced from approximately 300 individuals in 1996 to 40 individuals in 2010 as a result of habitat loss, bushfires and introduced predators. This drastic population decline resulted in a loss of two-thirds of the genetic diversity that was present in the population in 1996. The loss of genetic diversity puts small populations at further risk of extinction because of inbreeding between genetically similar individuals and their vulnerability to environmental change.

Mountain pygmy possum (*Burramys parvus*)

A high level of genetic diversity allows populations to adapt to new or changing environments. Low genetic diversity as seen when species or populations go through a genetic bottleneck, increases the risk of extinction.

Founder effect

The **founder effect** occurs when a small group of individuals is genetically isolated from a larger population, either by migration, new geographic barriers or habitat fragmentation. The smaller population only has a small portion of the alleles of the original population and therefore has lower genetic diversity. Like the bottleneck effect, there is increased inbreeding in the 'founder population' and a greater chance of loss of alleles due to genetic drift.

In the new environment, the environmental selection pressures on the founder population are likely to be different from those experienced by the original population. These differences in environmental selection pressures drive further changes in allele frequencies and may lead to divergence of the populations over time.

BIOFILE

Founder effect in Tasmania

Huntington's disease is an inherited disease that causes the degeneration of nerve cells in the brain. In Tasmania, many sufferers of Huntington's disease can trace their ancestry to a woman who migrated from Britain in the nineteenth century.

CASE STUDY **ANALYSIS**

Natural selection in the rock pocket mouse

FIGURE 11.1.7 Rock pocket mouse, wild type or light phenotype (*dd*), on light rock

Natural selection for colouration is seen in the American rock pocket mouse, which inhabits a region spanning northern Mexico, New Mexico and southern Arizona. The mice have two different hair colours. The majority of mice have light, sandy-coloured hair and inhabit light-coloured granite hills and desert sands (Figure 11.1.7). In some regions, where there are dark-coloured lava flows, dark-haired mice are found.

Predators of this mouse include rattlesnakes, foxes, coyotes and owls (Figure 11.1.8). Colouration of the mouse is an important protection from these visual predators, especially owls, which have a visual advantage from the air.

Researchers identified the *Mc1r* gene as the genetic basis for the colour difference. *Mc1r* codes for melanocortin 1 receptor and controls skin and hair pigment production in mammals. Rock pocket mice have two alleles for the *Mc1r* gene: *D* and *d*. Mice with genotype *dd* have light-coloured hair. This is considered the wild type genotype. Mice with genotypes *DD* and *Dd* have dark-coloured hair, called a melanic phenotype. Researchers sequenced the alleles and identified the mutations that result in four amino acid differences in the proteins produced. Compared to the wild type *d* allele, the *D* allele produces a more active protein, resulting in more melanin pigment production.

FIGURE 11.1.8 Owls are major predators of the rock pocket mouse. They can see a light mouse on dark rock even when hunting at night.

Field observations showed a strong association between mouse phenotype and the rock colour of their habitat. The evidence indicates selection for dark phenotype in mice inhabiting the dark lava flows.

The researchers hypothesised that positive natural selection is the mechanism for the presence of two mouse hair colour phenotypes in the granite and lava rock habits. They sampled animals from six locations—three granite and three lava flow areas—along a transect of 35 km through a national wildlife refuge in Arizona. The mice were trapped, measurements taken and data recorded by field researchers, then they were released. The pooled data are shown in Table 11.1.1.

TABLE 11.1.1 Data collected regarding mice that were trapped by researchers

Ground type and colour	*N*	Phenotype (hair colour)	Number of mice	Proportion of population (%)
granite/sand light brown (3 sites)	168	light hair	120	
		dark hair	48	
lava fields dark grey/ black (3 sites)	57	light hair	3	
		dark hair	54	

Analysis

1 Copy and complete Table 11.1.1 by calculating the proportion of each phenotype at each location. Give your answers to one decimal place.

2 Describe the association between the melanic phenotype and ground colour.

3 Table 11.1.2 shows the genotypes of mice collected from the dark lava rock habitat, known as the Pinacate lava flow site.

TABLE 11.1.2 Genotype–phenotype associations between *Mclr* alleles (*D*,*d*) and hair colour in at the Pinacate site

		Genotype		
		DD	*Dd*	*dd*
mouse phenotype	light	0	0	12
	dark	11	6	0

Use this equation to calculate the following:

$$\text{allele frequency} = \frac{2(\text{number of homozygotes}) + (\text{number of heterozygotes})}{2(\text{total number of individuals})} \times 100$$

a frequency of the *D* allele in light and dark mice at this site

b frequency of the *d* allele in light and dark mice at this site

4 State whether the *Mc1r* mutant allele *D* correlates with mouse hair colour.

5 Identify the most significant selection pressure(s) or selecting agent(s) acting on the mouse population. Describe how the selection pressure leads to the allele frequency seen at the Pinacate lava flow site.

11.1 Review

OA ✓✓

SUMMARY

- Natural variation exists between individuals of the same species because many genes have multiple alleles that are present in the population in different frequencies.
- Allele frequencies are a measure of how common a particular allele is in the gene pool of a population.
- Evolution is the change in the genetic composition (allele frequencies) of populations over time due to mutation, natural selection, gene flow and genetic drift.
- New alleles, genes and chromosomes are created or modified through mutation.
- Mutations may have a beneficial effect, a harmful effect or no effect at all on the individual.
- Natural selection is the influence of environmental selection pressures on allele frequencies in a population.
- Environmental selection pressures act on phenotypic variation in a population—individuals with phenotypes well suited to their environment will have a greater chance of surviving and reproducing.
- Alleles associated with the successful phenotypes are more likely to persist in the gene pool and increase in frequency over time.
- Gene flow is the movement of alleles into and out of a gene pool. It can occur when different populations interbreed or individuals migrate between populations.
- Genetic drift is the random change in allele frequencies in a population due to chance events. Genetic drift has a greater impact on the allele frequencies of small populations because they often have fewer alleles than large populations. Genetic drift can occur when a population decreases for a period of time (a bottleneck effect) or in small founding populations (the founder effect).

KEY QUESTIONS

Knowledge and understanding

1 What are the four main processes that drive changes in allele frequencies in a population?

2 Define natural selection.

3 Describe two ways that gene flow can occur between populations.

4 Explain why genetic drift may have a greater impact on a small population than a large one.

5 Explain the notion that low genetic diversity in Tasmanian devils relates to an increased risk of extinction due to factors such as devil facial tumour disease.

Analysis

6 In a population of 150 individuals, 80 are homozygous for purple flowers, 40 are homozygous for white flowers and the remaining 30 individuals are heterozygous. Calculate the allele frequency for the purple flower allele.

11.2 Selective breeding

Evolution through natural selection is an ongoing and, as the name implies, natural process. In addition, humans have been manipulating allele frequencies in the gene pools of populations for thousands of years through deliberate selection of particular individuals. The process by which humans decide which individuals may breed and leave offspring to the next generation is called **selective breeding** (also known as **artificial selection**). In this section you will learn about some of the potential advantages and disadvantages of selective breeding.

FIGURE 11.2.1 Dairy cows, such as these Holstein Friesians, are selectively bred for their high milk yields.

> Selective breeding is the process by which animals or plants with desired characteristics are bred together to produce offspring that will also show these characteristics.

INCREASING THE FREQUENCY OF DESIRED TRAITS

Selective breeding has led to improved agricultural crops and the domestication of animals for food or other uses (Figure 11.2.1). Darwin, for example, kept and bred pigeons, gathering information from stockbreeders. While developing his theory of natural selection, he observed the success of such selective breeding in producing new types of pigeons.

Through selective breeding, humans choose individual organisms with desirable traits and deliberately interbreed them to increase the allele frequency of those desired traits in the gene pool. This allows certain extreme forms to reproduce while preventing others from reproducing.

There are four basic steps that apply to all forms of selective breeding:

1 Determine the desired trait.
2 Interbreed parents that have the desired trait.
3 Select the offspring with the best form of the trait and interbreed these offspring.
4 Continue this process until the population reliably reproduces the desired trait.

All modern crops and livestock were developed by genetic manipulation of plant and animal species through this process of selective breeding. However, new molecular technologies (e.g. genetic engineering) are being used to alter the characteristics of organisms in a more targeted and specific way, and more quickly than by traditional selective breeding. Genetic engineering can also allow the exchange of genes between organisms that normally cannot interbreed. New forms of plants and animals developed in this way are referred to as genetically modified organisms (GMOs). You learnt about GMOs in Chapter 4.

Selective breeding in plants

Most selective breeding of plants is done to produce higher-quality food. Typically, seeds are collected from the individuals with the largest or most numerous grains, fruits, nuts or other part of the plant that will be eaten. Those seeds are planted and the new generation of plants is cross-pollinated with other individuals with similar traits. The resulting plants produce larger, more nutritious or more aesthetically pleasing food products.

Maize, or corn (*Zea mays*), is one of the most widely grown crops in the world. It is thought that maize was selectively bred from a wild grass of the genus *Teosinte*. Modern maize has significantly larger cobs with many more rows of much larger kernels compared to the ancestral *Teosinte*. The higher-yielding modern maize provides more food for people than the ancestral form (Figure 11.2.2).

Many other food crops, such as tomatoes, potatoes and bread wheats, have also been modified by selective breeding to have higher yields, as well as greater resistance to common diseases.

FIGURE 11.2.2 These varieties of maize, grown in Mexico, are among some of the first that were cultivated from the wild grass *Teosinte* thousands of years ago.

Selective breeding in animals

Just as crops have been selectively bred for desired traits, so too have many animal species. In agriculture, sheep have been selected for the quality and quantity of the wool they grow, dairy cows have been selected for the milk they produce, and beef cattle for their muscle mass.

CASE STUDY

Selective breeding of edible Australian plants

For thousands of years Indigenous Australians have interacted with and influenced the life cycle, form and structure of food plants through selective breeding. Recently, there has been renewed interest in the traditional bush plant foods of Australia's First Peoples. Some bush plant foods have become widely used for cooking, often as flavourings, such as lemon myrtle and finger lime. The macadamia nut, native to Queensland and selectively bred to improve nut size and flavour, has become popular all over the world. Bush tucker foods are much more varied than these examples and scientific research into their potential has only just begun.

Bush tucker

One staple food of Indigenous Australians living in desert environments is the maloga bean (*Vigna lanceolata*). It has an edible root and small beans that can be eaten raw. Another related species is *Vigna radiata*, the wild mung bean. Indigenous Australians did not traditionally make use of the bean's small black seeds, but in more recent times the wild mung been has been selectively bred with other plants to produce the cultivated mung bean.

The cultivated mung bean has a seed that can be green or black, and is more than double the size of the wild form (Figure 11.2.3).

The cultivated plant also grows upright, instead of being a wiry creeper.

FIGURE 11.2.3 Seeds of wild mung bean (top) and the larger green seeds of the cultivated variety (bottom)

Conservation and use of wild genes

Native wild plants are of interest to modern plant breeders. Wild plants are a source of genes that may be used for crop improvement. For example, crosses between wild and cultivated mung beans may produce hybrid forms best adapted to Australian soils and climates.

To conserve the genetic diversity in wild populations for future use, samples of native species that are the wild relatives of agricultural crops are collected and stored in a seed bank.

The CSIRO Centre for Plant Biodiversity Research has established a significant collection of another type of native pea, *Glycine*, which is related to the cultivated soybean. An Australian species of *Glycine* has resistance to the leaf rust fungus, a trait needed for protecting soybean.

Functional genomics

Functional genomics aims to identify genes that determine particular functions. For example, it could be used to determine the genes that allow plants to grow under drier conditions, in saline soils or with resistance to fungal diseases. Once a gene is identified, collections of wild plants can be screened for natural variants of the desired phenotype (for example, drought resistance). Any new and useful alleles discovered by genetic screening can be incorporated into existing agricultural variants through plant breeding by cross-pollination. Also, once identified, unwanted genes can be switched off and desired genes from other species can be introduced through genetic engineering.

WS 30

WS 31

When a species has a variety of traits, different traits may be useful in different situations. A single wild species can be the original source of a great variety of different breeds. For example, it is widely accepted that all domestic dogs were selectively bred from a wolf species. Today there are hundreds of domesticated dog breeds, some of which would be unlikely to survive in the wild (Figure 11.2.4). Examples include soft-mouthed, strong swimming dogs such as Labradors, which were bred for duck hunting, and sheepdogs, bred for their intelligence. Humans have also selectively bred a wide variety of traits in chickens, horses and many other domestic animals to produce gene pools that consistently produce the desired traits.

FIGURE 11.2.4 Dogs have been selectively bred for particular traits, such as size, coat colour, speed, protectiveness, strength and endurance.

POTENTIAL NEGATIVE EFFECTS OF SELECTIVE BREEDING

Most selective breeding requires similar individuals to interbreed. Although this increases the allele frequency of the desired trait, it also decreases the frequency of all other alleles for this trait. This reduces the genetic diversity within the gene pool and increases the number of homozygotes in a population. Furthermore, genes do not exist in isolation, but are carried on chromosomes with other characteristics. Gene linkage means that selecting for one allele may result in the selection of a number of other traits; for example, 'fluffy' cocker spaniels get sore eyes due to extra lashes on the inside of the eyelid. The success of selective breeding of both plants and animals may be limited by the presence of undesirable linked alleles (Figure 11.2.5).

> Genes that are located close together on a chromosome are said to be 'linked'. This is referred to as 'gene linkage'.

FIGURE 11.2.5 Farmers must select wheat varieties with high grain production, as well as strong, moderate-length stems. Some varieties of wheat have long stems that cannot support the weight of the grains. This causes the plants to fall over, making it difficult to harvest the grain.

Reduced resistance to environmental change

A population with low genetic diversity is one in which all the individuals are very similar. As long as the alleles in the gene pool have a high adaptive value for the environmental conditions, the species will persist. However, should the environmental conditions and the resulting selection pressures change, it is unlikely that the same alleles will still have the same adaptive values. A single disease could potentially wipe out entire populations if none of the individuals are resistant.

Reduced biodiversity

Selectively bred species are also replacing wild varieties, reducing the number and variability of species (**biodiversity**) used in agriculture. Low biodiversity, combined with low genetic diversity within the selectively bred populations, puts global food security at great risk. This risk is increased when there are additional environmental selection pressures on populations, such as climate change. The loss of a crop species such as sorghum, maize or rice could easily lead to mass starvation, if there were no wild varieties to fall back on. This possibility has led to the construction of food arks and seed banks around the world, where the seeds of both heirloom and modern varieties are stored. Svalbard Global Seed Vault is funded by the Norwegian government and stores seeds for more than 4000 essential food crops from around the world, including valuable seed stock from Australia.

Increased chance of genetic abnormalities

Selective breeding can also increase genetic abnormalities. For example, many purebred dogs are born with conditions that can cause health problems including hip dysplasia, deafness and an increased risk of cancers, heart diseases and neurological diseases. Other less detrimental traits, such as an underbite caused by malformation of the lower jaw, are rare in wild dog populations but relatively common in domesticated breeds (Figure 11.2.6).

Many of these problems are recessive conditions. This means that an individual needs two copies of the same allele for the condition to be present. Inbreeding and small gene pools greatly increase the frequency of particular alleles. This, in turn, increases the likelihood of homozygous recessive conditions.

FIGURE 11.2.6 An underbite is common in many domestic dog breeds but is not often seen in wild wolves, dingoes or foxes.

CASE STUDY ANALYSIS

Domestication of wild animals

Silver fox

Silver foxes are a natural colour variant of the red fox (*Vulpes vulpes*) and range from blue-grey to black in colour (Figure 11.2.7). They were prized for their unusual colour and hunted for their fur.

In 1959, Russian scientist Dmitri K. Belyaev began an experiment in which he selectively bred silver foxes for 'tameness' (Figure 11.2.8). He found that within eight to ten generations the foxes showed clear signs of domestication, wagging their tails when people approached. However, they no longer resembled their wild ancestors. Instead, the domestic foxes had floppy ears, short or curly tails and their fur had changed considerably in colour and texture. The genes related to 'tameness' are carried on chromosomes with many other characteristics.

The selection of this one trait affected the inheritance of other alleles. By selectively breeding the tamest foxes, the allele frequencies for other traits changed, resulting in domesticated foxes with different phenotypes to the original population.

FIGURE 11.2.7 These foxes are both red foxes, *Vulpes vulpes*. (a) The silver colour is a natural colour variant of (b) the red fox.

Musk ox

The musk ox (*Ovibos moschatus*) (Figure 11.2.9) is a large Arctic mammal, prized for its thick, wool-like fleece called qiviut. Qiviut is a prized luxury item in North America and musk ox meat is considered a lean alternative to beef. Unregulated hunting of the musk ox led to the near extinction of the species in the late 1800s. Conservation efforts have allowed the species to survive. Hunting restrictions were introduced and musk ox from the surviving populations were relocated to repopulate regions where the animals had died out.

In the 1950s, the Musk Ox Farm Project was set up in Alaska in an attempt to domesticate the animals. Thirty-three individuals were captured from wild populations and selectively bred for domestication. Several other musk ox farms appeared after this time, greatly reducing the reliance on hunting. The domesticated individuals have been kept as livestock to create sustainable farms, in which qiviut is combed out of the living adults.

FIGURE 11.2.8 Dmitri Belyaev with his partially domesticated foxes

FIGURE 11.2.9 Musk ox (*Ovibos moschatus*)

In many regions of the Arctic Circle, domesticated musk ox were released into the wild where the native populations had been hunted to extinction. In the 1970s, one farm in Northern Quebec closed due to poor profits and 54 musk ox were released into the wild. Slowly, the native population of musk ox has increased to over 1000 adults, all descending from domesticated individuals.

Analysis

1 What were the reasons for attempting to domesticate the musk ox and silver fox via selective breeding?

2 Compare the success of each program.

3 What danger does environmental change, such as global warming, present to populations that have been selectively bred, such as the musk ox population that has been returned to the wild?

11.2 Review

OA ✓✓

SUMMARY

- Allele frequencies can change as a result of natural selection or artificial selection.
- Selective breeding is the traditional form of artificial selection. In selective breeding, humans select desired traits and interbreed organisms with these traits.
- There are four basic steps that apply to all forms of selective breeding:
  - Determine the desired trait.
  - Interbreed parents who show the desired trait.
  - Select the offspring with the best form of the trait and interbreed these offspring.
  - Continue this process until the population reliably reproduces the desired trait.
- Modern molecular technologies have allowed for faster development of genetically modified organisms (GMOs) and for the transfer of DNA between species that do not naturally interbreed.
- Agricultural plants are typically bred for high yield and high resistance to common diseases. Animals are often bred for high quality traits and products (such as wool and milk), or for personality traits (such as tameness in pets).
- Selectively bred populations tend to have low genetic diversity, meaning that:
  - they are more susceptible to environmental change
  - biodiversity may be reduced if selectively bred populations replace wild populations and varieties
  - an increase in the incidence of genetic abnormalities can occur.

KEY QUESTIONS

Knowledge and understanding

1 Identify the steps of artificial selection that would lead to the production of large corn cobs.

2 Outline how selective breeding has changed with the development of new technology.

3 Discuss the case for and against the manipulation of plant and animal breeding in agriculture. Use some specific examples in your answer.

4 Explain why some conditions, such as hip dysplasia in dogs, are more common in selectively bred populations than in wild populations.

Analysis

5 A typical example of a breeding program in agriculture is for increased egg production in chickens. At the start of a particular breeding program, the average number of eggs per hen per year in a flock was 125. Hens that produced the most eggs per year were chosen as the female parents of the next generation. Roosters used in the program were the offspring of high-yielding hens. The average number of eggs per hen per year increased from 125 to 230 over 15 years, but the rate of increase was slower in subsequent years.

a Explain how the breeding program is an example of selective breeding by humans.

b What possible reasons might account for the decreasing rate of increase in egg production over time?

c What are the possible negative effects on hens of increasing their egg production?

11.3 Changes in the genetic composition of bacteria and viruses

Like all species, bacteria and viruses undergo changes in the genetic composition of their populations over time. Often these changes are slow, but sometimes they can be rapid and this causes problems for health authorities trying to contain and control the spread of pathogens.

As you learnt in Chapter 9, adaptive immunity to a pathogen is developed when memory B and T cells are formed after contact with the pathogen's antigens. However, if the pathogen's antigens change then immunity will be lost because the host's memory B and T cells will no longer recognise the antigens.

When immunological memory to a particular pathogen is formed it usually involves the formation of many different versions of the B and T memory cells, reflecting the variety of different antigens present on the pathogen. This means that if there are only changes in some antigens then a level of immunity will be retained, although it will be less effective than previously. If major changes occur to the antigens, then immunological memory may be lost completely.

In this section you will learn about how some bacteria have developed resistance to antibiotics in use today. You will also learn about how viruses become a threat to human health through obtaining genes from viruses usually found in other animals. You will explore how health authorities and governments respond to these altered or new pathogens in order to maintain the health of human populations and limit the spread of new pathogens.

BACTERIAL RESISTANCE TO ANTIBIOTICS

> Host immunity is reduced and may be lost as bacterial antigens change because of mutations in their DNA.

Bacteria are prokaryotes, having a single circular chromosome. They have simple metabolisms and lack some of the DNA check and repair mechanisms present in eukaryotic cells. Therefore, the DNA of bacteria is likely to accumulate mutations more quickly than the DNA of eukaryotes; however, such changes tend to be relatively slow. As the DNA of a bacterium changes, the antigens on the bacterium's surface also change. Individuals who have previously developed immunity to this strain of bacteria will no longer have immunity as their immune system will not recognise the new bacterial antigens. Because of these changes over time, scientists may need to modify the composition of vaccines in order to take these changes into account.

A more significant problem that arises in the fight against bacterial pathogens is the development of strains of bacteria which are resistant to the treatments currently available.

Bacterial infections are generally treated with a class of chemicals called antibiotics. These are chemicals which disrupt the normal metabolic activities and/or the reproduction of bacteria (Figure 11.3.1).

> Antibiotic resistance is the ability of bacteria to survive in the presence of antibiotics.

The discovery of the first of these chemicals, penicillin, revolutionised the treatment of bacterial infections but now the usefulness of antibiotics is under threat as strains of bacteria that are resistant to the effects of the medications are evolving. The ability of bacteria to survive in the presence of an antibiotic that they were once susceptible to is termed **antibiotic resistance**.

Bacteria can resist antibiotics in a variety of ways (Figure 11.3.2). Bacteria may reduce the intake of the drug into the cell, alter the target molecule to which the drug attaches, pump the drug out of the cell or enzymatically deactivate the drug. If these resistance properties are present in members of the bacterial population, the bacteria possessing them will survive the drug treatment and go on to become the dominant population.

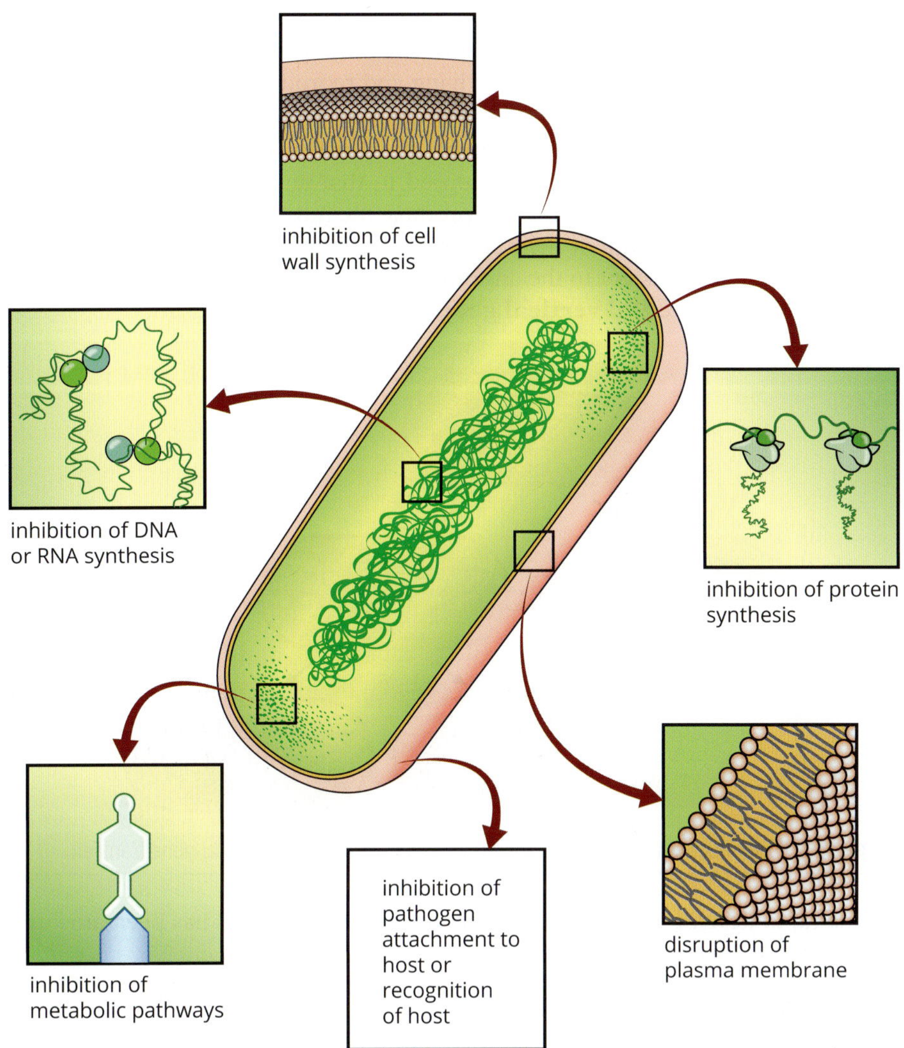

FIGURE 11.3.1 Mechanisms of antibiotic action

> Antibiotics work in a variety of ways to inhibit bacterial metabolism and reproduction.

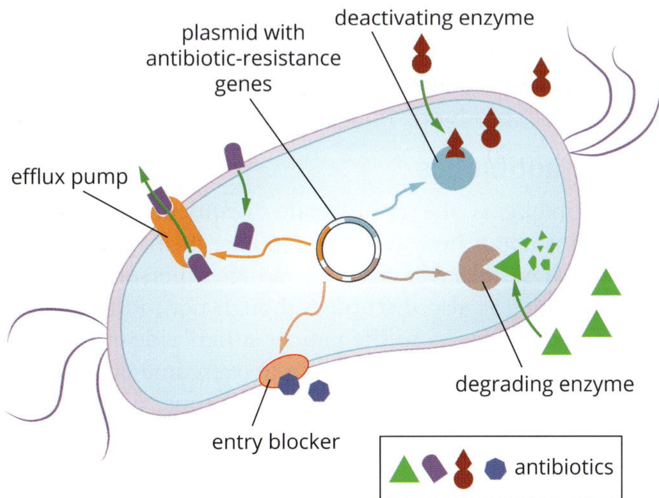

FIGURE 11.3.2 Mechanisms of bacterial resistance to antibiotics

Vertical gene transfer is the passing of genetic information from parents to offspring. Antibiotic-resistant bacteria pass their resistance to their offspring through vertical gene transfer.

In Section 11.1, you learnt about how populations change over time as a result of natural selection. Natural selection also acts on populations of bacteria—some individuals have a better chance of surviving and reproducing than others and the most biologically fit individuals produce the most offspring.

Some individual bacteria have a natural resistance to the effects of a particular antibiotic. If a colony of bacteria is exposed to that antibiotic, then the resistant individuals will survive the longest and produce the most offspring. It is then likely that their offspring will inherit their alleles for antibiotic resistance, increasing the number of bacteria resistant to the antibiotic. The passing of genetic material from parent to offspring is called **vertical gene transfer** (Figure 11.3.3).

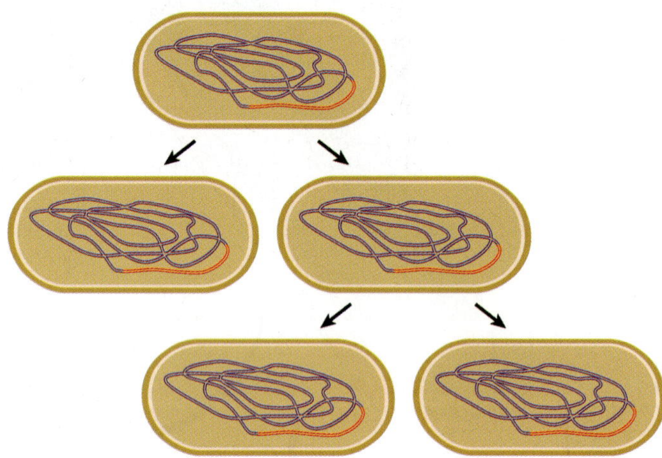

FIGURE 11.3.3 Parents pass their genetic material to their offspring via vertical gene transfer.

The rapid development of bacterial resistance to antibiotics has been attributed to two factors – overuse and improper use.

Overuse of antibiotics

Widespread use of antibiotics in animal agriculture has exposed many bacterial colonies to this strong selection pressure. Antibiotics are used to reduce rates of infection, and hence increase production, in the farming of sheep, cattle, pigs, poultry, fish and even shellfish such as scallops and oysters.

Excessive and improper use of antibiotics have created a strong selection pressure, resulting in increased numbers of antibiotic resistant bacterial strains.

Along with overuse in farming, people also overuse antibiotics to fight human infections. Prescribing of antibiotics to treat very minor infections, which would clear up on their own, or even viral infections, for which antibiotics are ineffective, have been common over the last decades. This is problematic because every time an antibiotic is used there is a chance of selecting resistant bacteria. The resistant individuals then go on to produce whole colonies of resistant bacteria.

Improper use of antibiotics

Even when an antibiotic is the appropriate treatment, patients often do not follow the instructions about how and when to take them. Antibiotics disrupt cell functioning—this is how they kill bacteria. When a person takes the antibiotic their own cell functioning is also disrupted, though not to the same extent as the functioning of the bacterial cells, so the patient suffers side effects such as nausea and diarrhoea. Consequently, as people begin to overcome the infection, they stop taking their medication to avoid the side effects. Improper use of antibiotics results in natural selection of the bacteria that are resistant to the antibiotic.

Horizontal gene transfer

Natural selection through the improper use of antibiotics is not the only way bacterial colonies gain resistance. Bacteria also have the ability to swap genes with each other via **horizontal gene transfer** (Figure 11.3.4). Many bacteria contain small, circular DNA molecules, known as plasmids. Plasmids may contain genes that give the bacteria resistance to one or more antibiotics. A bacterium can possess several copies of the same plasmid and share these plasmids with other bacteria of the same or closely related species, thereby passing genes for antibiotic resistance to other individuals.

Horizontal gene transfer is the passing of genetic material between individuals other than through reproduction. Bacteria do this through the transfer of plasmids.

Plasmids are small circular pieces of DNA found in bacteria. A single bacterial cell can contain several plasmids.

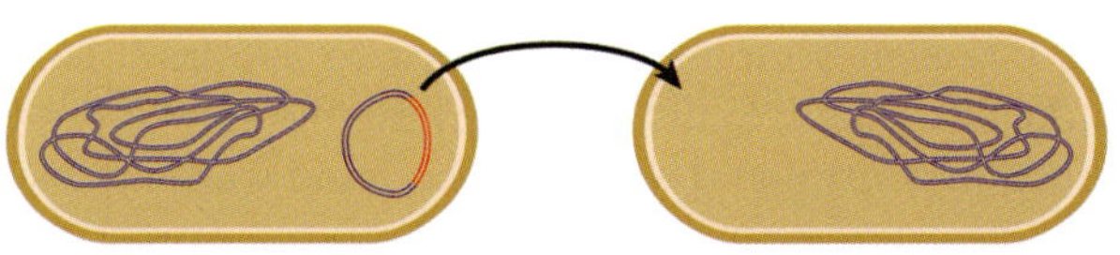

FIGURE 11.3.4 Horizontal gene transfer allows bacteria to swap genetic material.

Challenges in treating bacterial disease

Over time bacteria have become resistant to more and more antibiotics. As a result, many antibiotics are no longer effective. Those that remain effective are much more toxic to humans. We have now seen the rise of multidrug-resistant tuberculosis (MDR-TB) and methicillin-resistant *Staphylococcus aureus* (MRSA).

Multidrug-resistant tuberculosis

The rise in resistant bacteria among those infected with the bacterium that causes tuberculosis (TB), *Mycobacterium tuberculosis*, is creating an emerging problem for health authorities across the world. Many individuals are contracting a strain of TB that is resistant to the antibiotic rifampicin, the first option for TB treatment (known as a first-line drug). Of even greater concern is that new strains of TB with resistance to second-line drugs have also appeared (Figure 11.3.5).

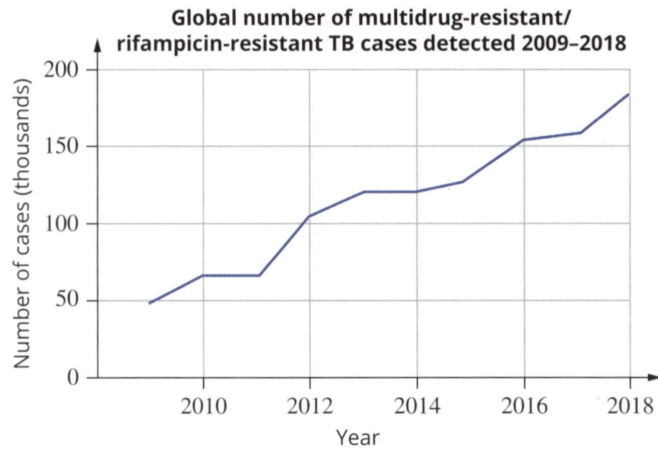

FIGURE 11.3.5 The rise in the number of cases of multidrug-resistant/rifampicin-resistant tuberculosis (MDR/RR-TB) has been monitored since 2009.

The rise of these strains of TB is of particular concern in eastern Europe and parts of Asia due to TB's very high incidence in those areas. Studies undertaken in 2004 and published in 2006 put the global incidence of antibiotic-resistant TB at 2.7% of all TB infections. By 2016, the World Health Organization (WHO) had raised that figure to 4.1%. This seems like a small proportion but when one considers that the number of new TB infections was nearly 9 000 000 in 2002 when surveillance began, these percentages represent a large number of people and a significant drain on health systems to treat patients and prevent the spread of the pathogen. In the worst affected countries, the incidence of antibiotic-resistant TB is as high as 18% of all new infections.

Methicillin-resistant Staphylococcus aureus

Of greater concern at the moment to Australian health officials is methicillin-resistant *Staphylococcus aureus* (MRSA), especially in Australian hospitals. *S. aureus* is responsible for many wound infections following operations or accidents. The first case of MRSA was identified in Britain in 1961; thereafter, its incidence slowly rose across the world.

Across Australia in 2012, 391 cases were identified. The increase in mortality and the increased effort and money required to treat patients with MRSA meant that health authorities needed to institute a response to the disease. Following the introduction of multiple new initiatives the number of MRSA infections has ceased to rise and there has even been a small decrease to 290 in 2016–17. Some of the most important health initiatives aimed at slowing the spread of the disease , such as enhanced infection control and more regular hand washing, have clearly had at least some effect.

It was initially thought that only people in hospital carried MRSA; however, it is now known that approximately one-third of all people carry *S. aureus* on their skin and that for around 2% of people this is MRSA. On the skin *S. aureus* causes no problems. Disease only occurs when the bacteria gets into the body. At this stage even the most resistant strains of MRSA are still treatable with the antibiotic vancomycin, but recently some vancomycin-resistant individuals have been identified in bacterial colonies.

The search for new antibiotics continues and promising candidates have been found. However, it takes many years after candidates have been identified to develop an antibiotic and ensure that it is safe for human use.

VIRAL ANTIGENIC DRIFT AND SHIFT

After a host has been infected by a virus, the host's immune system usually recognises the specific antigens on the surface of the virus. Once these antigens have been identified, the host can mount an immune response to prevent or minimise the severity of future infections. However, as in bacteria, genetic changes can lead to changes to the antigens on the surface of a virus. Antigenic drift and shift are two processes by which viruses change their antigens.

Antigenic drift

During replication, viruses naturally accumulate mutations. Some of these mutations will result in changes to the antigens on the viruses' surface. This gradual process of antigen change is known as **antigenic drift** (Figure 11.3.6a). Usually, the changes in antigen structure are so minor that the host's immune system will recognise the virus if a similar virus has infected the host on a previous occasion. Over a long period of time, multiple episodes of antigenic drift can result in significant changes in the viral antigens, effectively creating a new virus.

> Antigenic drift is the slow change in viral antigens due to the gradual accumulation of mutations in the genetic material of the virus.

Antigenic shift

Antigenic shift (Figure 11.3.6b), by contrast, is a much more abrupt change in the genome of a virus due to the re-assortment of genes from different viral strains, resulting in significantly different antigens on the surface of the virus. One source of antigenic shift is individuals who are simultaneously infected by two different strains of a particular virus. In this situation the viruses can swap blocks of genetic material, giving them new characteristics.

> Antigenic shift is a sudden and significant change in viral antigens due to the merging of genetic material from different viruses.

Many cases where antigenic shift occurs are **zoonotic** in origin. This means the viruses originated in another species of animal and then acquired the ability to infect humans.

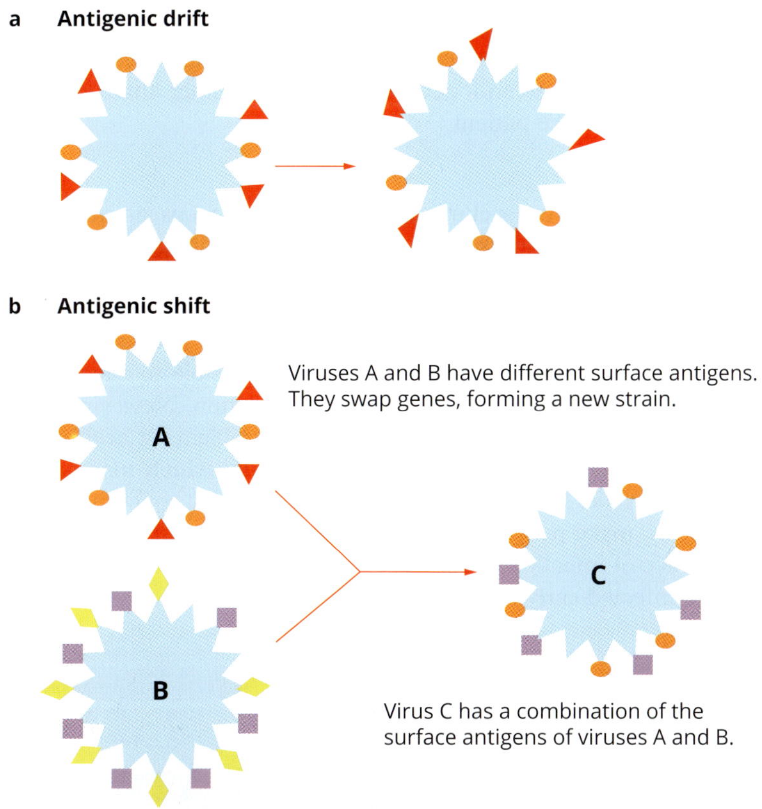

a **Antigenic drift**

b **Antigenic shift**

A

Viruses A and B have different surface antigens. They swap genes, forming a new strain.

C

B

Virus C has a combination of the surface antigens of viruses A and B.

FIGURE 11.3.6 (a) Antigenic drift is a gradual process that occurs as a result of the accumulation of mutations, whereas (b) antigenic shift usually occurs as a result of gene swapping between viral strains and can result in significant changes to the virus.

Antigenic shift can greatly increase the virulence of viruses. Antigenic shift was responsible for the worst influenza epidemic of all time, the 'Spanish flu'. The 'Spanish flu' broke out in 1918 as World War I was ending. Initially, as people were focused on the war, the flu did not attract much attention. However, by the end of the pandemic, an estimated one-third of the world's population had been infected and approximately 50 million people (possibly up to 100 million) had died, representing a mortality rate of 10–20%. Some patients died directly from the influenza virus and others from secondary infections, such as pneumonia, contracted while in a weakened state. While there is no absolute certainty, genetic analysis of samples of the virus suggests that the 'Spanish flu' was caused by an avian (bird-infecting) virus that gained the ability to pass from bird to human through antigenic shift. This type of antigenic shift can occur when humans and animals live in close proximity. A typical pathway for antigenic shift is shown in Figure 11.3.7.

'Spanish flu' is one example of antigenic shift that probably occurred as a result of horizontal gene transfer between bird and human flu viruses.

A pandemic is an outbreak of a disease that covers a wide geographical area.

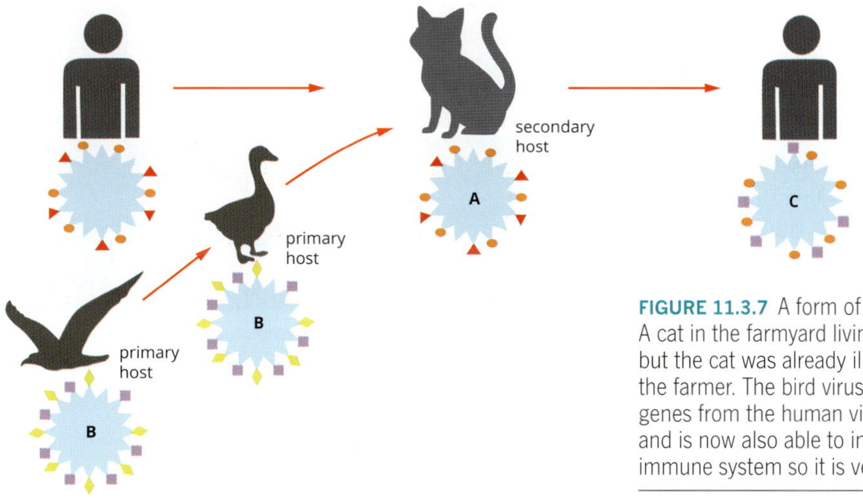

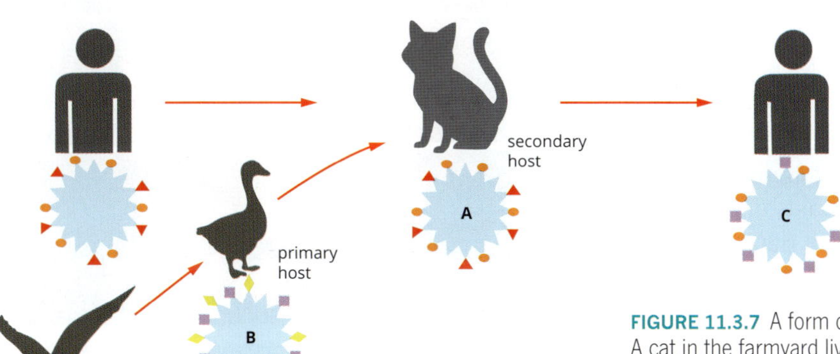

FIGURE 11.3.7 A form of influenza common to birds is passed to domestic ducks. A cat in the farmyard living in close proximity to the infected duck catches the virus but the cat was already ill with a human influenza virus that it had caught from the farmer. The bird virus, which was previously unable to infect a human, obtains genes from the human virus in the cat. The bird virus has undergone antigenic shift and is now also able to infect humans. This is a completely new virus to the human immune system so it is very virulent because humans have no level of immunity.

Challenges in treating viral disease

Treatment of viruses is highly problematic. Public health initiatives generally focus on vaccination, containment and treating symptoms rather than attempting to eradicate the infection in the patient.

Vaccination

> Vaccines contain antigens from the pathogen that trigger the immune system to develop antibodies and memory B and T cells. Memory cells protect against future infections.

The most important efforts to eliminate viral diseases focus on vaccinating the population. By presenting the immune system with the antigens from the virus, memory cells can be made, ensuring that even if people do come in contact with the virus they will not become ill or spread the virus to others. Vaccines, however, can only be made for known viruses that have been studied and had their antigens replicated in the laboratory. It is not possible to vaccinate against a virus until after it has been observed in the population. New viruses such as SARS-CoV-2 (COVID-19) are always going to be a problem for health authorities as the development of new vaccines takes time, during which many people can become ill and even die. Influenza viruses are of particular concern because they can spread rapidly, make people very ill and overwhelm health systems, especially as the second wave of infection usually occurs among health professionals treating those who were infected early in the epidemic.

CASE STUDY

Australia's COVID-19 vaccine research

To infect a cell, a virus must insert its genetic material into a host cell. Some viruses do this by making a hole in the host cell's plasma membrane and entering through it. All coronaviruses, including COVID-19, use this method.

Coronaviruses are named after the crown-like projections on their surface ('corona' means 'crown' in Latin) (Figure 11.3.8). These surface projections are corkscrew-shaped proteins, called spike proteins, which uncoil like a spring to pierce the host cell's plasma membrane. These proteins are the obvious antigens to use to create a vaccine; but they easily uncoil and once the shape has changed, any antibodies formed will be ineffective.

A group of researchers from the University of Queensland (UQ) has devised a new approach to this problem, called a 'molecular clamp'. This is a protein that stops the coronavirus proteins from uncoiling, so that when they are introduced to the immune system in a vaccine, the shape can be recognised, and the appropriate antibodies generated. The molecular clamp uses two fragments of a protein called glycoprotein 41 (gp41). This protein is found in the human immunodeficiency virus (HIV), but the protein alone cannot replicate or infect cells. Unfortunately, the presence of gp41 in the UQ vaccine caused some vaccine trial participants to return false-positive HIV tests. Follow-up tests confirmed that the participants did not have HIV. While the UQ vaccine is safe, the interference with HIV testing led to the decision to stop trials of the vaccine. Although the UQ vaccine will not be used against COVID-19, the molecular clamp technology is an important innovation that holds promise for future vaccines.

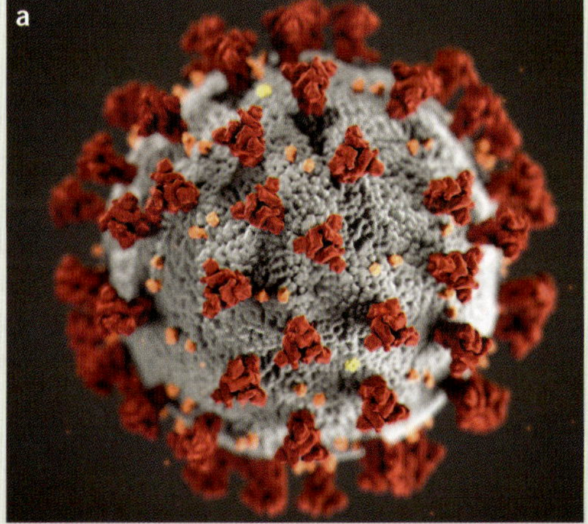

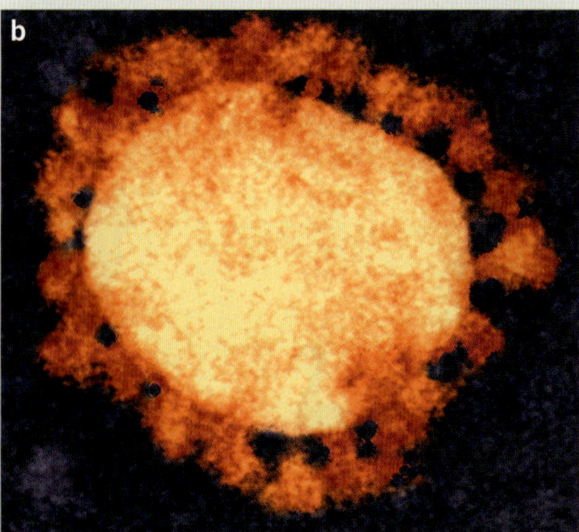

FIGURE 11.3.8 (a) Model of the virus that causes COVID-19. (b) Coloured transmission electron micrograph of the virus, showing the crown-like projections that give coronaviruses their name

Containing disease

Quarantine of infected individuals is probably the most effective method to stop widespread transmission of disease to others. In some cases, this is relatively easy to institute as patients become ill quickly and so are quickly identified and isolated. Some viruses, however, have a significant time lag between infection and symptoms appearing. If the patient is infectious before symptoms appear, they can spread the infection to many others before they are quarantined. Such diseases are especially problematic, as was seen with the recent outbreaks of COVID-19. In the absence of vaccines or medications, spread was contained through physical distancing, mask wearing, border closures and strict quarantine of anyone suspected of infection. New diseases such as COVID-19 are especially challenging in today's mobile world. An individual can contract a disease on one continent today and spread it to another continent tomorrow. In such a world, pandemics of some diseases are almost inevitable so there is a need for effective treatments as well.

Viral medications

While most bacterial infections can be treated with antibiotics, very few specific treatments for viruses exist and those that do are highly specific to the particular virus concerned. The usual medical response to a virus is to treat the symptoms and to try to keep the patient alive while their immune system overcomes the infection. This can put significant strain on the health system if there are many people needing care.

CASE STUDY

Drug therapy for a virus

Zanamivir, sold as Relenza®, was designed to target the influenza A strains which cause the winter influenza outbreaks each year. Like all viruses influenza A takes over a cell in the host's body and makes that cell produce more viruses. Once the viral particles (virions) have been produced they move to the outside of the plasma membrane where they are attached to proteins on the cell's surface using a protein called haemagglutinin. The virus must then cut the attachment protein so it can escape and infect a new cell. The virus uses an enzyme, called neuraminidase, to make this cut.

Relenza is a drug which attaches to the active site of neuraminidase and blocks its action. Figure 11.3.9 shows how Relenza stops the virus infecting more cells. This slows the reproduction of the virus within the host's body and gives the immune system more time to develop the appropriate antibodies.

The challenge with the use of drugs such as Relenza is first that they are very expensive to design and test. Second, as they are highly specific in nature (designed to fit in the active site of a particular enzyme), they can become quite ineffective if there is a genetic change in the virus and the drug is no longer a shape to fit or if the virus acquires the ability to circumvent the effects of the drug through mutation or horizontal gene transfer.

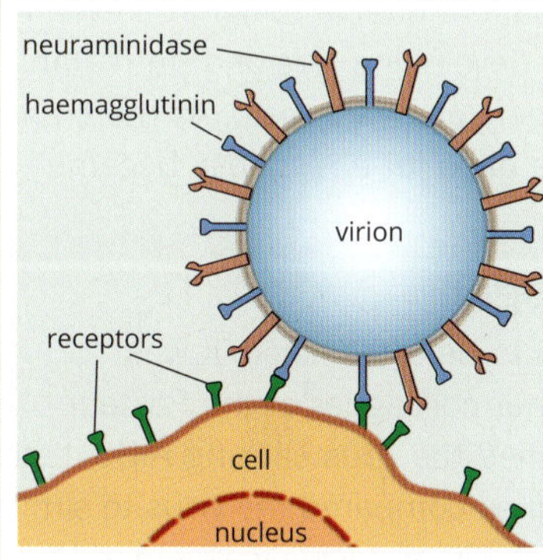

1 New virion being released from a cell is bound by haemagglutinin (blue cup) to a receptor (green).

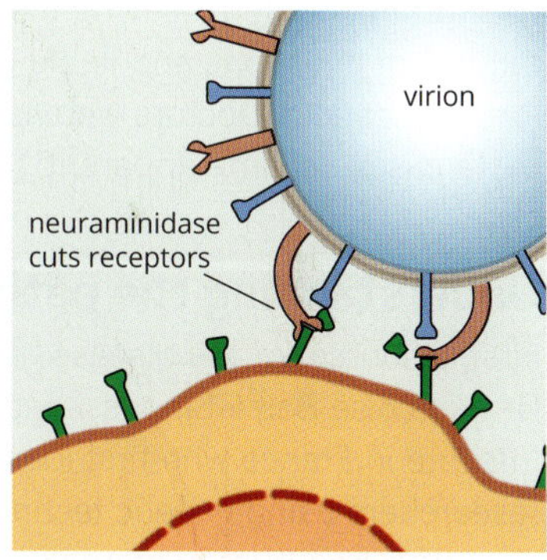

2 Neuraminidase cuts the receptors to let the virus escape and infect another cell.

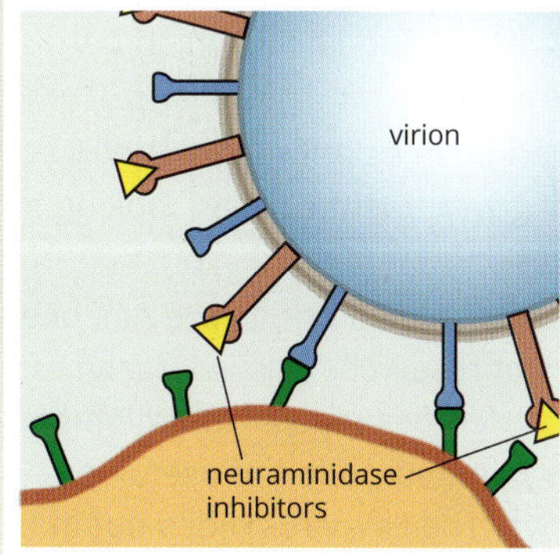

3 Relenza (yellow) blocks the active site of neuraminidase.

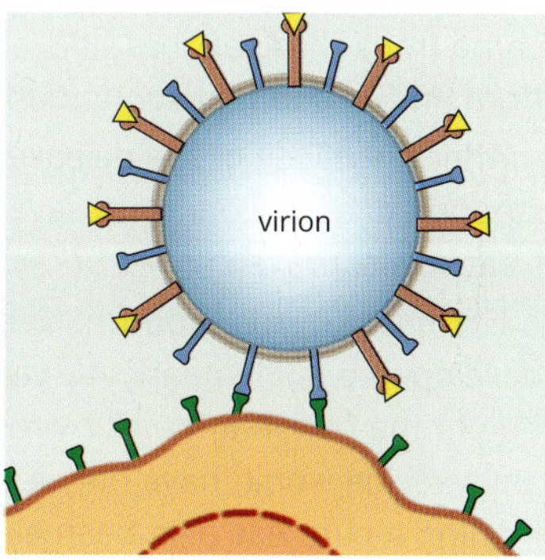

4 Now the virus cannot escape from the cell surface.

FIGURE 11.3.9 Mechanism of Relenza action

CASE STUDY ANALYSIS

Human immunodeficiency virus (HIV)

Human immunodeficiency virus (HIV) and the response of the world's governments, health authorities and medical researchers illustrate how an emerging disease is dealt with. Of course, the strength of the response is dependent on the seriousness of the disease, how easily it spreads and, sadly, the affluence of the people it affects.

Historical evidence suggests that HIV first appeared in humans in Africa, in around 1920. From then until the 1970s few cases were documented, and they were only identified later when research to find the origin of the virus began. Occasional cases were identified through the early 1970s but by the end of that decade it was recognised that HIV infections and the resulting disease, acquired immune deficiency syndrome (AIDS), had reached epidemic proportions. By 1980, AIDS had spread across the world and cases were appearing on five continents. It was now a pandemic.

Until well into the 1980s, infection with HIV was an effective death sentence. In 1981, 337 people were identified with symptoms that later were shown to be AIDS; of that group, 130 had died before the end of the year. The virus basically shut down the immune system, leaving the patient unable to fight off even the mildest of infections and liable to developing rare and untreatable forms of cancer. Doctors were largely unable to help. As more and more cases became apparent, the need for a response became clear.

Understanding the pathogen

First, a source for the disease had to be established. It was Dr Françoise Barré-Sinoussi and her colleagues at the Pasteur Institute in France who first identified the virus causing AIDS. Later researchers using genetic techniques compared the virus to similar viruses in non-human animals and it is now widely accepted that HIV is a variant of SIV (simian immunodeficiency virus) which crossed to humans. More genetic studies of the strains of the virus in humans show that there are two separate strains: HIV-1, which came from chimpanzees and/or gorillas, and HIV-2, which derives from an SIV strain in sooty mangabeys.

After the cause of the disease had been established the WHO stepped in and, along with several governments, began looking for treatments, preventatives and cures.

HIV has proved stubborn. Neither a vaccine nor a cure has been developed despite nearly 40 years of research. This is largely due to the fact that the HIV virus undergoes regular mutation so any vaccine would have only a very limited period of usefulness. Finding a cure has also been elusive. The HIV virus hides out in helper T cells in the immune system. You learnt about these cells in Chapter 9. The hijacking of the helper T cells explains why HIV has such serious effects on the patient.

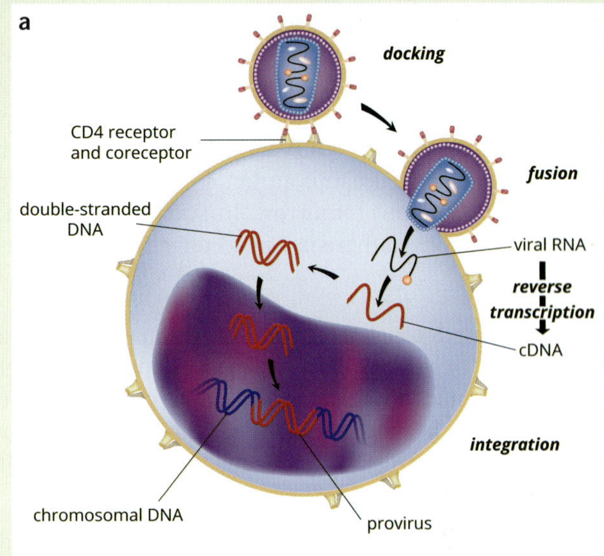

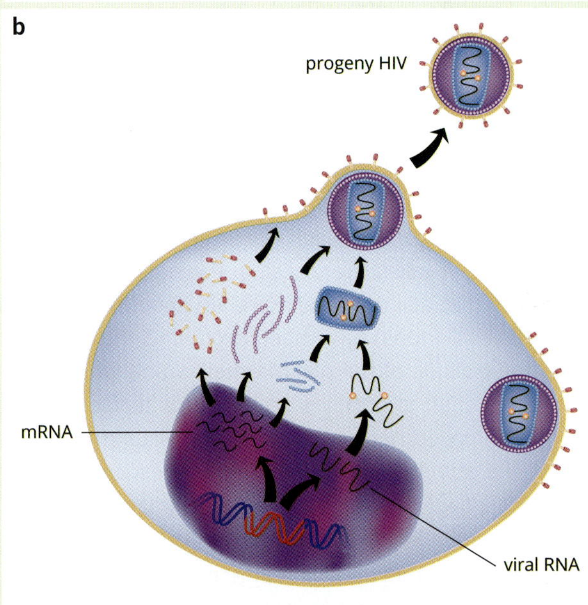

FIGURE 11.3.10 (a) HIV docks with a cell using a CD4 receptor. It then uses a second receptor called CCR5 to trick the cell into allowing it entry. The virus then releases its RNA and uses an enzyme, reverse transcriptase, to make a DNA copy. Finally, it uses another enzyme, integrase, to insert its DNA into the chromosomes of the host cell. (b) The cell is then forced to make various proteases which can build new virus particles. The newly made viruses bud off the surface of the host cell ready to infect more cells.

While preventatives and cures have not been discovered, many treatments have been developed and more become available each year.

To treat a new disease, a detailed understanding of the metabolism and reproduction of the pathogen must be acquired. Throughout the 1980s and 1990s intensive research into HIV occurred in laboratories across the world, resulting in a good understanding of the virus and how it worked. It was discovered that HIV is a retrovirus. That is, it uses RNA as its genetic material. Researchers eventually established how the virus enters a cell, how it integrates its genetic material, the processes it forces the cell to use to construct new viral particles and how the new viral particles escape from the cell ready to infect new cells. These processes are summarised in Figure 11.3.10.

Treatment

As a result of the intensive research undertaken across the world, drugs have been developed which target each stage of HIV. The classes of drugs that have been invented can be seen in Table 11.3.1, along with their modes of action, and their sites of activity are illustrated in Figure 11.3.11.

TABLE 11.3.1 HIV drugs and what they do

Drugs	Target
non-nucleoside reverse transcriptase inhibitors	block the action of reverse transcriptase so that the viral RNA cannot be copied into DNA
protease inhibitors	block the formation of the proteins necessary to form the new viral particles
fusion inhibitors	stop the virus from tricking the cell into granting entry
CCR5 antagonists	block the receptor CCR5 so that HIV cannot use it to enter the cell
post-attachment inhibitors	block the co-receptors that HIV uses and so deny it entry into the cell
integrase strand transfer inhibitors	block the activity of integrase so the viral DNA cannot integrate into the cell's chromosomes and thus the virus is prevented from making new copies of itself

As a result of these new drugs, fewer people are dying of AIDS (Figure 11.3.12) and many are living longer, healthier lives with the disease (Figure 11.3.13).

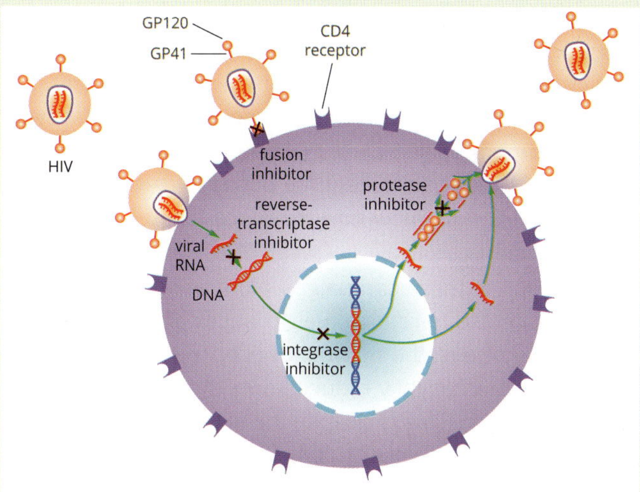

FIGURE 11.3.11 Antiretroviral drugs aimed at HIV: fusion inhibitor, reverse transcriptase inhibitor, integrase inhibitor and protease inhibitor

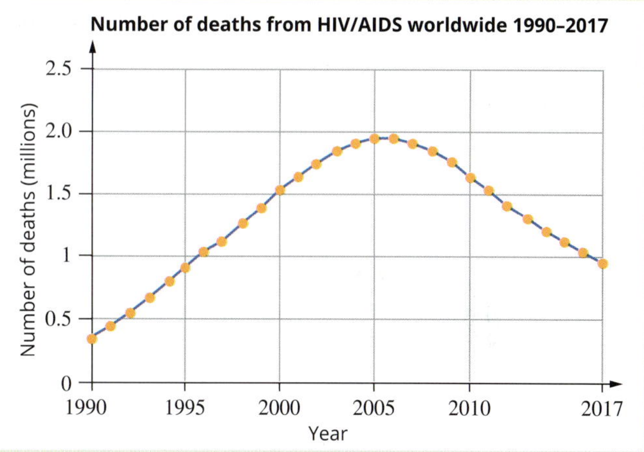

FIGURE 11.3.12 Number of deaths worldwide attributed to HIV/AIDS

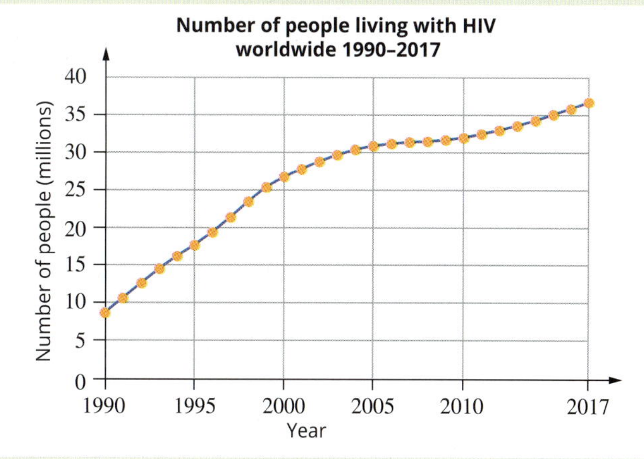

FIGURE 11.3.13 Number of people worldwide living with HIV

continued over page

Public health initiatives

While medical approaches have reduced the mortality of those who contract HIV, there have been other initiatives instituted by governments across the world to reduce the epidemic and contain the disease. The most effective of these programs has been massive public education campaigns, designed to help the public to understand the disease and to know how to reduce their chances of contracting it.

As a result of these campaigns, along with drug therapy of those infected to reduce their chances of passing on the virus to others, the number of people being infected by HIV has dramatically reduced over the last decades (Figure 11.3.14). This is despite the increase in the number of people who are living with HIV.

While there is currently no cure and no vaccine, the epidemic caused by HIV is now fairly well contained, especially in developed nations. Access to the drugs used to supress HIV is still more difficult in developing countries due to the sparsity of their health facilities and the cost of the medications. Organisations such as WHO aim to have all patients with HIV under treatment in the future in order to further reduce the spread of HIV.

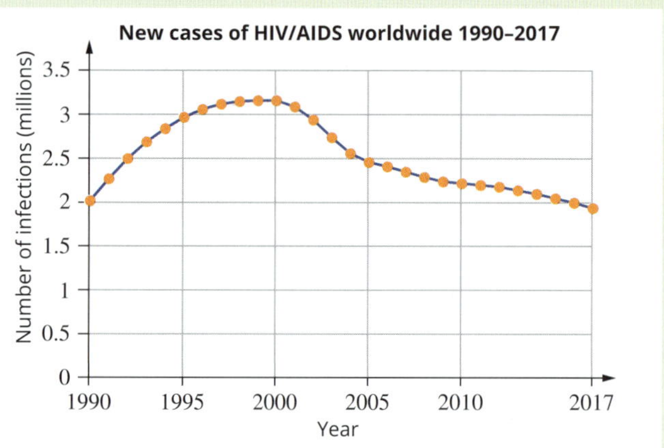

FIGURE 11.3.14 The number of people being infected by HIV continues to decrease despite there being more people worldwide with the infection.

Analysis

1 Propose why scientists think that it will be impossible to permanently eradicate HIV infections in humans.

2 Referring to the graphs above, explain why the number of people with HIV is increasing even though the number of new infections is decreasing.

DEALING WITH AN EMERGING DISEASE—A SUMMARY

The approach to dealing with new or newly virulent diseases involves two basic pathways, medical and public health, with both of the paths involving more than one aspect. Figure 11.3.15 summarises the pathways involved.

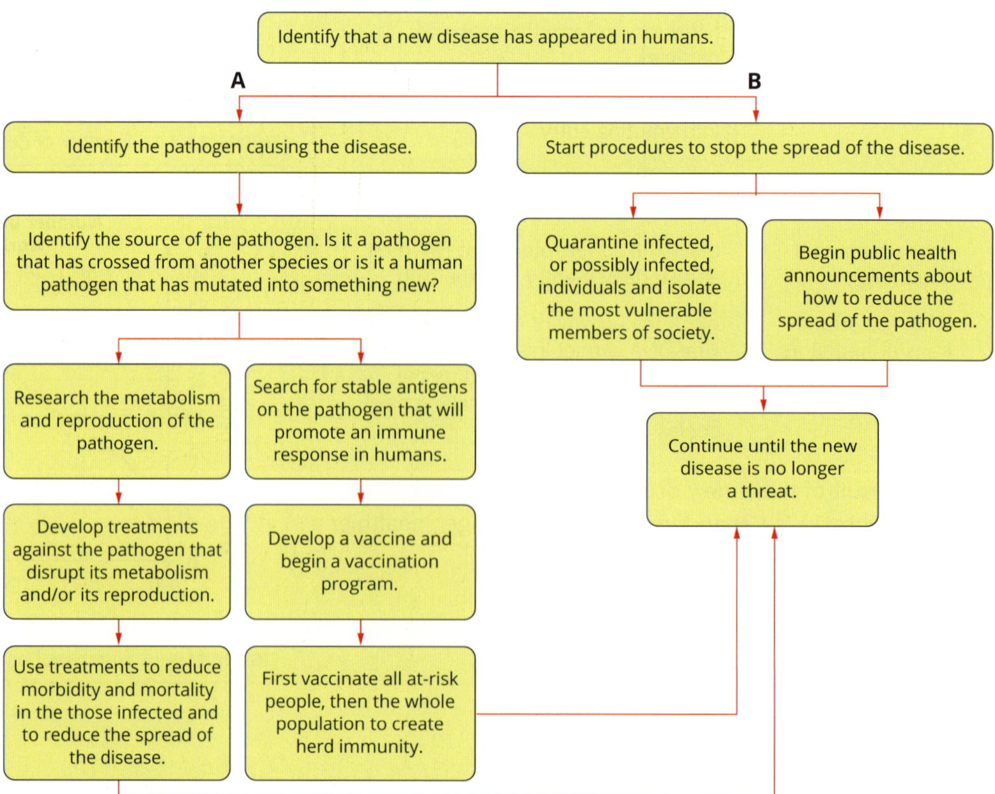

FIGURE 11.3.15 Medical (A) and public health (B) pathways for dealing with emerging infectious disease

11.3 Review

OA ✓✓

SUMMARY

- Both bacteria and viruses possess antigens which must be recognised by the immune system for an immune response to occur.
- Genetic mutations can change a pathogen's antigens and they may no longer be recognised by the host's immune system, leading to a loss of immunity.
- Inappropriate use of antibiotics selects antibiotic-resistant strains of bacteria.
- Antibiotic-resistant bacteria create a challenge for health systems as treatment becomes more difficult.
- Bacteria can obtain resistance to antibiotics through horizontal gene transfer.
- Multidrug-resistant tuberculosis (MDR-TB), caused by *Mycobacterium tuberculosis*, and methicillin-resistant *Staphylococcus aureus* (MRSA) are two bacterial strains that are providing challenges to modern health systems.
- Viruses undergo antigenic change due to the slow accumulation of mutations, called antigenic drift.
- Viruses that infect humans can also merge with animal viruses to create new/novel viruses in a process called antigenic shift.
- In the absence of a vaccine viruses can be treated using specially designed antiviral drugs, such as those designed to treat HIV.
- Creation of antiviral drugs requires a detailed knowledge of the structure and functioning of the virus.
- Quarantine of affected individuals is an important method of reducing the spread of novel viruses.
- Emergent diseases are treated using a combination of medical interventions, such as vaccines and medication, and the spread of disease is prevented using public health initiatives, such as quarantine and education about how to reduce infections.

KEY QUESTIONS

Knowledge and understanding

1 Explain the difference between antigenic drift and antigenic shift.

2 Describe the role of secondary viral hosts such as pigs in creating novel viruses which infect humans.

3 What is meant by the term 'antibiotic resistance'?

4 Why does developing drugs to treat viruses take such a long time?

5 State an important challenge to the development of vaccines against viruses such as HIV.

Analysis

6 Influenza viruses are named according to the variants of two proteins that they contain: haemagglutinin (H), which the virus uses to infect the cell, and neuraminidase (N), which it uses to leave the cell. One variant of influenza is seasonal H1N1. This is a common variant of influenza that is covered by the annual vaccine. In 2007 a novel variant of H1N1 appeared. This was a new subtype with some new characteristics. It was named pandemic H1N1 by health authorities. Some testing of individuals to assess their antibody levels was undertaken in 2007–2008. An antibody titre is determined by taking the patient's serum, diluting it and exposing it to the antigen to be tested. The titre is the maximum dilution which still gives a positive response. A higher value represents a stronger immune response.

Refer to the table below to answer these questions.

a **i** Which group has the strongest antibody response to the pandemic H1N1 before vaccination?

ii Propose why this group has the strongest response.

b Explain whether the data supports the proposition that the pandemic H1N1 is a novel/new viral strain.

c Using the data and your knowledge of the immune response, explain why older adults could benefit from a second dose of vaccine against seasonal influenza.

Antibody titre before and after seasonal flu vaccine, 2007–2008

Age group	Antibody titre before administration of seasonal flu vaccine		Antibody titre after administration of seasonal flu vaccine	
	seasonal H1N1	pandemic H1N1	seasonal H1N1	pandemic H1N1
children	42	10	574	12
younger adults	46	25	578	54
older adults (60+)	31	92	143	97

Chapter review

KEY TERMS

adaptation
adaptive value
allele
allele frequency
antibiotic resistance
antigenic drift
antigenic shift
artificial selection
biodiversity
biological fitness
bottleneck effect
deoxyribonucleic acid (DNA)
evolution
founder effect
gamete
gene flow
gene pool
genetic diversity
genetic drift
genotype
germline mutation
heterozygote
homozygote
horizontal gene transfer
mutagen
mutation
natural selection
phenotype
reproductive success
selection pressure
selective breeding
somatic mutation
trait
vertical gene transfer
zoonotic

REVIEW QUESTIONS

Knowledge and understanding

1 Recall the name given to an individual that has two different alleles for the one gene.

2 Which of the following is the source of new alleles?
A gene flow
B natural selection
C genetic drift
D mutation

3 Define environmental selection pressure.

4 Describe three key factors that contribute to natural selection.

5 **a** Recent studies of human and chimpanzee genomes have shown that populations of chimpanzees living near each other show greater genetic diversity than human populations spread on different continents. Explain how this supports the hypothesis that the human population experienced a genetic bottleneck about 75 000 years ago when Mt Toba, a supervolcano in Sumatra, erupted.

b The genetic bottleneck is one hypothesis to explain the lack of human genetic diversity. Another is that human genetic diversity is low because we are the result of a series of founder populations. Explain why the descendants of founder populations have low genetic diversity.

6 Compare gene flow and genetic drift as mechanisms for changing allele frequencies in populations.

7 Consider the following four populations:
1 a large population experiencing large environmental changes
2 a small population experiencing large environmental changes
3 a large population in a stable environment
4 a small population in a stable environment

What is the expected rate of evolution, from fastest to slowest, in these populations?
A 3, 4, 1, 2
B 4, 2, 3, 1
C 1, 2, 3, 4
D 2, 1, 4, 3

8 Compare natural and artificial selection.

9 Selective breeding:
A reduces biodiversity
B reduces resistance to environmental change
C increases genetic abnormalities
D all of the above

10 How do antigenic drift and shift present a problem for the immune system?

Application and analysis

11 A study involving 23 832 individuals in Lagos, Nigeria, examined the incidence of the blood groups A, B, AB and O in the population. The data on the incidence of the various alleles is shown in the table below.

Incidence of blood groups A, B, AB and O, Lagos, Nigeria

Blood group	Genotype	Number of individuals
O	I^OI^O	12700
A	I^AI^A	467
	I^AI^O	4871
B	I^BI^B	403
	I^BI^O	4523
AB	I^AI^B	868

Use the following formula to calculate the allele frequencies in the population.

$$\text{allele frequency} = \frac{2(\text{number of homozygotes}) + (\text{number of heterozygotes})}{2(\text{total number of individuals})} \times 100$$

- **a** What is the frequency of the I^O allele?
- **b** What is the frequency of the I^A allele?
- **c** What is the frequency of the I^B allele?

12 View the diagram of a hypothetical beetle population at right. Each population started with one phenotype for colour—population 1 had all purple individuals, population 2 had all light pink individuals and population 3 had all red individuals.

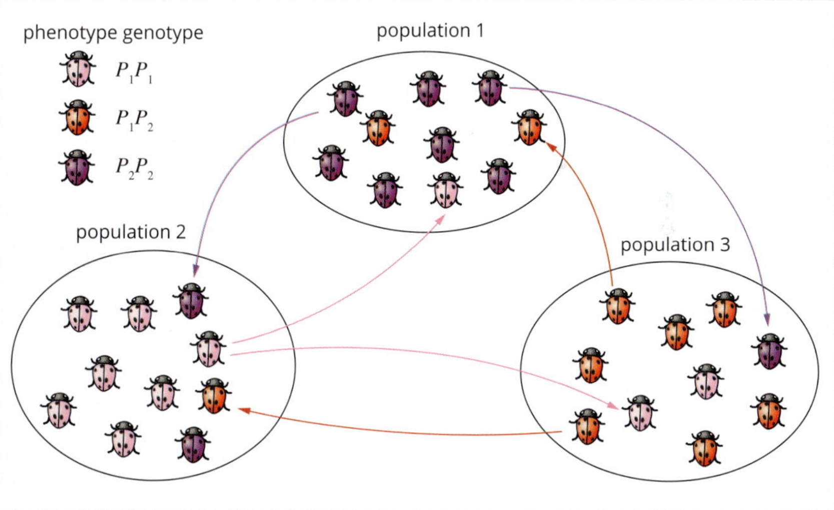

- **a** Name the process that has occurred to result in the phenotype distribution shown in the diagram.
- **b** Describe the importance of this process in natural selection.
- **c** Identify the population with the highest frequency of the P_1 allele. Show your calculation for this population.
- **d** Predict the consequence of environmental change leading to negative selection of the P_2 allele in population 2. Refer to phenotype, genotype and allele frequency in your answer.
- **e** A flood completely isolated population 3 from the others and 75% of the population died. After the flood subsided, environmental conditions stabilised and returned to the previous conditions. Over time, the predominant phenotype was dark. Name the process and describe the key points.

13 You are studying two populations of butterflies. Population 1 has 78 individuals and 14 of them display a rare colour variation. Population 2 has 237 individuals and three of them display the rare colour variation. Using the terms 'genetic drift' and 'bottleneck effect', discuss how the populations' allele frequencies are likely to be affected by a natural disaster.

14 The varying degrees of the sickle-cell disease are determined by the following combinations of alleles in individuals.

Normal haemoglobin	Sickle-cell trait	Sickle-cell disease
Hb^AHb^A	Hb^AHb^S	Hb^SHb^S

Heterozygotes show greater resistance to the mosquito-borne parasite *Plasmodium falciparum*, which causes malaria, than do individuals with normal haemoglobin.

a Identify the process that caused the Hb^S allele to appear.

b Identify the process resulting in the increased heterozygote frequency in regions where malaria is endemic.

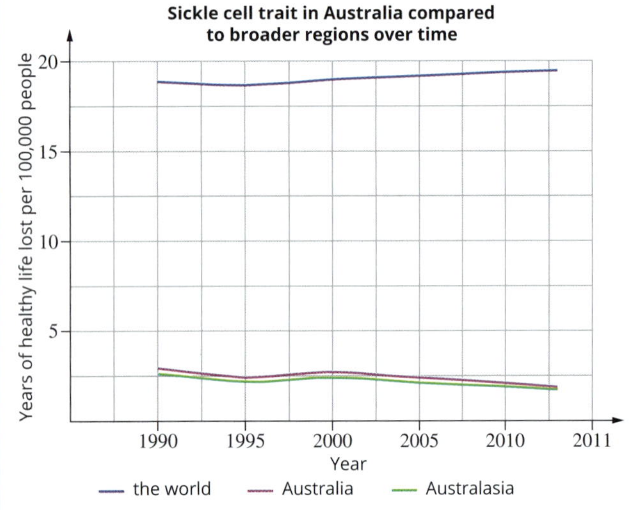

c Describe qualitatively the prevalence of the Hb^S allele in Australia compared to the global prevalence.

15 Red-bellied black snakes are predators of cane toads. Describe the kind of evidence that would indicate biological evolution of red-bellied black snakes in relation to cane toads, rather than short-term physiological change or learned behaviours.

16 Rabbit calicivirus is a disease that was introduced into the Australian mainland in 1995 and Tasmania in 1997. The purpose of the introduction was to reduce the number of wild rabbits as they had reached plague levels and were causing land degradation. Initially, millions of rabbits died but by 2005 numbers were again rising as the rabbits developed resistance to the virus.

a Explain how resistance to calicivirus has developed. Ensure you refer to allele frequencies in your answer.

b Is the action of calicivirus on the Australian rabbit population an example of artificial or natural selection? Explain your reasoning.

c The Department of Primary Industries introduced a new and more virulent strain of calicivirus into Tasmania in March 2017. Explain if this is likely to be more successful in eradicating the rabbit from Tasmania.

17 Cancer is not usually transmissible between individuals but at least one known exception to this exists. Devil facial tumour disease is passed between individual Tasmanian devils (*Sarcophilus harrisii*) when they bite each other. Tasmanian devils are aggressive and frequently bite each other when competing for food or even while mating. The facial tumour eventually makes eating impossible and the devil dies from starvation. Ecologists fear that the devil facial tumour will result in the extinction of the Tasmanian devil, so considerable research has been done to try to solve the problem. Two promising lines of research are being investigated.

It has been shown that less aggressive devils are much less likely to be infected with the devil facial tumour and that members of the population of Tasmanian devils in north-west Tasmania have what appears to be some level of genetic resistance to the disease. It has been suggested that selective breeding of devils for either resistance, low aggression or both traits could help to stop the Tasmanian devil becoming extinct.

a Explain how selective breeding in a captive population of Tasmanian devils could be used to help increase the survival chances for the species.

b Discuss any possible negative consequences for the gene pool of the Tasmanian devil from such a program.

18 a Viruses have the potential to cause a pandemic. Propose two characteristics that a virus needs to possess in order to cause a pandemic disease.

b How might these conditions arise?

19 A drug company is testing a chemical recently extracted from a fungus that they believe has the potential to be developed into a new antibiotic for use in humans. They take 500 identical agar plates; half of them have plain nutrient agar and the other half have nutrient agar and the newly discovered chemical. Bacteria from the same colony are grown on the plates. The plates are incubated at 37°C for 24 hours and are then examined for bacterial growth.

a State the independent variable in this experiment.

b State two controlled variables.

c i After 24 hours, the plates are examined. The plain nutrient agar plates are covered with bacteria. Most of the plates with the new chemical have no bacterial growth; however, one of the plates has a single colony growing. How has this colony survived when all the other bacteria have died?

ii Does this result suggest that the development of this antibiotic should be abandoned?

20 Between 2000 and 2004 the incidence of multidrug-resistant tuberculosis (MDR-TB) was monitored across the world. The data collected is shown in the table below.

Incidence of multidrug-resistant tuberculosis (MDR-TB), worldwide, 2002–2004

Region	Number of people tested	Number of people testing positive for MDR-TB
North America, Australia, Japan and Western Europe	2499	821
Latin America	985	543
Eastern Europe and Russia	1153	406
Africa and the Middle East	665	156
Asia (excluding Japan)	12330	1572

a For each region calculate the percentage of people tested who had a strain of TB that is resistant to treatment.

b i Based on the data, which region had the highest incidence of MDR-TB?

ii Explain why percentages needed to be calculated before analysing the data.

OA ✓✓

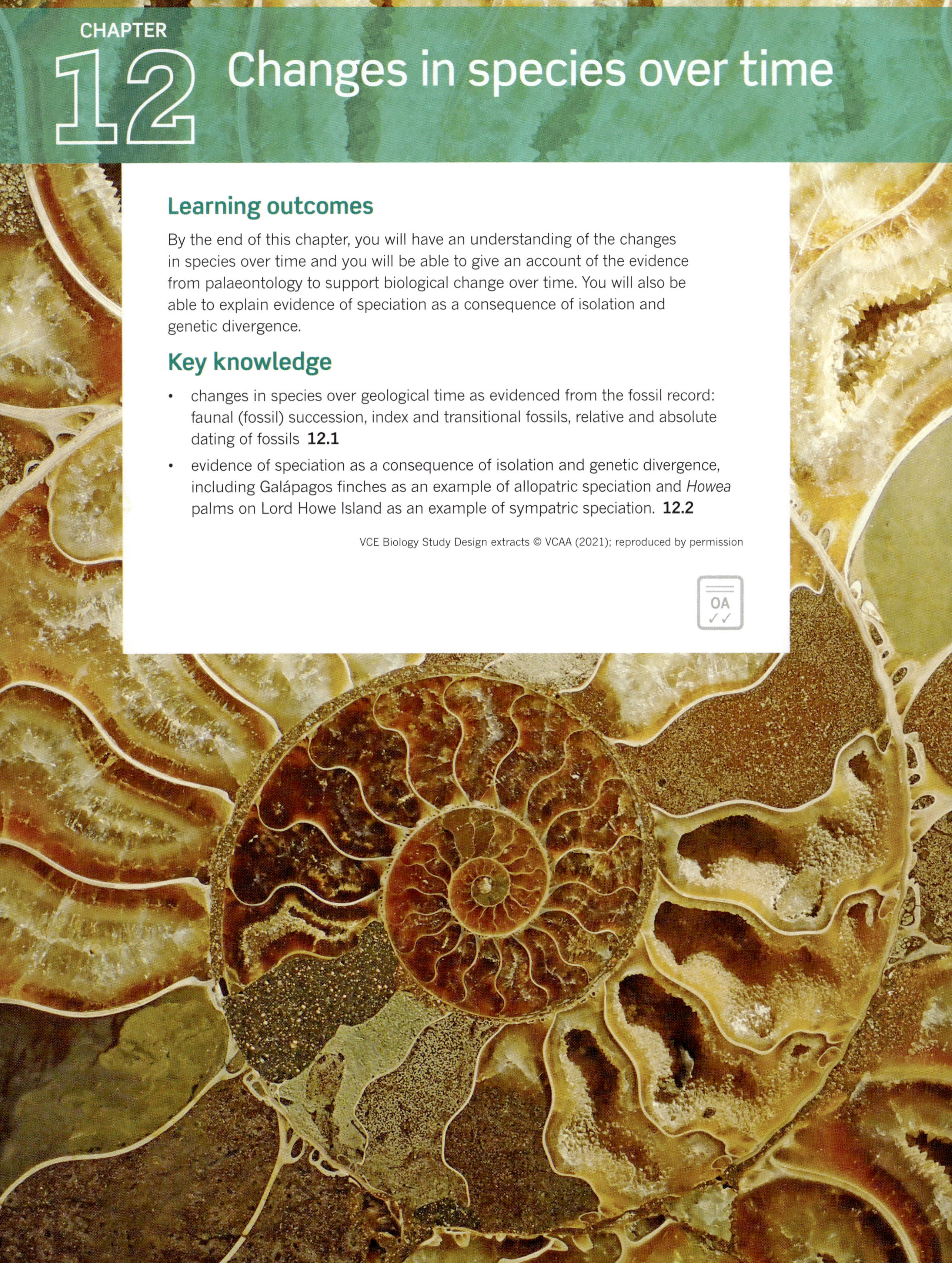

CHAPTER 12

Changes in species over time

Learning outcomes

By the end of this chapter, you will have an understanding of the changes in species over time and you will be able to give an account of the evidence from palaeontology to support biological change over time. You will also be able to explain evidence of speciation as a consequence of isolation and genetic divergence.

Key knowledge

- changes in species over geological time as evidenced from the fossil record: faunal (fossil) succession, index and transitional fossils, relative and absolute dating of fossils **12.1**
- evidence of speciation as a consequence of isolation and genetic divergence, including Galápagos finches as an example of allopatric speciation and *Howea* palms on Lord Howe Island as an example of sympatric speciation. **12.2**

OA ✓✓

12.1 The fossil record

The conditions on Earth have always determined the variety of living organisms that can exist. Earth's atmosphere has changed remarkably over time. The early Earth, which formed 4.6 billion years ago, was volcanically very active and the atmosphere was very different from that of today (Figure 12.1.1). Conditions allowed the formation of biomolecules, and cellular life evolved, with the earliest prokaryotes experiencing an atmosphere lacking in oxygen. Climates changed over time, and there have been multiple ice ages and hot, dry periods. The movement of continents has greatly affected the distribution of seas and area of land, with major consequences for organisms.

In this section, you will learn about the significant changes in species over geological time as evidenced from the fossil record.

FIGURE 12.1.1 When Earth formed 4.6 billion years ago from a hot mix of gases and solids, it had almost no atmosphere.

FIGURE 12.1.2 (a) Geological layers can be seen at Fossil Cove, Tasmania, Australia. (b) Ammonite fossils embedded in rock

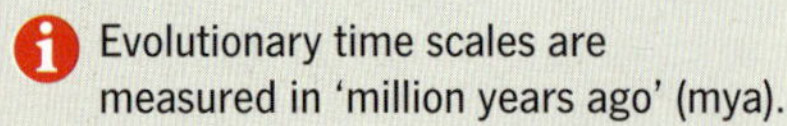

Evolutionary time scales are measured in 'million years ago' (mya).

GEOLOGICAL TIME SCALE OF EARTH

The history of Earth and evolving life can be chronologically followed using the **geological time scale**, which covers events that have occurred on Earth from its formation to the present time.

The geological time scale is constructed using the order of rocks laid down in a sedimentary rock sequence (a relative time scale in which the oldest rocks are at the bottom), and the fossilised remains of ancient animals and plants within the rock strata (Figure 12.1.2). Today geologists also use techniques such as radiometric dating to directly determine the age of rocks.

The geological time scale is divided into many subdivisions. The largest of these subdivisions is the **eon**. Eons are subdivided into **eras**, which are further subdivided into **periods**, and into still smaller subdivisions called **epochs** (Table 12.1.1).

TABLE 12.1.1 Geological time scale in millions of years ago (mya)

Eon	Era	Period	Epoch	Age (mya)	Plant life	Animal life
Phanerozoic eon	**Cenozoic**	Quaternary	Holocene	0.01	• modern plants	• evolution of humans
			Pleistocene	2.58		
		Neogene	Pliocene	5.33	• angiosperms dominate forests and grasslands	• mammals diversify, including primates • whales appear in oceans
			Miocene	23.03		
		Palaeogene	Oligocene	33.9	• angiosperms continue to dominate	• many primate groups appear
			Eocene	56.0	• angiosperms continue to dominate	• mammals continue to diversify
			Palaeocene	66.0	• angiosperms continue to dominate	• mammals, birds and pollinating insects diversify
	Mesozoic	Cretaceous		145.0	• angiosperms become dominant	• dinosaurs become extinct • mammals diversify or further develop • birds diversify • first primates
		Jurassic		201.3	• conifers abundant, first angiosperms	• age of reptiles, some flying reptiles • first birds
		Triassic		252.2	• conifer trees dominate forests	• first mammals • first dinosaurs • reptiles dominate land • amphibians decline
	Palaeozoic	Permian		298.9	• early seed plants develop, including cycads and early conifers	• reptiles diversify • familiar insects develop • many land vertebrates • many marine invertebrates become extinct
		Carboniferous		358.9	• first large swamp forests of vascular land plants	• insects become more common • first reptiles
		Devonian		419.2	• tree-like vascular land plants, including lycopods; ferns appear	• fishes and coral reefs common
		Silurian		443.8	• first small vascular land plants, many algae	• many coral reefs, shells • first animals on the land—amphibians and invertebrates
		Ordovician		485.4	• types of large algae	• many invertebrates • first vertebrates—fishes
		Cambrian		541.0	• more types of algae appear	• animals with bodies protected by shells • first fishes appear
Proterozoic eon		Ediacaran		635	• some algae	• soft-bodied animals • a few fossils of animals with jelly-like bodies found from this period
				2500	• first animal traces • multicellular life develops in shallow warm seas • fossils rarely found from this time period, due to the age of the rocks and the soft fragile bodies of these organisms	
Archaean eon				4000	• bacteria (prokaryotes) abundant • fossilised and living stromatolites from this eon are still found on Earth today • oldest known sedimentary rocks, and oldest 'fossil' remains—chemical traces of living things	
Hadean time				4600	• solidification of the Earth from a ball of molten rock	

Note: The Proterozoic eon, Archaean eon and Hadean time are collectively known as Precambrian time. Hadean time is not a geological era/eon/period.

EVIDENCE FOR EVOLUTION

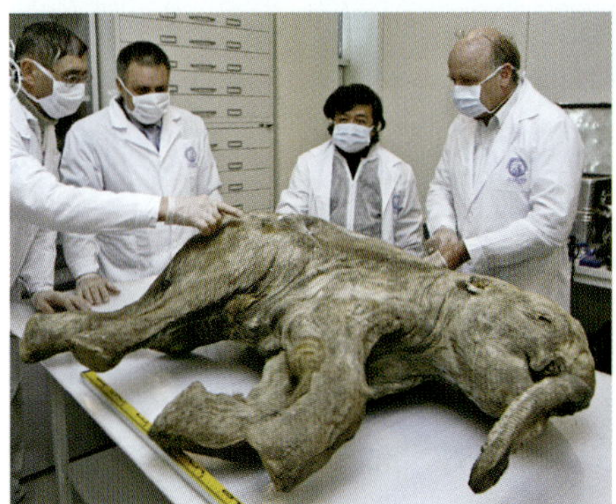

FIGURE 12.1.3 Palaeontologists studying 'Lyuba', one of the best-preserved woolly mammoths (*Mammuthus primigenius*). Lyuba was a female calf that died about 40 000 years ago at the age of about one month. Woolly mammoths became extinct about 10 000 years ago.

Evolution is a process of change. The modern theory of evolution states that all living organisms share a common origin that dates back to 3.8–4.1 billion years ago. The earliest organisms were bacteria and, over a long period of time, very different groups of organisms diverged from these early forms of life. Some groups became extinct, while others evolved to become the types of organisms that we see today. Evidence for biological change over time can be found in the fossil record and by comparing organisms' features (structural morphology) and DNA sequences (molecular homology). You will learn about structural and molecular features as evidence of evolution in Chapter 13.

FOSSILS

Palaeontology involves the study of ancient life represented by **fossils** (Figure 12.1.3). Fossils are the **preserved remains**, impressions or traces of organisms found in rocks, amber (fossilised tree sap), coal deposits, ice or soil (Figure 12.1.4). The **fossil record** refers to the total number of fossils that have been discovered, and it provides evidence of the evolution of living organisms through geological time. Fossils tell palaeontologists about the kinds of organisms that lived in the past, what they looked like, and where and when they lived, allowing them to develop an evolutionary time scale.

FIGURE 12.1.4 Macrophotograph of a fossilised midge insect (family: Chironomidae) found embedded in Baltic amber. The insect is related to present-day non-biting midges, and more distantly to mosquitoes. It is about 40 million years old and from the Late Eocene epoch.

Fossilisation process

Fossilisation is the preservation of the hardened remains or traces of organisms in rocks.

The chances of an organism becoming fossilised after death are small. Soft-bodied organisms are unlikely to be preserved, because soft body parts decay readily or are subject to predation and scavenging. Fossilised remains are usually hard structures that are not easily destroyed or are slow to decompose, such as bone, shell, wood, leaves, pollen and spores.

Fossilisation is more likely to occur when an organism is buried by sediments. This reduces the chance of decay, due to lack of oxygen for decomposer microorganisms, and hides the organism from scavengers. When sediments of sand, silt or mud in the sea, a lake or slow-flowing stream accumulate over the organism, the organism is preserved. The weight of many layers of sediments squeezes out the water between the particles of sand, silt or mud. As the deposit deepens, the temperature increases and soft sediments become solid rock—sandstone, siltstone, mudstone or shale (a mixture of clay and silt) (Figure 12.1.5).

Sediments accumulate in bodies of water such as seas, estuaries and lakes, hence a large proportion of fossils are found where ancient bodies of water existed. Fossils of shells formed in this way in Tasmania. In the Carboniferous and Early Permian periods (about 280 million years ago), a marine gulf formed the Tasmanian Basin. The basin filled with mud and silt washed down from glaciated uplands. The sediments formed layers of mudstone, siltstone, sandstone and some limestone. The basin later became a larger plain, with lakes and freshwater streams that deposited more sediment layers. Fossils of shells, fishes (including lungfish) and amphibians have been found in siltstone and sandstone at Mt La Perouse in Tasmania (Figure 12.1.6).

> The fossil record provides evidence of changes in species over geological time.

> Fossilisation is a rare event, so our understanding of organisms from the past is limited.

Organisms on land are less likely to be preserved than those that live in aquatic environments. For example, plants that grow along river banks or the edge of swamps, where sediments can trap leaves, fruits and seeds, are more likely to be fossilised than plants that grow only on rocky outcrops where water bodies are not present. Delicate plant parts such as flowers are rarely fossilised, although some are preserved by being buried rapidly, for example by ash from an erupting volcano. The fossil record preserves only certain sorts and parts of organisms under certain environmental conditions, which consequently limits evidence of past life and our understanding of it.

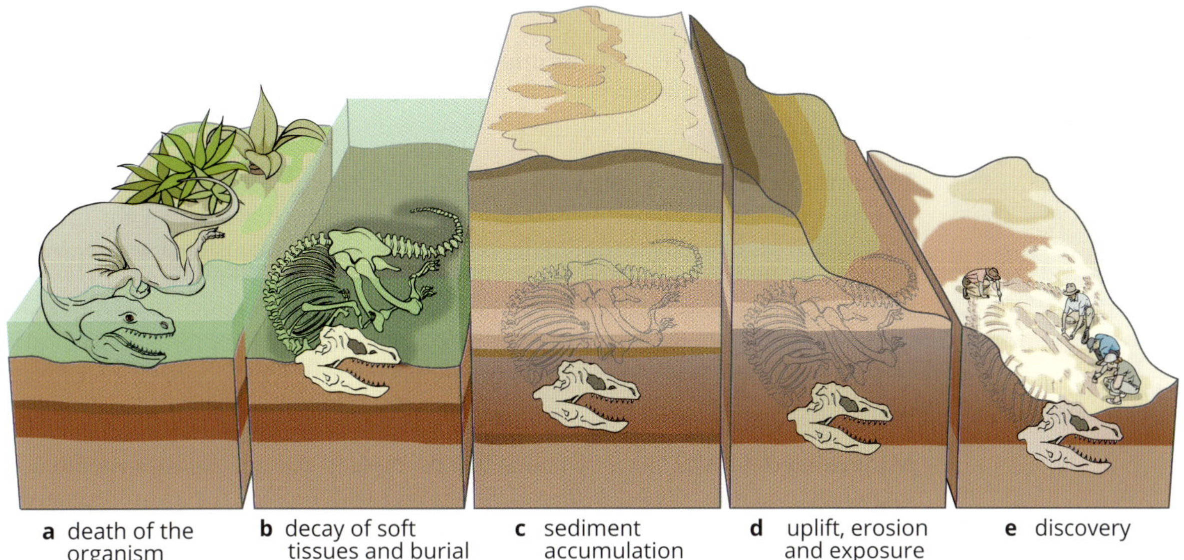

FIGURE 12.1.5 The sequence of events leading to the formation and subsequent discovery of a fossil. (a) The organism dies and (b) is buried. (c) Sediments accumulate and solidify to rock. (d) Subsequent uplift, erosion and exposure (e) lead to its discovery.

FIGURE 12.1.6 Fossil shells, fishes (including lungfish) and amphibians about 280 million years old have been found at Mt La Perouse in Tasmania.

CASE STUDY

The oldest stegosaur

In 2019 the remains of a new genus and species of stegosaur were discovered in Morocco (Figure 12.1.7), suggesting stegosaurs are older and were more geographically widespread than previously believed. The new species, named *Adratiklit boulahfa,* is the first stegosaur to be found in North Africa and is dated to the middle Jurassic period, about 168 million years ago. Most stegosaur fossils date to the late Jurassic period, from 155 to 150 million years ago, and are found in Europe and the United States. *A. boulahfa* places the stegosaurs further back in time than palaeontologists previously thought—about 100 million years earlier than *Tyrannosaurus rex.*

FIGURE 12.1.7 Complete stegosaur skeleton in the Natural History Museum in London. Remains of a new stegosaur genus and species were discovered in Morocco in 2019.

Types of fossils

The four main types of fossils are impression fossils, mineralised fossils, trace fossils and mummified organisms.

Impression fossils

Impression fossils are left when the entire organism decays but the shape or impression of the external or internal surface remains (Figure 12.1.8). In some rocks, such as limestone, the fossils retain their three-dimensional shape, but in rocks (e.g. shales) or coal deposits that are physically compressed, fossils are flattened. Impression fossils include the internal surface of a shell, tree trunks and plant leaves. If the vacant space of the mould is later filled with foreign material, a three-dimensional 'sculpture' of the organism is formed. This is called a **cast fossil**.

FIGURE 12.1.8 Sandstone block split open to show a rare impression fossil of a terminal bud from an *Equisetum* horsetail plant from the Triassic period

Mineralised fossils

Mineralised fossils occur when minerals replace the spaces in structures of organisms, such as bones. Minerals may eventually replace the entire organism, leaving a replica of the original fossil (e.g. petrified wood). This process is known as **mineralisation** (or petrification). Minerals can include opal, pyrite and silica (Figure 12.1.9).

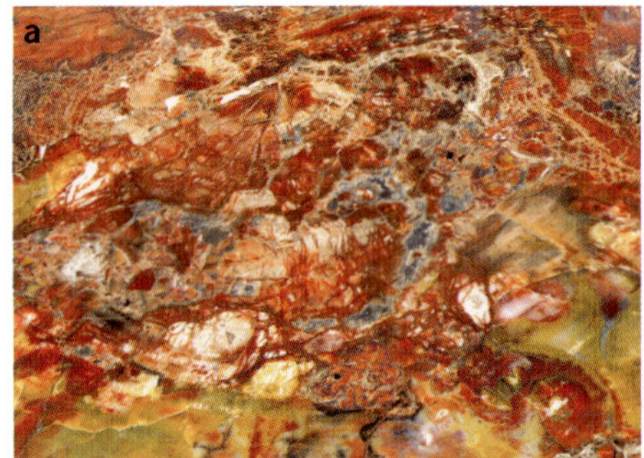

FIGURE 12.1.9 Examples of mineralised fossils. (a) Petrified wood from the Petrified Forest, Arizona, USA. This fossil is from the Late Triassic, when the forest was rapidly buried under volcanic ash. (b) Rock containing a fossil of a *Phareodus* sp. fish from the Eocene epoch. (c) Coloured scanning electron micrograph (SEM) of fossilised diatoms, single-celled planktonic algae. Diatoms have a wall of silica that provides protection and support, which is readily fossilised. These diatoms are from the Miocene epoch.

Trace fossils

Trace fossils (also known as ichnofossils) are the preserved evidence of an animal's activity or behaviour, without containing parts of the organism. Transient impressions or footprints would be considered trace fossils, as would casts of burrows or even coprolites (fossilised faeces) (Figure 12.1.10).

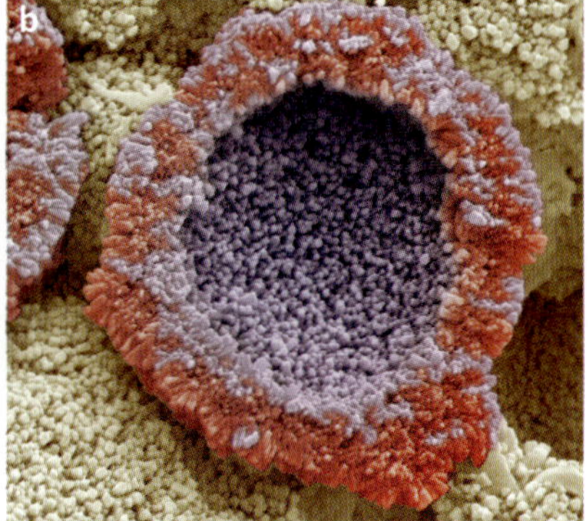

FIGURE 12.1.10 Examples of trace fossils. (a) Fossil worm burrows (*Arthrophycus* sp.) from the early Silurian. (b) Coloured SEM of a section through a coprolite (fossilised faeces) from a dinosaur

Mummified organisms

Mummified organisms are those that have been trapped in a substance under conditions that reduce decay and experience little change. When organisms are preserved relatively intact and normal decomposition has not occurred, they provide a useful record for paleontologists. In order for mummification to occur, the body of a dead organism must not be eaten by scavengers and must then either be frozen or rapidly buried. Both freezing temperatures and a lack of oxygen prevent microbial decay. Mummified organisms become exposed and can be discovered when there is uplift of the Earth's surface, erosion or glacial melting.

Examples of mummified organisms include insects trapped in amber, leaves that still contain carbon (Figure 12.1.11), and animals frozen in ice or trapped in a peat bog (known as 'bog body') (Figure 12.1.12). Fossil mummified animals, including humans, can have hair and skin preserved in a dehydrated state, while limbs and occasionally entire bodies are preserved in peat bogs and tar pits.

FIGURE 12.1.12 Mummified body of Tollund Man, dated 220–40 BCE. This well-preserved body of an adult man was discovered in 1950 in a bog at Tollund Mose in Jutland, Denmark.

FIGURE 12.1.11 This leaf from the coal deposits in Anglesea, Victoria, is a mummified fossil and still contains carbon. It was preserved in layers of mud sandwiched between the layers of coal. Many of the fossils are rainforest plants related to those that survive today in the wet tropics of Queensland. The environment at Anglesea must have once been wet and warm.

BIOFILE

Fossil footprints

Trace fossils of footprints are formed when an organism steps into soft mud. The impression is then covered with loose sand so that the footprint is filled. The sand in the footprint eventually consolidates and is compacted into sandstone. Finally, when the rock is split open along the bedding surface, the original footprint is revealed.

Fossils appear when rock slowly forms around objects buried in mud. As the rock forms, the shape and anatomy of buried animals and plants can be preserved, including tracks such as these footprints.

Transitional fossils

Transitional fossils exhibit features that are intermediate between ancestral and descendant groups, indicating a progression from one form to another.

Transitional fossils provide evidence for an intermediate evolutionary form between an ancestor and its descendants. A transitional fossil exhibits common traits between the ancestral group and descendant group, often providing a link between related species that appear to be very different. For example, fossils of *Archaeopteryx* (Figure 12.1.13) confirm the transition from dinosaurs to modern birds. *Archaeopteryx* is a genus of bird-like dinosaurs that possesses similarities to non-avian dinosaurs, such as a long, feathered tail and small teeth. However, unlike non-avian dinosaurs, *Archaeopteryx* also has flight feathers and wings.

BIOFILE

Dinosaur Cove

Dinosaur Cove, on the southern coast of Victoria, is famous for its fossil deposits. An ancient stream flowed through the site 106 million years ago, depositing soft sand and mud, which turned to rock. Dinosaur bones were trapped in these sediments.

Dr Tom Rich and Dr Pat Vickers-Rich found and described small, bipedal dinosaurs (hypsilophodontids). At the time that these dinosaurs lived near the cove, Australia was further south and connected to Antarctica. Although not frigidly cold, winter was a long period of darkness. The hypsilophodontids had large optic lobes in their brains, meaning that they could probably see well in the dark.

Cretaceous bird tracks on a slab of sandstone found at Dinosaur Cove, southern Victoria, Australia

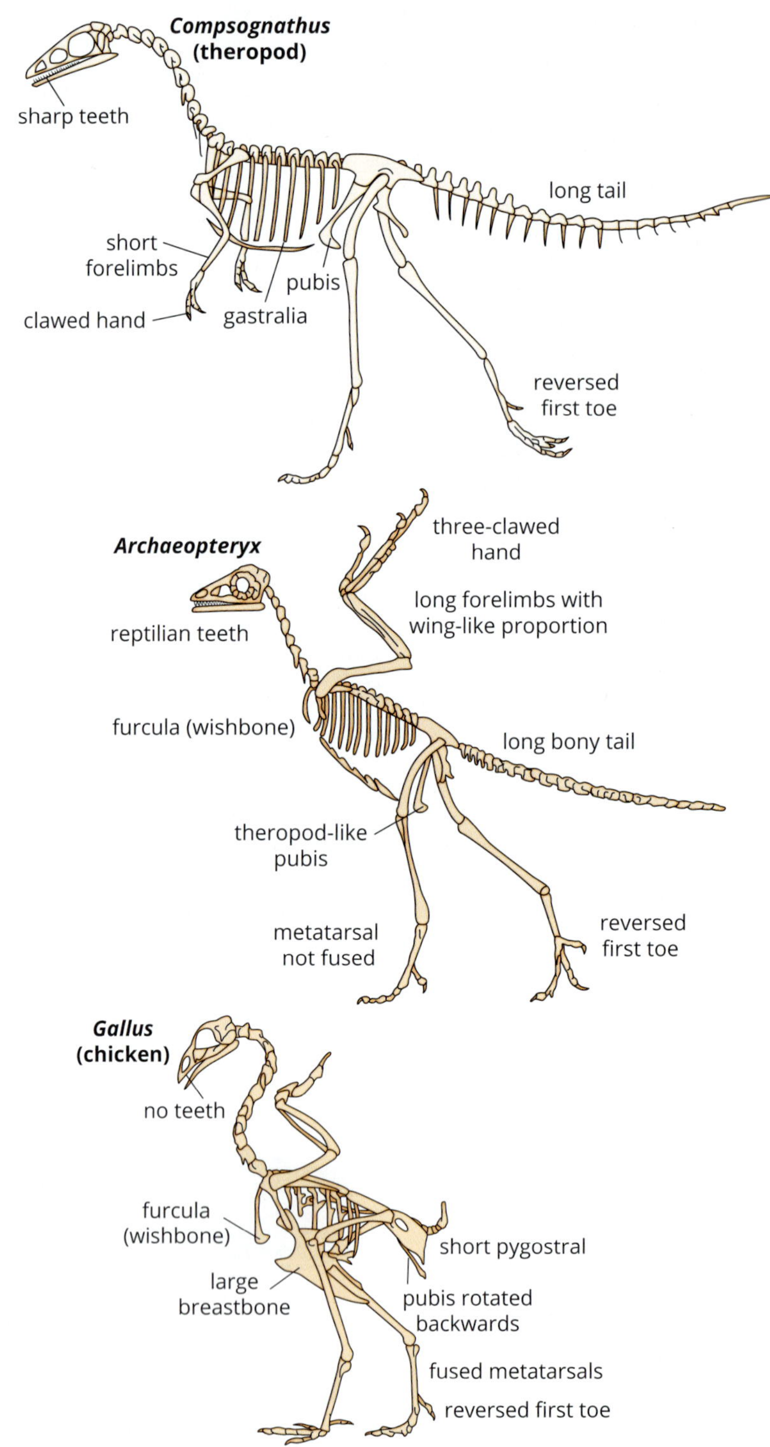

FIGURE 12.1.13 The skeletal structure of *Compsognathus* (theropod), *Archaeopteryx* and *Gallus* (chicken), showing the structures that are similar between the three organisms

Dating fossils

The age of a fossil is almost as important as its physical details because it gives a time scale of evolution. The age of a fossil can be determined by relative dating or by absolute dating methods.

Relative dating

Relative dating is based on stratigraphy. **Stratigraphy** is the study of the relative positions of the rock **strata (singular: stratum)**, or layers, some of which contain fossils. The lowest stratum is the oldest and the upper strata are progressively younger. The age of a fossil is estimated relative to the known age of the layers of rock above and below the layer in which the fossil is found (Figure 12.1.14). For example, if a layer containing fossils lies below rock that is dated at 200 million years old, then the fossils must be at least that age or older. Relative dating can be difficult in areas where rock layers have been eroded away, or where rocks have been buckled, moved or reburied, altering the original sequence of strata.

Relative dating estimates the age of an object relative to another by comparing their positions in rock layers.

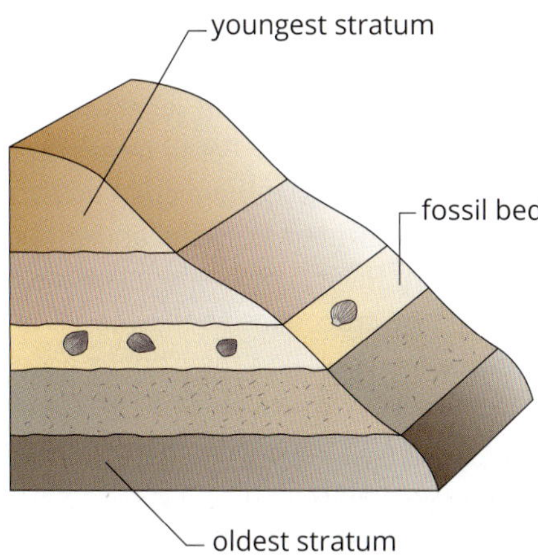

FIGURE 12.1.14 Relative dating assumes that rock strata (layers) are laid down in the order of the formation—the bottom stratum is the oldest, and the top is the youngest.

Faunal succession is a fundamental principle of stratigraphy that states that the fossils contained in sedimentary rock strata succeed one another in a predictable order, even when they are found in different places. The age of strata from different geographic regions can be estimated by comparing the fossils within the rock layers. When the same sequence of fossils is found at another location, it is likely that the rock strata containing them are from the same geological time.

An **index fossil** (sometimes known as an indicator fossil) is a fossil used to define and identify geological periods. Sometimes the only way to age a fossil bed is by using index fossils together with stratigraphy. Index fossils are commonly found fossils from similar sites for which an absolute age has been determined. For example, in Europe the same type of ammonoids (extinct molluscs) are found in different regions. A species of ammonoid fossil is called an index fossil because it indicates that the rocks at each locality are of similar age (Figure 12.1.15).

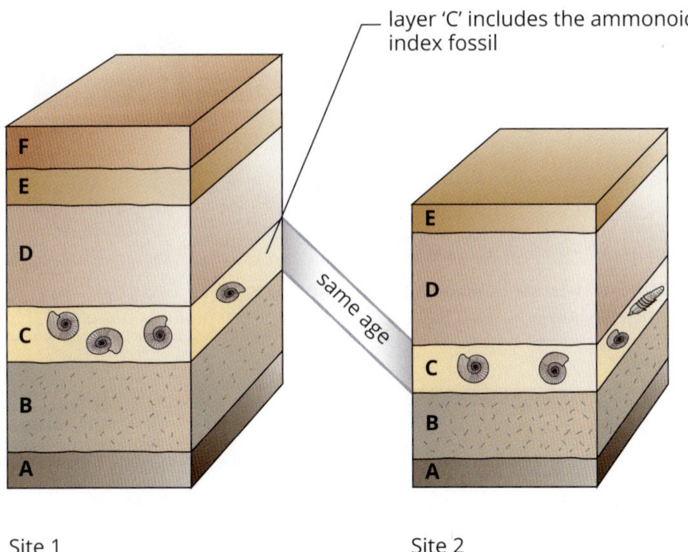

FIGURE 12.1.15 Stratigraphic comparison of sites in different parts of the world provides evidence of the relative age of particular strata. The ammonoid (mollusc) fossils of known age at site 1 are the same as at site 2, so the two strata are assumed to be the same age (even though site 1 has an additional, younger layer, F, at the top). The ammonoid is an index fossil for the age of all the other fossils in the fossil layer at site 2.

The principle of faunal succession states that fossils succeed each other in a predictable order, even when found in different geographic regions.

An index fossil is a fossil that is characteristic of a particular time period. They are abundant, widespread and distinctive, and from a narrow span of geological time.

CASE STUDY

Two fossil sites of inland Australia

Riversleigh in north-west Queensland

Riversleigh and the nearby Gregory River contain one of the great fossil sites in the world. Extinct animals are preserved in limestone of various ages, dating back to more than 30 million years (Oligocene period). The fossils at this site reveal the story of ancient rainforest animals and plants that once lived in inland Australia. Fossil remains of the extinct marsupial *Diprotodon optatum* have been uncovered from the ancient bed of the Gregory River. These animals were probably killed by crocodiles as they came to drink from the river, and their bones accumulated as fossils in layers of sands and gravels (Figure 12.1.16). These particular fossils are approximately 1 million years old (from the Late Pleistocene).

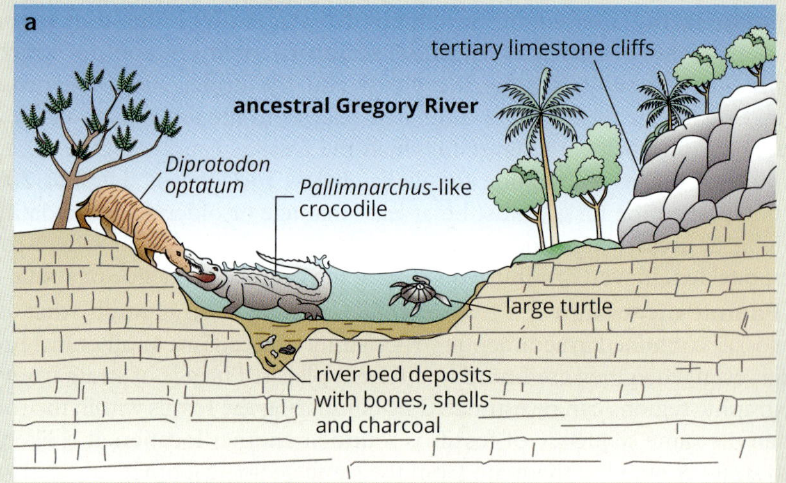

FIGURE 12.1.16 (a) Formation of fossils in the ancestral Gregory River, about 1 million years ago and (b) giant wombat *Diprotodon optatum*

Koonwarra in Gippsland, Victoria

At Koonwarra in Gippsland, Victoria, scientists have uncovered a great diversity of organisms trapped in the bed of an ancient lake (Figure 12.1.17). The fossils are preserved in mudstone 115 million years old. Many of the fossils are fishes and the fossil site is named the 'Koonwarra fish beds'. One plant fossil is a leaf of *Ginkgo*. Today, the *Ginkgo* (*G. biloba*) is native to China, but trees of this genus were probably once widespread in the world, as evidenced by the 115 million-year-old fossil from Koonwarra.

FIGURE 12.1.17 (a) Fossil fish and (b) *Ginkgo* leaves, 115 million years old, found at Koonwarra, Victoria

Absolute dating

Absolute dating provides a more precise estimate of age, although it does not mean that it provides an exact date. Radiometric dating, thermoluminescence and electron spin resonance are all methods of absolute dating that are used to determine the age of fossils.

Absolute dating provides a numerical age or date range of an object using techniques such as radiometric dating and thermoluminescence.

- **Radiometric dating** is a quantitative technique used to determine the proportion of particular radioactive elements, **isotopes**, within rocks around the fossil or sometimes within the fossil. Radioactive elements decay into different forms (e.g. uranium to lead, rubidium to strontium) at rates that are constant for a particular element. The rate of decay of the element is independent of the nature of the rocks or the environmental conditions to which they are exposed, so they act as accurate clocks. The **half-life** of a radioactive element is the time taken for half the element to decay, and can be used to calculate the age of the rock in which it is contained (Figure 12.1.18).

  Particular isotopes are used depending on the time scale involved. For example, the half-life of radioactive carbon-14 (^{14}C) allows an estimate of the age of carbon-bearing materials to be calculated up to about 58000–62000 years of age. The method of radioactive carbon dating is limited to samples not older than 60000 years, because by that age there is very little ^{14}C left. For samples older than 60000 years, potassium-40, which is found in volcanic rock, can be used. As the volcanic rock cools, its potassium-40 decays into argon-40 with a half-life of 1.25-billion-years.

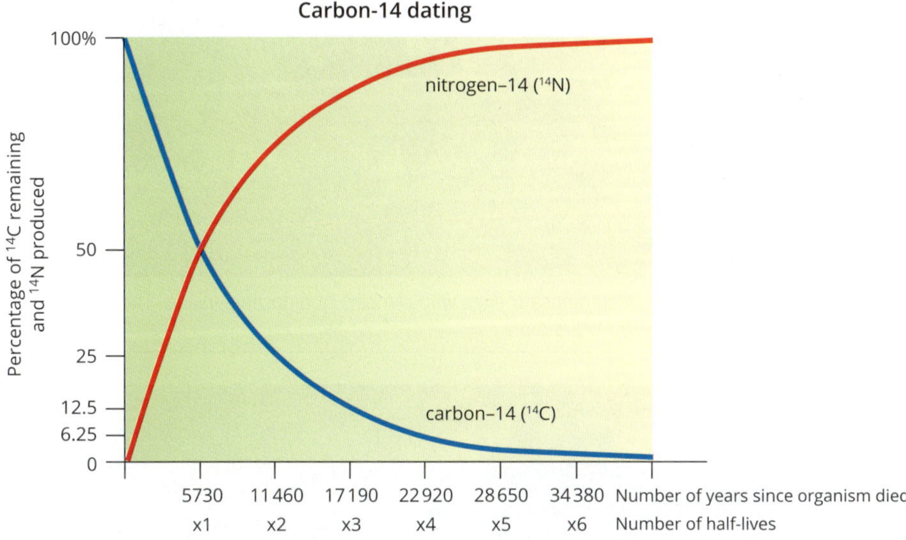

FIGURE 12.1.18 Carbon-14 (^{14}C) dating. The age of a fossil can be determined by measuring the proportion (percentage) of ^{14}C to carbon-12 (^{12}C) in a sample. When the fossilised organism was alive, its ^{14}C to ^{12}C ratio was constant (the same as the atmosphere). From the time the organism died the amount of ^{14}C reduced because it decays at a known rate to nitrogen-14. The ^{14}C decays by half every 5730 years (its half-life), as shown in the graph.

- **Thermoluminescence** is a technique that can be used to date objects such as pottery, cooking hearths and fire-treated tools that are up to 500000 years old, older than is possible with radiocarbon dating. Thermoluminescence is the emission of light from a mineral when it is heated. The amount of light emitted is proportional to the amount of radiation an object has absorbed—the older the object, the more light it emits. The intensity of the light can be calibrated to reveal how much time has passed since the object was last heated or burnt in a fire. This technique is used to date artifacts related to human evolution.

- **Electron spin resonance (ESR)** is used to date calcium carbonate in limestone, coral, fossil teeth, molluscs and egg shells. Palaeoanthropologists have used ESR mostly to date samples from the last 300 000 years. Unlike thermoluminescence dating, the sample is not destroyed with the ESR method, which allows samples to be dated more than once.

Information from fossils

Fossils provide an indication of the appearance and structure of an organism, but other information can also be gained or inferred from examining fossils. For example, animal fossils have been found with young in the uterus or inside eggs or guarding eggs (Figure 12.1.19). If young are fossilised next to adults, it is likely that the animal parented the young for a period (Figure 12.1.20). If large numbers of organisms are fossilised together, it could be surmised that they lived in herds. The contents of the animal's last meal may even be preserved in the stomach region of the fossil.

FIGURE 12.1.19 A nest of fossilised dinosaur eggs with remains of dinosaurs inside. The eggs are estimated to be at least 65 million years old.

FIGURE 12.1.20 This dinosaur nest with 15 one-year-old *Protoceratops andrewsi* discovered in the Djadokhta formation in the Gobi Desert, central Asia, suggests these animals were growing together with some sort of parental care.

CASE STUDY **ANALYSIS**

Fossils and the environment: Using fossil teeth

In one study, the enamel of fossil teeth deposited during the period of the Miocene/Pliocene boundary (between 6 and 8 million years ago) from a large range of species across the globe were analysed to determine their ratio of two isotopes of carbon: carbon-12 and carbon-13 (^{13}C). These two isotopes are incorporated during photosynthesis to a different extent by plants using C_3 and C_4 photosynthetic pathways: C_4 plants have a lower proportion of ^{13}C. The carbon is utilised by, and incorporated into, animals that eat the plants.

In the C_3 photosynthetic pathway, carbon dioxide is initially incorporated into 3-phosphoglyceric acid, which contains three carbon atoms. In the C_4 pathway, carbon dioxide is initially incorporated into oxaloacetate, which contains four carbon atoms. Ultimately, both pathways result in the production of glucose. Eighty-five per cent of modern plants use the C_3 pathway. This group includes most cereal crops, like rice, wheat, soybeans and peas, some grasses and most tree species. About 3% of modern plants, including a few crops such as sugar cane, sorghum and maize, some grasses and most sedges, use the C_4 pathway.

Analysis of modern plants shows that plants using the C_3 pathway contain a greater proportion of ^{13}C than plants using the C_4 pathway.

Fossil teeth from the period from 8 to 6 million years ago contain gradually decreasing levels of ^{13}C. This indicates that the animals had an increasing proportion of C_4 plants in their diet. C_4 plants photosynthese more efficiently than C_3 in lower carbon dioxide concentrations. Using these two observations, scientists have postulated that this gradual change in diet of the animals indicates a decreasing concentration of atmospheric carbon dioxide. This is one hypothesis about conditions in the past.

The shift to C_4 metabolism is also correlated with falling mean atmospheric temperatures, as C_3 metabolism is more efficient at higher temperatures (like those found in the tropics today).

As with all methods of determining past environments using fossil evidence, this method should not be used on its own. Confidence in conclusions comes from using a variety of different methods and obtaining a range of data that supports the same hypothesis.

Analysis

1 Identify at least two assumptions that the conclusions of this study rely upon.

2 Discuss the possible effect of a rise in carbon dioxide levels on the sugar cane industry in Queensland.

3 Contrast conditions 8 million years ago with those 6 million years ago.

4 From the data related to ^{13}C, infer how competition between C_3 and C_4 plants altered their abundance through the period from 6 to 8 million years ago.

12.1 Review

OA ✓✓

SUMMARY

- The fossil record is the record of the occurrence and evolution of living organisms through geological time as inferred from fossils.
- Fossils are the preserved remains, impressions or traces of organisms found in rocks, amber (fossilised tree sap), coal deposits, ice or soil.
- There are four main types of fossils:
  - impression fossils
  - mineralised fossils
  - trace fossils (ichnofossils)
  - mummified organisms.
- Fossilisation involves:
  1. death of the organism
  2. burial of the organism by sediments
  3. the weight of many layers of sediments squeezing out water between the particles of sand, silt or mud
  4. soft sediments become solid rock—sandstone, siltstone, mudstone or shale (a mixture of clay and silt) as the deposit deepens, and pressure and temperature increase.
- Dating of fossils can be determined by:
  - relative dating based on stratigraphy and faunal succession, which places the age of a fossil according to, or relative to, the known age of layers or strata of rock above and below the layer of rock in which the fossil is found
  - using index fossils, which are commonly found fossils from similar sites for which an absolute age has been determined
  - absolute dating using radiometric dating, thermoluminescence or electron spin resonance.
- Transitional fossils are any fossilised organisms that show intermediate traits and evidence of major change, such as animals moving from aquatic to terrestrial habitats.
- The fossil record is not a complete record of all past life because the chance of a fossil forming is small. Often only hard parts of organisms are preserved and only under certain environmental conditions.

KEY QUESTIONS

Knowledge and understanding

1. What types of organisms or parts of organisms are most likely to be fossilised? Why?
2. Explain the difference between an impression fossil, a cast fossil and a trace fossil.
3. Define stratigraphy.
4. What is an index or indicator fossil?
5. Explain the difference between relative dating and absolute dating of fossils.

Analysis

6. The half-life of ^{14}C is 5730 years.
   - **a** If a fossil sample originally included 1.0 g of ^{14}C, how much would be left after 5730 years?
   - **b** How much would be left after 11 460 years?
   - **c** How many years would it take for the amount to be 0.125 g?

12.2 Speciation

Evolution is the change in the genetic composition of populations over time. This can be observed as changes in allele frequencies and phenotypes in a population over time. As you learnt in Chapter 11, natural selection is a driving force of evolution. It causes alleles to increase or decrease in frequency, depending on their adaptive value in their environment. New species can evolve in response to changes in environmental conditions, as a result of chromosome duplications or after populations become isolated and accumulate genetic differences (i.e. mutations) over time (Figure 12.2.1). In this section, you will learn about the mechanisms that lead to speciation.

Evolution is the change in the genetic composition of a population during successive generations, which may result in the development of new species.

FIGURE 12.2.1 (a) Medium ground finch (*Geospiza fortis*), (b) warbler finch (*Certhidea olivacea*) and (c) common cactus finch (*Geospiza scandens intermedia*) are species of finches found on the Galápagos Islands. These species evolved from a common ancestor but each has different physical characteristics.

DEFINING SPECIES

A **species** is defined as a group of individuals that are genetically similar enough to interbreed and produce **viable offspring** that are able to reproduce (Figure 12.2.2). A species can also be thought of as a gene pool that is isolated from the gene pools of other species. While this definition of species fits most groups of organisms, there are exceptions. For instance, organisms that reproduce asexually, particularly single-celled organisms, can be difficult to categorise into discrete species. Also, species covering a wide geographic range may vary subtly over that range. Adjacent populations may freely interbreed, but the two extreme ends of the range may be genetically incompatible. The current definition of species is also problematic because species change over time.

A species is a group of organisms that can interbreed to produce viable, fertile offspring.

Viable offspring are offspring that are fertile and able to survive and breed the next generation.

FIGURE 12.2.2 Koalas are able to interbreed to produce viable fertile offspring and thus are considered to be the same species.

FIGURE 12.2.3 (a) Australia's southern boobook is closely related to (b) the New Zealand morepork. The two species are separated by a geographical barrier.

An ecological niche is the role and position a species has in its environment.

GENETIC ISOLATION

Genetic isolation (also called reproductive isolation) occurs when alleles are no longer exchanged between populations. The mechanisms of genetic isolation are classified into two main types:

- before reproduction—prezygotic isolation
- after reproduction—postzygotic isolation.

Prezygotic isolating mechanisms

Prezygotic isolating mechanisms are those that typically prevent individuals from different populations from interbreeding; in other words, they prevent fertilisation or mating from occurring in the first place. Less common forms of prezygotic isolating mechanisms work after breeding takes place; these prevent gametes from fusing and forming a zygote. There is a variety of prezygotic isolating mechanisms that work to prevent interbreeding at different stages.

Geographical (spatial) isolation

Geographical isolation occurs when populations are separated by physical and geographical barriers, such as oceans, deserts, mountain ranges and glaciers. For example, the southern boobook (*Ninox boobook*) (Figure 12.2.3a) is an Australian owl that is genetically distinct from the New Zealand owl, morepork (*Ninox novaeseelandiae*) (Figure 12.2.3b). One reason that they are genetically isolated is that the Tasman Sea separates them.

Ecological isolation (or niche partitioning)

Populations occupy different **ecological niches** within the same ecosystem. For example, *Eucalyptus baxteri* (brown stringybark) and *Eucalyptus verrucata* (Mt Abrupt stringybark) are closely related species that grow side by side in the Grampians, in Victoria. *E. verrucata* grows on upper slopes on rocky sites and *E. baxteri* occurs on lower slopes in deeper soils. This **ecological isolation** means that the two species are usually reproductively isolated, but sometimes their flowering times overlap and neighbouring trees will interbreed. Their offspring are fertile but are generally found only along the border (ecotone) between the two species. The obvious boundary between species can be seen in Figure 12.2.4.

FIGURE 12.2.4 Brown stringybark (*Eucalyptus baxteri*) (seen in the background of this image) and Mt Abrupt stringybark (*Eucalyptus verrucata*) (seen in the foreground) grow in the same ecosystem but in different niches. The different ecological requirements and flowering times of these two species mean that their populations are mostly reproductively isolated.

Temporal isolation

Temporal isolation (temporal means relating to time) occurs when the breeding cycles or active times of populations do not overlap. For example, a nocturnal animal is unlikely to breed with a diurnal one. Likewise, many similar plant species will flower at slightly different times of the year, preventing cross-pollination.

Behavioural isolation

Behavioural isolation occurs when behaviours such as mating calls and courtship rituals are different. This isolating mechanism is only possible in animals. An example is mate attraction to different types of vocal signals, such as bird songs or frog calls, which are unique to species. Behavioural isolation is often the result of sexual selection (see below).

Structural or morphological isolation

Morphological isolation occurs when two populations live close enough to interact, but are unable to interbreed due to physical differences. For example, if the reproductive organs of different species are physically incompatible then individuals will be unable to mate. This can also occur if the individuals are different sizes. For example, a sparrow could not breed with an albatross. For more-similar species, even slight differences can prevent mating, such as the different breeding pheromones produced by different moth species.

Gamete mortality

In **gamete mortality**, egg and sperm fail to fuse in fertilisation. For example, the sperm of one species may not be able to 'find' the egg of another without the appropriate signalling molecules, or the conditions of the female reproductive tract of one species may not sustain the sperm of another species. Pollen may not germinate on the style of the flower of another species because of a chemical barrier preventing sperm from reaching an egg.

Sexual selection

Sexual selection is a form of natural selection in which mates are chosen based on specific traits, such as antler size, tail feather length or colourful plumage. The different appearance of males and females of some species is known as **sexual dimorphism** and it is a result of sexual selection of particular traits over many generations. Most animals exhibit some level of sexual selection in which at least one biological sex selects their mate based on specific traits. Although it may appear that mates are chosen on the basis of an irrelevant characteristic, the chosen traits are often indicators of good health, strength and fitness or high adaptive value. The alleles of these beneficial traits may then be inherited by offspring. Sexual selection is very common in birds. For example, barn swallows (*Hirundo rustica*) select mates on the basis of the length of tail streamers, which indicate health and fitness.

Animals may compete with members of their own sex for mates of the opposite sex. Animals such as sea lions, antelope and kangaroos come into direct physical conflict over mates, usually resulting in a single male winning the right to mate with a large number of females. These conflicts ensure that the individuals with the 'fittest' phenotypes are the ones that are most likely to produce more and healthier offspring after mating. In this way, the 'fitter' alleles are more likely to be inherited by the next generation and increase in frequency in the gene pool over time.

BIOFILE

Sexual selection in bowerbirds

The bowerbird selects a mate on the basis of the showiness, structure and colour of the bower it builds from collected objects. Satin bowerbirds (*Ptilonorhynchus violaceus*) are found in wet forests in the southeast of Australia and females are attracted to males that build a bower using blue objects (Figure 12.2.5a). Great bowerbirds (*Chlamydera nuchalis*) occur in the tropical ecosystems of northern Australia and females are attracted to males that build bowers using white, green and red objects (Figure 12.2.5b). Although the species are geographically isolated, their difference in mate attraction is an example of sexual selection and behavioural prezygotic isolation.

FIGURE 12.2.5 (a) Female satin bowerbirds (*Ptilonorhynchus violaceus*) are attracted to mates, like this male bird, that build bowers using blue objects, while (b) female great bowerbirds (*Chlamydera nuchalis*) are attracted to mates that build bowers using white, green and red objects.

Postzygotic isolating mechanisms

FIGURE 12.2.6 The zonkey is a hybrid individual resulting from the interbreeding of a zebra and a donkey.

Postzygotic isolating mechanisms are those that typically prevent a zygote of two different species from developing into a fertile adult. The offspring resulting from interbreeding between individuals from different species are called **hybrids**. An example of a hybrid is a zonkey—the offspring of a zebra and a donkey (Figure 12.2.6).

Hybrid inviability is a mechanism of reproductive isolation in which the sperm from one species does successfully fertilise the egg of another species to form a hybrid zygote, but the hybrid zygote has unmatched (non-homologous) chromosomes. As a result, normal embryonic development cannot proceed because of the lack of homologous chromosome pairs in the zygote. The zygote does not typically survive long.

Sometimes the zygote survives and undergoes cell division but the offspring does not develop fully and will not reach adulthood. This is known as **reduced hybrid viability**. Most hybrids that do develop into adulthood are sterile; that is, they are incapable of producing offspring themselves (Figure 12.2.7). **Hybrid sterility** usually results from problems during gamete formation.

FIGURE 12.2.7 (a) Eastern grey kangaroos (*Macropus giganteus*) were crossed with a very similar but separate species, (b) the western grey kangaroo (*Macropus fuliginosus*). (c) The resulting male joeys were all sterile.

ALLOPATRIC SPECIATION

 Speciation is the evolution of new species from an ancestral species.

Speciation is the evolution of new species from an ancestral species. There are different evolutionary mechanisms that lead to speciation, but the most common form of speciation is allopatric speciation.

Allopatric speciation occurs when a population becomes divided by a geographical barrier. The spatial isolation prevents individuals of the separated sub-populations from interbreeding.

Over time, different environmental selection pressures and genetic drift drive change in the allele frequencies of the two sub-populations. Eventually, the two sub-populations may diverge genetically and morphologically, to the point where they can no longer successfully interbreed, should they come into contact again; that is, they may become two distinct species (Figure 12.2.8).

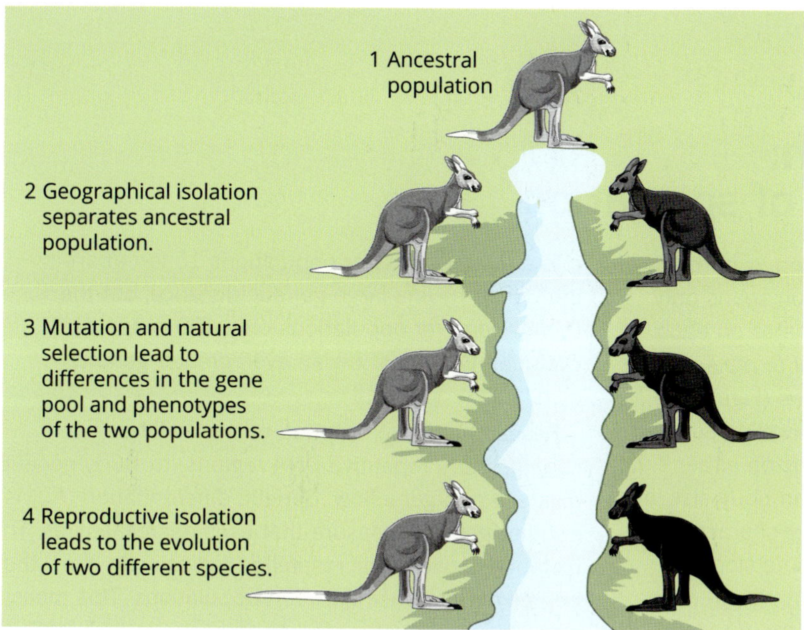

FIGURE 12.2.8 Geographical isolation leads to allopatric speciation.

> Allopatric speciation occurs after populations are geographically isolated.

Australia can be divided into a number of distinct regions of biodiversity that are isolated geographically by water, mountain ranges and deserts. For example, the southwest corner of Western Australia has a very high number of **endemic** species (species that are only found in one region) due to its geographical isolation by the central arid zone. Tasmania also has many endemic species as a result of its separation from the Australian mainland by Bass Strait (Figure 12.2.9).

FIGURE 12.2.9 Tasmania has a high number of unique species that are not found on the mainland, such as the leatherwood (*Eucryphia lucida*).

BIOFILE

Ligers

A liger is the hybrid offspring of a male lion ($2n = 38$) and a female tiger ($2n = 38$). Ligers also have a diploid number ($2n$) of 38. However, despite having the same number of chromosomes, because of differences in the genes in the two species, meiosis is rarely successful in ligers. Ligers typically experience hybrid sterility or hybrid breakdown.

All ligers have been bred in captivity because the natural ranges of tigers and lions do not overlap. The two species are naturally geographically isolated in the wild.

This liger is the hybrid offspring of a Siberian tiger and a lion.

CASE STUDY **ANALYSIS**

Isolation within a canyon: *Drosophila* on the verge of speciation

A canyon in Israel, aptly called Evolution Canyon, is home to the fruit fly *Drosophila melanogaster*. The canyon spans only 100–400 m but it has very different microclimates on each side (Figure 12.2.10a). Different populations of *D. melanogaster* inhabit the north-facing and south-facing slopes (Figure 12.2.10b). Although the flies are capable of travelling across the canyon, they maintain separate populations on either side of the canyon and exhibit marked differences in body size, and tolerance to heat and desiccation, among other characteristics. They also prefer to mate with flies from the same slope.

Researchers have been studying adaptation in these populations. One measure is genetic distance (*D*). If two populations have many alleles in common, they have a small genetic distance, but if the populations have few alleles in common then there is a greater genetic distance, indicating genetic divergence (Table 12.2.1). Two identical populations have a D value close to zero. A *D* value > 0.15 represents a large genetic distance.

The *D. melanogaster* populations 15 km apart in the Congo have a large genetic distance, but the *D. melanogaster* populations only 400 m apart in Evolution Canyon show an even greater genetic distance.

Researchers also mapped microsatellites (non-coding, short, repetitive DNA regions) to study genetic diversity and gene flow. Genetic differentiation (F_{ST}) is another genetic measure that uses microsatellites. The distance between microsatellite regions can show the degree of difference between populations. This measure also indicates whether the populations reproduced with each other in the past. A large, interbreeding population has little to no genetic differentiation and therefore an F_{ST} value of zero or close to zero. When populations become isolated and gene flow declines, genetic distance and genetic differentiation of the gene pools increase (Table 12.2.2).

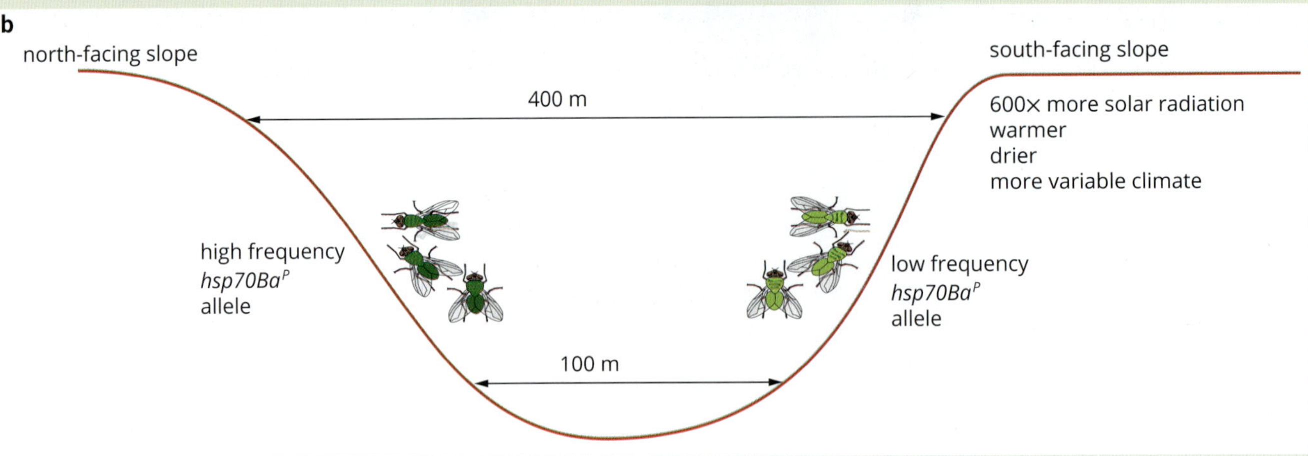

FIGURE 12.2.10 *Drosophila melanogaster* in Evolution Canyon are a model population for studying evolution. (a) Evolution Canyon in Israel. (b) *Drosophila melanogaster* populations occupy different slopes of the canyon, which have different microclimates.

F_{ST} values are on a scale of 0–1:

0 means low differentiation (populations freely interbreed)

> 0.25 means a high degree of differentiation

> 0.5 means very high degree of differentiation

1 means populations are completely separate with no gene flow.

Tables 12.2.1 and 12.2.2 show the results of these analyses for the Evolution Canyon populations and other *Drosophila* species or populations.

TABLE 12.2.1 Genetic distance between *Drosophila* populations or species

Population or species comparisons	Genetic distance (*D*)
D. melanogaster in Evolution Canyon (north-facing slope vs south-facing slope)	0.566
D. melanogaster vs *D. simulans*	0.5–0.6
D. melanogaster (two populations, 15 km apart, Congo)	0.44

Table 12.2.1 shows that the genetic distance between the Evolution Canyon populations of *D. melanogaster* (*D* = 0.566) is similar to the genetic distance between the different species *D. melanogaster* and *D. simulans*. Despite being only 400 m apart, they show greater genetic distance than two populations separated by 15 km in the Congo.

The Evolution Canyon *Drosophila* also had a much higher genetic differentiation score (F_{ST} = 0.361) than the same species in other locations, meaning they are more genetically isolated and have little gene flow with other *Drosophila* populations, including the ones across the canyon (Table 12.2.2). The *D* and F_{ST} data together suggest that the Evolution Canyon *Drosophila* populations may become separate species if the genetic isolation is maintained.

TABLE 12.2.2 Genetic differentiation in populations of *Drosophila melanogaster*

Populations	Microsatellite differentiation (F_{ST})
Evolution Canyon (north-facing slope versus south-facing slope)	0.361
California, Zimbabwe, Kenya	0.121

The researchers also analysed gene sequences. One locus of interest is a regulatory region for the heat shock protein (hsp) 70Ba, which has a normal allele and a variant allele. The north-facing slope flies had an almost 30-fold higher frequency of the variant allele, indicating very limited gene flow between the populations. Therefore, although these populations occupy the same canyon, they are genetically isolated. The researchers conclude that, while some gene flow occurs, these populations are close to being separate species. They propose the microclimates on the north and south slopes of the canyon are driving adaptation of the populations and speciation.

Analysis

1 In the *Drosophila* populations, the *hsp70Ba* gene has a variant allele named $hsp70Ba^P$. For simplicity, we will call the normal allele *A* and the variant allele *P*. The table below lists the number of individuals analysed and their genotype.

Genotype data for *hsp70Ba* locus in *Drosophila* in Evolution Canyon

Population	Number of individuals analysed	Homozygous (*PP*)	Heterozygous (*AP*)	Homozygous (*AA*)	Frequency of *P* allele (%)
north-facing slope	96	5	55	36	
south-facing slope	124	0	3	121	

a Copy and complete the table. Calculate the frequency of the *P* allele in each population.

b Calculate the ratio of this allele in the north-facing slope and south-facing slope populations.

c Identify the evidence leading the researchers to conclude that 'gene flow is low but not zero'.

d The researchers conclude that, 'The *D. melanogaster* case presented here, however, is unique not only in detecting massive fine-scale genetic differentiation in a very mobile organism but in revealing genetic changes associated with developing speciation driven by adaptation to local environments'. Infer the specific type of speciation the researchers identified. Justify your choice.

2 Evaluate the influences of geographical isolation and gene flow in the speciation of *Drosophila* flies inhabiting the north-facing and south-facing slopes of Evolution Canyon. Use the genotype data from the *hsp70Ba* locus in the table above and your understanding of the principles of speciation.

CASE STUDY

Allopatric speciation in the coast banksia

The coast banksia (*Banksia integrifolia*) is one of the largest and most common banksias on Australia's east coast (Figure 12.2.11a).

It grows as far north as Mackay in Queensland and as far south as Port Phillip Bay in Victoria (Figure 12.2.11b). Over this extensive geographical range, populations vary in the shape of their fruits and particularly in their leaves.

Four forms have been identified. Plant taxonomists have recognised three of these forms as varieties or subspecies of *Banksia integrifolia*. The fourth form, *Banksia aquilonia*, was originally identified as another subspecies but is now recognised as a related but distinct species. Geographically, *B. aquilonia* is separated from the *B. integrifolia* subspecies by more than 200 km.

A molecular study using a DNA fingerprinting technique (called amplified fragment length polymorphism or AFLP) confirmed that the three *B. integrifolia* subspecies are genetically distinct from one another and different also from *B. aquilonia*. The technique also confirmed that plants with a leaf shape intermediate between subspecies are the result of gene flow between populations. The distributions and characteristics of the species and subspecies of coast banksias are summarised in Table 12.2.3.

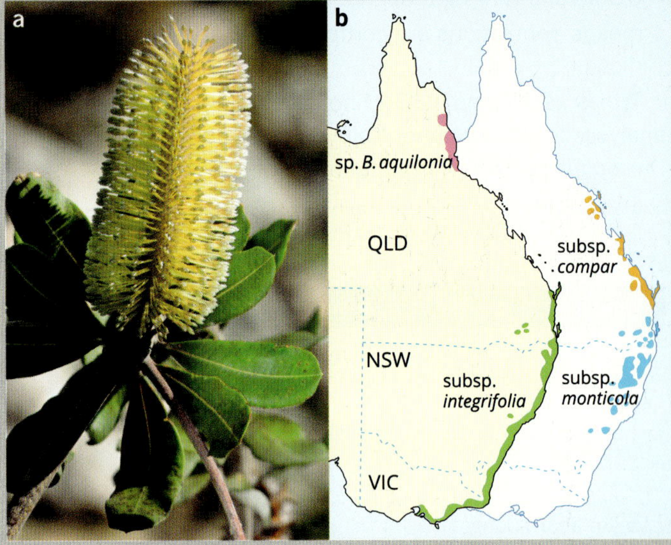

FIGURE 12.2.11 (a) Coast banksia, *Banksia integrifolia*. (b) Distribution of the different forms of coast banksia

TABLE 12.2.3 Distribution and characteristics of species and subspecies of coastal banksias

Species and subspecies	Distribution	Form
B. integrifolia subsp. *compar*	scattered distribution, southern Queensland and north-eastern NSW, coastal	large glossy adult leaves; seedling leaves elliptical, small straight side teeth
B. integrifolia subsp. *monticola*	south-eastern Queensland and north-eastern NSW, montane regions above 650 m in altitude	narrow adult leaves; seedling leaves obovate (widest at the top), large teeth with curved sides
B. integrifolia subsp. *integrifolia*	southern Queensland, NSW, Victoria, coastal	leaves shorter than *B. integrifolia monticola*; seedling leaves obovate, curved
B. aquilonia	north-eastern Queensland, coastal	long narrow adult leaves; largest fruits, fringe of stiff hairs on the midrib, on the underside of the leaf

Darwin's Galápagos finches

Galápagos Island finches studied by Charles Darwin are a typical example of allopatric speciation. Darwin collected specimens of finches from different Galápagos Islands when he sailed on the voyage of HMS *Beagle* (1831–36). The Galápagos Archipelago is an isolated group of volcanic islands about 1000 km off the coast of South America in the eastern Pacific (Figure 12.2.12).

FIGURE 12.2.12 Galápagos Islands, Ecuador

Darwin noticed that finches from different islands had unique morphology and specialised features. Between the islands, finches differed in beak shape, body size and feeding behaviour (Figure 12.2.13). When Darwin returned to England, scientists informed him that each specimen was a different species. Each species appeared to have evolved differently to meet the conditions on the island on which they live.

Darwin's finches display a relationship between morphology and environmental selection pressures, in a form of allopatric speciation called adaptive radiation. **Adaptive radiation** is the rapid evolution of a large number of related species from a single common ancestor. In this case, the availability of different food resources was the environmental selection pressure that resulted in adaptive radiation. Adaptive radiation is a type of **divergent evolution**, which is the evolution of new species from a common ancestor.

> Divergent evolution is the evolution of species from a common ancestor. In contrast, convergent evolution is the independent evolution of similar features in unrelated groups of organisms.

FIGURE 12.2.13 Darwin's Galápagos finches are an example of allopatric speciation. Finch populations on different islands evolved in geographical isolation from a common ancestor and became different species.

The Galápagos finch genome sequence provides researchers with information about the process of divergent evolution. The finch species have diverged from the same ancestral species, which established populations on the isolated islands millions of years ago. Over time, the finch populations adapted to the different environmental selection pressures on each island and became phenotypically and genetically distinct. Geographical isolation prevented gene flow between the islands and allopatric speciation occurred.

SYMPATRIC SPECIATION

Sympatric speciation occurs when populations with overlapping geographic ranges become reproductively isolated.

Speciation can occur even when geographical barriers do not isolate populations. **Sympatric speciation** occurs when populations of a species that share the same geographic range become reproductively isolated from each other. Sympatric speciation is more common in plants than in animals, and there are multiple ways in which it can occur.

Most habitats are made up of microhabitats, small areas with highly specific environmental conditions. When part of a population occupies a microhabitat, the difference in environmental selection pressures can be enough to drive sympatric speciation. In this situation the two populations become genetically isolated by temporal or behavioural isolating mechanisms.

Sympatric speciation in plants is more commonly caused through **polyploidy** (having more than two sets of chromosomes). Polyploidy is caused by abnormal cell division which results in a change in the number of chromosome sets (e.g. from diploid, two sets of chromosomes, to tetraploid, four sets of chromosomes). If a mutation causes polyploidy in an individual, the polyploid plant can still reproduce successfully through self-pollination or vegetative reproduction. However, a polyploid plant is unlikely to successfully reproduce with its diploid counterparts and so sympatric speciation by polyploidy is instantaneous.

Sympatric speciation on Lord Howe Island

FIGURE 12.2.14 Lord Howe Island, a small isolated volcanic island approximately 600 km east of Australia

Lord Howe Island is a very small isolated volcanic island approximately 600 km east of Australia (Figure 12.2.14). The island was formed by volcanic activity 7 million years ago and its isolation from other land masses has led to the evolution of unique species that are found nowhere else. Almost half of the island's plant species are endemic to the island, with five endemic genera.

While the island's isolation explains the uniqueness of its flora and fauna, it is the divergence of species on Lord Howe Island that is of great interest to researchers. Because the island is so small, it is considered impossible to have true geographical isolation between the populations on the island. Therefore, other forms of isolation or unusual pathways to speciation may be at work.

Researchers have found two related species of swaying palms (*Howea forsteriana* and *Howea belmoreana*) living side by side on Lord Howe Island. The researchers used molecular analyses to determine genetic relatedness and variation between the island's plant species. They found that the swaying palm species diverged much more recently than the island's formation by volcanic activity 6.4–6.9 million years ago. It is estimated that the palms' common ancestor arrived from mainland Australia as long as 4.5–5.5 million years ago and that *H. forsteriana* diverged from an ancestor of *H. belmoreana* more recently.

The researchers concluded that sympatric speciation was driven by flowering times of the palm trees, which differed on different soils. Ancestors of the two species may have adapted to different soil types, which began the speciation process.

12.2 Review

OA ✓✓

SUMMARY

- A species is a group of individuals that can produce viable, fertile offspring through interbreeding and has a gene pool that is isolated from the gene pools of other species.
- Populations may evolve into different species through various genetic isolating mechanisms.
- Isolating mechanisms can be prezygotic, preventing fertilisation of gametes of different species, or postzygotic, preventing production of viable offspring after fertilisation.
- Prezygotic isolating mechanisms include:
  - geographical isolation
  - ecological isolation
  - temporal isolation
  - morphological isolation
  - behavioural isolation (animals only)
  - gamete mortality.
- Postzygotic isolating mechanisms include:
  - hybrid inviability
  - reduced hybrid viability
  - hybrid sterility
  - hybrid breakdown.
- Speciation is the evolution of new species from an ancestral species. The new species are genetically different enough from the ancestral species that they can no longer produce viable offspring should they interbreed.
- New species can evolve in response to changes in environmental conditions or after populations become isolated and accumulate genetic differences over time.
- The most common form of speciation is allopatric speciation, in which a population becomes divided by a geographical barrier. In this type of speciation, spatial isolation is the mechanism preventing gene flow, leading to genetic isolation.
- Sympatric speciation occurs without geographical isolation. Populations with overlapping distributions are reproductively isolated by genetic (e.g. polyploidy), behavioural, temporal or ecological barriers.

KEY QUESTIONS

Knowledge and understanding

1 Define the following terms:
- **a** species
- **b** speciation

2 Apart from genetic isolation, name two other factors that contribute to speciation.

3 Name the type of prezygotic isolating mechanism that prevents gene flow between populations during allopatric speciation. Give two examples of conditions that can result in allopatric speciation.

4 Compare temporal and spatial isolation mechanisms.

Analysis

5 Speciation can take tens to hundreds of thousands of years or it can happen very rapidly. An example of rapid speciation has occurred over the last 150 years in the United States. In the early 1900s three species of related wildflowers from the genus *Tragopogon* (*T. dubius*, *T. pratensis* and *T. porrifolius*) were introduced into the country from Europe. In all three species, $n = 12$. Initially these were separate populations, but eventually their ranges began to overlap, interactions between the individuals of different species occurred and hybrids were formed. These hybrids were sterile.
- **a** Explain why the hybrids were sterile.
- **b** In the 1950s scientists noticed that two varieties of the 'hybrids' (now called *T. miscellus* and *T. mirus*) were reproducing sexually. Each of the variants could reproduce sexually with individuals of the same variant but not with the sterile hybrids or any of the original three species. Examination of the chromosomes of these plants showed that they had double the number of chromosomes of the sterile hybrids.
  - **i** How might these hybrids have acquired the ability to reproduce sexually?
  - **ii** Why are they now classified as new species?
  - **iii** How many chromosomes would these new species possess?

Chapter review

OA ✓✓

12

KEY TERMS

absolute dating
adaptive radiation
allopatric speciation
behavioural isolation
cast fossil
divergent evolution
ecological isolation
ecological niche
electron spin resonance (ESR)
endemic
eon
epoch
era
evolution
faunal succession
fossil
fossil record
fossilisation
gamete mortality
genetic isolation
geographical isolation
geological time scale
half-life
hybrid
hybrid inviability
hybrid sterility
impression fossil
index fossil
isotope
mineralisation
mineralised fossil
morphological isolation
mummified organism
palaeontology
period
polyploidy
postzygotic isolating mechanism
preserved remains
prezygotic isolating mechanism
radiometric dating
reduced hybrid viability
relative dating
sexual dimorphism
sexual selection
speciation
species
stratigraphy
stratum (pl. strata)
sympatric speciation
temporal isolation
thermoluminescence
trace fossil
transitional fossil
viable offspring

REVIEW QUESTIONS

Knowledge and understanding

1 What are fossils?

2 Suggest some reasons why the fossil record is incomplete.

3 Describe some of the different ways in which organisms may be preserved relatively intact.

4 Describe the following techniques used to date fossils and rock. Outline the applications or limitations of each.
- **a** relative dating
- **b** indicator fossils
- **c** absolute dating using radiocarbon dating

5 Which of the following statements about carbon-14 (^{14}C) dating is incorrect?
- **A** It measures the rate of decay from carbon-12 to ^{14}C.
- **B** It requires organic matter to be present in the fossil.
- **C** It is limited to dating fossils that are less than about 50 000 years old.
- **D** It is an absolute measure of dating fossils.

6 **a** For each of the following fossils identify whether ^{14}C dating would be a suitable method to obtain an absolute date for the fossil:
- **i** an insect preserved in amber that formed sometime during the Palaeogene
- **ii** a dinosaur footprint
- **iii** a hand axe used by a member of *Homo neanderthalensis* during the last days before they became extinct
- **iv** an Egyptian mummy

b For each fossil for which ^{14}C dating is not suitable, explain why.

7 How might gene flow occur between two populations?

8 Why is genetic isolation an important step in speciation?

9 Apart from genetic isolation, what two other factors are required for speciation to occur?

10 In hybrid inviability, why does the zygote fail to develop?

11 Australia's southern boobook (*Ninox boobook*) (Figure 12.2.3a on page 406) is closely related to the New Zealand morepork (*Ninox novaeseelandiae*) (Figure 12.2.3b on page 406). The two species are separated by a geographical barrier. If you attempted to crossbreed a morepork owl with a southern boobook owl, what could you assume about their offspring?

12 What mechanism(s) is likely to maintain reproductive isolation between two different but related species of frogs living in the same marsh?
- **A** The frogs breed at different times of the year.
- **B** The mating calls of the frogs are different and they only respond to their own call.
- **C** Hybrids between the two frogs are sterile.
- **D** All of the above are possible mechanisms.

13 If a hybrid offspring was formed between a red kangaroo (*Macropus rufus*, $n = 10$) and a grey kangaroo (*Macropus giganteus*, $2n = 16$), what would be the diploid number of the offspring?
- **A** 18
- **B** 36
- **C** 13
- **D** 28

14 Give some examples of conditions that can result in allopatric speciation.

15 The coyote (*Canis latrans*) and the grey wolf (*Canis lupus*) are closely related species and both have a diploid number of 78. The ranges of the two animals overlap and hybrids have been identified in the wild. Breeding experiments in which female coyotes were crossed with male wolves resulted in viable offspring but showed reduced fertility, with two out of three pregnancies failing. Hybrids, however, are able to reproduce with a lower success rate than pure breeds of either species. Based on this information, identify a term that this situation is most closely related to.

Application and analysis

Use the following diagram to answer questions 16 and 17.

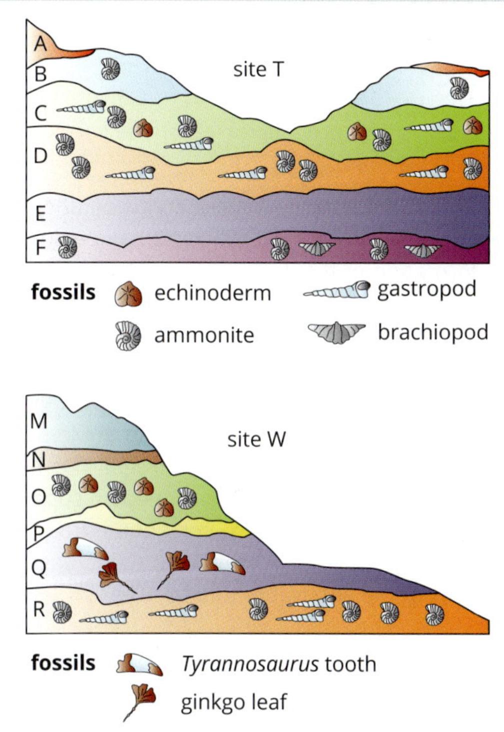

16 Fossils are found in sedimentary rocks. Barring major tectonic upheavals, it can be assumed that the layers have been laid down in order, with the oldest on the bottom.

One method of dating the layers of rock is to use indicator fossils (also called index fossils).

a **i** Which of the fossils observed at site T has the potential to be a good indicator fossil?

ii Explain your choice.

b Consider the two sites.

i Which layer is the most ancient?

ii How do you know?

c Stratum E at site T is sedimentary but there is no evidence that it contains any fossils. Why might there have been no fossils found in that stratum?

17 A method of absolute dating used for rock strata is radioactive dating. This method can only be used for igneous rocks. In order to date rocks by this method, the amounts of a radioactive material in the rock and the decay products are measured. First the half-life for the radioactive material must be calculated.

a **i** What is a half-life?

ii One radioactive material (the parent) and its decay product (the daughter) used for this sort of dating is shown in the graph below. What is another radioactive material and its daughter?

iii A decay curve is generated for the radioactive material in order to determine the age of the rock. Use the graph to determine the half-life of potassium-40 (^{40}K).

b The rocks in strata N and P at site W were determined to be igneous. They were both analysed to determine the percentage of ^{40}K left in the rock.

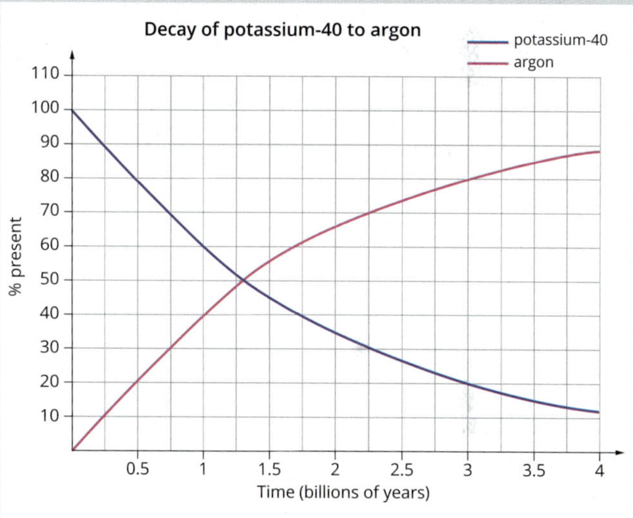

Use the graph to determine the age of the two strata if the rocks of:

i stratum N contain 90% ^{40}K

ii stratum P contain 75% ^{40}K.

c What is the likely age of stratum O?

18 Carbon-14 (^{14}C) decays to nitrogen-14 with a half-life of approximately 5730 years. If a sample of material contained 10 000 atoms of ^{14}C 30 000 years ago, what is the approximate number of ^{14}C atoms that it will contain today?

A 5000

B 1250

C 312

D 156

19 Even though not all dogs can interbreed with each other, currently all domestic dogs are considered to be a single species, *Canis lupus familiaris*. This is because alleles can be spread between different breeds by dogs of mixed breed. If a situation were to occur in which all breeds of dog died out (perhaps due to a disease) except for Jack Russell terriers and Irish wolfhounds (pictured below), should the two breeds still be considered the same species? Justify your position.

Irish wolfhound (left) and Jack Russell terrier (right)

20 One iconic Australian animal is the corroboree frog. There are two species: the southern corroboree frog (*Pseudophryne corroboree*) and the northern corroboree frog (*Pseudophryne pengilleyi*), both pictured below. They are quite small, measuring 2.5–3 cm in length. The two species are closely related but still have distinct differences in colour, mating calls and skin chemistry. Both breed in damp marshy areas and both species are seriously endangered.

(a) Southern corroboree frog (*Pseudophryne corroboree*) and (b) northern corroboree frog (*Pseudophryne pengilleyi*)

As seen in the map below, the ranges of the two species do not overlap.

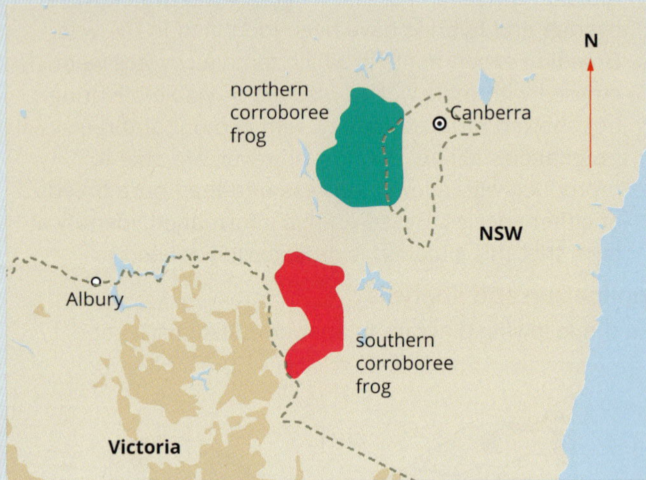

Distribution of the northern and southern corroboree frogs

a The two species have a recent common ancestor. Describe the processes that could have resulted in the formation of the two species of frog.

b The northern corroboree frog is divided into two genetically distinct populations, both of which contain few individuals. How might this affect the viability of these populations over the next decade?

c There is a captive breeding program at Melbourne Zoo for the northern corroboree frog. It has been suggested that individuals from the two separate populations of the northern corroboree frog should be interbred. How would this help to increase the chances of the survival of this species?

OA
✓✓

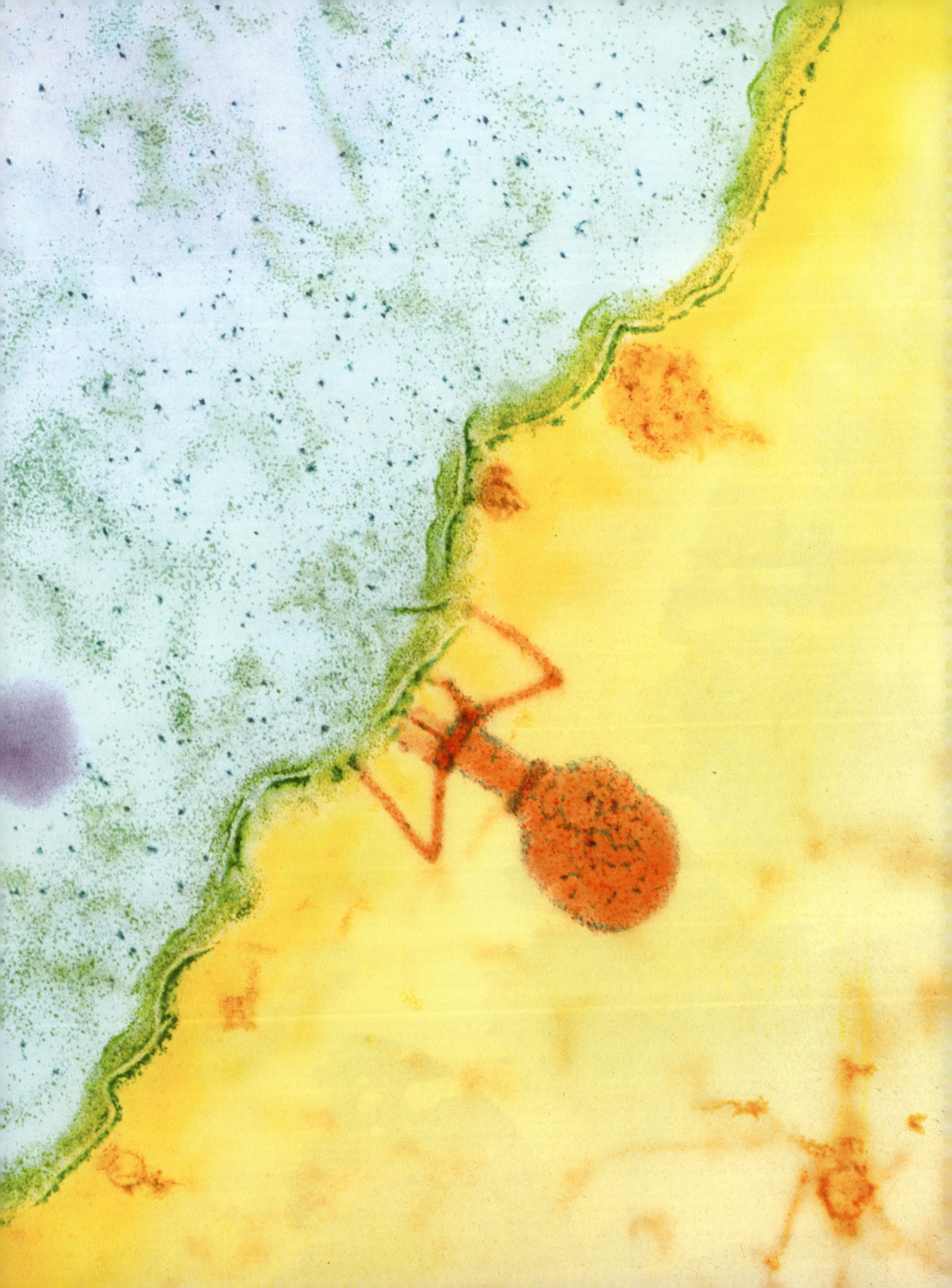

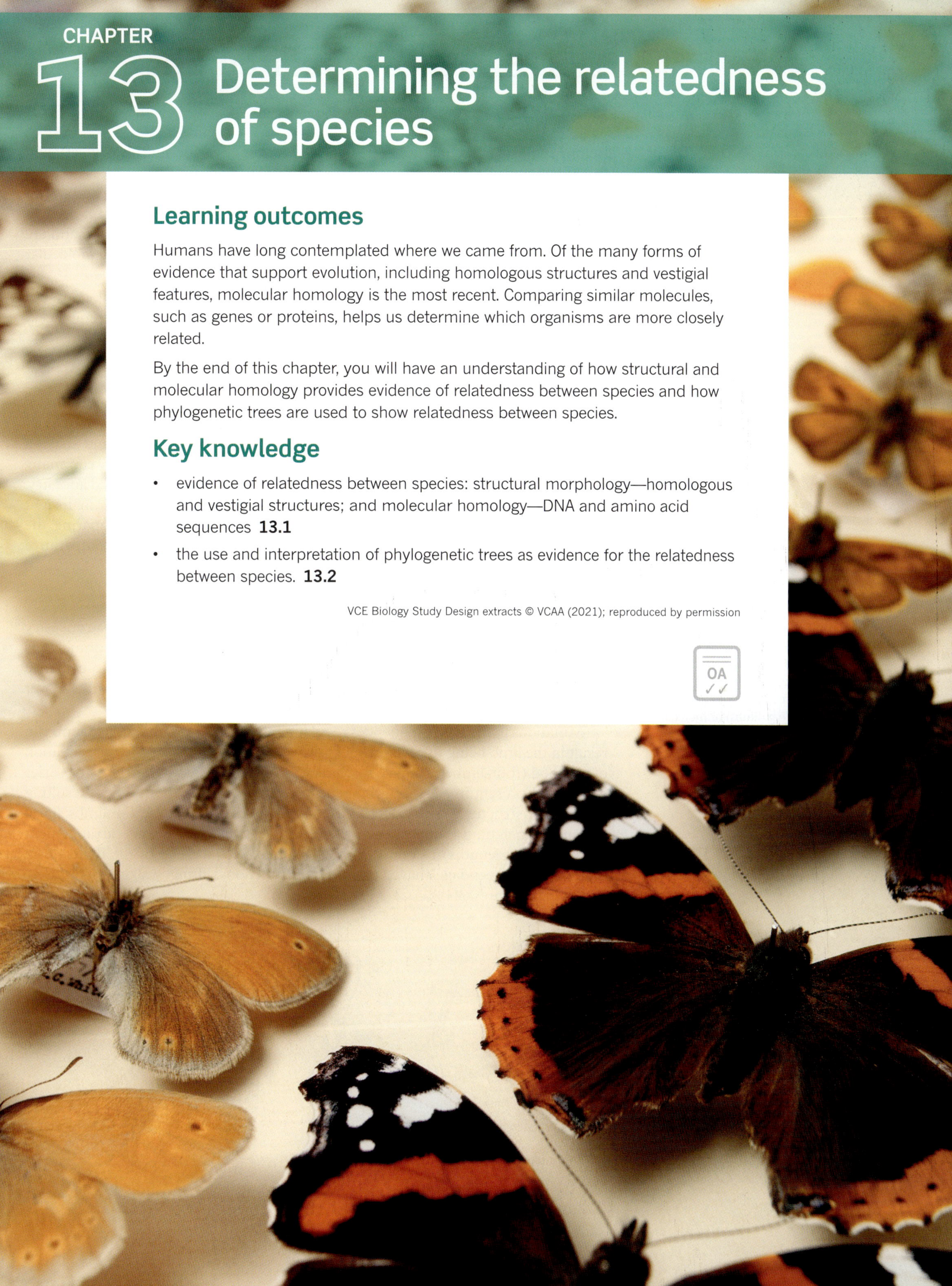

CHAPTER 13

Determining the relatedness of species

Learning outcomes

Humans have long contemplated where we came from. Of the many forms of evidence that support evolution, including homologous structures and vestigial features, molecular homology is the most recent. Comparing similar molecules, such as genes or proteins, helps us determine which organisms are more closely related.

By the end of this chapter, you will have an understanding of how structural and molecular homology provides evidence of relatedness between species and how phylogenetic trees are used to show relatedness between species.

Key knowledge

- evidence of relatedness between species: structural morphology—homologous and vestigial structures; and molecular homology—DNA and amino acid sequences **13.1**
- the use and interpretation of phylogenetic trees as evidence for the relatedness between species. **13.2**

OA ✓✓

13.1 Evidence of relatedness between species

Before molecular techniques were available, structural (or morphological) and functional similarities were the main evidence used to determine relatedness between species. Humans, chimpanzees, gorillas and orangutans each have forelimbs with hands and five digits (fingers). These morphological similarities can be used as evidence to support the theory that humans are related to these great apes and are descended from a common ancestor. We can now also use evidence from molecular analysis to gain insight into evolutionary relationships.

In this section, you will learn how structural morphology (i.e. homologous and vestigial structures) and molecular homology (i.e. DNA and amino acid sequences) are used as evidence of relatedness between species.

STRUCTURAL MORPHOLOGY

Structural morphology refers to the physical features of an organism, including external features, such as body coverings, and internal features, such as bones. If you compare the features of a human and a chimpanzee, you can see a striking resemblance in many structures. Similarities can also be observed between many other species. These similarities are not a coincidence, but a result of shared ancestry. Studying the morphology of species, or their body structures, gives an insight into the evolutionary relationships between species. The field of comparing the structures of organisms is referred to as comparative morphology or comparative anatomy.

Homologous structures

Homology refers to features that are similar in different organisms due to a shared evolutionary history (common ancestry). Homologous features may be structural (e.g. limbs) or molecular (e.g. DNA sequences).

Features of organisms that have a fundamental similarity based on common ancestry are called **homologous features** (or homologous structures). Often homologous features evolve different functions, but their similar structures provide evidence that the organisms shared a **common ancestor** from which they diverged over time. This is known as **divergent evolution**.

Mutations in the DNA sequences regulating the length of the bones in a limb can result in the limb being used in different ways in related species. Close examination of tetrapod (four-limbed animal) forelimbs, for example, shows that the same series of bones is present in each, but the genetic sequence has been modified, resulting in different structures with different functions. For example, the forelimbs of all mammals, including humans, cats, whales and bats, show the same arrangement (with different lengths) of bones from the shoulder to the tips of the digits, even though these appendages have very different functions: lifting, walking, swimming and flying (Figure 13.1.1).

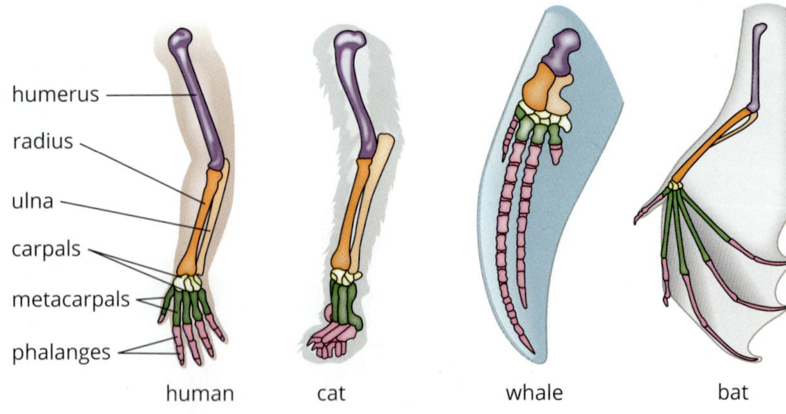

FIGURE 13.1.1 Even though they have become adapted for different functions, the forelimbs of all mammals are constructed from the same basic skeletal elements: one large bone (purple), attached to two smaller bones (orange and tan), attached to several small bones (yellow), attached to several metacarpals (green), attached to approximately five digits, each of which is composed of phalanges (pink).

Homologous features are evident in all groups of organisms. For example, the seeds of cycads, ginkgo, conifer trees and flowering plants (angiosperms) show a variety of shapes and sizes but have the same basic structure. The seeds of most conifers are winged and blown about by the wind, whereas *Acacia* seeds lack a wing but have a tough outer coat and a coloured nutritious appendage to attract ants that disperse the seeds. Despite the variation, these plants all reproduce by seeds, and it can be argued that they have evolved from a common ancestral group (Figure 13.1.2).

FIGURE 13.1.2 Homologous structures in plants. (a) The seed of a fir tree has a large wing that allows it to be carried by the wind. (b) *Acacia* seeds are small and easily carried by ants. (c) Dandelion seeds have a light parachute attached that easily catches the wind.

Vestigial structures

Some organisms possess structures that seem to have little or no function. These structures are often remnants of organs that had a function in an ancestral species but have become reduced in size over time and have ceased to be used. Such structures are referred to as **vestigial structures** or vestigial organs and they provide further evidence of divergent evolution from a common ancestor. Examples of vestigial structures include pelvic bones in whales and pythons (Figure 13.1.3); the coccyx, ear muscles, wisdom teeth and inner eyelid in humans; and the reduced eyes of certain blind cavefish and salamanders. Structures such as the wings of flightless birds can be considered vestigial for flight, but in many cases, such as that of the ostrich, the reduced wings provide a new function of temperature regulation.

Analogous structures

Anatomical similarities can also be seen in organisms that do not share a recent common ancestor. These features are not evidence of relatedness between species but indicate that the organisms have experienced and adapted to similar environmental conditions. A shark and a dolphin, for example, have a similar shape, which is suited to living and swimming in water. Despite an overall similarity of appearance due to living in similar environments and experiencing similar environmental selection pressures, sharks and dolphins have very different DNA sequences, indicating that they are not closely related. Similar features that are the result of independent evolutionary paths are known as **analogous features**. The evolution of similar features in unrelated groups of organisms is known as **convergent evolution**.

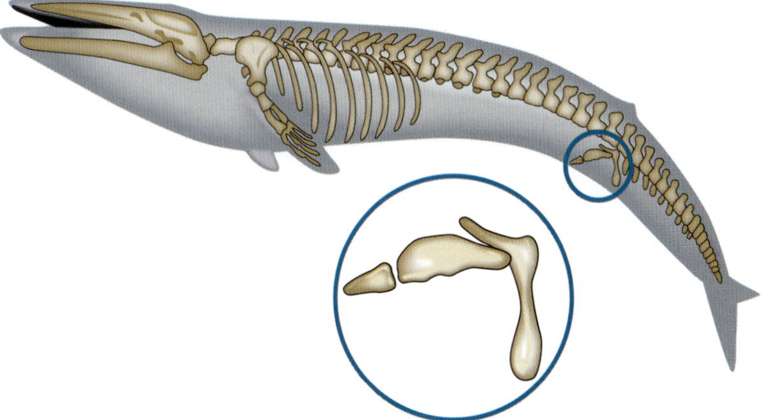

FIGURE 13.1.3 The skeleton of a baleen whale, a representative of the group of mammals that includes the largest living species, contains pelvic bones. These bones resemble those of other mammals, but are underdeveloped in the whale and have no apparent function.

Analogous features have a similar structure and function but have evolved independently due to unrelated species experiencing similar environmental selection pressures (e.g. bird and bat wings).

MOLECULAR HOMOLOGY

Molecular homology is the similarity in the molecular characters (e.g. DNA and proteins) of organisms due to common ancestry. Today it is possible to sequence DNA and compare individual genes or whole genomes. Comparing genomic features to determine evolutionary relatedness is called comparative genomics.

If species have a very similar set of proteins, chromosomes or DNA sequences it is interpreted as evidence that they share a recent common ancestor. Analysis of molecular characters, such as DNA and amino acid sequences, can provide insight into the evolutionary history of species.

> 'Recent', in evolutionary terms, may be hundreds of thousands, or even millions, of years.

DNA sequences

All living organisms on Earth once shared a common ancestor. If two populations become isolated from each other, they will accumulate different mutations in their DNA. As time passes, the sequence of nucleotides in their DNA becomes more different and what was once similar DNA gradually diverges (Figure 13.1.4). The more mutations that accumulate in the DNA sequences between two species, the more time will have passed since the two species diverged from their common ancestor. For example, there are more differences between the DNA sequence coding for the cartilage protein of a frog and a dog, than between the DNA sequence coding for the cartilage protein of a frog and a toad. This is because a dog and a frog have a more distant common ancestor than a frog and a toad and have therefore had more time to accumulate genetic changes.

> Mutations occur regularly as a species evolves—like 'molecular clockwork'. As time passes, mutations accumulate in DNA, resulting in genetic differentiation and divergence of species.

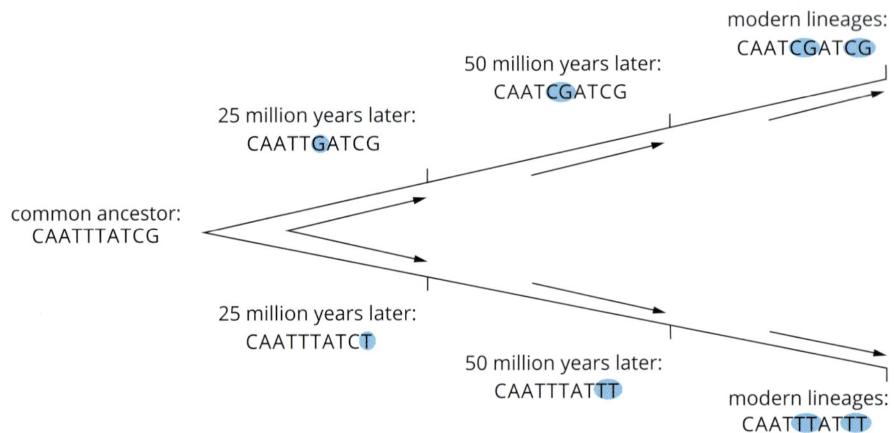

FIGURE 13.1.4 When species diverge, they start accumulating different mutations in their nucleotide sequences. The more time that passes, the more mutations they accumulate.

Changes in nucleotide sequences are caused by mutations. When a cell copies its DNA, it may make errors. Usually these errors are repaired before mitosis occurs. Occasionally these errors are not repaired and become a permanent part of the genome. This is a mutation. If these mutations occur within the germline cells (gametes), then they can be passed on to the next generation (see Chapter 11).

Amino acid sequences

As two species diverge from a common ancestor, they accumulate different mutations in their DNA and start accumulating differences in the amino acid sequences of their proteins. The more time that has passed since the two species diverged from the common ancestor, the more differences there are between their amino acid sequences.

Sometimes **point mutations**, such as nucleotide substitutions, insertions or deletions, in the DNA sequence may not cause a difference in the amino acid sequence. This is because the genetic code is degenerate, which means more than one codon codes for the same amino acid; for example, GUU, GUC, GUA and GUG all code for the amino acid valine. Consequently, differences in amino acids accumulate more slowly than differences in DNA. You learnt about the degeneracy of the genetic code in Chapter 3. Even when a point mutation leads to a change in an amino acid, the mutation may not lead to a change in phenotype.

The order of the nucleotides in the DNA indicates the evolutionary relationship between species more accurately than the amino acid sequence. All mammals, for example, produce milk containing the protein casein, suggesting that all mammal species have the same gene for this protein. However, if the DNA sequence of this gene is compared, slight differences can be observed. For this reason, and because it is now technically easier and cheaper to analyse nucleic acids than proteins, DNA comparisons are the preferred type of data.

Table 13.1.1 shows the number of amino acid differences in the cytochrome *c* molecule between humans and selected organisms, thereby comparing the relatedness of humans to each of these species. For example, there are 51 amino acid differences in the protein sequences of cytochrome *c* between yeast and humans. This indicates a low level of relatedness.

Not all DNA mutations lead to changes in amino acids and proteins. This is because there is more than one DNA triplet code for each amino acid.

Point mutations occur when a single nucleotide base is changed, inserted or deleted from DNA or RNA.

TABLE 13.1.1 Number of amino acid differences in the cytochrome *c* molecule of different organisms

	Human	Rhesus monkey	Horse	Donkey	Sheep	Dog	Yeast
Human	0						
Rhesus monkey	1	0					
Horse	12	11	0				
Donkey	11	10	1	0			
Sheep	10	9	3	2	0		
Dog	11	10	6	5	3	0	
Yeast	51	51	51	50	50	49	0

CASE STUDY ANALYSIS

The case of the whale and the hippo

A multiple sequence alignment of the casein milk protein gene in seven species is illustrated in Figure 13.1.5. Table 13.1.2 shows that, in this gene, there are three nucleotide differences between the whale and hippopotamus and 11 nucleotide differences between the whale and the camel.

The result of this gene analysis confirms that the whale is most closely related to the hippopotamus. This single nucleotide change shared by the whale and hippopotamus place them together in a clade (a group of organisms that includes an ancestor and all the ancestor's descendants).

FIGURE 13.1.5 Section of the sequence alignment. Blue letters indicate unchanged nucleotides across all species. Red letters represent nucleotides that help biologists organise organisms into clades. X in the genetic code represents an unknown nucleotide. The single nucleotide change at position 166 identifies the whale and hippopotamus as a clade.

TABLE 13.1.2 Number of base changes between paired sequences determined from the multiple sequence alignment of a milk protein gene in the whale and six hoofed species

	Camel	Peccary	Pig	Hippo	Whale	Deer	Cow
Camel	–						
Peccary	15	–					
Pig	13	6	–				
Hippo	12	13	10	–			
Whale	11	12	11	3	–		
Deer	18	18	14	10	11	–	
Cow	14	15	13	7	9	5	–

Analysis

1. Explain how the casein gene sequence changes the following statement: 'Whales are more closely related to the camel than to other hoofed animals'.
2. Describe some of the limitations of the molecular analysis shown in this example.
3. Suggest additional information that would make the molecular analysis stronger.

CASE STUDY **ANALYSIS**

Chromosomes in common

Homologies between chromosomes of different species can also be used as evidence of relatedness between species. To analyse condensed chromosomes, they are first stained with a dye called Giemsa stain. Some regions of the DNA take up more stain than other regions and appear as dark horizontal bands (Figure 13.1.6). The pattern of bands reflects the structure of the chromosome. An image of the stained chromosomes is called a karyotype.

Chromosomes can be identified by three features:

- their length
- the location of the centromere
- the dark sections of the chromosome (banding).

Organisms that shared a recent common ancestor will have many chromosomes in common. A common ancestor is evident when comparing length and banding of human and chimpanzee chromosomes. Although humans have 46 chromosomes and chimpanzees 48, there are many similarities in the banding of the chromosomes of the two species (Figure 13.1.7). The difference in the number of chromosomes can be accounted for by comparing human chromosome 2 with two smaller chimpanzee chromosomes, which have the same banding pattern.

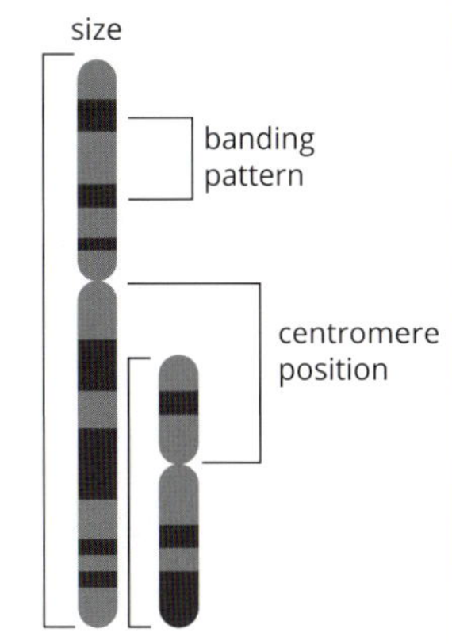

FIGURE 13.1.6 When preparing a karyotype, scientists compare the number and size of the chromosome, the banding pattern and the position of the centromere.

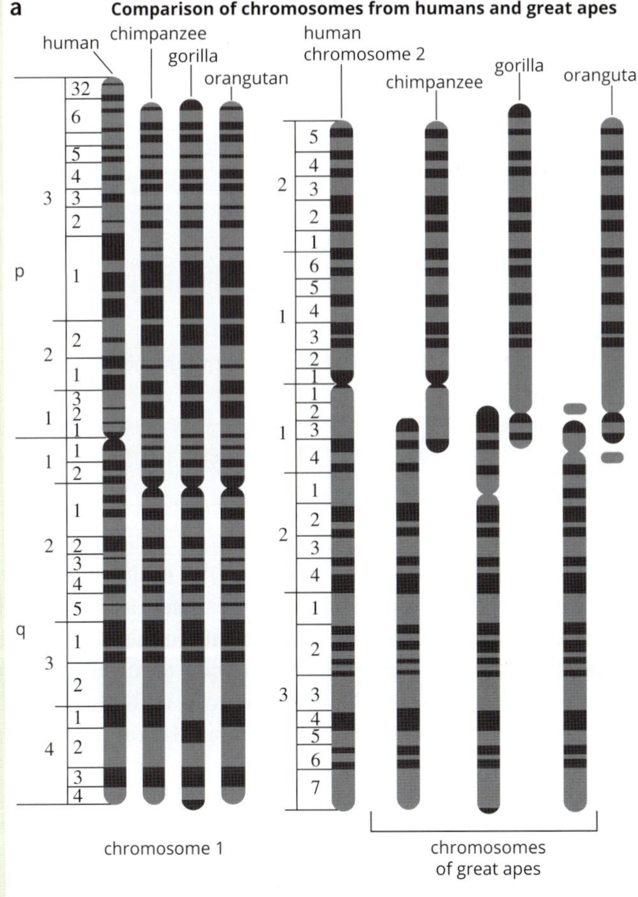

FIGURE 13.1.7 (a) A comparison between the karyotypes of (b) humans, (b) chimpanzees, gorillas and orangutans reveals many similarities between these species.

Analysis

1 What are images of stained chromosomes called?

2 **a** Which great ape is most closely related to humans?

b Explain how you reached this conclusion.

3 **a** Which great ape is most distantly related to humans?

b What does this tell us about the evolution of these species?

Molecular clocks

A molecular clock is a technique that uses the rate of accumulation of mutations in DNA to calculate how long ago organisms diverged from one another. The molecular clock hypothesis is the basis of this technique and was first proposed in the 1960s by Emile Zuckerkandl and Linus Pauling. This hypothesis states that changes in DNA and proteins are constant over evolutionary time and across different lineages.

The change in DNA over time is also known as the **mutation rate** and can be expressed as the number of nucleotide changes that occur every million years. The molecular clock hypothesis can be applied by calculating the rate of mutation of a region of DNA, along with the number of differences between the DNA of two organisms, and using this information to estimate how long ago they diverged. The more unique mutations each species accumulates, the more time has passed since they shared a common ancestor.

For example, Figure 13.1.8 illustrates the evolutionary relationships between three butterfly species. Each mutation in the DNA sequences used to construct this evolutionary tree is represented by a red line. The time scale (millions of years) over which the divergence occurred is to the left of the tree. The mutation rate can be calculated from the number of mutations that have accumulated over the evolutionary time period. The most recent common ancestors are represented by the nodes (branching points).

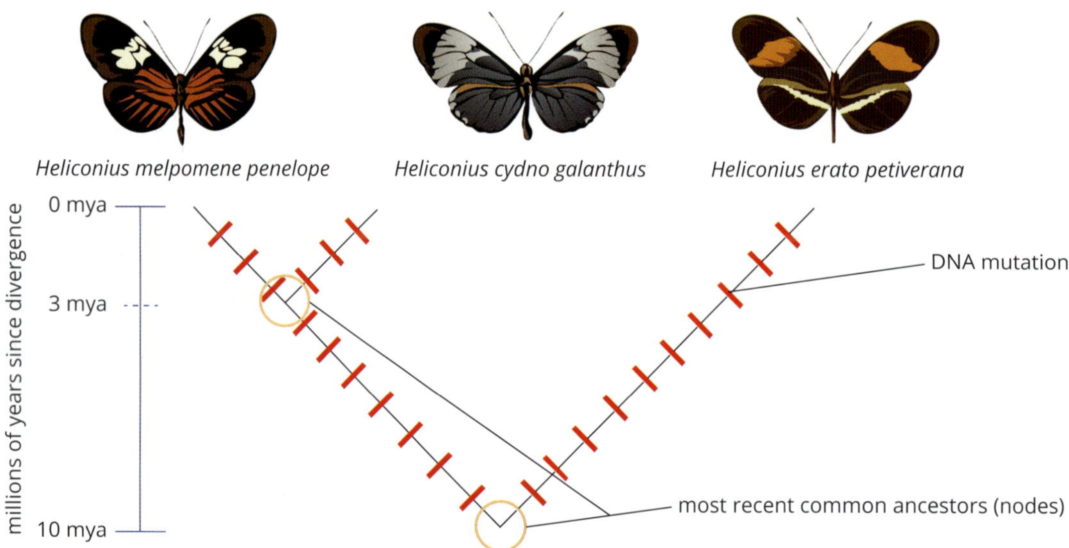

FIGURE 13.1.8 The evolutionary relationship between three butterfly species. Each red line on the tree represents a change in the DNA sequence (mutation).

In order to estimate how long ago two lineages diverged, the molecular clock is calibrated using evidence from the fossil record. Techniques such as radiometric dating and stratigraphy (the study of rock layers) are used to date fossils (Chapter 12, pages 399–402). The molecular clock for a particular gene can then be calibrated by comparing the number of differences in DNA sequences with the dates of evolutionary branch points known from the fossil record of similar organisms.

Limitations of molecular clocks

The molecular clock is a useful tool for understanding the evolutionary history of species but it does have some limitations. One of its major limitations is the assumption that the rate of genetic change is constant and therefore accurately represents evolutionary time (Figure 13.1.9).

In order for genetic changes to occur at a constant rate, those changes (mutations) need to be neutral or not affected by natural selection. This needs to be considered when applying a molecular clock to genetic data. Any DNA regions that code for the phenotype of the organism (i.e. its structure or function) are under natural selection and will be influenced by environmental selection pressures. Therefore the mutation rate of proteins and protein-coding DNA (genes) will not be constant.

Some sections of DNA mutate more frequently than others: that is, at a faster rate. This means there are different molecular clocks within an organism, ticking at different speeds in different regions of DNA. Genes that are essential to an organism's survival, such as those that code for cytochrome *c*, very rarely accumulate mutations and the gene sequence is therefore highly conserved (mostly unchanged) throughout evolution.

The molecular clock is also limited when looking at very recent or ancient timescales. When looking at recent timescales, it is less likely that enough time has passed to generate evolutionarily meaningful fixed differences in the sequences of different populations. Instead, alternative alleles that may be present in both populations will lead to an overestimation of evolutionary distance. Over ancient timescales, single sites in the sequence will have changed multiple times. This is known as saturation. Because we can only know of those changes that can be observed today, a molecular clock will underestimate the divergence that has occurred.

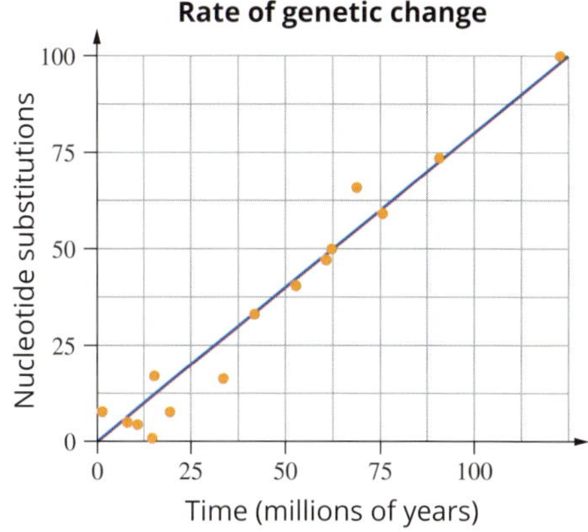

FIGURE 13.1.9 Constant rate of genetic change (nucleotide substitution) over time (millions of years) assumed under the molecular clock hypothesis

Mitochondrial DNA as a molecular clock

Genetic material is not just found in the nucleus of a cell. Mitochondria found in eukaryotic cells have their own genome (mitochondrial DNA or mtDNA). In humans, mtDNA contains 37 genes that code for 2 ribosomal RNAs, 22 transfer RNAs and 13 proteins.

MtDNA is unique in that it is passed through the maternal line of sexually reproducing organisms; that is, from mothers to their offspring (Figure 13.1.10). A father's mtDNA is not passed on to his offspring.

Mutations in mtDNA accumulate over time just as they do in nuclear DNA. However, because mtDNA does not have the same repair mechanisms as nuclear DNA, the rate of mutation in mtDNA is usually higher than in nuclear DNA (there are also some highly conserved regions). For this reason, mtDNA can be used as a molecular clock in relatively closely related species, while nuclear DNA is used to compare older lineages. An added advantage of using mtDNA is that it is easier to obtain high yields of DNA because most cells contain many mitochondria.

i Mitochondrial DNA is inherited via the maternal line, while nuclear DNA is passed on from both parents (biparentally inherited).

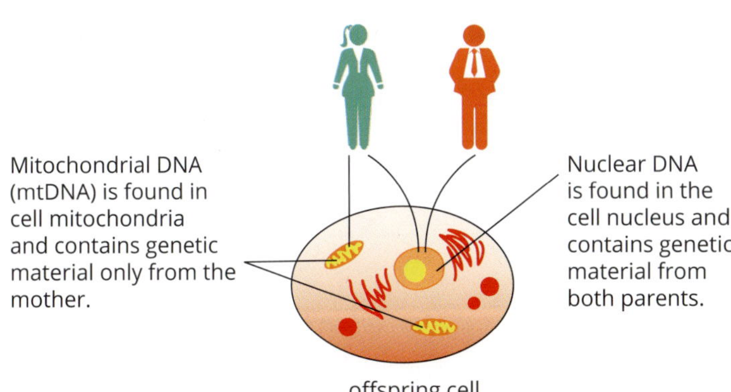

FIGURE 13.1.10 When a sperm and egg fuse, the sperm contributes nuclear DNA only. The egg contributes nuclear DNA as well as the mtDNA of the mitochondria in its cytoplasm. Therefore, mtDNA is only inherited through the maternal line, while nuclear DNA is inherited from both parents.

BIOFILE

Mitochondrial Eve

Mitochondrial Eve is the most recent common female ancestor of all present day humans. As mitochondrial DNA (mtDNA) is inherited from the mother (maternally inherited), there is no recombination, making it possible to trace unbroken maternal lines. In 1987, researchers from the University of California, Berkeley, discovered that the mtDNA of all living humans originated from one woman who lived in Africa approximately 140 000 to 200 000 years ago.

Although Mitochondrial Eve represents an unbroken female lineage from one woman until the present day, she was not the only female alive at the time or the earliest female. Other women who came before her or were alive at the same time have not had a continuous female lineage that persisted to the present day. This is because if a woman doesn't have daughters who also have daughters to pass on her mtDNA, her mitochondrial lineage will die out. It is for this reason that the most recent common ancestor (matrilineal or otherwise) will continually shift over time as lineages end and others carry on.

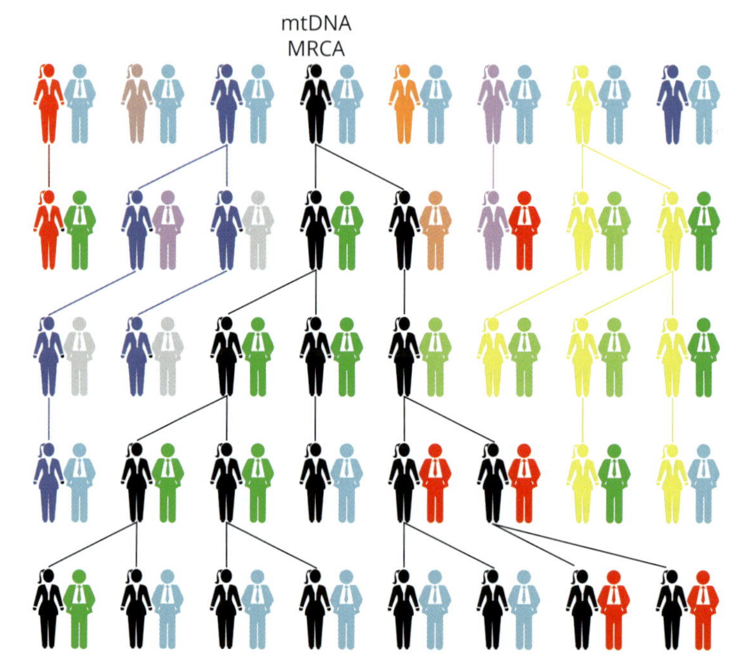

Tracing mitochondrial lineages over five generations. The coloured figures represent extinct mitochondrial lineages, while the black figures represent the female lines that are directly descended from the most recent common ancestor (MRCA).

CASE STUDY

Ancient DNA from extinct giant kangaroo and wallaby

Ancient DNA has been extracted from the bones of a giant short-faced kangaroo (*Simosthenurus occidentalis*) and a giant wallaby (*Protemnodon anak*) that inhabited Australia about 40 000–50 000 years ago (Figure 13.1.11). This research was conducted by scientists from the University of Adelaide's Australian Centre for Ancient DNA (ACAD) and provides us with new insight into the evolutionary past of Australia's megafauna (Figure 13.1.12).

Remains of the giant short-faced kangaroo and giant wallaby were discovered inside a cave in Mount Cripps, Tasmania. The cool, dry conditions of the cave preserved the remains, allowing the researchers to extract the DNA from the ancient bones and reconstruct part of the mitochondrial genomes of both species. This is the first glimpse into the DNA of the Australian megafauna—previous attempts to sequence genetic material were unsuccessful due to the poor preservation of the specimens. Well-preserved megafauna specimens are rarely found in Australia due to the harsh climate and age of the remains (39 000–52 000 years). The specimens in this study are the oldest Australian remains from which DNA has been sequenced.

Information from the mitochondrial genome of the giant wallaby revealed that it is a close relative of the living genus *Macropus*, which includes grey and red kangaroos and the tammar wallaby. DNA evidence confirmed that the short-faced kangaroo belongs to the extinct subfamily Sthenurinae. The short-faced kangaroo has no living descendants. Its closest living relative is the endangered banded hare-wallaby, the only surviving member of the ancient subfamily Lagostrophinae. The banded hare-wallaby is now restricted to islands off the coast of Western Australia and is the last representative of a once-diverse lineage of ancient kangaroos.

Ancient DNA gives us insight into the past, but it also strengthens our understanding of the evolutionary history and biology of living species. This knowledge has applications in conservation and environmental management, as we learn from the past to predict the future.

FIGURE 13.1.11 Short-faced kangaroo (*Simosthenurus occidentalis*). Fossil remains of this species were discovered in a cave in Tasmania.

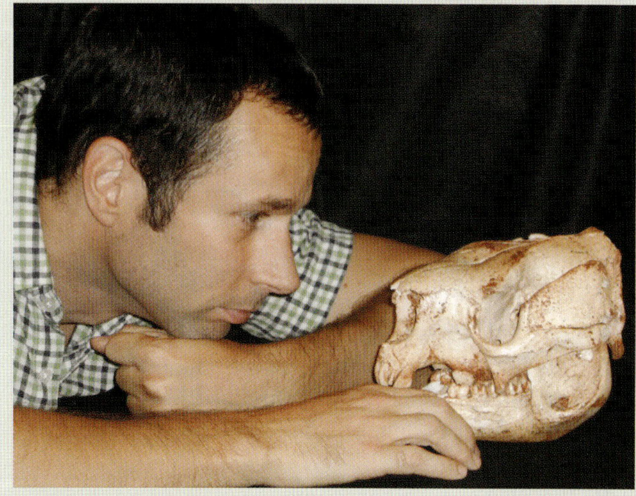

FIGURE 13.1.12 Researcher Dr Bastien Llamas with a skull of the extinct giant short-faced kangaroo (*Simosthenurus occidentalis*). Scientists extracted and sequenced ancient DNA from this species to understand its evolutionary relationship with living and extinct fauna.

13.1 Review

OA ✓✓

SUMMARY

- Structural morphology (also known as comparative morphology or comparative anatomy) is the study of the form and structure of organisms.
- Homologous features are those that are similar due to a shared evolutionary history. These features are evidence that the organisms share a common ancestor.
- Homologous features may be structural (e.g. limbs) or molecular (e.g. DNA or amino acid sequences).
- Analogous features have a similar structure and function but have evolved independently due to unrelated species experiencing similar environmental selection pressures.
- A vestigial structure is a reduced structure with no apparent function, but is evidence of an evolutionary relationship.
- Molecular homology is the similarity in the molecular characters (e.g. DNA and proteins) of organisms due to common ancestry.
- If two species have a similar set of proteins or DNA sequences, it is evidence that they shared a recent common ancestor.
- Mutations accumulate throughout the genome over time.
- A mutation to a DNA sequence may not necessarily cause a change to the amino acid sequence because the genetic code is degenerate. However, as more differences in the nucleotide sequence of DNA occur due to mutations, more differences in the amino acid sequence occur.
- Some genes accumulate mutations faster than others.
- The more mutations accumulated in the DNA sequences of two species, the more time has passed since they shared a common ancestor. This is the principle of a molecular clock.
- Mitochondrial DNA (mtDNA) is passed through the maternal line, while nuclear DNA is passed through the maternal and paternal lines.
- MtDNA does not have the same repair mechanisms as nuclear DNA. This means mutations can accumulate at a faster rate than in nuclear DNA, making mtDNA a useful molecular clock for species that diverged recently in evolutionary time.

KEY QUESTIONS

Knowledge and understanding

1 Which of the following species has accumulated the most mutations from the initial sequence: CAATTATCG?

A C A T T T A T C G
B C A A A T A A C G
C C T A T T T A C G
D C A T T T G T G C

2 Which of the following is correct?

A The number of differences in the amino acid sequence of a polypeptide would be identical to the number of differences in the DNA sequence of the coding gene.
B The DNA code is degenerate.
C A single change in the DNA will always cause a change in the amino acid sequence.
D The DNA code is grouped into codons.

3 Offspring share the same mitochondrial DNA as their:

A father
B paternal grandmother
C mother
D maternal grandfather

Analysis

4 **a** What are homologous features?

b Use an example of an animal and an example of a plant to explain the significance of homologous features in terms of evolutionary relationships.

5 The table below shows the number of differences in the nucleotide sequence between each of the organisms. List the organisms in order from the one with the most distant common ancestor to the horse to the one with the most recent common ancestor to the horse.

Human	0							
Monkey	1	0						
Dog	13	12	0					
Horse	17	16	10	0				
Donkey	16	15	8	1	0			
Pig	13	12	4	5	4	0		
Rabbit	12	11	6	11	10	6	0	
Yeast	66	65	66	68	67	67	67	0
	Human	**Monkey**	**Dog**	**Horse**	**Donkey**	**Pig**	**Rabbit**	**Yeast**

Use the evolutionary tree below to answer questions 6 and 7. Each red line on the tree represents a change in the DNA sequence (mutation).

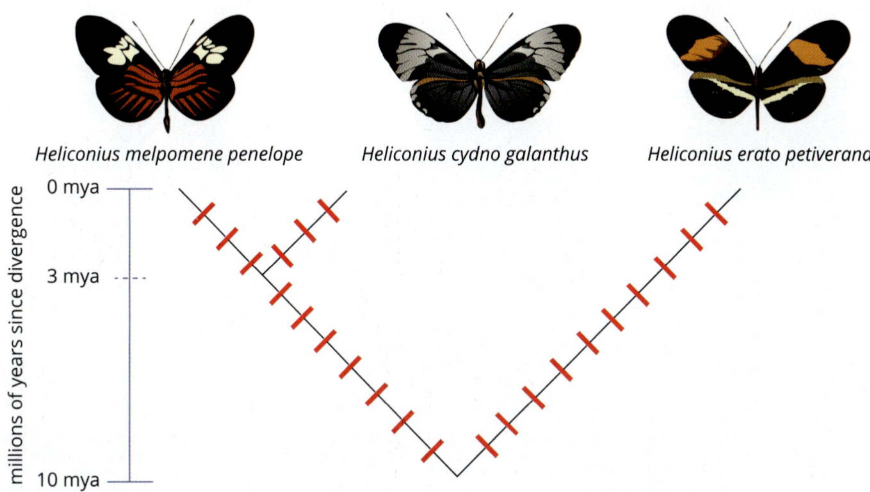

6 **a** Which two species of butterfly diverged most recently?

b Circle the point on the tree where they diverged from their common ancestor.

c Approximately how long ago did these two species diverge from one another?

7 **a** How many mutations has *Heliconius erato petiverana* accumulated since diverging from the most recent common ancestor of the other butterfly species?

b Over what period of time did this occur?

c What is the mutation rate (per million years) of the sequence of DNA used to construct this evolutionary tree?

13.2 Phylogenetic trees

Phylogenetic trees (or phylogenies, also known as evolutionary trees) are branching diagrams that depict the evolutionary relationships between different groups of organisms. They are constructed using homologous features, both morphological and molecular, to reveal the history of common ancestry between groups of organisms. As information regarding the true evolutionary history of an organism is mostly unknown, scientists use evidence from the morphology and DNA or RNA sequences of living species to reconstruct their evolutionary past. A phylogenetic tree represents an evolutionary hypothesis. A published phylogenetic tree represents a hyphothesis only, because the data used to develop the tree can create more than one possible tree. Often, the phylogenetic tree (or trees) published is the one that most closely aligns with current understanding or the one that provides the simplest interpretation of evolutionary relationships. As well as enhancing understanding of evolutionary relationships, phylogenetic trees provide useful classification data.

Today, most phylogenetic trees are built using DNA or RNA sequence data, and organisms are grouped on the basis of the similarity of their nucleotide sequences. Using this technology, we have gained remarkable insight into the evolutionary history of life on Earth. Phylogenetic trees based on molecular characters (DNA or RNA nucleotides) can be used to compare any organism, even if they seem to have very few characteristics in common (Figure 13.2.1). This is because the molecular characters on which the phylogenetic tree is based are the same for all organisms; however, the different mutation rates of different DNA or RNA regions needs to be taken into account when constructing phylogenetic trees. For this reason, phylogenetic trees are usually constructed using sequence alignments of the same gene in different organisms.

FIGURE 13.2.1 Phylogenetic trees based on molecular characters can be used to compare any organisms, even if they seem to have very few characteristics in common.

In this section, you will learn about the different types of phylogenetic trees and how they are applied to gain an understanding of evolutionary relationships.

BIOLOGICAL CLASSIFICATION

Phylogeny is the evolutionary history of lineages as they diverge from a common ancestor over time. The Swedish naturalist Carl Linnaeus (1707–1778) established the modern biological system of classification. In this system, now known as the Linnaean system of classification, organisms are organised into a hierarchy of groups (taxa; singular **taxon**) reflecting their evolutionary relationships: domains, kingdoms, phyla, subphyla, classes, orders, families, genera and species (Figure 13.2.2). The science of classifying organisms based on their shared characteristics and the evolutionary relationships inferred from them is called **taxonomy**.

A taxon (plural taxa) is a group of organisms that form an evolutionary unit.

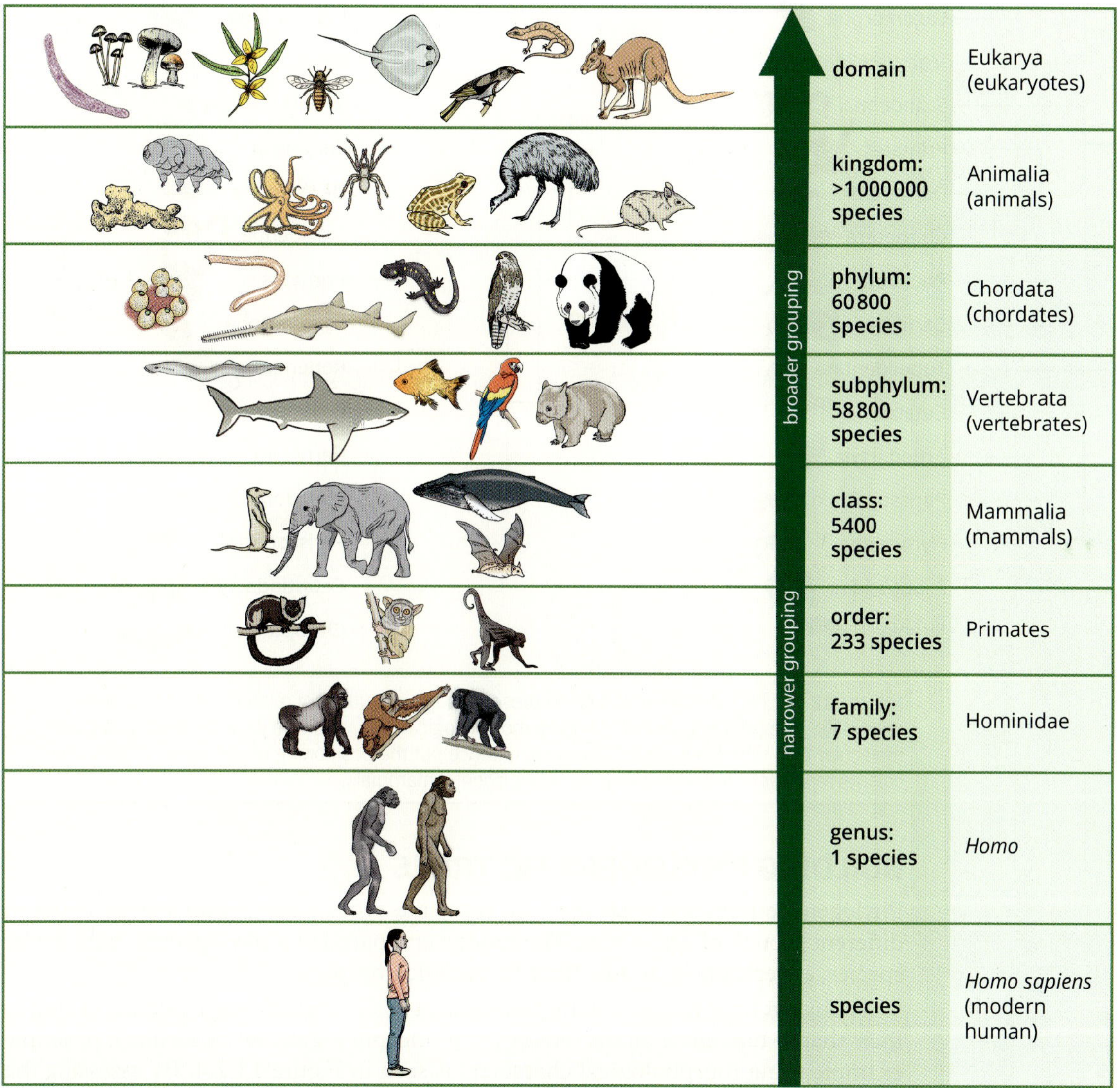

FIGURE 13.2.2 Categories in the Linnaean system of classification, which is still used today to classify organisms on the basis of their evolutionary relationships. In this figure, the biological classification of humans (*Homo sapiens*) is used as an example.

As you learnt in the previous section, comparison of DNA sequences allows the discovery of evolutionary relationships that were previously unknown, and the re-classification of organisms into more accurate taxonomic groups. Molecular data, combined with morphological and ecological information, continues to strengthen our understanding of evolution and refine the system of biological classification (Figure 13.2.3).

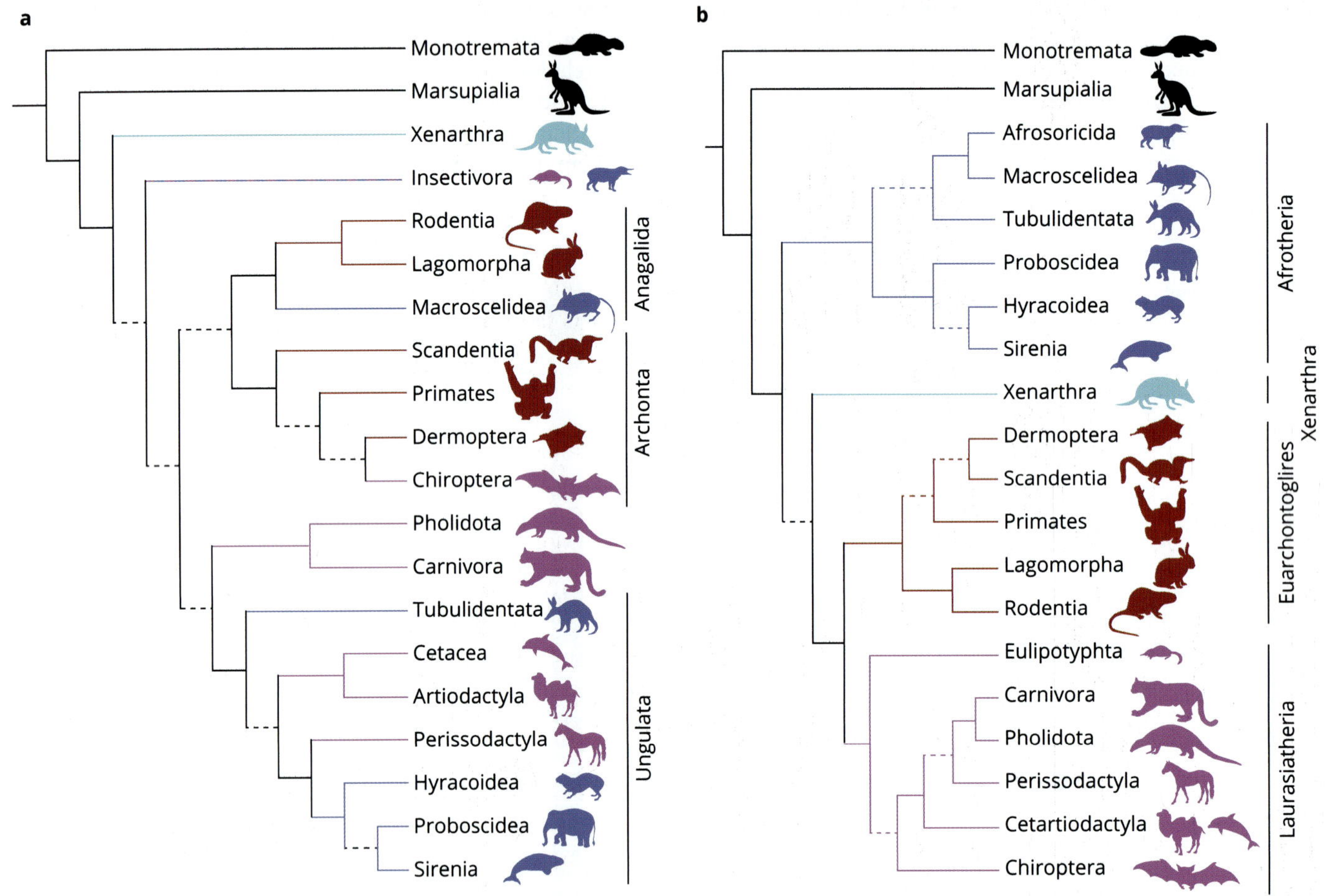

FIGURE 13.2.3 Two phylogenetic trees representing the evolutionary relationships of placental mammals. Tree (a) was constructed using morphological data, and tree (b) was constructed using molecular data. The trees are similar in appearance but the positioning of some taxa differs. The colours represent the four major groups of placental mammals.

BUILDING PHYLOGENETIC TREES

Phylogenetic trees are diagrams that show the evolutionary relationships between different groups of organisms. The groups compared in a phylogenetic tree can be species, genera, phyla or any other taxonomic group.

A phylogenetic tree is built by placing taxa in a branching sequence, according to their shared biological characteristics (e.g. morphological or molecular). A simple example using morphological characters is seen in Figure 13.2.4. By assessing the characters that different organisms share, the evolutionary relationships between taxa can be hypothesised. Starting with the most shared character, which is assumed to be the most ancestral, taxa are added to the tree sequentially, ending with the least shared character at the top of the tree (Figure 13.2.4). Each branch in the tree represents a change in character state from the last common ancestor. The greater the number of differences between taxa, the greater the distance between them in the tree, reflecting their evolutionary relationships (Figures 13.2.4 and 13.2.5). Most phylogenetic trees are now built using computational methods to generate more complex trees from large datasets.

a Character table

characters \ taxa	seastar (outgroup)	lamprey	shark	frog	chicken	dog
hair	0	0	0	0	0	1
amniotic egg	0	0	0	0	1	1
walking legs	0	0	0	1	1	1
jaws	0	0	1	1	1	1
vertebral column (backbone)	0	1	1	1	1	1

b Phylogenetic tree

seastar (outgroup)
lamprey
shark
frog
chicken
dog
hair
amniotic egg
walking legs
jaws
vertebral column

FIGURE 13.2.4 Building a simple phylogenetic tree using morphological characters. (a) A character table lists different morphological characters and their presence (1) or absence (0) in each taxon is indicated. (b) Each taxon is sequentially added to the tree according to the presence of shared characters, which are assumed to reflect when they appeared in evolutionary history.

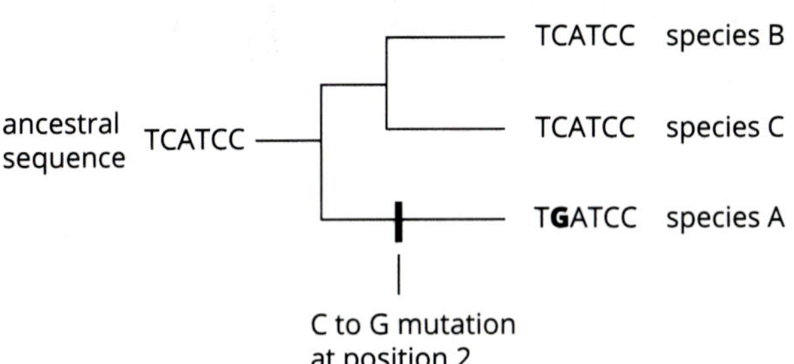

FIGURE 13.2.5 Phylogenetic trees can be built using DNA sequence data to estimate the evolutionary distance between taxa. DNA mutations (in this case the ancestral nucleotide C has been substituted for G, highlighted bold) are represented by branches in the tree; the greater the number of nucleotide differences between taxa, the greater the distance between them in the phylogenetic tree.

Parts of phylogenetic trees

As the name suggests, a phylogenetic tree is shaped like a tree with branches and leaves that extend from the **root** or ancestral **lineage** (Figure 13.2.6). Each line on the tree is called a **branch** and represents the evolutionary path from a common ancestor. The end of each branch contains a scientific name and is called a **leaf**. The point at which two branches diverge is called a **node** (or branch point) and represents the last common ancestor that the diverging groups shared. An **outgroup** species is sometimes included to assess the evolutionary relationships of those in the ingroup (the taxa of focus) relative to more distantly related taxa.

A lineage is all the species that are descendants of a common ancestor.

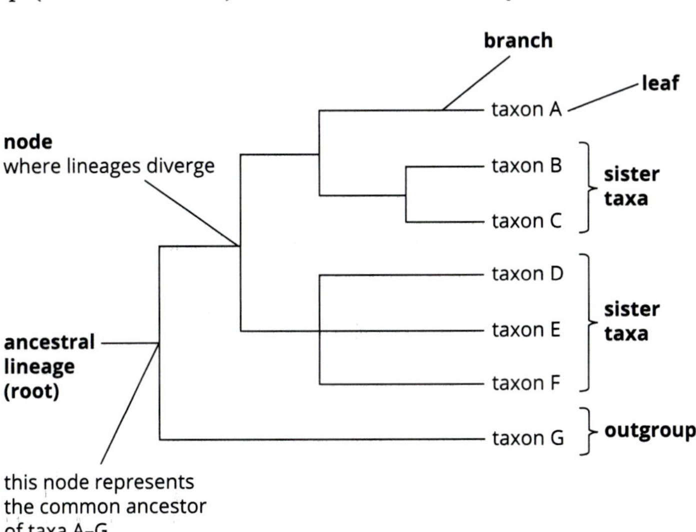

FIGURE 13.2.6 The components of a phylogenetic tree, with taxa diverging from the root. Branches represent the evolutionary paths and nodes represent the divergence points.

Pairs of taxa grouped together are called **sister taxa** and are the most closely related relative to other taxa in the tree. The most closely related (most recently diverged) taxa have the shortest branch lengths between them. In Figure 13.2.6, you can see that the closest relative of taxon C is taxon B, as that is the closest taxon following the branch from taxon C. Because no other taxa are more closely related in the tree, taxa B and C are sister taxa. The next closest relative to taxon C is taxon A. Even though taxon D is listed under taxon C and they appear close to each other, if you follow the branch lengths between them, you can see that they are actually quite distantly related.

A clade includes an ancestor and all of its descendants.

Each section of the phylogenetic tree is called a **clade**. A clade is a group of organisms that includes an ancestor and all the descendants of that ancestor (also called a monophyletic group; shown in Figure 13.2.7). The order in which clades diverged from their common ancestor is represented by the order of the branching points (or nodes), with the oldest branching point closest to the root of the tree (the bottom in vertical trees; the left side in horizontal trees) and the most recent branching point closest to the tips of the tree (the top in vertical trees; the right side in horizontal trees; shown in Figure 13.2.8).

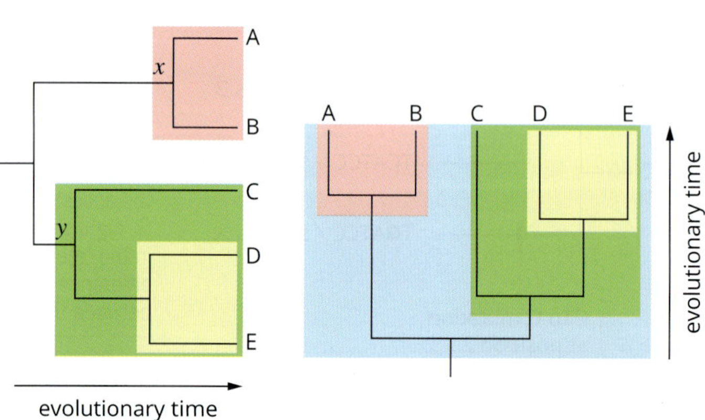

FIGURE 13.2.7 Phylogenetic trees consist of groups of taxa called clades. A clade includes a common ancestor and all its descendants. Each coloured square in these two trees represents a clade.

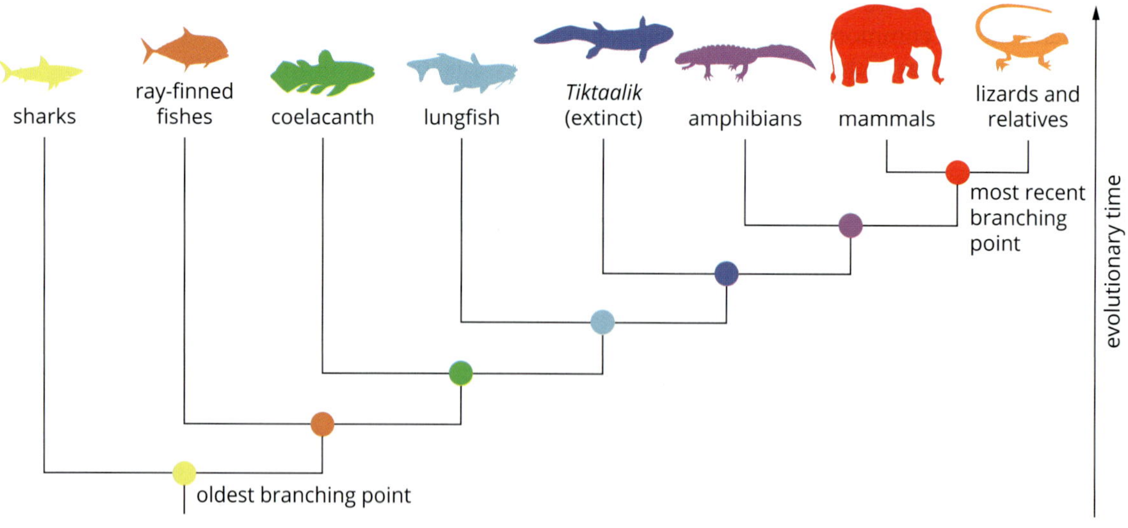

FIGURE 13.2.8 The order of the branching points (or nodes), from root to the tips of the tree (bottom to top in a vertical tree; left to right in a horizontal tree), represents the order in which clades diverged from one another.

There are three different ways in which taxa can be grouped within a phylogenetic tree (Figure 13.2.9):

- **Monophyletic groups** (one tribe) include a common ancestor and all of its descendants (a clade). This grouping is the only taxonomically viable group and is the basis of evolutionary biology. (A group that does not form a monophyletic group is not a taxonomic group.) All the members of a monophyletic group can be removed from the tree with a single 'cut'.
- Paraphyletic groups (beside or surrounding the tribe) include a common ancestor and only some of its descendants. Although not taxonomically accurate groupings, paraphyletic groups are useful for describing subsets of evolutionary groups. For example, dinosaurs are a paraphyletic group. Although birds are also dinosaurs (both groups descended from a common ancestor), dinosaurs and reptiles are almost always referred to as a separate evolutionary group from birds.
- Polyphyletic groups (many tribes) include multiple descendants but do not include a common ancestor. This grouping is rarely used, but may group taxa on the basis of shared characteristics, such as the ability of birds and mammals to maintain their body temperature.

A monophyletic group contains a common ancestor and all its descendants. This group is also called a clade and is the only true taxonomic group.

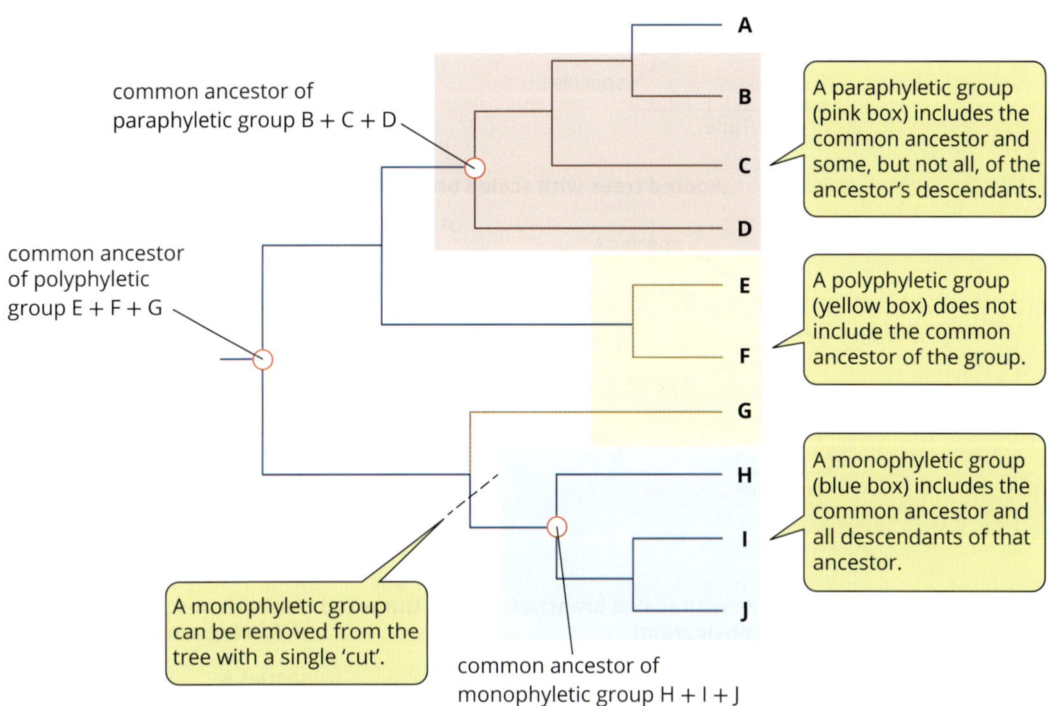

FIGURE 13.2.9 Three types of groupings within phylogenetic trees: monophyletic, paraphyletic and polyphyletic. Monophyletic groups are the only true taxonomic groups.

Phylogenetic trees can be drawn in many different ways: vertical or horizontal; rectangular or circular; rooted or unrooted; diagonal or horizontal branches; and scaled or unscaled branches. Trees that look quite different can represent the same evolutionary relationships.

Different forms of phylogenetic trees

Phylogenetic trees come in two main forms: rooted trees (Figure 13.2.10a–d) and unrooted trees (Figure 13.2.10e, f). Both these tree types can also have scaled or unscaled branches. These trees are referred to as **cladograms** if they are unscaled (Figure 13.2.10a, b) and **phylograms** if they are scaled (Figure 13.2.10c, d). The branch lengths of a cladogram are not proportional to the amount of evolutionary divergence (nucleotide changes) between taxa.

Phylogenetic trees may be presented horizontally (Figure 13.2.10a–d) or vertically, and branches may be diagonal (Figure 13.2.10a, c) or square (Figure 13.2.10b, d). These different tree types and configurations may appear quite different even when representing the same evolutionary relationships.

Rooted tree with unscaled branches (cladogram)

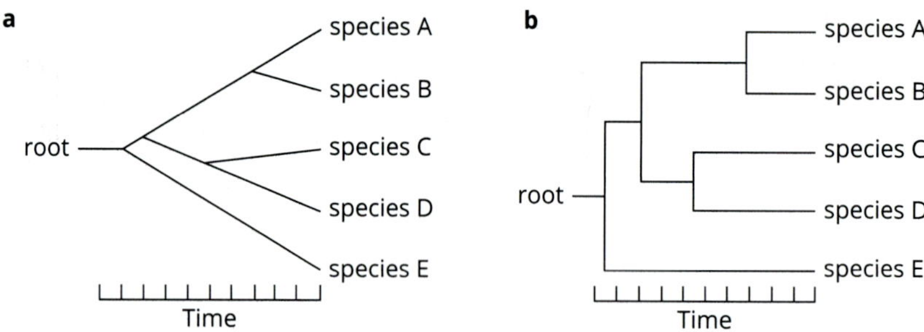

Rooted trees with scaled branches (phylogram)

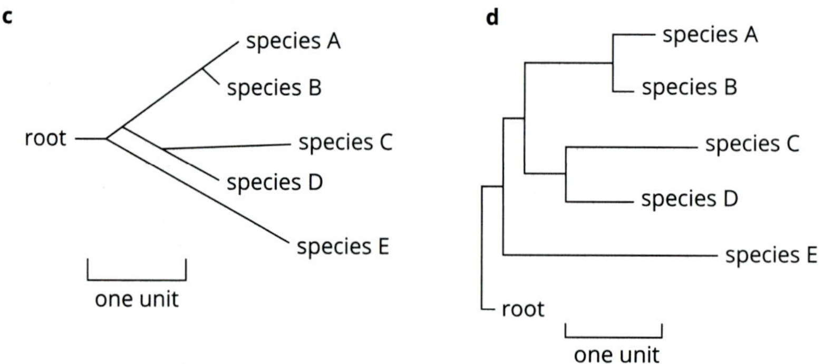

e Unrooted tree with scaled branches (phylogram)

f Unrooted tree with unscaled branches (cladogram)

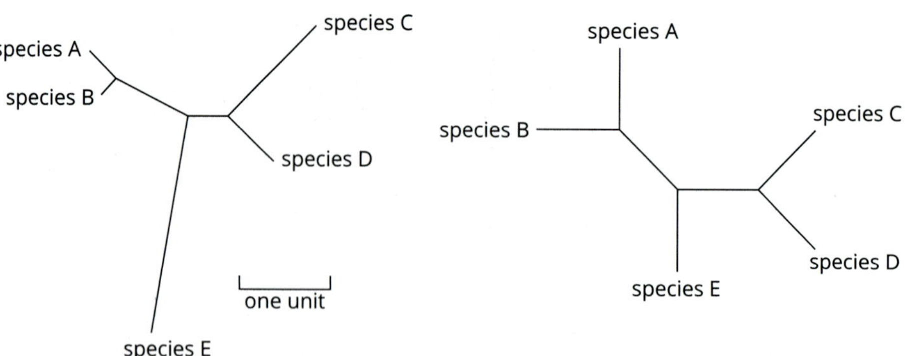

FIGURE 13.2.10 Phylogenetic trees can be depicted in many different ways to show the same evolutionary relationships. Two different ways of drawing rooted trees are with (a, b) unscaled branches (cladograms) and (c, d) scaled branches (phylograms). The unscaled trees in (a) and (b) have been calibrated to a timescale. Two unrooted phylogenetic trees, (e) one with scaled branches and (f) the other with unscaled branches, are also shown. All of these trees present the same evolutionary relationships in different ways.

Rooted trees

Rooted trees are drawn with the first branches coming from the base or 'trunk' of the tree. These may be drawn vertically (Figure 13.2.11), horizontally or in a circular format. The root represents the hypothesised common ancestor of all taxa in the tree and the branches represent the evolutionary path of those taxa over time. In comparison, unrooted trees depict the evolutionary relationships between taxa in the tree but do not show their evolutionary path from a common ancestor.

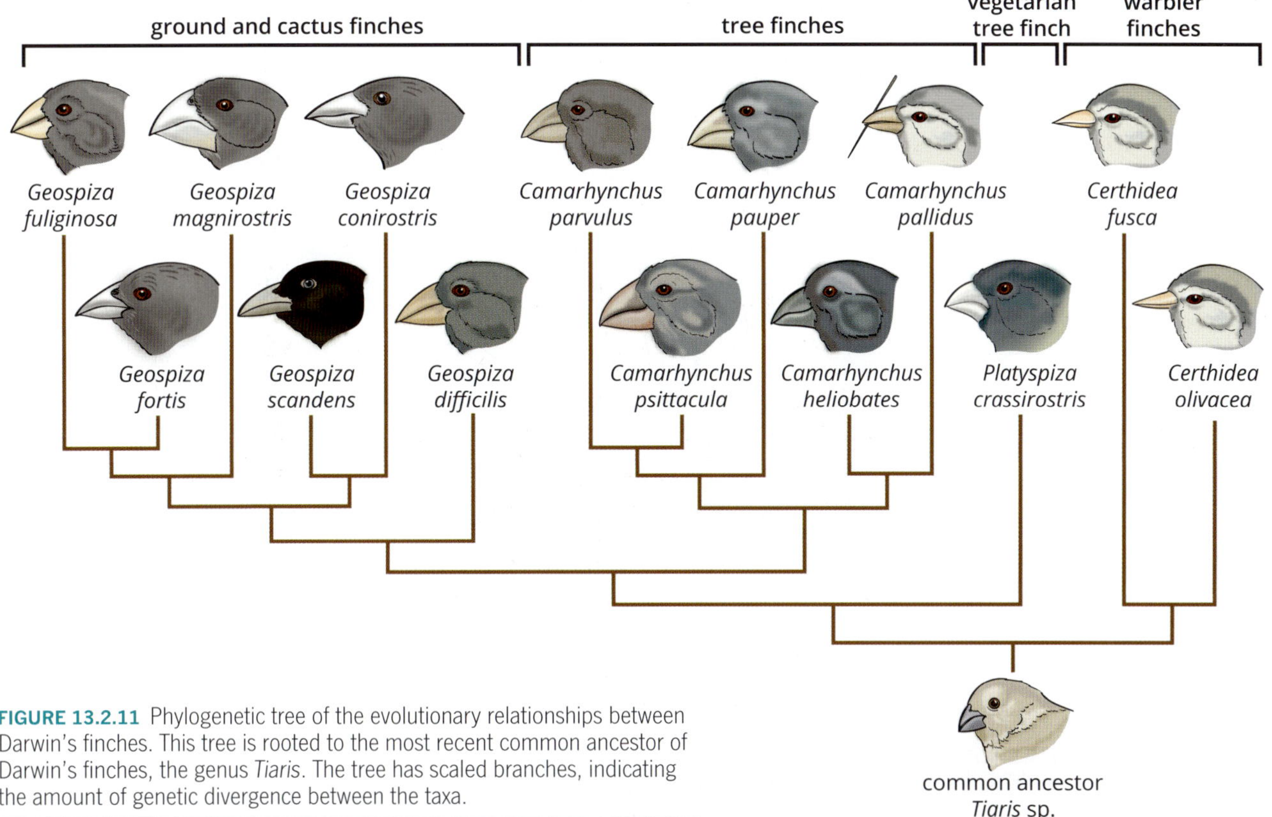

FIGURE 13.2.11 Phylogenetic tree of the evolutionary relationships between Darwin's finches. This tree is rooted to the most recent common ancestor of Darwin's finches, the genus *Tiaris*. The tree has scaled branches, indicating the amount of genetic divergence between the taxa.

In order to root the tree to a common ancestor, an organism that is known to be related to the main group of interest is included in the phylogenetic analysis as an outgroup. The outgroup (warbler finches in Figure 13.2.11) provides a comparison point to assess where the main group of organisms (ingroup) sits in relation to other closely related taxa. The outgroup must be related to the ingroup in order to make meaningful comparisons, but not be more closely related than organisms in the ingroup are to one another. An outgroup can be selected by aligning DNA sequences to determine how closely related they are.

In phylograms the lengths of the branches connecting two organisms indicates the amount of genetic divergence between them. The time that has passed since the organisms shared a common ancestor can also be represented by the branch length, by applying a molecular clock (Section 13.1).

Polytomies

Occasionally there will be nodes with more than two descendent lineages (like a garden rake). These nodes are called **polytomies** (Figure 13.2.12). Polytomies occur where there is not enough information to distinguish the order of evolution or when rapid speciation has occurred after adaptive radiation in a new environment (Chapter 12, page 413). If rapid speciation has occurred, then all the daughter lineages will be closely related.

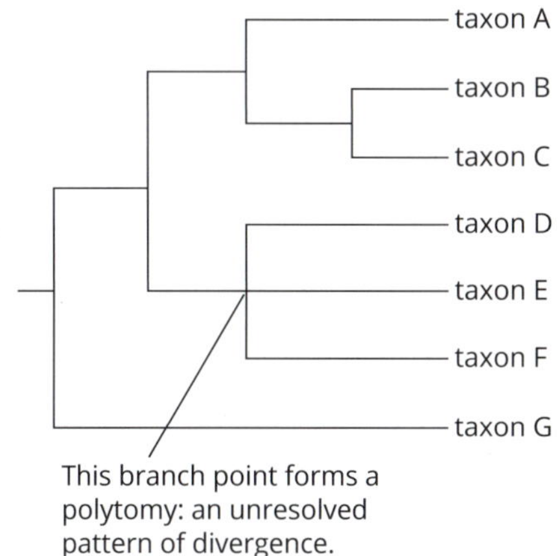

FIGURE 13.2.12 Phylogenetic tree with a polytomy between taxa D, E and F. Polytomies are formed when the evolutionary relationships between three or more taxa cannot be resolved due to lack of information or rapid speciation events.

Unrooted phylogenetic trees

Unrooted phylogenetic trees do not include an ancestral root. Because the ancestor is not defined in unrooted trees, they often have a radial layout. These trees only indicate the relationship between the different leaf nodes without indicating which node is the most ancestral. Like the rooted phylogram, unrooted trees can also have scaled branches where the distance between each leaf (along the branch lengths) is an indication of the amount of genetic divergence between the two groups of organisms. A greater distance suggests more time has passed since they shared a common ancestor (Figure 13.2.13).

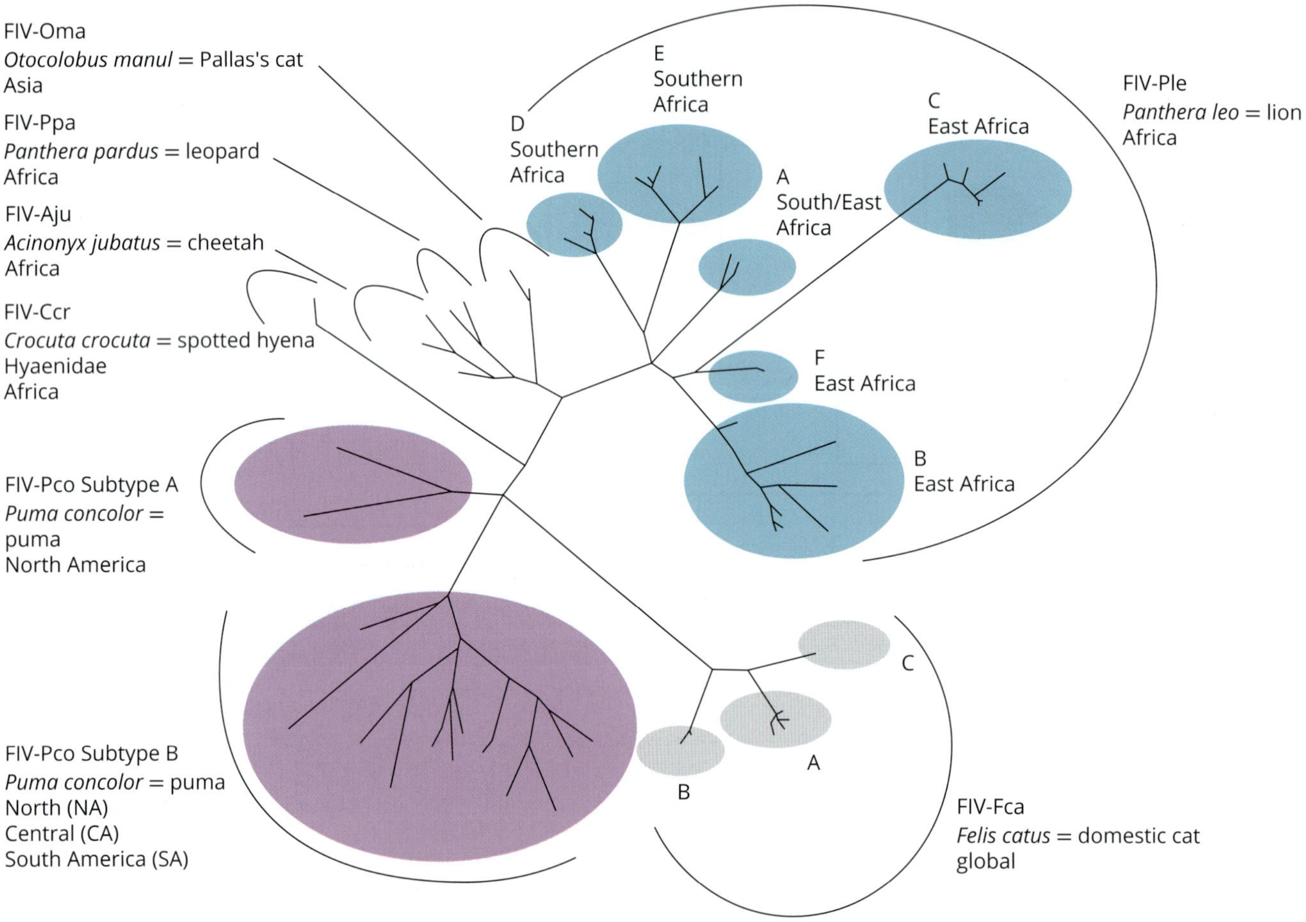

FIGURE 13.2.13 Unrooted phylogenetic tree representing evolutionary relationships between subtypes of the feline immunodeficiency virus (FIV) from seven carnivore species. The common ancestor has not been defined in this tree and so, without a root, it has a radial configuration. Branches are scaled to represent the genetic distance between the virus subtypes.

CASE STUDY **ANALYSIS**

The evolutionary uncertainty of ray-finned fish

Phylogenetic modelling of fish species reveals that there is uncertainty around the evolutionary relationships of a group of ray-finned fish known as the Percomorphs (highlighted in yellow in Figure 13.2.14). You might have noticed that the section of the phylogenetic tree (clade) with the Percomorphs looks different from the rest of the tree. Instead of each node (branch point) having two lineages, the Percomorph clade looks like a rake, with many lineages. This feature of a phylogenetic tree is called a polytomy.

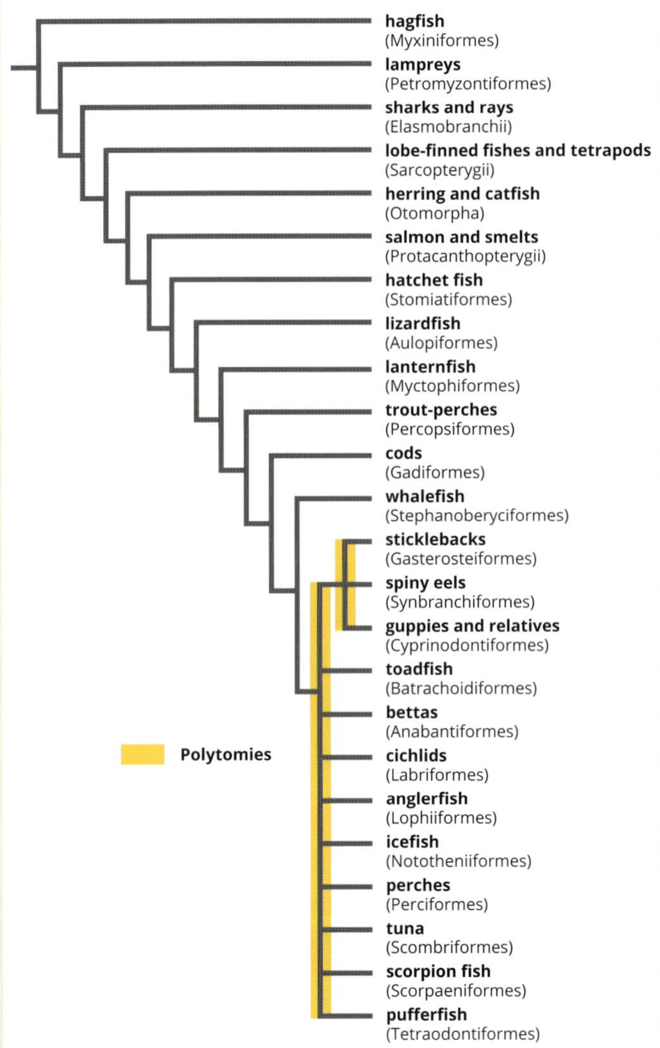

FIGURE 13.2.14 Polytomies in this phylogenetic tree (highlighted in yellow) indicate that the evolutionary relationships of Percomorph fish are uncertain.

FIGURE 13.2.15 (a) Spiny eel and (b) scorpion fish

In this tree, there are two polytomies (highlighted in yellow), one with three lineages (sticklebacks, spiny eels and guppies) and the other with nine lineages (toadfish, bettas, cichlids, anglerfish, icefish, perches, tuna, scorpion fish and pufferfish, Figure 13.2.15). Polytomies indicate that there is not enough information to resolve the evolutionary relationships of lineages. The tree tells us that these lineages are closely related but we can't be sure of the order in which they diverged from one another. From the phylogenetic tree in Figure 13.2.14, we know that sticklebacks, spiny eels and guppies are closely related, but we can't tell which is most closely related to which. Further study and additional data can help scientists resolve polytomies and gain more certainty about evolutionary relationships.

Analysis

1. Using Figure 13.2.14, identify which group the Percomorphs are most closely related to.
2. **a** Are sticklebacks more closely related to spiny eels or toadfish?

   b Explain your answer.
3. What other data could help scientists learn more about Percomorph evolution?

13.2 Review

SUMMARY

- Phylogenetic trees show the evolutionary relationship between different groups of organisms based on morphological and molecular homology.
- The branches and nodes of phylogenetic trees indicate common ancestry between organisms.
- A clade is a group of organisms that includes an ancestor and all the descendants of that ancestor.
- Groups in phylogenetic trees can be described as monophyletic, paraphyletic or polyphyletic depending on their evolutionary relationships.
- Monophyletic groups are the only taxonomically viable group because they contain a common ancestor and all its descendants (a clade).
- Rooted phylogenetic trees can be used to indicate the length of time that has passed since organisms shared a common ancestor.
- Unrooted phylogenetic trees do not include an ancestral root, and only indicate the relationship between the different leaf nodes.

KEY QUESTIONS

Knowledge and understanding

1 What is a phylogenetic tree and what does it represent?

2 List four forms of phylogenetic tree.

3 Why is an outgroup included in some phylogenetic trees?

Analysis

4 From the phylogenetic tree below, which animal is most closely related to the pig?

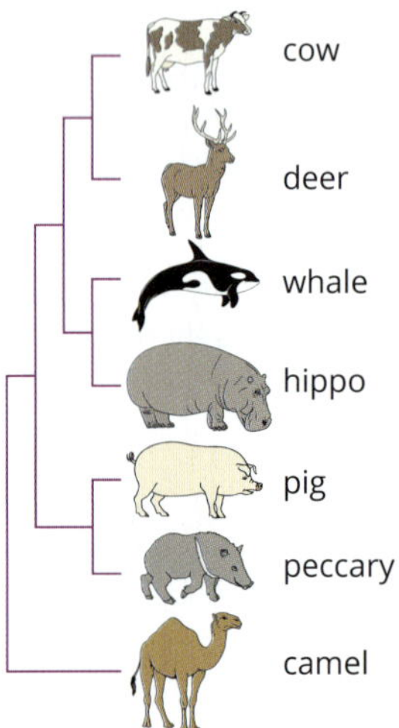

5 Which species is most closely related to species A?

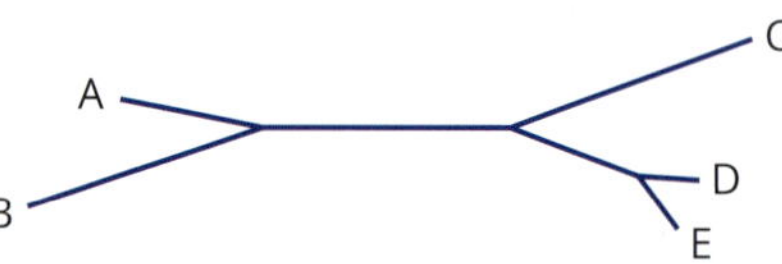

6 Number (1–7) the nodes of the phylogenetic tree in the order in which they diverged.

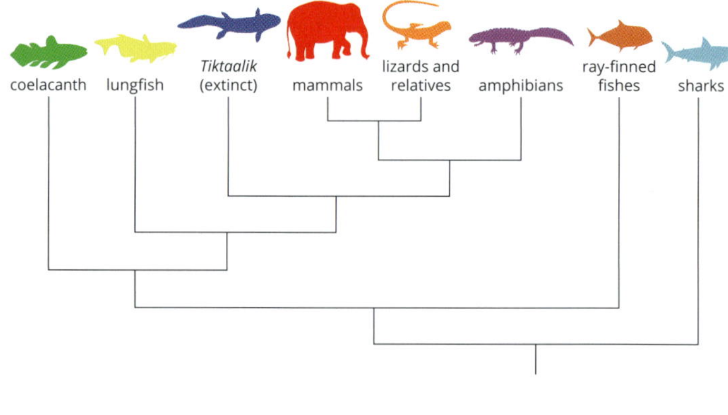

Chapter review

13

KEY TERMS

analogous feature
branch
clade
cladogram
common ancestor
convergent evolution
divergent evolution
homologous features
leaf
lineage
molecular homology
monophyletic group
mutation rate
node
outgroup
phylogenetic tree
phylogeny
phylogram
point mutation
polytomy
root
sister taxa
structural morphology
taxon (plural taxa)
taxonomy
vestigial structure

REVIEW QUESTIONS

Knowledge and understanding

1 What are vestigial structures? Give an example.

2 Identify the following as analogous or homologous features.
  - **a** the wings of butterflies and the wings of birds
  - **b** the flippers of whales and the fins of fishes
  - **c** the arms of humans and the flippers of seals

3 Explain why mutations in the DNA of two species can represent the time since they shared a common ancestor.

4 Explain why point mutations may not always result in a change in phenotype.

5 Explain why mitochondrial DNA (mtDNA) is used instead of nuclear DNA to differentiate closely related species.

6 Which graph would best illustrate the assumed rate of the molecular clock?

A

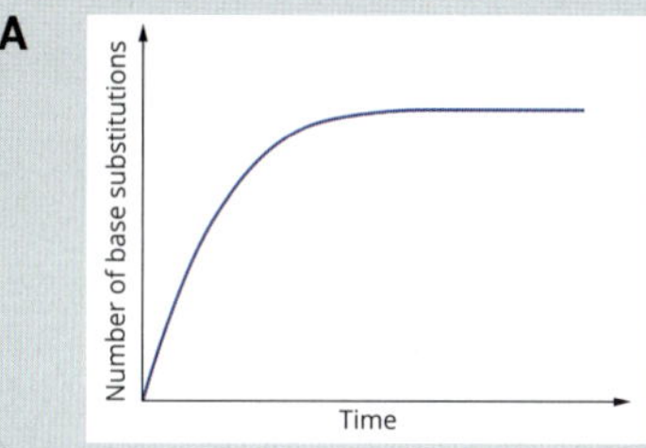

B

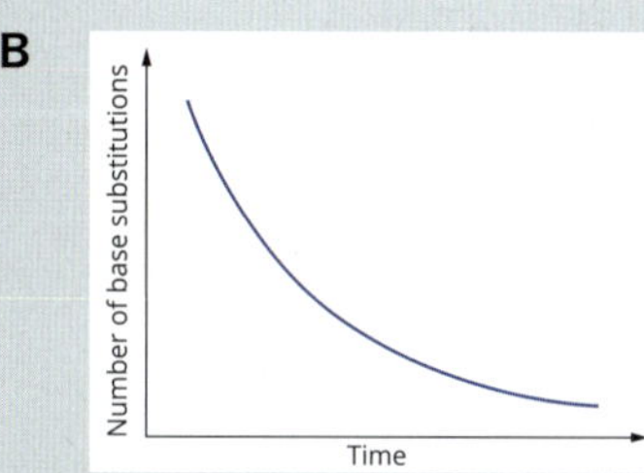

C

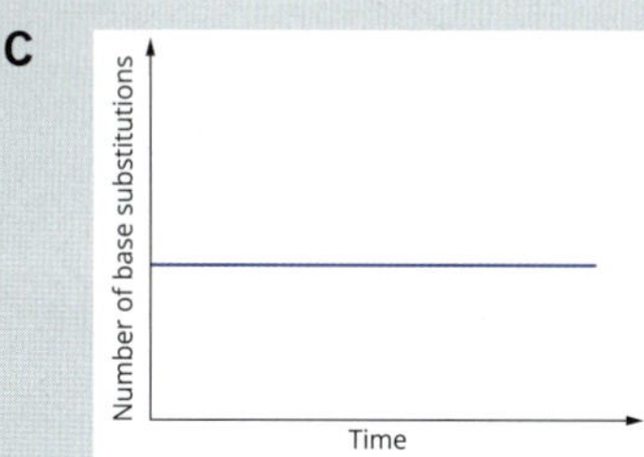

D

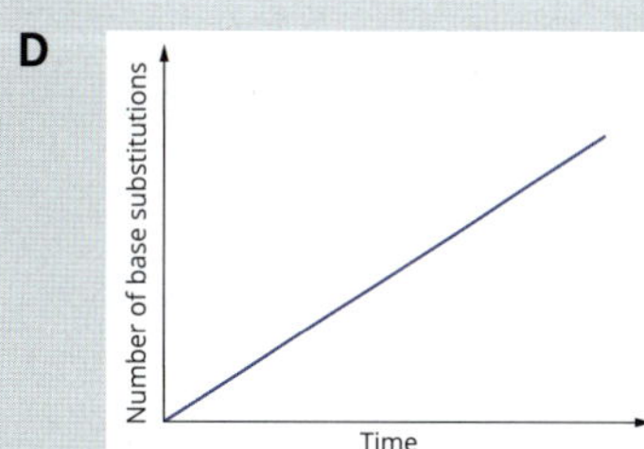

7 Which phylogenetic grouping (monophyletic, paraphyletic or polyphyletic) is the only taxonomically viable group and why?

8 It has been established that human mitochondrial DNA is maternally inherited. Using this understanding, scientists have postulated that all living humans are related to a single female (Mitochondrial Eve), who lived between 140 and 200 thousand years ago. There is a general misconception that this means that all humans are descended from only one female. Explain why this interpretation of the situation is incorrect.

Application and analysis

9 Which of the following statements best describes the phylogenetic tree shown below?

A scaled and rooted
B unscaled and rooted
C scaled and unrooted
D unscaled and unrooted

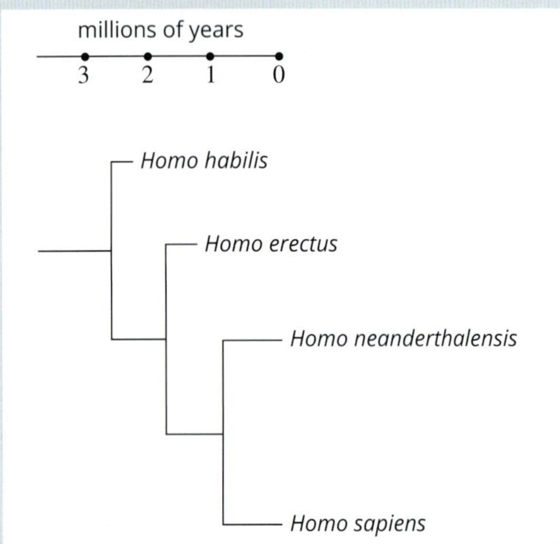

10 According to the phylogenetic tree of selected primates shown below, which of the following statements is true?

A Humans are more closely related to chimpanzees than they are to bonobos.
B The most recent common ancestor of gorillas and bonobos lived about 15 million years ago.
C The most recent common ancestor of bonobos and chimpanzees lived about 3 million years ago.
D Humans and orangutans last shared a common ancestor around 10 million years ago.

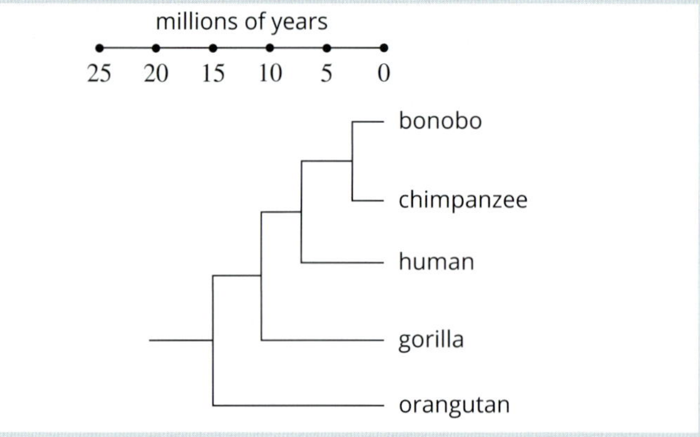

11 Four plants have similar leaf shape, but plant 4 has a different colour and was proposed to be distantly related to the others. A section of DNA from each species is aligned.

a Copy and complete the table for base changes in the sequence alignment.
b From this data, infer which two species are most distantly related. Justify your answer.

Multiple alignment
P1 AGGCCAAGCCATAGCTGTCC
P2 AGGCAAAGACATACCTGACC
P3 AGGCCAAGACATAGCTGTCC
P4 AGGCAAAGACATACCTGTCC

	Plant 1	Plant 2	Plant 3	Plant 4
Plant 1	–			
Plant 2		–		
Plant 3			–	
Plant 4				–

12 Compare the three sequences below.

monkey ATGCACACCTCCATTATA
gorilla ATGTACGCTACCATAACC
human ATGCACACTACTATAACC
*** ** * * ** *

a Complete the table of number of base changes.

	Monkey	Gorilla	Human
Monkey	–		
Gorilla		–	
Human			–

b Complete the unrooted phylogenetic tree for these three species based on the data provided.

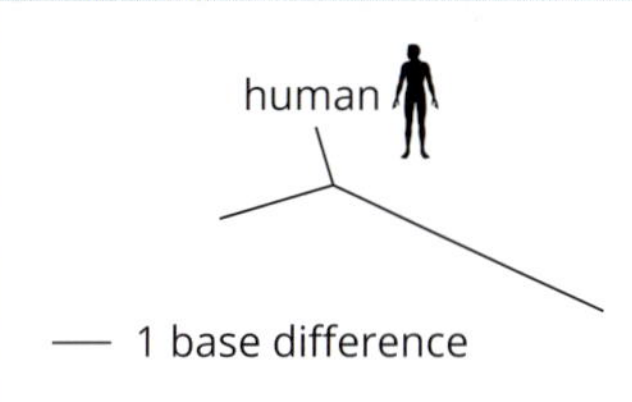

c Determine which species is more closely related to the gorilla from the provided data.

13 Hearing is quite a complex process. In humans and other mammals, it involves an outer membrane (the tympanic membrane or eardrum) and three bones (the incus, malleus and stapes), also known as the auditory ossicles. The ear bones can transfer vibrations from the tympanic membrane to an inner fluid-filled chamber called the cochlea. The cochlea is lined with sensors that detect the vibrations in the fluid and transfer the information, via nerve signals, to the brain. The mammalian incus and malleus evolved from parts of the jaw of fish and the stapes evolved from the hyomandibula bone. In fish, this bone helps to support the gills.

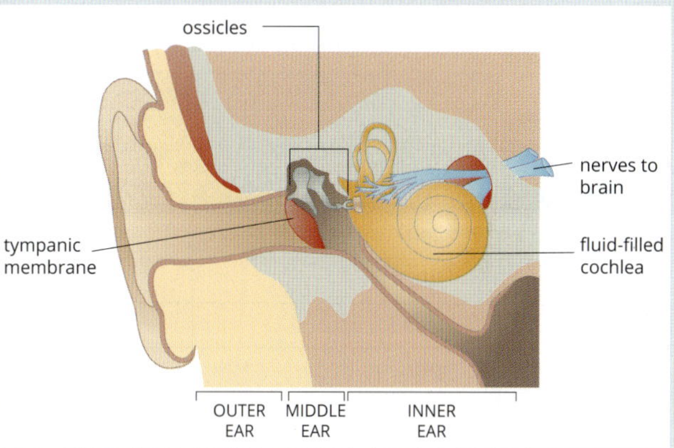

Reptiles have a single bone, the columella, which evolved from the hyomandibula bone. The columella transfers the sound from the tympanic membrane to the fluid-filled inner ear.

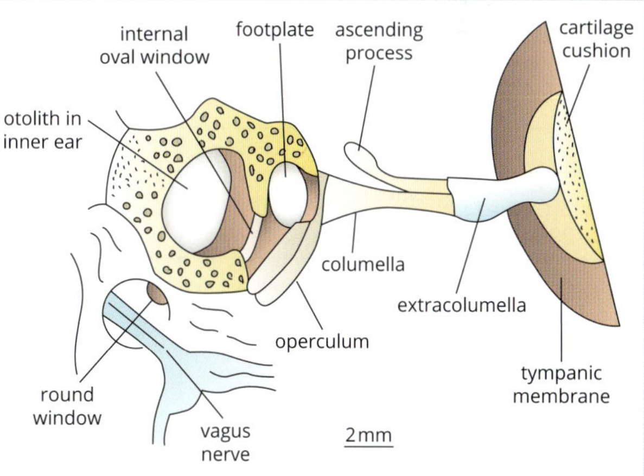

Explain whether the bones of the middle ear in mammals and reptiles are homologous or analogous.

14 a DNA sequences provide evidence of the evolution of organisms. Justify this statement.

b Explain how scientists can use DNA sequences to reconstruct the evolutionary history of species.

c Draw a phylogenetic tree to support your answer.

15 The tyrosine-related protein 1 gene (*TYRP1*) is a gene that codes for an enzyme that is involved in the production of melanin in the skin, hair and eyes in humans. The gene is also found in many other vertebrates where it also codes for melanin production. Consider the table below, which shows the amino acid sequence for part of the protein. The first column shows the sequence in humans. The other columns show the differences between the sequence in humans and the other species. Empty boxes mean that the amino acid is the same as that in a human.

a **i** Based only on the data given, which organism is most closely related to humans? Justify your opinion.

ii Based only on the data given, which organism is most distantly related to humans? Justify your opinion.

b *Gallus gallus* is a chicken and *Takifugu rubripes* is a pufferfish. Most evolutionary data indicates that chickens have a more recent common ancestor with humans than do pufferfish, but a comparison of the TYRP1 protein of these two organisms and the human protein indicates that pufferfish are more closely related than chickens to humans. Why wouldn't scientists rewrite the evolutionary tree based on this data?

	Species							
Amino acid position	***Homo sapiens* (human)**	***Capra hircus* (goat)**	***Canis lupus familiaris* (dog)**	***Ovis aries* (sheep)**	***Mus musculus* (mouse)**	***Gallus gallus* (chicken)**	***Xenopus laevis* (frog)**	***Takifugu rubripes* (pufferfish)**
	leu						ser	ala
	ile						leu	
	ser							
	phe		leu			gln	ser	ala
280	asn							
	ser							
	val					ile		ile
	phe							
	ser							
	gln					thr		arg
	trp							
	arg							
	val							
	val					leu		
290	cys							
	asp			glu				
	ser						phe	
	leu					ile	val	val
	glu							
	asp				glu			
	try							
	asp						glu	
	thr					ser	ser	
	leu							
300	gly							
	thr							
	leu					ile	ile	val
	cys							
	asn							

16 The diagram below is a cladogram illustrating the evolutionary relationships between various groups of organisms.

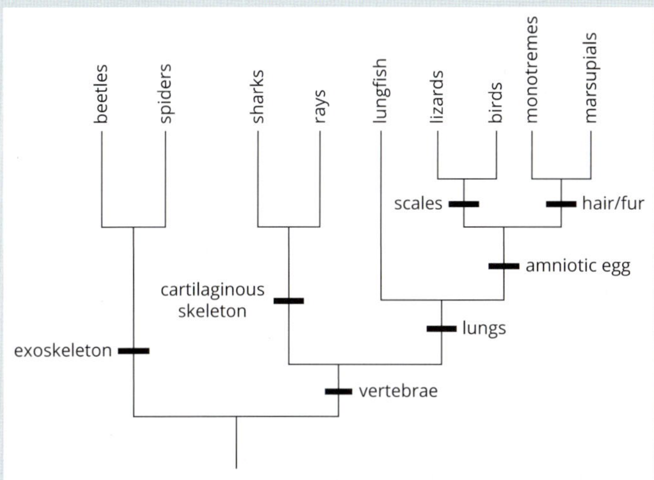

a What extra information could be drawn from the diagram if it had been drawn as a phylogram?

b **i** Are lungfish more closely related to rays or lizards? Explain.

ii What characteristic arose in the common ancestor of lizards, birds, monotremes and marsupials and is thus present in only that clade today?

c For each group, identify whether it is an example of a monophyletic, paraphyletic or polyphyletic grouping:

i lungfish, sharks and rays

ii lizards, birds, monotremes and marsupials

iii birds, lizards and beetles.

17 The ratite birds have long interested palaeontologists and biologists as they provide something of a mystery. Members of the ratite group (which includes emus and ostriches) are unable to fly. The mystery was how, if ratites never flew, did they become distributed across the world.

The first hypothesis suggested for their distribution was that they had a common ancestor that lived on the southern supercontinent, Gondwana. This resulted in the formation of many ratite species when Gondwana broke apart over a period of 85 million years.

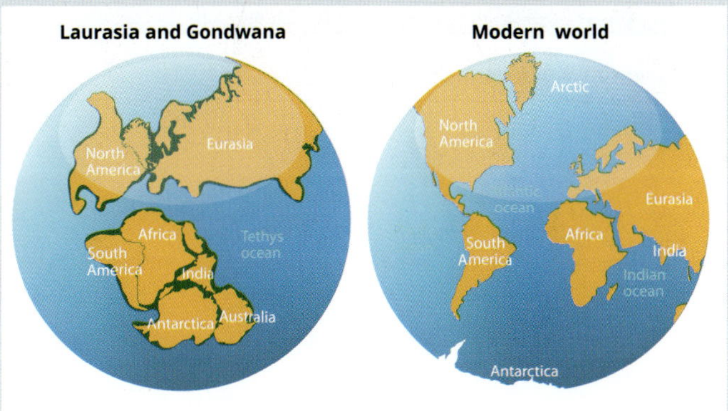

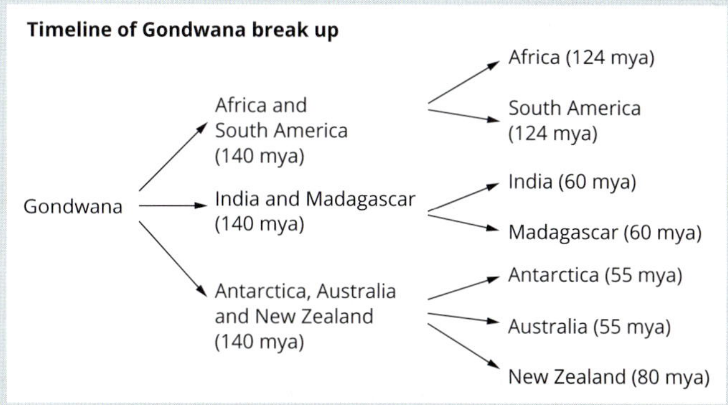

a Explain how the breakup of Gondwana could have resulted in the formation of species as different as the ostrich and the kiwi.

b The phylogenetic tree for the ratites was hypothesised, on the basis of their morphology, to be as shown below.

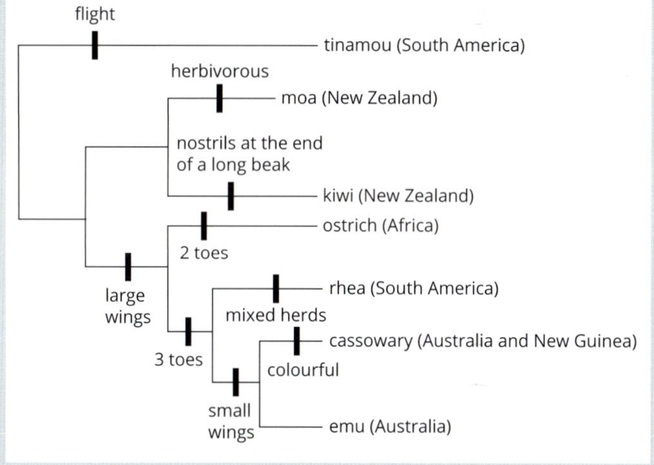

i Which group could be considered to be an outgroup in the tree above?

ii Consider the phylogenetic tree above and the timeline of the continents breaking away from Gondwana. Identify one piece of evidence from the tree that supports the hypothesis that the ratites developed from populations that became isolated when Gondwana broke apart.

iii Identify one piece of evidence in the phylogenetic tree above that refutes the hypothesis that ratites developed as a result of the breakup of Gondwana.

Question 16 continues over page

c One extinct member of the ratite group that has not been included in the phylogenetic tree is the elephant bird of Madagascar. Madagascar is an island found off the east coast of Africa. There are no ratites in India and no evidence that they ever lived there, so one hypothesis is that Madagascar was colonised from Africa by rafting. (An organism gets trapped on a log or something similar and is transported across the ocean to an island.)

- **i** Why is it relevant that there are no ratites in India?
- **ii** Considering all of the data presented so far, which of the ratites in the phylogenetic tree on the previous page should be most closely related to the elephant bird?

d In the last few years DNA analysis has advanced to the stage that it has become possible to sequence DNA from fossils of the moa, which has been extinct for around 600 years, and the elephant bird, which has been extinct for around 250 years. The results of these analyses along with the analysis of the extant (living) species of ratites led to a complete revision of the ratite phylogenetic tree, along with the development of an entirely new hypothesis to explain their distribution.

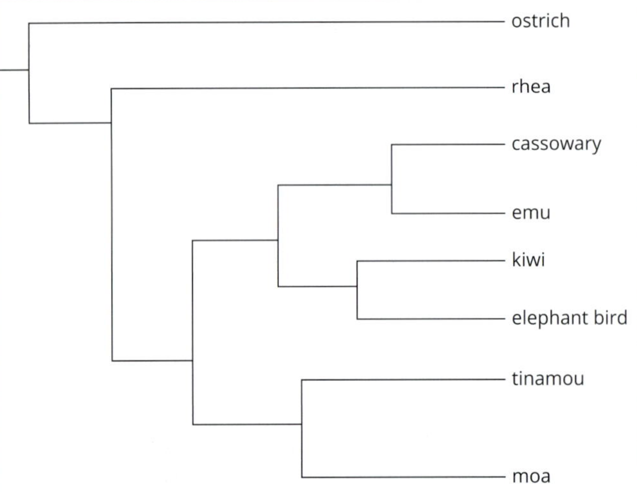

- **i** Explain how evidence from the DNA analysis disagrees with the theory that the ratite distribution occurred due to the separation of Gondwanan populations of an ancestral flightless ratite bird.
- **ii** It has now been hypothesised that the ratite ancestors were all flighted birds whose ancestral populations flew to where they are found today and that the loss of flight occurred independently on several different occasions in this group. Explain the type of evolution that this represents.

18 In the past, the primary method taxonomists used to work out the evolutionary relationships between organisms was to investigate the morphology of the organisms. Important characteristics were compared and scored in a table called a character matrix. Observe the insects below.

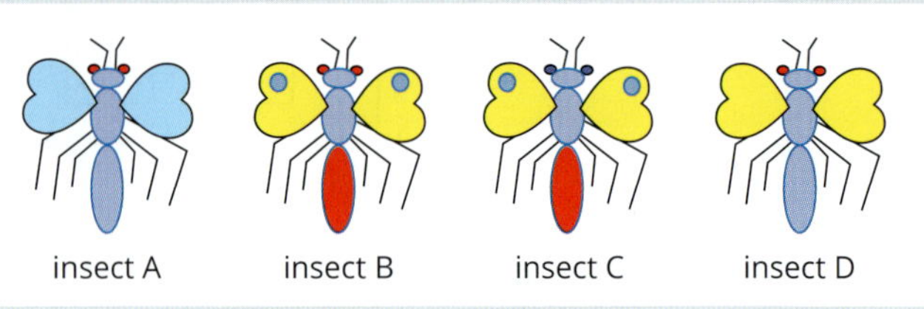

a Complete the character matrix for the insects. Use '0' for ancestral traits and '1' for derived traits. Assume that insect A is an outgroup for these species.

Trait	A	B	C	D
yellow wings				
wing spots				
red abdomen				
blue eyes				

b Draw a cladogram to show the evolutionary relationships between the insects. Include significant traits on the diagram.

19 Scientists were studying the relationships between a number of species in the genus *Conus*. This is a genus in the phylum Mollusca. All members of the genus have cone-shaped shells.

Conus marmoreus shell

Cytochrome *c* oxidase I (COI) is an enzyme found in many species, including the four species of *Conus* below.

Species	Abbreviation
Conus marmoreus	M
Conus chaldeus	C
Conus omaria	O
Conus magnus	MG

The grids below show the base sequence for part of a gene coding for COI. The grids contain 21 triplet nucleotides of the COI gene of all four species and have been coloured to make observing and counting differences easier.

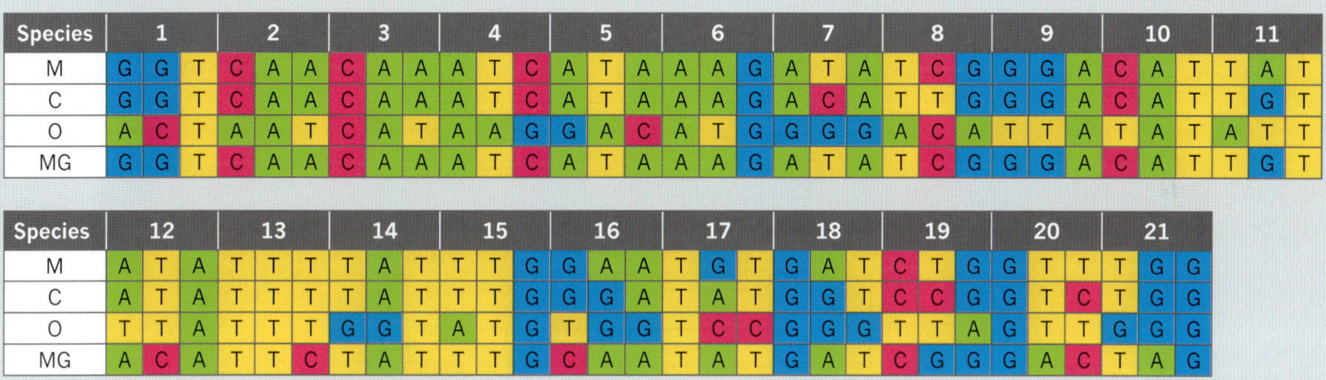

Species	1	2	3	4	5	6	7	8	9	10	11
M	G G T	C A A	C A A	A T C	A T A	A A G	A T A	T C G	G G A	C A T	T A T
C	G G T	C A A	C A A	A T C	A T A	A A G	A C A	T T G	G G A	C A T	T G T
O	A C T	A A T	C A T	A A G	G A C	A T G	G G G	A C A	T T A	T A T	A T T
MG	G G T	C A A	C A A	A T C	A T A	A A G	A T A	T C G	G G A	C A T	T G T

Species	12	13	14	15	16	17	18	19	20	21
M	A T A	T T T	T A T	T T G	G A A	T G T	G A T	C T G	G T T	T G G
C	A T A	T T T	T A T	T T G	G G A	T A T	G G T	C C G	G T C	T G G
O	T T A	T T T	G G T	A T G	T G G	T C C	G G G	T T A	G T T	G G G
MG	A C A	T T C	T A T	T T G	C A A	T A T	G A T	C G G	G A C	T A G

a Complete the following table by counting the number of nucleotide differences between pairs of the species. Each pair only needs to be counted once so shaded squares do not need to be filled.

Species	*Conus marmoreus* (M)	*Conus chaldeus* (C)	*Conus omaria* (O)	*Conus magnus* (MG)
Conus marmoreus (M)				
Conus chaldeus (C)				
Conus omaria (O)				
Conus magnus (MG)				

b Use the data obtained from the DNA sequences to generate a possible cladogram for these species.

20 A DNA sequence from three extant ratites (rhea, kiwi and cassowary) and the extinct elephant bird is shown below in a multiple sequence alignment. The DNA sequence is 40 nucleotides long.

```
rhea           AATAGTT-CATAAATATCCGGCACCATAAATCAGCGCCTA
kiwi           AATAGTC-CGTAAATATCTGGTTCCATAAATTAACGCAAA
cassowary      AATAGTT-CATAAATATCCGACCCCATAAATTAACGCGAA
elephant bird  AATAGTG-TTTAAATATCTGGAACCCAGACCACATACAAA
```

The researchers constructed a phylogenetic tree based on all of the mitochondria and nuclear DNA sequence data available from all extant and extinct ratites for which DNA is available. A simplified phylogenetic tree of the major taxa is shown below.

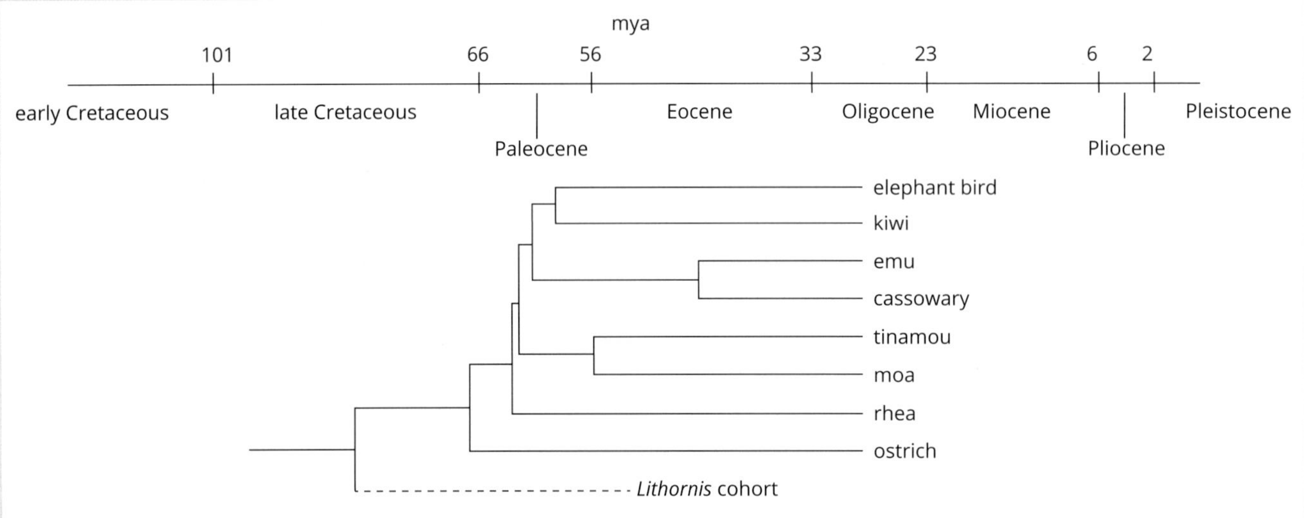

a Copy the table below into your workbook. Count the number of nucleotide differences for each pair of birds and record the numbers in the table.

	Rhea	Kiwi	Cassowary	Elephant bird
Rhea	–			
Kiwi		–		
Cassowary			–	
Elephant bird				–

b Based on your counts of this sequence, identify the living species that is the closest relative of the extinct elephant bird. Justify your answer.

c Explain whether your determination of the closest relative of the elephant bird from the DNA sequence alignment agrees with the phylogenetic tree pictured.

d Compare the result for the cassowary and rhea obtained from the short DNA sequence alignment and the phylogenetic tree produced from the mitochondrial and nuclear DNA sequence data. Identify any discrepancy or disagreement between these results and suggest a reason for any difference.

e State the time, in millions of years, and identify the geological period of the common ancestor of the elephant bird and kiwi from the phylogenetic tree.

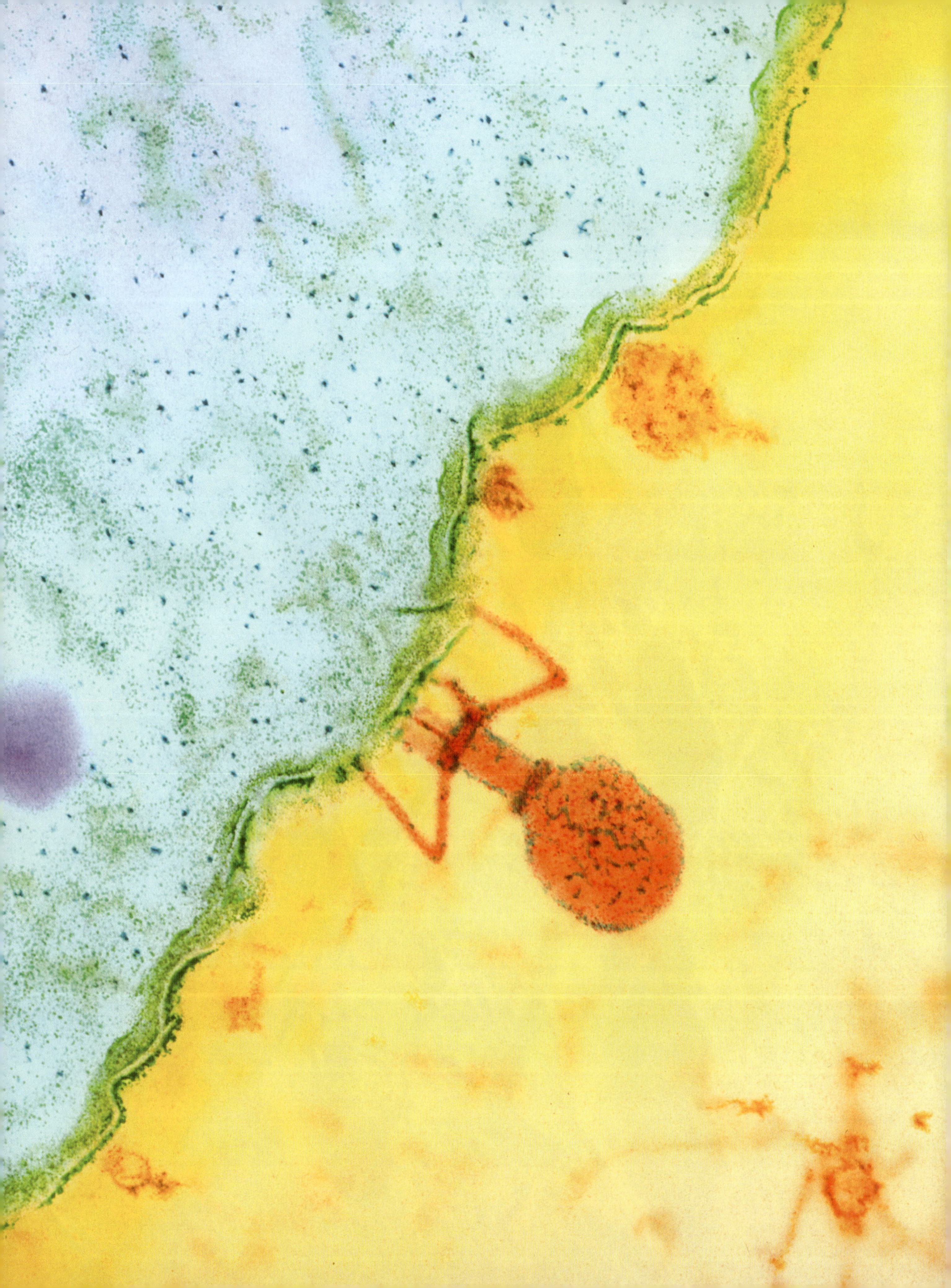

CHAPTER 14

Human change over time

Learning outcomes

The quest for answers about our origins and the evolutionary paths that led us to where we are today is more exciting than ever, with rapidly advancing DNA technology and the discovery of new fossils and archaeological artefacts. Our evolutionary story is a dynamic one, changing as new information comes to light or new ideas are found to have greater explanatory power.

In this chapter you will learn about what it is that makes us human: our shared and defining characteristics from other mammals, primates, hominoids and hominins. You will also examine the characteristics of some of our known ancestors and come to understand patterns in hominin evolution, from the genus *Australopithecus* to our own genus, *Homo.* You will also learn about the human fossil record and how it is used, along with DNA evidence, to understand the evolutionary history of modern humans and their migration around the world.

Key knowledge

- the shared characteristics that define mammals, primates, hominoids and hominins **14.1**
- evidence for major trends in hominin evolution from the genus *Australopithecus* to the genus *Homo*: changes in brain size and limb structure **14.1, 14.2**
- the human fossil record as an example of a classification scheme that is open to differing interpretations that are contested, refined or replaced when challenged by new evidence, including evidence for interbreeding between *Homo sapiens* and *Homo neanderthalensis* and evidence of new putative *Homo* species **14.1, 14.2**
- ways of using fossil and DNA evidence (mtDNA and whole genomes) to explain the migration of modern human populations around the world, including the migration of Aboriginal and Torres Strait Islander populations and their connection to Country and Place. **14.2**

14.1 Defining humans

Homo sapiens is the Latin term for our species. The term translates as 'wise man' and refers to anatomically and behaviourally modern humans. *H. sapiens* is one of the most widespread, adaptable and influential species to have ever existed. Under the Linnaean system of biological classification, *H. sapiens* is a eukaryote and a member of the animal kingdom.

Within the animal kingdom, humans are further classified as mammals, with the characteristics of body hair and the ability to suckle young (Figure 14.1.1). Humans are also classified as **primates** (order Primates), having a grasping hand, bicuspid teeth, a short nose and well-developed eyes and brain. Within the order Primates, humans belong to the same family (**Hominidae**) as the great apes, which includes orangutans, gorillas, chimpanzees and bonobos. All **hominids** lack a tail and have similar skeletal and skull features. As well as sharing many anatomical and behavioural features with all of the great apes, modern humans share 98.8% of their DNA with chimpanzees and bonobos, our closest living relatives. Although modern humans are very similar to chimpanzees and bonobos, we did not directly evolve from them or any other living primate. DNA and fossil evidence tells us that we last shared a common ancestor with chimpanzees and bonobos approximately 6–8 million years ago (Figure 14.1.2).

> Chimpanzees and bonobos are our closest living relatives but we did not directly evolve from them. Our last shared common ancestor existed approximately 6–8 million years ago.

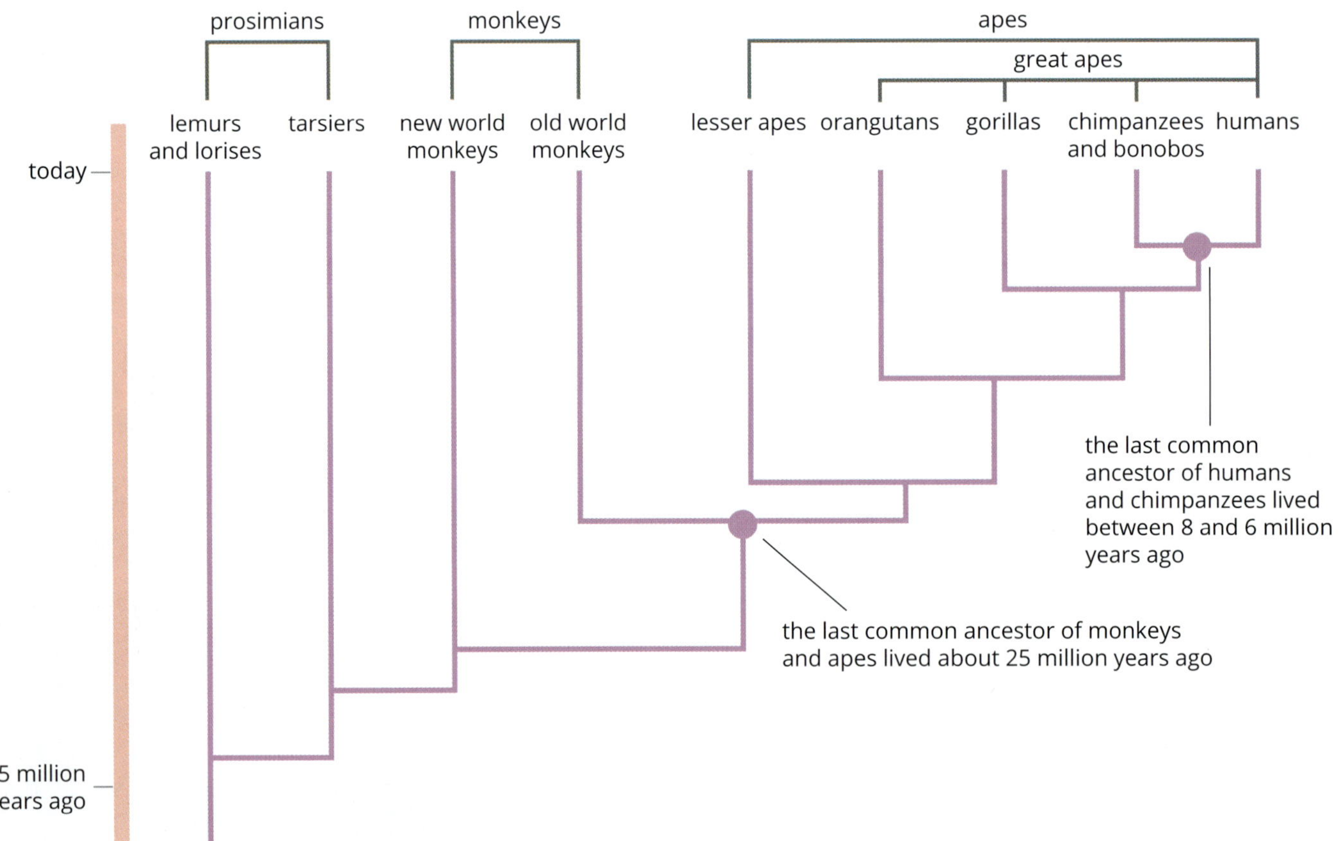

FIGURE 14.1.2 Evolutionary relationships of living primates

FIGURE 14.1.1 'Tree of life' depicting the evolutionary relationships of humans with other organisms

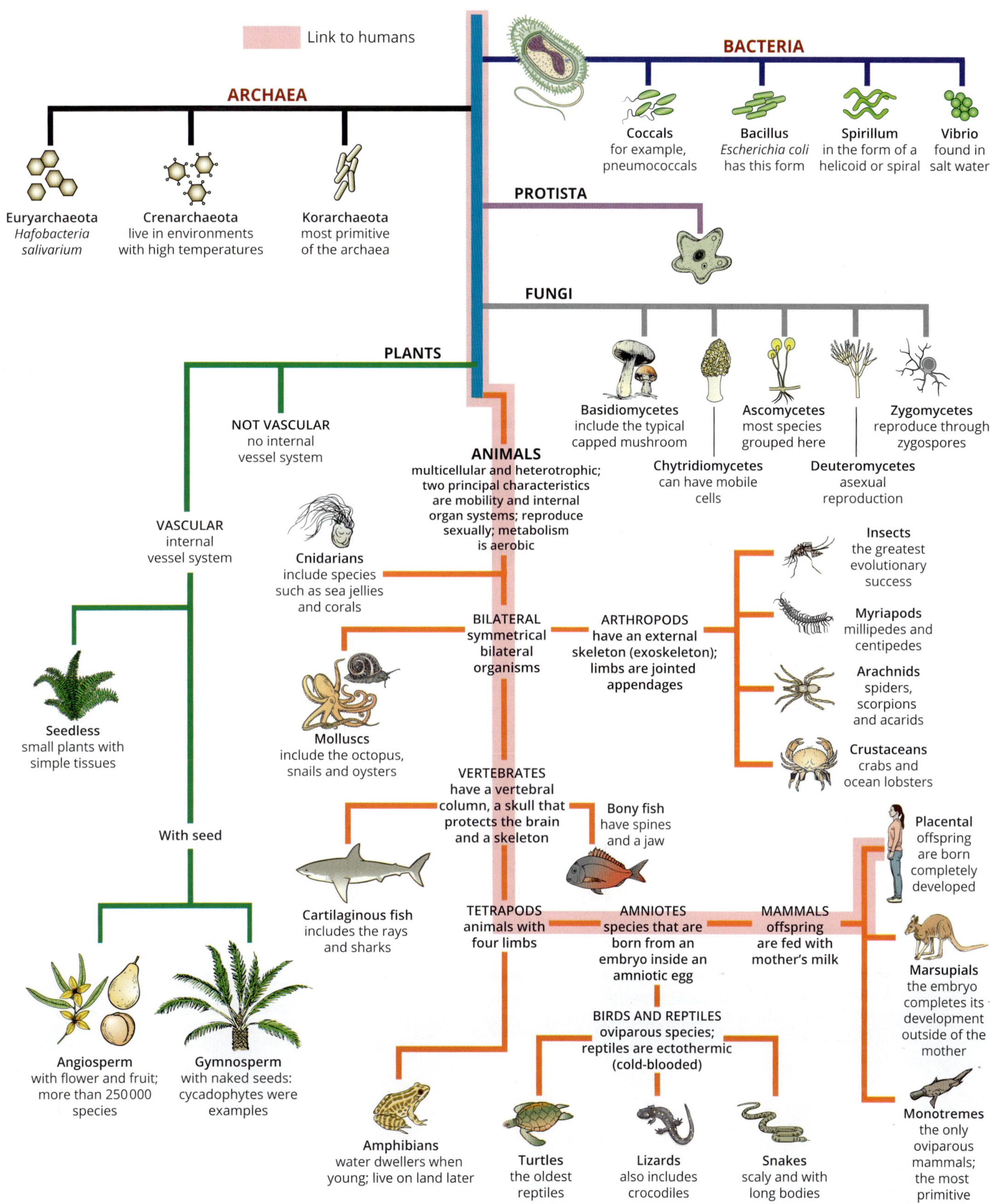

In taxonomy, a 'tribe' is the taxonomic rank above genus and below subfamily and family.

BIOFILE

Hominins, hominids and hominoids

The taxonomic classification of humans has changed over time, with some confusion arising in regard to the terms 'hominin', 'hominid' and 'hominoid.' Humans belong to the tribe Hominini and hence are hominins. Hominini also includes extinct *Homo* species, *Australopithecus*, *Paranthropus* and *Ardipithecus*. We are also part of a larger group, the family Hominidae, which includes orangutans, gorillas and chimpanzees, hence the term hominid. Finally, the superfamily Hominoidea includes humans, the great apes and the lesser apes (gibbons), hence the term hominoid.

The term **hominoid** refers to the superfamily **Hominoidea**, which includes humans, great apes (orangutans, gorillas, chimpanzees and bonobos) and the lesser apes (gibbons). Hominids, which include humans and great apes, belong to the family Hominidae. Hominins refers to members of the tribe **Hominini**, which includes modern humans, extinct human species and our **bipedal** ancestors (e.g. *Australopithecus*, *Paranthropus* and *Ardipithecus*). Within Hominini, humans belong to the ***Homo*** genus, with modern humans (*H. sapiens*) the only surviving species.

There are various fossilised **hominins** that have been discovered, all characterised by walking upright on two limbs (bipedalism). *Sahelanthropus* is the oldest known ancestor of modern humans, with skulls found in central Africa that date to approximately 7 million years ago. Some scientists think that *Sahelanthropus* may represent a common ancestor of humans and chimpanzees, but this is still debated in the scientific community. Species of the extinct genus *Australopithecus* are the closest relatives to the *Homo* genus. *Australopithecus* had a forward-jutting face, a small brow and a pronounced ridge above the eyes, but hands and teeth that were similar to ours.

Homo diverged from an australopithecine (*Australopithecus* species) approximately 2.8 million years ago. The genus *Homo* is characterised by a larger brain and includes 9–15 extinct species known to date (the identities of some fossil specimens are uncertain). Of the extinct species, *Homo habilis* is the oldest, found in Africa. A later species, *Homo erectus*, which had a larger brain than *H. habilis*, spread into South-East Asia about 1.6 million years ago, possibly surviving to about 300 000 years ago.

Fossil and DNA evidence is continuing to uncover new information and helping us to tell our story; but many questions remain concerning the origins of modern humans and fossil species.

TRENDS IN HUMAN EVOLUTION

The foramen magnum is the opening for the spinal cord through the skull from the brain. A foramen magnum positioned further towards the front of the skull indicates an upright stance and bipedal movement.

The fossil remains of our human ancestors demonstrate significant changes in physical, behavioural and cultural traits. Over time, our ancestors' arms got shorter, their legs longer, the pelvis modified and the **foramen magnum** moved forwards as they spent less time in trees and more time walking upright. Bodies became leaner and taller with less hair as climates became warmer and drier. As their diets moved away from plant material and towards a more omnivorous diet, teeth became smaller, and the increase in cooking of food further allowed for a reduction in jaw muscles. This led to an increased capacity for the skull to get larger and accommodate a larger brain (Figure 14.1.3). Culturally, the increased control our ancestors had over the environment enabled progressively less movement, more building of permanent shelters and the gradual development of technology, rituals and societal structures.

Today *H. sapiens* is one of the most successful species on Earth. Our ability to adapt readily to changing environments and the technology we have developed to support our survival has led us to become one of the most populous, widespread and influential species to ever exist on Earth.

FIGURE 14.1.3 The evolution of skull morphology in *Homo*. From left to right: *Australopithecus africanus* (2.1–3.3 million years ago), *H. habilis* (1.4–2.4 million years ago), *H. erectus* (0.14–1.89 million years ago), a modern human (*H. sapiens*) from Qafzeh in Israel (approximately 92 000 years old), and a Cro-Magnon human (*H. sapiens*) from France (approximately 22 000 years old)

CASE STUDY

Fossil evidence of modern humans

Our species, *H. sapiens*, has been around for approximately 200 000 years, with fossil and DNA evidence placing some of the earliest of our species in Africa. One of the oldest fossil records of anatomically modern humans was found in a region called Omo Kibish in Ethiopia and has been dated to 195 000 years ago (Figure 14.1.4). Two specimens, known as Omo I and Omo II, were recovered from this region between 1967 and 1974. Omo I has anatomically modern human morphology with some primitive traits, representing a transition from archaic *H. sapiens* to early modern *H. sapiens*. Omo II has more primitive features than Omo I, such as a sloping forehead and more robust build, leading researchers to suggest that it belonged to a population that was transitional between *Homo heidelbergensis* and *H. sapiens*. As both specimens have been dated to approximately 195 000 years ago, they may have co-existed. Omo Kibish and the surrounding regions are of great archaeological significance, as this area is the source of many important findings regarding the origin of our species.

For many years, Omo I was the oldest evidence of *H. sapiens*. However, in 2019 new technologies in this rapidly changing field of research were used to analyse fossils that are now thought to be older than Omo I. In the 1970s, anthropologists discovered two skull fossils in a cave in southern Greece. It was not until 2019 that modern technologies allowed scientists to reconstruct these fragments to form their probable original shape, and to date the fragments. When the back of one of the skulls was reconstructed, it matched the shape of the back of a *H. sapiens* skull. Radiometric dating estimated that the skull fragments are 210 000 years old. If confirmed, the fossils from Greece could pre-date the Omo Kibish fossils and become the earliest fossil record of anatomically modern *H. sapiens*.

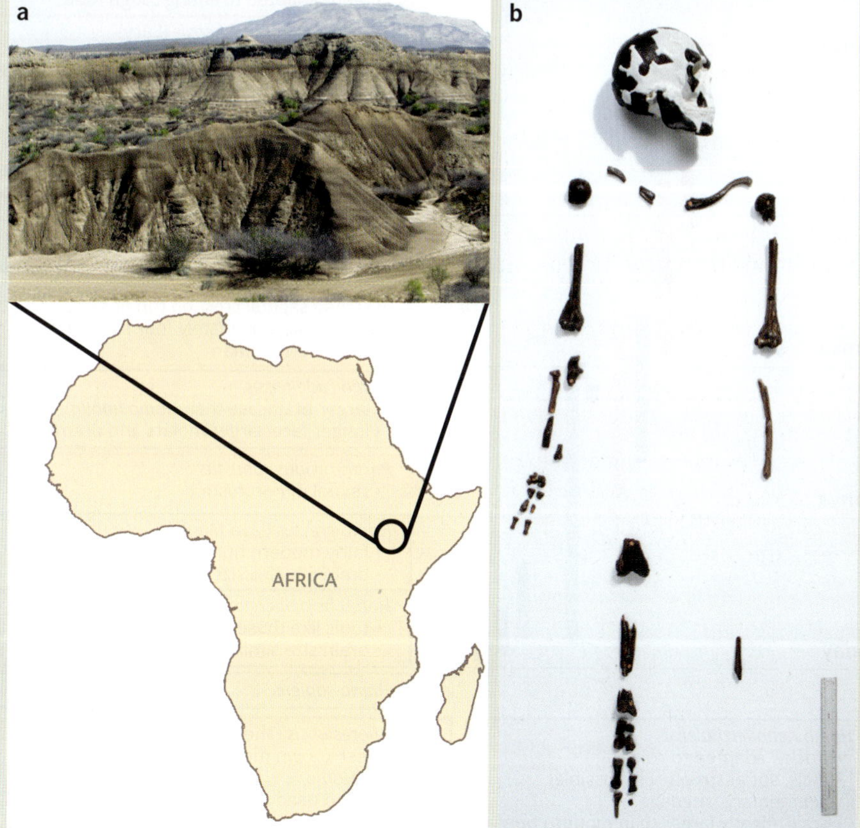

FIGURE 14.1.4 (a) The Omo Kibish formation in Ethiopia, Africa, is the site of one of the earliest known *H. sapiens* fossils. These fossils date back to 195 000 years ago. (b) Omo I was discovered in Omo Kibish in 1967, and at that time was identified as the oldest anatomically modern *H. sapiens* specimen.

HOMO SAPIENS: MODERN HUMANS

Time range: present day–200 000 years ago
Geographic range: worldwide

Limb structure

> Bipedal means to walk upright on two limbs. Bipedalism is the defining feature of hominins.

Modern *H. sapiens* can generally be characterised by a leaner, more agile (but less strong) bipedal build than our predecessors. Our pelvis is narrower and deeper than that of our ancestors, and our skulls have a high braincase and short base, with rounding at the back indicating reduced neck muscles. Our brow ridge is much flatter than that of our predecessors, and we have squarer eye sockets, smaller faces and reduced canine teeth (Figure 14.1.5).

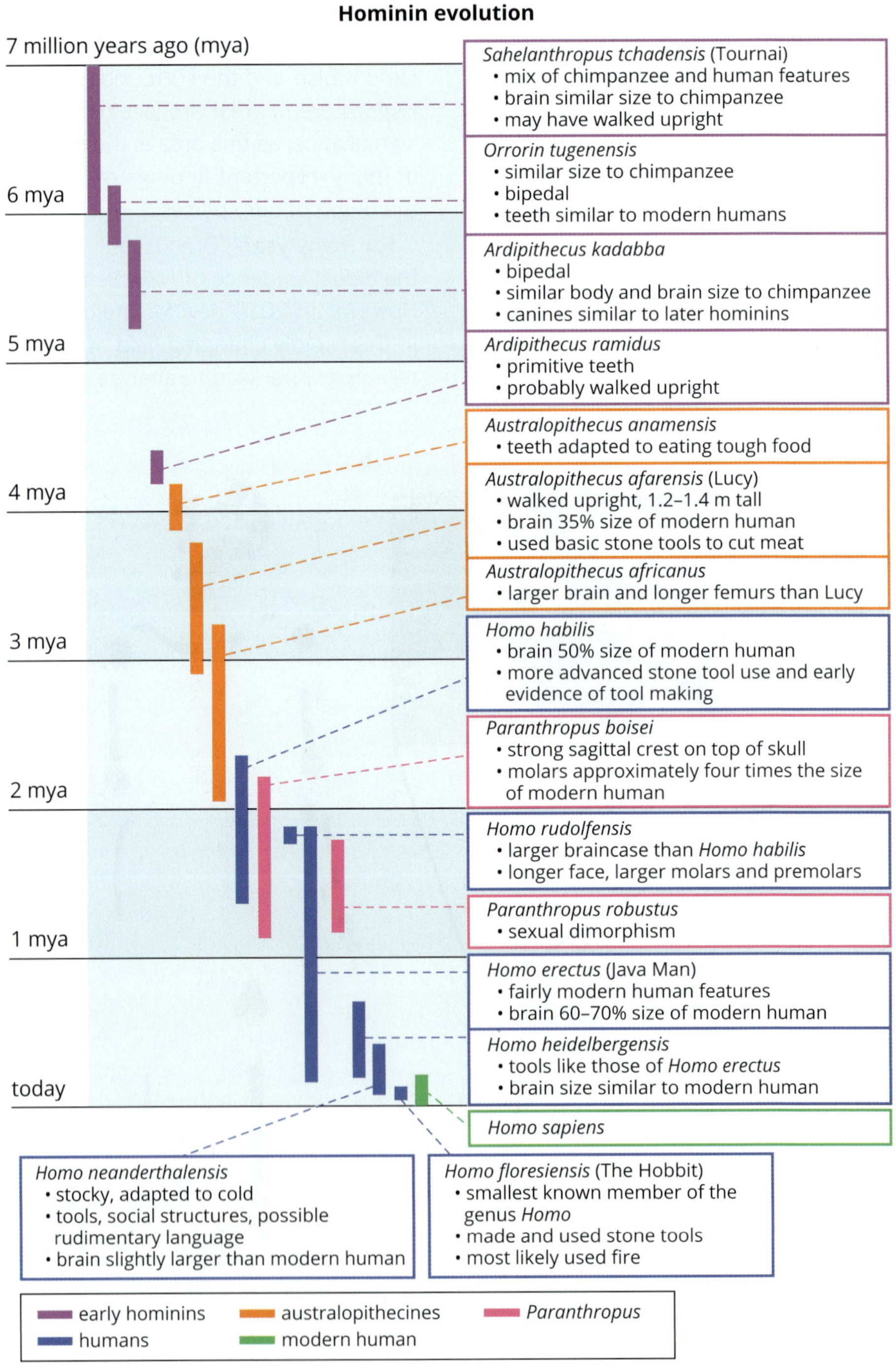

FIGURE 14.1.5 Evolution of the physical and behavioural characteristics of hominin species over the last 7 million years

H. sapiens have a complex larynx, which has allowed the development of remarkably complex languages. Our hands have a well-developed precision grip, which permits fine motor control rather than simply having opposable thumbs. The limbs of *H. sapiens* show further development for bipedal locomotion and a reduction in traits suited for arboreal living compared to our earlier ancestors who, while bipedal, also spent more time in trees (Figure 14.1.6).

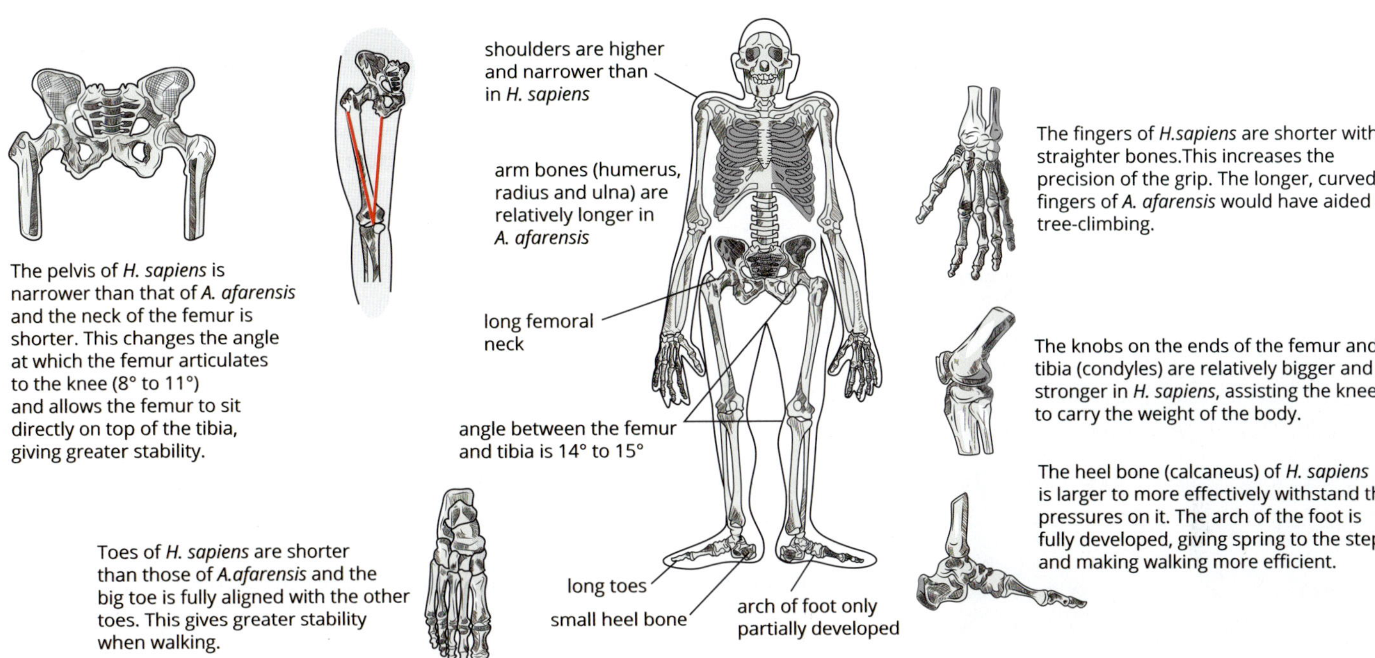

FIGURE 14.1.6 Limb development from *Australopithecus* to *Homo sapiens*

Over time, the relative lengths of the limbs also showed change, with the arms becoming relatively shorter and the legs relatively longer. The increased length of the leg bones resulted in an overall increase in height. *Australopithecus afarensis* had an average height of 1.5 m for males and 1.05 m for females while modern humans have an average height of 1.7 m.

Brain size

H. sapiens has one of the largest, most complex brains of all the hominin species, allowing for the development of sophisticated tools, innovations, social structures and cultures. There has been a trend throughout hominin evolution for larger braincases, with brain size almost tripling in the last two million years. The braincase capacity of *H. habilis* was approximately 600 cm^3, while *Homo neanderthalensis* had a braincase capacity of 1600 cm^3, the largest known brain size of all hominins. The average braincase capacity of present-day humans is about 1100–1300 cm^3, the second largest of the hominins. Although the braincase capacity of *H. neanderthalensis* was larger than that of *H. sapiens*, researchers have suggested that Neanderthal brains were specialised for vision and movement, with less brain volume devoted to social interactions and problem-solving.

The complex brains of *H. sapiens* enable much higher levels of communication and interaction between individuals. Our brains also allow for a much better understanding of, and consequent manipulation of, the living and non-living components of our environment compared to any other species. Complex behavioural traits are characteristics that distinguish modern *H. sapiens* from earlier populations. The behaviour of modern *H. sapiens* is characterised by the ability to plan, use abstract thinking, ritual and symbolism (e.g. in art, ornamentation and music), capture large prey and make advanced use of tools. The behavioural complexity of *H. sapiens* led to the development of diverse cultures around the world.

GENUS *HOMO*

Fossils that have been classified as belonging to the genus *Homo*, the genus to which we belong, are **gracile**; that is, they have a relatively lightweight skeleton and a more upright bipedal stance (slender build). *Homo* fossils all have a relatively large braincase and reduced jaw size.

There are currently 10–16 species that are recognised within the genus *Homo*; the reason for this range is that some are thought to be **subspecies** or not distinct enough to warrant recognition as separate species. The number and evolutionary relationships of species changes as new evidence from fossils and DNA is uncovered.

Fossil and DNA evidence indicate that different *Homo* species existed at the same time in the same regions. There is evidence to suggest that some species interbred and gave rise to hybrids.

Homo floresiensis

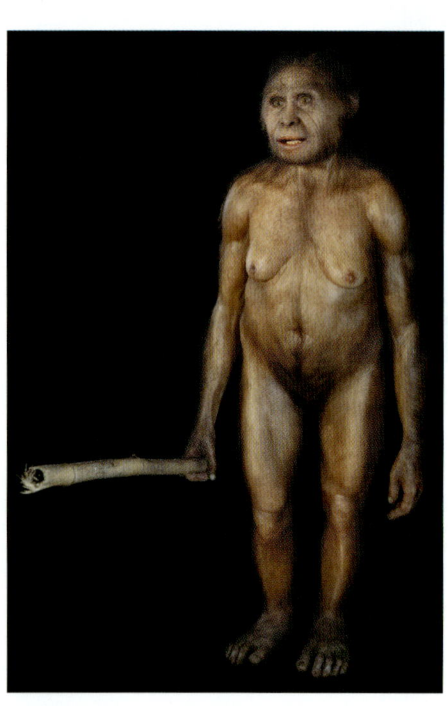

FIGURE 14.1.7 Reconstruction of *H. floresiensis* based on skeletal remains found in Indonesia in 2003

Time range: 17 000–95 000 years ago

Geographic range: Indonesia

Nicknamed 'the hobbit', *Homo floresiensis* is one of the most recent additions to the human family tree. It **coexisted** with *H. sapiens* throughout its entire known time range and was the most recent *Homo* species to become extinct. Discovered in Indonesia on the island of Flores in 2003, *H. floresiensis* was the smallest known member of the genus *Homo* (Figure 14.1.7). With a stature of just over 1 m tall, and a small brain, oversized teeth and feet, this human relative may have suffered from 'island dwarfism'—an evolutionary process that results from isolation on an island with limited resources and environmental selection pressures from predation.

H. floresiensis made and used stone tools, successfully hunted pygmy elephants, coped with giant Komodo dragons, and most likely used fire.

Homo denisova

Time range: 41 000–125 000 years ago (still unconfirmed)

Geographic range: Russia to South-East Asia

Homo denisova (Denisovan or Denisova hominin) was first identified from a finger bone and two teeth that were discovered in Denisova cave in Siberia, Russia. The finger bone that was discovered belonged to a young female who lived approximately 30 000–50 000 years ago. The bone was broad and robust, suggesting that the species had a very robust build, possibly similar to that of the Neanderthals.

DNA from the fossils was well preserved due to the cold, dry conditions of the cave in which the bones and teeth were found. Analysis of mitochondrial DNA (mtDNA) from the finger bone revealed that Denisovans were closely related to Neanderthals and modern humans but genetically distinct.

Fossils from Neanderthals and modern humans have also been found in Denisova cave and evidence from molecular studies suggests that **interbreeding** occurred between Denisovans and both these species. Mitochondrial DNA shows that the Denisovans may have interbred with Neanderthals, with 17% similarity between Denisovan and Neanderthal genomes. Nuclear DNA studies have also shown that 3–5% of the DNA of present-day modern humans from Melanesian and Aboriginal Australian populations is shared with Denisovans, indicating that Denisovans also interbred with modern humans (Figure 14.1.8).

With such little evidence available, much about this species is still yet to be discovered.

> Evidence from DNA and fossils has shown that *Homo sapiens*, *H. neanderthalensis* and *H. denisova* all coexisted and interbred, leaving genetic evidence of these extinct species in the DNA of modern humans.

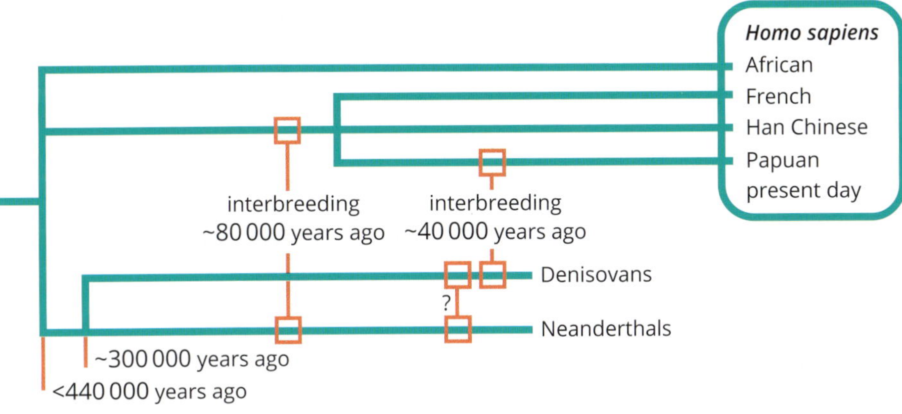

FIGURE 14.1.8 Evolutionary relationship of Denisovans with *H. sapiens* and Neanderthals

Homo neanderthalensis

Time range: 40 000–400 000 years ago

Geographic range: Europe and south-western to central Asia

H. neanderthalensis (Neanderthal) is our closest extinct human relative (Figure 14.1.9). Neanderthals differed from *H. sapiens* in a variety of ways: they had large faces with angled cheek bones, large noses for coping with cold, dry air, and chunkier, shorter builds suited to colder climates.

Neanderthals did not use complex tools but they did use fire and lived with family groups in shelters. It is thought that they had complex social structures and were possibly the first human species to have language. Neanderthals were the first species to wear clothes and jewellery, have burial rituals and display symbolic behaviour. Their brains were just as big as ours, relative to body size, and sometimes even larger.

Modern humans share approximately 1–4% of chromosomal DNA with Neanderthals, suggesting that interbreeding occurred between these two species after modern humans left Africa.

FIGURE 14.1.9 Reconstruction of a Neanderthal family based on fossil evidence. It is thought that Neanderthals lived in small family groups and had complex social structures and language.

CASE STUDY **ANALYSIS**

Homo naledi

In 2013, a discovery of ancient skeletons in a cave in South Africa changed the story of human evolution and gave rise to a new species of hominin: *Homo naledi*. Partial skeletons of 15 specimens were found in the Rising Star Cave near Johannesburg, South Africa. *H. naledi* has similar morphological features to modern humans, but also has many primitive features such as a small skull, and hands and arms specialised for climbing (Figure 14.1.10). Researchers initially had difficulty dating the remains, as the bones of *H. naledi* were found scattered on the floor of the cave and buried in shallow sediment. Based on some of the primitive physical features, several anthropologists believed that the remains could be 2 million years old.

In 2017, researchers used a variety of advanced dating methods to determine that the remains were approximately 250 000 years old, far younger than initially thought. This dating, however, supports the modern physical traits present in *H. naledi*, as well as the body positions. The odd positioning of the bones, which were piled together on the floor of the remote cave chamber, has led researchers to believe that *H. naledi* may have deliberately placed the bodies there, using the cave as a burial chamber. As this complex behaviour evolved much later in *Homo* evolution, it supports the research suggesting an age of 250 000 years old rather than 2 million years old.

The discovery of *H. naledi* represents one of the greatest palaeoarchaeological finds in history, adding pieces to the puzzle of human evolution while raising even more questions that we are yet to answer.

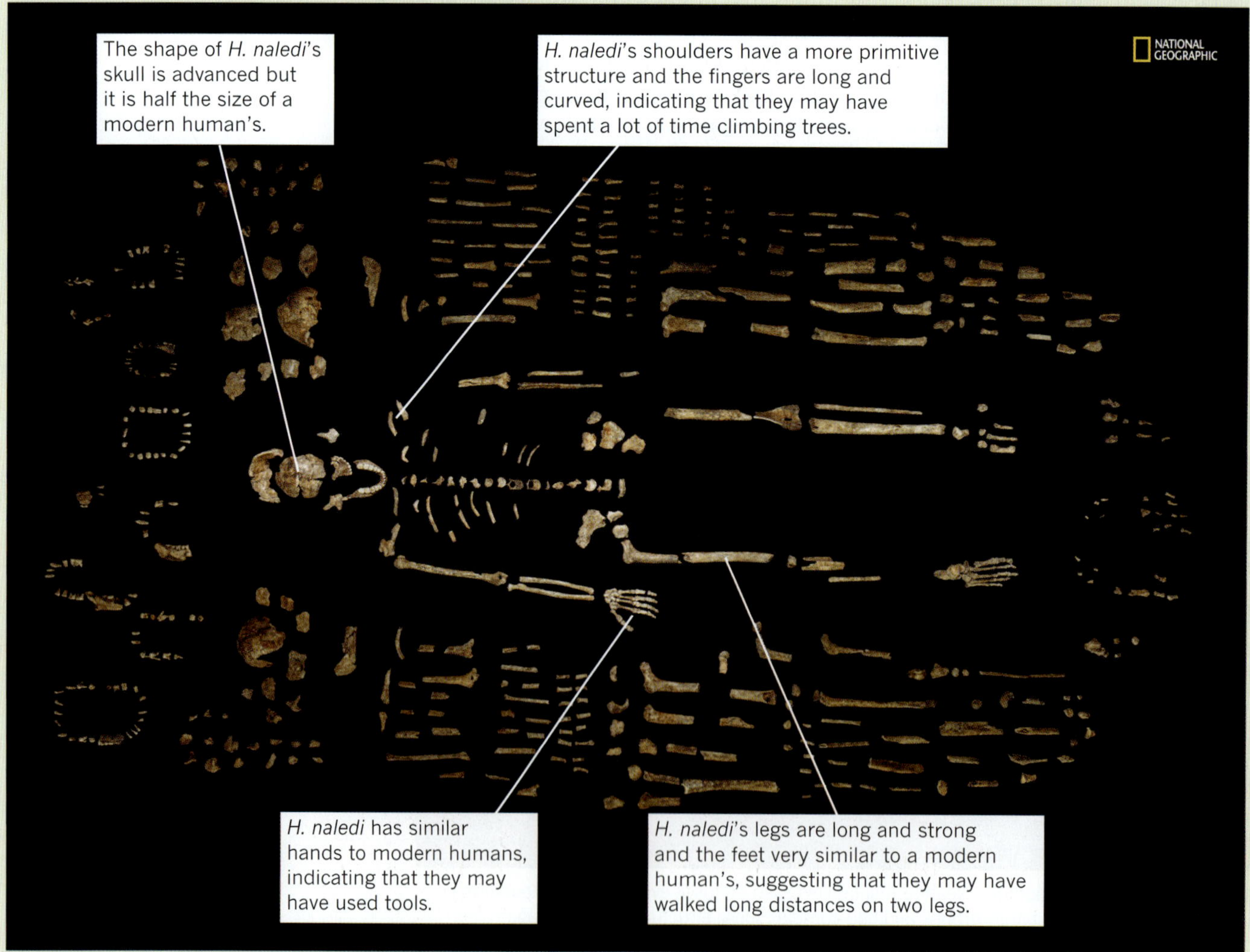

FIGURE 14.1.10 Partial skeleton of the recently discovered hominin species, *H. naledi*. The specimen was found in a South African cave in 2013.

Analysis

H. naledi has been dated at 250 000 years old. It was originally believed to be much older, with some scientists suggesting it was 2 million years old.

1 Mark both of these dates on the hominin timeline below.

2 Which hominins coexisted with *H. naledi* 250 000 years ago?

3 Which hominin species would have coexisted with *H. naledi* if they were alive 2 million years ago, as initially thought?

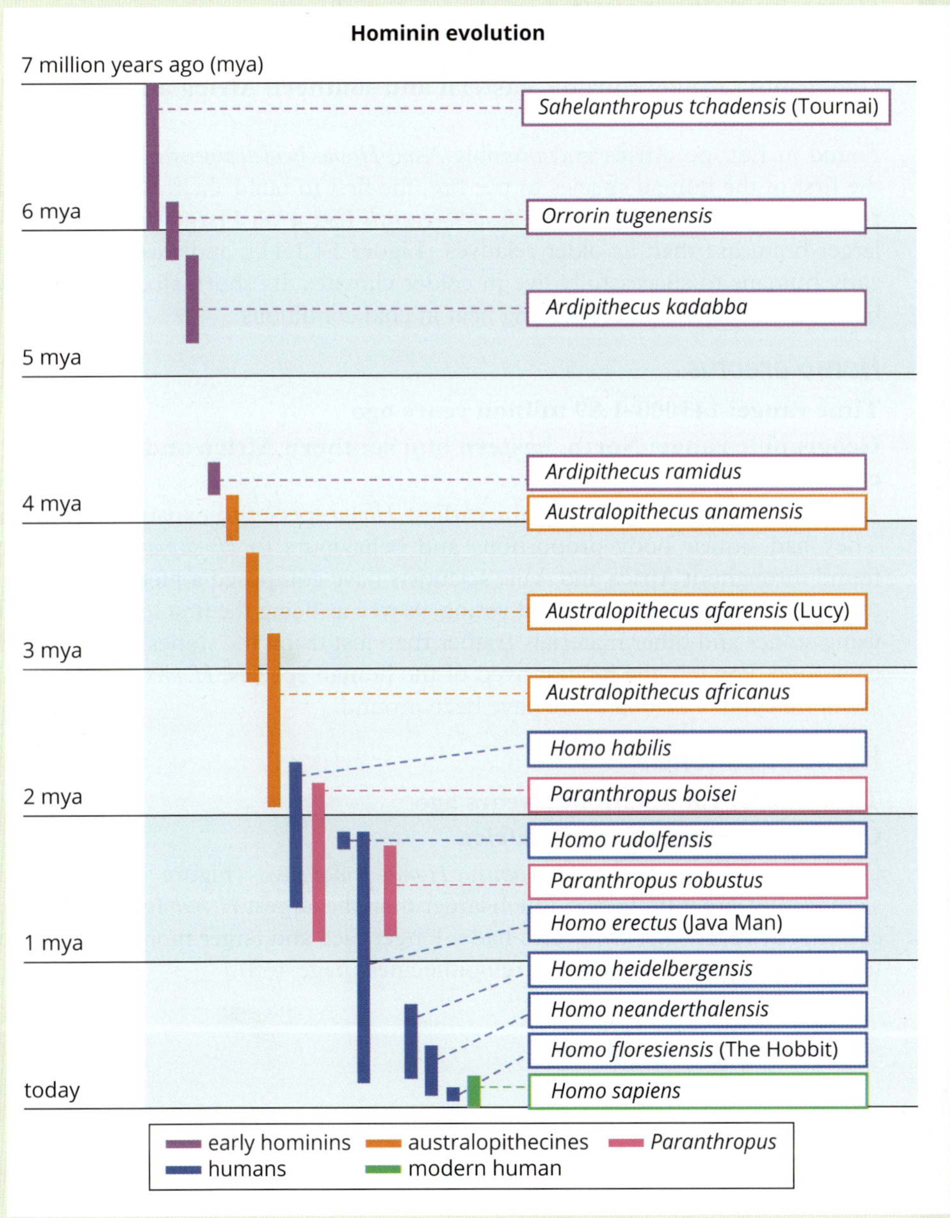

Homo luzonensis

Time range: 50 000–80 000 years ago

Geographic range: Philippines

In 2007 in the Callao Cave in the Philippines, a team of researchers uncovered a nearly complete foot bone that appeared to be human in among some animal bones. Initially it was believed that the bone belonged to a small *H. sapiens*, however more evidence was required. By 2015, 12 more bones had been discovered (two more toe bones, seven teeth, two finger bones and part of a femur) and the species name *Homo luzonensis* had been assigned. It is believed the remains uncovered so far come from at least three individuals. While little is known about this newly discovered *Homo* species, including whether there were possible interactions with *H. floresiensis* or *H. denisova*, there was evidence in the Callao cave to suggest that they were skilled toolmakers and hunters.

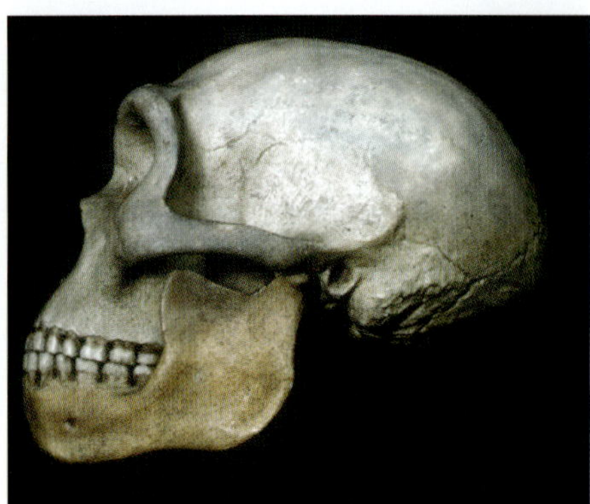

FIGURE 14.1.11 Skull of *H. heidelbergensis*, displaying its characteristic prominent brow ridge

Homo heidelbergensis

Time range: 200 000–700 000 years ago

Geographic range: Europe, eastern and southern Africa and possibly China

Found in Europe, Africa and possibly Asia, *Homo heidelbergensis* is thought to be the first of the human species to use fire, the first to build shelters and the first to routinely hunt large animals. *H. heidelbergensis* had a very large brow ridge and a larger braincase than its older relatives (Figure 14.1.11), and was the first of the early humans to successfully live in colder climates. Its short, stocky build would have been beneficial in conserving heat in cold conditions.

FIGURE 14.1.12 Model of *H. erectus* based on fossil evidence

Homo erectus

Time range: 143 000–1.89 million years ago

Geographic range: North, eastern and southern Africa and west and east Asia

H. erectus (Figure 14.1.12) was the earliest *Homo* species to expand out of Africa. They had similar body proportions and behaviours to *H. sapiens*. From fossils found throughout Africa and Asia, scientists have composed a picture of a species that looked after the sick, old and young, as well as being the first to construct tools using stones and other materials (rather than just using the stones as tools) and to cook food. Possibly the longest lived of the human species, *H. erectus* survived for about nine times as long as we have been around.

FIGURE 14.1.13 Reconstruction of *H. rudolfensis* from fossil evidence

Homo rudolfensis

Time range: 1.8–1.9 million years ago

Geographic range: Eastern Africa

Originally thought to be *H. habilis*, *Homo rudolfensis* (Figure 14.1.13) had a significantly larger braincase, much larger than the largest *H. habilis* skull. Found in eastern Africa, *H. rudolfensis* also had a longer face and larger molar and premolar teeth, and was more like the australopithecines (page 468).

Homo habilis

Time range: 1.4–2.4 million years ago

Geographic range: Eastern and southern Africa

Homo habilis was found only in Africa, and is the earliest known *Homo* species. It had a much larger braincase than its predecessors and has been associated with some of the earliest tools. *H. habilis* was so-named to represent what was thought to be the first evidence of making stone tools. Although the name stands, the role of 'handy man' has since been taken by an earlier species. *H. habilis* commonly used tools to break open bones and feed on the bone marrow. *H. habilis* had more ape-like features than later species, with longer arms, and jaws that projected forwards.

GENUS *PARANTHROPUS*

Time range: 1.2–2.7 million years ago

Geographic range: South-eastern Africa

Fossils of the genus *Paranthropus* are found in south-eastern Africa and are characterised by their large teeth and powerful jaws. They were bipedal and stood at around 1.3–1.4 m tall with a muscular build. *Paranthropus* species were prevalent at the time when some species of the *Homo* genus existed, but it is thought that *Paranthropus* species were more specialised and less adaptable than *Homo* species. This lack of adaptability in changing environments may have led to their extinction.

- The face shape of *Paranthropus robustus* (~1.2–1.8 million years ago) was wide and angular to accommodate the powerful jaw muscles. They also demonstrated significant sexual dimorphism, with males being taller and heavier than females.
- *Paranthropus boisei* (~1.2–2.3 million years ago) had an even wider face than *P. robustus*, as well as a strong sagittal crest on top of the skull (Figure 14.1.14). This species had very large molars, approximately four times the size of a modern human's, with the thickest dental enamel of any known early human.
- *Paranthropus aethiopicus* (~2.3–2.7 million years ago) possessed similar features for chewing, but few fossils have been found to ascertain other characteristics.

In this context, the symbol ~ means approximately.

The sagittal crest is a ridge of bone running lengthwise along the midline of the top of the skull. The presence of this ridge of bone indicates that the jaw muscles are exceptionally strong.

FIGURE 14.1.14 Digital reconstruction of *P. boisei* from skull specimens

FIGURE 14.1.15 Reconstruction of 'Lucy', *Australopithecus afarensis*

FIGURE 14.1.16 Reconstruction of *Sahelanthropus tchadensis*, the oldest known fossil leading to the lineage of *Homo*

GENUS *AUSTRALOPITHECUS*

Time range: 2–4.2 million years ago

Geographic range: Eastern and southern Africa

Australopithecine skeletons, all found in eastern Africa, show that members of this genus walked upright on a regular basis but still climbed trees. The genus *Homo* is related to *Australopithecus*.

- *A. afarensis* (~2.95–3.85 million years ago) is better known as 'Lucy's species' due to the key fossil named Lucy, found in Hadar, eastern Africa (Figure 14.1.15). More than 300 individuals have been identified from fossils, making it the best-known of the early hominin species, as well as being one of the longest-lived, having survived more than 900 000 years. *A. afarensis* had both ape and human features: a flat nose, a projecting chin, a small braincase, long arms and long, curved fingers adapted for climbing trees.
- *Australopithecus garhi* (~2.5 million years ago) and *A. africanus* (~2.1–3.3 million years ago) had larger brains than Lucy and longer femurs, suggesting they took longer strides when walking upright.
- *Australopithecus anamensis* (~3.9–4.2 million years ago) was also similar to Lucy, but was likely the size of a chimpanzee.

BEFORE *AUSTRALOPITHECUS*

The earliest hominins predating the australopithecines were significant in standing upright (bipedalism). Bipedalism is thought to be most likely a development related to a savannah environment in Africa that emerged due to drier, warmer climates. The large, open areas of the savannah favoured those who could move quickly over the land, see above the tall grasses, and were not reliant on trees.

- *Ardipithecus ramidus* (~4.4 million years ago) demonstrated features of bipedalism and tree climbing, with little sexual dimorphism evident from the teeth.
- *Ardipithecus kadabba* (~5.2–5.8 million years ago) is known from only a very small sample of skull, tooth and other partial skeletal remains. These fossils are able to provide a lot of information, demonstrating bipedalism, body and brain sizes similar to those of chimpanzees, and canines similar to those of later hominins.
- *Orrorin tugenensis* (~5.8–6.2 million years ago) was also similar in size to chimpanzees, was able to walk on two legs (but it is unknown how often it did this), and possessed small teeth with thick enamel, similar to those of modern humans.
- *Sahelanthropus tchadensis* (~6–7 million years ago) is the oldest fossil species of hominins to be considered ancestral to the lineage that eventually led to *Homo* (Figure 14.1.16). Discovered in west-central Africa, this species is the oldest to possess a foramen magnum (opening for the spinal cord through the skull) that is located further forwards than in apes or any other primates except humans. This position of the foramen magnum demonstrates an upright stance that is indicative of time spent on two legs. All this is known from just nine cranial fossils.

14.1 Review

OA ✓✓

SUMMARY

- Modern humans (*Homo sapiens*) are the only living species belonging to the genus *Homo*.
- Humans belong to the tribe Hominini and are called hominins. Hominins include modern humans, extinct species of *Homo* and our bipedal ancestors (e.g. *Australopithecus*, *Paranthropus* and *Ardipithecus*).
- Modern humans are also part of the family Hominidae, called hominids, which includes the great apes. The superfamily Hominoidea includes humans, the great apes and the lesser apes, and its members are all called hominoids.
- Our closest living relatives are chimpanzees and bonobos, but we did not directly evolve from them or any other living primate. Our last shared common ancestor existed around 6–8 million years ago.
- The oldest fossil of anatomically modern humans have been dated to 195 000 years ago (Omo Kibish) and 210 000 years ago (southern Greece).
- There has been a trend throughout hominin evolution for larger braincases, with brain size tripling in the last two million years. While *H. neanderthalensis* had a slightly larger cranial capacity than *H. sapiens*, it is thought their brains were more specialised for vision and movement.
- Limbs continued to develop as hominins became more highly adapted to bipedal locomotion.
- Our closest extinct relative is *H. neanderthalensis*. *H. sapiens* coexisted with Neanderthals and there is evidence of interbreeding between the two species. DNA and fossil evidence also suggests that both these species coexisted and interbred with *H. denisova*.

KEY QUESTIONS

Knowledge and understanding

1 Use the terms below to show how *H. sapiens* is classified under the Linnaean system of classification.

Chordata *sapiens* Hominidae Animalia
Primates Eukarya Mammalia *Homo*

Domain: ______

Kingdom: ______

Phylum: ______

Class: ______

Order: ______

Family: ______

Genus: ______

Species: ______

2 When did humans last share a common ancestor with chimpanzees?

3 Which two hominin species were thought to have interbred with *H. sapiens*?

A *H. habilis* and *H. denisova*
B *H. neanderthalensis* and *H. denisova*
C *H. denisova* and *H. heidelbergensis*
D *H. floresiensis* and *H. habilis*

Analysis

4 Add the following terms to the diagram below to accurately show the relationship between hominins, hominoids and hominids:

bonobos	Hominoidea	lesser apes
hominids	*Homo* species	gorillas
chimpanzees	Hominini	orangutans
Hominidae	hominoids	hominins
Australopithecus	*Paranthropus*	*Ardipithecus*

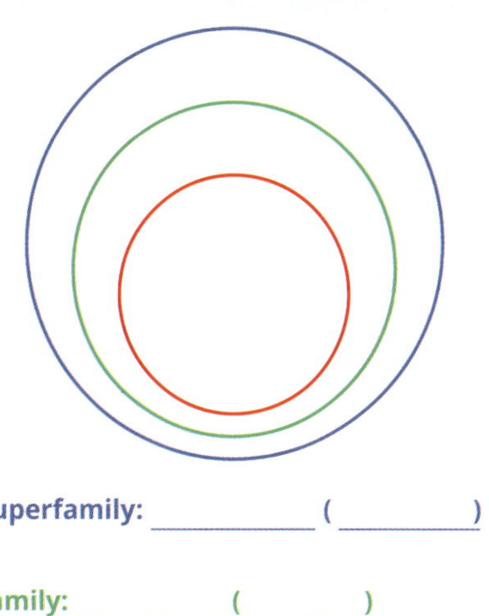

superfamily: ______ (______)

family: ______ (______)

tribe: ______ (______)

14.2 Evidence of human evolution and migration

Our knowledge of the human fossil record and human evolution are changing as new information and evidence comes to light.

Putative evidence is evidence that is generally considered or reputed to be accurate.

The human fossil record is an example of a classification scheme that is open to interpretations that are contested, refined or replaced when new evidence challenges them or when a new model has greater explanatory power. This has occurred numerous times, sometimes with little impact and at other times completely overturning our idea of our ancestry. With new discoveries of fossils and **archaeological** artefacts, along with rapid advances in DNA technology and bioinformatics, our family tree is being discovered in more detail (Figure 14.2.1), and our understanding of our relationships continues to grow. Most recently, evidence has been discovered that supports the theory that *H. sapiens* and *H. neanderthalensis* interbred. A new fossil species of *Homo*, *H. floresiensis*, was discovered in 2004 in Indonesia; another, *H. denisova*, was discovered in Russia in 2008. In 2015 palaeoanthropologists described yet another member of the genus *Homo*: *H. naledi* from South Africa (see Section 14.1). **Palaeoanthropology** is a branch of anthropology that involves the study of fossil hominins, and contributes to our knowledge of human evolution.

FIGURE 14.2.1 Models of our hominin ancestors based on reconstructions of fossils. Back row, from left to right: *Homo ergaster*, two male Neanderthals (*H. neanderthalensis*) carrying dead animals with a Neanderthal female and child in between, a Cro-Magnon (*H. sapiens*) hunter throwing a spear; middle row, left to right: a female *H. habilis*, a male *Homo georgicus* wearing fur, *A. africanus*, *P. boisei*, *S. tchadensis*; front row, left to right: *H. georgicus* (female with throwing stone), Lucy and Lucien (*A. afarensis*)

CULTURAL EVOLUTION

One of the world's foremost evolutionary biologists, Richard Dawkins, famously declared, 'What lies at the heart of every living thing is not a fire, not warm breath, not a "spark of life". It is information, words, instructions.' What sets modern humans apart from all other species is our cultural evolution.

Culture is the accumulated knowledge that is passed on from generation to generation and evolves with changing environments, information and time.

Culture is the accumulated knowledge passed on to the next generation by verbal, written or symbolic communication. A very important step for the human species was the development of language and the ability to record information. The study of the relationships of different languages, together with information from skeletal remains, artefacts and DNA, provides evidence of the evolution of the different cultures of *H. sapiens* living today in different geographic regions.

As humans became increasingly able to communicate their discoveries from one generation to the next, their culture also changed over time. This is called cultural evolution. As culture is passed on, it is modified by each generation and is subject to influences such as new discoveries about the environment. Like gene flow, the transmission of cultural information can also be interrupted by geographic barriers: over time cultures that were geographically isolated evolved independently and became increasingly different, resulting in the cultural diversity in *H. sapiens* that we see today.

The most significant time for early cultural development of *H. sapiens* is thought to have occurred approximately 50 000 years ago (although evidence of earlier culture is beginning to emerge). It is around this time that rituals, burials, clothes and more complex hunting techniques become evident in the fossil record (Figure 14.2.2).

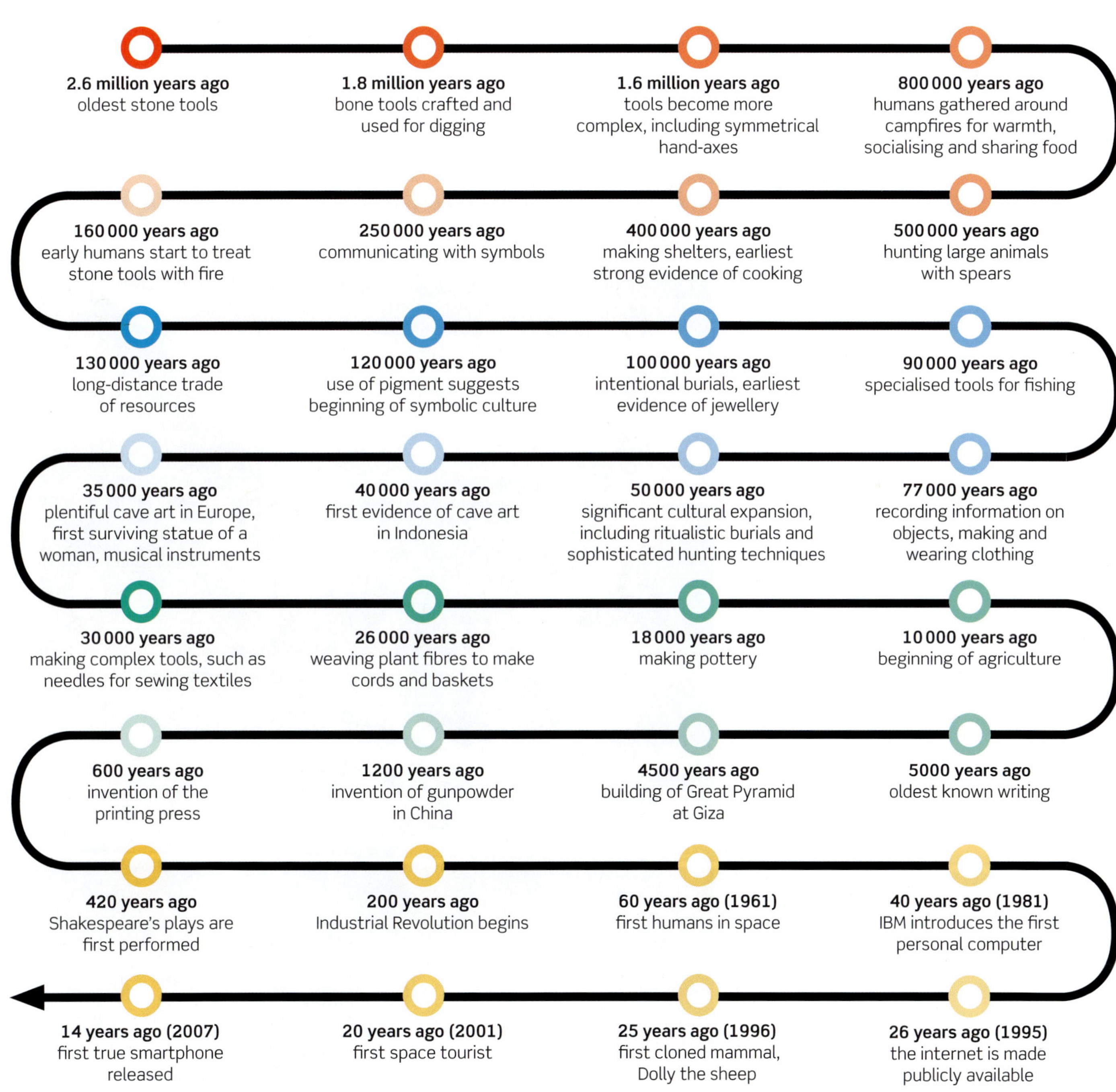

FIGURE 14.2.2 Timeline of the major events in the cultural evolution of *Homo*

CASE STUDY

Lake Mungo

Lake Mungo is a dry lake located in south-western New South Wales, Australia. It is part of a World Heritage–listed region because of the many significant archaeological remains that have been found there. The fossilised remains of three humans have been discovered in the sediment deposits of the lake, dated to between 25 000 and 50 000 years ago. These fossil specimens are known as Mungo Lady (or Lake Mungo 1), Mungo Man (or Lake Mungo 3) (Figure 14.2.3) and Lake Mungo 2.

The remains of Mungo Lady and Mungo Man have been dated to 40 000 years old, which makes them the second oldest anatomically modern humans to be found east of India. There is evidence that Mungo Lady's body was cremated and covered in ochre (mineral pigment), making her one of the earliest examples of cremation and burial rituals in the world. Mungo Man was found with his hands placed in his lap, also indicating ritualistic practices in placing the body for burial.

Lake Mungo is a significant place for Indigenous Australians and the story of the migration of modern humans throughout the world.

FIGURE 14.2.3 Mungo Man is one of the oldest anatomically modern human specimens found in Australia, dated to approximately 40 000 years ago.

Tool use

Tool making was once thought to be a defining characteristic of the genus *Homo*, but basic stone tool use has now been found associated with *Australopithecus*. Australopithecines created some of the earliest tools, using sticks and rocks to help them capture and eat small prey. *H. habilis* made more sophisticated tools, striking flakes off either side of stones. Such tools have been dated as being up to 2.6 million years old. There is evidence of more advanced craftsmanship of stone tools, from flake tools to axes, by *H. erectus* and *H. habilis* approximately 1.6 million years ago.

Throughout evolutionary time, and with the increasing brain size of *Homo* species, tools became more complex and their uses more varied. The earliest evidence of the emergence of modern behaviour and culture in *Homo* is from tools found in Africa from approximately 300 000–400 000 years ago.

Tool use has had an important role in shaping culture in *H. sapiens*. Some of the earliest examples of complex tools that represent modern human culture have been found in Border Cave in South Africa. These tools have been dated to 44 000 years ago and show evidence of complex behaviours and cultural practices, such as using weapons for large prey capture and beads for adornment (Figure 14.2.4). Blades, bone tools, specialised weapons for hunting, and evidence of long-distance travel and trade signify the beginnings of cultural development in humans.

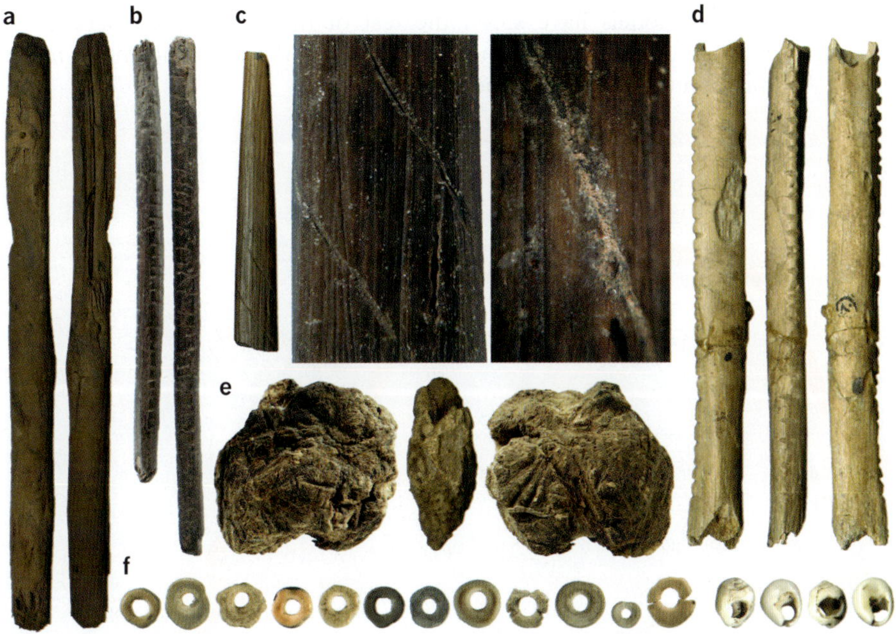

FIGURE 14.2.4 Tools found in Border Cave in South Africa. The artefacts recovered from the cave included (a) digging sticks, (b) poison applicators, (c) an arrow point made of bone, (d) bones with notches carved into them, (e) beeswax mixed with resin and (f) beads made from ostrich eggs and marine shells.

The sophisticated use of tools was crucial to the evolution of human culture for hunting, adornment, ritual, clothes making, agriculture, writing, building and many other important advances. The invention and manipulation of tools by modern humans has resulted in incredible feats in engineering and demonstrates the remarkable ingenuity of our species.

BIOFILE

Cro-Magnon: European early modern humans

Archaeological deposits in Europe from the Upper Palaeolithic period (approximately 10 000 to 40 000 years ago) tell the story of the first early modern *H. sapiens* in Europe. These people are called Cro-Magnon or European early modern humans. Cro-Magnon had a robust build, a short, wide face, a prominent chin and a slightly larger cranial capacity than people of today. Skeletons have been found with evidence of long-term injuries, indicating that Cro-Magnon may have cared for sick and injured group members. It is thought that Cro-Magnon may have constructed semi-permanent dwellings, with evidence from a series of huts made from mammoth bones that date back approximately 15 000 years in a Ukrainian village. Evidence of their culture has been found in sophisticated tools, jewellery, clothing, cave paintings and the use of ochre for colouring objects.

Reconstruction of a Cro-Magnon man painting on a cave wall

THE ORIGIN OF MODERN HUMANS

H. sapiens have spread throughout the world, but their origins are still debated. Palaeoanthropologists conclude that *H. sapiens* first evolved in Africa and that much of human evolution occurred on that continent, given the evidence of fossil hominins in Africa from as far back as 6 million years ago. Advances in the analysis of fossil and DNA evidence (including mtDNA and whole genome sequencing) has enabled hypotheses about the origins of humans to be tested directly. This has facilitated numerous breakthroughs in our understanding of human evolutionary history. Our interpretation of the evolutionary history and adaptation of humans is continually being transformed and revised by analyses of this new data.

Scientists currently agree that there are 10–16 or more fossil species of *Homo*, but there is disagreement surrounding how, when and where each species evolved, how they are related and whether they died out or evolved into a new species. Fossil evidence puts *Homo* in Africa from 2.5 million years ago; but some populations of *H. erectus* migrated northwards, and arrived in Asia and South-East Asia 1.8–2 million years ago, in Europe 1–1.5 million years ago and other parts of the world much later. Our species, *H. sapiens*, is thought to have left Africa between 60 000 and 125 000 years ago, arriving in Europe and Asia 35 000–45 000 years ago, Australia 50 000–65 000 years ago and the Americas 12 000–20 000 years ago (Figure 14.2.5), but exact migration dates are speculative. Of the theories that abound, three models have stood the test of time and much debate: the Multiregional evolution (continuity) model, the Out of Africa (replacement) model and the Assimilation (partial replacement) model.

FIGURE 14.2.5 Journey of modern humans throughout the world. Fossil and DNA evidence supports the theory that all modern humans originated in Africa and began to migrate to other parts of the world between 60 000 and 125 000 years ago.

Multiregional evolution (continuity) model

Several scenarios have been proposed that include the interbreeding of *Homo sapiens* with **archaic** human species in Africa prior to migration. The **Multiregional evolution (continuity) model** proposes significant migration of *H. erectus* across Africa, Asia and Europe for the last 1.8 million years. It is argued that isolation of the populations resulted in the divergence of gene pools, traits and behaviour, but occasional contact ensured some gene flow was maintained and led to concurrent evolution of all groups as one recognisable species, *H. sapiens* (Figure 14.2.6a). There is speculation that interbreeding could have occurred on any number of occasions but that, over time, genetic drift has led to the DNA of the other species being lost from the genome of *H. sapiens*.

The three dominant theories of present-day human origins all support an African origin but have different views on the pattern of migration and interbreeding that occurred in other parts of the world.

Supporting evidence

- Physical variations in modern *H. sapiens* indicate a relatively long time period since migration from an African ancestor.
- Similarities exist between modern humans and fossils of extinct species found in the same region.
- Living humans show little genetic diversity, consistent with regular gene flow across all regions.

Out of Africa (replacement) model

The **Out of Africa (replacement) model** suggests that after *H. erectus* left Africa, populations became isolated and diverged into different species (e.g. *H. neanderthalensis* and *H. heidelbergensis*). All living modern humans evolved from a single common ancestor in Africa about 200 000 years ago. Migrations of *H. sapiens* from Africa occurred some time between 60 000 and 125 000 years ago, although stone artefacts found in the Arabian Desert suggest it may have been as early as 160 000 years ago.

As modern humans (*H. sapiens*) spread throughout the world, they displaced all other human species (Figure 14.2.6b); *H. heidelbergensis* was displaced in Africa and Europe, *H. erectus* was displaced in Asia and *H. neanderthalensis* was displaced in Europe. Evidence also points towards several migration events, the earliest about 130 000 years ago along the southern coastlines into Australia and Papua New Guinea. There was a later migration about 50 000 years ago along the Nile River Valley northwards into Europe and Asia, and possible additional migration events between and after these events.

The most extreme version of this model suggests competition between *H. sapiens* and other human species without interbreeding, although analysis of the Neanderthal genome in 2010 supports theories that include some interbreeding, at least between *H. neanderthalensis* and *H. sapiens*.

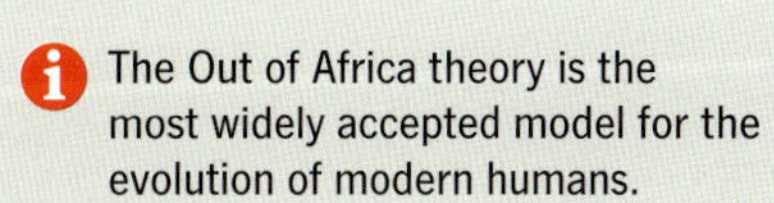

The Out of Africa theory is the most widely accepted model for the evolution of modern humans.

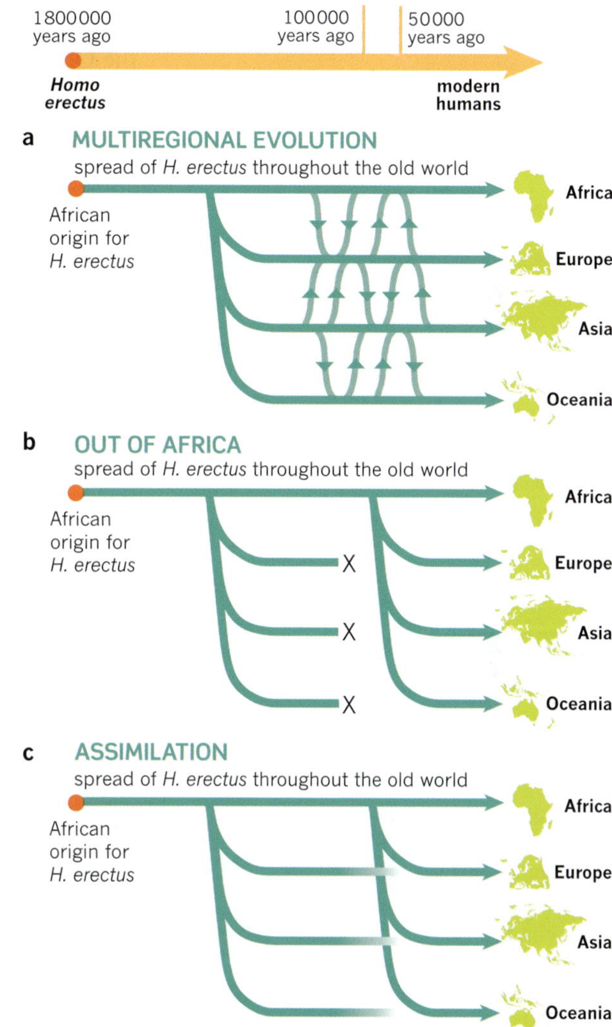

FIGURE 14.2.6 The three dominant theories of the origin of all living humans: (a) the Multiregional evolution model, (b) the Out of Africa model and (c) the Assimilation model. All three theories support an African origin of modern humans but vary in the pattern of migration and interbreeding that occurred in other parts of the world.

Supporting evidence

- The oldest fossil evidence of *H. sapiens* has been found in Africa.
- Living *H. sapiens* show little genetic diversity, suggesting a relatively recent emergence.
- Analysis of the DNA of living humans can be used to map the movement of humans and the approximate timeline on the basis of the estimated rates of mutation. This analysis points to Africa as the point of origin.
- DNA analysis of more than 1000 unrelated humans from different populations shows that present-day humans belong to three genetic groups: Africans, Eurasians (people native to Europe, the Middle East and southwest Asia) and East Asians (people native to Japan, South-East Asia, the Americas and Oceania). Differences between these groups have been found to be mostly due to genetic drift during periods of isolation. The African group has the most genetic diversity, indicating that they are the source population, which supports the Out of Africa theory.

Assimilation (partial replacement) model

The **Assimilation (partial replacement) model** is a newer hypothesis that combines elements of the Out of Africa and Multiregional evolution models. This model proposes that all modern humans had an African origin and when people migrated out of Africa there was occasional interbreeding with archaic humans who were already living in other parts of the world, resulting in hybrid populations (assimilation) (Figure 14.2.6c on page 475).

Supporting evidence

- The abrupt appearance of modern humans in Europe approximately 40 000–45 000 years ago (Cro-Magnon people) suggests that modern humans originated in Africa and then migrated to Europe and Asia.
- DNA and fossil evidence supports the theory of interbreeding between Neanderthals and modern humans. It is proposed that hybrid populations of Neanderthals and Cro-Magnon may have evolved into modern Europeans.
- Skeletal remains that have characteristics of both archaic and modern humans have been found.
- DNA evidence shows that interbreeding between *Homo* species from Asia, Europe and Africa has been occurring over the past 600 000 years.

HUMAN MIGRATION TO AUSTRALIA

There is ongoing debate about how and when humans first migrated to Australia. Ongoing research at the Moyjil site in south-west Victoria suggests human activity in the area 120 000 years ago. Indigenous Australians believe they have inhabited the continent since the beginning of time. However, from a Western scientific perspective, the oldest archaeological evidence places humans in northern Australia approximately 65 000 years ago and genomic analyses of Indigenous Australians indicate that humans have continuously occupied Australia for at least 50 000 years. Developments in genetic technology, dating techniques and new archaeological discoveries continue to improve the Western scientific understanding of the journey of human populations to and across Australia.

BIOFILE

Oldest human outside of Africa

Australian scientists from Griffith University, in conjunction with teams from Greece and Germany, reconstructed and dated a skull, known as Apidima 1, which was found in Apidima Cave in southern Greece in the 1970s. The scientists determined that Apidima 1 was *H. sapiens* and was at least 210 000 years old. The skull was found with stone tools and has a combination of modern human and more primitive *Homo* features. The age of this skull supports the theory that modern humans spread out from Africa on multiple occasions, rather than all at once.

Sahul

At the time of the first migration events, Australia was part of a larger continent known as Sahul. Sahul included what is known today as Australia and New Guinea. Two main migration routes from Asia to Sahul have been proposed by researchers—one via Timor into what is now the north-west coast of Australia, and the other via the west of New Guinea (Figure 14.2.7).

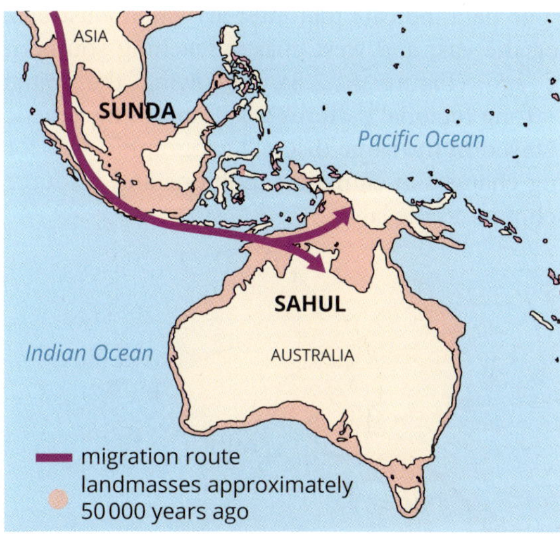

FIGURE 14.2.7 Possible migration routes travelled by humans from South-East Asia to Sahul (present-day Australia and New Guinea). Two likely routes have been proposed: the more southerly route is via Timor and the northerly route is via New Guinea.

Although land bridges appeared between South-East Asia and Sahul during times of low sea levels, voyagers still had to navigate vast stretches of ocean. The migration from Asia to Sahul is evidence of one of the earliest seafaring journeys and is considered one of the most remarkable achievements of early humans.

Migration routes

Researchers have used mathematical modelling to determine the most likely migration route that humans travelled to reach Sahul. Factors such as survival rate, longevity, fertility and climatic conditions were included in the modelling as well as the size of the groups that was required to survive the journeys and establish viable populations. The data suggests that the migration to Sahul was planned and relied on sophisticated watercraft technology and complex knowledge of navigation and open-ocean voyages. The researchers hypothesise that humans first arrived in Sahul by sailing from South-East Asia and island-hopping to western New Guinea (also known as Papua). Humans may have arrived in Sahul in a single, large migration event, consisting of at least 1000 people, or there may have been multiple smaller migration events of around 100 people over approximately 700 years.

Whole genome analyses of Indigenous peoples from Australia and New Guinea shows evidence that their ancestors left Africa approximately 70 000–75 000 years ago, reaching Sahul approximately 50 000 years ago. The genomic data indicates that Indigenous peoples from Australia and New Guinea are closely related and that their populations separated approximately 35 000–40 000 years ago. This data supports the theory of human migration from South-East Asia to New Guinea and into Australia. It is likely that the ancestors of Indigenous Australians first arrived at what is now the north coast of Australia, and then smaller groups migrated to other coastal regions and then inland into the harsher desert climate.

BIOFILE

Songlines

Songlines are travel routes used by Indigenous Australians to describe the environment and teach stories of cultural significance. These will often follow a Creation Story and can span across Australia. Following a Songline and learning the stories associated with it could tell an Indigenous Australian person about the lore of the land, and important cultural teachings.

From a Western scientific perspective, it is believed that the Songlines follow the lines of human migration.

Connection to Country and Place

Researchers investigating the genetic and geographic structure of Indigenous Australian populations have found evidence for a strong connection between Aboriginal populations and discrete geographic areas, supporting the important cultural connection to Country and Place held by Indigenous Australians today. The researchers used whole mitochondrial genomes extracted from historical hair samples of Aboriginal Australians to reconstruct the genetic and geographic history of Indigenous Australians. The data indicate that after arrival in Australia, populations expanded rapidly along the east and west coasts, reaching southern Australia by 45 000–49 000 years ago (Figure 14.2.8). Following the initial population expansion, evidence of strong regional patterns was found. The genetic data shows that small populations stayed in the same discrete geographic area for tens of thousands of years, enduring changes in climatic conditions and resource availability, indicating a strong attachment to the land.

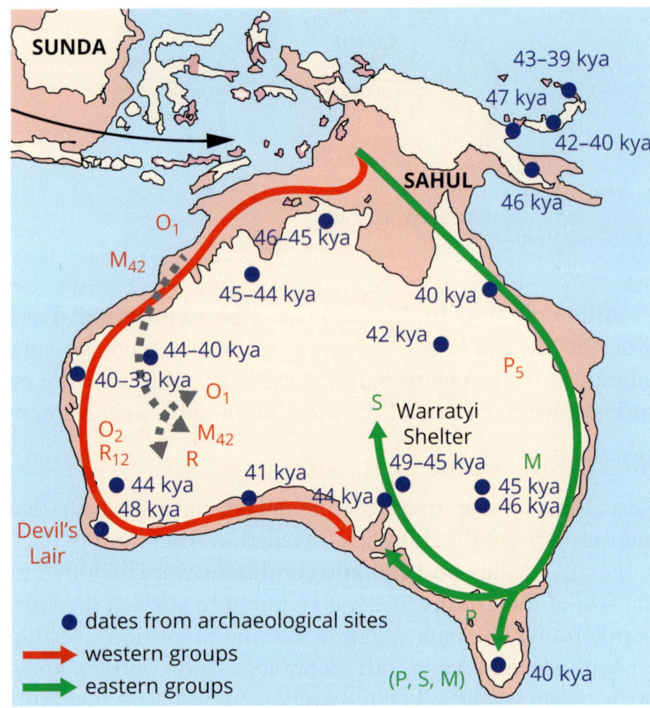

FIGURE 14.2.8 Western scientific representation of the migration and settlement of people throughout Australia following the first arrival of humans over 50 000 years ago. The blue dots indicate dates from archaeological sites and the arrows indicate the movement of genetic groups based on mitochondrial DNA sequences (O and R are western groups; P, S and M are eastern groups).

Evidence from fossils and genetic analyses matches many of the Songlines and traditional stories told by Indigenous Australians. The areas identified by Western science as migration routes and new settlement points are culturally significant for Indigenous Australians, as the relationship between people and Country is powerful. The belief that Indigenous peoples are caregivers of the land, rather than owners, means that all living things and land on Country have value and should be respected. While this does mean that it can be hard to obtain fossil evidence and explore some cultural sites, this connection to Country and Place is valuable and must be respected. Indigenous storytelling provides a rich source of information about Indigenous peoples' histories, reducing the need to disturb important cultural sites.

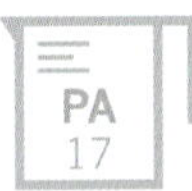

14.2 Review

OA ✓✓

SUMMARY

- The human fossil record is open to interpretations that are contested, refined or replaced when new evidence challenges them or when a new model has greater explanatory power.
- Tools were (and still are) crucial to the evolution of human culture for hunting, adornment, ritual, clothes making, agriculture, writing, building and many other important advances.
- *Homo sapiens* spread to all corners of the Earth, but their origins are still debated.
- There are three main theories to explain the origins of *H. sapiens*:
  - The Out of Africa theory suggests that all living modern humans evolved from a single common ancestor in Africa about 200 000 years ago and spread throughout the rest of the world, replacing other hominin species.
  - The Multiregional theory proposes significant migration of *H. erectus* across Africa, Asia and Europe, with interbreeding between various populations and concurrent evolution of all groups into *H. sapiens*.
  - The Assimilation theory proposes that all living humans had an African origin and migrated out of Africa, occasionally interbreeding with archaic humans, resulting in hybrid populations (assimilation).
- There is debate about when Australia was first inhabited by humans due to differences in fossil and DNA evidence.
  - Fossil evidence indicates that humans first occupied Australia at least 65 000 years ago.
  - DNA evidence indicates that Indigenous Australians have continuously occupied Australia for at least 50 000 years.
- At the time of the first migration events, Australia was part of a larger continent, known as Sahul, which also included New Guinea.
- Evidence suggests the most likely migration route from Asia into present-day Australia was via island-hopping into western New Guinea.
- Migration into Australia may have occurred in a single large event or multiple smaller events.
- Genetic data indicates that Indigenous peoples from Australia and New Guinea are closely related and that their populations separated approximately 35 000–40 000 years ago.
- Connection to Country and Place is an important part of Aboriginal and Torres Strait Islander cultures. Evidence of this connection has been found with strong genetic links to geographic areas.
- Evidence from fossils and genetic analyses matches many of the Songlines and traditional stories told by Indigenous Australians.

KEY QUESTIONS

Knowledge and understanding

1 Give three reasons why our understanding of human evolution might change.

2 Describe how brain size has changed throughout hominin evolution. Include at least three hominin species within your description.

Analysis

3 How did modern humans get to Australia? Outline the evidence of migration routes to Australia.

4 In your own words explain the key differences between the three models of the origins of modern humans. Explain which model you most agree with and why.

Chapter review

OA ✓✓

14

KEY TERMS

archaeological
archaic
Assimilation (partial replacement) model
bipedal
coexisted
foramen magnum
gracile
hominid
Hominidae
hominin
Hominini
hominoid
Hominoidea
Homo
interbreeding
Multiregional evolution (continuity) model
Out of Africa (replacement) model
palaeoanthropology
primate
subspecies

REVIEW QUESTIONS

Knowledge and understanding

1 The family Hominidae includes which groups of primates?
   A modern humans, extinct *Homo* species, *Australopithecus*, *Paranthropus* and *Ardipithecus*.
   B lemurs, gibbons, gorillas, chimpanzees, orangutans and humans
   C orangutans, gorillas, chimpanzees and humans
   D only humans

2 Humans are mammals, specifically hominins, which include primates. List two defining features of:
   a mammals
   b primates
   c hominins

3 State the taxonomic class, order, superfamily, family and tribe that *Homo sapiens* belongs to.

4 **a** Define hominoid.
   b Define hominin.

5 Examine the table below. For each feature of humans, place an X in the relevant box to identify whether the feature is common to all primates, hominoids or is only observed in hominins.

Feature	Primates	Hominoids	Hominins
binocular vision			
foramen magnum in more central position			
opposable thumbs			
S-shaped spine			
teeth of four types			
grasping hands			
skeletal flexibility (e.g. shoulder rotation)			
broad chest			
no tail			
brains larger than 800 cm^3			
flat faces			
tool use			
fingernails and toenails			
large head and short neck to femur			
long period of juvenile development			

6 Describe three significant physical changes *H. sapiens* has undergone throughout evolution.

7 Explain how the trend for increasing cranial capacity in hominins has influenced the evolution of these species.

8 Consider the evolutionary tree below, which includes the genera *Homo*, *Paranthropus*, *Australopithecus* and *Sahelanthropus*.

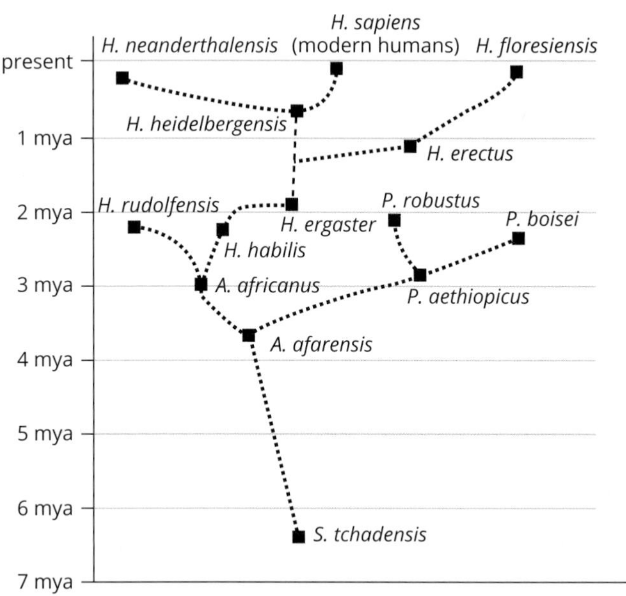

According to the evolutionary tree, which of the following statements is incorrect?

A *H. erectus* gave rise to modern humans.

B *H. ergaster* is a direct descendant of *H. habilis*.

C *A. afarensis* lived approximately 3.7 million years ago.

D The most recent common ancestor of *H. erectus* and *H. heidelbergensis* lived about 1.5 million years ago.

9 The hominin fossil record is fragmentary and confusing and there is a number of species that have been identified from very limited fossil evidence. Such fossils are classified by their structures into the genus that has other better-studied fossils to which they are most similar both in structure and time. Some of the less well-known hominin species are listed. Which of these is likely to be the most ancient?

A *Paranthropus aethiopicus*

B *Australopithecus sediba*

C *Ardipithecus kaddaba*

D *Homo naledi*

10 Sections of DNA from two archaic species of human, Neanderthals and Denisovans, have been sequenced and compared to the genome of modern humans. This research has shown that Neanderthals in Europe and the Middle East and Denisovans in Asia interbred with modern humans.

Modern humans descended from European populations have Neanderthal DNA in their genomes. Only populations descended from native Australians, Melanesians and some native South-East Asian groups, such as the Manobo of the Philippines, have Denisovan DNA.

a There is a significant number of scientists who argue that the classification of Neanderthals and Denisovans should be *Homo sapiens denisova* and *Homo sapiens neanderthalensis*. Why might they be suggesting this?

b Mitochondrial DNA from Neanderthals and modern humans have both been fully sequenced. Modern human mtDNA shows no relationship between Neanderthals and modern humans. How does this support the idea that they are separate species?

11 A group of archaeologists discovered fossilised remains of what appears to be a member of the genus *Homo*. To help them classify the specimen, identify one feature that is unique to each of the following taxonomic groups:

a hominini

b *Homo*

c anatomically modern *Homo sapiens*

12 What are the three main theories of the origin of present-day humans?

13 The diagram below shows one model of human spread.

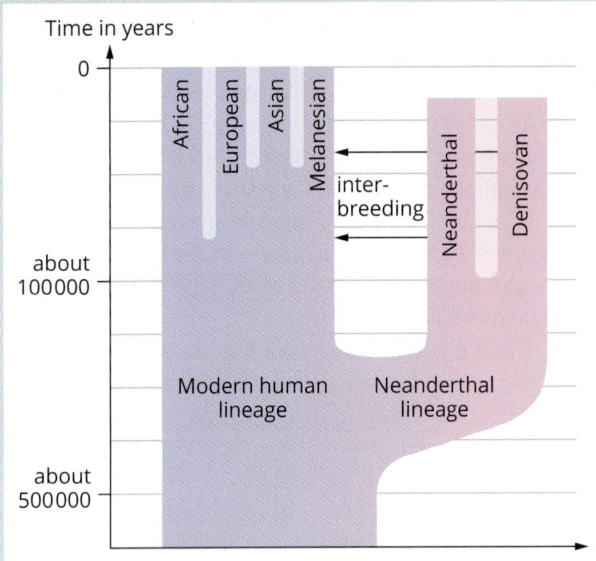

The model represented in this diagram is closest to:

A the Assimilation model

B the Out of Africa model

C the Multiregional evolution model

D none of the above

14 Researchers have used mitochondrial DNA from historical hair samples of Indigenous Australians to reconstruct their genetic and geographic history.

a Describe how this data improves our understanding of the migration of humans across Australia.

b The data from the mitochondrial DNA analysis provides an understanding of human migration across Australia from a Western scientific perspective. How does this understanding align with Aboriginal and Torres Strait Islander peoples' connection to County and Place?

Application and analysis

15 Identify three challenges that researchers face in developing a complete classification scheme for hominins.

16 Look carefully at the three skulls shown below.

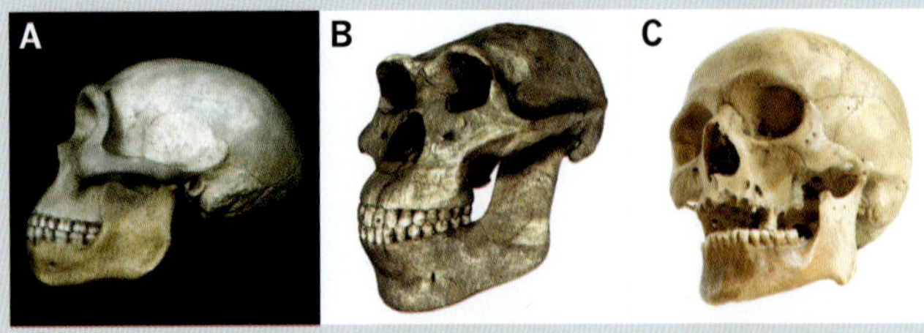

a List the skulls in order from oldest to youngest.

b Describe the features of the skull that support your contention.

17 Bipedalism is one defining feature of modern humans and their ancestors. The fossil record leading to modern humans shows a series of developments in skeletal structure that make bipedal locomotion more efficient. What was the reason for this development?

18 The discovery of well-preserved fossil hominins allows body weight to be estimated. The table below shows the geological dates and the estimated body and brain sizes of eight species of hominin.

Species	Dates (mya)	Body weight (kg)		Brain volume (cm^3)
		Male	Female	
Australopithecus afarensis	2.9–3.8	45	29	384
Australopithecus africanus	2.0–3.0	41	30	420
Paranthropus aethiopicus	2.3–2.7	–	–	399
Paranthropus boisei	1.2–2.3	49	34	488
Paranthropus robustus	1.2–1.8	40	32	502
Homo habilis	1.4–2.4	52	32	597
Homo erectus (early)	1.5–1.8	58	52	804
Homo erectus (late)	0.3–0.5	60	55	980
Homo sapiens	0.0–0.2	58	49	1350

(Source: Adapted from McHenry, H. M. (1994). Tempo and mode in human evolution. *Proceedings of the National Academy of Sciences of the United States of America*, (91): 6780–6786. http://www.pnas.org/content/91/15/6780.full.pdf)

a Compare the body weight of *Australopithecus* with that of *Homo*.

b Based on the results in the table, suggest which species is the most distance ancestor of *H. sapiens*. Give a reason for your answer.

c Suggest one reason why there is no data on body mass for *P. aethiopicus*.

19 Anthropologists have various theories about the evolutionary tree that lead to *H. sapiens*. The two such evolutionary trees shown below include members of the genera *Homo*, *Paranthropus*, *Australopithecus* and *Sahelanthropus*.

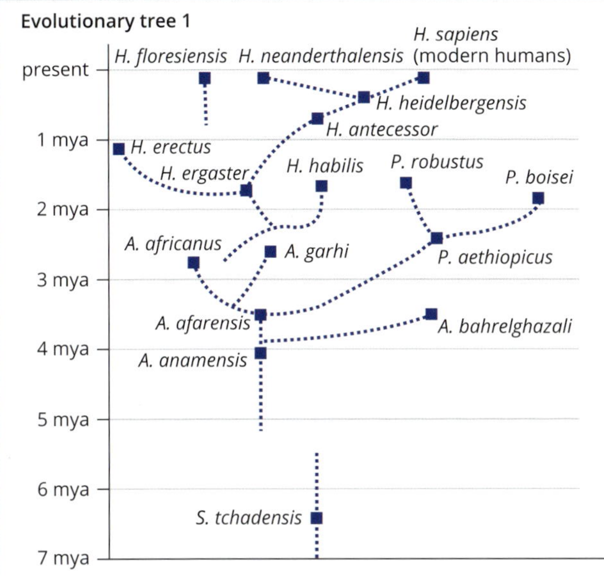

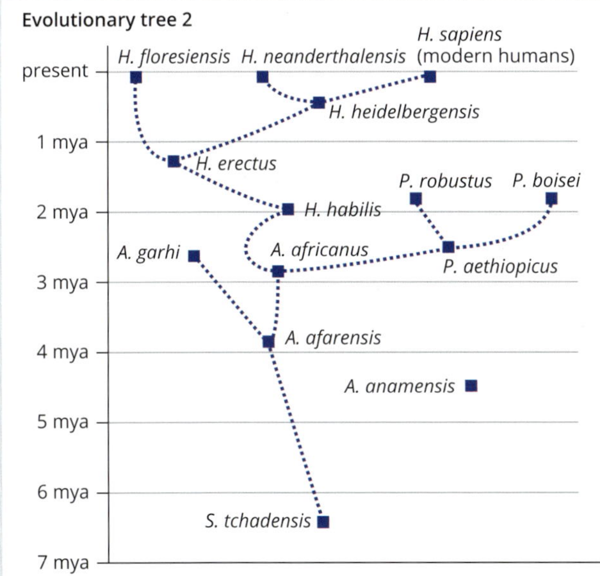

a Identify one point of agreement and one point of difference between the two evolutionary trees.

b Why are anthropologists unable to agree on a single view of human evolution?

20 Ancient fossils are unlikely to contain usable DNA; anthropologists can ony use their morphology to classify them. In 2003 a new fossil hominin was discovered on the island of Flores in Indonesia. One view of this fossil is that it represents a previously undescribed species of hominin that was a direct descendant of *H. erectus*. This new species has been given the name *H. floresiensis*. The skulls of *H. floresiensis* and *H. erectus* are shown below.

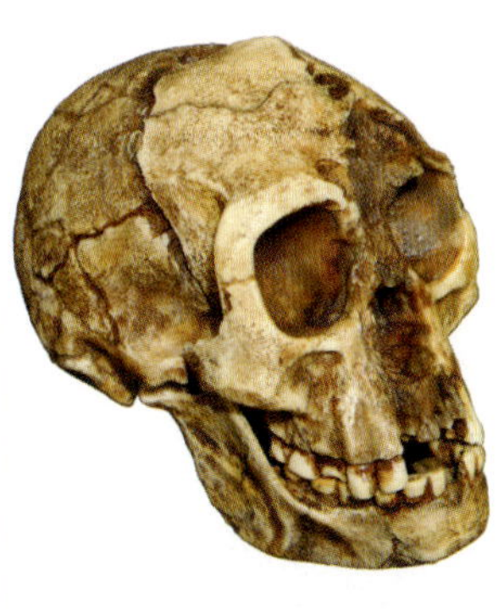
Homo floresiensis

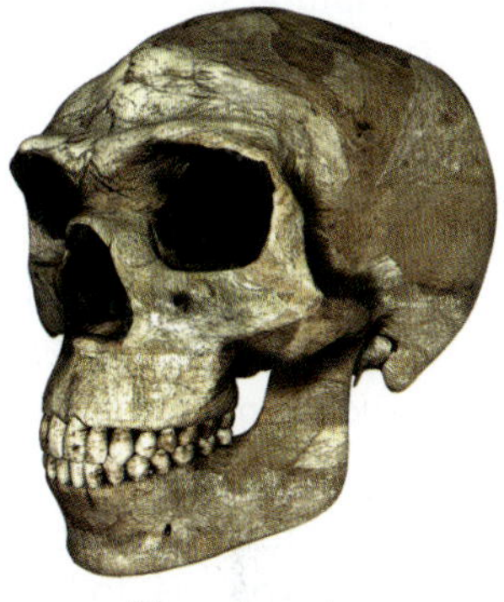
Homo erectus

a i Describe two features of the skulls that indicate that these two species are closely related.

ii Some anthropologists argue that *H. floresiensis* is not a new species but is a variant of *H. sapiens*. Describe one feature of the skull that would support this idea.

b Anthropologists examined the skull of *H. floresiensis* and inferred that it has a significantly smaller brain than *H. sapiens*.

i What is the difference between an observation and an inference?

ii What observations about the skull did the anthropologists make that led them to infer that *H. floresiensis* had a smaller brain than *H. sapiens*?

OA ✓✓

UNIT 4 • Area of Study 2

REVIEW QUESTIONS

How are species related over time?

Multiple-choice questions

1 Which of the following statements most correctly relates evolution and natural selection?

A Natural selection blocks evolution.
B Evolution leads to natural selection.
C Natural selection is a mechanism for evolution.
D Natural selection is the mechanism for evolution.

2 Selective breeding programs for plants often result in polyploidy. Select the situation that would be most likely to result in polyploidy.

A gamete with *n* chromosomes
B gametic cell with 2*n* chromosomes
C pollen or ovules with *n* chromosomes
D somatic cell with 2*n* chromosomes

3 There are two wild species of banana, *Musa acuminata* and *Musa balbisiana*. Today there are many varieties of this popular fruit. These original wild varieties have 22 chromosomes. The most popular variety of bananas grown in Australia today is the Cavendish. It accounts for more than 95% of all production. Cavendish bananas have been around since before 1850. The Cavendish banana has a chromosome count of 33. The development of this variety of banana is most likely due to:

A polyploidy
B hybridisation between *Musa acuminata* and *Musa balbisiana*
C genetic engineering
D self-fertilisation by a member of *Musa acuminate*

4 Stratigraphic correlation is used to work out the relative ages of sedimentary strata at different sites. Strata containing the same index fossils are assumed to be the same age. It is also assumed that lower layers are older than upper layers.

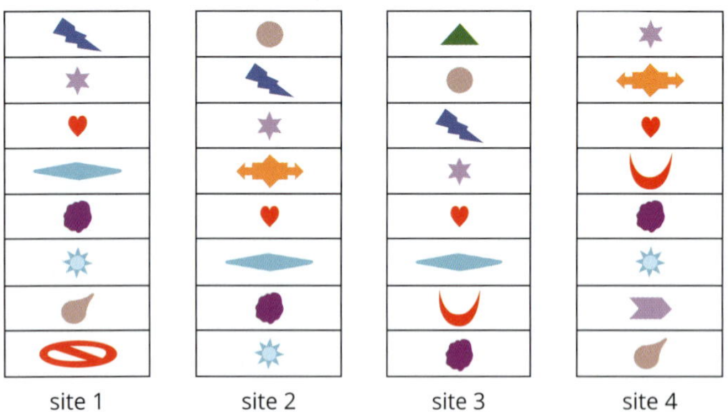

Consider the four sites illustrated above. Which one of the following statements is accurate?

A Site 3 contains the youngest stratum and site 4 the oldest.
B Site 1 contains both the youngest and the oldest strata.
C All strata at the same depth are the same age.
D Site 1 contains the oldest stratum and site 3 the youngest.

5 Radioactive dating methods are frequently used to give absolute dates to fossils. One radioactive dating method used to date fossils directly uses the carbon-14 isotope. The graph below shows the decay of carbon-14 to nitrogen-14.

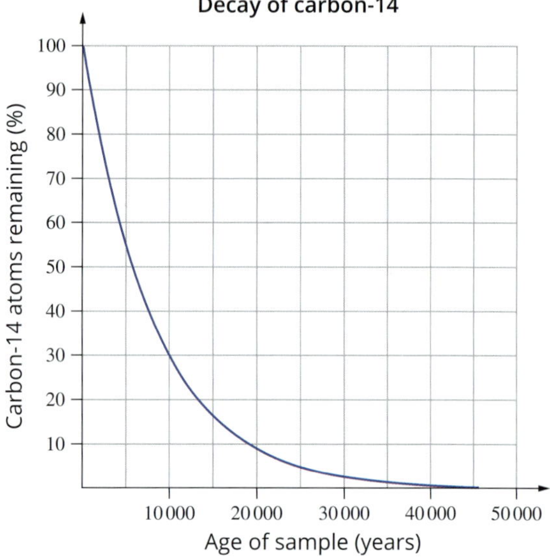

A fossil is found in which the percentage of carbon-14 remaining is 60%. What is the approximate age of the fossil?

A 10000 years
B 8000 years
C 6000 years
D 4000 years

6 The family Felidae includes many species of felines. Three species in this family are *Catopuma badia* (species A), *Catopuma temminckii* (species B) and *Profelis aurata* (species C). Together these are known as golden cats. An outgroup of this clade is *Panthera leo* (species D).

Assuming that these species are classified correctly, which of these phylogenetic trees most accurately shows the relationships between the species?

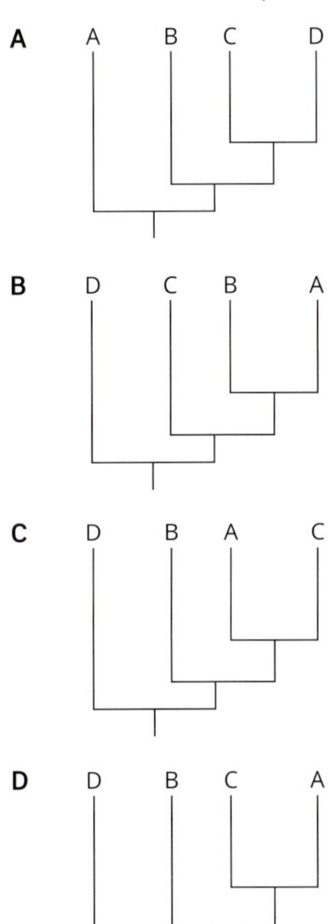

Short-answer questions

7 **a** Outline the three conditions required for natural selection to occur.

b Identify an example of evolution where natural selection is the mechanism for change in a species.

8 The genus *Equus* includes several species, both extinct and extant (still living). The extant species are Przewalski's horse (*Equus przewalskii*) and the domestic horse (*Equus caballus*). The last wild member of *Equus przewalskii* was seen in 1965 and shortly after it was declared extinct in the wild. Back in 1945, 13 individuals were being held in captivity. A conservation and breeding program based on nine members of the captive population has been so successful that they were reintroduced into their natural habitat, the steppes of Mongolia. The wild population in Mongolia now numbers over 400 and the captive population exceeds 1500 but Przewalski's horse is still listed as endangered. All current members of *Equus przewalskii* are descendants of the nine members from the 1945 captive population.

a The population of Przewalski's horse from which the current population descended was quite small in number. Clarify if this is a result of a population bottleneck or a founder event.

b Despite the considerable increase in the numbers of Przewalski's horse, it is still considered to be endangered. Discuss why this is the case.

c The nearest living relative of Przewalski's horse is the domestic horse. Przewalski's horses have 66 chromosomes and domestic horses have 64. Despite this difference, matings between domestic horses and Przewalski's horses produce fertile offspring.

i Suggest how the change in chromosome numbers may have come about.

ii Przewalski's horse and the modern domestic horse are very closely related. Recall the standard definition of a species.

iii Discuss whether Przewalski's horse and the domestic horse are separate species or, as in the alternative view of some taxonomists, they are subspecies of a horse species called *Equus ferus*.

9 Wheat (*Triticum* spp.) is one of the world's major food crops. A disease that results in severe crop losses is wheat leaf rust, caused by three species of fungus of the genus *Puccinia*. It can cause crop losses of up to 20% and substantial loss of income for wheat producers. This has resulted in significant research, such as identifying resistant strains of wheat and using selective breeding to enhance the offspring of these strains. One strain, called Norm, has strong resistance to leaf rust fungus.

- **a** Research shows around 46 different genes are associated with resistance to the *Puccinia* leaf rust fungus. Scientists have used selective breeding to increase the frequency of the alleles conferring most resistance in wheat plants. Selective breeding for particular traits can have unexpected effects on the phenotypes of target organisms. Explain why.
- **b** Identification of the Norm strain as highly resistant to leaf rust means that many farmers are deciding to plant this variety of wheat. Some scientists consider exclusive use of one strain to be a very dangerous practice that could lead to worldwide famine. Do you agree with these scientists? Explain why or why not.
- **c** Are these resistant populations of wheat likely to maintain their resistance to leaf rust fungus over the long term? Justify your answer.

10 **a** Recall the basic structure of a prokaryotic bacterial cell by drawing a simple labelled diagram.

- **b** Using your knowledge of evolution by the mechanism of natural selection, summarise how pathogenic bacteria may develop antibiotic resistance. Your answer should account for bacteria using asexual reproduction, which would usually limit the introduction of new genetic diversity into a population.
- **c** Bacteria are known to transfer genetic material from cell to cell in two different ways.
  - **i** Name each transfer method and compare their essential features.
  - **ii** Discuss why the bacterial transfer of genetic material poses a challenge for human health care.

11 The fossil record is an important source of evidence of evolution. Shown below is a series of fossil skeletons showing the evolution of modern whales.

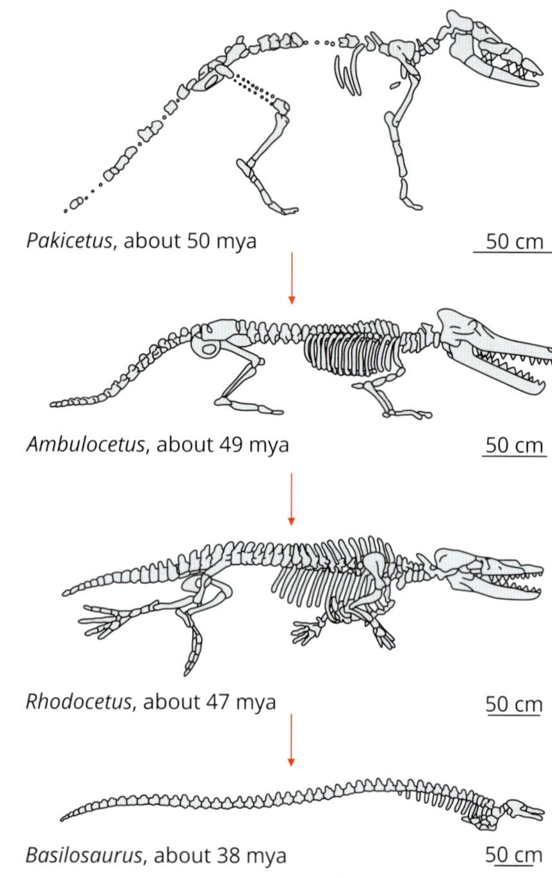

- **a** Explain how this group of fossils provides evidence of evolution.
- **b** Describe how the age of the *Ambulocetus* fossil would have been determined.
- **c** *Dorudon* was another whale ancestor. Its skeleton is shown below. *Dorudon* was a contemporary of one of the above organisms. Which of the fossil series is this most likely to be? Provide reasoning to support your answer.

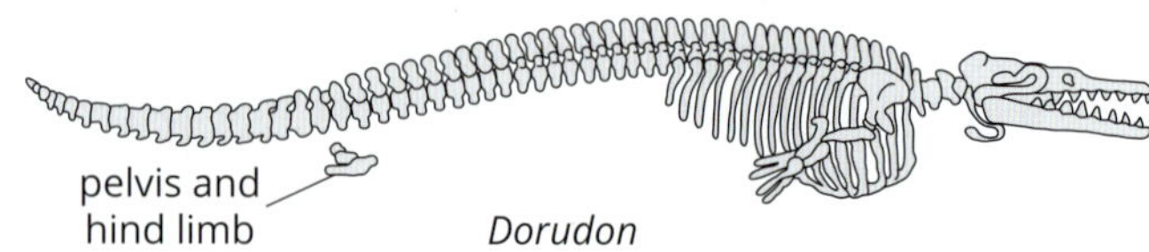

12 On Lord Howe Island, off the east coast of Australia, researchers have found two related species of palms (*Howea forsteriana* and *Howea belmoreana*) living side by side. Because the island is so small, it is considered impossible to have true geographical isolation between the natural endemic populations on the island. Their research showed that the species diverged much more recently than the island's volcanic origin. Therefore, another form of isolation must have led to the speciation of the palms.

Curly palms (*Howea belmoreana*) on Lord Howe Island

a Name the type of speciation that occurs by an isolating mechanism other than a geographic barrier.

b Suggest one possible isolating mechanism that could have led to speciation of the Lord Howe Island palms.

13 *Lymnaea* is a genus of mollusc that has snail-like shells. Members of *Lymnaea* have shells that coil either left or right, as shown in the diagram below. In most individuals, development of this trait is purely genetic, with offspring showing a phenotype that is identical to the maternal phenotype. Occasionally, and seemingly at random, an environmental factor influences the outcome and an individual with a genetically right-coiling shell grows a shell that coils left, and vice versa.

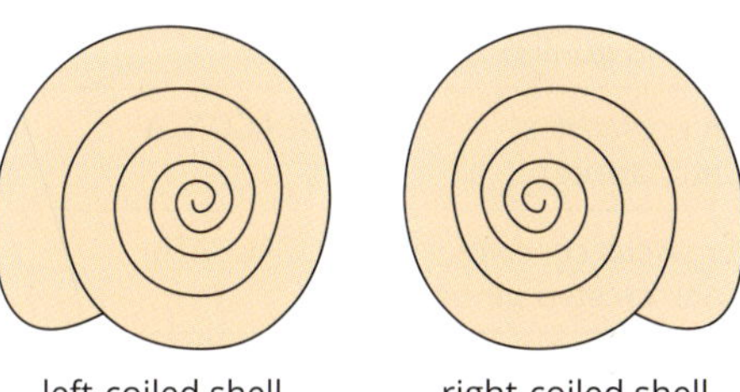

a As a result of physical incompatibility, individuals with shells that coil in the opposite direction are unable to mate.

  i If the environmental factor were to disappear, the two populations would find mating very difficult. What name is given to this sort of isolation?

  ii Without the occasional environmental effect on shell growth, this incompatibility could lead to speciation of the two groups. Explain how this could occur.

b Why is speciation unlikely as long as the environmental effect continues to create individuals that are genetically of one form but phenotypically the other form?

14 The diagram illustrates a sequence of DNA from a chicken, a quail and a turkey. Asterisks (*) indicate that all three species have the same base at that position. Dashes indicate insertions and/or deletions.

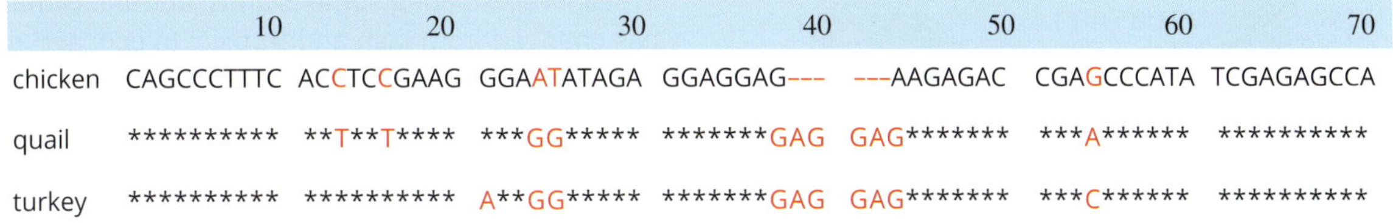

	10	20	30	40	50	60	70
chicken	CAGCCCTTTC	ACCTCCGAAG	GGAATATAGA	GGAGGAG---	---AAGAGAC	CGAGCCCATA	TCGAGAGCCA
quail	**********	**T**T****	***GG*****	*******GAG	GAG*******	***A******	**********
turkey	**********	**********	A**GG*****	*******GAG	GAG*******	***C******	**********

a Based on this sequence, infer which two of the organisms are most closely related. Justify your inference.

b Draw a phylogenetic tree of the three birds based on the data.

15 A mitochondrial DNA (mtDNA) sequence from five primate species is shown below.

Species	mtDNA sequence
Homo sapiens (human)	ACACCATA
Pan paniscus (bonobo)	ACACCATA
Gorilla gorilla (lowland gorilla)	CCACCACA
Pan troglodytes (chimpanzee)	CCACCACA
Nomascus concolor (black crested gibbon)	CCACCATA

a Define mtDNA.

b **i** Identify the pairs of primates that are most closely related.

ii Explain why analysis of mtDNA is useful as evidence for relatedness between species.

c mtDNA is also used to trace the historical migration patterns of early humans around the world. Summarise the general path of migration as indicated by current evidence, including the arrival of the earliest human populations in Australia.

16 While on his famous trip on the HMS *Beagle*, Darwin observed finches on the Galápagos Islands. These islands contain 13 different species. On Cocos Island there is a fourteenth species of finch that has been shown to be related to the Galápagos finches. The Galápagos finches are hypothesised to be most closely related to the tanager finches of Central and South America. The map below shows the relationship between the islands and part of the mainland of South America.

The hypothesised relationships between the various finches is shown in the phylogenetic tree. Note: Vegetarian finches are also known as herbivorous finches.

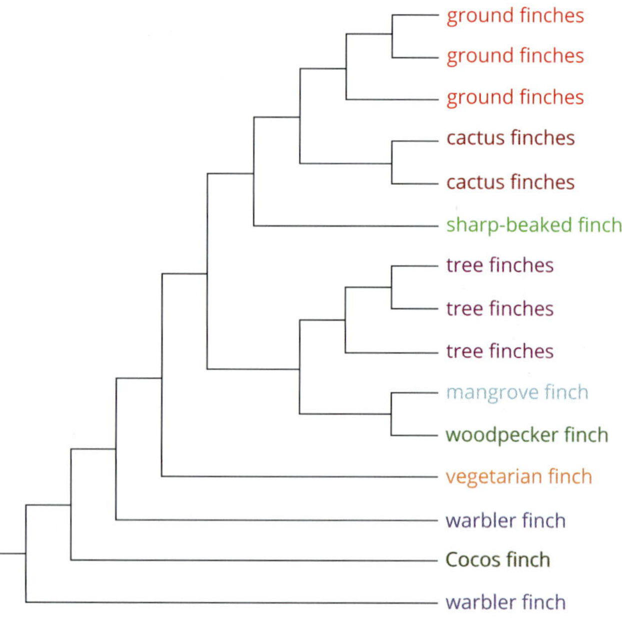

Genetic sequences show that finches with similar feeding styles tend to be closely related.

a According to the phylogenetic tree shown, which of the Galápagos groups is most closely related to the Cocos finches?

b The finches of the Galápagos Islands show considerable genetic diversity. What does this suggest about the size of the founder population?

c It is thought that the first finches arrived on the Galápagos about 2 million years ago. Since that time the environment has changed considerably. There has been volcanic activity, which added 14 islands to the chain. There have also been changes in sea level, isolating some islands, and the climate has changed from lush and tropical to cool and dry. How might the changes in the Galápagos Islands have resulted in speciation of the original population of finches to arrive there?

17 Birds and mammals share a common ancestor. The ancestor of these two classes was a terrestrial vertebrate. Today some species in both taxa spend large amounts of time in an aquatic environment. Two such species are penguins and seals. The hypothesised evolution of both of these groups is shown in the phylogenetic tree below.

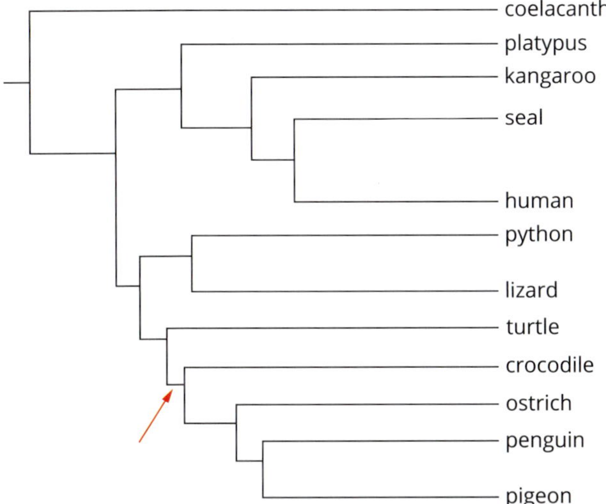

The last common ancestor of penguins and seals was a vertebrate that lived some time between 305 and 340 million years ago. Evidence suggests that, like modern reptiles, this organism was ectothermic.

a Penguins and seals are both endothermic. What kind of evolution has resulted in the two groups having this characteristic? Explain.

b Propose why there is uncertainty about when the last common ancestor of penguins and seals lived.

c Seals and penguins both have forelimbs modified as flippers. These flippers can be considered to be both homologous and analogous. Explain how this can be the case.

d On the phylogenetic tree above, the red arrow points to a common ancestor called an archosaur, which palaeontologists believe gave rise to a number of later lineages. Identify all of the members of the monophyletic group that includes the archosaurs.

e Identify a polyphyletic group from the tree. Explain how this group is polyphyletic.

18 *PAX-6* is a master regulator gene found in a large variety of animals. *PAX-6* encodes a transcription factor that binds to DNA at two sites, thereby controlling development of sensory organs, especially the eyes. This gene has been highly conserved over time. It is so similar in most organisms with eyes that the human gene when inserted into the genome of otherwise sightless fruit flies (*Drosophila melanogaster*) results in the formation of eyes.

Clarify the phrase used in this question 'this gene has been highly conserved over time'.

19 Taxonomists use the features of skeletons, and for hominins, associated cultural materials, to develop cladograms and phylogenetic trees for species to show how they may be related. In order to develop their cladogram, the scientist will first look at the various traits of the group of organisms and develop a character matrix. Such a matrix is shown below for one group of hominins.

Species	Opposable big toe	Large brow ridge	Recognisable tool making	Relatively small teeth	Taller than 1.5 m	Cranial capacity normally greater than 1200 cm^3	Sophisticated ritual burial practices
1	1	1	0	0	0	0	0
2	0	1	0	0	0	0	0
3	0	1	1	0	0	0	0
4	0	1	1	1	0	0	0
5	0	0	1	1	1	1	0
6	0	0	1	1	1	1	1

a Using the information in the matrix, develop a cladogram for the species shown.

b The species used to develop this matrix are: *Homo neanderthalensis*, *Australopithecus* sp., *Homo habilis*, *Homo sapiens*, *Ardipithecus* sp. and *Homo erectus*. *Ardipithecus* preceded *Australopithecus* and had an opposable big toe, an adaptation for climbing trees. Using your knowledge of hominin evolution, match each numbered species in the matrix to its correct name.

c Consider a hypothetical group of anthropologists working in China who have found a group of fossilised hominin bones. As is common, some of the bones are missing and some show signs of animal activity, such as teeth marks. Why would many fossilised bones show signs of animal activity?

continued over page

d The bones found by the researchers had a number of significant features. The cranial capacity was around 600 cm^3. The length of the femur was around 30% shorter than the femur of a modern human. The brow ridges were pronounced and the zygomatic arches (cheek bones) were extremely large. The teeth were large and a number were distinctly pointed. The foramen magnum was positioned more forwards than in chimpanzees and gorillas but very slightly more to the rear of the skull than in the australopithecines. The arms bones were significantly shorter than the leg bones. No bones from the feet or ankles were found.

- **i** If you were presented with these bones, where would you put the hominin on your cladogram? Give reasons for your decision.
- **ii** What other inferences could you make about this hominin?

This statement relates to questions 20 and 21.

Indigenous Australians believe that they have inhabited the land since time began. Questions 20 and 21 are based on the Western scientific understanding of human migration into and around Australia.

20 Some evidence of hominin habitation from 120 000 years ago exists in Australia. Given our current understanding of modern human migration across the world, discuss if the hominins present were likely to be members of the species *Homo sapiens*.

21 Haplogroup analysis is an important method of studying human spread throughout the world. Haplogroups are groups of people who share a common ancestor. Female haplogroups are determined by examining mitochondrial DNA and male haplogroups can be determined using the Y chromosome. In the map below, each haplogroup is denoted by a capital letter.

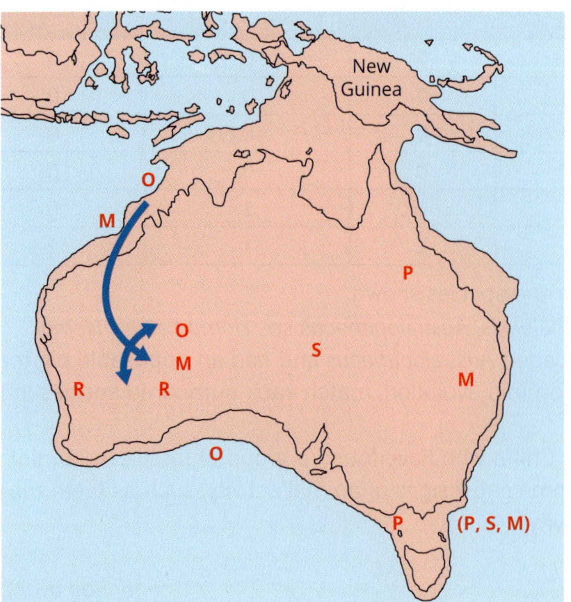

a Account for why female lineages are determined using mitochondria and male lineages are determined using the Y chromosome.

Using hair samples collected from 111 individuals during the early twentieth century, and examining mitochondrial DNA, the haplogroups of Indigenous Australians from pre-European arrival were determined and their migration into and around Australia was hypothesised. The distribution of these haplogroups is shown in the map below.

b Considering just the data shown, deduce which of the haplogroups M, O, P, R or S is the most ancient.

c Evidence from Asia shows that both the R and M groups were present in India earlier than 55 000 years ago. Suggest why the R haplogroup is only seen on the west coast of Australia.

d Using the data, draw arrows to show the most likely route(s) of migration through Australia after humans first arrived in the north of Australia.

e Evidence from studying the mitochondrial DNA suggests that after an initial rapid population spread across Australia around 50 000 to 60 000 years ago, the populations became relatively fixed in position. Populations moved around, but only in a fairly confined area. This situation started around 45 000 years ago. Suggest why this supports the strong connection between Indigenous groups and Country.

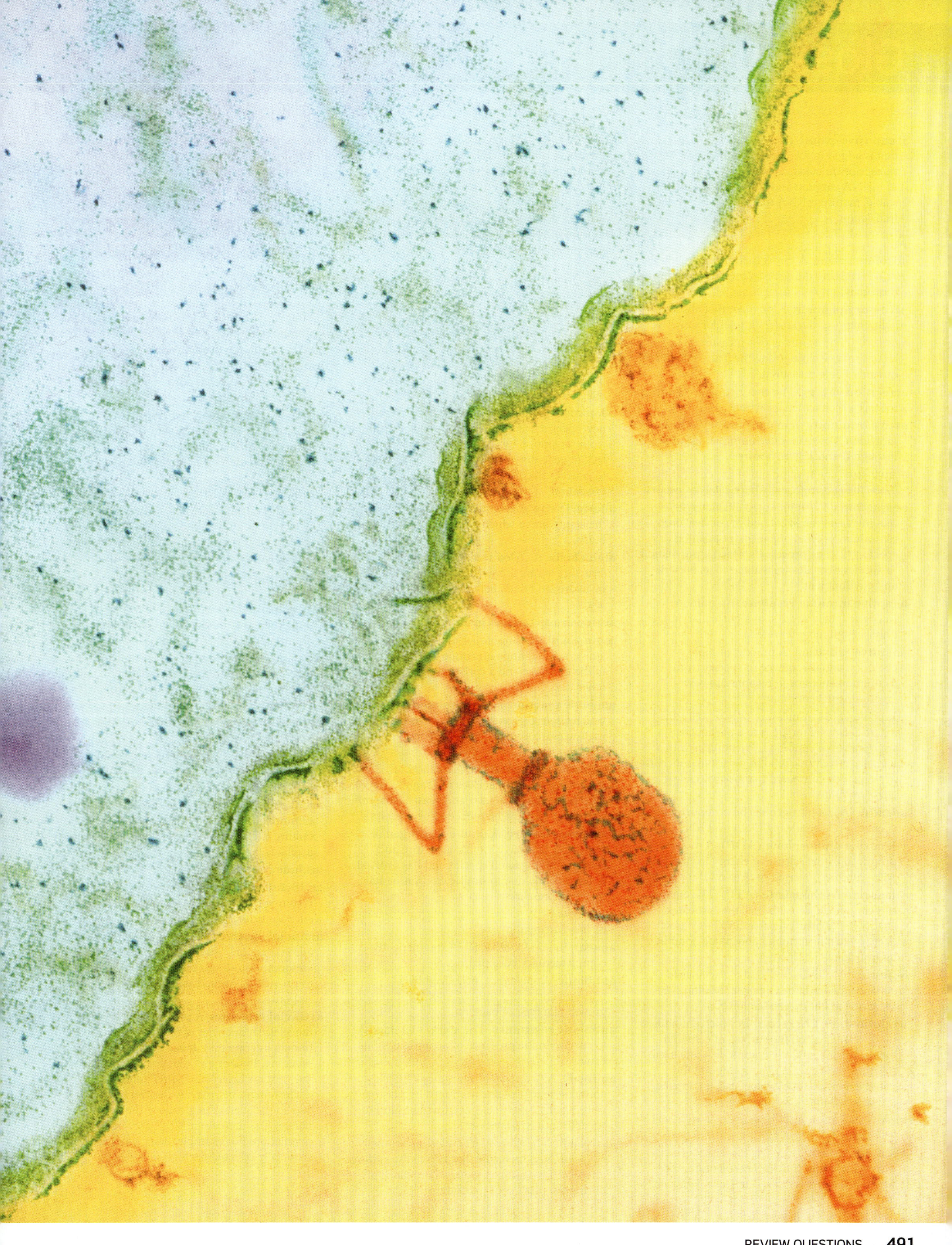

Glossary

5' cap (five-prime cap) A special nucleotide that is added to the 5' end of primary transcripts in eukaryotes. The process is known as mRNA capping and it functions to make stable, mature mRNA that is ready to undergo translation.

A

absolute dating A direct quantitative method of determining the age of a rock or object using radioactivity.

accessory pigment A pigment that absorbs light energy and transfers it to chlorophyll α.

accuracy The accuracy of a measurement relates to how close it is to the 'true' value of the quantity being measured.

activation energy The energy that is required to start a biochemical reaction.

active immunity Immunity that involves an individual's own adaptive immune response, through B and T lymphocytes.

active site The specific site of the enzyme that binds the substrate and where catalysis occurs.

adaptation (1) An inherited characteristic that increases the likelihood of survival and reproduction of an organism, population or species. (2) The process by which a population or species becomes well-suited to its lifestyle and environment.

adaptive immune response An immune response that is specific to a particular antigen; only present in vertebrates.

adaptive radiation A process of rapid evolution and divergence of species from a common ancestor in response to new environmental conditions.

adaptive value A measure of how well suited a particular phenotype is to a particular environmental condition. Phenotypes with a high adaptive value tend to persist and increase in frequency compared to those with a low adaptive value.

adenine (A) A nitrogenous base (a purine) that occurs in nucleotides of DNA and RNA.

adenosine diphosphate (ADP) A molecule produced by the release of energy from ATP. ADP can be converted back to ATP for re-use.

adenosine triphosphate (ATP) The energy-carrying molecule of the cell that provides energy for cellular processes. ATP releases energy when its terminal phosphate bond is hydrolysed.

adjuvant A substance added to vaccines to enhance the immune response, resulting in increased antibody production.

agglutination The process in which antibodies bind to antigens on the surface of cells and form antigen–antibody complexes that clump together and activate phagocytes and the complement cascade, which leads to antigen/cell destruction.

aim A statement describing in detail what will be investigated.

alkaloid An often toxic chemical produced by plants, including caffeine and nicotine, and which helps protect the plant from pathogens.

allele A different form of a gene. Alleles of a gene have different DNA sequences, sometimes resulting in different forms of a phenotype. For example, the gene that codes for hairline shape in humans has two alleles—straight and widow's peak.

allele frequency The relative proportion of a particular allele in a gene pool. It is typically presented as a decimal or percentage of the allele of that gene in the gene pool.

allergen An antigen that elicits an allergic response.

allergic response The rapid and vigorous overreaction of the immune system to antigens called allergens. Allergic reactions involve the production of IgE by B lymphocytes and the release of histamine by mast cells.

allopatric speciation The evolution of two new species from an ancestral species resulting from separation by a geographical barrier. The lack of gene flow between populations leads to genetic divergence over time, resulting in speciation.

allosteric site A site on an enzyme other than the active site to which an effector molecule binds.

alpha helix A coiled secondary protein structure within a polypeptide chain stabilised by hydrogen bonds between adjacent amino acids.

amine group (amino group) A -NH_2 group.

amino acid The monomer of polypeptides. All amino acids contain an amine group at one end of the molecule and a carboxyl group at the other end.

anabolic reaction A biochemical reaction in which larger molecules are made from smaller molecules; requires an input of energy to build new bonds.

anaerobic fermentation A method of producing energy that does not require oxygen, used by animals, bacteria and yeasts. Anaerobic fermentation must be preceded by glycolysis.

analogous feature A feature (e.g. organ or structure) that has a similar structure and function in unrelated species and has evolved independently due to similar environmental selection pressures.

anaphylaxis A severe allergic reaction that can be life-threatening.

anneal To join DNA or RNA fragments by complementary base pairing.

antibiotic A substance, produced by a microorganism or synthesised, that inhibits the growth of a type of bacteria.

antibiotic resistance The ability of a microbe to survive in the presence of an antimicrobial drug.

antibody Also known as immunoglobulins, antibodies are proteins produced by plasma cells that are highly selective for, and bind to, specific antigen molecules

anticodon The three nucleotides on a transfer RNA (tRNA) molecule that join to the codons on mRNA by complementary base pairing during the process of translation.

antigen A substance that reacts with antibodies and T lymphocyte receptors; antigens that induce an immune response are immunogens.

antigen–antibody complex A specific chemical interaction between an antibody (immunoglobulin) molecule and an antigen molecule.

antigen presentation The presentation of antigens by antigen-presenting cells.

antigen-presenting cell (APC) A cell that uses MHC-II on its surface to present foreign antigens to helper T cells to elicit an adaptive immune response. Examples include dendritic cells and macrophages.

antigenic drift The slow change to antigens on the surface of a virus due to the accumulation of genetic mutations.

antigenic shift An abrupt change in the genetic code of a virus due to re-assortment of genes from different viral strains, resulting in significantly different antigens on the surface of the virus.

antigenic variation The mechanism of changing surface antigens, usually to avoid detection or an immune attack. Employed by certain protozoans such as *Plasmodium* spp.

antimicrobial drug A compound that inhibits the growth of microorganisms, including bacteria, fungi, protists and viruses.

antimicrobial resistance The ability of a microbe to survive in the presence of an antimicrobial drug.

antiparallel Running in opposite directions, with one strand in the 5' to 3' direction and the other in the 3' to 5' direction (referring to the two strands in DNA molecules).

antiserum A serum containing specific antibodies.

antiviral A drug that inhibits replication of a virus by blocking entry or exit from the cell, or blocking viral replication enzymes.

archaeological Relating to the scientific study of human history or prehistory and their culture from the analysis of fossil remains and artefacts.

archaic Very old or ancient.

artificial active immunity Active immunity resulting from the administration of antigens, such as through vaccination.

artificial passive immunity The administration, usually by injection, of antibodies produced by another organism to provide an immediate, specific immune response.

artificial selection A process of changing the allele frequencies of a population through human intervention. It is a form of selective breeding. Phenotypes that are selected for may not necessarily be better suited to the environment, but may be desired by humans and so the alleles that produce the desired phenotypes increase in frequency in the population. For example, dairy cows are artificially selected for high milk yields.

Assimilation (partial replacement) model A model to explain the origins of present day humans. This model proposes that all living humans had an African origin and when they migrated out of Africa, there was occasional interbreeding with archaic humans that were already living in other parts of the world, resulting in hybrid populations.

autoimmune disease Any disease in which there is a failure of tolerance and an adaptive immune response is directed against a self-antigen, causing T lymphocytes to attack tissues directly and B lymphocytes to produce antibodies against the self-antigen. Autoimmune diseases can be organ-specific or generalised.

autotroph A living organism capable of synthesising all of its own food by photosynthesis or chemosynthesis.

B

B cell see *B lymphocyte*

B cell receptor (BCR) A molecule found on the surface of a B lymphocyte that detects specific antigens.

B lymphocyte (or B cell) Lymphocytes that when stimulated produce large quantities of antibodies specific to a particular antigen. They are responsible for the humoral immune response and include both memory and plasma cells.

bacterial competence The ability of a bacterial cell to alter its genome by taking in DNA from other cells or the environment.

bacterial transformation The incorporation of DNA from another organism into a bacterial cell.

bacteriophage A virus that infects bacteria.

bacterium (plural bacteria) All prokaryotes not members of the domain Archaea.

base One of five nitrogenous chemicals present in the nucleotides of nucleic acids (DNA or RNA). The five bases are adenine, guanine, cytosine, thymine (DNA only) and uracil (RNA only).

basophil A type of white blood cell (leukocyte) that releases histamine during an allergic reaction.

behavioural isolation Genetic isolation that results from populations displaying different behaviours, such as mating calls and courtship rituals.

benign tumour A mass of abnormal (but not cancerous) cells. They are not cancerous because they do not invade nearby tissue or spread throughout the body through a process of metastasis.

beta-pleated sheet A secondary protein structure stabilised by hydrogen bonds between different regions of a polypeptide chain that create pleat-like formations.

biochemical pathway (metabolic pathway) A sequence of biochemical reactions, each catalysed by a specific enzyme in which the product of one reaction becomes the substrate of the next.

biodiversity Biological diversity; the variety of all life forms, the genes that they contain and the ecosystems of which they are a part.

biofuel A fuel that can be produced from crops or other organic material. Examples of biofuels are ethanol from the fermentation of sugars, methane from digestion in animals, and biogas from plant and animal wastes.

biological fitness An organism's ability to survive and reproduce in its natural environment.

biomass The mass of living matter per unit area (e.g. kg/m^2), or the equivalent amount of chemical energy bound in the mass of tissue (e.g. kJ/m^2). Biomass measurements may be for total biomass, or for the biomass of a particular group of organisms such as plants.

biomolecule A molecule involved in the maintenance or metabolism of living organisms.

biosecurity Measures to detect, respond rapidly to and recover from pests and diseases, including introduced species, to protect agricultural production and wildlife biodiversity.

bipedal Describes an animal that walks on two legs.

bispecific monoclonal antibody (bispecific mAb) A monoclonal antibody (mAb) constructed through molecular technology to have two different binding sites: one for a cancer cell and one for an immune cell. Bispecific mAbs 'identify' and 'deliver' cancer cells to the immune system.

blunt-end restriction enzyme A restriction enzyme that leaves clean-cut ends because it cuts both strands of the DNA molecule at the same location within the recognition site.

bottleneck effect The impact of substantial population reduction on the smaller remaining population. The remaining population has reduced genetic diversity and is more likely to lose alleles through genetic drift and experience inbreeding.

branch A line on a phylogenetic tree that represents the evolutionary path from a common ancestor (lineage).

bundle sheath cell A cell in the leaf or stem of a vascular plant that forms part of the layer of cells that surrounds the vascular tissue.

C

C_3 plant A plant that uses the most common form of photosynthesis in plants, in which the first carbon-containing product is a three-carbon compound.

C_4 plant A plant that fixes a four-carbon molecule during photosynthesis rather than the more common three-carbon molecule.

Calvin cycle A cyclic biochemical pathway in the light-independent reactions of photosynthesis, in which carbon from carbon dioxide becomes fixed in the synthesis of carbohydrate.

CAM plant A plant that uses crassulacean acid metabolism (CAM), a form of photosynthesis that occurs in many plants growing in hot, dry environments. Stomata open at night to take in carbon dioxide, which is incorporated into malate. During the day the stomata are closed to reduce transpiration, and malate is metabolised to release oxaloacetate.

cancer vaccine Vaccines made up of cancer cells, parts of cells, or pure antigens that stimulate the immune system to prevent or fight cancer cells.

carbon fixation The incorporation of carbon into organic compounds by living organisms.

carboxyl group A -COOH group containing a carbonyl and hydroxyl group.

carcinogen A substance that damages cell DNA. A carcinogen can be physical, chemical or biological.

cast fossil A three-dimensional 'sculpture' of an organism formed by materials such as silica or phosphate filling the vacant space in an impression or fossil mould.

catabolic reaction A biochemical reaction in which there is a breakdown of macromolecules into smaller molecules; releases energy.

catalyse To increase the rate of a reaction.

catalyst A substance that increases the rate of a chemical reaction but is not consumed in the reaction. An enzyme is a catalyst.

catalytic power The ability or potential of an enzyme to increase the rate of a biochemical reaction compared to the reaction occurring without the enzyme present.

cell-mediated immunity An immune response that is mediated by T lymphocytes. Compare with *humoral immunity*.

cellular pathogen A cellular organism that is a source of non-self antigens and causes disease in a host organism. Bacteria, protozoa, oomycetes, fungi, several types of worms and arthropods are types of cellular pathogens.

cellular respiration (1) Generally—the complete breakdown of glucose to provide energy in cells. (2) Specifically—the second aerobic stage that occurs in the mitochondria and produces 36–38 molecules of ATP per molecule of glucose.

chain of infection The series of events that occur during the transmission of an infectious agent, which includes reservoir, portal of exit, mode of transmission, portal of entry, susceptible host.

chemical barrier A chemical product or mechanism that is one of the first-line defences of an organism's immune system. Examples include the lysozyme enzymes in saliva.

chemical energy Energy stored in the chemical bonds that join atoms together in molecules, and which can be released by breaking the bonds apart.

chemical group A group of covalently linked atoms, such as an amino group or hydroxyl group, that has a characteristic chemical behaviour.

chemokine A cytokine that attracts white blood cells to the site of infection.

chimeric monoclonal antibody (chimeric mAb) A monoclonal antibody made of mouse and human molecular components.

chlorophyll The green pigment found in chloroplasts in plants, and within some prokaryotic cells, that absorbs light energy for photosynthesis.

chloroplast An organelle that uses light energy, carbon dioxide and water to produce glucose. Site of photosynthesis.

***cis* face** The side of the Golgi apparatus facing the nucleus.

cisterna (pl. cisternae) The flattened sac-like membranes found in the Golgi apparatus and endoplasmic reticulum.

clade A group of organisms that includes an ancestor and all descendants of that ancestor.

cladogram A branching diagram representing the evolutionary relationships between taxa. The branches of a cladogram are scaled (the lengths of the branches do not represent evolutionary distance).

clonal selection The theory that in a group of lymphocytes, a specific antigen will activate only the lymphocyte that has a receptor that specifically recognises it. This lymphocyte will proliferate into clones of itself.

coding strand The strand of DNA that has the same nucleotide base sequence as the mRNA strand produced by transcription (uracil in the mRNA in place of thymine in the DNA).

codon The basic unit of the genetic code. A sequence of three nucleotides on mRNA that codes for a particular amino acid, or indicates the beginning or end of translation.

coenzyme A small organic molecule that combines with an enzyme and is necessary for its activity.

coexisted Existed at the same time and place.

cofactor A chemical component such as a metal ion or coenzyme that is required for the proper function of a protein.

common ancestor An organism from which two or more species diverged. Also known as a shared ancestor.

competitive inhibition A type of enzyme inhibition that occurs when the shape of the inhibitor is similar to the shape of the substrate that normally binds to the active sites of an enzyme. Due to their similar shapes, the inhibitor is able to bind to the active site of the enzyme and block the substrate from binding to the site.

complement protein A protein that is able to kill foreign cells by lysis. There are more than 30 different complement proteins that are activated in response to antigen–antibody complexes, antigens and carbohydrates on the surfaces of some bacteria and parasites.

complementary base pairing The pairing in DNA and RNA molecules of the nitrogenous bases between two strands. In DNA adenine always pairs with thymine, and cytosine always pairs with guanine.

complementary DNA (cDNA) Double-stranded DNA that contains no introns; copied from mRNA by the enzyme reverse transcriptase.

conclusion An evidence-based statement that is developed from the analysis of results from a scientific investigation. A conclusion summarises the findings of an investigation and explains the extent to which the hypothesis or research question was addressed.

conformational change A change in the spatial (three-dimensional) arrangement of atoms in a macromolecule such as a protein or nucleic acid.

conjugated monoclonal antibody (conjugated mAb) A monoclonal antibody (mAb) that has been attached to a drug, toxin or radioactive particle in order to deliver the treatment specifically to cancer cells.

conjugated protein A protein that contains a non-protein (prosthetic) group.

constant region The region of antibody molecules that remains the same and interacts with receptors on the body's cells.

constitutive gene A gene or protein that is always expressed or active.

continuous variable A variable that can have any number value within a given range.

control group The experimental conditions of the control group are identical to those of the experimental group, except that the variable of interest (the independent variable) is also kept constant.

controlled variable A variable that is kept constant during an investigation.

convergent evolution The evolution of similar features in unrelated groups of organisms.

corepressor A molecule that binds to and activates a repressor transcription factor.

crassulacean acid metabolism (CAM) A form of photosynthesis that occurs in many plants growing in hot, dry environments. Stomata open at night to take in carbon dioxide, which is incorporated into malate. During the day the stomata are closed to reduce transpiration, and malate is metabolised to release carbon dioxide, which is then used by cells.

CRISPR-Cas9 A gene editing technique that uses guide RNA (gRNA) to guide a Cas9 enzyme to a target DNA site where the enzyme cuts the DNA. CRISPR stands for 'clustered regularly interspaced short palindromic repeats', which means segments of DNA with short repetitive sequences that are interspersed with unique DNA sequences.

cyanogenic glycoside A compound produced by plants that breaks down to form hydrogen cyanide, a compound extremely toxic to eukaryotic cells. These compounds assist the plant to resist infection by pathogens and damage by herbivores.

cytokine One of a group of peptides and proteins released from cells that are important in cell signalling, particularly between cells of the immune system.

cytoplasm The contents of a cell, enclosed by the plasma membrane, including the fluid (cytosol) and all organelles except the nucleus.

cytosine (C) A nitrogenous base (a pyrimidine) that occurs in nucleotides of DNA and RNA.

cytosol The fluid inside a cell in which the cell's organelles, proteins and other structures are suspended.

cytotoxic A substance or process that is toxic to a cell, and can cause death of that cell.

cytotoxic T cell A T lymphocyte that is stimulated by cytokines to bind to antigen–MHC I complexes on infected host cells and release cytotoxic compounds that destroy the infected cells.

D

defensin Molecule that is active against bacteria, fungi and certain viruses.

degenerate More than one codon may code for a particular amino acid.

denature (noun denaturation) To irreversibly change the tertiary structure of a protein, as a result, for example, of heating the protein above a critical temperature.

dendritic cell A type of antigen-presenting cell.

deoxyribonucleic acid (DNA) A double-stranded nucleic acid that contains the genetic code in its sequence of bases. DNA is found in all organisms, and most viruses, in chromosomes, as well as in mitochondria and chloroplasts.

deoxyribose The five-carbon sugar molecule found in DNA. Deoxyribose is derived from ribose but lacks an oxygen molecule; it has a hydrogen atom rather than a hydroxyl group.

dependent variable A variable that may change in response to a change in the independent variable, and is measured or observed.

differentiation The modification of the structure and function of a cell that occurs during its development.

digestive enzyme inhibitor An enzyme or lectin that blocks normal digestion of starch by insects.

discrete variable A separate or distinct value that can be counted.

disease Disorder in the structure and function of an organism.

divergent evolution The evolution of two or more different species from a common ancestral species.

DNA see *deoxyribonucleic acid*

DNA amplification The process of creating millions of identical copies of a DNA sample using the polymerase chain reaction (PCR).

DNA ladder DNA standards; a set of DNA molecules of known size used as a 'molecular ruler' on gel electrophoresis to determine the size of other DNA molecules.

DNA ligase An enzyme that joins together fragments of DNA by forming a phosphodiester bond between the 3'-hydroxyl and 5'-phosphate of adjacent nucleotides.

DNA polymerase An enzyme that catalyses the formation of polymers of DNA by linking nucleotides into a chain by complementary base pairing with a template strand.

DNA profiling A technique used to produce an individual's unique pattern of DNA bands on a gel. Produced by analysing short tandem repeat (STR) regions of the genome.

DNA sequencing A technique to determine the sequence of bases in DNA. DNA sequencing can be used to determine relationships between individuals of a species and for determining the entire genome of an organism.

DNA thermocycler The machine used in the polymerase chain reaction that alters the temperature in pre-programmed steps.

double helix The double-stranded, coiled structure of a DNA molecule.

E

ecological isolation Genetic isolation that results from populations occupying different ecological niches. Also known as niche partitioning.

ecological niche The distribution and ecological role of a species in its environment; how it meets its needs for food and shelter, how it survives, and how it reproduces.

electron A negatively charged subatomic particle that usually occupies the orbit surrounding the nucleus of an atom.

electron spin resonance (ESR) A method of absolute dating that is used to date calcium carbonate in samples from the last 300 000 years, such as limestone, coral, fossil teeth, molluscs and egg shells.

electron transport chain A chain or interconnected series of enzymes and cytochromes embedded in the inner mitochondrial membrane.

emerging infectious disease A disease that is newly identified and has increased in incidence over recent years, or that may increase in the near future.

endemic A term used to describe a species as being native to a specific area or region. For example, koalas are endemic to Australia.

endergonic reaction A chemical reaction that requires the input of energy.

endocytosis The movement of material into a cell by enclosing it in the plasma membrane, which then pinches off to form a vesicle within the cell. Endocytosis includes phagocytosis (the entry of solids) and pinocytosis (the entry of liquids).

endonuclease An enzyme, also called a restriction enzyme, that occurs naturally in bacteria and can cut DNA at a particular site (a recognition site); used in genetic engineering.

enzyme A protein molecule that acts as a biological catalyst. Enzymes speed up rates of reactions that would otherwise take place much more slowly. Their action is often specific to only one type of reaction.

enzyme-linked immunosorbent assay (ELISA) A method to detect the presence of antigen in a sample, or antibodies in blood or serum.

enzyme–substrate complex The complex that forms when an enzyme binds to a substrate.

eon One of several subdivisions of geological time enabling cross-referencing of rocks and geological events from place to place. Eons are the largest subdivisions.

eosinophil A type of leukocyte that predominates in parasitic infections and contains granules that are stained by eosin dye.

epidemic A sudden increase in the number of cases of a disease above what is normally expected in that population in that area.

epoch One of several subdivisions of geological time enabling cross-referencing of rocks and geological events from place to place. Epochs are the smallest subdivisions.

era One of several subdivisions of geological time enabling cross-referencing of rocks and geological events from place to place. Eons are larger subdivisions than eras; eras may be divided into periods and epochs.

evolution A change in the inheritable traits of a population (or species) over successive generations. See *genetic drift* and *natural selection*.

exergonic reaction A chemical reaction in which energy is released.

exocytosis The movement of materials out of a cell via a vesicle. The vesicle fuses with the plasma membrane, and the vesicle contents are released out of the cell.

exon The region of a gene that codes for a protein.

experimental group A group in an experiment for which controlled (fixed) variables are kept constant, a single experimental (independent) variable is changed and the dependent variable is measured to determine any effect of the change.

exponential relationship Variables that are exponentially proportional to each other will produce a curved trend line when graphed.

extracellular pathogen A pathogen that does not invade cells but lives and reproduces in the extracellular environment.

F

faunal succession The stratigraphic principle that the fossils contained in sedimentary rock strata succeed one another in a predictable order, even when they are found in different places.

feedback inhibition Occurs when a product produced late in a biochemical pathway acts as the inhibitor of an enzyme acting earlier in the pathway.

fever An increase in body temperature that results from the regulated body temperature set point in the hypothalamus of the brain being set to a higher level by inflammatory cytokines, to slow the replication of bacteria and improve the adaptive immune response.

fibrous protein A type of protein that forms long fibres and provides structural support to cells and tissues.

flavin adenine dinucleotide (FAD) A coenzyme that functions as an electron carrier in metabolic reactions.

foramen magnum The hole at the base of the skull through which the spinal cord connects to the brain.

fossil The preserved remains, impressions or traces of organisms found in rocks, amber (fossilised tree sap), ice or soil.

fossil record The record of the evolution of organisms through geological time based on information from fossils.

fossilisation The process of preservation of the hardened remains, impressions or traces of organisms in rocks.

founder effect Occurs when a small group of individuals is genetically isolated from a larger population. The smaller population only has a small portion of the alleles of the original population and usually has lower genetic diversity.

fungus (pl. fungi) Non-phototrophic eukaryotes that have rigid cell walls made from chitin; includes moulds, yeasts, mushrooms and toadstools.

G

gamete A haploid cell capable of fusion with another haploid cell to form a zygote. In vertebrates the gametes are sperm and egg cells.

gamete mortality Genetic isolation that results from incompatibility between gametes (egg and sperm) and a failure to fuse at fertilisation.

gel electrophoresis A technique used for separating fragments of DNA, or different proteins, based on their molecular weight (or length). Fragments migrate through a gel at rates that are dependent on their length and charge.

gene A specific sequence of nucleotides that codes for a particular protein or RNA molecule. It is the unit of heredity.

gene cloning The production of identical copies of a gene.

gene editing The modification of genes by removal, substitution or alteration by mutation, without necessarily introducing a foreign gene.

gene expression The process that leads to the transformation of the information stored in a gene into a functional gene product (usually a protein or RNA molecule).

gene flow The movement of alleles between individuals of different populations; includes the dispersal of pollen and seeds in plants.

gene pool All of the alleles in a population.

gene regulation Processes that control gene expression, turning genes on or off.

genetic code The linear sequence of three nucleotides in DNA or RNA that determines, or codes for, the sequence of amino acids in a protein.

genetic diversity The variety of genes or alleles in a population or species.

genetic drift Random changes to allele frequencies in a gene pool as the result of a chance event. This has a more significant impact on smaller populations, as the chance death of one individual could eliminate an allele from the gene pool.

genetic isolation The prevention of gene flow between two populations.

genetic transformation The process of cells incorporating foreign DNA. This process can be natural or artificial.

genetically modified organism (GMO) An organism with a genetic modification (GM) made by transfer of specific genes from another organism (transgenic) or by gene editing techniques.

genome The complete set of genes or DNA in an organism.

genotype The genetic composition of an individual. Contrast with phenotype.

geographical isolation Genetic isolation that results from the separation of populations by physical and geographical barriers.

geological time scale The time scale of events that have occurred on Earth from its formation to the present time.

germline mutation A mutation that can affect gamete formation and can therefore be inherited by offspring.

globular protein A type of protein that is folded and coiled to form a compact spherical shape. It has a tertiary or quaternary structure specific to its function; for example, enzymes.

glucose A common six-carbon monosaccharide carbohydrate, or hexose. It is the product of photosynthesis and the substrate for respiration.

glycolysis The biochemical pathway in which glucose is broken down to pyruvate. The first stage of cellular respiration.

Golgi apparatus An organelle composed of a stack of cisternae in which proteins are assembled and then packaged in vesicles for exocytosis. Also known as Golgi body, Golgi complex.

gracile Slender, thin build.

grana (singular granum) A stack of flattened discs composed of the thylakoid membranes found in chloroplasts.

granulocyte A type of white blood cell containing granules (sacs filled with enzymes that digest pathogenic microorganisms). Neutrophils, basophils and eosinophils are granulocytes.

guanine (G) A nitrogenous base (a purine) that occurs in nucleotides of DNA and RNA.

guide RNA (gRNA) A short, synthetic RNA composed of a scaffold sequence necessary for the binding of the Cas9 enzyme. Guide RNA guides the Cas9 enzyme to DNA target sites and the enzyme cuts the DNA in the gene editing technique CRISPR-Cas9.

H

half-life The time taken for half a radioactive element to decay. The half-life of an element can be used to calculate the age of the rock in which it is contained.

heavy chain The polypeptide chain that forms the 'stem' of a Y-shaped antibody molecule.

helper T cell The helper T lymphocytes that bind to antigen-MHC II complexes on antigen-presenting cells and activate B lymphocytes to secrete antibodies, macrophages to phagocytose, and cytotoxic T cells to kill infected cells.

herd immunity A phenomenon in which vaccination of a large proportion of a population provides protection from a pathogen to non-immune or non-vaccinated individuals.

heterotroph An organism that must obtain nutrients from other organisms.

heterozygote (adj. heterozygous) A diploid individual with different alleles for a particular gene.

histamine An organic compound involved in inflammatory responses and allergic reactions, which causes surface blood vessels to dilate and become more permeable to immune cells and fluids. Common hay fever symptoms such as runny nose and eyes and sneezing are the result of histamine action and are aimed at flushing out allergens.

hominid A member of the family Hominidae, which includes humans, chimpanzees, gorillas and orangutans.

Hominidae The taxonomic family of primates, which includes humans, chimpanzees, gorillas and orangutans.

hominin A member of the tribe Hominini, which includes modern humans, extinct *Homo* species and our bipedal ancestors (e.g. *Australopithecus*, *Paranthropus* and *Ardipithecus*).

Hominini The taxonomic tribe that includes modern humans, extinct human species and our bipedal ancestors (e.g. *Australopithecus*, *Paranthropus* and *Ardipithecus*).

hominoid A member of the superfamily Hominoidea, which includes humans, great apes (orangutans, gorillas, chimpanzees and bonobos) and lesser apes (gibbons).

Hominoidea The taxonomic superfamily that includes humans, great apes (orangutans, gorillas, chimpanzees and bonobos) and lesser apes (gibbons).

Homo The genus of anthropoid mammals of which humans (*Homo sapiens*) are the only living species; includes numerous extinct species and subspecies, such as Neanderthals (*H. neanderthalensis*), *H. erectus*, *H. ergaster* and *H. habilis*.

homologous features Features that have a common evolutionary origin, which is evident in the underlying fundamental similarities in their structure. Homologous features are found in different organisms and may have evolved different functions (e.g. a human hand and a bat wing) as a result of divergent evolution from a common ancestor. DNA sequences or proteins can also be homologous.

homozygote (adj. homozygous) A diploid individual with two identical alleles at a particular genetic locus.

horizontal gene transfer The transfer of genes between cells, sometimes of different species, such as transfer of plasmids between bacteria. Contrasts with inheritance of genes from parent to daughter cell through cell division (vertical gene transfer).

human leukocyte antigen (HLA) Another name for the major histocompatibility complex in humans.

human monoclonal antibody (human mAb) A monoclonal antibody that is comprised entirely of human molecular components.

humanised monoclonal antibody (humanised mAb) A monoclonal antibody that is mostly comprised of human molecular components.

humoral immunity An immune response involving B lymphocytes that produce specific antibodies against foreign antigens.

hybrid The result of mixing, through sexual reproduction, two individuals of different breeds, varieties, species or genera.

hybrid inviability A form of postzygotic isolation between different species where any fertilised hybrid zygotes do not develop properly and do not reach birth/germination.

hybrid sterility A form of postzygotic isolation between different species where hybrid offspring that may survive to reproductive maturity are incapable of reproducing.

hybridoma The product of the fusion of an immortal cell line with a B lymphocyte to produce an immortal B lymphocyte; used in the production of antibodies.

hydrolytic enzyme An enzyme, found in plants, that breaks down the cell walls of fungi, oomycetes and bacteria and the exoskeletons of insects. Such enzymes are produced in greater quantities when the plant is under attack.

hypothesis A possible explanation to a research question that can be used to make predictions that can often be tested experimentally.

I

immortal cell line A cell line that can continually undergo division without the mutations that would normally occur as a cell ages, and can therefore be cultured for long periods.

immunogen An antigen that elicits an immune response.

immunoglobulin (Ig) Alternate name for an antibody; a type of protein produced by B lymphocytes in an immune response to the presence of a particular antigen, to which the immunoglobulin binds.

immunological memory The ability of lymphocytes of the adaptive immune system to 'remember' antigens after primary exposure, and to mount a larger and more rapid response when exposed to the same antigen again.

immunotherapy Any treatment that harnesses the immune system of the patient to fight diseases; for example, monoclonal antibody therapy.

impression fossil A type of fossil where the impression of the external or internal surface of the organism is preserved.

in vitro Occurring in a culture dish, test tube or other location outside the living organism (compared to 'in vivo', a process taking place in a living organism).

inactivated vaccine A vaccine made from an inactivated (killed) form of pathogens. Inactivation destroys the pathogen's ability to replicate, but keeps it 'intact' so it can be recognised by the immune system.

incidence In epidemiology, the rate of occurrence of new cases of a medical condition in a population over a given time period. Incidence data can be used to determine the risk of contracting a disease in a given population.

independent variable The variable that is altered during an experiment to test its effect on another variable (the dependent variable). Also called the experimental variable.

index fossil A fossil that is used to define and identify geological periods.

induced To promote or activate. When a gene is induced its transcription is activated.

induced-fit model A model that states that when a substrate binds to the active site of an enzyme, a change in shape (or conformational change) of the enzyme's active site occurs.

inducer A molecule that regulates gene expression.

infectious disease A medical condition or disease that is caused by pathogens (infectious agents). Infectious diseases are transmissable between hosts.

inflammation (or inflammatory response) A protective response, triggered by damaged tissue or invading pathogens, that leads to increased blood flow and migration of white blood cells to the site of damage/infection. It results in heat, pain, swelling, redness and loss of function.

innate immune response An immune response that non-specifically and rapidly protects against a wide variety of pathogens using innate immune cells such as leukocytes (e.g. phagocytes) and defensive molecules such as complement proteins. The innate immune response is triggered when the physical, chemical and microbiological barriers that provide resistance to infection are breached.

innate immunity Immunity that non-specifically protects against a wide variety of pathogens. It consists of physical, chemical and microbiological barriers that provide resistance to infection, and an innate response to infection that involves leukocytes such as phagocytes and defensive molecules such as complement proteins.

interbreeding The mating of two different species.

interferon A type of cytokine important in antiviral immunity. Interferons are produced by virus-infected cells to inhibit viral replication by resulting in the transcription of antiviral genes and the expression of antiviral proteins; they have a lesser role in bacterial and parasitic immune responses. They regulate the immune response in a number of ways, such as enhancing T lymphocyte activity.

intracellular pathogen A pathogen that invades cells and requires a host's cells to survive and/or reproduce.

intron A section of DNA that does not code for proteins and is spliced during mRNA processing in eukaryotes.

inverse relationship A mathematical relationship in which one variable increases as the other decreases.

irreversible inhibition Enzyme inhibition in which an inhibitor binds strongly to an enzyme with covalent bonds, permanently reducing or stopping the enzyme's function.

isotope One of two or more atoms that have the same atomic number (the same number of protons) but a different number of neutrons; for example, carbon-12 and carbon-14.

K

Krebs cycle A cyclic biochemical pathway that metabolises acetyl coenzyme A, producing carbon dioxide, ATP, NADH and $FADH_2$. The Krebs cycle is one of the three major biochemical pathways that constitute aerobic respiration.

L

***lacZ* gene** A gene in the *lac* operon that codes for beta galactosidase; used in recombinant plasmids for detecting transformed bacteria.

leaf The branch tip on a phylogenetic tree where the scientific name of the taxon is.

leukocyte A white blood cell; includes phagocytes and lymphocytes.

ligase An enzyme that joins together two molecules or fragments of molecules.

ligation The process of joining two fragments of DNA using a DNA ligase enzyme.

light chain A short polypeptide chain that forms the 'arms' of a Y-shaped antibody molecule.

light-dependent reaction The reactions in photosynthesis in which light captured by chlorophyll is used to split water to produce oxygen, and ATP and NADPH for use in the light-independent reactions.

light-independent reaction The reactions in photosynthesis in which the energy in ATP and NADPH from the light-dependent reactions is used to fix carbon into carbohydrates. These reactions are part of a biochemical pathway called the Calvin cycle.

light saturation curve A plateau-shaped graph of the rate of photosynthesis versus light intensity. The point at which the curve plateaus is the point at which a further increase in light intensity brings about no further increase in photosynthesis.

lineage All the species that are descendants of a common ancestor.

linear relationship A mathematical relationship in which variables are directly proportional to each other and produce a straight trend line when graphed.

live attenuated vaccine A vaccine that uses a weakened form of the disease-causing agent to stimulate an immune response, but which doesn't cause disease.

loaded coenzyme The form of a coenzyme that has a proton, electron or chemical group to donate.

lock-and-key model A model that states that an active site of an enzyme and a substrate fit together like a key into a lock.

lymph A colourless fluid that contains white blood cells, bathes tissues, and travels through the lymphatic system, draining into the bloodstream.

lymphatic system The body system that transports immune cells including antigen-presenting cells throughout the body, and is where antigen recognition by lymphocytes occurs; important for adaptive immune responses in mammals.

lymphocyte A type of leukocyte involved in adaptive immune responses; includes B and T lymphocytes.

lysis The destruction of a cell, usually by rupturing the plasma membrane.

lysozyme An antibacterial enzyme present in body secretions such as saliva and tears. It disrupts the bacterial cell wall.

M

macrophage A type of large white blood cell that is responsible for engulfing and digesting foreign matter in the body, as well as damaged cells or the remnants of apoptosis.

major histocompatibility complex (MHC) A complex of genes that code for proteins on the surface of cells that are involved in antigen presentation to T lymphocytes. MHC proteins are also known as human leukocyte antigens.

malignant tumour A mass of cancer cells that can invade nearby tissue and spread throughout the body through a process of metastasis.

mast cell An immune cell containing granules of histamine. This cell mediates allergic responses by binding IgE–allergen complexes and releasing histamines.

mean The average value of a set of values, calculated by dividing the sum of the values by the number of values.

median The value in the middle of an ordered list of values.

memory B cell B lymphocytes activated against a specific antigen that remain in the lymphoid tissues for a long time, and permit a faster and more effective secondary immune response if the same antigen is encountered again.

memory T cell T lymphocytes activated against a specific antigen that remain in the lymphoid tissues for a long time, and permit a faster and more effective secondary immune response if the same antigen is encountered again.

meniscus The curved upper surface of liquid in a tube, caused by surface tension. A meniscus can be concave (as in water in a glass tube) or convex (as in mercury in a thermometer).

mesophyll cell A cell that is found within the thin-walled, loosely packed photosynthetic plant tissue that forms most of the interior of leaves.

messenger RNA (mRNA) A type of RNA molecule transcribed from DNA in the nucleus, which passes into the cytoplasm and binds to a ribosome. At the ribosome, mRNA is translated into a polypeptide.

metabolism The total of all chemical processes that take place in an organism.

metastasis The process by which cancer cells break away from the original (or primary) tumour, travel through the blood and lymph vessels, and form secondary tumours at other locations.

method The specific steps taken to collect data during a scientific investigation. Also known as the procedure.

methodology A brief description of the general approach taken in a scientific investigation. Examples of scientific investigation methodologies are controlled experiments, fieldwork, literature reviews, modelling and simulation.

microbiological barrier A microbiological agent that is one of the first-line defences of an organism's innate immune system, such as the non-pathogenic bacteria (microbiota) found on the skin and in parts of the digestive and excretory systems.

microbiota Microorganisms that colonise particular sites; normal microbiota do not usually cause disease.

microflora Bacteria and microscopic fungi and algae that colonise a particular site. Microflora are a subset of microbiota.

microsatellite A short repeated sequence of nucleotides found at a defined locus on a chromosome. The number of repeats varies between individuals and so these are useful in DNA profiling.

mineralisation A process of fossilisation in which minerals replace the spaces in the structures of organisms, such as bones.

mineralised fossil Fossil in which minerals replace the spaces in the structure of the organism such as bone. Minerals may eventually replace the entire organism, leaving a replica of the original fossil.

mitochondrion (pl. mitochondria) An organelle in eukaryotic cells consisting of folded membrane structures called cristae. It is where the Krebs cycle and the electron transport chain in aerobic respiration occur.

mode The value that appears most often in a data set.

mode of transmission The different ways an infectious agent can be transmitted from its natural reservoir to a susceptible host.

molecular homology A DNA or protein molecule (or their precise sequence) shared by two species or other taxa because of common ancestry.

monoclonal antibody (mAb) An antibody produced by a single clone of B lymphocytes grown in culture. The antibodies produced by the clone are identical and specific to the same antigen.

monomer A smaller subunit of a larger unit (called a polymer); examples include amino acids and nucleotides.

monophyletic group A taxonomic group containing a common ancestor and all its descendants.

morphological isolation Genetic isolation that results from incompatibility between physical structures, such as reproductive organs.

Multiregional evolution (continuity) model A model used to explain the origins of modern humans. This theory proposes significant migration of *Homo erectus* across Africa, Asia and Europe for the last 1.8 million years. Isolation between the populations resulted in the divergence of biology and behaviour, but occasional contact ensured gene flow was maintained and led to concurrent evolution of all groups into *Homo sapiens*.

mummified organism A type of fossil in which the organism is fully preserved and may include features such as skin, fur and organs.

mutagen Any substance or condition that causes mutation. Some chemicals and radiation are common types of mutagens.

mutation A permanent change in a genetic sequence, including changes to the nucleotide sequence of DNA or chromosomal arrangement. Mutations can have a beneficial effect, a harmful effect or no effect at all on the survival ability of the individual.

mutation rate The rate at which genetic mutations occur over time.

myeloma cell Myeloma is a cancer of B lymphocytes. Myeloma cells are B lymphocytes that proliferate uncontrollably.

N

natural active immunity Active immunity induced as a result of survival of a natural infection.

natural killer cell (or NK cell) A type of lymphocyte that is involved in the innate immune response. NK cells recognise and destroy tumour and virus-infected cells.

natural passive immunity The passive transfer of antibodies from mother to fetus through the placenta prior to birth, and from mother to baby through breastfeeding.

natural selection A mechanism of evolution. Phenotypes that are well suited to the environment are more likely to survive and reproduce and pass their alleles on to the next generation.

neutralisation The binding of neutralising antibodies to toxins or antigens on the surface of pathogens that inhibits their action or ability to enter cells.

neutrophil A type of leukocyte for which phagocytosis is a major role. Neutrophils also release defensins to destroy pathogens and cytokines to recruit and activate other leukocytes in response to infection. Neutrophils are the most common leukocyte in mammals, accounting for 60–70% of all white blood cells.

nicotinamide adenine dinucleotide (NAD^+) A coenzyme that functions as an electron carrier during cellular respiration.

nicotinamide adenine dinucleotide phosphate (NADPH) A coenzyme that functions as an electron carrier during photosynthesis.

node The point at which two branches in a phylogenetic tree diverge (also known as branch point). The node represents the last common ancestor that the two diverging taxa shared.

non-cellular pathogen A non-cellular, non-living agent that causes disease. Non-cellular pathogens include viruses, viroids and prions.

non-competitive inhibition A type of enzyme inhibition that occurs when an inhibitor binds to an enzyme site (an allosteric site) other than the active site. Binding to the allosteric site either changes the shape (or conformation) of the enzyme such that the substrate cannot bind to its active site or it prevents a catalytic reaction from proceeding even if the substrate is bound.

non-self antigen An antigen that does not belong to an organism's own cells.

nucleic acid The genetic material of all organisms (DNA and RNA) that controls cellular activities, and is made up of monomer units called nucleotides.

nucleotide A monomer, or building block, of the nucleic acids DNA and RNA. Consists of a phosphate, a sugar and a nitrogenous base.

O

observation Closely monitoring something or someone.

oomycete A fungus-like pathogen of plants with branching hyphae (haustoria) that penetrates living cells and absorb nutrients, or releases enzymes that digest cytoplasm into molecules that can be absorbed.

operator region The segment of DNA that is the binding site of the transcription factor.

operon A unit of DNA under the regulation of a single promoter that codes for several proteins.

Out of Africa (replacement) model A model to explain the origins of modern humans. This model proposes that all living modern humans evolved from a single common ancestor in Africa about 200 000 years ago and as they spread throughout the world they displaced all other human species.

outgroup A taxonomic group that is closely related to the other groups (ingroups) but less closely related than any single one of the ingroups is to each other. It has a common ancestor with the ingroups that is older than the common ancestor of the ingroups. It is included in phylogenetic trees for comparison to the group of interest.

outliers Readings that lie a long way from other results are sometimes called outliers. Repeating readings may be useful in further examining an outlier.

P

palaeoanthropology A branch of anthropology that involves the study of fossil hominins, contributing to our knowledge of human evolution.

palaeontology The study of ancient life preserved as fossils in rocks and ancient sediments.

palindromic sequence A DNA or RNA sequence that is the same when read from 5' to 3' on one strand and 5' to 3' on the complementary strand.

pandemic An epidemic that has spread over several countries or continents, usually affecting a large number of people.

passive immunity The immunity provided by the transfer of antibodies produced in another organism.

pathogen An organism that can produce disease in another organism; includes many micro-organisms and parasites. Also known as an infectious agent.

pathogen-associated molecular pattern (PAMP) A specific molecule typical of certain pathogens found on the surface of their cell wall or plasma membrane, which can be recognised by receptors found in animal cells and which causes an immune response in the animal. Each subsequent infection by the same pathogen elicits the same response.

pattern recognition receptor (PRR) A receptor that recognises molecular patterns common to various microbes, allowing early detection of infection and rapid activation of the host's immune cells.

PCR see *polymerase chain reaction*

peer-reviewed Describes information that other scientists have checked and have agreed is appropriate for publication.

peptide A polypeptide that consists of fewer than 50 amino acids.

peptide bond A covalent chemical bond linking amino acids in a polypeptide chain.

period One of several subdivisions of geological time enabling cross-referencing of rocks and geologic events from place to place. Eons and eras are larger subdivisions than periods while periods themselves may be divided into epochs.

personal errors Include mistakes or miscalculations.

phagocyte A cell capable of engulfing pathogens or foreign particles to destroy them.

phagocytosis The engulfment of solid materials in which the plasma membrane surrounds the material, forming a vacuole (phagosome) and allowing the substance into the cell.

phenolic A protective chemical found in plants; examples include tannins and phytoalexins. Phenolics bind to salivary proteins and enzymes thereby de-activating them. They cause the death of the pathogen through inadequate energy intake.

phenotype The observable trait; expression of a genotype in an individual for a particular trait. The dominance of the alleles and the environmental conditions influence the phenotype of an individual. For example, nutrient availability may influence pigment synthesis in flower petals or hair follicles in animals.

phosphate group One of the three basic units that makes up a nucleotide, and forms part of the sugar–phosphate backbone of RNA and DNA.

phospholipid An essential component of the plasma membrane, comprised of a hydrophilic phosphate head and a hydrophobic fatty acid tail.

photorespiration The process in which plants absorb oxygen and release carbon dioxide, the opposite process to photosynthesis.

photosynthesis The process used by plants, algae and some prokaryotes in which nutrients are produced from carbon dioxide, water and light energy.

phylogenetic tree A diagram that represents the evolutionary relationships between different species, but does not show evolutionary distance. It may be rooted or unrooted.

phylogeny The evolutionary relationships of organisms, usually represented by a branching tree diagram (phylogenetic tree).

phylogram A branching diagram representing the evolutionary relationships between taxa. The branches of a phylogram are scaled (i.e. the lengths of the branches represent evolutionary distance).

physical barrier A physical entity that is one of the first-line defences of an organism's immune system, such as skin or bark.

pigment A substance that absorbs light energy.

placebo An inactive treatment that provides no therapeutic value and is used with a control group in an investigation.

plasma cell An activated B lymphocyte that produces large quantities of the same type of antibody.

plasmid Small, circular pieces of double-stranded DNA found in bacterial cells. Plasmids replicate independently of the bacteria's chromosomal DNA and are used in genetic engineering for creating recombinant DNA.

point mutation A type of gene mutation that typically only affects a single nucleotide. Types of point mutations include substitution and frameshift mutations.

poly-A tail A long tail of adenine (A) nucleotides (100–250) that is added to the end of mRNA during processing. This increases the stability of the mRNA.

polymerase A group of enzymes that catalyse the formation of polymers, in particular the formation of nucleic acid polymers by complementary base pairing with a template strand.

polymerase chain reaction (PCR) A laboratory technique used to amplify (make millions of copies of) a piece of DNA in a short period of time.

polymorphism Genetic variation within a population. The least common allele has to have a frequency in a population of 1% or more to be considered polymorphism rather than mutation.

polynucleotide A polymer of nucleotides joined together through a condensation polymerisation reaction. Can refer to DNA or RNA.

polypeptide chain A polymer of amino acids joined by peptide bonds through a condensation polymerisation reaction.

polyploidy A condition in which every cell of an organism has more than two sets of each chromosome. It is denoted xn, where x is the number of copies of chromosomes, such as $3n$ (triploid) or $6n$ (hexaploid) wheat varieties.

polytomy A node in a phylogenetic tree that indicates three or more lineages evolving from a common ancestor.

portal of entry The manner in which an infectious agent enters a host.

portal of exit The path that a pathogen uses to leaves a host.

postzygotic isolating mechanism A process that stops gene flow between different species by causing reproductive failures after fertilisation. Common forms of postzygotic isolating mechanisms are hybrid inviability, reduced hybrid viability, hybrid sterility and hybrid breakdown.

precipitation (in immunity) One of the mechanisms used by antibodies to interfere with the function of pathogens; by binding to soluble antigens, antibodies cause them to become insoluble and precipitate out of solution.

precision Refers to how closely a set of measurement values agree with each other. Precision gives no indication of how close the measurements are to the true value and is therefore a separate consideration to accuracy.

preserved remains Biological material that has remained intact after the death of an organism due to conditions that have prevented decomposition.

prevalence In epidemiology, the proportion of cases of a medical condition in a population at any given time.

prezygotic isolating mechanism A process that stops gene flow between different species by preventing fertilisation. Common forms of prezygotic isolating mechanism include spatial, temporal, ecological, structural and behavioural isolation.

primary data The measurements or observations that you collect during your investigation.

primary immune response The immune response to an antigen that has been encountered for the first time.

primary lymphoid organ and tissue A major organ or tissue of the lymphatic system: the bone marrow and the thymus.

primary source A source that includes first-hand information, such as the results of an original experiment.

primary structure The linear sequence of amino acids in the polypeptide chain of a protein.

primate A member of the order Primates, which includes humans, great apes (orangutans, gorillas, chimpanzees and bonobos), lesser apes (gibbons), monkeys (Old World monkeys and New World monkeys) and prosimians (lemurs, tarsiers and lorises).

primer A short strand of DNA or RNA that is able to bind or anneal to single-stranded DNA to create a region where DNA polymerase can join and initiate DNA synthesis.

principle A principle is usually more specific than a theory. See *theory*.

prion A small protein particle that, when its shape is altered due to mutation, causes protein aggregation and is toxic to neurons. Prions are the cause of the spongiform encephalopathy diseases BSE in cattle and CJD in humans.

processed data Data that has been mathematically manipulated in some way.

promoter region An upstream region of a gene (a specific DNA sequence) to which RNA polymerase attaches, initiating transcription.

prosthetic group A non-protein compound that is involved in protein structure or function. A protein with a prosthetic group is known as a conjugated protein.

protease inhibitor A chemical produced by plants to protect against pathogens and herbivores which blocks the activity of proteases used for digestion by the invader.

protein An organic compound consisting of one or more long chains of amino acids connected by peptide bonds, and that has a distinct and varied three-dimensional structure.

protein secretory pathway The pathway taken by proteins from production to secretion from the cell. The protein is synthesised and modified in the endoplasmic reticulum, may then be modified in the Golgi apparatus, and is transported to the plasma membrane inside a vesicle. The vesicle merges with the plasma membrane and the protein is expelled from the cell.

protein synthesis The production of a protein through the processes of gene expression which, in eukaryotes, comprises transcription, RNA processing and translation.

proteome The entire set of protein products of the genome.

proteomics The study of proteomes, including the structure, function and interactions of proteins.

proton A positively charged subatomic particle that forms part of the nucleus of an atom.

protozoan A unicellular, eukaryotic organism that may have multiple stages in a complete life cycle and may replicate within the cells of its host.

provisional data Data that is subject to revision.

purine A nitrogenous base that has a double ring structure (e.g. adenine and guanine). Each purine base pairs with a specific pyrimidine base (cytosine, thymine, uracil).

pyrimidine A nitrogenous base that has a single ring structure (e.g. cytosine, thymine and uracil). Each pyrimidine base pairs with a specific purine base (adenine and guanine).

Q

qualitative data Data collected about categorical variables.

quantitative data Data collected about numeric variables.

quaternary structure Two or more polypeptide chains joined as a single functional protein.

R

radioactive Emitting radiation (a form of energy from the nucleus of an unstable atom).

radiometric dating A method of absolute dating that is used to calculate the age of rocks and minerals using radioactive isotopes.

random coil A polymer conformation with the monomers orientated randomly. Adjacent monomers are bonded together.

random errors Affect the precision of a measurement and are present in all measurements except for measurements involving counting. Random errors are unpredictable variations in the measurement process and result in a spread of readings. The effect of random errors can be reduced by making more or repeated measurements and calculating a new mean and/or by refining the measurement method or technique.

random selection A form of sampling in which subjects are randomly selected to participate in a study.

range The difference between the highest and lowest values in a set of data.

raw data The data you record in your logbook.

reactant A chemical that is used up in a biochemical pathway. Enzymes act on reactants to form a final product. Also known as a substrate.

recognition site The short sequence of DNA bases recognised and cut by a restriction enzyme; also called a restriction site.

recombinant DNA DNA that has been genetically engineered by joining fragments of DNA from two or more different organisms.

recombinant DNA technology Technology that combines DNA molecules from two or more sources in cell cultures in the laboratory to create a new DNA molecule or genetic sequence.

recombinant plasmid A plasmid containing a foreign gene that has been inserted by the use of restriction enzymes and DNA ligase.

reduced hybrid viability A form of postzygotic isolation in which offspring of different species (hybrids) survive to birth or germination but do not reach reproductive maturity and are therefore incapable of reproducing.

regulatory gene A gene that codes for transcription factors (which in turn control gene expression at the transcription stage).

relative dating A method of dating geological deposits based on the relative order of layers (strata) and, if present, the fossils within those layers. It is assumed that the deepest layer is the oldest and the uppermost layer is the youngest.

repeat trials Collecting multiple data sets by performing an experiment again after the initial test.

repeatability The closeness of the agreement between the results of successive measurements of the same quantity being measured, carried out under the same conditions of measurement. These conditions include the same measurement procedure, the same observer, the same measuring instrument used under the same conditions, the same location, and repetition over a short period of time.

replication (1) The mechanism by which DNA can be copied. (2) Experimentation carried out on duplicate sets at the same time.

reporter gene A gene that allows detection of gene expression in genetic engineering. Examples are the *lacZ* gene and genes for fluorescent proteins.

repressed Describes a gene that is inhibited and cannot be transcribed.

repressible operon An operon that is turned off by the presence of a particular molecule.

repressor protein A protein that binds to DNA or RNA to inhibit gene expression.

reproducibility The closeness of the agreement between the results of measurements of the same quantity being measured, carried out under changed conditions of measurement. These different conditions include a different method of measurement, different observer, different measuring instrument, different location, different conditions of use, and different time.

reproductive success A measure of how many offspring an individual produces, reflecting the likelihood of their alleles being passed on to the next generation.

research question A statement that defines what is being investigated.

reservoir The habitat in which an infectious agent normally reproduces/replicates.

restriction enzyme A type of enzyme, also called an endonuclease, that occurs naturally in bacteria and can cut DNA at a particular site (a recognition site); used in genetic engineering.

reverse transcriptase A type of polymerase enzyme used by retroviruses to copy their RNA genome into DNA; used in genetic engineering to copy messenger RNA (mRNA) into complementary DNA (cDNA).

reversible inhibition A type of enzyme inhibition in which an inhibitor binds weakly to an enzyme in a temporary manner with hydrogen or hydrophobic bonds.

ribonucleic acid (RNA) A nucleic acid that is a single strand made up of a sequence of ribose sugars and bases (adenine, cytosine, guanine and uracil) linked by phosphodiester bonds. There are three forms: messenger RNA (mRNA), ribosomal RNA (rRNA) and transfer RNA (tRNA).

ribose A five-carbon sugar molecule found in RNA as a component of RNA nucleotides.

ribosomal RNA (rRNA) A nucleic acid synthesised in the nucleolus that forms part of a ribosome.

ribosome A small non–membrane bound organelle, made of RNA and protein. Ribosomes are often attached to rough endoplasmic reticulum and are the site of translation in the process of protein synthesis.

risk assessment A systematic way of identifying the potential risks associated with an activity.

RNA see *ribonucleic acid*

RNA ligase A ligase enzyme that joins together fragments of RNA.

RNA polymerase An enzyme that catalyses the synthesis of RNA, using an existing strand of DNA as a template.

RNA processing The removal of introns from the primary transcript produced in transcription. The exons are joined to form mRNA, ready for translation. This stage of gene expression occurs only in eukaryotes.

root The part of a phylogenetic tree that represents the common ancestor of all the taxa in the tree.

rough endoplasmic reticulum (RER) A large organelle comprised of layers of cisternae with ribosomes studding the surface. It is the site of protein synthesis and modification.

Rubisco An enzyme involved in the photosynthesis and photorespiration pathways in plants.

S

safety data sheet (SDS) A document that contains important information about the possible hazards in using a substance and how the substance should be handled and stored.

saponin A soap-like compound present in plants that breaks down lipids and that disrupts the plasma membranes of potential pathogens.

saturation point The maximum rate of an enzyme-catalysed reaction when all active sites are occupied and no further rate increase can be achieved by increasing substrate concentration.

scientific method The experimental approach to the study of science that involves formulating a hypothesis, designing and performing an experiment to test the hypothesis, and analysing whether the results support or refute the hypothesis.

secondary data Data you have not collected yourself.

secondary immune response The immune response to an antigen that has previously been encountered and which elicited a primary immune response. The process activates memory cells and so is faster and more effective that the primary response.

secondary lymphoid organ and tissue An organ or tissue of the lymphatic system in which an adaptive immune response is initiated: the lymph nodes, spleen, tonsils, adenoids and appendix.

secondary source A resource that interprets primary documents, written after the event by a person who was not a witness to the event.

secondary structure The folding or coiling of the polypeptide chains in proteins due to hydrogen bonds. The main forms are the alpha helix structure, beta-pleated sheets and random coils.

secretory protein A protein synthesised for export out of the cell.

secretory vesicle A vesicle that buds from the Golgi apparatus and contains material that is to be secreted out of the cell by exocytosis.

selection pressure An environmental factor that affects the survival and reproductive success of an individual based on their phenotype.

selective breeding The artificial selection of individuals with desired traits to be interbred. Humans selectively breed both plant and animal species, creating specific strains or breeds.

self-antigen An organism's own antigen, which is normally tolerated (does not elicit an immune response).

self-tolerance The inability of the adaptive immune system to respond to self-antigen.

serum The fluid portion of blood that remains after blood cells and material involved in blood clotting have been removed.

sexual dimorphism The occurrence of marked differences in the physical appearance of males and females of the same species (in addition to differences in sexual organs).

sexual selection The difference in the ability of individuals to acquire mates. Individuals that possess the desired characteristic are more likely to mate and pass on their desired alleles to the next generation. The desired trait is often an indication of overall health and fitness and of other alleles of high adaptive value.

short tandem repeat (STR) A region of non-coding DNA with 4–6 base pair repeated sequences. Used in DNA profiling.

sister taxa A pair of taxa grouped together in a phylogenetic tree (the closest relative of any given taxon).

solar energy The energy from sunlight.

somatic mutation A mutation that occurs in somatic, or non-gamete, cells of an organism. These types of mutations may affect the individual, but cannot be passed on to offspring. Cancer is a form of somatic mutation.

speciation The formation of new species following a lineage splitting event. Speciation may result from geographic, anatomical, physiological or behavioural barriers to breeding, leading to divergence over evolutionary time, or may be rapid as a result of adaptive radiation.

species A group of individuals that are able to interbreed to produce viable, fertile offspring.

specificity The ability to recognise and respond exclusively to specific antigens.

spliceosome An enzyme that removes the introns from the primary transcript to create mRNA during RNA processing (in eukaryotes).

splicing The removal of introns from the primary transcript produced in transcription. The exons are joined to form mRNA, ready for translation.

start codon A codon that indicates where translation should begin in messenger RNA (mRNA). The most common start codon is AUG.

sticky-end restriction enzyme A type of restriction enzyme that makes a staggered cut in DNA to leave fragments with overhanging (or 'sticky') ends. The exposed bases of these sticky ends are then able to form complementary base pairs with nucleotides of other DNA molecules that have been cut with the same restriction enzyme.

stoma (pl. stomata) A pore structure bordered by two guard cells found in plants (often more abundant on the underside of leaves) that allows exchange of gases between the outside and inside of the leaf. The stomata open to allow carbon dioxide to enter and oxygen to be released, and close to reduce water loss.

stop codon A codon that indicates where transcription should stop in messenger RNA (mRNA). The most common stop codons are UAG, UAA and UGA.

stratigraphy The study of the relative positions of layers of rock (strata), some of which contain fossils. The lowest stratum is the oldest and upper strata are progressively younger.

stratum (pl. strata) Rock layers.

stroma (pl. stromata) The fluid matrix part of a chloroplast, in which the light-independent reactions occur.

structural gene A gene that codes for proteins and RNA molecules that are not involved in gene regulation (e.g. enzymes).

structural morphology The study of the form and structure of organisms.

subspecies Populations within a species that show genetic differences across a geographic range.

substrate A chemical that is used up in a biochemical pathway. Enzymes act on substrate molecules to form a final product. Also known as a reactant.

subunit vaccine A vaccine that contains one or more antigens that stimulate an adaptive immune response.

susceptible host A host organism that does not have immunity to an infectious agent.

sympatric speciation The evolution of two new species from an ancestral species that occurs when populations that share the same geographic range become reproductively isolated from each other. Sympatric speciation is more common in plants than in animals.

systematic errors Affect the accuracy of a measurement. Systematic errors cause readings to differ from the true value by a consistent amount each time a measurement is made, so that all the readings are shifted in one direction from the true value. The accuracy of measurements subject to systematic errors cannot be improved by repeating those measurements.

T

T cell see *T lymphocyte*

T cell receptor (TCR) A molecule found on the surface of T lymphocytes that is responsible for recognising fragments of antigen as peptides bound to major histocompatibility complex (MHC) proteins. It is made up of two polypeptide chains that have a variable and a constant region and only one antigen-binding site.

T lymphocyte (or T cell) A type of lymphocyte that originates in the bone marrow and matures in the thymus, and is responsible for cell-mediated immune responses. See *cytotoxic T lymphocyte* and *helper T lymphocyte*.

***Taq* polymerase** A type of heat-resistant DNA polymerase that is widely used in PCR.

target DNA A particular region of a DNA molecule that a scientist intends to study or manipulate.

TATA box A name given to a common sequence of bases in eukaryotic genes, TATAAA, that codes for the promoter region.

taxon (pl. taxa) A biological group classified on the basis of their shared characteristics and evolutionary relationships.

taxonomy The science of the classification of organisms into hierarchical groupings based on their shared characteristics and evolutionary relationships.

template strand A strand of DNA or RNA used as a template for building a complementary strand of a precise nucleotide sequence.

temporal isolation Genetic isolation that results from the separation of populations by timing of activity or breeding cycles.

terpene A chemical produced by plants that is highly toxic to fungi and which is also toxic to many insects. In insects, terpenes mimic certain hormones, causing disruptions to their life cycle. Pyrethrins and phytoectysones are included in this group.

tertiary structure The structure in proteins created by further folding as a result of bonds forming between the R groups of the amino acids, leading to greater stability than the folding in secondary structures.

theory When, after many experiments, a hypothesis has been supported by all the results so far, it is referred to as a theory or principle.

thermodynamics The study of all forms of energy exchange.

thermoluminescence A method of absolute dating that is used to date objects that are less than 500 000 years old. Thermoluminescence uses the intensity of light emitted from a mineral when it is heated to determine the amount of radiation it has absorbed—the older the object, the more light it emits.

thylakoid lamella (pl. thylakoid lamellae) The sheet-like thylakoid membranes between the grana in a chloroplast.

thylakoid membrane A system of interconnected membranes inside a chloroplast. Thylakoids are the site of light-dependent reactions in photosynthesis.

thymine (T) A nitrogenous base (pyrimidine) found in the nucleotides of DNA.

toxoid vaccine A type of non-recombinant subunit vaccine that uses toxins inactivated by formalin to stimulate an adaptive immune response.

trace fossil Preserved evidence of an animal's activity or behaviour, such as footprints, that does not contain parts of the organism.

trait A particular characteristic or feature of an organism.

***trans* face** The side of the Golgi apparatus facing the plasma membrane.

transcription The process by which a base sequence in DNA is used to produce a base sequence in RNA.

transcription factor Proteins that control gene expression at the transcription stage by binding to DNA sequences close to the promoter region of a gene or to the RNA polymerase to induce or repress the expression of specific genes.

transfer RNA (tRNA) An RNA molecule that brings a specific amino acid to a ribosome so it can be joined to other amino acids during translation.

transgene A gene or genetic material that has been transferred naturally or by genetic engineering techniques from one organism to another.

transgenic organism A genetically modified organism (GMO) that has had a gene from another species inserted into its genome.

transitional fossil A fossil that exhibits features that are intermediate between ancestral and descendant groups, indicating a progression from one form to another.

translation The process in which the base sequence of an mRNA molecule is used to produce the amino acid sequence of a polypeptide.

transport vesicle A vesicle that buds from the rough endoplasmic reticulum and contains materials that are to be transported to the Golgi apparatus.

triplet A sequence of three nucleotides in DNA that carries the genetic information for the sequence of amino acids in a protein. Each triplet usually codes for one amino acid.

***trp* operon** A group of genes that are transcribed together to synthesise tryptophan in many bacteria including *E. coli*. It is a repressible operon that is switched off when tryptophan is present in the cell. Also known as the tryptophan operon.

***trp* repressor** A DNA-binding protein that binds to the operator site of the *trp* operon, inhibiting the transcription of the genes in it. It is only active when tryptophan is present as a corepressor.

trpR A regulator gene that codes for the *trp* repressor.

true value The value, or range of values, that would be found if the quantity could be measured perfectly.

tryptophan An amino acid that cannot be made by many animals, including humans, but can be synthesised by many bacteria, including *E. coli*.

tumour An abnormal growth of cells resulting from uncontrolled cell division or failure of programmed cell death. It may be benign or malignant.

U

uncertainty The uncertainty of the result of a measurement reflects the lack of exact knowledge of the value of the quantity being measured.

unloaded coenzyme The form of a coenzyme that is free to accept a proton, electron or chemical group.

uracil (U) A nitrogenous base (purine) found in the nucleotides of RNA. It forms a base pair with adenine.

V

vaccination The technique of artificially inducing an adaptive immune response by administering (usually by injection) a vaccine usually made of altered, weakened or killed microorganisms, or inactivated forms of toxins or antigens.

vaccine A substance used to induce artificial active immunity.

validity A measurement is said to be valid if it measures what it is supposed to be measuring. An experiment is said to be valid if it investigates what it sets out and/or claims to investigate.

variable A factor or condition that can change during your experiment.

variable R group A side chain found on an amino acid.

variable region The region of an antibody molecule that varies between different antibodies and allows them to interact with different antigens.

vector (1) In infectious disease: an object or organism that transfers a parasite from one host to another. (2) In molecular biology: a vehicle used to transfer foreign DNA into a cell (e.g. a plasmid, virus or liposome).

vertical gene transfer The transfer of genetic material from parents to offspring.

vesicle A small organelle consisting of a membrane filled with fluid. Vesicles are often involved in transport within the cell, but may have other functions.

vestigial structure A remnant structure that has lost all or most of its original function in an organism in the course of evolution.

viable offspring Members of the next generation who survive to maturity and are able to reproduce successfully.

viroid An infectious agent of plants that is a type of self-cleaving RNA enzyme (or ribozyme); composed of short, circular strands of RNA that lack a protein coat.

virulence (adj. virulent) The ability of a pathogen to cause disease.

virus An infectious agent composed of genetic material (DNA or RNA) enclosed in a protein coat, and sometimes also a lipoprotein envelope; is only able to multiply in a host cell.

WXYZ

zoonotic Infectious disease transmitted from non-human animal to human.

Index

Bold page numbers refer to where glossary terms are bolded in the text.

C

D

E

F

G

H

I

J

K

L

M

T

U

V

W

X

Y

Z

Attributions

We thank the following for their contributions to our text book:
The following abbreviations are used in this list: t = top, b = bottom, l = left, r = right, c = centre.

COVER: **Science Photo Library:** Alfred Pasieka.

123RF: Akhararat Wathanasing, p. 24cr; alila, pp. 120br, 382cr, 382tr; alycan, p. 341bl; Anna Yakimova, p. 248cb; Anton Lebedev, pp. 76bl, 76cl, 76tl, 94br; blueringmedia, pp. 70bl, 70br, 70tl; designua, pp. 296; 296tl, 326br, 69bl, 70cl, 83cl, 83cr, 90cl; Deyan Georgiev, p. 27tl; Geoffrey Whiteway, pp. 29bl, p. 305cr; gozzoli, p. 369tr; Hans Christiansson, p. 248cr; hypedesk, p. 74cr; Ivan Mikhaylov, p. 39br; Jarun Ontakrai, p. 265tr; Jeff Grabert, p. 372cr; Juan Gartner, p. 267tr; kasto, p. 27br; Kevin Griffin, p. 443tr; lculig, p. 75cl; Leah-Anne Thompson, p. 410c; Luca Alarico Sperotti, p. 450br; luk Cox, p. 264tl; miloszg, p. 268cl; molekuul, p. 363cl;ninglu, p. 212bl; pan demin, p. 434bl; petervick167, p. 406cl; prill, p. 423tr; reddogs, pp. 371br, 371tr; rioblanco, p. 333cr; Roberto Biasini, p. 68br; Sebastian Kaulitzki, p. 350tl; snapgalleria, p. 70cr; tempusfugit, p. 362tl; Valerii Kirsanov, p. 372tr; Vit Krajicek, p. 84r; vladislav ivantsov, p. 482br; zuzanaa, p. 269cr.

AAP: AP Photo, pp. 464c, 472br; Aqua Bounty Farms, p. 164bl.

Alamy Stock Photo: Agencja Fotograficzna Caro, p. 134cr; Auscape International Pty Ltd, pp. 366cl, 414cl; B Christopher, p. 468cl; blickwinkel, p. 423tl; BSIP SA, pp. 280bl, 316tl, 319bl; Chris Mattison, p. 408c; David Tipling Photo Library, p. 407cr; Encyclopaedia Britannica Inc., p. 413bl; Friedrich Saurer, pp. 482bc, 483cl; IanDagnall Computing, p. 143cr; Kaiser/Agencja Fotograficzna Caro, p. 331br; Martin Shields, p. 153br; MShieldsPhotos, p. 29bc; Nature's Geometry by NanoArt, p. 268bl; Rick & Nora Bowers, p. 367cr; Rick & Nora Bowers, p. 367tr; rob3000, p. 104br; Robert Wyatt, p. 407br; Sabena Jane Blackbird, p. 483cr; Science Picture Co, p. 466cl; Science Photo Library, pp. 9cr, 12tl, 25cr, 25bl, 30bc, 45tl, 78br, 78bc, 121br, 132cl, 150cr, 150c, 161cr, 239t, 240ct, 243tr, 278bl, 280tl, 292cr, 292cl, 317cr, 396cr, 396bl, 397b, 408tr; Simon Murrell/Cultura Creative (RF), pp. 420–421; Stephen Barnes/Business, p. 16c; The Natural History Museum, p. 459c; Tony French, p. 395br.

ARC Centre of Excellence for Mathematical & Statistical Frontiers: p. 478.

ATOR (Arc-team Open Research): p. 467b.

Backwell, Lucinda & d'Errico, Francesco: pp. 473cl, 50.

Banwell, Tom: p. 332tc.

BioMed Central: p. 165cr.

Cengage Learning Australia: Solomon & Martin, Biology 10th Edition by Linda Berg, Charles Martin, Diana Martin and Eldra Solomon, 2014, pp. 155c, 319.

Centers for Disease Control and Prevention (CDC): Alissa Eckher, MS, Dan Higgins, MAM, pp. 360bl, 380cr; Based on data from CDC, p. 389tr; 'Weekly US Map: Influenza Summary Update, week ending January 06 2018'. https://www.cdc.gov/flu/weekly/usmap.htm, p. 331l.

China Daily: p. 402cr.

Commonwealth Scientific and Industrial Research Organisation (CSIRO): Electron Microscopy Unit AAHL, p. 346cl; CSIRO Australia, p. 380br.

Cotton Australia: Photograph by Greg Kauter, courtesy of Cotton Australia, p. 163tr.

Department of Health: Adapted from 'Vaccine preventable diseases in Australia, 2005 to 2007' by Clayton Chiu, Aditi Dey, Han Wang, Robert Menzies, Shelley Deeks, Deepika Mahajan, Communicable Diseases Intelligence, 34 Supp:S1-S167, p. S65, 2010, p. 51tc.

Dr. J Liberato, Dr. M Suzuki & Dr. A Drenth, p. 341bc;

Elsevier: 'Molecules consolidate the placental mammal tree' by Mark S. Springer, Michael J. Stanhope, Ole Madsen, Wilfried W. de Jong, p. 436c; 'Pax 6: Mastering eye morphogenesis and eye evolution' by Walter J Gehring, Kazuho Ikea. Trends in Genetics, Vol. 15, Iss. 9, 1999, p. 129bc.

Fleagle, John: p. 459cl.

Fotolia: elenabsl, p. 196br; juanjo tugores, p. 167tr; KAR, p. 40cl; san724, p. 212cl; vgenia sh, p. 234bl.

Getty Images: Adam Pretty, p. 388br; Bethany Clarke/Stringer, p. 341br; BSIP, p. 108tl; Ed Reschke, p. 44bl; Encyclopaedia Britannica, pp. 71cl, 71cr; Ralph Orlowski, p. 338tr.

Genome Research Limited: Licensed under Creative Commons CC-BY 4.0, p. 163b.

ImageFolk (previously Amanda/Corbis) p. 79tr, p. 83tl; Anthony Bannister/Gallo Images, p. 111cr; Bettmann, p. 4c; CDC/PHIL, p. 348tl; Dr. Alvin Telser/Visuals Unlimited, p. 280cr; Hero Images, p. 132tl; John Cancalosi/National Geographic Creative, p. 396cl; Laguna Design, pp. 82tr, 89c; 120tl; Michael Rosenfeld/Science Faction, p. 3br; Nigel Cattlin/Visuals Unlimited, p. 220br; p. 272cr; Regis Bossu/Sygma, p. 466bl; Science Photo Library/Juan Gaertner, p. 295br; Science Photo Library/Steve Gschmeissner, pp. 90bl, 280cl; Tkachev Andrei/ITAR-TASS, p. 394tl; Tom Till/SuperStock, p. 396tl.

Institute for Health Metrics and Evaluation: Based on data from Institute for Health Metrics and Evaluation, pp. 383br, 383cr, 384tr.

Kohout, Michele: p. 269cl.

Low, Tim: p. 370ct.

McGraw-Hill Australia Pty Ltd: McGraw-Hill Education Australia, p. 296tr.

MDPI: 'Emerging Viruses in the Felidae: Shifting Paradigms' by Stephen J. O'Brien, Jennifer L. Troyer, Meredith A. Brown, Warren E. Johnson, Agostinho Antunes, Melody E. Roelke and Jill Pecon-Slattery, Viruses, 4(2), 236–257; doi:10.3390/v4020236, 2012, p. 442c.

Museum Victoria: Based on data from The Human Story by C. Lockwood 2008, p. 50cr.

National Center for Biotechnology Information: 'Multilocus Species Trees Show the Recent Adaptive Radiation of the Mimetic Heliconius Butterflies' by Kryzysztof M Kozak, Niklas Wahlberg, Andrew F E Neild, Kanchon K Dasmahaptra, James Mallet, Chris D Jiggins. Systematic Biology, 64: 505–524, 10.1093/sysbio/syv007, 2015, pp. 428c, 433c.

National Geographic Society: O. Louis Mazzatenta, p. 402br.

National Human Genome Research Institute: pp. 103br, 115bl, 116tr.

National Oceanic & Atmospheric Administration (NOAA): Kyle Carothers, p. 229br.

Natural History Museum Picture Library: Michael R Long/The Trustees of the Natural History Museum, London, p. 431tr.

New England Journal of Medicine: Based on data from The New England Journal of Medicine, p. 385b.

Newman, Hans: p. 28tl.

NSW Office of Environment and Heritage: David Hunter, pp. 418bc, 418bl.

Oxford Designers & Illustrators Ltd: pp. 26tl, 38cl, 212tl, 235tl, 265bl.

Pearson Education Australia: Campbell Biology 10th Edition by Jane B. Reece, Michael L. Cain, Steven A. Wasserman, Lisa A. Urry, 2014, pp. 83, 88bc.

Pearson Education Inc: Human Anatomy & Physiology 10th Edition by Elaine N. Marieb, Katja Hoehn, Figure 21.15, 2013, p. 299.

Pearson Education Ltd: p. 308.

Phoenix, Dave: p. 205cl.

RCSB Protein Data Bank (PDB): 'Molecule of the Month: Inspiring a Molecular View of Biology' feature by H.M. Berman, Shuchismita Dutta, Stephen K. Burley, David S. Goodsel, Maria Voigt, Christine

Zardecki, Nucleic Acids Research, 28: 235–242; www.rcsb.org, 2015, p. 113cr.

Safe Work South Australia: Commonwealth of Australia © 2020, p. 19t.

Science Photo Library (SPL): pp. 26br, 29tr, 35tr, 186br, 187tr, 242cl, 263, 271cr, 293tr; A. Dowsett, Health Protection Agency, p. 341tl; Alfred Pasieka, pp. 9tr, 92cr, 96tc, 96tl, 96tr; AMI Images, p. 274cl; AMI Images/Dartmouth college - Louisa Howard, p. 336br, 340cr; Bildagentur-Online/Tschanz, p. 332tr; Biozentrum, University of Basel, p. 262; CDC, p. 341cl; Chris Butler, p. 392cr; Clouds Hill Imaging Ltd, p. 270br; CNRI, p. 255tr; Dan Dunkley, pp. 54tl, 137tr; David M. Phillips/Science Photo Library, p. 66; David Parker, pp. 288–289; Dennis Kunkel Microscopy, p. 67; Doncaster and Bassetlaw Hospitals, p. 306bl; Dr David Furness, p. 72tl; Dr Gopal Murti, pp. 130–131; Dr Kari Lounatmaa, p. 215tl; Dr Linda Stannard, UCT, pp. 314–315; Dr P Marazzi, p. 159br; Dr Tony Brain, p. 269br; Dr. Kari Lounatmaa, p. 73tl; Elizabeth Daynes, p. 468tl; Equinox graphics, p. 186tl; E.R. Degginger, pp. 466tl, 482bl; Eye of Science, p. 111tr; Gilles Mermet, p. 396tl; Gunilla Elam, pp. 97tr, 316br; Ingo Arndt / Nature Picture Library, p. 361; ISM, p. 238cl; J. G. Paren, p. 196tr; J.C. Revy, ISM, p. 145tc; Jacopin/BSIP, p. 148tl; James Gathany/CDC, p. 145br; James King-Holmes, pp. 145tc, 204tr, 271bl; Javier Trueba/MSF, p. 18bl; John Durham, pp. 4cr, 208–209; John Reader, pp. 454–455; Juergen Berger, p. 270cl; Kallista Images/Custom Medical Stock Photo, pp. 79br, 92c; K. R. Porter, p. 180cb; Laguna Design, pp. 11bl, 200tr; Louise Hughes, p. 180c; Marty Snyderman, p. 211cr; Martyn F Chillmaid, pp. 26bc, 28bl, 28cl; Massimo Brega, The Lighthouse, p. 161c; Mauricio Anton, p. 400cr; Maurizio de Angelis, p. 320tl; Mauro Fermariello, p. 30cr; Medimage, pp. 90tl, 238tl; Microscape, pp. 180bl, 180ct; Nancy Kedersha, p. 280br; Pascal Goetgheluck, p. 458bc; Peter Menzel, p. 133tc; Philippe Plailly, pp. 145cr, 369br; P. Plailly/E.Daynes, pp. 462cl, 463b, 470c; Power and Syred, p. 78cl; Prof P.M. Motta, S. Makabe & T. Naguro, p. 73tc; Professors P Motta & T Naguro, pp. 72bc, 72bl, 72tc, 72tr; Ramon Andrade 3DCIENCIA, p. 148cl; Revi, ISM, p. 265br; Rosenfeld Images Ltd., p. 166br; SCI-COMM Studios, p. 32cl; Science Stock Photography, p. 396c; Scubazoo, p. 423tc; Sherman Thomson, p. 211br; Silkeborg Museum, Denmark/ Munoz-Yague, p. 397cl; Sinclair Stammers, p. 165tr; Sputnik, p. 372bl; Stephen Ausmus/US Department of Agriculture, p. 30cl; Steve Gschmeissner, pp. 72br, 232–233, 268tl, 274bc, 274tl, 279br, 280cr, 290tl, 292tl, 218tl; Susumu Nishinaga, pp. 184–185; Tek Image, p. 30tc; Tony Camacho, p. 408cr; Trevor Clifford Photography, p. 27bl; Volker Steter/Nordstar - 4 Million Years of Man, p. 473cr; Wladimir Bulgar, p. 158bl.

Science Source/Photo Researchers Inc: Omikron, pp. 98-99; Ralph Jr, p. 337cl.

Shutterstock: Alexander Raths, p. 152br; Alila Medical Images, p. 279tr; Alila Medical Media, pp. 266c, 268br, 324c, 376t, 377tl; Ashley Whitworth, p. 487cl; attem, p. 24tr; Australis Photography, p. 321br; Bildagentur Zoonar GmbH, p. 270bc; Bitcyte, p. 81br; BMCL, p. 405br; Bork, p. 27tr; Catalin Petolea, p. 277cr; catolla, p. 4br; Christian Mueller, p. 329tr; CkyBe (Adapted by Pearson Education Australia), p. 360cr; Crevis, p. 434cl; D. Kucharski K. Kucharska, p. 270tc; David Lade, pp. 392cl, 409cr; Deman, p. 427c; Designua, p. 302cr, 449bl; Dmitry Naumov, p. 305tr; Dmussman, p. 418tl; ellepigrafica, pp. 279c, 284bl; extender_01, p. 83tr; Fon Hodes, p. 217cr; Georgios Kollidas, p. 350c; gillmar, p. 371cr; Gypsytwitcher, p. 406tl; igorstevanovic, p. 144cr; iofoto, p. 293c; Ipatov, p. 409cb; Jacek Fulawka, p. 434br; Jamen Percy, p. 434cr; Jamilia Marini, p. 91tc; Jane Rix, p. 392bl; Janelle Lugge, p. 364cl; Jens Goepfert, p. 25tl; Jianghaistudio, p. 443tc; Juan Gaertner, p. 301tr; KentaStudio, p. 24br; Kjuuurs, p. 408cl; kongsak sumano, p. 405tc; Lebendkulturen.de, p. 23br; LightField Studios, p. 36tl; magnetix, p. 88tc; Matusciac Alexandru, p. 167bc; Miles Aways Photography, p. 249tr; Mizina Oksana, p. 245br; Monkey Business Images, p. 18cl, 362c; Mor Ben Zion, p. 412tc; Natalia van D, p. 390–391; Natalie Jean, p. 223cr; NH, p. 247cb; photoinnovation, p. 211tr; photong, p. 27tc; Richard Griffin, p. 214bl; Rob Byron, p. 195tr; Ronsmith, p. 36cl; RTimages, p. 23cl; Ryan M. Bolton, p. 405tl; Science Photo Library, p. 187c; Sebastian Kaulitzki, p. 309tr; Serg Zastavkin, p. 372br; Sozaijiten, p. 223tr; Stephen van Horn, p. 222cl; Stubblefield Photography, p. 405tr; toeytoey, p. 280tr; Vector Tradition (Adapted by Pearson Education Australia), p. 461tc; vetpathologist, p. 198cr; vitstudio, p. 349br; Vladimir Mulder, pp. 1–2; Y.F.Wong, p. 427cr; zimowa, p. 79cl.

Sinaeur Associates Inc: p. 439c.

Smith, Suzi (www.diagnosisdiet.com): p. 192cr.

South Australian Museum: © The Museum Board of South Australia. Photo by Michael Lee, p. 431br.

Sozaijiten: p. 217cb.

Tait, Alan: p. 398bl.

The United Nations Economic Commission for Europe (UNECE): p. 19.

The University of Melbourne: Dr. Andrew Drinnan/School of Botany, p. 400br; Pauline Ladiges, pp. 397cr, 400bl, 406br.

Victorian Curriculum and Assessment Authority (VCAA): Selected VCE extracts from the VCE Biology Study Design (2022-2026) are copyright Victorian Curriculum and Assessment Authority (VCAA), reproduced by permission. VCE® is a registered trademark of the VCAA. The VCAA does not endorse this product and makes no warranties regarding the correctness or accuracy of its content. To the extent permitted by law, the VCAA excludes all liability for any loss or damage suffered or incurred as a result of accessing, using or relying on the content. Past VCE exams, current VCE Study Designs and related content can be accessed directly at www.vcaa.vic.edu.au

Western Australian Museum: p. 3tr.

Wikimedia Commons: Alexbateman, p. 32bc; Thomas Splettstoesser, p. 383tr.

World Health Organisation (WHO): pp. 333tl, 342bc.

W. W. Norton & Company, Inc: Molecular Biology of the Cell 4th Edition by Bruce Alberts, Alexander Johnson, Julian Lewis, Martin Raff, Keith Roberts and Peter Walter, used by permission of W.W. Norton & Company, Inc., 2002, pp. 75bc, 313tc.